Motion and Time Study

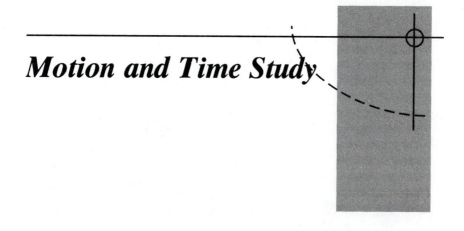

Motion and Time Study

BENJAMIN W. NIEBEL
Professor Emeritus of Industrial Engineering
The Pennsylvania State University
Niebel & Associates

Ninth Edition

IRWIN

Burr Ridge, Illinois
Boston, Massachusetts
Sydney, Australia

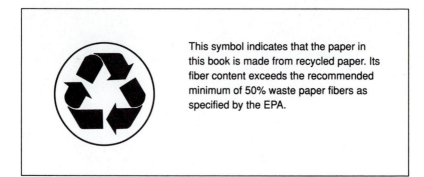

This symbol indicates that the paper in this book is made from recycled paper. Its fiber content exceeds the recommended minimum of 50% waste paper fibers as specified by the EPA.

Sponsoring editor: Richard T. Hercher, Jr.
Senior editorial coordinator: Shelley McDonald
Marketing manager: Robb Linsky
Project editor: Karen J. Nelson
Production manager: Bette K. Ittersagen
Art manager: Kim Meriwether
Compositor: Bi-Comp, Inc.
Typeface: 10/12 Times Roman
Printer: R. R. Donnelley & Sons Company

Library of Congress Cataloging-in-Publication Data

Niebel, Benjamin W.
    Motion and time study / Benjamin W. Niebel.—9th ed.
        p.   cm.
    Includes bibliographical references and index.
    ISBN 0-256-09248-6
    1. Motion study.   2. Time study.   I. Title.
T60.7.N54   1993
658.5'42—dc20   92–15051

Printed in the United States of America

2 3 4 5 6 7 8 9 0   DOC   9 8 7 6 5 4

Preface

As we approach the 21st Century, with a widely expanded market and manufacturing capability both in Europe and the south Pacific rim, both the opportunities and the need for technical competence are growing dramatically. American companies are feeling the pressure of global competition as never before. Ten years ago, foreign competition was centered in only a few industries—electronics and automotive in particular. But today this competition is both industrywide and worldwide. Almost every industry, business, and service organization is restructuring itself to operate more effectively in an increasingly competitive world. Each segment of these organizations is increasing the intensity of its cost reduction and quality improvement efforts. Cost effectiveness and product reliability without excess capacity are the keys to successful activity in all areas of business, industry, and government. And cost effectiveness with improved quality under restricted plant capacity is the end result of methods engineering, equitable time standards, and improved employee motivation through the introduction of modern management reward systems. These tools are the keys to productivity improvement in any business, industry, or service organization, whether it be in a bank, a hospital, a department store, a railroad, or the postal system. Furthermore, success in a given product line or service leads to new products and innovations. It is this accumulation of successes that drives hiring and the growth of an economy. These are the tools that the Japanese have used so effectively in connection with their lean production concepts. This concept emphasizes methods engineering and the employment of teams of multiskilled workers at all levels of the organization and uses highly flexible increased automated facilities to produce volumes of product in large variety. The tremendous success of the Japanese auto, camera, art of electronics,

electronics, and appliance industries has been attributed to their lean production concept.

The ninth edition has been written for several reasons. First, it aims to emphasize the importance of continued quality improvement and product reliability based upon simplified design in connection with methods engineering work and management reward systems. Second, this edition continues to focus on the extensive use of personal computers to establish standards and computer-aided design and engineering for conceptualizing possibilities and evaluating costs. Computers produce standards from fundamental motion data up to 50 percent faster than manual methods. Third, this edition introduces new material that has proved successful, particularly in the establishment of indirect labor standards such as maintenance. Currently, approximately 55 percent of the work force in this country is involved in indirect labor activities, much of which is clerical. Fourth, this edition further broadens the application of motion and time study into service areas such as railroads, hospital administration and maintenance, the U.S. Postal Service, retirement centers, and business management. Fifth, this ninth edition updates existing material and examples that may have become obsolete because of technological change and inflation.

This edition contains new material describing the MODAPTS procedure of systems. Also new basic motion times developed under the Micro and Macro Motion Analyses Systems have been included. Expanded information in the use of videotape, human factors considerations, and productivity sharing programs have been included. In addition, each chapter ends with pertinent reference material and problems. Applicable chapters provide summaries of recommended videotapes and computer software.

This edition has the same objectives as the first eight—to provide a practical, up-to-date college text describing engineering methods, time study, and wage payment, and to give practicing labor and management analysts an authentic source of reference material.

Suggestions received from people in the universities, colleges, technical institutes, industries, and labor organizations who regularly use this text have helped materially in the preparation of this ninth edition. The reviews by Dulio Furtado of Louisiana Tech University, K.N. Balasubramanian of California Polytechnic State University, and Marvin M. Kuers of Texas A & M University were most helpful.

The author wishes to acknowledge the constructive criticisms of Emory Enscore of The Pennsylvania State University, who reviewed sections of this edition. Also, I acknowledge the help received from Dick Ward of the Material Handling Institute in connection with new material provided in the areas of material handling and plant layout. The new material on Micro and Macro Motion Analyses Systems provided by

Clifford M. Sellie of Standards, International, was most useful. Also, the help of Robert M. Wygandt in connection with providing information relative to the use of MODAPTS greatly expands and improves the coverage of basic motion times. Finally, I acknowledge, with considerable gratitude, the assistance of my wife, who produced the final manuscript.

Benjamin W. Niebel

Contents

Delays. Extra Allowances: *Clean Workstation and Oil Machine. Power Feed Machine Time Allowance.* Application of Allowances: *Typical Allowance Values.* Summary.

Methods, Time Study, and Wage Payment Today

THE IMPORTANCE OF PRODUCTIVITY

The only way a business or enterprise can grow and increase its profitability is by increasing its productivity. Productivity improvement refers to the increase in output per work-hour or time expended. The United States has long enjoyed having the world's highest productivity. Over the last 100 years, this country has increased in productivity approximately 4 percent per year. However, in the last decade, the U.S. rate of productivity improvement has been exceeded by that of Japan, Korea, and Germany, and it has been challenged by Italy and France.

The fundamental tool whose use results in increased productivity is methods, time study (frequently referred to as work measurement), and wage payment. Of the total cost of the typical metal products manufacturing enterprise, 12 percent is direct labor, 45 percent is direct material and 43 percent is overhead. All aspects of a business or industry—sales, finance, production, engineering, cost, maintenance, and management—provide fertile areas for the application of methods, time study, and sound wage payment. Too often, people consider only the production function when applying methods, standards, and wage payment. Important as the production function is, other aspects of the enterprise also contribute substantially to the cost of operation and are equally valid areas for the application of cost improvement techniques. In sales, for example, modern information retrieval methods usually result in more reliable information, leading to greater sales at less cost; product quotas for specific territories provide a base or standard that individual salespeople endeavor to exceed; and payment for results always results in above-standard performance.

Today most U.S. businesses and industries are by necessity restructuring themselves in order to operate more effectively in an increasingly competitive world. They are addressing cost reduction and quality improvement through productivity improvement with more intensity than ever before. They are also examining critically all business components that do not contribute to their profitability.

Since the field of production within manufacturing industries utilizes the greatest number of young men and women in methods, time study, and wage payment work, this text will treat that field in more detail than any other. However, examples from other areas of the manufacturing industry—such as maintenance, transportation, sales, and management—as well as the service industry will be provided.

The areas of opportunity existing in production for students enrolled in engineering, industrial management, business administration, industrial psychology, and labor-management relations are: (1) work measurement, (2) work methods, (3) production engineering, (4) manufacturing analysis and control, (5) facilities planning, (6) wage administration, (7) safety, (8) production and inventory control, and (9) quality control. Other position areas, such as personnel or industrial relations, cost, and budgeting, are closely related to, and dependent on, the production group. These areas of opportunity are not confined to manufacturing industries. They exist and are equally important in such enterprises as department stores, hotels, educational institutions, hospitals, banks, airlines, insurance offices, military service centers, government agencies, and retirement complexes. Today, in the United States, only about 20 percent of the total labor force is employed in manufacturing industries. The remaining 80 percent is engaged in service industries or staff-related positions. As this country becomes more service industry oriented, the philosophies and techniques of methods, time study, and wage payment must be utilized in the service sector. Wherever people, materials, and facilities interact to obtain some objective, productivity can be improved through the intelligent application of methods, time study, and wage payment.

The production section of an industry may well be called its heart; once the activity of this section is interrupted, the whole industry ceases to be productive. The production department includes methods engineering, time study, and wage payment activity, offering the young technical graduate one of the most satisfying fields of endeavor.

It is in the production department that material to produce is requisitioned and controlled; the sequence of operations, inspections, and methods is determined; tools are ordered; time values are assigned; work is scheduled, dispatched, and followed up; and customers are kept satisfied with quality products delivered on time. Training in this field demonstrates how production is accomplished, where it is done, when it is performed, and how long it takes to do. A background including such training will prove invaluable, whether one's ultimate objective is in sales, production, or cost.

If the production department is considered the heart of an industrial enterprise, the *methods, time study, and wage payment* activity is the heart of the production group. Here, more than in any other place, people determine whether a product is going to be produced on a competitive basis. Here is where they use initiative and ingenuity to develop efficient tooling, worker and machine relationships, and workstations on new jobs in advance of production, thus assuring that the product will stand the test of stiff competition. Here is where they continually use creativity to improve existing methods and product to help assure the company of leadership in its product line. In this activity good labor relations may be maintained through establishing fair labor standards, or may be impeded by setting one inequitable rate.

Methods, time study, and wage payment offer real challenges. Industries with competent engineers, business administrators, industrial relations personnel, specially trained supervisors, and psychologists carrying out methods, time study, and wage payment techniques are inevitably better able to meet competition and better equipped to operate profitably.

The objective of the manufacturing manager is to produce a quality product, on schedule, at the lowest possible cost, with a minimum of capital investment and a maximum of employee satisfaction. The realiability and quality control manager focuses his or her objectives so as to maintain engineering specifications and satisfy customers with the product's quality level and reliability over its expected life. The production control manager is principally interested in establishing and maintaining production schedules with due regard for both customer needs and the favorable economics obtainable with careful scheduling. The manager of methods, time study, and wage payment is mostly concerned with combining the lowest possible production cost and maximum employee satisfaction. The maintenance manager is primarily concerned with minimizing facility downtime because of unscheduled breakdowns and repairs. Figure 1–1 illustrates the relationship of the manager of the methods, time study, and wage payment department to the staff and line departments under the general manager.

THE SCOPE OF METHODS ENGINEERING AND TIME STUDY

The field of methods engineering and time study includes designing, creating, and selecting the best manufacturing methods, processes, tools, equipment, and skills to manufacture a product after working drawings have been released by the product engineering section. When the best method interfaces with the best skills available, an efficient worker-machine relationship exists. Once the complete method has been established, the responsibility of determining the time required to produce the product falls within the scope of this work. Also included is the

FIGURE 1–1

Typical organization chart showing the influence of methods, time study, and wage payment on the operation of the enterprise

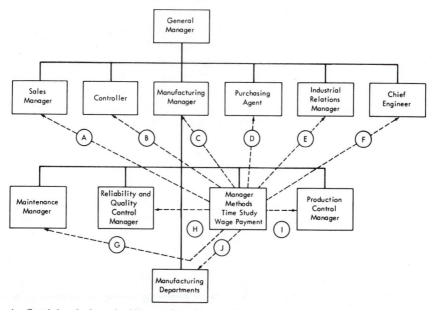

A—Cost is largely determined by manufacturing methods.
B—Time standards are the bases of standard costs.
C—Standards (direct and indirect) provide the bases for measuring the performance of production departments.
D—Time is a common denominator for comparing competitive equipment and supplies.
E—Good labor relations are maintained with equitable standards and fair base rates.
F—Methods and processes strongly influence product designs.
G—Standards provide the bases for preventive maintenance.
H—Standards enforce quality.
 I—Scheduling is based on time standards.
J—Methods and standards provide how the work is to be done and how long it will take.

responsibility of following through to see that predetermined standards are met, that workers are adequately compensated for their output, skills, responsibilities, and experience, and that they have a feeling of satisfaction from the work that they do.

This procedure includes defining the problem related to expected cost, breaking the job down into operations, analyzing each operation to determine the most economical manufacturing procedures for the quantity involved with due regard for operator safety and job interest, applying proper time values, and then following through to assure that the prescribed method is put into operation. Figure 1–2 illustrates the opportunities for reducing manufacturing time through the application of methods engineering and time study.

FIGURE 1–2
Opportunities for savings through the applications of methods engineering and time study.

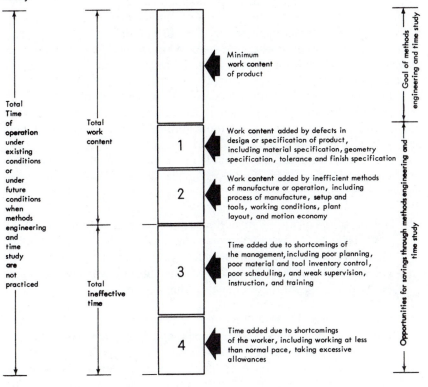

Methods Engineering

The terms *operation analysis, work simplification,* and *methods engineering* are frequently used synonymously. In most cases the person is referring to a technique for increasing the production per unit of time and, consequently, reducing the unit cost. However, *methods engineering,* as defined in this text, entails analysis work at two different times during the history of a product. Initially, the methods engineer is responsible for designing and developing the various work centers where the product will be produced. Second, he or she continually restudies the work centers to find a better way to produce the product and/or improve its quality. The more thorough the methods study made during the planning stages, the less the necessity for additional methods studies during the life of the product.

Methods engineering implies the utilization of technological capability. Primarily because of methods engineering, improvements in productivity are a never-ending procedure. The productivity differential resulting from technological innovation can be of such magnitude that developed countries will always be able to maintain competitiveness with low-wage developing countries. Research and development leading to new technology is essential to methods engineering. The 10 countries with the highest R&D expenditures per workers, as reported in *United Nations Industrial Development Organization,* are: United States, Switzerland, Sweden, Netherlands, West Germany, Norway, France, Israel, Belgium, and Japan. Certainly these countries are among the leaders in productivity. As long as they continue to emphasize research and development, methods engineering through technological innovation will be instrumental in their ability to provide goods and services at a high level.

Methods engineers use a systematic procedure to develop a work center, to produce a product, or to provide a service. This procedure is outlined below.

1. *Select the project.* Typically, projects selected represent either new products or existing products that have a high cost of manufacture and a low profit. Also, products that currently experience difficulty in maintaining quality and are having problems meeting competition are logical method engineering projects.
2. *Get the facts.* Assemble all the important facts relating to the product or service. These include drawings and specifications, quantity requirements, delivery requirements, and projections about the anticipated life of the product or service.
3. *Present the facts.* Once all important information has been acquired, record it in an orderly form for study and analysis. The development of process charts at this point is very helpful.
4. *Make an analysis.* Utilize the primary approaches to operations analysis and the principles of motion study to decide which alternative will produce the best product or service. These primary approaches include: purpose of operation, design of part, tolerances and specifications, materials, process of manufacture, setup and tools, working conditions, material handling, plant layout, and principles of motion economy.
5. *Develop the ideal method.* Select the best procedure for each operation, inspection, and transportation by considering the various constraints associated with each alternative.
6. *Presenting the method.* Explain the proposed method in detail to those responsible for its operation and maintenance.
7. *Install the method.* Consider all details of the work center to assure that the proposed method will provide the results anticipated.
8. *Develop a job analysis.* Make a job analysis of the installed method to assure the operator or operators are adequately selected, trained, and rewarded.

9. *Establish time standards.* Establish a fair and equitable standard for the installed method.
10. *Follow up the method.* At regular intervals, make an audit of the installed method to determine if the anticipated productivity and quality is being realized, if costs were correctly projected, and if further improvements can be made.

When methods studies are made to improve the existing method of operation, experience has shown that to achieve the maximum returns, analysts should follow a systematic procedure similar to that advocated for designing the initial work center. Westinghouse Electric Corporation's Operation Analysis program advocates these steps to assure the most favorable results:

1. Make a preliminary survey.
2. Determine the extent of analysis justified.
3. Develop process charts.
4. Investigate the approaches to operation analysis.
5. Make motion study when justified.
6. Compare the old and the new methods.
7. Present the new method.
8. Check the installation of the new method.
9. Correct time values.
10. Follow up the new method.

Actually, methods engineering includes all of these steps.

Methods engineering can be defined as the systematic close scrutiny of all direct and indirect operations to find improvements making work easier to perform and allowing work to be done in less time with less investment per unit. Thus, the real objective of methods engineering is profit improvement.

Time Study

Time study is often referred to as work measurement. It involves the technique of establishing an allowed time standard to perform a given task, based on measurement of the work content of the prescribed method, with due allowance for fatigue and for personal and unavoidable delays. Time study analysts use several techniques to establish a standard: a stopwatch time study, computerized data collection, standard data, fundamental motion data, work sampling, and estimates based on historical data. Each of these techniques has application under certain conditions. Time study analysts must know when it is best to use a certain technique and then use that technique judiciously and correctly.

The functions of time study analysts and methods engineers are closely allied. Although the objectives of the two differ, good time study analysts are good methods engineers, since their positions include methods engineering as a basic component.

To be assured that the prescribed method is the best, time study engineers frequently assume the roles of methods engineers. In small industries these two activities are often handled by the same individual. Establishing time values is a step in the systematic procedure of developing new work centers and improving methods in existing work centers.

Today, in order to position your firm as a world class competitor, there must be implementation of performance measurement systems to meet the demands of just-in-time quality control and time-compressed management.

Wage Payment

The wage payment function, similarly, is closely associated with the time study and methods sections of the production activity. In many companies, and particularly in smaller enterprises, the wage payment activity is performed by the same group responsible for the methods and standards work. In general, the wage payment activity is performed in concert with those responsible for conducting job analyses and job evaluations, so that these closely related activities function smoothly.

Job analysis refers to the procedure for making a thorough appraisal of each position and recording details of the work so that the job can be evaluated.

Job evaluation is a technique for equitably determining the relative worth of the different work assignments within an organization. This technique establishes fair base rates for different work assignments. In general, job evaluation methodologies consider what the employee brings to the job in the form of education, experience, and special skills, and what physical and mental effort the job requires. A third important factor is the amount of responsibility the job requires.

Because of the nature of a given enterprise, it may have two, or even three, entirely different wage payment plans in effect (daywork, piecework, group incentives). The administration of these plans falls on the wage payment group.

Production control, plant layout, purchasing, cost accounting and control, and process and product design are additional areas closely related to both the methods and standards functions. All of these areas depend on time and cost data, facts, and procedures of operation from the methods and standards department to operate effectively. These relationships are briefly discussed in Chapter 24.

Objectives of Methods, Time Study, and Wage Payment

The principal objectives of methods, time study, and wage payment are to increase productivity and product reliability and lower unit cost, thus

TABLE 1–1

Methods (operations analysis)	Work measurement (time study)	Wage payment
Chapters 1, 2, 3, 4, 5, 6, 7, 8, 9, 10, 21, 24	Chapters 1, 2, 10, 12, 13, 14, 15, 16, 17, 18, 19, 20, 21, 22, 23, 24	Chapters 11, 25, 26

allowing more quality goods and/or services to be produced for more people. The ability to produce more for less will result in more jobs for more people for a greater number of hours per year. Only through the intelligent application of the principles of methods, time study, and wage payment can producers of goods and services increase while, at the same time, the purchasing potential of all consumers grows. Through these principles unemployment and relief rolls can be minimized, thus reducing the spiraling cost of economic support to nonproducers.

Corollaries that apply to the principal objective are to:

1. Minimize the time required to perform tasks.
2. Continually improve the quality and the reliability of products and services.
3. Conserve resources and minimize cost by specifying the most appropriate direct and indirect materials for the production of goods and services.
4. Produce with a concern for the availability of power.
5. Maximize the safety, health, and well-being of all employees.
6. Produce with an increasing concern to protect our environment.
7. Follow a humane program of management that results in job interest and satisfaction for each employee.

Table 1–1 reflects where the three areas: methods, work measurement (time study), and wage payment are discussed in detail. Some chapters such as 1 and 2 give some attention to all three areas. Typically, chapters 1 through 10 can be covered in one three-credit course, and chapters 11 through 26 in a separate three-credit course for which the first course should be a prerequisite. Thus, this text has been designed to cover six credits of course work.

TEXT QUESTIONS

1. What is another name for time study?
2. Which job opportunities exist in the general field of production?
3. Explain the scope of methods engineering.
4. What activities are considered the key links to the production group within a manufacturing enterprise?

5. Define the terms *operations analysis, work simplification,* and *methods engineering.*
6. List the 10 steps in applying methods engineering.
7. Which steps have the Westinghouse Electric Corporation advocated to assure real savings during a methods improvement program?
8. What is the function of the time study department?
9. Is it possible for one enterprise to have more than one type of wage payment plan? Explain.
10. What is the principal objective of methods engineering?
11. What four broad opportunities are there for savings through methods and time study?
12. Explain in detail what a *work center* encompasses.
13. What is the function of job evaluation?
14. How is job evaluation dependent on job analysis?
15. Which three considerations constitute a successful job evaluation plan?
16. What is your understanding of the terms *just-in-time, total quality control,* and *time-compressed management?*

GENERAL QUESTIONS

1. Explain why underdeveloped countries with labor costs only a fraction of developed countries have difficulty competing in the manufacturing sector for such products as motor cars and appliances.
2. Describe the *lean production concept* used by Japanese industry.
3. How do well-organized methods, time study, and wage payment procedures benefit the company?
4. Show the relationships between time study and methods engineering. Explain each fully.
5. Discuss the reasoning behind the statement, "A good time study analyst is a good methods engineer."
6. Comment on the general responsibility of the wage payment group.
7. Why does the purchasing department need data and information from the methods, time study, and wage payment department? Give several examples.
8. Why is job evaluation a part of the wage payment function?
9. What is meant by fundamental motion data?
10. Based on your reading of Chapter 1, what percentage savings do you estimate is possible in a hospital that has never practiced methods engineering and time study? Do you feel that this is a realistic estimate?
11. Explain why in a typical metal manufacturing plant only 15 percent of total cost is direct labor cost.
12. Contact the following service industries in your community and find out how professional competence in methods, standards, and wage payment is used:
    A hospital.
    A high school.

A post office.
A police department.
A bus service.

13. In addition to the 10 listed in the chapter, which countries do you feel are in the top 20 in the world as far as R&D contribution per worker is concerned? Why?

PROBLEMS

1. In the XYZ hospital, management has been charging $1,000 per day for a semiprivate room. An analysis of present costs reveals the following:

Direct labor (including fringe benefits)......	$20 per hour
Materials...............................	$15.25 per patient/day
Indirect costs...........................	$16 per 100 sq. ft./day
Hospital room occupancy................	80 percent

The typical patient uses 12.2 hours of direct labor per 24 hours. The average semiprivate room is 14' × 20'.

Was the $1000 per day (charge per person) an adequate figure? What dollar charge would you recommend? Explain how you would initiate a cost reduction and quality improvement program.

2. In the Dorben Department Store, the local union and management agreed to use work measurement standards and base rates determined by job evaluation. After the job evaluation was made, the following job classes and money rates were established, based on the job points noted:

Job class	Job points	Money base rate per hour
A................	100	$ 5.25
B................	250	6.00
C................	400	7.10
D................	550	8.40
E................	700	10.00
F................	850	11.40

Because of the wide point range associated with each job class, union representatives asked for the establishment of five additional job classes based on these job points: 175, 325, 475, 625, 775.

What money rates per hour should be assigned to the five new job classes? (Hint: The original plan plotting points against money was parabolic.)

3. Gains in both productivity and quality have long been a goal of the Dorben Manufacturing Company. Since reduction in failure cost directly increases productivity, quality improvement is recognized as a key contributor to these gains. Dorben management has established a goal of reducing failure costs by

50 percent in the next two years. Explain the strategy that you would recommend to achieve the established goal.

SELECTED REFERENCES

Camp, Robert C. "Benchmarking: The Search For Industries Best Practices—Part III", *Quality Progress,* March 1989, p. 76.

Feorene, O. J. "Organization and Administration of Industrial Engineering." Chapter 1.2 in *Handbook of Industrial Engineering,* ed. Gavriel Salvendy, New York: John Wiley & Sons, 1982.

Konz, Stephan. *Work Design.* Columbus, Ohio: Grid, 1983.

Nadler, Gerald "The Role and Scope of Industrial Engineering." In *Handbook of Industrial Engineering,* 2nd ed., edited by Gavriel Salvendy, Chap. 1. New York: John Wiley & Sons, 1992.

United Nations Industrial Development Organization. *Industry in the 1980's Structural Change and Interdependence.* United Nations, 1985.

The Development of Motion and Time Study

2

THE WORK OF TAYLOR

Frederick W. Taylor is generally conceded to be the father of modern time study in this country. However, time studies were conducted in Europe many years before Taylor's time. In 1760 Jean Rodolphe Perronet, a French engineer, made extensive time studies on the manufacture of No. 6 common pins, and arrived at a standard of 494 per hour (2.0243 hrs./1000). Sixty years later, an English economist, Charles W. Babbage, conducted time studies on the manufacture of No. 11 common pins. These studies determined that one pound (5,546 pins) should be produced in 7.6892 hours (1.3864 hrs./1000).[1]

Taylor began his time study work in 1881 while associated with the Midvale Steel Company in Philadelphia. After 12 years' work, he evolved a system based on the "task." Taylor proposed that the work of each employee be planned out by the management at least one day in advance. Workers were to receive complete written instructions describing their tasks in detail and noting the means to accomplish them. Each job was to have a standard time determined by time studies made by experts. This time was to be based on the work possibilities of a first-rate worker who, after being instructed, was able to do the work regularly. In the timing process, Taylor advocated breaking up the work assignment into small divisions of effort known as "elements." Experts were to time these individually, and use their collective values to determine the allowed time of the task.

In June 1895 Taylor presented his findings and recommendations at a

---

[1] Charles Babbage, *On the Economy of Machinery and Manufactures,* 1832.

13

Detroit meeting of the American Society of Mechanical Engineers (ASME). His paper was received without enthusiasm because many of the engineers interpreted his findings to be a new piece rate system rather than a technique for analyzing work and improving methods.

The distaste for piecework that prevailed in the minds of these engineers can well be appreciated. At that time piecework standards were established by supervisors' estimates; at best, these were far from being accurate or consistent. Both management and employees were rightfully skeptical of piece rates based on the supervisor's guess. Management looked on the rates with doubt, in view of the possibility that bosses would make conservative estimates to protect the performance of their departments. Because of unfortunate past experiences, workers were concerned over any rate established merely by judgment and guess, since these rates vitally affected their earnings.

Then in June 1903, Taylor presented his famous paper, "Shop Management," at the Saratoga meeting of the ASME. In that paper he proposed these elements of scientific management:

Time study, with the implements and methods for properly making it.

Functional, or divided responsibilities for supervisors, with its superiority to the old-fashioned single boss.

Standardization of all tools and implements used in the plant, and also of the acts or movements of workers for each class of work.

The desirability of a planning room or department.

The "exception principle" in management.

Use of slide rules and similar timesaving implements.

Instruction cards for workers.

The task idea in management, accompanied by a large bonus for the successful performance of the task.

The "differential rate."

Mnemonic systems for classifying manufactured products as well as the implements used in manufacturing.

A routing system.

A modern cost system.

Taylor's "Shop Management" technique was well received by many factory managers, and with modifications it resulted in many satisfactory installations.[2]

In 1898, while at the Bethlehem Steel Company (he had resigned his

---

[2] In 1917, C. Bertrand Thompson reported on the record of 113 plants that had installed "scientific management." Of these, 59 considered their installations completely successful, 20 partly successful, and 34 failures. C. Bertrand Thompson, *The Taylor System of Scientific Management* (Chicago: A. W. Shaw, 1917).

post at Midvale), Taylor carried out an experiment that came to be one of the most celebrated demonstrations of his principles. The tale is remembered as the "pig-iron story." Details of this research are recorded in Taylor's book, *The Principles of Scientific Management*. In this story, he brings out that by establishing the correct method along with financial incentive, workers carrying 92-pound pigs of iron up a ramp onto a freight car would be able to increase their productivity from an average of 12.5 long tons per day to between 47 and 48 long tons per day. This work was performed with an increase in daily rate of $1.15 to $1.85. Taylor claimed that workmen performed at the higher rate "without bringing on a strike among the men, without any quarrel with the men and were happier and better contented."

His studies were made on a man referred to as Schmidt (Taylor's fictitious name for the man) and described as "a little Pennsylvania Dutchman who had been observed to trot back home for a mile or so after his work in the evening, about as fresh as when he came trotting down to work in the morning." Taylor's studies indicated that a pig-iron handler should be under load only 43 percent of the time and entirely free from the load the remaining 57 percent. But while under load he should never be standing still, as had often been the case at Bethlehem, because merely holding the 90-pound pig caused nearly as much fatigue as walking with it. Forced rest periods at regular intervals, permitted adequate recovery from fatigue so that pig iron could be loaded during the entire work day at about the same pace.

Another of Taylor's Bethlehem Steel experiments that gained fame was referred to as "the science of shoveling." Workers who shoveled at Bethlehem owned their own shovels and would use the same one for any job—lifting heavy iron ore to lifting light rice coal. After considerable study, Taylor designed shovels to fit the different loads. For example, shovels for iron ore were given short handles; those for the light rice coal were given broad scooped shape and longer handles.

Another of Taylor's well-known contributions was the discovery of the Taylor-White process of heat treatment for tool steel. Studying self-hardening steels, he developed a means of hardening a chrome-tungsten steel alloy without rendering it brittle, by heating it close to its melting point. The resulting "high speed steel" more than doubled machine cutting productivity and remains in use today all over the world. Later, he developed the Taylor equation for cutting metal. Not as well known as his engineering contributions is the fact that in 1881 he was a U.S. tennis doubles champion. Here he used an odd-looking racket he had designed with a spoon curved handle.

In the early 1900s the country was going through an unprecedented inflationary period. The word *efficiency* became passé, and most businesses and industries were looking for new ideas that would improve their performance. The railroad industry also felt the need to substantially increase shipping rates to cover general cost increases. Louis Brandeis, who at that time represented the eastern business associations, contended that the railroads did not deserve or, in fact, need the increase because they had been remiss in not introducing the new "science of management" into their industry. Brandeis claimed that the railroad companies could save $1 million a day by introducing the techniques advocated by Taylor. Thus, Brandeis and the Eastern Rate Case (as the hearing came to be known) first introduced Taylor's concepts as "scientific management."

At this time many people without the qualifications of Taylor, Barth, Merrick, and other early pioneers, were eager to make names for themselves in this new field. They established themselves as "efficiency experts" and endeavored to install scientific management programs in industry. Soon they encountered a natural resistance to change from employees, and since they were not equipped to handle problems of human relations, they met with great difficulty. Anxious to make a good showing and equipped with only a pseudoscientific knowledge, they generally established rates that were too difficult to meet. Situations became so acute that some managers were obliged to discontinue the whole program in order to continue operation.

In other instances, factory managers would allow the establishment of time standards by the supervisors, but this was seldom satisfactory. Once standards were established, many factory managers of that time, interested primarily in the reduction of labor costs, would unscrupulously cut rates if some employee made what the employer felt was too much money. The result was harder work at the same, and sometimes less, take-home pay. Naturally, violent worker reaction resulted.

These developments spread in spite of the many favorable installations started by Taylor. At the Watertown Arsenal, labor objected to the new time study system to such an extent that in 1910 the Interstate Commerce Commission started an investigation of time study. Several derogatory reports on the subject influenced Congress in 1913 to add a rider to the government appropriation bill, stipulating that no part of the appropriation should be made available for the pay of any person engaged in time study work. This restriction applied to the government-operated plants where government funds were used to pay the employees.

The Military Establishment Appropriation Act, 1947 (Public Law 515, 79th Congress), and the Navy Department Appropriation Act, 1947 (Public Law 492, 79th Congress), provide as follows:

Sec. 2. No part of the appropriation made in this Act shall be available for the salary or pay of any officer, manager, superintendent, foreman or other person having charge of the work of any employee of the United States Government while making or causing to be made with a stopwatch, or other time-measuring device, a time study of any job of any such employee between the starting and completion thereof, or of the movements of any such employee while engaged upon such work; nor shall any part of the appropriation made in this Act be available to pay any premiums or bonus or cash reward to any employee in addition to his regular wages, except as may be otherwise authorized in this Act.

Finally, in July 1947, the House of Representatives passed a bill that allowed the War Department to use time study. In 1949, the prohibition against using stopwatches was dropped from appropriation language, so that now no restriction of time study practice prevails. It is of interest that today the use of the stopwatch is prohibited by some unions in some railroad repair facilities.

MOTION STUDY AND THE WORK OF THE GILBRETHS

Frank B. Gilbreth was the founder of the modern motion study technique. This may be defined as the study of the body motions used in performing an operation to improve the operation by eliminating unnecessary motions and simplifying necessary motions, and then establishing the most favorable motion sequence for maximum efficiency.

Gilbreth originally introduced his ideas and philosophies into the bricklayer's trade where he was employed. After introducing methods improvements through motion study, including an adjustable scaffold that he had invented and operator training, he was able to increase the average number of bricks laid to 350 per worker per hour. Prior to Gilbreth's studies, 120 bricks per hour was considered a satisfactory rate of performance for a bricklayer.

More than anyone else, Gilbreth and his wife, Lillian, were responsible for industry's recognition of the importance of a minute study of body motions to increase production, reduce fatigue, and instruct operators in the best method of performing an operation.

With assistance from his wife, Gilbreth also developed the technique of filming motions to study them. In industry, this technique is known as micromotion study. The study of movements through the aid of the slow motion moving picture is by no means confined to industrial applications. The world of sports has found it invaluable as a training tool to show the development of form and skill.

In addition, the Gilbreths developed the cyclegraphic and chronocyclegraphic analysis techniques for studying the motion paths made by an operator. The cyclegraphic method involves attaching a small electric light bulb to the finger or hand or part of the body being studied and then photographing the motion while the operator is performing the operation. The resulting picture gives a permanent record of the motion pattern employed and can be analyzed for possible improvement.

The chronocyclegraph is similar to the cyclegraph, but its electric circuit is interrupted regularly, causing the light to flash. Thus, instead of showing solid lines of the motion patterns, the resulting photograph shows short dashes of light spaced in proportion to the speed of the body motion being photographed. Consequently, with the chronocyclegraph it is possible to compute velocity, acceleration, and deceleration as well as study body motions.

EARLY CONTEMPORARIES

Carl G. Barth, an associate of Frederick W. Taylor, developed a production slide rule for determining the most efficient combination of speeds and feeds for cutting metals of various hardnesses, considering the depth of cut, size of tool, and life of the tool.

Barth is also noted for the work he did in determining allowances. He investigated the number of foot-pounds of work a worker could do in a day. He then developed a rule that equated a certain push or pull on a worker's arms with the amount of weight that worker could handle for a certain percentage of the day.

Harrington Emerson applied scientific methods to work on the Santa Fe Railroad and wrote a book, *Twelve Principles of Efficiency,* in which he made an effort to inform management of procedures for efficient operation. He reorganized the company, integrated its shop procedures, installed standard costs and a bonus plan, and transferred its accounting work to Hollerith tabulating machines. This effort resulted in annual savings in excess of $1.5 million.

It was Emerson who coined the term *efficiency engineering.* His ideal was efficiency everywhere and in everything. His philosophy of efficiency as a basis for operation in all fields first appeared in 1908 in *Engineering Magazine.* When, in 1911, Emerson enlarged his ideas in *Twelve Principles of Efficiency,* this volume was perhaps the most comprehensive guide to good management. His first principle of management that he emphasized was "people work most effectively when they have clearly defined goals."

In 1917 Henry Laurence Gantt developed simple graphs that would measure performance while visually showing projected schedules. This production control tool was enthusiastically adopted by the shipbuilding industry during World War I. This tool made it possible for the first time to compare actual performance against the original plan, and to adjust daily schedules in accordance with capacity, backlog, and customer requirements.

Gantt is also known for his invention of the task and bonus wage system. This was developed in 1901, after he spent six years as Taylor's right-hand man at the Midvale and Bethlehem Steel companies. Gantt's wage payment system rewarded workers for above-standard performance and eliminated any penalty for failure. Perhaps more important than the deviation from Taylor's advocacy of penalizing the below-standard operator was offering the boss a bonus for every worker who performed above standard. Gantt made it obvious that scientific management could and should be much more than an inhuman "speedup" of labor. Gantt proclaimed, "We do not approve of foremen 'cussing' their men, but we do approve of their showing the men how the work is to be done." Thus, Gantt is known today for his emphasis on the human factors, a closer study of job difficulties, and the importance of leadership.

Morris L. Cooke, former director of the Department of Public Works in Philadelphia, made an effort to bring the principles of scientific management into city governments. In 1940 Cooke and Philip Murray, a past president of the CIO, published *Organized Labor and Production,* in which they brought out that the goal of both labor and management is "optimum productivity." This they defined as "the highest possible balanced output of goods and services that management and labor skills can produce, equitably shared and consistent with a rational conservation of human and physical resources."

After Taylor retired, Dwight V. Merrick started a study of unit times; these were published in the *American Machinist,* edited by L. P. Alford. Merrick, with the assistance of Carl Barth, developed a technique for determining allowances on a rational basis. Merrick is also known for his multiple piece rate wage payment plan based on three graded piece rates.

Motion and time study received added stimulus during World War II when Franklin D. Roosevelt, through the Department of Labor, advocated establishing standards, from which increases in production resulted. On November 11, 1945, Regional War Labor Board III (for Pennsylvania, southern New Jersey, Maryland, Delaware, and the District of Columbia) issued a memorandum stating the policy of the War Labor Board on incentive proposals. Sections I, II, and IV are reproduced since they contain matter pertinent to standards and wage incentives.

I—General Considerations Applicable to All Incentive Proposals

1. The expected effect of an incentive plan should be an increase in the present production per man-hour without increasing the unit labor cost in the plant, department, or job affected.
2. The proposal should not be in effect merely as a means of giving a general increase in wages, nor should it result in wage decreases.
3. The plan should offer more pay only for more output.
4. If a union has bargaining rights for workers affected, the plan in all of its details should be collectively bargained.
5. No incentive wage plan should be proposed as a substitute for carrying out the responsibilities of both the management and employees.
6. No incentive plan should be put into operation, even if the money is held rather than advanced to the workers, until the approval of the War Labor Board has been secured.

II—*Establishing Incentive Rates for a Specific Production Operation*

When incentive rates are proposed for a specific production job or job classification the following principles apply:

1. Where feasible the operation should be *time studied* carefully to set the production standard. Results of this time study in as much detail as possible should appear in the application. If a time study is impracticable, the application should show why.

2. If a time study is impracticable, the production standard may be based upon *records of past production* provided: *(a)* The records for an appropriate production period are submitted with the application; *(b)* The applicant establishes that the period is representative and that the present product, methods, prospective volume of work, and work force are comparable to those existing during the particular period of past production; *(c)* Exceptionally high or low production figures for short periods are satisfactorily explained in the application.

3. The production standard should be a quantity of output *higher than has been attained previously* by the average worker, or at least, higher than has been attained customarily. If the production standard is below an amount previously attained at some one or more times, the application should explain fully the reasons therefor.

IV—*Plant-wide Incentive Plans*

Because plant-wide incentive plans are comparatively new to American industry and because the effects of such plans on worker efficiency and production are difficult to predict, the Regional Board is not adopting a position on them at this time, but will consider each case on its individual merits. The general considerations set forth in I above are applicable here.

ORGANIZATIONS

Since 1911 there has been an organized effort to keep industry abreast of the latest developments in the techniques inaugurated by Taylor and Gilbreth. Technical organizations have contributed much toward bringing the science of time study, motion study, work simplification, and methods engineering up to present-day standards.

In 1911 the Conference of Scientific Management, under the leadership of Morris L. Cooke and Harlow S. Persons, was started at the Amos Tuck School of Administration and Finance at Dartmouth College.

In 1912 the Society to Promote the Science of Management was organized. It was renamed the Taylor Society in 1915.

The Society of Industrial Engineers was organized in 1917 by those interested in production methods.

American Management Associations (AMA) traces its origins back to 1913, when a group of training managers formed the National Association of Corporate Schools. This society later merged with the National Association of Employment Managers and in 1923 broadened its mission and adopted the AMA name. Since then AMA has rounded out its coverage of the management field by forming 12 functional divisions for its more than 80,000 members. Courses, seminars, and publications on corporate goal setting, productivity management, and standards of performance for executives are among the offerings of the AMA General Management Division. Its Manufacturing Division has courses and publications on productivity improvement, work measurement, wage incentives, and industrial engineering. Its General and Administrative Services Division includes programs on work simplification and clerical standards. Together with the American Society of Mechanical Engineers, AMA presents annually the Gantt Memorial Medal for the most distinguished contribution to industrial management as a service to the community.

The Society for the Advancement of Management (SAM) was organized in 1936 by the merging of the Society of Industrial Engineers and the Taylor Society. This organization has continued to emphasize the importance of time study and methods and wage payment. Industry has used SAM's time study rating films over a long period of years. It annually offers the Taylor key for the outstanding contribution to the advancement of the art and science of management as conceived by Frederick W. Taylor. Also awarded annually is the Gilbreth medal for noteworthy achievement in the field of motion, skill, and fatigue study. In 1972 SAM combined forces with AMA, while maintaining its separate identity and its network of local chapters.

The Institute of Industrial Engineers has grown rapidly since its founding at Columbus, Ohio, on September 9, 1948. At that time, it was named the American Institute of Industrial Engineers. Later to give more visibility to its international character, it became the Institute of Industrial Engineers (IIE). The purposes of IIE are to maintain the practice of industrial engineering on a professional level; to foster a high degree of integrity among the members of the industrial engineering profession; to encourage and assist education and research in areas of interest to industrial engineers; to promote the interchange of ideas and information among members of the industrial engineering profession; to serve the public interest by identifying persons qualified to practice as industrial engineers; and to promote the professional registration of industrial engineers. IIE's Work Measurement and Methods Engineering Division keeps the membership up-to-date on all facets of this area of work. This division annually gives the Phil Carroll achievement award, established in memory of the division's first director. The criteria for the award specifically state that the recipient's contribution to the profession must apply to work measurement and/or methods engineering.

PRESENT TRENDS

Time and motion study has steadily improved since the 1920s; today it is recognized as a necessary tool for the effective operation of business and industry. Practitioners of the art and science of time and motion study have come to realize the necessity of considering the human element. No longer is the cut-and-dried procedure so characteristic of efficiency experts acceptable. Today, through employee testing and training, they consider that individuals differ in performance potential. Now they recognize that such factors as sex, age, health and well-being, physical size and strength, aptitude, training attitudes, job satisfaction, and response to motivation have a direct bearing on output. Furthermore, present-day analysts recognize that workers object, and rightfully so, to being treated as machines. Workers dislike and fear a purely scientific approach to methods, work measurement, and wage incentives. They inherently dislike any change from their present way of operation. This psychological reaction is not characteristic of factory workers only, but is the normal reaction of all people. Management frequently rejects worthwhile methods innovations because of its reluctance to change. In fact, in my experience, management has been harder to sell on new ideas than any other group within the plant. After all, it is responsible for the existing methods, and it frequently defends those methods regardless of the potential savings through change.

Workers tend to fear methods and time study, for they see that this results in an increase in productivity. To them, this means but one thing: less work and consequently less pay. They must be sold on the fact that they, as consumers, benefit from lower costs, and that broader markets result from lower costs, meaning more work for more people for more weeks of the year.

Some fear of time study today is without a doubt due to unpleasant experiences with efficiency experts. To many workers, motion and time study is synonymous with the *speedup* or the *stretch-out*. These terms denote using incentives to spur employees to higher levels of output, followed by establishing new levels as normal production, thus forcing the workers to still greater exertions to maintain even their previous earning power. In years past, undoubtedly some shortsighted and unscrupulous managements did resort to this practice.

Even today, some unions oppose the establishment of standards by measurement, the development of hourly base rates by job evaluation, and the application of incentive wage payment. It is the belief of these unions that the time allowed to perform a task and the amount that an employee should be paid represent issues that should be resolved by collective bargaining arrangements.

Today's practitioners of motion and time study must use the "humane" approach. They must be well versed in the study of human behavior and accomplished in the art of communication. They must be good listeners at all times, indicating that they respect the ideas and thinking of others, particularly the worker at the bench. They must give credit where credit is due. In fact, they should habitually give the other person credit, even if there is some question of that person deserving it. Also, practitioners of motion and time study should always remember to use the questioning attitude as emphasized by the Gilbreths, Taylor, and the other pioneers in this field. The idea that there is "always a better way" needs to be continually pursued in the development of new methods that improve the productivity, quality, delivery, worker safety, and well being.

In their industrial engineering curricula a great number of colleges and universities are teaching the principles, techniques, and philosophies of this field. Most labor unions are training their representatives in the results and uses of motion and time study. Managements of both small and large industries are embarking on mass training programs, realizing the potentialities of a well-formulated program utilizing this tool.

Industry, business, and government are in agreement that the untapped potential for increasing productivity is the best hope for dealing with inflation and competition. And the principal key to increased productivity is a continuing application of the principles of methods, standards, and wage payment. Only in this way can more output from people and machines be realized. American labor expects and has the bargaining strength to get a continuing increase in wage levels. American government has pledged itself to an increasingly paternalistic philosophy of providing for the disadvantaged—housing for the poor, medical care for the aged, jobs for the minorities, and so on. In order to accommodate the spiraling costs of labor and government taxes and still stay in business, we must get more from our productive elements—people and machines.

Today, progressive industries are extending the methods engineering tool of ergonomics/human factors for use in designing jobs, workplaces, equipment, and products. These efforts have improved productivity throughout organizations, increased employees' health and safety, and created a more satisfied work force. Countless instances have proved that greater productivity can be achieved by removing unnecessary effort and demands from the workplace.

The extended application of methods, standards, and wage payment to all combinations of people, materials, and machines is assured. For example, today military equipment contractors and subcontractors are finding an increasing pressure to document direct labor standards as a result of MIL-STD 1567 (USAF), work measurement, released June 30, 1975 and superseded by MIL-STD 1567A released March 11, 1983, and revised on January 30, 1987. Any firm awarded a contract exceeding $1 million is subject to MIL-STD 1567A, which requires a work measurement plan and procedures, a plan to establish and maintain engineered standards of known accuracy and traceability, a plan for methods improvement in conjunction with standards, and a plan for use of standards as an input to budgeting, estimating, planning, and performance evaluation.

On May 3, 1986, MIL-STD 1567A Work Measurement Guidance Appendix was finalized. The military standard requires the application of a disciplined work measurement program as a management tool to improve productivity on those contracts to which it is applied. This document establishes criteria that must be fulfilled by the contractor's work measurement activity and guidelines in the application of work measurement techniques to help assure cost effective equipment and systems.

The military standard defines Type I engineered labor standards as those developed from a recognized technique: time study, standard data, predetermined time system, or a combination thereof, to derive at least 90 percent of the normal time associated with the labor effort covered by the standard. All Type I standards must reflect an accuracy of plus or minus 10 percent with a 90 percent or greater confidence at the operation level. Type I standards also are required to: (1) include documentation of an operations analysis, (2) provide a record of standard practice or method followed when the standard was developed, (3) provide a record of the performance rating or leveling, (4) include a record of the standard time computation, including the allowances, and (5) include a record of observed or predetermined time system values utilized in determining the standard time. All other work measurement standards are defined as Type II and have no specified accuracy requirements. Appendix 4 provides MIL-STD-1567A with revised and superseded pages as of January 30, 1987.

While work measurement has concentrated on direct labor in the past, there has been increasing use of methods and standard development for indirect labor during the past few years. This trend will continue. The use of computerized techniques will also continue to grow. Several of the predetermined time systems (see Chapter 19) are fully computerized today. Notable among these are 4M, MOST, and WOCOM. Today many companies have developed time study and work sampling software. Typically the programs use electronic data collectors for compiling the study.

Table 2–1 shows a time ladder that illustrates the progress made in connection with work methods and time standards.

TABLE 2–1

Year	Important events
1760	Perronet makes time studies on No. 6 common pins.
1776	Adam Smith publishes *The Wealth of Nations*.
1820	Charles W. Babbage makes time studies on No. 11 common pins.
1832	Charles W. Babbage publishes *On the Economy of Machinery and Manufactures*.
1881	Frederick W. Taylor begins his time study work.
1895	Taylor presents his findings to ASME. He publishes his paper "A Piece Rate System."
1901	Henry L. Gantt develops the task and bonus wage system.
1903	Taylor presents paper on Shop Management to ASME.
1906	Taylor publishes paper "On the Art of Cutting Metals."

TABLE 2–1 *(concluded)*

Year	Important events
1909	Frank B. Gilbreth publishes "Bricklaying System."
1910	The term *scientific management* was coined by Louis D. Brandeis at a meeting in the home of H. L. Gantt. Interstate Commerce Commission starts an investigation of time study. Gilbreth publishes "Motion Study." Gantt publishes "Work, Wages, and Profits."
1911	Conference of Scientific Management sponsored by the Amos Tuck School of Administration and Finance at Dartmouth College. Taylor publishes paper on "The Principles of Scientific Management." Harrington Emerson publishes "Efficiency as a Basis for Operation and Wages."
1912	Society to Promote the Science of Management is organized. Emerson estimates $1 million per day can be saved if eastern railroads apply scientific management. Gilbreth publishes "Primer of Scientific Management."
1913	Emerson publishes "The Twelve Principles of Efficiency." Congress adds rider to government appropriation bill stipulating that no part of this appropriation should be made available for the pay of any person engaged in time study work. Henry Ford unveils the first moving assembly line in Detroit.
1914	Professor Robert Hoxie publishes "Scientific Management and Labor." Ford Motor Company introduces the $5 day.
1915	Taylor Society formed to replace the Society to Promote the Science of Management.
1916	Gantt publishes "Industrial Leadership."
1917	Frank B. and Lillian M. Gilbreth publish "Applied Motion Study."
1923	American Management Associations formed.
1927	Elton Mayo begins Hawthorne experiments at Western Electric Company's plant in Hawthorne, Illinois.
1933	Ralph M. Barnes receives the first Ph.D. granted in the United States in the field of Industrial Engineering from Cornell University. His thesis leads to the publication of "Motion and Time Study."
1936	Society for the Advancement of Management organized.
1940	Morris L. Cooke and Philip Murray publish "Organized Labor and Production."
1945	Department of Labor advocates establishing standards to improve productivity of supplies for the war effort.
1947	Bill passed allowing the War Department to use time study.
1948	Founding of the Institute of Industrial Engineers in Columbus, Ohio. Eiji Toyoda and Taichi Ohno at Toyota Motor Company pioneered the concept of lean production.
1949	Prohibition against using stopwatches dropped from appropriation language.
1972	Society for the Advancement of Management combines with the American Management Associations.
1975	MIL-STD 1567 (USAF), Work Measurement released.
1983	MIL-STD 1567A, Work Measurement released.
1986	MIL-STD 1567A, Work Measurement Guidance Appendix finalized.

TEXT QUESTIONS

1. What was the first principle of sound management that Harrington Emerson emphasized?
2. Where were time studies originally made and who conducted them?
3. Explain Frederick W. Taylor's principle of functional foremanship.
4. What effect has Congress had on time study?
5. What is meant by motion study, and who is generally conceded to be the founder of the motion study technique?
6. What is Carl G. Barth primarily noted for in the production end of industry?
7. Which organizations are concerned with advancing the ideas of Taylor and the Gilbreths?
8. Was the skepticism of management and labor toward rates established by "efficiency experts" understandable? Why or why not?
9. What psychological reaction is characteristic of workers when methods changes are suggested?
10. Explain the importance of the humanistic approach in methods and time study work.
11. What contributions in the area of motion study were made by the Gilbreths?
12. What was Emerson's philosophy of efficiency?
13. What was Gantt's contribution to motion and time study?
14. Who was Louis Brandeis, and what part did he play in the introduction of scientific management?
15. How did the Military Establishment Appropriation Act, 1947, slow the development of work measurement methodology?
16. Who was the first director of the Work Measurement and Methods Engineering Division of the Institute of Industrial Engineers?
17. What is the impact of MIL-STD 1567 on work measurement?
18. Who was Adam Smith? Why were his writings important in connection with motion and time study? (See Table 2–1 and consult your local library.)
19. Why did the AIIE change its name to IIE?

GENERAL QUESTIONS

1. Identify the concept of *lean production.*
2. How is time and motion study used in industry today?
3. Why are labor unions training their representatives in time and motion study techniques?
4. Explain the function of the AMA as compared to that of the IIE.
5. Why was the government restriction on the use of stopwatches carried through the World War II years?
6. Interview a group of five industrial workers and obtain their opinions on the necessity of motion and time study.
7. How would you rank in order of importance the following management techniques: material handling, work measurement, ergonomics, systems analysis, incentive application, and method study? Justify your ranking.

PROBLEMS

1. The industrial engineer of a local township is considering the location of a hospital on a plot 3.2 miles from borough A and 5.8 miles from borough B. If borough A is 4.7 miles from borough B, find the angle at which straight roads joining the two boroughs with the hospital will intersect.

    The population of borough A is three times that of borough B, and a second alternative is to place the new hospital 2.3 miles from borough A and 7.4 miles from borough B. If the cost of a hospital visit is computed at 0.20 per vehicle-mile, and hospital visitors are known to be 25 percent of the population per year, which location would be the most favorable from the standpoint of visitation cost?

2. By requiring the implementation of MIL-STD 1567A, the government projects a 15-percent saving on a particular contract. The manufacturer that obtained the contract for $10 million (this includes an estimated 10 percent gross profit) estimates that it will need to employ two additional industrial engineers and two technical assistants to provide the requirements outlined in MIL-STD 1567A. It feels these personnel costs will be approximately 120,000 per year including fringe benefits. The life of the contract is five years. Without the use of these technical people, the company estimates it would have had to quote a figure of $12 million (this includes technical outside assistance) for this contract in order to realize the same gross profit. However, at this higher quotation there would have been only a 20 percent chance that the company would have been able to get the contract. Did this manufacturer make the best decision? Explain in detail.

SELECTED REFERENCES

Barnes, Ralph M. *Motion and Time Study: Design and Measurement of Work.* 7th ed. New York: John Wiley & Sons, 1980.

Eastman Kodak Co., Human Factors Section. *Ergonomic Design for People at Work.* New York: Van Nostrand Reinhold, 1983.

Mundell, Marvin E. *Motion and Time Study: Improving Productivity.* 5th ed. Englewood Cliffs, N.J.: Prentice-Hall, 1978.

Niebel, Benjamin W. *A History of Industrial Engineering at Penn State.* University Park, Pa.: University Press, 1992.

Saunders, Byron W. "The Industrial Engineering Profession." In *Handbook of Industrial Engineering,* ed. Gavriel Salvendy. New York: John Wiley & Sons, 1982.

Graphic Tools of the Methods Analyst

3

Regardless of how methods work is being used—to design a new work center or to improve one already in operation—it is helpful to present in clear, logical form the factual information related to the process. The first step in methods work is to gather all the necessary facts related to the operation or process. Pertinent information—such as the quantity to be produced, delivery schedules, operational times, facilities, machine capacities, special materials, and special tools—may have an important bearing on the solution of the problem. This chapter will discuss the techniques that best present the factual data. Once the facts are presented clearly and accurately, they are examined critically so that the most practical, economic, and effective method can be installed.

Every craftsperson uses tools to facilitate performance. Just as the machinist has micrometers and calipers, and the patternmaker has chisels and plane, so does the methods analyst use tools to do a better job in a shorter time. One of the most important tools of the methods engineer is the process chart. A process chart is defined as a graphic presentation of any manufacturing or business process. Methods engineers usually use eight different process charts, each of which has specific applications. These are:

1. The operation process chart.
2. The flow process chart.
3. The flow diagram.
4. The worker and machine process chart.
5. The gang process chart.
6. The operator process chart.
7. The travel chart.
8. The PERT chart.

The operation, flow, and PERT charts and the flow diagram are used principally to present problems. Usually a problem cannot be adequately solved unless it is properly presented. Consequently, these charts are discussed at this time. Worker and machine, gang, and operator process charts are discussed in Chapter 6. The travel chart is discussed in Chapter 5. All four of these charts usually are constructed in conjunction with operations analysis, which is explained in Chapters 4 and 5.

THE OPERATION PROCESS CHART

The operation process chart shows the chronological sequence of all operations, inspections, time allowances, and materials used in a manufacturing or business process—from the arrival of raw material to the packaging of the finished product. It depicts the entrance of all components and subassemblies to the main assembly. Just as a blueprint gives at a glance such design details as fits, tolerances, and specifications, so does an operation process chart give manufacturing and/or business details at a glance.

Before we can improve a design, we must get a drawing of the product as it is presently designed. Likewise, before we can improve a manufacturing process, it is well to construct an operation process chart so that we understand the problem fully and can determine what areas afford the most possibilities for improvement. The operation process chart effectively states the problem, and if a problem cannot be stated, it usually cannot be solved. Information needed to construct the operation process chart is obtained from direct observation and measurement. It is important that the exact starting and ending points of the operation under study be clearly identified.

Constructing The Operation Process Chart

In constructing the operation process chart, two symbols are used: a small circle, usually 3/8 inch in diameter, denotes an operation, and a small square, usually 3/8 inch on a side, denotes an inspection.

An operation takes place when the part being studied is intentionally transformed or when it is being studied or planned prior to performing productive work on it. Some analysts prefer to separate manual operations from those for paper work. Manual operations usually are related to direct labor while paperwork operations frequently are a portion of indirect or expense costs. Different classes of operations can be coded as shown in Figure 3–3 on page 36.

An inspection takes place when the part is being examined to determine its conformity to standard.

Before beginning the actual construction of the operation process chart, analysts should identify it with a title placed at the top of the paper: Operation Process Chart. Usually the identifying information, the part

FIGURE 3-1
Conventional practice to show
that no juncture occurs when
vertical flow line crosses horizontal material line

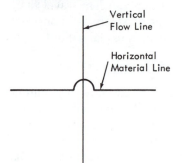

number, drawing number, process description, present or proposed method, date, and name of the person doing the charting, follows. Sometimes additional information may be added to identify completely the subject being charted. This may include such items as the chart number, plant, building, and department.

Vertical lines indicate the general flow of the process as work is being accomplished, while horizontal lines feeding into the vertical flow lines indicate material, either purchased or on which work has been performed during the process. Thus, parts can be shown as entering a vertical line for assembly or leaving a vertical line for disassembly. Materials that are disassembled or extracted are represnted by horizontal material lines drawn to the right of the vertical flow line, while assembly materials are shown as horizontal lines drawn to the left of the vertical flow line. In general, the operation process chart should be constructed so that vertical flow lines and horizontal material lines do not cross. If for some reason it becomes necessary to cross a vertical and a horizontal line, the conventional practice to show that no juncture occurs is to draw a small semicircle in the horizontal line at the point where the vertical line crosses it (see Figure 3-1).

Time values should be assigned to each operation and inspection. Often, these values are not available (especially in the case of inspections), so analysts make estimates of the times needed to perform various events. In such instances, analysts must go to the work floor and take some time measurements. Methods analysts, more than anyone else, recognize that time is money; consequently, time information must be included on the operation process chart. A typical completed operation process chart appears in Figure 3-2.

FIGURE 3–2

Operation process chart illustrating manufacture of telephone stands

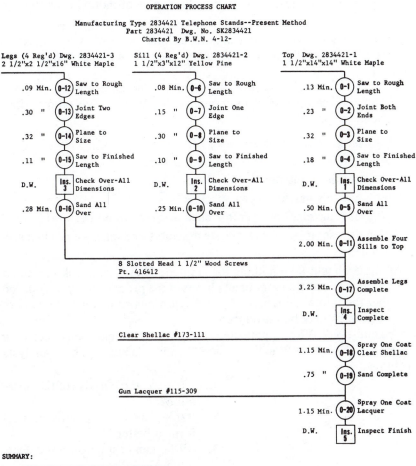

OPERATION PROCESS CHART

Manufacturing Type 2834421 Telephone Stands--Present Method
Part 2834421 Dwg. No. SK2834421
Charted By B.W.N. 4-12-

SUMMARY:

Event	Number	Time
Operations	20	17.58 minutes
Inspections	5	Day work

Using The Operation Process Chart

After completing an operation process chart, analysts should review each operation and inspection from the standpoint of the primary approaches to operation analysis (see Chapters 4 and 5). In particular, the following approaches apply when studying an operation process chart:

1. The purpose of the operation.
2. The design of the part.
3. Tolerances and specifications.
4. Materials.

5. The process of manufacture.   8. Material handling.
6. Setup and tools.   9. Plant layout.
7. Working conditions.   10. Principles of motion economy.

Analysts should adopt a questioning attitude on how each of these 10 criteria influences the time (cost), quality, and output of the product under study.

The most important question that analysts should ask when studying the events on the operation process chart is "Why?" Typical questions that should be asked are:

"Why is this operation necessary?"

"Why is this operation performed in this manner?"

"Why are these tolerances this close?"

"Why has this material been specified?"

"Why has this class of operator been assigned to do the work?"

"Would closer tolerances facilitate assembly and improve product reliability?"

Analysts should take nothing for granted. They should ask these and other pertinent questions about all phases of the process and then proceed to gather the information to answer the questions so that a better way of doing the work may be introduced.

The question "Why?" immediately suggests other questions, including "What?" "How?" "Who?" "Where?" and "When?" Thus, analysts might ask:

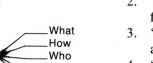

1. "*What* is the purpose of the operation?"
2. "*How* can the operation be performed better?"
3. "Who can best perform the operation?"
4. "*Where* could the operation be performed at a lower cost or improved quality?"
5. "*When* should the operation be performed to give the least amount of material handling?"

For example, in the operation process chart shown in Figure 3–2, analysts might ask the questions listed in Table 3–1 to determine the practicability of the methods improvements indicated. Answering these questions helps initiate the elimination, combining, and the simplification of operations.

By answering such questions, analysts become aware of other questions that may lead to improvement. Ideas seem to generate ideas, and

TABLE 3-1

Question	Method improvement
1. Can fixed lengths of 1½″ × 14″ white maple be purchased at no extra square footage cost?	Eliminate waste ends from lengths that are not multiples of 14″.
2. Can purchased maple boards be secured with edges smooth and parallel?	Eliminate jointing of ends (operation 2).
3. Can boards be purchased to thickness size and have at least one side planed smooth? If so, how much extra will this cost?	Eliminate planing to size.
4. Why cannot two boards be stacked and sawed into 14″ sections simultaneously?	Reduce time of 0.18 (operation 4).
5. What percentage of rejects do we have at the first inspection station?	If the percentage is low, perhaps this inspection can be eliminated.
6. Why should the top of the table be sanded all over?	Eliminate sanding of one side of top and reduce time (operation 5).
7. Can fixed lengths of 1½″ × 3″ yellow pine be purchased at no extra square footage cost?	Eliminate waste ends from lengths that are not multiples of 12″.
8. Can purchased yellow pine boards be secured with edges smooth and parallel?	Eliminate jointing of one edge.
9. Can sill boards be purchased to thickness size and have one side planed smooth? If so, how much extra will this cost?	Eliminate planing to size.
10. Why cannot two or more boards be stacked and sawed into 14″ sections simultaneously?	Reduce time of 0.10 (operation 9).
11. What percentage of rejects do we have at the first inspection of the sills?	If the percentage is low, perhaps this inspection can be eliminated.
12. Why is it necessary to sand the sills all over?	Eliminate some sanding and reduce time (operation 10).
13. Can fixed lengths of 2½″ × 2½″ white maple be purchased at no extra square footage cost?	Eliminate waste ends from lengths that are not multiples of 16″.
14. Can a smaller size than 2½″ × 2½″ be used?	Reduce material cost.
15. Can purchased white maple boards be secured with edges smooth and parallel?	Eliminate jointing of edges.
16. Can leg boards be purchased to thickness size and have sides planed smooth? If so, how much extra will this cost?	Eliminate planing to size.
17. Why cannot two or more boards be stacked and sawed into 14″ sections simultaneously?	Reduce time (operation 15).
18. What percentage of rejects do we have at the first inspection of the legs?	If the percentage is low, perhaps this inspection can be eliminated.
19. Why is it necessary to sand the legs all over?	Eliminate some sanding and reduce time (operation 16).
20. Could a fixture facilitate assembly of the sills to the top?	Reduce assembly time (operation 11).
21. Can a sampling inspection be used on the first inspection of the assembly?	Reduce inspection time (operation 4).
22. Is it necessary to sand after one coat of shellac?	Eliminate operation 19.

experienced analysts always arrive at several possibilities for improvement. Analysts must keep an open mind so that previous disappointments do not discourage the trial of new ideas.

The completed operation process chart helps analysts visualize the present method with all its details, so that new and better procedures may be devised. It shows analysts what effect a change on a given operation will have on the preceding and subsequent operations. The mere construction of the operation process chart inevitably suggests possibilities for improvement to alert analysts. It is not unusual to realize a 30 percent reduction in performance time utilizing the principles of operations analysis in conjunction with the operation process chart.

The operation process chart indicates the general flow of all components entering into a product, and since each step is shown in its proper chronological sequence, the chart is in itself an ideal plant layout. Consequently, methods analysts, plant layout engineers, and persons in related fields find this tool extremely helpful in making new layouts and in improving existing ones.

The operation process chart is an aid in promoting and explaining a proposed method. Since it gives so much information so clearly, it provides an ideal comparison between two competing solutions. This important tool:

1. Identifies all operations, inspections, materials, moves, storages, and delays involved in making a part or completing a process.
2. All events are shown in their particular sequence.
3. The chart clearly shows the relationship between parts and the complexity of fabrication of each part.
4. It distinguishes between produced and purchased parts.
5. It provides information as to the number of employees utilized and the time required to perform each operation and inspection.

THE FLOW PROCESS CHART

In general, the flow process chart contains considerably more detail than does the operation process chart. Consequently, it is not adapted to complicated assemblies as a whole. It is used primarily on one component of an assembly or a system at a time to effect maximum savings in manufacturing or in the procedures applicable to a particular component or sequence of work. The flow process chart is especially valuable in recording hidden costs, such as distances traveled, delays, and temporary storages. Once these nonproductive periods are highlighted, analysts can take steps for improvement.

In addition to recording operations and inspections, the flow process chart shows all the moves and delays in storage encountered by an item as it goes through the plant. Along with the operation and inspection symbols used in operation process charts, flow process charts use several other symbols. A small arrow signifies a transportation, which can be defined as the moving of an object from one place to another, except when the movement takes place during the normal course of an operation or an inspection. A large capital D indicates a delay. A delay occurs when a part is not permitted to be immediately processed at the next workstation. An equilateral triangle standing on its vertex signifies a storage, which occurs when a part is held and protected against unauthorized removal. When it becomes necessary to show a combined activity, such as one operator performing both an operation and an inspection at a workstation, then the identifying symbol is a square ⅜ inch on a side with a small circle ⅜ inch

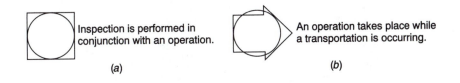

Inspection is performed in conjunction with an operation.

(a)

An operation takes place while a transportation is occurring.

(b)

in diameter inscribed within it. Figure 3–3 illustrates how one company uses process chart symbols to identify industrial activity.

Two types of flowcharts are in general use: product or material (see Figure 3–4) and operative or person (see Figure 3–5). While the product chart provides all details of the events that take place involving a product or a material, the operative flowchart details how a person performs an operation sequence.

Constructing The Flow Process Chart

Like the operation process chart, the flow process chart should be properly identified with a title appearing at the top: Flow Process Chart. The identifying information usually includes the part number, drawing number, process description, present or proposed method, date, and name of the person doing the charting.

Sometimes additional data is valuable to completely identify the job being charted. This may include the plant, building, or department; chart number; quantity; and cost information.

Since the flowchart represents only one item rather than an assembly, a neat-appearing chart can be constructed by starting on the top central section of the paper. First, draw a horizontal material line; over this write the part number and the description, as well as the material from which the part is processed. Second, draw a short vertical flow line (about ¼ inch) to the first event symbol, which may be an arrow indicating a transportation from the storeroom. Recorded just to the right of the transportation symbol is a brief description of the move, such as "moved to cutoff saw by material handler." Listed immediately under this is the type of

FIGURE 3–3
Examples of process chart symbols

OPERATION ⭕ A large circle indicates an operation, such as ➡	Drive nail	Mix	Drill hole
Paperwork operation to create a record or set of papers ➡	Type letter	Make out repair order	Initiate broken tool record
Paperwork operation to add information to a record ➡	Post piece count	Update balance of stores	Post production control schedule
TRANSPORTATION An arrow indicates a transportation, such as ➡	Move material by truck	Move material by conveyor	Move material by carrying (messenger)
STORAGE A triangle indicates a storage, such as ➡	Raw material in bulk storage	Finished stock stacked on pallets	Protective filing of documents
DELAY A large capital D indicates a delay, such as ➡	Wait for elevator	Material in truck or on floor at bench waiting to be processed	Papers waiting to be filed
INSPECTION A square indicates an inspection such as ➡	Examine material for quality or quantity	Read steam gauge on boiler	Examine printed form for information

material handling equipment used, if any. For example, "hand two-wheeled truck" or "gasoline-powered fork truck" would identify the equipment used. Recorded to the left of the symbol is the time required to perform the event, and about one inch still farther to the left, the distance in feet or meters moved.

FIGURE 3–4
Flow process chart (product)

SUBJECT CHARTED ___Shower head face___ CHART NO ___1128___

DRAWING NO ___BA-14782___ PART NO ___B-14782-2___ CHART OF METHOD ___Present___

CHART BEGINS ___Bar stock storage___ CHARTED BY ___E. Dunnick___

CHART ENDS ___Assembly Department Storeroom___ DATE ___9-7___ SHEET ___1___ OF ___2___

DIST IN FEET	UNIT TIME IN MIN	CHART SYMBOLS	PROCESS DESCRIPTION	DIST IN FEET	UNIT TIME IN MIN	CHART SYMBOLS	PROCESS DESCRIPTION
		1	In bar stock storage until requisitioned.		60	6	Waiting press operator.
20	.02	1	Bars loaded on truck upon rec'ipt of requisition.	100		5	To Bliss 74 1/2 press #16 by operator.
600	.05	1	Extruded rod to air saw #72.		.075	6	Pierce 6 holes.
15	.02	2	Bars removed from truck and stored on rack near machine.		120	7	Waiting drill press operator.
	120	1	Waiting for operation to begin.	50		6	To drill press by operator.
	.077	3	Saw slug on air saw.		.334	7	Rough ream and chamfer L. & G. drill press #19.
	30	2	Waiting for move--man.		30	8	Waiting drill press operator.
70	.03	2	Slugs to Nat. Maxi-press #8.	20		7	To Avery drill press #21 by operator.
	15	3	Awaiting forging operation.		.152	8	Drill three 13/64" holes Avery press #21.
	.234	1	Forge (3-man operation) and inspect.		20	9	Awaiting turret lathe operator.
	10	4	Waiting for press operator.	60		8	To turret lathe section by operator.
30		3	To press by operator.		.522	9	Turn stem and face #3 W. & S.
	.061	4	Trim Flash Bliss 74 1/2 press #16.		60	10	Awaiting turret lathe operator.
	30	5	Waiting for pickling operator.	30		9	To adjacent turret lathe operator.
100		4	To pickling tanks by operator.		.648	10	From outside diameter and face back.
	.007	5	Pickle (HCL tank).		15	11	Waiting press operator.

This charting procedure is continued by recording all operations, inspections, moves, delays, permanent storages, and temporary storages that occur during the processing of the part. All events are numbered chronologically for reference purposes, using a separate series for each event class.

FIGURE 3–4 *(concluded)*

SUBJECT CHARTED Shower head face ___ CHART NO. 1128
DRAWING NO. BA-14782 ___ PART NO. B-14782-2 ___ CHART OF METHOD Present
CHART BEGINS Bar stock storage ___ CHARTED BY E. Dupnick
CHART ENDS Assembly Department Storeroom ___ DATE 9-7 SHEET 2 OF 2

DIST. IN FEET	UNIT TIME IN MIN.	CHART SYMBOLS	PROCESS DESCRIPTION	DIST. IN FEET	UNIT TIME IN MIN.	CHART SYMBOLS	PROCESS DESCRIPTION
60		10	To Bliss 20B press by operator.				
	.097	11	Stamp identification Bliss 20B press #21.				
	15	12	Waiting next press operator.				
	.167	12	Broach six holes to size Bliss 20B press #22.				
	60	13	Waiting for move--man.				
350	.012	11	To inspection by move--man.				
	20	14	Waiting for inspector.				
	.05	1	Inspect complete (10% spot check).				
75		12	To storeroom by inspector.				
		2	Stored until requisitioned.				

SUMMARY

EVENT	NUMBER	TIME	DISTANCE
OPERATIONS	13	2.414 min.	
INSPECTIONS	1	.050 min.	
COMBINED ACTIVITY	1	.234 min.	
TRANSPORTATIONS	12		1580 ft.
STORAGES	2	undetermined	
DELAYS	14	605 min.	

The transportation symbol indicates the direction of the flow. Thus, when straight-line flow is taking place, the arrow points to the right side of the paper. When the process reverses or backtracks, the arrow points to the left. If a multifloor building is housing the process, an upward-pointing arrow indicates that the process is moving upward, and a downward-pointing arrow shows the flow of work descending.

FIGURE 3–5
Flow process chart (operative)

Field Inspection of LUX Reservoirs

OPERATION

| PAGE 1 OF 1 | PRESENT METHOD [X] | PROPOSED METHOD ☐ | DATE 4/17 |

| LOCATION Centre County, PA | | | | By Russ Ruhf | |

SUMMARY	OPERATION	OPERATION CREATE A RECORD	OPERATION ADD INFORMATION	TRANSPORTATION	STORAGE	DELAY	INSPECTION
TOTAL NO.	7	1	1	5		2	6
TOTAL DIST.				375 ft			
TOTAL TIME	8.30 min	1.00 min	0.40 min	4.35 min		14.0 min	5.15 min

EVENT	EVENT SYMBOL	TIME (MIN)	DIST (FT)	METHOD RECOMMENDATION
Leave vehicle, walk to front door, ring bell,	○ ⊙ ⊘ ⇨ ▽ D ☐	1.00	75	Call home in advance to reduce writing
wait, enter home.	○ ⊙ ⊘ ⇨ ▽ D ☐			delays.
Walk to field reservoir.	○ ⊙ ⊘ ⇨ ▽ D ☐	0.25	25	
Disconnect field reservoir from unit.	○ ⊙ ⊘ ⇨ ▽ D ☐	0.35		
Inspect for dents, cracks in shroud, cracked glass	○ ⊙ ⊘ ⇨ ▽ D ☐	1.25		This can be done while walking back
or missing hardware.	○ ⊙ ⊘ ⇨ ▽ D ☐			to vehicle.
Clean unit with approved cleaner and disinfectant.	○ ⊙ ⊘ ⇨ ▽ D ☐	2.25		This can be done more effectively at vehicle.
Return to vehicle with empty tank.	○ ⊙ ⊘ ⇨ ▽ D ☐	1.00	75	
Unlock vehicle, place empty tank in fixture,	○ ⊙ ⊘ ⇨ ▽ D ☐	1.75		
connect hardware.	○ ⊙ ⊘ ⇨ ▽ D ☐			
Open valve, begin fill.	○ ⊙ ⊘ ⇨ ▽ D ☐	0.25		
Wait for tank to fill.	○ ⊙ ⊘ ⇨ ▽ D ☐	12.00		Clean unit while being filled.
Check humidifier for proper function.	○ ⊙ ⊘ ⇨ ▽ D ☐	0.50		Eliminate—no need to do this twice.
Check pressure (indicator).	○ ⊙ ⊘ ⇨ ▽ D ☐	0.20		
Check reservoir contents (indicator).	○ ⊙ ⊘ ⇨ ▽ D ☐	0.20		
Return to patient with filled tank.	○ ⊙ ⊘ ⇨ ▽ D ☐	1.10	100	
Hook up filled tank.	○ ⊙ ⊘ ⇨ ▽ D ☐	1.00		
Check humidifier for proper function.	○ ⊙ ⊘ ⇨ ▽ D ☐	.75		
Wait for patient to remove nasal cannula	○ ⊙ ⊘ ⇨ ▽ D ☐	2.00		
or face mask.	○ ⊙ ⊘ ⇨ ▽ D ☐			
Install new nasal cannula or face mask.	○ ⊙ ⊘ ⇨ ▽ D ☐	2.50		
Check flows with patient.	○ ⊙ ⊘ ⇨ ▽ D ☐	2.25		
Affix a dated, initialed	○ ⊙ ⊘ ⇨ ▽ D ☐	1.00		Perform this while unit is being filled.
inspection sticker.	○ ⊙ ⊘ ⇨ ▽ D ☐			
Return to vehicle.	○ ⊙ ⊘ ⇨ ▽ D ☐	1.00	100	
Unlock vehicle.	○ ⊙ ⊘ ⇨ ▽ D ☐	.20		
Log the odometer reading on the gas	○ ⊙ ⊘ ⇨ ▽ D ☐	.40		Perform this while unit is being filled.
slips.	○ ⊙ ⊘ ⇨ ▽ D ☐			
	○ ⊙ ⊘ ⇨ ▽ D ☐			
	○ ⊙ ⊘ ⇨ ▽ D ☐			
	○ ⊙ ⊘ ⇨ ▽ D ☐			
	○ ⊙ ⊘ ⇨ ▽ D ☐			
	○ ⊙ ⊘ ⇨ ▽ D ☐			
	○ ⊙ ⊘ ⇨ ▽ D ☐			

To determine the distance moved, it is not necessary to measure each move accurately with a tape or a six-foot rule. Usually a sufficiently correct figure results by counting the number of columns that the material is moved past and then multiplying this number, less one, by the span. Moves of five feet or less are usually not recorded; however, they may be if analysts feel that they materially affect the overall cost of the method being plotted.

All delay and storage times must be included on the chart. It is not sufficient to indicate that a delay or storage takes place. Since the longer a part stays in storage or is delayed, the more cost it accumulates and the longer it takes for the customer to receive delivery, it is important to know how much time it spends at each delay or storage.

The most economical method of determining the duration of delays and storages is to mark several parts with chalk indicating the exact time they went into a storage or were delayed. Then the section is checked periodically to see when the parts marked are brought back into production. By taking a number of cases and recording the elapsed time and then averaging the results, analysts obtain sufficiently accurate time values.

Using The Flow Process Chart

The flow process chart, like the operation process chart, is not an end in itself, but merely a means to an end. This tool of analysis facilitates the elimination and/or reduction of hidden costs of a component. Since the flow chart clearly shows all transportations, delays, and storages, it provides the information that can lead to both the reduction of the quantity and the duration of these elements.

Once analysts have constructed the flow process chart, they should use the questioning approach based on the considerations primary to operation analysis. With the flow process chart, special consideration is given to:

1. Material handling.
2. Plant layout.
3. Delay time.
4. Storage time.

In all probability, analysts will already have constructed and analyzed an operation process chart of the assembly of which the part under study in the flowchart is a component. Flowcharts usually depict the components of the particular assembly thought to contain hidden costs. If the operations and inspections performed on the component have already been studied, then analysts would not spend a great deal of time restudying them when analyzing the flowchart. They would be more concerned with studying the distance parts must be moved from operation to operation, the material handling, and the delays that occur. Of course, if analysts constructed the flow process chart initially, then all of the primary approaches to operation analysis should be used for studying the events shown on it.

To eliminate or minimize delay and storage time to improve deliveries to customers, and to reduce cost, analysts should consider these check questions in studying the job:

1. How often does the full amount of material fail to be delivered to the operation?
2. What can be done to schedule materials so that they come in more even quantities?
3. What is the most efficient batch or lot size, or manufacturing quantity?
4. How can schedules be rearranged to provide longer runs?
5. What is the best sequence for scheduling orders to allow for the type of operation, tools required, colors, and so on?
6. What can be done to group similar operations so that they are performed at the same time?
7. How much downtime and overtime can be saved by improved scheduling?
8. What is the cause of emergency maintenance and rush orders?
9. How much delay and storage time can be saved by running certain products on certain days to make schedules more regular?
10. What alternate schedules can be developed to use materials most efficiently?
11. Would it be worthwhile to accumulate pickups, deliveries, and shipments?
12. What is the proper department to do the job so that it will be done with the same class of work and save a move, delay, or storage?
13. How much would be saved by doing the job on another shift? At another plant?
14. What is the best and most economical time to run tests and experiments?
15. What information is lacking on orders issued to the factory that may cause a delay or storage?
16. How much time is lost by changing shifts at different hours in related departments?
17. What are the frequent interruptions to the job, and how should they be eliminated?
18. How much time do employees lose by waiting for or failing to receive the proper instructions, drawings, and specifications?
19. How many holdups are caused by congested aisles?
20. What improvements can be made in locating doorways and aisles and in making aisles that reduce delays?

Specific check questions that analysts should use to shorten the distances traveled and reduce material handling time include the following:

1. Is product group technology being practiced to reduce the number of setups and to allow for larger production runs? (Product group technology is the classification of different products into similar geometric configurations and sizes to take advantage of the economics in manufacture brought about by large quantities.)
2. Can a facility be economically relocated to reduce the distances traveled?
3. What can be done to reduce handling of materials?
4. What is the correct equipment for handling materials?
5. How much time is lost in getting materials to and from the workstation?
6. Should product grouping be considered rather than process grouping?
7. How can the size of the unit of material handled be increased to reduce handling, scrap, and downtime?
8. How can elevator service be improved?
9. What can be done about runways and roadways to speed up transportation?
10. What is the proper position in which to place material to reduce handling by the operator?
11. What use can be made of gravity delivery?

A study of the completed flowchart (Figure 3–4) will familiarize analysts with all the pertinent details related to the direct ar. 1 indirect costs of a manufacturing process so that they can analyze these costs for improvement. Unless all the facts relating to a method are known, it is difficult to improve that method. Casual inspection of an operation does not provide the information needed to do a thorough job of methods improvement. Since distances are recorded on the flow process chart, the chart is exceptionally valuable in showing how the layout of a plant can be improved. Intelligent use of the flow process chart results in improvements.

THE FLOW DIAGRAM

Although the flow process chart gives most of the pertinent information relative to a manufacturing process, it does not show a pictorial plan of the flow of work. Sometimes this added information is helpful in developing a new method. For example, before a transportation can be shortened, it is necessary to see or visualize where room can be provided to add a facility so that the transportation distance can be diminished. Likewise, it is helpful to visualize potential temporary and permanent storage areas, inspection stations, and work points. The best way to provide this information is to take an existing drawing of the plant areas involved, and then

FIGURE 3–6

Flow diagram of the old layout of a group of operations on the Garand rifle.
(Shaded section of plant represents the total floor space needed for the revised
layout [Figure 3–7]. This represented a 40 percent savings in floor space.)

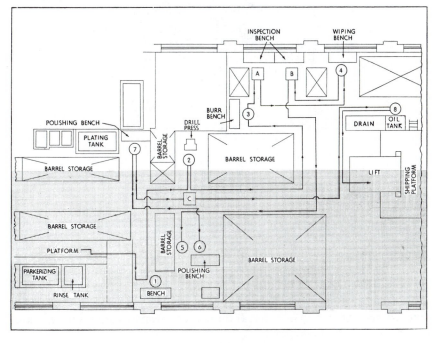

to sketch in the flow lines indicating the movement of the material from
one activity to the next. A pictorial representation of the layout of floors
and buildings showing the location of all activities on the flow process
chart is a flow diagram.

When constructing a flow diagram, analysts should identify each activity by symbols and numbers corresponding to those appearing on the flow
process chart. The directon of flow is indicated by placing small arrows
periodically along the flow lines. Different colors can indicate flow lines
for more than one part.

Figure 3–6 illustrates a flow diagram made in conjunction with a flow
process chart to improve the production of the Garand (M1) rifle at
Springfield Armory. This pictorial representation together with the flow
process chart resulted in savings that increased production from 500 rifle
barrels per shift to 3,600—with the same number of employees. Figure
3–7 illustrates the flow diagram of the revised layout.

The flow diagram is a helpful supplement to the flow process chart
because it indicates backtracking and areas of possible traffic congestion,
and facilitates making an ideal plant layout.

FIGURE 3–7
Flow diagram of the revised layout of a group of operations on the Garand rifle

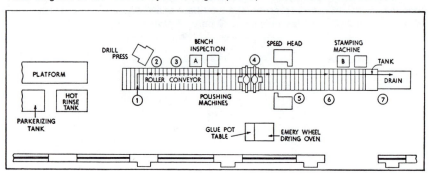

PERT CHARTING

PERT charting is a prognostic planning and control method that graphically portrays the optimum way to attain some predetermined objective, generally in terms of time. Usually methods analysts use PERT charting to improve scheduling from the standpoint of cost reduction and/or customer satisfaction.

In using PERT for scheduling, analysts generally provide two or three time estimates for each activity. If three time estimates are used, they are based on the following questions:

1. How much time is required to complete this activity if everything works out ideally (optimistic estimate)?
2. Under average conditions, what would be the most likely duration of this activity?
3. What is the time required to complete this activity if almost everything goes wrong (pessimistic estimate)?

With these estimates, a probability distribution of the time required to perform the activity can be made. See Table 3–2.

On the PERT chart, events—represented by nodes—are positions in time showing the start and/or completion of a particular operation or group of operations. Each operation or group of operations in a department is referred to as an activity and is identified as an arc on the PERT chart. Each arc has attached to it a number representing the time (days, weeks, months) needed to complete the activity. Activities that utilize no time or cost, yet are necessary to maintain a correct sequence, are referred to as dummy activities and are plotted as dotted lines.

The minimum time needed to complete the entire project would correspond to the longest path from the initial node to the final node. In Figure 3–8 the minimum time needed to complete the project would be the longest path from node 1 to node 12. The longest path is the critical path since

TABLE 3–2
Cost and time values to perform a variety of activities under
normal and emergency conditions

Activities	Normal		Emergency	
	Weeks	*Dollars*	*Weeks*	*Dollars*
A	4	4,000	2	6,000
B	2	1,200	1	2,500
C	3	3,600	2	4,800
D	1	1,000	0.5	1,800
E	5	6,000	3	8,000
F	4	3,200	3	5,000
G	3	3,000	2	5,000
H	0	0	0	0
I	6	7,200	4	8,400
J	2	1,600	1	2,000
K	5	3,000	3	4,000
L	3	3,000	2	4,000
M	4	1,600	3	2,000
N	1	700	1	700
O	4	4,400	2	6,000
P	2	1,600	1	2,400

FIGURE 3–8
Network showing critical path (heavy line). Code numbers within nodes signify
events. Connecting lines with directional arrows indicate operations that are
dependent on prerequisite operations. Time values on the connecting lines
represent normal duration in weeks. Hexagonals associated with events show the
earliest event time. Dotted circles associated with events show the latest event time.

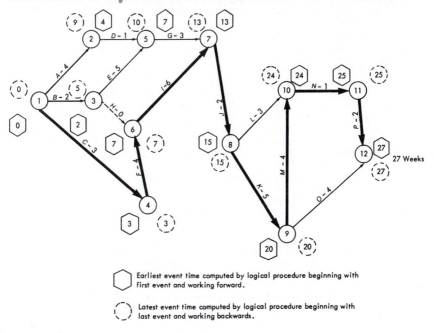

Earliest event time computed by logical procedure beginning with
first event and working forward.

Latest event time computed by logical procedure beginning with
last event and working backwards.

this path establishes the minimum project time. There is always at least one such path through any project. However, more than one path can reflect the minimum time needed. This is the meaning behind the concept of critical paths.

Activities not on the critical path have a certain time flexibility. This time flexibility, or freedom, is referred to as "float." The amount of float is computed by subtracting the normal time from the time available. Thus, the float is the amount of time that a noncritical activity can be lengthened without delaying the project's completion date.

Figure 3–8 illustrates an elementary network portraying the critical path. This path, identified by a heavy line, would involve a duration of 27 weeks. Several methods can be used to shorten the project's duration. The cost of the various alternatives can be estimated. For example, assume that the following cost table has been developed and that a linear relation exists between the time and the cost per week. The cost of various time alternatives can be readily computed:

27-week schedule–normal duration of project . . . .   cost = $22,500
26-week schedule–the least expensive way to gain one week would be
to reduce activity M or J by one week for an additional cost of $400 .   cost = $22,900
25-week schedule–the least expensive way to gain two weeks would be
to reduce activities M and J by one week each for an additional cost
of $800 . . . .   cost = $23,300
24-week schedule–the least expensive way to gain three weeks would
be to reduce activities M, J, and K by one week each for an
additional cost of $1,300. . . . .   cost = $23,800
23-week schedule–the least expensive way to gain four weeks would be
to reduce activities M and J by one week each and activity K by two
weeks for an additional cost of $1,800 . . . .   cost = $24,300
22-week schedule–the least expensive way to gain five weeks would be
to reduce activities M and J by one week each, activity K by two
weeks, and activity I by one week for an additional cost of $2,400 . . .   cost = $24,900
21-week schedule–the least expensive way to gain six weeks would be
to reduce activities M and J by one week each and activities K and I
by two weeks each for an additional cost of $3,000. . . . .   cost = $25,500
20-week schedule–the least expensive way to gain seven weeks would
be to reduce activities M, J, and P by one week each and activities
K and I by two weeks each for an additional cost of $3,800 . . . .   cost = $26,300
19-week schedule–the least expensive way to gain eight weeks would
be to reduce activiites M, J, P, and C by one week each and
activities K and I by two weeks each for an additional cost of $5,000   cost = $27,500
(Note that a second critical path is now developed through nodes 1,
3, 5, and 7.)
18-week schedule–the least expensive way to gain nine weeks would
be to reduce activities M, J, P, C, E, and F by one week each and
activities K and I by two weeks each for an additional cost of $7,800   cost = $30,300
(Note that by shortening the time to 18 weeks, we develop a second
critical path.)

SUMMARY

Operation and flow process charts, the flow diagram, and PERT are valuable tools for presenting and solving problems. Just as several types of tools are available for a particular job, so several designs of charts can help solve an engineering problem. However, in determining a specific solution, one chart usually has advantages over another. Analysts should understand the specific functions of each process chart and choose the correct one to solve a specific problem. In summary, their functions are as follows:

1. OPERATION PROCESS CHART. Used to analyze relations between operations. Good for studying operations and inspections on assemblies involving several components. Helpful for plant layout work.

2. FLOW PROCESS CHART. Used to analyze hidden or indirect costs, such as delay time, storage costs, and material handling costs. Best chart for complete analysis of the manufacture of one component part.

3. FLOW DIAGRAM. Used as a supplement to the flow process chart, especially where considerable floor space is involved in the process. Indicates backtracking and traffic congestion. Necessary tool in making revised plant layouts.

4. PERT CHART. Used as a project scheduling tool. Especially desirable for use in major projects involving six months or more.

The operation and flow process charts, the flow diagram, and PERT all have their place in developing improvements. Their correct use can aid in presenting the problem, solving the problem, selling the solution, and installing the solution. These charts can be valuable descriptive and communicative aids for understanding a process and its related activities. Thus, they are effective in presenting improved methods to management, training employees in the prescribed method, and focusing pertinent details in conjunction with plant layout work.

TEXT QUESTIONS

1. Explain how you would show material handling at the workstation when constructing a flow process chart of a complex machining operation done at a work center.
2. Who uses process charts?
3. What does the operation process chart show?
4. What symbols are used in constructing the operation process chart?
5. How are materials introduced into the general flow when constructing the operation process chart?
6. How does the flow process chart differ from the operation process chart?
7. What are the two types of flowcharts?
8. What is the principal purpose of the flow process chart?
9. What symbols are used in constructing the flow process chart?

10. When would you advocate using the flow diagram?
11. How can the flow of several different products be shown on the flow diagram?
12. Why are the operation and flow process charts merely a means to an end?
13. In the construction of the flow process chart, what method can be used to estimate distances moved?
14. How can delay times be determined in the construction of the flow process chart? Storage times?
15. Explain how the concept of PERT charting can save a company money.
16. What two flowchart symbols are used exclusively in the study of paper work?
17. Explain how disassembly operations are shown on the operation process chart.

GENERAL QUESTIONS

1. What are the limitations of the operation and flow process charts and of the flow diagram?
2. What relation is there between the flowchart and material handling? Between the flow diagram and plant bottlenecks? Between the operation process chart and material specifications?
3. What is the connection between effective plant layout and the operation process chart?
4. Distinguish between process and product grouping.
5. What is meant by group technology?
6. Why is the flow process chart a valuable tool in training new employees?
7. Explain how you would obtain "optimistic" and "pessimistic" times to be used in connection with a PERT chart that you are developing.
8. Why is it necessary to construct process charts from direct observation as opposed to information obtained from the foreman?

PROBLEM

1. Based on the following "emergency" cost table, what would be the minimum time to complete the project described by Figure 3–8, whose normal costs are shown in Table 3–2? What would be the added cost to complete the project within this time period?

	Emergency Schedule	
	Weeks	Dollars
A	2	$7,000
B	1	$2,500
C	2	$5,000
D	0.5	$2,000
E	4	$6,000
F	3	$5,000
G	2	$6,000
H	0	0
I	4	$7,600
J	1	$2,200
K	4	$4,500
L	2	$2,200
M	3	$3,000
N	1	$1,000
O	2	$6,000
P	1	$3,000

SELECTED REFERENCES

Apple, James M. and Rickles, Harvey V. "Material Handling and Storage." In *Production Handbook* edited by John A. White. New York: John Wiley & Sons, 1987.

Buffa, Elwood S. *Operations Management.* 6th ed. New York: John Wiley & Sons, 1980.

Francis, Richard L., and John A. White. *Facility Layout and Location: An Analytical Approach.* Englewood Cliffs, N.J.: Prentice-Hall, 1974.

Kadota, Takeja. "Charting Techniques." In *Handbook of Industrial Engineering,* ed. Gavriel Salvendy. New York: John Wiley & Sons, 1982.

Konz, Stephan. *Work Design.* Columbus, Ohio: Grid, 1983.

Mundel, M. E. *Motion and Time Study,* 5th ed. Englewood Cliffs: Prentice Hall, 1978.

Operation Analysis

Methods analysts use operation analysis to study all productive and nonproductive elements of an operation to improve them. Methods engineering in itself is concerned with devising methods to increase productivity per unit of time and reduce unit costs while maintaining or improving quality.

Operation analysis is just as effective in planning new work centers as it is in improving those already in operation. By using the questioning approach on all facets of the workstation, dependent workstations, and the design of the product, analysts can develop an efficient work center.

Since the improvement of existing operations is a continuing process in industry, this chapter deals primarily with that process. Naturally these principles are equally valid and important in planning new work centers. Investigating the approaches to operation analysis is the step immediately following the presentation of facts in the operation and/or flow process chart. This is the step in which analysis takes place and the various components of the proposed method crystallize.

With foreign competition becoming increasingly keen at the same time that labor and material costs are spiraling, operation analysis has become increasingly important. This procedure is never complete. Competition (both national and international) usually necessitates the continuing study of a given product so that the manufacturing processes and methods can be improved and a part of the gains passed on to the consumer in the form of a better product at a reduced selling price. Once this is done by a given plant, competitors invariably introduce similar improvement programs, and it is only a matter of time until they have produced a more salable product at a reduced price. This starts a new cycle in which the given plant reviews its operations and improves its manufacturing processes and methods again, necessitating improvements in competing plants.

Conditions in industry cannot be static; otherwise, bankruptcy would result. Progress in productivity and quality improvement is the only key to continued profitable operation.

Another basic economic law, that of supply and demand, must be considered when savings are effected. The volume of goods consumed is inversely proportional to the selling price. As improvements that inaugurate real savings are developed, the market is broadened through lower selling prices. This has been proved time and time again. It happened with electric lighting, refrigerators, radios, automobiles, television, VCRs, motorboats, hand-held calculators, computers, and many other products. As progress allowed reduction in costs, more people could afford to purchase these items. The increased volume permitted further economies, which again resulted in lower prices, and lower prices further increased the volume. Each time the selling price of any commodity is reduced by as little as 5 percent, the product is immediately brought within reach of more people's pocketbooks.

Experience has proved that practically all operations can be improved if sufficient study is given to them. Since the procedure of systematic analysis is equally effective in large and small industries, in job shops, and mass production, operation analysis is applicable in all areas of manufacturing, business, and government. When properly utilized, it develops a better method of doing the work by simplifying operational procedures and material handling, and by utilizing equipment more effectively. Thus, firms are able to increase output and reduce unit cost; to ensure quality and reduce defective workmanship; and to facilitate operator enthusiasm by improving working conditions, minimizing operator fatigue, and permitting higher operator earnings.

INTRODUCTION TO OPERATION ANALYSIS

Probably one of the most common attitudes of management is that its problems are unique. Consequently, it feels that any new method will be impractical for it. Actually, all work, whether clerical, maintenance, office, machine, assembly, or general labor, is much the same. The Gilbreths concluded that any work, whether productive or nonproductive, was done by using combinations of 17 basic motions that they called "therbligs." These basic movements along with their letter, graphic, and color designations are shown in Table 7–1. Regardless of what the operation is, its basic divisions are quite similar to others. For example, the elements of work in driving a car are much like those required to operate a turret lathe; the basic motions employed in dealing a bridge hand are almost identical with certain manual inspection and machine loading elements. The fact that all work is similar in many respects verifies the principle that if methods can be improved in one plant, opportunities exist for methods improvements in all plants.

   History repeats itself over and over in accounts of the unwillingness of people to accept something new. Iron ships, steamboats, automobiles, airplanes, radios, telephones, television, and computers are just a few examples of products that are commonplace today; yet at one time all of them were considered impractical by most people. Today, the average person still scoffs at the ideas of interplanetary travel and of atomic power plants for private vehicles.

   Anyone engaged in methods work is well aware of the natural, inherent resistance to change prevailing in people's minds, regardless of their level in an organization. Overcoming this resistance to change is one of the major obstacles in the path of an improvement program. Managers continually make statements to the effect that "it might have worked at the _____ plant, but our operation is quite different." Supervisors say, "It can't work here," and even the operator bluntly declares, "It won't work." Selling whole organizations on "It will work here" is a never-ending job for methods analysts. A person who is successful in this type of work never accepts anything as being right just because it exists today or has been done this way for years. Instead, analysts question, probe, investigate, and finally, after all angles have been considered, make a decision for the moment. These persons are always conscious that this method may be satisfactory today, but that it will not be tomorrow, for there is always a better way.

   To reduce the resistance to change characteristic of all people, methods analysts endeavor to establish an atmosphere of participation, understanding, and friendliness. They recognize the knowledge other people have of their jobs and solicit their assistance in making improvements. The analysts see to it that all communication channels are kept open and are utilized, so that all personnel affected by contemplated changes are kept informed. The analysts provide information freely, and openly discuss facts related to the investigation; thus, they obtain confidence rather than mistrust and suspicion. Above all else, they maintain an enthusiastic attitude toward improvement.

   In a recent study, the following factors were established as being the principal reasons that either stopped or slowed down continuous improvement activities.[1]

1. Lack of awareness of the program by all employees.
2. Lack of understanding of why and how it is being done.
3. Insufficient or ineffective training.
4. Inadequate planning before launching the program.
5. Lack of cooperation among functional areas.
6. Lack of coordination among functional areas by teams.

---

[1] William Winchell, "Avoiding Failure," *Continuous Quality Improvement,* (Dearborn, Michigan: Society of Manufacturing Engineers, 1981).

7. Resistance to change by middle management.
8. Lack of appropriate rewards or incentives.
9. Ineffective leadership skills for the changing culture.

Operation Analysis Procedure

As outlined in Chapter 1, the use of a systematic procedure is invaluable in producing real savings. The first step is to get all information related to the anticipated volume of the work. To determine how much time and effort should be devoted to improving the present method or planning the new job, analysts evaluate the expected volume, chance of repeat business, life of the job, chance for design changes, and labor content of the job. If the job promises to be quite active, a more detailed study is justified.

Once an estimate is made of the quantity, job life, and labor content, operations analysts then collect all factual manufacturing information. This information would include all operations, facilities used to perform the operations, and operational times; all moves or transportations, facilities used for transportation, and transportation distances; all inspections, inspection facilities, and inspection times; all storages, storage facilities, and time spent in storage; all vendor operations together with their prices; all drawings and quality and design specifications. After all this information affecting quality and cost is gathered, it must be presented in a form suitable for study. One of the most effective ways of doing this is through the flow process chart (see Chapter 3). This chart graphically presents all manufacturing information, much as an engineering drawing shows all design information.

Upon completing the flow process chart, analysts review the problem with thought toward improvement. Up to this point, they have merely stated the problem, as a necessary preliminary to solving it. One of the most common techniques used by methods people is to prepare a check sheet for asking questions about every activity shown on the flowchart. Typical questions are: Is this operation necessary? Can the operation be performed better another way? Can it be combined with some other operation? Are tolerances correct? Are they perhaps closer than necessary? Or perhaps they should be held closer? Can a more economical material be used? Can better material handling be incorporated?

Figure 4-1 is a typical checklist of pertinent questions. The figure also illustrates how this form was used to make a cost reduction study on an electric blanket control knob shaft. By redesigning the shaft so that it could be economically produced as a die casting rather than a screw machine part, factory cost was reduced from $68.75 per thousand pieces to $17.19 per thousand pieces.

Analysts ask and answer all related questions on the check sheet for all

FIGURE 4-1

Date ___9/15___   Dept. ___11___   Dwg. ___18-4612___   Sub. ___2___

Mould _____   Die _____   Style _____   Item ___2___

Pattern _____   Ins. Spec. ___C___   L. Spec. _____   Sub. _____

Part Description ___Blanket control knob shaft___

Operation ___Turn, groove, drill, tap, knurl, thread, cut-off___ Operator ___Blazer___

DETERMINE AND DESCRIBE

1. PURPOSE OF OPERATION

To form contours of 3/8" S.A.E. 1112 rod on automatic screw machine
to achieve drawing specifications.

2. COMPLETE LIST OF ALL OPERATIONS PERFORMED ON PART

No.	Description	Work Sta.	Dept.
1.	Turn, groove, drill, tap, knurl, thread, cut-off	B. & S.	11
2.	Burr	Bench	12
3.	Inspect 1%	Bench	18
4.			
5.			
6.			
7.			
8.			
9.			
10.			

3. INSPECTION REQUIREMENTS

a—Of previous oprn.

b—Of this oprn.   Yes.   Perhaps S.Q.C. will reduce amount of inspection.

DETAILS OF ANALYSIS

Can purpose be accomplished better otherwise? *Yes - by die casting*

Can oprn. being analyzed be eliminated? *No*

be combined with another? *No*

be performed during idle period of another? *Yes, by machine coupling.*

Is sequence of oprns. best possible? *Yes*

Should oprn. be done in another dept. to save cost or handling? *Perhaps can be purchased outside at a savings.*

Are tolerance, allowance, finish and other requirements necessary?

c—Of next oprn.

4. MATERIAL   Zinc base die cast metal would be less expensive.

Cutting compounds and other supply materials

5. MATERIAL HANDLING
a—Brought by 4 wheel truck to automatics
b—Removed by hand 2 wheel trucks
c—Handled at work station by

6. SET-UP (Accompany description with sketches if necessary)

This is satisfactory as being done.

a—**Tool Equipment**
Present

Suggestions   Redesign part to be made as zinc base die casting rather than S.A.E. 1112 screw machine part.

---

too costly?

suitable to purpose?

Consider size, suitability, straightness, and condition.

Can cheaper material be substituted?

Should crane, gravity conveyors, totepans, or special trucks be used?

Consider layout with respect to distance moved. *perhaps gravity to burring station.*

How are dwgs. and tools secured?
Can set-up be improved?
Trial pieces.
Machine Adjustments.

Tools

Suitable?
Provided?
Ratchet Tools
Power Tools
Spl. Purpose Tools
Jigs, Vises
Special Clamps
Fixtures
Multiple
Duplicate

FIGURE 4–1 (concluded)

7. CONSIDER THE FOLLOWING POSSIBILITIES.

1. Install gravity delivery chutes.
2. Use drop delivery
3. Compare methods if more than one operator is working on same job.
4. Provide correct chair for operator.
5. Improve jigs or fixtures by providing ejectors, quick-acting clamps, etc.
6. Use foot operated mechanisms.
7. Arrange for two handed operation.
8. Arrange tools and parts within normal working area.
9. Change layout to eliminate back tracking and to permit coupling of machines.
10. Utilize all improvements developed for other jobs.

8. WORKING CONDITIONS.

Generally satisfactory.

a—Other Conditions

9. METHOD (Accompany with sketches or Process Charts if necessary.)

a—Before Analysis and Motion Study.

RECOMMENDED ACTION

*yes, to accumulate
for tumbling.*

Light *o.k.*
Heat *o.k.*
Ventilation, Fumes *o.k.*
Drinking Fountains *o.k.*
Wash Rooms *o.k.*
Safety Aspects *o.k.*
Design of Part *o.k.*
Clerical Work Required (to fill
out time cards, etc.) *o.k.*
Probability of Delays *o.k.*
Probable Mfg. Quantities *o.k.*

Arrangement of Work Area

Placement of

Tools.

Materials.

Supplies.

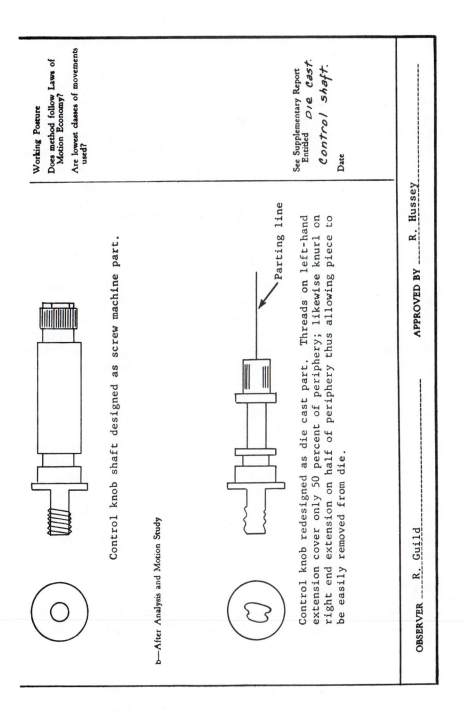

Control knob shaft designed as screw machine part.

b—After Analysis and Motion Study

Parting line

Control knob redesigned as die cast part. Threads on left-hand extension cover only 50 percent of periphery; likewise knurl on right end extension on half of periphery thus allowing piece to be easily removed from die.

Working Posture

Does method follow Laws of Motion Economy?

Are lowest classes of movements used?

See Supplementary Report Entitled *Die Cast Control shaft.*

Date

OBSERVER ___ R. Guild ___   APPROVED BY ___ R. Hussey ___

57

steps appearing on the flowchart. This procedure invariably leads to efficient ways of performing the work. As ideas develop, methods people record them immediately so that they will not be forgotten. And, they include sketches at this time. Usually analysts are surprised at the numerous inefficiencies that prevail and have little trouble in compiling many improvement possibilities. One improvement usually leads to another. Analysts must have open minds and creative ability to be successful in this type of work. The check sheet is also useful in giving methods training to factory foremen and superintendents. Thought-provoking questions, when intelligently used, help factory supervisors to develop constructive ideas. The check sheet serves as an outline, which can be referred to by the discussion leader handling the methods training.

TEN PRIMARY APPROACHES TO OPERATION ANALYSIS

When the 10 primary approaches to operation analysis are used in studying each individual operation, attention focuses on the items most likely to produce improvement (see pages 31 and 32). All these approaches are not applicable to each activity appearing on the flowchart, but usually more than one should be considered. The recommended method of analysis is to take each step in the present method individually and to analyze it with a specific approach toward improvement clearly in mind, considering all the key points of analysis. Then follow the same procedure on the succeeding operations, inspections, moves, storages, and so on, shown on the chart. After each element has been thus analyzed, consider the product being studied as a whole rather than in light of its elemental components, and reconsider all points of analysis looking for overall improvement possibilities. There are usually unlimited opportunities for methods improvement in every plant. To develop maximum savings, carefully study individual and collective operations as outlined. Wherever this procedure has been followed by competent engineers, beneficial results have always been realized. Each of the 10 primary approaches to operation analysis is discussed in detail.

Purpose of the Operation

Probably this is the most important of the 10 points of operation analysis used to improve an existing method or to plan a new job. An analyst's cardinal rule is to try to eliminate or combine an operation before improving it. In my experience, as much as 25 percent of the operations being performed by American industry can be eliminated if sufficient study is given to the design and process.

Far too much unnecessary work is done today. In many instances, the job or the process should not be simplified or improved, but eliminated entirely. Eliminating a job saves money on installing an improved method. No interruption or delay is caused while an improved method is being developed, tested, and installed. It is not necessary to train new operators for the new method. The problem of resistance to change is minimized when an unnecessary job or activity is eliminated. In connection with paperwork, before a form is developed for information transfer, analysts should always ask: "Is the form really needed?" The advance of today's computer-controlled systems should reduce the generation of forms and paperwork. The best way to simplify an operation is to devise some way to get the same or better results at no cost at all.

Unnecessary operations frequently result from improper planning when the job was first set up. Once a standard routine is established, it is difficult to make a change, even if such a change would eliminate a portion of the work and make the job easier. When new jobs are planned, the planner usually includes an extra operation if there is any possibility that the product would be rejected without the extra work. For example, if there is some question of whether to take two or three cuts in turning a steel shaft to maintain a 40-microinch finish, invariably the planner specifies three cuts, even though the proper maintenance of cutting tools, supplemented by ideal feeds and speeds, would allow the job to be done with two cuts. Likewise, if there is some question of the ability of a drill press to hold a 0.005-inch tolerance on a ¼-inch drilled hole, the planner calls for a reaming operation. Actually, the drilling operation would be adequate if all variable factors (speed, feed, cutting lubricant, drill size) were controlled.

Many times, an unnecessary operation may develop because of the improper performance of a previous operation. A second operation must be done to touch up or make acceptable the work done by the first operation. In one plant, for example, armatures were previously spray painted in a fixture, making it impossible to cover the bottom of the armature with paint because the fixture shielded the bottom from the spray blast. It was necessary to touch up the armature bottoms after spray painting. A study of the job resulted in a redesigned fixture that held the armature and still allowed complete coverage. The new fixture permitted seven armatures to be spray painted simultaneously, while the old method called for spray painting one at a time. Thus, by considering that an unnecessary operation may develop because of the improper performance of a previous operation it was possible to eliminate the touch-up operation (see Figure 4–2).

In manufacturing large gears, it was necessary to introduce a hand-scraping and lapping operation to remove waves in the teeth after they had been hobbed. An investigation disclosed that contraction and expansion brought about by temperature changes in the course of the day were responsible for the waviness in the teeth surface. By enclosing the whole unit and installing an air-conditioning system within the enclosure, proper temperature was maintained during the whole day. The waviness disappeared immediately, and it was no longer necessary to continue the hand-scraping and lapping operations.

FIGURE 4–2
A. Painted armature as removed from old fixture
and as removed from improved fixture

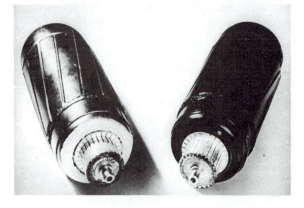

B. Armature in spray-painting fixture allowing complete
coverage of the armature bottom

Some unnecessary operations are introduced to facilitate an operation that follows. When wiring commutators, for example, each pair of wires was twisted to keep the correct pair together and to increase the strength of the wires. However, the correct pair of wires could be placed in the proper slot without twisting. Also, observation revealed that twisting caused the wires to be unequal in length, and that the uneven tension in the twisted wires weakened rather than strengthened them. This was discovered when someone asked: "Can a change in assembly eliminate the need for a previous operation?" (see Figure 4–3).

FIGURE 4–3
Twisting wires during commutator winding operation was found to be unnecessary
because the correct pair of wires could be placed in the proper slot without twisting

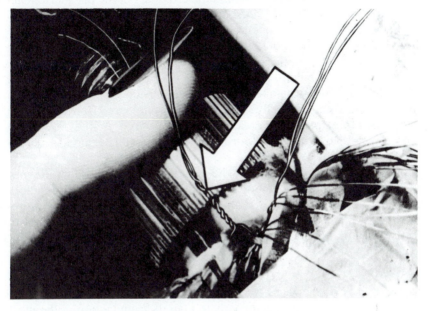

In endeavoring to eliminate operations, analysts should consider the
question: Is an additional operation justified by savings that it will effect
in a subsequent operation? For example, a brush holder was originally
planned so that two holes were drilled and tapped in each holder. One
hole was drilled and tapped in the bottom; another was drilled and tapped
in the top. The brush holder was assembled so that a holder with a tapped
hole in the bottom and a holder with a tapped hole in the top were used
alternately. Since all parts had tapped holes in both the bottom and the
top, it was merely a matter of positioning the parts for assembly. By
questioning the need for having both tapped holes in each piece, it was
discovered to be more economical to have two parts—one with the hole
tapped in the top, the other with the hole tapped in the bottom. The two
parts were then assembled alternately. One drilled and tapped hole was
eliminated from each brush holder because someone questioned the necessity of performing this operation.

Also, an unnecessary operation may develop to give the product
greater sales appeal. One company originally used cast brass nameplates
for various lines of products. Although the brass nameplate was attractive, the appearance could be kept attractive and the cost notably reduced
by using an etched steel nameplate instead.

To produce a smooth joint in electric fan guards, it was thought necessary to perform a coining operation where the ends of the wire cage were joined by butt welding. By slightly modifying the design, the joined wire ends were positioned directly behind the crossbars of the fan guard, making it impossible to see the joined ends. Since the butt-welded ends could not be seen, the coining operation was no longer necessary. This design made it difficult to reach the butt-welded portion, so there was no chance of scratching hands on the rough welded area.

To eliminate, combine, or shorten each operation, analysts should ask and answer the following question: "Can an outside supplier perform the operation more economically?" Ball bearings purchased from an outside vendor had to be packed in grease prior to assembly. A study of bearing vendors revealed that "sealed-for-life" bearings could be purchased from another supplier at lower cost.

The above examples highlight the desirability of establishing the purpose of each operation before endeavoring to improve the operation. Once the necessity of the operation has been determined, then the remaining nine approaches to operation analysis should be considered to determine how it can be improved.

Design of the Part

Methods engineers are often inclined to feel that once a design has been accepted their only recourse is to plan its manufacture in the most economical way. Granted, usually introducing even a slight change in design is difficult; still, a good methods analyst should review every design for possible improvements. Designs are not permanent—they can be changed; and if improvement is the result, and the activity of the job is significant, then the change should be made.

To improve the design, analysts should keep in mind the following pointers for lower cost designs on each component and each subassembly:

1. Reduce the number of parts, simplifying the design.
2. Reduce the number of operations and the length of travel in manufacturing by joining the parts better and by making the machining and assembly easier.
3. Utilize a better material.
4. Liberalize tolerances and rely for accuracy on "key" operations rather than on series of closely held limits.

The General Electric Company has summarized the ideas for developing minimum cost designs in Table 4–1.

These methods improvements resulted from considering a better material in an effort to improve the design:

One manufacturer always used cast-iron brackets on its motors. A methods analyst questioned this design, and this led to the redesign of the

TABLE 4-1

Castings
1. Eliminate dry sand (baked-sand) cores.
2. Minimize depth to obtain flatter castings.
3. Use minimum weight consistent with sufficient thickness to cast without chilling.
4. Choose simple forms.
5. Symmetrical forms produce uniform shrinkage.
6. Liberal radii—no sharp corners.
7. If surfaces are to be accurate with relation to each other, they should be in the same part of the pattern, if possible.
8. Locate parting lines so that they will not affect looks and utility, and need not be ground smooth.
9. Specify multiple patterns instead of single ones.
10. Metal patterns are preferable to wood.
11. Permanent molds instead of metal patterns.

Moldings
1. Eliminate inserts from parts.
2. Design molds with smallest number of parts.
3. Use simple shapes.
4. Locate flash lines so that the flash does not need to be filed and polished.
5. Minimum weight.

Punchings
1. Punched parts instead of molded, cast, machined, or fabricated parts.
2. "Nestable" punchings to economize on material.
3. Holes requiring accurate relation to each other to be made by the same die.
4. Design to use coil stock.
5. Punchings designed to have minimum sheared length and maximum die strength with fewest die moves.

Formed parts
1. Drawn parts instead of spun, welded, or forged parts.
2. Shallow draws if possible.
3. Liberal radii on corners.
4. Bent parts instead of drawn.
5. Parts formed of strip or wire instead of punched from sheet.

Fabricated parts
1. Self-tapping screws instead of standard screws.
2. Drive pins instead of standard screws.
3. Rivets instead of screws.
4. Hollow rivets instead of solid rivets.
5. Spot or projection welding instead of riveting.
6. Welding instead of brazing or soldering.
7. Use die castings or molded parts instead of fabricated construction requiring several parts.

Machined parts
1. Use rotary machining processes instead of shaping methods.
2. Use automatic or semiautomatic machining instead of hand-operated.
3. Reduce the number of shoulders.
4. Omit finishes where possible.
5. Use rough finish when satisfactory.
6. Dimension drawings from same point as used by factory in measuring and inspecting.
7. Use centerless grinding instead of between-center grinding.
8. Avoid tapers and formed contours.
9. Allow a radius or undercut at shoulders.

Screw machine parts
1. Eliminate second operation.
2. Use cold-rolled stock.
3. Design for header instead of screw machine.
4. Use rolled threads instead of cut threads.

Welded parts
1. Fabricated construction instead of castings or forgings.
2. Minimum sizes of welds.
3. Welds made in flat position rather than vertical or overhead.
4. Eliminate chamfering edges before welding.
5. Use "burnouts" (torch-cut contours) instead of machined contours.
6. Lay out parts to cut to best advantage from standard rectangular plates and avoid scrap.
7. Use intermittent instead of continuous weld.
8. Design for circular or straight-line welding to use automatic machines.

Treatments and finishes
1. Reduce baking time to minimum.
2. Use air drying instead of baking.
3. Use fewer or thinner coats.
4. Eliminate treatments and finishes entirely.

Assemblies
1. Make assemblies simple.
2. Make assemblies progressive.
3. Make only one assembly and eliminate trial assemblies.
4. Make component parts RIGHT in the first place so that fitting and adjusting will not be required in assembly.

This means that drawings must be correct, with proper tolerances, and that parts must be made according to drawings.

General
1. Reduce number of parts.
2. Reduce number of operations.
3. Reduce length of travel in manufacturing.

Source: Adapted from *American Machinist*, reference sheets, 12th ed. (New York: McGraw-Hill Publishing Co.).

bracket, making it from welded sheet steel. The new design was not only stronger, lighter, and more eye-appealing, but it was also less expensive to produce.

A similar design improvement was made in the construction of conduit boxes. Originally, they were built of cast iron, while the improved design, making a stronger, neater, lighter, and less expensive conduit box, was fabricated from sheet steel (see Figure 4–4).

In another instance, a brass cam switch used in control equipment was made as a brass die casting. By slightly altering the design, the less expensive process of extruding was utilized. The extruded sections were cut to the desired length to produce the cam switch (see Figure 4–5).

These improvements resulted from better joining of parts:

Attaching nuts were originally arc welded to transformer cases. This proved to be a slow, costly operation. Furthermore, the resulting design had an unsightly appearance due to the overflow and spatter from the arc welding process. By projection welding nuts to the transformer case, not only were time and money saved, but the resulting design also had considerably more sales appeal (see Figure 4–6).

A similar improvement occurred by changing from arc welding to spot welding to join mounting brackets to resistor tubes (see Figure 4–7).

Design simplification through the better joining of parts was also used in assembling terminal clips to their mating conductors. It had been the practice to turn the end of the clip up to form a socket. This socket was filled with solder, and the wire conductor was then tinned, inserted into the solder-filled socket, and held there until the solder solidified. By altering the design to call for resistance welding the clip to the wire conductor, both the forming and the dipping operations were eliminated (see Figure 4–8).

A motor cover thumbscrew was originally made of three components: head, pin, and screw. These components were assembled by joining the head to the screw with the pin. A much less costly thumbscrew was developed by redesigning the part for an automatic screw machine that was able to turn out the part complete with no secondary operations. A simplified design resulted in a less expensive part that still met all service and operating requirements (see Figure 4–9).

Design simplification can improve the process as well as the product. For example, Figure 4–10 illustrates a simple improvement in machine design that eliminated a division of work between operating and maintenance people.

The old bolted design required that a maintenance mechanic change the yarn guide because tools were required. The improved snap-in design, permits the operator to pull out the guide and replace it with a new one. This new design permits more timely changing of guides; without tools or coordination with another group, the operator can change a guide as soon

FIGURE 4–4
Improved design of
conduit box (new design
fabricated from sheet
steel)

FIGURE 4–5
A. Redesigned brass cam
switch allowing part to be
made from extrusion

B. New design shown cut
to length from section of
brass extrusion

FIGURE 4–6
Projection welding of nuts to surface provides neat-appearing and inexpensive method of joining

as it becomes defective. In addition, it takes less time to make the yarn guide change.

Just as there are opportunities to improve productivity through better product design, so are there similar opportunities in improving the design of forms used throughout an industry or business. Once a form is proved

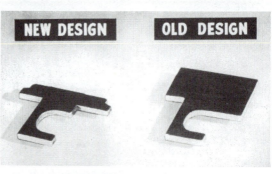
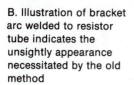

FIGURE 4–7
A. Mounting bracket changed so that spot welding rather than arc welding may be used

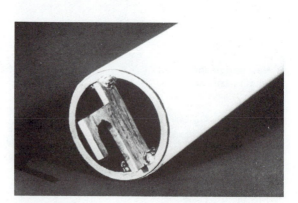

B. Illustration of bracket arc welded to resistor tube indicates the unsightly appearance necessitated by the old method

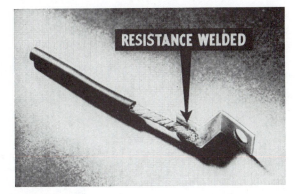

FIGURE 4–8
A. Old design required tinned and soldered connection

B. Improved design illustrating resistance-welded clip to wire conductor

necessary, then it should be studied to improve both the collection and flow of information. The following criteria apply to the development of forms:

1. Maintain simplicity in the form design, keeping the amount of necessary input information (writing, typewriter, word processor) at a minimum.
2. Provide ample space for each bit of information, allowing for different input methods.
3. Sequence input information in a logical pattern.
4. Color code the form to facilitate distribution and routing.
5. Provide adequate margins to accommodate standard filing facilities and procedures.
6. Confine forms for computer terminals to one page.

The foregoing examples are characteristic of the possibilities for improvement when the design of the part is investigated. It is always wise to check the design for improvement because design changes can be valu-

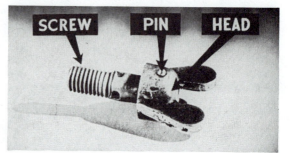

FIGURE 4–9
A. Old thumbscrew was in
three parts

B. Improved one-piece
design of motor cover
thumbscrew

able. To be able to recognize good design, methods engineers should have had some training and practical experience in this area. Good designs do not just happen; they are the result of broad experience and creative thinking, tempered with an appreciation of cost.

Tolerances and Specifications

The third of the 10 points of operation analysis concerns tolerances and specifications. Many times this point is considered in part when reviewing the design. This, however, is usually not adequate, and tolerances and specifications should be considered independently of the other approaches to operation analysis. Tolerances and specifications relate to the quality of the product and quality according to the American Society for Quality Control (ASQC) is the totality of features and characteristics of a product or service that bear on its ability to satisfy given needs.

Today, "geometric dimensioning and tolerancing" is a drawing language being used extensively by both manufacturing industries and government agencies as a means of specifying the geometrical configuration or shape of a part on an engineering drawing. This technique also provides information as to how the part should be inspected so as to assure the intent of the design. Figure 4–11 illustrates geometric tolerancing symbols. Datum identification applies to a point or points, line, plane, or

FIGURE 4–10
Old, bolted and new, snap-in design of rotating yarn guide

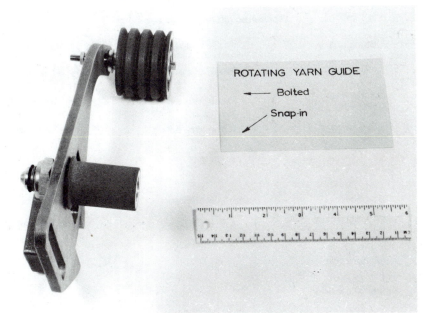

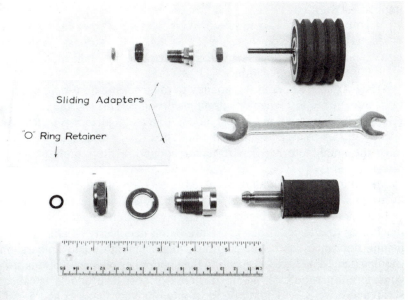

E. I. du Pont de Nemours & Co., Inc.

FIGURE 4–11
Geometric tolerancing symbols

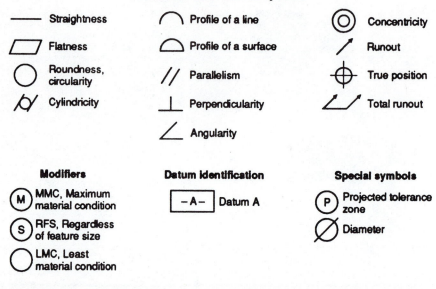

Geometric characteristic symbols

——	Straightness	⌒	Profile of a line	◎	Concentricity
▱	Flatness	⌓	Profile of a surface	∕	Runout
○	Roundness, circularity	∥	Parallelism	⊕	True position
⌭	Cylindricity	⊥	Perpendicularity	⌰	Total runout
		∠	Angularity		

Modifiers	**Datum identification**	**Special symbols**
Ⓜ MMC, Maximum material condition	|–A–| Datum A	Ⓟ Projected tolerance zone
Ⓢ RFS, Regardless of feature size		⌀ Diameter
○ LMC, Least material condition		

surface of an object. Geometric dimensioning takes place from datums that are considered to be exact. Modifiers are shown in order to clarify implied tolerances.

Thus, geometric tolerances provide the tolerance of basic geometric characteristics of which there are 11: straightness, flatness, perpendicularity, angularity, roundness, cylindricity, profile, parallelism, concentricity, runout, and true position.

Designers have a natural tendency to incorporate more rigid specifications than necessary when developing a product. This is brought about by (1) a lack of appreciation for cost, and (2) the feeling that it is necessary to specify closer tolerances and specifications than are actually needed to have the manufacturing departments produce to the required tolerance range.

Methods analysts should be versed in the aspects of cost and be fully aware of what unnecessarily close tolerances and/or rejects can do to the selling price. Figure 4–12 illustrates the relationship between the cost and machining tolerance. If designers are being needlessly "tight" in establishing tolerances and specifications, management should embark on a training program clearly presenting the economies of specifications. Also, consideration should be given to the extra cost of products because of scrap and/or rejects. Today, there is only one way that a concern can be competitive, that is, all parts in every product must be produced to the

FIGURE 4–12
Approximate relationship between cost and machining tolerance

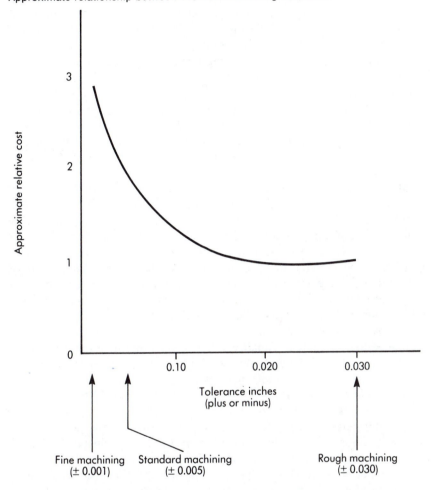

precise dimensions given on the drawings. Developing quality products in a manner that actually reduces costs is a major tenet of the approach to quality instituted by Dr. Genichi Taguchi, executive director of the American Supplier Institute of Dearborn, Michigan, and winner of the 1986 Willar D. Rockwell medal for excellence in technology, as well as four Deming prizes. Taguchi's approach is congruent to what this author recommends the methods analyst use. This involves combining engineering and statistical methods to achieve improvements in costs and quality by optimizing product design and manufacturing methods.

Analysts must be alert for too liberal, as well as too restrictive, specifications. Closing up a tolerance often facilitates an assembly operation or some other subsequent step. This may be economically sound, even though it increases the time required to perform the present operation. In this connection, analysts should recognize that the overall tolerance is equal to the square root of the sum of the squares of the individual tolerances comprised by the overall tolerance.

Beyond the principle of operating economies through correct tolerances and specifications is the consideration of establishing the ideal inspection procedure. Invariably, inspection is a verification of the quantity, quality, dimensions, and performance. Inspection in all these areas can usually be performed by numerous methods and techniques. One way is usually best, not only from the standpoint of quality control, but also from time and cost consideration. Methods analysts should question the present way with thought toward improvement.

Methods people must consider the possibilities for installing spot inspection, lot-by-lot inspection, or statistical quality control. Spot inspection is a periodic check to assure that established standards are being realized. For example, a nonprecision blanking and piercing operation set up on a punch press should have a spot inspection to assure the maintenance of size and the absence of burrs. As the die begins to wear or deficiencies in the material being worked begin to show up, the spot inspection would catch the trouble in time to make the necessary changes without generating an appreciable number of rejects.

Lot-by-lot inspection is a sampling procedure in which a sample is examined to determine the quality of the production run or lot. The size of the sample selected depends on the allowable percentage defective and the size of the production lot under check.

Statistical quality control is an analytic tool employed to control the desired quality level of the process.

If a 100 percent inspection is being done, consider the possibility of spot inspection or lot-by-lot inspection. A 100 percent inspection is the process of inspecting every unit of production and rejecting the defective units. Experience has shown that this type of inspection does not assure a perfect product. The monotony of screening tends to create fatigue, thus lowering operation attention. The inspector may pass some defective parts as well as reject good parts. Because a perfect product is not assured under 100 percent inspection, acceptable quality may be realized from the considerably more economical methods of either lot-by-lot or spot inspection.

Usually an elaborate quality control procedure is not justified if the product does not require close tolerances, if its quality is easily checked, and if the generation of defective work is unlikely.

Just as there are several mechanical methods for checking a 0.500 to 0.502 inch reamed hole, so there are several overall policy procedures that can be adopted as a means of control. Methods analysts must be alert and well grounded in various techniques to make sound recommendations for improvement.

In one shop a certain automatic polishing operation had a normal rejection quantity of 1 percent. Subjecting each lot of polished goods to 100 percent inspection would have been quite expensive. Management decided, at an appreciable saving, to consider 1 percent the allowable percentage defective, even though this quantity of defective material would go through to plating and finishing only to be thrown out in the final inspection before shipment.

Analysts must always be aware that the reputation of and demand for a company's product depends on the care taken in establishing correct specifications and in maintaining them. Once quality standards are established, no deviations should be permitted. In general, tolerances and specifications can be investigated by asking these three questions: "Are they absolutely correct?" "Are ideal inspection methods and inspection procedures being used?" and "Are modern quality control techniques being exercised?"

One manufacturer's drawings called for a 0.0005-inch tolerance on a shoulder ring for a DC motor shaft. The original specifications called for a 1.8105 to 1.8110-inches tolerance on the inside diameter. This close tolerance was thought necessary because the shoulder ring was shrunk on the motor shaft.

Investigation revealed that a 0.003-inch tolerance was adequate for the shrink fit. The drawing was immediately changed to specify a 1.809 to 1.812 inches inside diameter. A reaming operation was saved because someone questioned the absolute necessity of a close tolerance.

In another instance, it was possible to introduce an automatic control on an external cylindrical grinder. Formerly, the operator had to manually feed the grinding wheel to the required stop. Each piece had to be carefully inspected to assure the maintenance of tolerance on the outside diameter. With the automatic machine control, the feed is tripped and the piece released upon completion of the infeed. The automatic machine control freed the operator to do other work because some methods analyst endeavored to develop an ideal inspection procedure. (See Figure 4–13.)

In an automatic screw machine shop, workers were inspecting 100 percent of the parts coming off the machine because of the critical tolerance requirements. However, adequate quality control could be maintained by inspecting every sixth piece. This sampling procedure allowed one inspector to service three machines rather than one machine.

By investigating tolerances and specifications and taking action when desirable, the costs of inspection can be reduced, scrap minimized, repair costs diminished, and quality kept high.

Material

One of the first questions an engineer considers when designing a new product is, "What material shall I use?" The ability to choose the right material is based on the engineer's knowledge of materials. Since choos-

FIGURE 4–13
Automatic control attached to external cylindrical grinder

ing the correct material is difficult because of the great variety available, many times it is more practical to incorporate a better and more economical material into an existing design.

Methods analysts should consider these possibilities for the direct and indirect materials utilized in a process: (1) finding a less expensive material; (2) finding materials that are easier to process; (3) using materials more economically; (4) using salvage materials; (5) using supplies and tools more economically; (6) standardizing materials; (7) finding the best vendor from the standpoint of price and vendor stocking.

FINDING A LESS EXPENSIVE MATERIAL. Monthly publications available to all engineers summarize the approximate cost per pound of steel sheets, bars, and plates, and the cost of cast iron, cast steel, cast aluminum, cast bronze, thermoplastic and thermosetting resins, and other basic materials. These costs can be used as anchor points from which to judge the application of new materials. Industry is developing new processes for producing and refining materials continually. Thus, a material that was not competitive in price yesterday may be so today.

One company used Micarta spacer bars between windings of transformer coils. Separating the windings permitted the circulation of air between the windings. An investigation revealed that glass tubing could be substituted for the Micarta bars at a saving. Not only was the glass tubing less expensive, but it met service requirements better because the glass could withstand higher temperatures. Furthermore, the hollow tubing permitted more air circulation than did the solid Micarta bars (see Figure 4–14).

Another company also used a less expensive material that still met service requirements in the production of distribution transformers. Originally, a porcelain plate separated and held the wire leads coming out of the transformers. The company found a fullerboard plate stood up just as well in service, yet was considerably less expensive (see Figure 4–15).

Methods analysts should remember that such items as valves, relays, air cylinders, transformers, pipe fittings, bearings, couplings, chains, hinges, hardware, and motors can usually be purchased at less cost than they can be manufactured.

Today, many plastics are competing effectively with metals and wood. For example, Figure 4–16 illustrates a change in material in the manufacture of gasoline mechanical pump computer wheels; a 13-cent-per-unit

FIGURE 4–14
A. Rectangular Micarta bars were formerly used to separate windings of transformer coils

B. Use of glass tubing instead of bars provides substantial savings

FIGURE 4–15
A. Porcelain plate was
formerly used to hold wire
leads from distribution
transformers

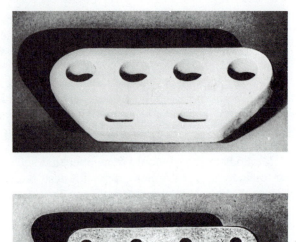

B. Fullerboard plate
makes economical and
effective substitute

savings was made as well as a $10,000-per-year savings in tool mainte-
nance by converting two parts of the assembly from steel to plastic.

The pawl had been made of sheet steel that had been blanked, tumbled,
and ground to tolerance. The new pawl was made by injection molding the
plastic Delrin. Similarly, the stamped and coined steel geer was changed
to a redesigned injection-molded thermoplastic. The new design of the
mechanical pump computer wheel assembly was not only 13 cents less
expensive to produce, but it proved to have greater reliability and be
maintenance free giving a longer service life.

FINDING A MATERIAL THAT IS EASIER TO PROCESS.   One material is
usually more readily processed than another. Referring to handbook data
on the physical properties usually helps analysts discern which material
will react most favorably to the processes it must be subjected to in its
conversion from raw material to finished product. For example, machin-
ability varies inversely with hardness, and hardness usually varies directly with strength.

By selecting a material that was easy to process, one methods analyst
produced real savings by changing the procedure for producing stainless
steel bearing shells. Originally, workers made them by drilling and boring
to size cut lengths of stainless steel bar stock. By specifying stainless steel
tubing as a material source, material was conserved, the rate of production was increased, and the cost of manufacture was reduced.

USING MATERIAL MORE ECONOMICALLY.   The possibility of using
material more economically is a fertile field for analysis. If the ratio of

FIGURE 4–16
Gasoline mechanical pump computer wheel. At left is shown the stamped and coined gear that was converted to a plastic gear, shown at right. The steel pawl shown assembled to the body (upper right) was converted to a Delrin plastic (bottom). The assembly is shown in the center.

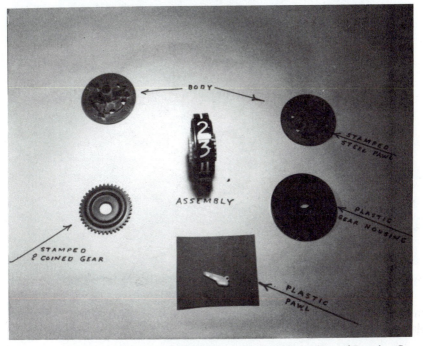

Courtesy of: Veeder Root Company Subsidiary of Danaher Co.

scrap material to that actually going into the product is high, then consider greater utilization.

At one time, the Procter & Gamble Company packed bars of Ivory soap into boxes with the opening on the largest face of the cardboard box. Figure 4–17 illustrates this operation. Flaps on the four edges of the open case fold over to form a double cover over that part of the box during packing. A methods study found that the bars could be packed satisfactorily, with no increase in time, by passing them through an open end. This permitted redesign of the case so that the flaps overlapped on the smallest face of the case, using approximately 15 percent less boxboard. Figure 4–18 shows the end packing. This change resulted in a substantial reduction in container cost.

Another example of economical use of material is in the compression molding of plastic parts. By preweighing the material put into the mold, only the exact amount of material required to fill out the cavity is used and excessive flash is eliminated.

FIGURE 4–17
Packing Ivory bars before methods change

In the production of stampings from sheet, if the skeleton seems to contain an undue amount of scrap material, consider going to the next higher standard width of material and utilizing a multiple die. If a multiple die is used, carefully arrange the cuts to assure maximum utilization of material. Figure 4–19 illustrates how the careful nesting of parts permits maximum utilization of flat stock.

Figure 4–20 illustrates efficient use of material. The bar at the top of the picture is fed automatically into the equipment where it is induction heated at one end. The heated end is then impacted twice to produce the final crankshaft shape shown. All operations are automatically performed in one machine.

SALVAGE MATERIALS.   Many times materials can be salvaged rather than sold as scrap. Sometimes by-products from an unworked portion or scrap section offer real possibilities for savings. One manufacturer of stainless steel cooling cabinets had sections of stainless steel four to eight

FIGURE 4–18
Packing Ivory bars after methods change (small end of box requires 15 percent less boxboard because of smaller overlapping flaps)

inches wide left as cuttings on the shear. An analysis brought out a by-product of electric light switch plate covers.

Another manufacturer, after salvaging the steel insert from defective bonded rubber ringer rolls, was able to utilize the cylindrical hollow rubber rolls as bumpers for protecting moored motor- and sailboats. Figure 4–21 illustrates how one manufacturer used pieces of circular plate scrap to produce needed components.

If it is not possible to develop a by-product, then segregate scrap materials for top scrap prices. Provide separate bins in the shop for tool steel, steel, brass, copper, and aluminum; and instruct chip-haulers and floor sweepers to keep the scrap segregated.

If large quantities are used, electric light bulbs can be salvaged. Store the brass socket in one area, and after breaking and disposing of the glass

FIGURE 4–19
Method of torch cutting heavy gear case side plates (note nesting of the point
and the heel for most effective use of plate)

General Electric Co.

bulb, remove the tungsten filament and store it separately for greatest
residual value.

Many companies save wooden boxes from incoming shipments, and
saw the boards to standard lengths for use in making smaller boxes for
outgoing shipments. This practice is always economical, and it is now
being followed by many large industries as well as by service maintenance
centers.

FULL USE OF SUPPLIES AND TOOLS.   Management should encourage
full use of all shop supplies. One manufacturer of dairy equipment introduced the policy that no new welding rod was to be distributed to workers
without the return of old tips under two inches long. The cost of welding
rods was reduced immediately by more than 15 percent.

Brazing or welding is usually the most economical way to repair expensive cutting tools, such as broaches, special form tools, and milling cutters. If it has been company practice to discard broken tools of this
nature, investigate the potential savings of a tool salvage program.

FIGURE 4–20
Crankshaft forging produced by upsetting one end of induction heated barstock

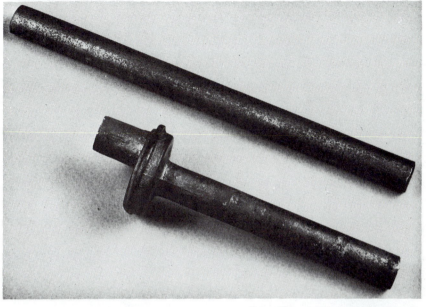

General Electric Co., Louisville, Kentucky

Analysts can also find a use for the unworn portion of grinding wheels, emery disks, and so forth, elsewhere in the plant. Such items as gloves and rags should not be discarded once they are soiled. Storing dirty items in containers to await laundering is less expensive than replacing them.

Waste of material benefits no one. Methods analysts can make a real contribution to their companies by preventing wastes which today claim about one fifth of our material.

STANDARDIZING MATERIALS. Methods analysts should always be alert to the possibility of standardizing materials. They must minimize the sizes, shapes, grades, and so on of each material utilized in the production and assembly of products. The following typical economies result from reductions in the sizes and grades of the materials employed: purchase orders are for larger amounts which are almost always less expensive; inventories are smaller since less material must be maintained as a reserve; fewer entries are made in stores records; fewer invoices need to be paid; fewer spaces are needed to house materials in the storeroom; sampling inspection reduces the total number of parts inspected; and fewer price quotations and purchase orders are prepared.

The standardization of materials, like other methods improvement techniques, is a continuing process. It requires the continual cooperation

FIGURE 4–21

A. Circular piece is scrap from motor frame head fabrication part (this scrap is used to produce gusset support pieces for locomotive platform)

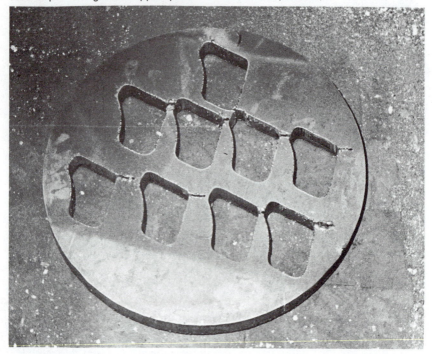

B. Airco No. 50 Travo-graph 8 torch burner cutting gusset support pieces from scrap blank

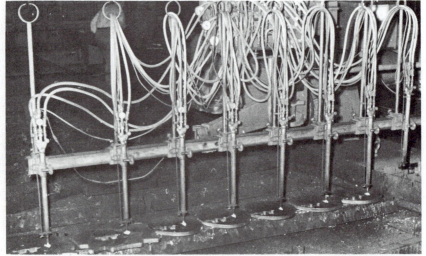

General Electric Co.

of members of the design, production planning, and purchasing departments.

FINDING THE BEST VENDOR FROM THE STANDPOINT OF PRICE AND STOCKING. For the vast majority of materials, supplies, and parts, one will find that there are numerous suppliers who will quote different prices, quality levels, delivery times, and willingness to hold inventories. It is the responsibility of the Purchasing Department to locate the most favorable supplier. However, the best supplier last year may not be best at the present time. It is also a good idea for the methods analyst to have the purchasing department rebid the highest cost materials, supplies, and parts to obtain better prices and superior quality and to increase the vendor stocking, where the vendors agree to hold inventories for their customers. It is not unusual for methods analysts to achieve a 10 percent reduction in cost of materials and a 15 percent reduction in inventories by regularly pursuing this approach through their purchasing departments.

Perhaps the most important reason for continued Japanese success in the manufacturing sector is the *keiretsu*. This is a form of business and manufacturing organization that links businesses together. It can be thought of as a web of interlocking relationships among manufacturers— often between a large manufacturer and its principal suppliers. Thus, in Japan such companies as Hitachi and Toyota and other international competitors are able to acquire parts for their products from regular suppliers who produce to the quality called for and are continually looking for improvement so as to provide better prices for the firms in their network. Alert purchasing departments often are able to create relationships with suppliers comparable to the so-called production keiretsu.

TEXT QUESTIONS

1. Explain how design simplification can be applied to the manufacturing process.
2. How is operation analysis related to methods engineering?
3. Does increased competition submerge the necessity for operation analysis? Explain.
4. Explain the relationship between market price and volume as related to production.
5. What factors did Bill Winchell list as reasons for slowing down improvement activities in an organization?
6. What is the therblig symbol and color designation for "rest to overcome fatigue"?
7. What is the major obstacle in the path of methods analysts?
8. How do unnecessary operations develop in an industry?
9. What has been the impact of the computer in connection with paperwork?
10. What four thoughts should analysts keep in mind to improve design?

11. Explain why it may be desirable to "tighten up" tolerances and specifications.
12. What is meant by lot-by-lot inspection?
13. When is an elaborate quality control procedure not justified?
14. What six points should be considered when endeavoring to reduce material cost?
15. Explain why corrugated metal sheet is more rigid than flat sheet of the same material.
16. How does a changing labor and equipment situation affect the cost of purchased components?
17. The finish tolerance on a shaft was changed from 0.004″ to 0.008″. How much cost improvement resulted from this change?
18. What overall tolerance would be applied to three components making up the overall dimension if component one had a tolerance of 0.002″; component two, 0.004″; and component three, 0.005″?
19. What six points should be remembered when designing forms?
20. What is meant by "geometric dimensioning and tolerancing"?
21. What is Dr. Genichi Taguchi's approach to quality of output?

GENERAL QUESTIONS

1. Formulate a checklist that would be helpful in improving operations.
2. Explain how the conservation of welding rod can result in 20 percent material savings.
3. Investigate the operations required to convert waste sulfur dioxide to usable sulfur.
4. Explain why this country has lost a large share of its steel business to Japan and Korea since the early 1970s. What needs to be done to regain the business lost?
5. Distinguish between roundness and concentricity. Between runout and cylindricity.
6. Identify several automobile components with which you are familiar that have been converted from metal to plastic in recent years.

PROBLEMS

1. A ceramic material is being considered as a possible mold material in conjunction with the die casting of 60–40 brass. A cylinder of the material 8 inches in diameter and 10 inches long was used to obtain a stress-strain relationship in compression. The material failed under a load of 265,000 pounds and a total strain of 0.012 inches. What was the material's fracture strength, percent contraction at fracture, and modulus of toughness?
2. The Dorben Company is designing a cast-iron part whose strength T is a known function of the carbon content C.

$$T = 2C^2 + \frac{3}{4}C - C^3 + k$$

To maximize strength, what carbon content should be specified?

3. To make a given part interchangeable, it was necessary to reduce the tolerance on the outside diameter from ±0.010 to ±0.005 at a resulting cost increase of 50 percent of the turning operation. The turning operation represented 20 percent of the total cost. By making the part interchangeable, the volume of this part could be increased by 30 percent. The increase in volume would permit production at 90 percent of the former cost. Should the methods engineer proceed with the tolerance change? Explain.

4. The following sketch illustrates the maximum material diameter and least material diameter of a shaft.

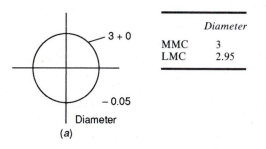

(a)

On the following sketch indicate the MMC and the LMC of the hole.

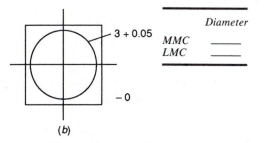

(b)

SELECTED REFERENCES

Bralla, James G. *Handbook of Product Design for Manufacturing*. New York: McGraw-Hill, 1986.

Brumbaugh, Philip S., and Russell G. Heikes. "Statistical Quality Control." In *Handbook of Industrial Engineering*, 2nd ed., edited by. Gavriel Salvendy, Chap. 87. New York: John Wiley & Sons, 1992.

Chang Tien-Chien, Wysk, Richard A., and Wang Hsu-Pin. *Computer Aided Manufacturing*. Englewood Cliffs: Prentice Hall, 1991.

DeMarle, David J., and M. Larry Shillito. "Value Engineering." In *Handbook of Industrial Engineering*, 2nd ed., edited by Gavriel Salvendy, Chap. 14. New York: John Wiley & Sons, 1992.

Drury, Colon G. "Inspection Performance." In *Handbook of Industrial Engineering*, 2nd ed., edited by Gavriel Salvendy, Chap. 88. New York: John Wiley & Sons, 1992.

Fallon, Carlos. *Value Analysis to Improve Productivity.* New York: John Wiley & Sons, 1971.

Kadota, Takeji, and Shigeyasu Sakamoto. "Methods Analysis and Design." In *Handbook of Industrial Engineering,* 2nd ed., edited by Gavriel Salvendy, Chap. 55. New York: John Wiley & Sons, 1992.

Mudge, Arthur E. *Value Engineering: A Systematic Approach.* New York: McGraw-Hill, 1971.

Niebel, Benjamin W. "Designing for Manufacturing." In *Handbook of Industrial Engineering,* ed. Gavriel Salvendy. New York: John Wiley & Sons, 1982.

Niebel, Benjamin W., and C. Richard Liu. "Designing for Manufacturing." In *Handbook of Industrial Engineering,* 2nd ed. edited by Gavriel Salvendy, Chap. 13. New York: John Wiley & Sons, 1992.

Taguchi, Genichi. *Introduction to Quality Engineering.* Tokyo, Japan: Asian Productivity Organization, 1986.

Trucks, H. E. *Designing for Economical Production.* Detroit: Society of Manufacturing Engineers, 1974.

Wick, Charles, and Raymond F. Veilleux. *Quality Control and Assembly,* vol. 4. Detroit: Society of Manufacturing Engineers, 1987.

SELECTED SOFTWARE

Nicks, J. E. Microcomputer Assisted Process Planning and Tolerance Control. Basic Programming Solutions for Manufacturing. Dearborn, Michigan: Society of Manufacturing Engineers, 1982. Source code owned by MiCAPP, Inc.

SELECTED VIDEOTAPES

Automated Inspection/Non-Destructive Testing. Manufacturing Insights Videotape Series. 1/2" VHS VT281-1368 & 3/4" U-Matic VT281U-1368. Dearborn, Mich.: Society of Manufacturing Engineers, 1988.

Design for Manufacturing. Manufacturing Insights Videotape Series. 1/2" VHS VT396-1368 & 3/4" U-Matic VT396U-1368. Dearborn, Mich.: Society of Manufacturing Engineers, 1990.

Layout Improvements for Just-In-Time. Manufacturing Insights Videotape Series. 1/2" VHS VT393-1368 & 3/4" U-Matic VT393U-1368. Dearborn, Mich.: Society of Manufacturing Engineers, 1990.

Total Quality Management. Manufacturing Insights Videotape Series. 1/2" VHS VT395-1368 & 3/4" U-Matic VT395U-1368. Dearborn, Mich.: Society of Manufacturing Engineers, 1990.

Operation Analysis (continued)

PROCESS OF MANUFACTURE

From the standpoint of improving the process of manufacture, investigate in four ways: (1) consider the possible effects on other operations when changing an operation; (2) mechanize manual operations; (3) utilize more efficient facilities on mechanical operations; and (4) operate mechanical facilities more efficiently.

Consider the Effects on Subsequent Operations Resulting from Changes in the Present Operation

Before changing any operation, consider detrimental effects that may result on subsequent operations down the line. Reducing the costs of one operation can result in higher costs for other operations. For example, the following change in the manufacture of AC field coils resulted in higher costs and was, therefore, not practical. The field coils were made of heavy copper bands which were formed and then insulated with mica tape. The mica tape was hand wrapped on the already coiled parts. Then the company decided to machine wrap the copper bands prior to coiling. This did not prove practical as the forming of the coils cracked the mica tape, necessitating time-consuming repairs prior to acceptance (see Figure 5–1).

The market for aluminum cylinder head castings is growing, and foundries are finding it cost effective to go from the steel mold casting process to the lost foam process. Lost foam is an investment casting procedure that uses an expendable pattern of polystyrene foam surrounded by a thin ceramic shell.

FIGURE 5–1
Machine-wrapped
copper bands prove
unsatisfactory in
manufacture of AC field
coils

A. Method of forming
heavy copper bands

B. Hand-wrapped field
coils

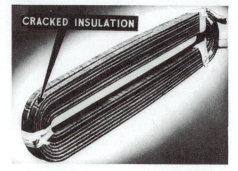

C. Cracked insulation in
coils machine wrapped
prior to forming

CRACKED INSULATION

Steel mold castings require considerable subsequent machining. However, the lost foam process reduces the amount of machining and also eliminates the sand disposal costs associated with investment casting.

After considering the reduction of machining costs in the subsequent operations, the lost foam process was introduced.

Rearranging operations often results in savings. The flange of a motor conduit box required four holes to be drilled—one in each corner. Also, the base had to be smooth and flat. Originally, the operator began by grinding the base, then drilling the four holes in a drill jig. The drilling operation threw up burrs which had to be removed in the next step. By rearranging operations so that the holes were drilled first and the base then ground, analysts eliminated the deburring operation. The base-grinding operation automatically removed the burrs.

Combining operations usually reduces costs. Formerly, a manufacturer fabricated the fan motor support and the outlet box of electric fans. After painting them separately, operators then riveted these parts together. By riveting the outlet box to the fan motor support prior to painting, analysts effected an appreciable savings in time on the painting operation.

Mechanize Manual Operations

Today any practicing method analyst should consider using special-purpose and automatic equipment and tooling, especially if production quantities are large. Notable among industry's latest offerings are program controlled, numerically controlled (NC), and computer controlled (CNC) machining and other equipment. These afford substantial savings in labor cost as well as these advantages: reduced work-in-process inventory, less damage of parts due to handling, less scrap, reduced floor space, and reduced throughput time for production. For example, one company experienced a 40 percent direct labor savings in the machining of a precision stainless steel component (see Figure 5–2). Prior to acquiring the CNC equipment, three machines with a total of eight cutting tools were used to produce the part. The change in method realized a 60-percent savings in floor space, a reduction in rejects, and lower in-process inventory.

Other automatic equipment includes: automatic screw machines; multiple-spindle drilling, boring, tapping, machines; index-table machine tools; automatic casting equipment combining automatic sand-mold making, pouring, shakeout, and grinding; and automatic painting and plating finishing equipment.

A company producing specialty windows was using manual methods to press rails over both ends of plate window glass that had been covered with a synthetic rubber wrap. The plates of glass were held in position by

FIGURE 5–2

Figure 5–2A illustrates the complete machining of a pulse dampner top plate. This precision stainless steel part is being machined on a CNC Bridgeport machining center using a CAM program. Seven tools are utilized from the 24-position carousel. A 40 percent savings resulted from changing the machining of this part from three separate operations on three different machines to this CNC equipment. (See Figures 5–2B, 5–2C, and 5–2D.)

A.

B.

C.

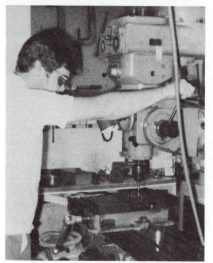

D.

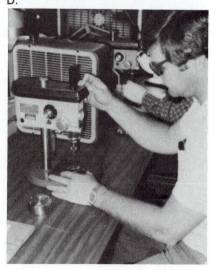

Courtesy of Scientific Systems Incorporated

FIGURE 5-3
Rough sketch of vertical glazing machine

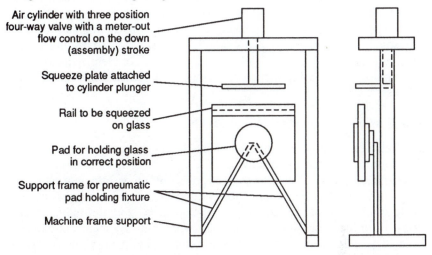

Air cylinder with three position
four-way valve with a meter-out
flow control on the down
(assembly) stroke

Squeeze plate attached
to cylinder plunger

Rail to be squeezed
on glass

Pad for holding glass
in correct position

Support frame for pneumatic
pad holding fixture

Machine frame support

Note: The flow control should be mounted to the cylinder to prevent the loss of control in case of hose breakage.
There may need to be a second cylinder and additional controls in order to move the squeeze plate back so as to provide clearance when rotating the glass to position to assemble the second rail.

two pads that were pneumatically squeezed together. The operator would pick up a rail and position it over the end of the window glass and then pick up a mallet and hammer the rail into position over the glass. The operation was slow and resulted in considerable operator carpal tunnel syndrome. Furthermore, scrap was high because of glass breakage brought on by pounding of the rails over the glass.

A new facility was designed that pneumatically squeezed rails on to the window glass over the synthetic rubber wrap (see Figure 5-3). Operators enthusiastically accepted the new facility because the work was much easier to perform; carpal tunnel syndrome disappeared, productivity increased, and glass breakage dropped to approximately zero.

Anytime heavy manual work is encountered, consider possible mechanization. To clean insulation and dried varnish from armature slots, one company resorted to tedious hand filing. Questioning this process of manufacture led to the development of an end mill placed in a power air drill. This not only took most of the physical effort out of the job, but allowed a considerably higher rate of production at a reduced cost.

The use of power assembly tools, such as power nut- and screwdrivers, electric or air hammers, and mechanical feeders, is often more economical than is the use of hand tools. Other options are automatic flame, laser, and other contour-cutting machines controlled by either optical or template tracing or computer memory.

The application of mechanization applies not only to process operations but also to paper work. For example, bar coding applications can be invaluable to alert operations analysts. Bar coding can rapidly and accurately enter a variety of data. Then, computers can manipulate the data into some desired objective, for example, counting and controlling inventory, routing specific items to or through a process, or identifying the state of completion and who is currently working on each item in a work-in-process.

Utilize More Efficient Mechanical Facilities

"Can a more efficient method of machining be used?" is a question that should be foremost in any analyst's mind. If an operation is done mechanically, there is always the possibility of a more efficient means of mechanization. Let us look at three examples. At one company, turbine blade roots were machined by performing three separate milling operations. Both the cycle time and the costs were high. By means of external broaching, all three surfaces were finished at once. A pronounced saving was the result.

Another company overlooked the possibility of utilizing press operation. This process is one of the fastest for forming and sizing operations. A stamped bracket had four holes that were drilled after the bracket was formed. By designing a die to pierce the holes, the work was performed in a fraction of the drilling time. Figure 5–4 illustrates a part made on a progressive die. Originally this part was made using three single station dies at considerably more direct labor time per piece.

The mechanization of work does not apply only to manual work. For example, one company in the food industry was checking the weight of

FIGURE 5–4
Output of a Progressive Die. Shows the completed part (at bottom) and the progressive fabrication steps. The material is fed from a coil into the press and automatically indexed through the die, supplying a completed part at the extreme end.

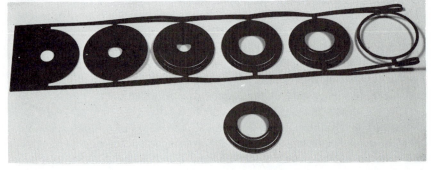

General Electric Co., Louisville, Kentucky

FIGURE 5–5
Charleston Chew candy bars being weighed on a shadowgraph balance

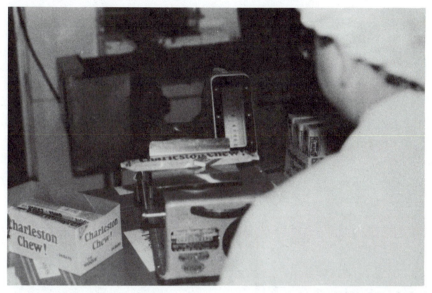

Courtesy of Nabisco Brands, Inc.

various product lines with a shadowgraph balance (see Figure 5–5). This equipment requires the operator to visually note the weight, record the weight on a form, and subsequently make arithmetic calculations utilizing the data. A methods engineering study resulted in the introduction of a statistical weight control system (Figure 5–6). Under the improved method, the operator weighs the product on a digital scale programmed to accept the product within a certain weight range. As the product is weighed, the weight information is transferred to a personal computer that compiles the information and prints the desired report.

Operate Mechanical Facilities More Efficiently

A good slogan for methods analysts to remember is "Design for two at a time." Usually multiple-die operation in presswork is more economical than single-stage operation. Again, multiple cavities in die-casting, molding, and similar processes are viable options when there is sufficient volume.

On machine operations analysts should be sure that proper feeds and speeds are used. They should investigate the grinding of cutting tools for maximum performance. They should check to see whether the cutting tools are properly mounted, whether the right lubricant is being used, and whether the machine tool is in good condition and is adequately maintained. Many machine tools are operated at a fraction of their possible

FIGURE 5–6

Charleston Chew candy bars being weighed on a digital scale programmed to accept a specific weight range. The weight information is transferred to the personal computer, shown in lower illustration. It compiles the data and generates the desired reports. The SAC system includes the personal computer, buffer units, and printer.

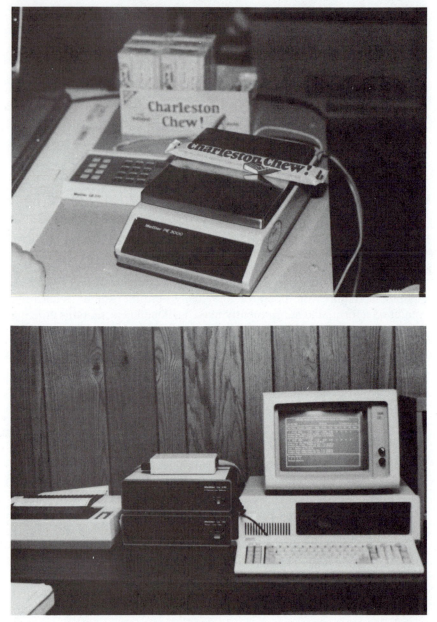

Courtesy of Nabisco Brands, Inc.

output. Endeavoring to operate mechanical facilities more efficiently always pays dividends.

SETUP AND TOOLS

One of the most important elements applying to all forms of work holders, tools, and setups is the economic one. The amount of "tooling up" that proves most advantageous depends on (1) the quantity to be produced, (2) the chances for repeat business, (3) the labor involved, (4) delivery requirements, and (5) the capital required.

The most prevalent mistake made by planners and toolmakers is to tie up money in fixtures that may show a large saving when in use but are seldom used. For example, a saving of 10 percent in direct labor cost on a job in constant use would probably justify greater expense in tools than an 80 or 90 percent saving on a small job that appears on the production schedule only a few times a year. The economic advantage of lower labor costs is the controlling factor in the determination of the tooling; consequently, jigs and fixtures may be desirable even where only small quantities are involved. Other considerations, such as improved interchangeability, increased accuracy, or reduction of labor trouble, may provide the dominant reason for elaborate tooling, although this is usually not the case.

For example, the production engineer in a machining department has under study a job being machined in the shop. He has devised two alternate methods involving different tooling for this job. Data on the present and proposed methods follow. Which method would be more economical in view of the activity? The base pay rate is $4.80 per hour. The estimated activity is 10,000 pieces per year. The fixtures are capitalized and depreciated in five years.

Method	Time value (minutes)	Fixture cost	Tool cost	Average tool life
Present method	3.50 each	None	$ 6	10,000 pieces
Alternate 1	2.80 each	$300	20	20,000 pieces
Alternate 2	1.85 each	600	35	5,000 pieces

A cost analysis of the above reveals that a unit total cost of $0.1670 represented by Alternate 2 is the most economical for the quantity anticipated. The elements of cost entering into this total are as follows:

Method	Unit direct labor cost	Unit fixture cost	Unit tool cost	Unit total cost	Annual cost
Present method.............	$0.280	None	$0.0006	$0.2806	$2,806
Alternate 1................	0.224	$0.006	0.0010	0.2310	2,310
Alternate 2................	0.148	0.012	0.0070	0.1670	1,670

A crossover (break-even) chart, as illustrated in Figure 5–7, is a helpful tool for deciding which method to use for given quantity requirements.

Once the needed amount of tooling has been determined, or if tooling already exists, once the ideal amount needed has been determined, then these specific points should be considered to produce the most favorable designs:

Can the fixture be used to produce other similar designs to advantage?

Will the fixture be similar to some other that has been used to advantage? If so, how can you improve on it?

Has the part undergone any previous operations? If so, can you use any of these points or surfaces to locate or master from?

Can any stock hardware be used for making the fixture?

Can the part be quickly placed in the fixture?

Can the part be quickly removed from the fixture?

FIGURE 5–7

A crossover (break-even) chart. This chart illustrates total costs for various methods and quantities. Note that the old method is the best for quantities up to about 7,400 per year, alternate method number one for quantities between 7,400 and 9,200 per year, and alternate method number two for quantities above 9,200 per year.

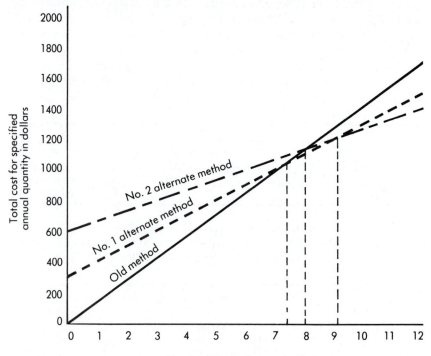

Is the part held firmly so that it cannot work loose, spring, or chatter while the cut is being made? (The cut should be against the solid part of the fixture and not against the clamp.)

Can the output be increased by placing more than one part in the fixture?

Can the chips be readily removed from the fixture?

Are the clamps on the fixture strong enough to prevent them from buckling when they are tightened down on the work?

Remember when using cams or wedges for binding or clamping the work, that through vibration or chatter they are apt to come loose and cause damage.

Must any special wrenches be designed to go with the fixture?

Is the work adequately supported so that the clamping force will not bend or distort it?

Can a gage be designed, or hardened pins added, to help the operator set the milling cutters or check up on the work?

Must special milling cutters, arbors, or collars be designed to go with the fixture?

Is there plenty of clearance for the arbor collars to pass over the work without striking?

If the fixture is of the rotary type, have you designed an accurate indexing arrangement?

Can the fixture be used on a standard rotary indexing head?

Can the fixture be made to handle more than one operation?

Have you, in designing the fixture, brought the work as close to the table of the miller as possible?

Can the part be milled in a standard vise by making up a set of special jaws, thus doing away with an expensive fixture?

If the part is to be milled at an angle, could the fixture be simplified by using a standard adjustable milling angle?

Can lugs be cast on the part to be machined to enable you to hold it?

How many different-sized wrenches must the operator have to tighten all clamps? Why will one not do?

Can the work be gaged in the fixture? Can a snap gage be used?

Can you use jack pins to help support the work while it is being milled?

Have you placed springs under all clamps?

Are all steel contact points, clamps, etc., hardened?

What kind or class of jigs are you going to design? Will any of the standard jig designs shown help you?

What takes the thrust of the drill? Can you use any jack pins or screws to support the work while it is being drilled?

Can you use a double or triple thread on the screw that holds the work in the jig, so that it will take fewer turns to get the screw out of the way to remove the part more quickly?

Have you made a note on the drawing, or have you stamped all loose parts with a symbol indicating the jig that they were made for, so that lost or misplaced parts can be returned to the jig when found?

Are all necessary corners rounded?

Can the toolmaker make the jig?

Are your drill bushings so long that it will be necessary to make up extension drills?

Are the legs on the jig long enough to allow the drill, the reamer, or the pilot of the reamer to pass through the part a reasonable distance without striking the table of the drill press?

Are all clamps located in such a way as to resist or help resist the pressure of the drill?

Has the drill press the necessary speeds for drilling and reaming all holes?

Must the drill press have a tapping attachment?

Drilling and reaming several small holes and only one large one in the jig is not practical since quicker results can be obtained by drilling the small holes on a small drill press, while having only one large one would require the jig to be used on a large machine. The questions then arise, Is it cheaper to drill the large hole in another jig, and will the result of doing so be accurate enough?

Is the jig too heavy to handle?

Is the jig identified with both a location number for storing and a part number that identifies the part or parts that the jig helps produce?

Setup ties in very closely with tooling because tooling of a job invariably determines the setup and teardown time. When we speak of setup time, we usually include such items as punching in on the job; procuring instructions, drawings, tools, and material; preparing workstations so that production can begin in the prescribed manner (setting up tools; adjusting stops; setting feeds, speeds, and depth of cut; and so on); tearing down the setup; and returning tools to the crib.

Setup operations are extremely important in the job shop when production runs tend to be small. Even if this type of shop had modern facilities and high effort were put forth, it would still be difficult to meet competition if setups were long because of poor planning and inefficient tooling. When the ratio of setup time to production-run time is high, a methods analyst can usually develop possibilities for setup and tool improvement. A notable possibility here is to design and develop a system of group technology.

FIGURE 5–8
Subdivision of a system grouping for group technology

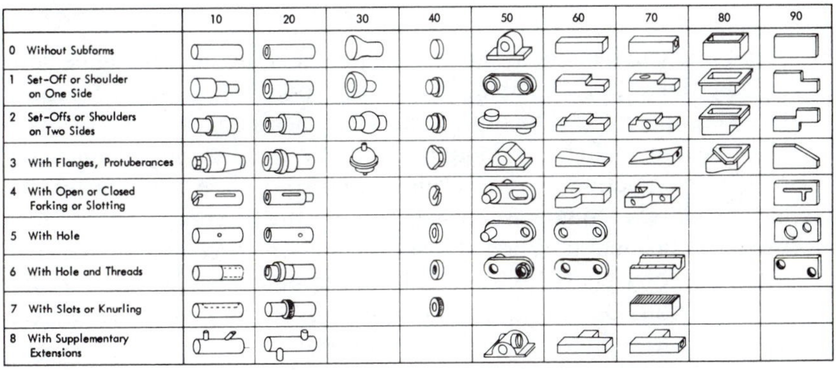

	10	20	30	40	50	60	70	80	90
0 Without Subforms									
1 Set–Off or Shoulder on One Side									
2 Set–Offs or Shoulders on Two Sides									
3 With Flanges, Protuberances									
4 With Open or Closed Forking or Slotting									
5 With Hole									
6 With Hole and Threads									
7 With Slots or Knurling									
8 With Supplementary Extensions									

The essence of group technology is to classify the various components entering into a company's products so that parts similar in shape and processing sequence are numerically identified. Parts belonging to the same family group, such as rings, sleeves, discs, and collars, are scheduled for production over the same interval of time on a line of general purpose facilities arranged in the operation sequence optimum to most conditions. Since both the size and the shape of the family of parts involve considerable variability, the line of facilities usually is equipped with universal-type, quick-acting jigs and fixtures.

The resulting line can mean greater output, less setup time, greater machine utilization, less material handling, shorter cycle time, and better cost improvement. The design and development of universal-type jigs and fixtures means that less equipment is required, with the added advantage of reduction of such hidden costs as tool storage and obsolescence.

For example, Figure 5–8 illustrates a subdivision of a system grouping by nine classes of parts. Note the similarity of parts within a given vertical column. If we were machining a shaft with external threads and bored partially through at one end, the part would be identified as Class 206.

To develop better methods, analysts should investigate the setup and tools in these three ways: (1) reduce setup time by better planning, methods, and production control; (2) design tooling to utilize the full capacity of the machine; and (3) introduce more efficient tooling.

Reduce Set-up Time by Better Planning Methods and Production Control

Just-in-time (JIT) techniques that had become so popular in recent years emphasize decreasing set-up times to the minimum by eliminating or simplifying them. For example, often a significant portion of set-up time can be eliminated by assuring raw materials are within specifications, that tools are sharp, and that fixtures are available and in good condition. Several points that the analyst should always consider in reducing set-up time include:

1. Work that can be done while the equipment is running should be done at that time. For example, pre-setting tools for numerical control (NC) equipment can be done while the machine is running.
2. Use the most efficient clamping. Usually quick-acting clamps that employ cam action, levers, wedges, and so on, are much faster and can provide adequate force and usually are a good alternative to threaded fasteners. When threaded fasteners need to be used (for clamping force), "C" washers or slotted holes can be used so that nuts and bolts do not have to be removed from the machine and can be reused to reduce set-up time on the next job.
3. Eliminate machine base adjustment. By redesigning part fixtures and through pre-set tooling, spacers or guide-blocks adjustment to the table position may be eliminated.
4. Use templates or block gages to make quick adjustments of machine stops.

The time spent in requisitioning tools and materials, in preparing the workstation for actual production, and in cleaning up the workstation and returning the tools to the tool crib, is usually included in set-up time. Since this time is often difficult to control, it is the portion of the work that is usually performed least efficiently. Effective production control can often reduce this time. Making the dispatch section responsible for seeing that the tools, gages, instructions, and materials are provided at the correct time, and that the tools are returned to their respective cribs after the job has been completed, eliminates the need for the operator to leave the work area. The operator has to perform only the actual setting up and tearing down of the machine. The clerical and routine function of providing drawings, instructions, and tools can be performed by those more familiar with this type of work. Thus, large numbers of requisitions for these requirements can be performed simultaneously, and set-up time minimized. Here again, group technology can be advantageous.

Another function of production control that should be carefully reviewed for possible improvement is scheduling. Considerable set-up time can be saved by scheduling like jobs in sequence. For example, if a one-inch round bar stock job is scheduled to a No. 3 Warner and Swasey turret lathe, from the standpoint of set-up time, it would be economical to have other one-inch round bar stock work scheduled to immediately follow the first one-inch job. This would eliminate collet-changing time, and would probably minimize the tool changes in the square and hex turrets.

Duplicate cutting tools should be available for the operators rather than making them responsible for sharpening their own tools. When operators get new tools, the dull ones should be turned in to the tool crib attendant and replaced with sharp ones. The benefits of standardization cannot be realized when tool sharpening is the responsibility of the operators.

To minimize downtime, each operator should have a constant backlog of work. There should never be a question about what each operator's next work assignment will be. A technique frequently used to keep the work load apparent to the operator, supervisor, and superintendent is to provide a board over each production facility with three wire clips or pockets to receive work orders. The first clip contains all work orders scheduled ahead; the second clip holds the orders currently being worked on; and the last clip holds the completed orders. While issuing work orders, the dispatcher places them in the work-ahead station. At the same time, the dispatcher picks up all completed job tickets from the work-completed station and delivers them to the scheduling department for recording. This system assures operators of perpetual loads in front of them and makes it unnecessary for them to go to the supervisor for their next work assignments.

Making a record of difficult recurring setups can save considerable set-up time when repeat business is received. Perhaps the simplest and yet most effective way to compile a record of a setup is to take a photograph of the setup once it is complete. The photograph should be either stapled to and filed with the production operation card, or placed in a plastic envelope and attached to the tooling prior to storage in the tool crib.

Design Tooling to Utilize the Full Capacity of the Machine

In designing tooling to utilize the full capacity of the machine, analysts should ask: "Can the work be held to permit all machining operations in one setup?" A careful review of many jobs brings out possibilities for multiple cuts, thus utilizing a greater share of the machine's capacity. For example, it was possible to change a milling setup for a toggle lever so that the six faces were milled simultaneously by five cutters. The old setup required that the job be done in three steps. The part had to be placed in a separate fixture three different times. The new setup reduced the total machining time and increased the accuracy of the relationship between the six machined faces.

Figure 5–9 illustrates the use of high die cutting of paper on a Hobbs Mfg. Co. platen press. This work station allows the operator to die cut 300 pages with every closing of the press. The operator positions the die and then activates the machine cycle by pressing a control button. The work table automatically moves to position under the upper platen, the press closes in a period of less than four seconds, and then the press opens for

FIGURE 5-9

High dinking die being used to cut to size advertising literature in stacks of several hundred. Work performed on a Hobbs Mfg. Co. platen press. Upper picture shows operator at workstation. Lower shows dinking die on stack of literature preparatory to cutting.

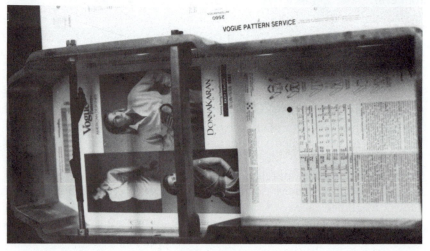

Courtesy of: Butterick Co., Inc.

the operator to remove the die-cut paper pages. By using the full capacity of the press, maximum output is obtained.

Analysts should also consider positioning one part while another is being machined. This opportunity exists on many milling machine jobs where it is possible to conventional mill on one stroke of the table and climb mill on the return stroke. While the operator is loading a fixture at one end of the machine table, a similar fixture is holding a piece being machined by power feed. As the table of the machine returns, the operator removes the first piece from the machine and reloads the fixture. While this internal work is taking place, the machine is cutting the piece in the second fixture.

In view of the ever-increasing cost of energy, it is important to utilize the most economical equipment to do the job. Several years ago, the cost of energy was such an insignificant proportion of total cost that little attention was given to utilization of the full capacity of machines. I have seen literally thousands of operations where but a fraction of machine capacity was utilized, with a resulting waste of electric power. In the metal trades industry today, the cost of power is over 2.5 percent of total cost, with strong indications that the present cost of power will increase by at least 50 percent in the next decade. It is highly probable that careful planning to utilize a large proportion of the capacity of the machine selected to do the work can effect a 50 percent savings in power usage in many of our plants.

Figure 5–10 illustrates the relationship between both efficiency and power factor to percentage of rated full load employed. If the percent of full load is increased from 25 percent to 50 percent, there would be approximately an 11 percent increase in efficiency.

About three quarters of all electric energy is used for motors in American industry. Energy-efficient AC motors provide between 2 and 4 percent more efficient operation than standard motors. See Figure 5–11. Also, energy efficient motors run cooler than standard motors thus giving longer service life. Energy-efficient AC motors work well on continuous installations, such as compressors, pumps, fans, and blowers. Standard motors with a lower initial cost usually are more cost effective on intermittent use motors.

For example, an analyst considers replacing a 25 HP motor that is 10 years old with a new energy-efficient motor. The motor will operate an estimated 6,000 hours per year at 91 percent of full load. The estimated annual savings for this motor based on a 3 percent improvement in efficiency and a cost of power of $0.05/kwh is:

Power cost/yr. (standard motor) = HP × 0.746 × F × H × C × 1/E

where

HP = Horsepower of motor

FIGURE 5–10
Relation between efficiency and power factor to percentage of full load. Power
factor is the true power (may be measured by a wattmeter) divided by the apparent
power (E × I, in an AC circuit carrying a current of I amperes at a potential
difference of E volts).

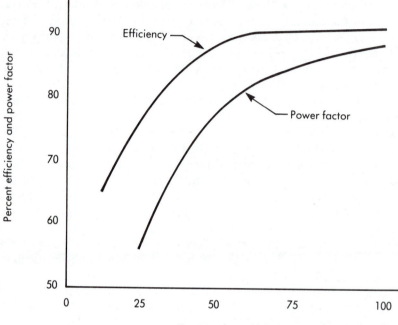

Percent of rated full load

0.746 conversion factor (horsepower to watts)

F = Percent of full load
H = Annual hours of operation
E = Efficiency

Power cost/yr. (standard motor) = 25 × 0.746 × 0.91 × 6000
× 0.05/0.88 = $5,786.00
Power cost/yr. (energy-efficient motor) = 25 × 0.746 × 0.91 × 6000
× 0.05/0.91 = $5,595.00

This indicates a $191 power savings per year. The analyst needs to
compare this annual savings with the extra cost of the energy-efficient
motor. If the extra cost can be saved within a three-year period, the
analyst should proceed.

FIGURE 5–11

Relation between motor size and efficiency of standard and energy-efficient motors

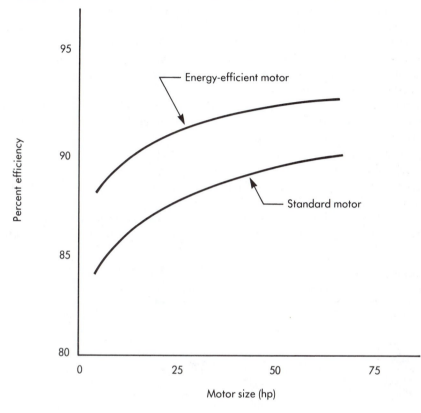

Introduce More Efficient Tooling

Just as new processing techniques are continually being developed, new and more efficient tooling should be considered. Coated cutting tools have dramatically improved the critical wear resistance-breakage resistance combination. For example, TiC-coated tools have provided a 50 to 100 percent increase in speed over uncoated carbide where each have the same breakage resistance.

With the recent development of the low-temperature physical vapor deposition (PVD) process, which is more suitable than the chemical vapor deposition (CVD) process for coating HSS_s and other tool steels and does not require subsequent heat treatment of the tools, coated tool steels are providing dramatic advantages over noncoated tool steels. Advantages include harder surfaces thus reducing abrasive wear, excellent adhesion to the substrates, low coefficient of friction with most work piece materials, chemical inertness, and resistance to elevated temperatures. Table 5–1 illustrates the increased tool life that can be attainted with coated cutting tools. Then too, the use of aluminum oxide-based ceramics are permitting extremely high cutting speeds. They are capable of machining both steel and cast irons in the range of 1,000 to 2,000 surface feet per minute, which is often double that recommended when using carbides. Whisker reinforced ceramic inserts (powder blend containing approximately 35 percent silicon carbide whiskers in an alumina matrix) are able to machine nickel-based super alloys several times faster than previously possible with traditional carbide inserts.

Recently, on-going research by Kennametal has developed Kyon 3000, a technologically advanced, second generation silicon nitride cutting insert material that is demonstrating advantages in high-speed (1000–4000 sfm) and high-feed (0.010–0.040 ipr) applications.

Of course, carbide tools usually offer savings over high-speed steel tools on many jobs. For example, one company realized a 60 percent

TABLE 5–1
Increased tool life attained with coated cutting tools

Cutting Tool				Workpieces Machined before Resharpening	
Name	HSS, AISI Type	Coating	Workpiece Material	Uncoated	Coated
End mill	M7	TiN	1022 Steel, R_C35	325	1200
End mill	M7	TiN	6061-T6 Aluminum Alloy	166	1500
End mill	M3	TiN	7075T Aluminum Alloy	9	53
Gear hob	M2	TiN	8620 Steel	40	80
Broach Insert	M3	TiN	303 Stainless Steel	100,000	300,000
Broach	M2	TiN	48% Nickel alloy	200	3400
Broach	M2	TiN	410 Stainless Steel	10,000–12,000	31,000
Pipe tap	M2	TiN	Gray iron	3000	9000
Tap	M2	TiN	1050 Steel, R_C30–33	60–70	750–800
Form tool	T15	TiC	1045 Steel	5000	23,000
Form tool	T15	TiN	303 Stainless Steel	1840	5890
Cutoff tool	M2	TiC-TiN	Low-carbon steel	150	1000
Drill	M7	TiN	Low-carbon steel	1000	4000
Drill	M7	TiN	Titanium alloy 662 Layered with D6AC Tool steel, R_C48–50	9	86

Source: Courtesy of Scientific Coatings Inc.

saving by changing the milling operation of a magnesium casting. Formerly, the base was milled complete in two operations, using high-speed steel milling cutters. An analysis resulted in employing three carbide-tipped fly cutters mounted in a special holder to mill parts complete. Faster feeds and speeds were possible, and surface finish was not impaired.

Often savings can be achieved by investigating tool geometries. Each setup has different requirements that can be achieved only by designing an engineered system optimizing feed range for chip control, cutting forces, and edge strength. For example, single-sided low force geometries may be designed to provide both good chip control and force reduction. Here high positive rake angles are grouped to reduce the chip thickness ratio providing a low cutting force and cutting temperature. Figure 5–12 illustrates orthogonal chip formation.

Selection of the correct drill is important. For example, the old method for drilling a hole ½ inch in diameter and 7 inches deep in a hard steel shaft with a high-speed twist drill resulted in considerable difficulty. It was necessary to back the drill out about 20 times to remove the chips and cool the drill. It was not possible to get the coolant to the cutting edge because of the depth of the hole. By utilizing a carbide-tipped V-shaped

FIGURE 5–12
Orthogonal chip formation

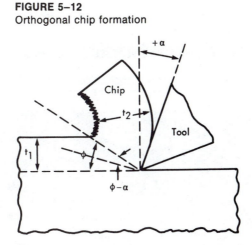

Note:

$$\frac{t_2}{t_1} = \frac{\cos(\phi - \alpha)}{\sin \phi}$$

where

t_1 = chip thickness before cutting
t_2 = chip thickness after cutting
ϕ = shear angle
α = rake angle

gun drill with a hole in the center through which the coolant was pumped, an operator drilled the hole in about one third the time (see Figure 5–13).

Again, it was possible to greatly reduce the costs of die maintenance by using carbide as a die-cutting edge material. To produce a certain stamping, a die with tool steel cutting edges had to be resharpened after every 50,000 pieces. A study of die-cutting edges resulted in an improved die with edges of tungsten carbide. Now, more than 600,000 pieces are produced before resharpening is necessary (see Figure 5–14).

While introducing more efficient tooling, develop better methods for holding the work. Be sure that the work is held so that it can be positioned and removed quickly. For example Figure 5–15 illustrates a hinge support assembly made up of three components held together with three rivets (see A). In the method prior to an operations analysis study operators loaded these components into a fixture by hand and then activated the riveter by palm push buttons (see B). The new method (see C) introduced a dial indexing table with automatic stations to load the roller, assemble three rivets, and eject the completed assembly. The loading of the metal parts continued as a manual operation but there was a 280 percent increase in productivity. Another benefit of the new method was an increase in quality. The machine is designed to automatically check for the

FIGURE 5–13
Carbide-tipped V-shaped gun drill with hole through center for coolant answers deep-hole drilling problem

FIGURE 5–14
More than 600,000 pieces are produced on this die before resharpening is necessary

presence of all parts in the assembly, thus assuring that the completed assembly is correct.

Finally, management should provide proper hand tools. Many hand tools, such as screwdrivers, are designed for a specific task. A screwdriver that is efficient under one set of conditions may be very inefficient

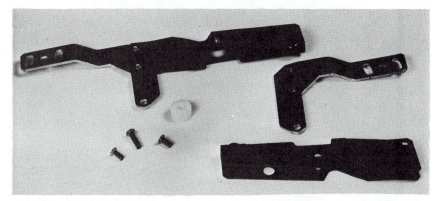

A. Parts assembled with rivets   *General Electric Co., Louisville, Kentucky*

FIGURE 5–15 (*concluded*)

B. Components manually loaded into fixture and riveter activated with push bottons

C. Improved methods of holding work and assembly—280 percent improvement in productivity

under another set. Analysts should specify to see that the very most effective hand tools are used.

WORKING CONDITIONS

Methods analysts should provide good, safe, comfortable working conditions. Experience has proved conclusively that plants with good working conditions outproduce those maintaining poor conditions. The economic return from investment in an improved working environment is usually significant. Ideal working conditions improve the safety record, reduce absenteeism, tardiness, and labor turnover, raise employee morale, and improve public relations, in addition to increasing production.

Some common ways to improve working conditions are as follows:

1. Improve lighting.
2. Control temperature.
3. Provide adequate ventilation.
4. Control sound.
5. Promote orderliness, cleanliness, and good housekeeping.
6. Arrange for the immediate disposal of irritating and harmful dusts, fumes, gases, vapors, and fogs.
7. Provide guards at nip points and at points of power transmission.
8. Provide personal protective equipment.
9. Sponsor and enforce a well-formulated first-aid program.

Improve Lighting

The intensity of light required depends primarily on what operations are being performed in an area. A toolmaker or an inspector requires greater intensity of light than personnel in a store-room. Analysts also need to consider glare, the quality of the light, the location of the light source, contrasts in color and brightness, and flickering and shadows. Some ways to obtain good lighting are as follows:

1. Reduce glare by installing a large number of light sources to give the total required light output.
2. Enclose filament-type bulbs in opalescent bowls to reduce glare by spreading the light output over a greater surface.
3. Produce a satisfactory approximation to white light for most uses by using a filament-type bulb or a single white fluorescent unit. (White light, or the composition of average sunlight, is generally considered ideal.)
4. Eliminate all shadows by providing the correct level of illumination at all points of the workstation. In view of power costs, identify areas with too much illumination, as well as areas with inadequate illumination.
5. Utilize the most efficient lighting that provides the quality and

FIGURE 5–16
Cutaway view of fluorescent light bulbs. These
bulbs are interchangeable with regular
incandescent light bulbs. At 13 to 15 watts, the
globe or cylindrically shaped bulbs provide
equivalent light while using about 75 percent less
energy and have an average life span of 9,000
hours. This service life is typically 10 times longer
than the standard incandescent's life expectancy
of about 900 hours. The initial cost of these
fluorescent bulbs is about 30 times the cost of
the incandescent counterpart.

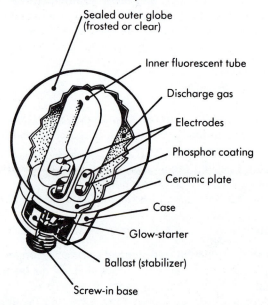

quantity of light desired at the work site. For example, fluorescent
bulbs, designed to replace 50 or 60 watt incandescents, create a bright
equivalent lighting effect for 75 percent less energy. See Figure 5–16.

Control Temperature

The human body endeavors to maintain a constant temperature of
about 98° F. When the body is exposed to unusually high temperatures,
large amounts of perspiration evaporate from the skin. During the perspiration process, sodium chloride is carried through the pores of the skin
and is left on the skin surface as a residue when evaporation takes place.
This represents a direct loss to the system and may create a disturbance to
the normal balance of fluids in the body. The result is heat fatigue and heat
cramps, with an accompanying slowdown in production. The performance of a very good operator deteriorates as rapidly as the performance

of the average and less-than-average operator. In clerical operations, such as typing and word processing, output not only suffers, but errors also increase.

Conversely, detailed time studies have repeatedly brought to attention the loss in production when working conditions are unduly cold. Temperature should be controlled so that it will be between 65 and 75° F the year round. If this level can be maintained, losses and slowdowns from heat fatigue, heat cramps, and lack of manipulative dexterity are kept to a minimum.

Provide Adequate Ventilation

Ventilation also plays an important role in the control of accidents and operator fatigue. Disagreeable fumes, gases, dusts, and odors cause fatigue that definitely reduces the physical efficiency of the worker and often creates a mental tension. Laboratory findings indicate that the depressing influence of poor ventilation is associated with air movement, as well as with temperature and humidity.

When humidity increases, evaporative cooling decreases rapidly, thus reducing the ability of the body to dissipate heat. Under these conditions high heart rates, high body temperatures, and slow recovery after work result in pronounced fatigue.

The New York State Commission on Ventilation disclosed that in heavy manual labor 15 percent less work was done at 75° F with 50 percent relative humidity than was done at 68° F with the same humidity, and that 28 percent less work was done at 86° F with 80 percent relative humidity. It also brought out that 9 percent less work was accomplished in stagnant air than in fresh air when the temperature and humidity remained constant. Further experiment showed a reduced work capacity of 17 percent at 75° F, and of 37 percent at 86° F, as compared with work done at 68° F.

Similar experiments made by the American Society of Heating and Ventilating Engineers showed that similar gains in production, safety, and employee morale follow when ideal ventilation is introduced to the production floor.

Control Sound

Both loud and monotonous noises are conducive to worker fatigue. Constant and intermittent noise also tends to excite the worker emotionally, resulting in loss of temper and difficulty in doing precision work. Quarrels and poor conduct by workers can often be attributed to disturbing noises. Tests have proved that irritating noise levels heighten the pulse rate and blood pressure and result in irregularities in heart rhythm. The nervous system of the body strains to overcome the effect of noise, resulting in neurasthenic states.

Promote Orderliness, Cleanliness, and Good Housekeeping

A good industrial housekeeping program will (1) diminish fire hazards, (2) reduce accidents, (3) conserve floor space, and (4) improve employee morale.

Industrial accident statistics indicate that a large percentage of accidents are the result of poor housekeeping. Many times the expression "A place for everything and everything in its place" has been cited as the basis for orderliness. This is true; methods analysts should be sure that a place is provided for everything and, if necessary, follow through with supervision to see that everything is in its place.

When the general layout of a plant shows management's and supervision's desire to have orderliness, cleanliness, and good housekeeping, then the employees follow the examples set for them, and practice good housekeeping themselves.

Arrange for the Disposal of Irritating and Harmful Dusts, Fumes, Gases, Vapors, and Fogs

Dusts, fumes, gases, vapors, and fogs generated by various industrial processes constitute one of the major dangers encountered by workers. The following classification of dusts, prepared by the National Safety Council, gives an indication of the problem:

1. Irritating dusts, such as metal and rock dusts.
2. Corrosive dusts, such as those from soda and lime.
3. Poisonous dusts, such as those from lead, arsenic, or mercury.
4. Dusts from fur, feathers, and hair may carry disease germs which may infect the worker.

All of these dangers can be eliminated by employing suitable methods, such as exhaust systems, complete enclosing of the process, wet or absorbing techniques, and complete protection of the operator through personal respiratory protective equipment. Probably the most effective measure for controlling dust and fumes is through local exhaust systems in which a hood collects the substance to be removed right at the point of generation. A fan draws the contaminated air through metal pipes or ducts to some properly provided place for filtering and disposal. The diameter of the exhaust pipe is an important detail in a satisfactory installation. Generally, the larger the exhaust pipe, the more costly the initial installation. However, larger pipes allow increased efficiency of the motive system since they require less power for the exhaust of air (see Table 5–2). And, as already pointed out, the conservation of power is an important consideration today.

Many times the dust or fumes collected can be used. Perhaps wood dust can be used as a source of heat, as can blast furnace gases and other gases containing carbon monoxide. One manufacturer in the lead industry

TABLE 5–2

Diameter of buffing wheels	Maximum grinding surface (sq. in.)	Minimum diameter of pipe (in.)
6 in. or less, not over 1 in. thick..............................	19	3
7 to 9 in., inclusive, not over 1½ in. thick	43	3½
10 to 16 in., inclusive, not over 2 in. thick	101	4
17 to 19 in., inclusive, not over 3 in. thick	180	4½
20 to 24 in., inclusive, not over 4 in. thick	302	5
25 to 30 in., inclusive, not over 5 in. thick	472	6

Source: National Safety Council Safe Practices pamphlet no. 32.

claims that it salvages enough lead annually to more than pay for the operation of its exhaust system.

Provide Guards at Nip Points and at Points of Power Transmission

In view of the Williams-Steiger Occupational Safety and Health Act, (OSHA) the employer is legally responsible for installing proper safeguards to protect all employees. Guards must be designed correctly if they are to provide protection and still not hinder production. The general requisites of good guards are to:

1. Effectively protect the employee.
2. Permit normal operation of the facility at a pace equal to or greater than that used prior to the guard installation. (Employees who know they are safe will tend to produce at a more effective pace.)
3. Permit normal maintenance to the facility.

Since the machine tool builder is in an ideal position to provide satisfactory guards, it is usually advantageous to purchase the necessary guards at the same time that the machines are procured. Although many homemade guards are doing an excellent job, they are not usually as efficient, attractive, or inexpensive as purchased units.

Production workers are fully aware that unguarded machinery is hazardous. If they are exposed to such working conditions, there is a natural tendency to execute at low effort as a precautionary measure. Also, employers who fail to spend any reasonable amount to remove visible hazards cannot expect full cooperation from their employees. See Figure 5–17.

Consider the Use of Robots

Today manufacturers can use robots in those work centers where danger to the worker prevails because of the nature of the process. For example, in the die-casting process there can be considerable danger due to splashing of hot metal during that portion of the cycle when the molten

FIGURE 5–17
Work center showing press operations. Note that each person has two hand buttons spread well apart (36 inches) so that the operator must have both hands in a safe position when the press starts. Activated palm buttons are the number one safety device for protecting an operator while loading and unloading fabrication presses.

metal is injected into the die cavity. One of the original applications for the development of robots was die casting.

In one company a five-axis robot developed by Unimation, Inc. serves a 600-ton microprocessor-controlled die-casting machine. In the operation, the robot moves into position when the die opens, grasps the casting by its slug, and clears it from the cavity. At the same time, it initiates automatic die-lubrication sprays. The robot displays the casting to infrared scanners, then signals the die-casting machine to accept another shot. The casting is deposited by the robot on an output station for trimming. Here an operator, remote from the die-casting machine, safely trims the casting preparatory to subsequent secondary operations.

Japanese motor vehicle manufacturers have placed particular emphasis on the use of robots in welding. For example, at Nissan Motors 95 percent of the welds on vehicles are made by robots; Mitsubishi Motors reported about 70 percent of its welding is performed by robots. In these companies robot downtime averages less than 1 percent.

Press operations represent another class of dangerous work where

FIGURE 5–18
Robot feeding a punch press

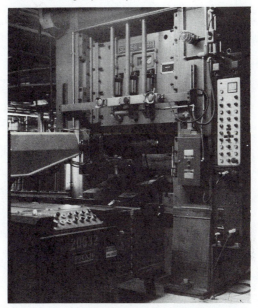

General Electric Co., Louisville, Kentucky

robots are utilized effectively. Figure 5–18 shows a robot feeding a fabricating punch press.

In addition to hazardous work centers, it is advantageous today to consider the use of robots in other manufacturing areas. For example, assembly areas should be considered since this type of work typically has a high direct labor cost—in some cases accounting for as much as half of the manufacturing cost of a product. The principal advantage of integrating a modern robot in the assembly process is its inherent flexibility. That is its ability to assemble multiple products on a single system and to be reprogrammed to handle various tasks with part variations. In addition, robotic assembly can provide consistently repeatable quality with predictable product output with little capital obsolescence.

A robot's typical life is approximately 10 years. If well maintained and if used for moving small payloads, the life can be extended up to 15 years. Consequently, a robot's depreciation cost can be relatively low. The reader should recognize that if a given robot's size and configuration are adequate, it can be used in a variety of operations. For example: load a die-casting facility, load a quenching tank, load and unload a board drop hammer forging operation, load a plate glass washing operation, and so on. In theory, a robot of the correct size and configuration can be programmed to do any job.

FIGURE 5-19
This is an illustration of a few common industrial robot applications. One welding
robot is shown (a), but typically a number of robots would be used along an
automotive assembly line. In diecasting application (b), a robot unloads diecasting
machines, performs quench operations, and loads material into a press. Production
machining line is used for producing cam housings (c). Assembly line (d) uses a
combination of robots, parts feeders, and human operators.

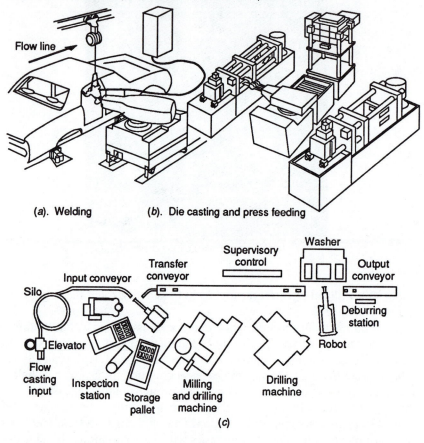

(*a*). Welding   (*b*). Die casting and press feeding

(*c*)

Provide Personal Protective Equipment

Because of the nature of the operation and/or economic considerations, changing methods, equipment, or tools does not always eliminate
certain hazards. When this is the case, operators can often be fully protected by personal protective equipment. Representative personal protective equipment would include goggles, face shields, helmets, aprons, jack-
ets, trousers, leggings, gloves, shoes, and respiratory equipment.

To assure that operating personnel conscientiously use protective
equipment, companies should either furnish it to employees at cost or at

FIGURE 5-19 (concluded)

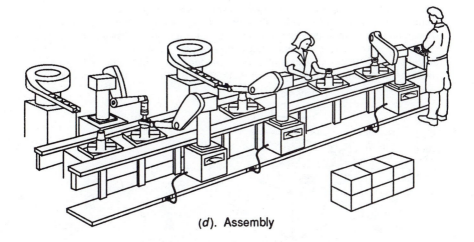

(d). Assembly

no expense. The policy of having the company absorb completely the cost of personal protective equipment is becoming more and more common. Innumerable cases can be cited where personal protective equipment has saved an eye, a hand, a foot, or a life. For example, one steel company reported that 20 fatalities were prevented in one year by the enforced wearing of company-provided helmets. A northwest lumber company reported within a 20-day period the use of protective hats prevented six serious head injuries.

A study recently conducted by the U.S. Bureau of Labor Statistics shows that three out of five workers who suffered eye injuries or chemical burns were not wearing eye protection at the time of the accident. In 75 percent of those cases, the workers thought protective eye wear was not required in that particular situation. Eye protection equipment is not expensive and should always be available. Employees must be taught the importance of utilizing the protective equipment specified and develop the attitude that they will not deviate from the prescribed use of such equipment. Compliance should be a condition of employment.

Sponsor and Enforce a Well-Formulated First-Aid Program

The most advanced program in industrial safety will never be able to completely eliminate all accidents and injuries. To adequately care for the injuries that do occur, a well-formulated first-aid program is essential. This will include training and publicity, so that the employees will be fully aware of the danger of infection and of the ease of avoiding infection through first aid. Also, a complete procedure to be followed in case of injury must be arranged, with proper instructions to all supervisory

FIGURE 5–20
This illustration shows several commercially available robots. Unit (a) has six axes of articulation. Lifting is accompanied with pivoting action of the turret.
General-purpose robot (b) also has six axes of freedom. Termination of the arm can be equipped with various types of power tools or gripping devices. Another type is shown mounted on an overhead bridge crane to perform welding at different work stations (c). Painting operations of a robot painter (d) are programmed by an operator. One control establishes arm motion and another the paint sequence.

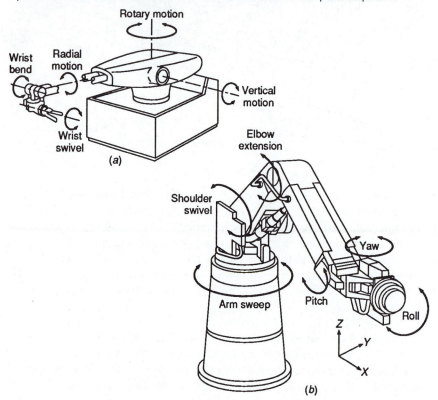

levels. A well-equipped first-aid room must be provided to care for injured and ill employees until professional medical aid is available.

OSHA

The Occupational Safety and Health Act of 1970 was passed by Congress "to assure so far as possible every working man and woman in the Nation safe and healthful working conditions and to preserve our human resources." Under the act, the Occupational Safety and Health Administration (OSHA) was created to:

1. Encourage employers and employees to reduce workplace hazards and to implement new or improved existing safety and health programs.

FIGURE 5-20 (*concluded*)

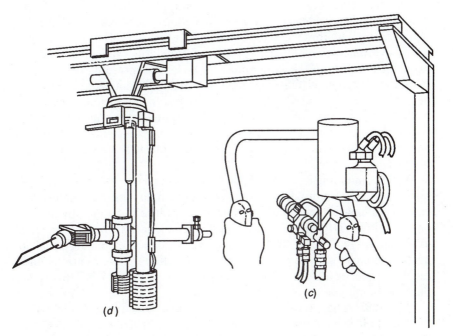

(c)

(d)

2. Establish "separate but dependent responsibilities and rights" for employers and employees for the achievement of better safety and health conditions.
3. Maintain a reporting and recordkeeping system to monitor job-related injuries and illnesses.
4. Develop mandatory job safety and health standards and enforce them effectively.
5. Provide for the development, analysis, evaluation, and approval of state occupational safety and health programs.

Since the act can intimately affect the design of the workplace, methods analysts should be knowledgeable regarding details of this act. The general duty clause of the act states that each employer "must furnish a place of employment which is free from recognized hazards that cause or are likely to cause death or serious physical harm to employees." Furthermore, the act brings out that it is the employers' responsibility to become familiar with standards applicable to their establishments and to ensure that employees have and use personal protective gear and equipment for safety.

OSHA standards fall into four categories: general industry, maritime, construction, and agriculture. All OSHA standards are published in the *Federal Register* which is available in most public libraries. Annual subscriptions can be obtained from the Superintendent of Documents, U.S.

Government Printing Office, Washington, D.C. 20402. OSHA can begin standards-setting procedures on its own initiative or on the basis of petitions from the Secretary of Health, Education, and Welfare (HEW), the National Institute for Occupational Safety and Health (NIOSH), state and local governments, nationally recognized standards-producing organizations such as the ASME, and employer or labor representatives.

Of these groups, NIOSH, an agency of HEW, is quite active in recommendations for standards. It conducts research on various safety and health problems and provides considerable technical assistance to OSHA. Especially important is the investigation of toxic substances by NIOSH and its development of criteria for the use of such substances in the workplace.

While conducting its research, NIOSH may make workplace investigations, gather testimony from employers and employees, and require that employers measure and report employee exposure to potentially hazardous materials. NIOSH may also require employers to provide medical examinations and tests to determine the incidence of occupational illnesses among employees. When such examinations and tests are required by NIOSH for research purposes, they may be paid for by NIOSH rather than the employer.

Once OSHA has developed plans to propose, amend, or delete a standard, it publishes these intentions in the *Federal Register* as an "Advance Notice of Proposed Rulemaking" or later as a "Notice of Proposed Rulemaking." No decision on a permanent standard is ever reached without due consideration of arguments and data received from the public in written submissions and at hearings. Any person who may be adversely affected by a final standard may file a petition for judicial review of the standard with the U.S. Circuit Court of Appeals for the circuit in which that person lives or establishes a principal place of business.

At times employers ask OSHA for a variance from a standard or regulation if they cannot fully comply by the effective date, due to shortages or unavailability of materials, equipment, or personnel. This is quite in order. Also, if employers can prove that their facilities or methods of operation protect their employees "at least as effectively as" that required by OSHA, then a variance may be obtained.

OSHA may allow a temporary variance until the shortage or unavailability of materials, equipment, or personnel is relieved. Meanwhile employers must demonstrate to OSHA that they are taking all available steps to safeguard employees and that they have enforced an effective program for complying with the standard regulation as soon as possible. A temporary variance may be granted for up to one year. It is renewable twice, each time for six months.

Prior to the OSHA no centralized and systematic method existed for monitoring occupational safety and health problems. With OSHA standards came the first basis for consistent, nationwide procedures.

Employers of 11 or more employees now must maintain records of occupational injuries and illnesses as they occur. Employers with 10 or fewer employees are exempt from keeping such records unless they are selected by the Bureau of Labor Statistics (BLS) to participate in periodic statistical surveys.

An occupational injury has been defined as "any injury such as a cut, fracture, sprain or amputation which results from a work-related accident or from exposure involving a single incident in the work environment." An occupational illness is "any abnormal condition or disorder, other than one resulting from an occupational injury, caused by exposure to environmental factors associated with employment." Occupational illnesses include acute and chronic illnesses which may be caused by inhalation, absorption, ingestion, or direct contact with toxic substances or harmful agents.

The act requires that all occupational illnesses be recorded regardless of severity and that all occupational injuries be recorded if they result in death, loss of one or more workdays, restriction in motion or ability to do the work that had been done, loss of consciousness, transfer to another job, or medical treatment other than first aid.

To enforce its standards, OSHA is authorized to conduct workplace inspections. Consequently, every establishment covered by the act is subject to inspection by OSHA compliance safety and health officers. The act states that "upon presenting appropriate credentials to the owner, operator, or agent in charge," an OSHA compliance officer is authorized to:

Enter without delay and at reasonable times any factory, plant, establishment, construction site or other areas, workplace, or environment where work is performed by an employee of an employer

and to

Inspect and investigate during regular working hours, and at other reasonable times, and within reasonable limits and in a reasonable manner, any such place of employment and all pertinent conditions, structures, machines, apparatus, devices, equipment and materials therein, and to question privately any such employer, owner, operator, agent or employee.

OSHA inspections, with few exceptions, are concluded without advance notice. In fact, alerting an employer in advance of an OSHA inspection can bring a fine of up to $1,000 and/or a six-month jail term.

Those special circumstances where OSHA may give notice of inspection to an employer include those organizations where:

1. Imminently dangerous situations exist which require correction as soon as possible.
2. Some inspections necessitate special preparation or must take place after regular business hours.

3. Some inspections where prior notice assures that the employer and employee representatives or other personnel will be present.
4. Those situations where the OSHA area director determines that advance notice would produce a more thorough or more effective inspection.

Upon inspection, if an imminently dangerous situation is found, the compliance officer asks the employer to voluntarily abate the hazard and to remove endangered employees from exposure. Notice of the imminent danger also must be posted. Before the OSHA inspector leaves the workplace, he or she will advise all affected employees of the hazard.

Industries are selected for inspection on the basis of criteria such as death, injury, and illness incidence rates, or employee exposure to toxic substances. Special emphasis may be regional or national in scope, depending on the distribution of the workplaces involved.

At the time of the inspection, the employer is asked to select an employer representative to accompany the compliance officer during the inspection. An authorized employee representative also is given the opportunity to attend the opening conference and to accompany the compliance officer during the inspection. In those plants having a union, the union ordinarily designates the employee representative to accompany the compliance officer. Under no circumstances may the employer select the employee representative for the inspection. The act does not require an employee representative for each inspection; however, where there is no authorized employee representative, the compliance officer must consult with a reasonable number of employees concerning safety and health matters in the workplace.

After the inspection tour, a closing conference is held between the compliance officer and the employer or the employer representative. Subsequently the compliance officer reports findings to the OSHA office, and the area director determines what citations, if any, will be issued, and what penalties, if any, will be proposed.

Citations inform the employer and employees of the regulations and standards alleged to have been violated, and of the proposed time set for their abatement. The employer will receive citations and notices of proposed penalties by certified mail. The employer must post a copy of each citation at or near the place where a violation has occurred, for three days or until the violation is abated, whichever period is longer.

The compliance officer has authority to issue citations at the worksite, following the closing conference. In order to do so, he or she must first discuss each apparent violation with the area director and receive approval to issue citations.

Five types of violations which may be cited and the penalties which may be imposed are:

1. OTHER THAN SERIOUS VIOLATION. This type of violation has a direct relationship to job safety and health, but probably would not cause

death or serious physical harm. A proposed penalty of up to $1,000 for each violation is discretionary. A penalty for an other than serious violation may be adjusted downward by as much as 80 percent, depending on the employer's good faith (demonstrated efforts to comply with the act), history of previous violations, and size of business. When the adjusted penalty amounts to less than $50, no penalty is proposed.

2. SERIOUS VIOLATION. This is a violation in which there is substantial probability that death or serious harm could result, stemming from a hazard about which the employer knew, or should have known. A mandatory proposed penalty ranging from $300 to $1,000 for each violation is assessed. A penalty for a serious violation may be adjusted downward by as much as 80 percent, based on the employer's good faith, history of previous violations, and size of business.

3. IMMINENT DANGER. This is a situation in which there is reasonable certainty that a danger exists that can be expected to cause death or serious physical harm immediately or before the danger can be eliminated through normal enforcement procedures. An imminent danger may be cited and penalized as a serious violation.

4. WILLFUL VIOLATION. This is a violation that the employer intentionally and knowingly commits. The employer either knows that his or her actions constitute a violation, or is aware that a hazardous condition exists and has made no reasonable effort to eliminate it. Penalties of up to $10,000 may be proposed for each willful violation. A proposed penalty for a willful violation may be adjusted downward, depending on the size of business and its history of previous violations. If an employer is convicted of a willful violation that has resulted in the death of an employee, the offense is punishable by a court-imposed fine of not more than $10,000, or by imprisonment for up to six months, or both. A second conviction doubles these maximum penalties.

5. REPEATED VIOLATION. A repeated violation occurs when a violation of any standard, regulation, rule, or order is reinspected and another violation of the same previously cited section is found. A citation for a repeated violation is not necessarily issued for violations involving the same piece of equipment or location. If on reinspection, a violation of the previously cited standard, regulation, rule, or order is found, but it involves another piece of equipment and/or different location in the establishment or worksite, it may be considered a repeated violation. Each repeated violation can bring a fine of up to $10,000.

Other violations for which citations and proposed penalties may be issued are as follows:

1. Falsifying records, reports, or applications on conviction can bring a fine of $10,000 and six months in jail.
2. Violating the posting requirements can bring a civil penalty of up to $1,000.
3. Assaulting a compliance officer, or otherwise resisting, opposing,

ng, or interfering with a compliance officer in the perform-
is or her duties is a criminal offense and can bring a fine of
than $5,000 and imprisonment for not more than three years.
, correct a violation can bring a civil penalty of up to $1,000
for each day the violation continues beyond the prescribed abatement
date.

Congress has authorized OSHA to provide a free on-site consultation
service for employers in all fifty states. This service is available today on
request, and priority is given to smaller businesses, which are generally
less able to afford private sector consultation. These consultants help
employers identify hazardous conditions and determine corrective measures.

The following responsibilities of the employer under OSHA are of
importance to methods analysts:

1. To provide a workplace which is free from recognized hazards that
    are causing or are likely to cause death or serious physical harm to
    employees, and which complies with standards, rules, and regula-
    tions issued under the act.
2. To examine workplace conditions to make sure that they conform to
    applicable standards.
3. To make sure employees have and use safe tools and equipment and
    that such equipment is properly maintained.
4. To update operating procedures and to communicate them so that
    employees follow safety and health requirements.

The seventh primary approach to operations analysis, working conditions, is never overlooked by competent analysts. This chapter has given
only a basic coverage of the physical environment. Chapter 9 presents a
much more detailed discussion of the physical environment, as well as the
physiological and psychological factors related to both the operator and
the work force. You may want to review Chapter 9 at this time.

MATERIAL HANDLING

Material handling includes motion, time, place, quantity, and space.
First, material handling must assure that parts, raw materials, in-process
materials, finished products, and supplies are moved periodically from
location to location. Second, since each operation in the process requires
materials and supplies at a particular point in time, material handling
assures that no production process or customer is hampered by either
early or late arrival of materials. Third, material handling must assure that
personnel deliver materials to the correct place. Fourth, material handling
must assure that materials be delivered at each location without damage
in the proper quantity. Finally, material handling must consider storage
space—both temporary and dormant.

Thus, good material handling provides for the delivery of an adequate inventory of material at the proper time and in the proper condition to the point of use at the least total cost. Good material handling must act in concert with good material management. Thus, when analysts consider the eighth primary approach to operation analysis, they should consider the following as an integrated system: inventory control, purchasing policy, receiving, shipping, inspection, storage, traffic control, pickup and delivery, layout, and facilities.

The tangible and intangible benefits of material handling can be reduced to four major objectives, as outlined by the American Material Handling Society: These are:

1. Reduction of handling costs.
   a. Reduction of labor costs.
   b. Reduction of material costs.
   c. Reduction of overhead costs.
2. Increase of capacity.
   a. Increase of production.
   b. Increase of storage capacity.
   c. Improved layout.
3. Improvement in working conditions.
   a. Increase in safety.
   b. Reduction of fatigue.
   c. Improved personnel comforts.
4. Better distribution.
   a. Improvement in handling system.
   b. Improvement in routing facilities.
   c. Strategic location of storage facilities.
   d. Improvement in user service.
   e. Increase in availability of product.

A study conducted by the Material Handling Institute brought out that between 30 and 85 percent of the cost of bringing a product to market is associated with material handling. Axiomatically, the best-handled part is the least manually handled part. Whether the distance of moves is large or small, these moves should be scrutinized. By considering the following six points, you can reduce the time and energy spent in handling material.

1. Reduce the time spent in picking up material.
2. Reduce material handling by using mechanical equipment.
3. Reduce material handling by using mechanized or automated equipment.
4. Make better use of existing handling facilities.
5. Handle material with greater care.
6. Consider the application of bar coding for inventory and related applications.

A good example of the application of these six points is the evolution of warehousing so that the former storage center has become an automated distribution center. Today, the automated warehouse has computer control of material movement, as well as information flow through data processing. In this type of automated warehouse, receiving, transporting, storing, retrieving, and inventory control are treated as an integrated function.

Reduce the Time Spent in Picking Up Material

Many people think of material handling as only transportation, and neglect to consider positioning at the workstation. This is equally important, and since it is often overlooked, it may offer even greater opportunities for savings than does transportation. Reducing the time spent in picking up material minimizes tiring, costly manual handling at the machine or the workplace. It gives the operator a chance to do the job faster and with less fatigue and greater safety.

Consider the possibility of avoiding loose piling on the floor. Perhaps the material can be stacked directly on pallets or skids after being processed at the workstation. This can result in a substantial reduction of terminal transportation time (the time material handling equipment stands idle while loading and unloading take place). In fact, a good axiom to remember is, if terminal time in the transportation of any material is long, then an improved material handling installation is needed. A good example of reducing terminal time to a minimum is the utilization of an electro-magnet on a crane to lift ferrous loads. The crane is tied up a negligible amount of time at the terminal points. Usually some type of conveyor or mechanical fingers can bring material to the workstation and thus avoid or reduce the time spent in picking up the material. Often plants install gravity conveyors, used in conjunction with the automatic removal of finished parts, thus minimizing material handling at the workstation.

In one plant that manufactured leaf springs for trucks, an early operation was to shear to length spring steel bar stock. The lengths of steel after being sheared fell into a large welded steel tote box. Upon completing a run (the size of the run depended upon the order size and usually was small—fewer than 200 pieces), the operator picked up each leaf (weighing about 20 lbs.) and carefully stacked it preparatory to being moved by fork truck to the next operation. Of course, while the operator was doing the stacking, the shear was idle. Furthermore, the stacking required more operator time than the shearing. By arranging a guide in connection with a gravity chute, the parts upon being sheared fell into a stacked alignment in the tote box. This method improvement eliminated the time spent on picking up the heavy material and stacking it.

Several types of positioning equipment are available to reduce the time spent in handling material to and from the workstation. Hydraulic tables can place sheets of material at the proper feeding height for shears, presses, brakes, and other machines, as well as transport and bring to position dies and other heavy tools. The portable elevator or positioning truck is an elevating mechanism supported on a truck mast and base. The hoisting unit is either motorized or has a winch mechanism. Another well-known positioning aid is the welding positioner. The latest models provide for powered rotation and the elevation of materials to permit downhand welding motions.

With regard to the time spent in picking up material, analysts should ask the following questions: "Can loose piling on the floor be avoided?" "Can material be weighed without picking it up?" "Can a conveyor be used to avoid picking it up?" "Is the unit load as large as practical?" See Figures 5–21 and 5–22 for examples of typical handling equipment.

Interfaces between different types of handling and storage equipment should be studied so as to provide efficient arrangements. For example, the following sketch of order picking arrangements shows how materials can be removed from reserve or staging storage by using a man-aboard order picking vehicle, left, or manually, right. A lift truck can be used to replenish pallet racks. After the required items are removed from the flow rack, they are sent to order accumulation and packaging operations by conveyor.

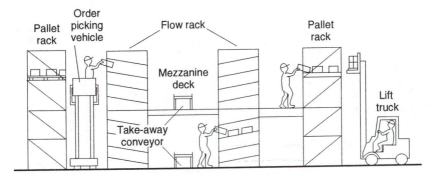

Reduce Material Handling by Using Mechanical Equipment

Mechanizing the handling of material usually reduces labor costs, reduces damage of handled materials, improves safety, alleviates fatigue, and increases production. However, care must be exercised on the proper selection of equipment and methods. Standardization of equipment is important because it simplifies operator training, provides interchangeability of equipment, and requires less stocking of repair parts. Compare the energy consumption of competitive material handling systems when preparing economic alternates. Similarly, consider any adverse effects on the environment when comparing alternative material handling systems.

FIGURE 5–21
Typical handling equipment used in industry today

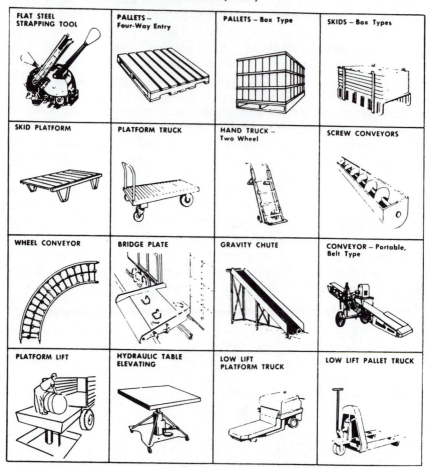

FLAT STEEL STRAPPING TOOL	PALLETS – Four-Way Entry	PALLETS – Box Type	SKIDS – Box Types
SKID PLATFORM	PLATFORM TRUCK	HAND TRUCK – Two Wheel	SCREW CONVEYORS
WHEEL CONVEYOR	BRIDGE PLATE	GRAVITY CHUTE	CONVEYOR – Portable, Belt Type
PLATFORM LIFT	HYDRAULIC TABLE ELEVATING	LOW LIFT PLATFORM TRUCK	LOW LIFT PALLET TRUCK

The Material Handling Institute

The savings possible through the mechanization of material handling equipment are typified by the following examples. One plant installed a monorail hoist over two workplaces and a paint basin. Formerly, 25 tons of tools in process were transported weekly by hand. The return of the monorail installation investment was estimated at 200 percent in the first year. In another instance, a 100 percent investment return was realized by moving one heavy cutoff saw close to another cutoff saw, and then furnishing a jib crane to service the two saws.

At the outset of the 360 program, IBM considered several methods for populating panels with the Solid Logic Technology circuit cards used in the computers manufactured at its Endicott plant.

FIGURE 5–22
Typical handling equipment used in industry today

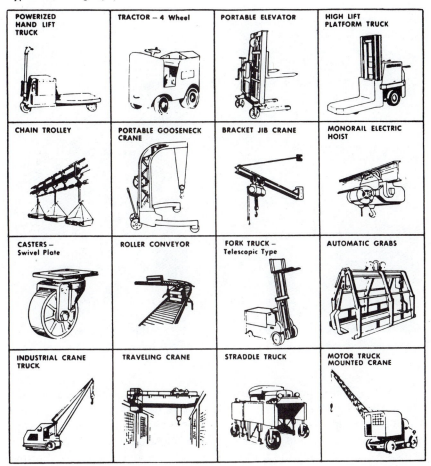

The Material Handling Institute

Under the previous method, the circuit cards were stored in containers on shelving units in a storage crib area. The operator would go to the storage crib, select the correct cards required for a specific panel based on its "plug" list, return to the workbench, and then proceed to insert the circuit cards into the panel in accordance with the plug list.

The improved method resulted in the installation of two automated vertical storage machines (see Figures 5–23 and 5–24). Both machines were in an area under one of the "sawtooth" sections of the building roof, which literally uses storage space above the normal ceiling height. Each automated vertical storage unit has 10 carriers with four pullout drawers

Motion and Time Study

FIGURE 5–23
Automated vertical storage machine used in conjunction with the assembly of
computer panels

per carrier; 20 stop positions were created so that two drawers are available at each stop position. The carriers move up and around in a system
which is a compressed version of a Ferris wheel. With 20 possible stop
positions on call, the unit always selects the closest route—either forward
or backward—to bring the proper drawers to the opening in the shortest
possible time.

The assembly station is arranged at the opening level, where an operator can select cards and place them in the proper position in the panel.
From this seated position, the operator dials the correct stop. The drawer
is pulled forward into a position exposing the needed cards, and the
proper card is withdrawn and placed in the panel as designated on the
"plug" list.

The improved method has reduced by approximately 50 percent the
storage area required in valuable manufacturing floor space, has improved
workstation layout, and has substantially reduced populating errors by
minimizing operator handling, decision making, and fatigue.

In considering the possible advantages of using mechanical and/or
automated equipment, the following questions should be asked: "Can
heavy material be handled more easily with mechanical and/or automated

FIGURE 5–24
Work area of vertical storage machine used in the assembly of computer panels

equipment?" "Can parts be handled more quickly with mechanical and/or automated equipment?" and "Can material be stacked higher using mechanical and/or automated equipment?"

Often an Automated Guided Vehicle (AGV) can replace a driver. AGVs are successfully used in a variety of applications. Perhaps the most familiar is for mail delivery. Typically these are not programmed but follow a magnetic or optical guide for a planned route. Stops are made at specific locations for a predetermined period giving an employee adequate time for unloading and/or loading. By pressing a "hold" button and then pressing a "start" button at the conclusion of the loading/unloading operation, the operator can lengthen the dwell period at each stop.

AGVs can be programmed to go to any location over more than one path. They are equipped with sensing and control instrumentation to avoid collisions with other vehicles. When using guide path equipment, material handling cost varies little with distance.

The following sketch illustrates common mechanical and automated transportation alternatives in warehouses. This includes lift trucks, conveyors, towlines, tractor train, driverless tractor train, and automatic guided vehicles.

Common warehouse transportation alternatives

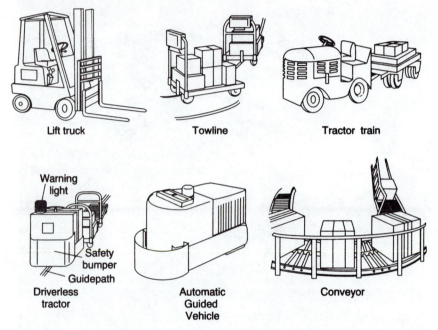

Lift truck   Towline   Tractor train

Warning light

Safety bumper

Guidepath

Driverless   Automatic   Conveyor
tractor   Guided
                              Vehicle

Make Better Use of Existing Handling Facilities

To assure the greatest return from material handling equipment, it must be used effectively. Thus, both the methods and the equipment should be sufficiently flexible so that a variety of material handling can be accomplished under variable conditions. By palletizing material in temporary and permanent storage, greater quantities can be transported faster than when material is stored without the use of pallets. United Wallpaper, Inc., realized the following gross direct labor savings through palletizing and mechanical material handling:

1. Reduced labor cost of warehousing finished stock 66 percent.
2. Reduced labor cost of assembling and shipping finished goods orders 30 to 65 percent.
3. Reduced labor cost of unloading and warehousing raw stock 80 percent.

FIGURE 5–25
Typical forklift truck

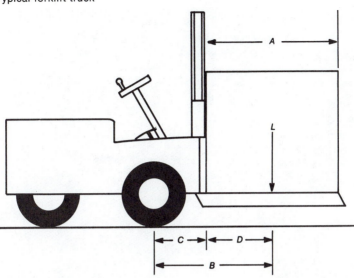

4. Reduced labor cost of unloading and warehousing other raw materials 40 percent.[1]

The maximum net load that can be safely handled can usually be easily computed. For example, one computes the inch-pound rating of fork trucks by multiplying the distance in inches from the center of the front axle to the center of the load (see Figure 5–25). If the distance from the center of the front axle to the front end of the fork truck is 18 inches, and the length of the pallet is 60 inches, then the maximum gross weight that a 200,000 inch-pounds fork truck should handle would be:

$$\frac{\text{Inch-pounds}}{B} = \frac{200,000}{18 + {}^{60}\!/_2} = 4,167 \text{ pounds}$$

By planning the pallet size to make full use of the equipment, a greater return can be realized from the material handling equipment.

Sometimes material can be handled in larger or more convenient units by designing special racks. When this is done, the compartments, hooks, pins, or supports for holding the work should be in multiples of 10 for ease of counting during processing and final inspection.

If any material handling equipment is used only part of the time, consider the possibilities for putting it to use a greater share of the time. By

---

[1] Philip F. Cannon, "Palletizing Makes New Warehousing System Really Work," *Factory Management and Maintenance*, Plant Operation Library no. 134.

FIGURE 5–26
Portable lift provides greater use of Hubbard tank in a physical therapy department

relocating production facilities or adapting material handling equipment to diversified areas of work, companies achieve greater utilization.

Here analysts should ask and answer the following questions: "Can the material be handled in larger or more convenient units?" "Can auxiliary handling equipment facilitate service at the workstation?" and "Can piece counting be combined with material handling?"

Handle Material with Greater Care

Industrial surveys indicate that approximately 40 percent of plant accidents happen during material handling operations. Of these, 25 percent are caused by lifting and shifting material. By exercising care in handling material, and using mechanical mechanisms for material handling, employees can reduce fatigue and accidents. Records prove that the safe factory is also an efficient factory. Although it is factual knowledge that the greater the amount of mechanized material handling, the safer the factory, analysts must make handling equipment safer, too. Safety guards at points of power transmission, safe operating practices, good lighting, and good housekeeping are essential to make material handling equipment safer. Workers should install and operate all material handling equipment in a manner compatible with existing safety codes.

Better handling reduces product damage. If the number of reject parts is at all significant in handling parts between workstations, then this area should be investigated. Usually damage to parts in handling can be kept to a minimum if specially designed racks or trays are fabricated to hold the parts immediately after processing. For example, one manufacturer of aircraft engine parts incurred a sizable number of damaged external threads on one component that was stored in metal tote pans after the completion of each operation. When two-wheeled hand trucks moved the tote pans filled with parts to the next workstation, the machined forgings bumped against one another and against the sides of the metal pan to such an extent that they became badly damaged. Someone investigated the cause of the rejects and suggested making wooden racks with individual compartments to support the machined forgings. This prevented the parts from bumping against one another or the metal tote pan. Production runs were also more easily controlled because of the faster counting of parts and rejects.

In a city hospital a portable mechanized lift permitted much greater application of a Hubbard tank in connection with physical therapy treatment (see Figure 5-26). With this controllable material handling equipment, patients can be comfortably immersed in the tank while in either a sitting or a prone position.

Two typical questions that should be asked are: "Can the material be handled with greater safety?" and "Can product damage be reduced by better handling?"

Consider Bar Coding for Inventory and Related Applications

Certainly the majority of technical people have some familiarity with bar coding and bar code scanning. Bar coding has shortened queues at grocery and department store checkout lines. Those black bars and white spaces represent 10 digits that identify both the item and the manufacturer. Once this Universal Product Code (UPC) is scanned by a reader at the checkout counter, the decoded data is sent to a computer that records timely information on labor productivity, inventory status, and sales. These five reasons justify the use of bar coding by methods and standards analysts for inventory and related applications:

1. *Accuracy.* Typically less than one error in 3.4 million characters is representative performance.[2] This compares favorably with the 2 to 5 percent error that is characteristic of keyboard data entry.

---

[2] U.S. Army LOGMARS study.

2. *Performance.* A bar code scanner enters data three to four times faster than typical keyboard entry.
3. *Acceptance.* Most employees enjoy using the scanning wand. Inevitably they prefer using a wand to keyboard entry.
4. *Low cost.* Since bar codes are printed on packages and containers, the cost of adding this identification is extremely low.
5. *Portability.* An operator can carry a bar code scanner into any area of the plant to determine inventories, status of orders, etc.

Bar coding is useful for receiving, warehousing, job tracking, labor reporting, tool crib control, shipping, failure reporting and quality assurance, tracking, production control, and scheduling. For example, the typical storage bin label provides the following information: part description, size, packing quantity, department number, storage number, basic stock level, and order point. Considerable time can be saved by using a scanning wand to gather all this detail from a bin for inventory reordering.

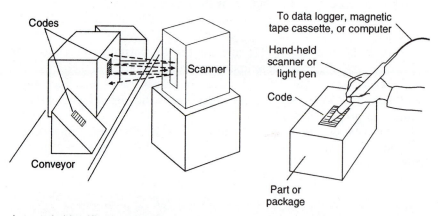

Automatic identification systems can include the use of hand-held scanner (right) and/or scanner mounted at a conveyor line.

Some practical applications reported by Accu-Sort Systems, Inc. include the automatic control of conveyor systems, diverting material to the location where it is needed; and providing material handlers with clear, concise instruction about where to take materials, automatically verifying that the proper material is handled.[3] By incorporating bar coding with programmable controllers and automatic packaging equipment, on-line real time verification of packing labels with container contents avoids costly product recalls. These few examples illustrate the potential of bar coding.

―――――――
[3] Society of Manufacturing Engineers, *Applying Industrial Bar Coding* (Dearborn, Mich.: Society of Manufacturing Engineers, 1985), pp. 133–38.

Some methods analysts may want to design bar code labels. They should realize that today standard code formats exist within the U.S. Department of Defense, the automotive industry, the health care industry, the railroad industry, and retail and wholesale distribution.

Summary: Material Handling

The goal

Analysts should always be on the alert to eliminate inefficient handling of material. The analyst should also understand the extent of what material handling involves. This may best be understood by the definition of material handling as developed by The Material Handling Alliance. This states: "Material handling is the movement, storage, control, and protection of materials and products, throughout the process of their manufacture and distribution". The following fundamental principles for doing a better job in material handling should be considered.

The above five points are based upon 20 principles of material handling which should be practiced by the methods analyst. These are:

1. *Planning principle.* Plan all materials and storage activities to obtain maximum overall efficiency.
2. *Systems principle.* Integrate as many handling activities as is practical into a coordinated system of operations, covering vendor, receiving, storage, production, inspection, packaging, warehousing, shipping, transportation, and customer.
3. *Material flow principle.* Provide an operation sequence and equipment layout optimizing material flow.
4. *Simplification principle.* Simplify handling by reducing, eliminating, or combining unnecessary measurements and/or equipment.
5. *Gravity principle.* Use gravity to move material wherever practical.
6. *Space utilization principle.* Make optimum use of building cube.
7. *Unit size principle.* Increase the quantity, size, or weight of unit loads or flow rate.
8. *Mechanization principle.* Mechanize handling operations.
9. *Automation principle.* Provide automation to include production, handling, and storage functions.
10. *Equipment selection principle.* In selecting handling equipment, consider all aspects of the material handled—the movement and the method to be used.
11. *Standardization principle.* Standardize handling methods as well as types and sizes of handling equipment.
12. *Adaptability principle.* Use methods and equipment that can best perform a variety of tasks and applications where special purpose equipment is not justified.
13. *Dead weight principle.* Reduce ratio of dead weight of mobile handling equipment to load carried.

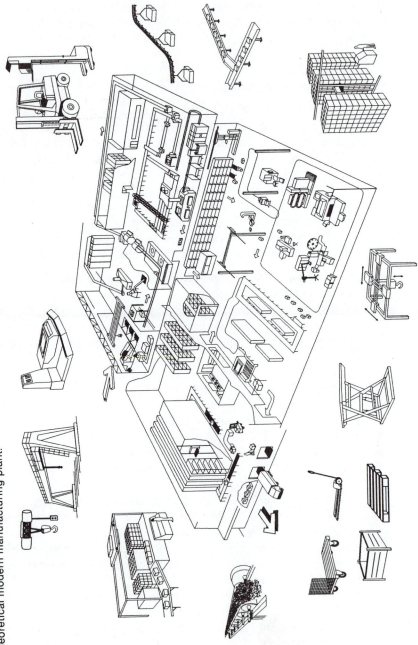

FIGURE 5–27
Illustration of the flow of material and a variety of material handling methods: manual, mechanical, and automated through a theoretical modern manufacturing plant.

14. *Utilization principle*. Plan for optimum use of handling equipment and manpower.
15. *Maintenance principle*. Plan for preventive maintenance and scheduled repairs of all handling equipment.
16. *Obsolescence principle*. Replace obsolete handling methods and equipment when more efficient methods or equipment will improve operations.
17. *Control principle*. Use material handling activities to improve control of production, inventory, and order handling.
18. *Capacity principle*. Use handling equipment to help achieve desired production capacity.
19. *Performance principle*. Determine effectiveness of handling performance in terms of expense per unit handled.
20. *Safety principle*. Provide suitable methods and equipment for safe handling.

It may be well to reiterate that the predominant principle to be kept in mind is that the less a material is handled, the better it is handled.

PLANT LAYOUT

The principal objective of effective plant layout is to develop a production system that permits the manufacture of the desired number of products of the desired quality at the least cost. Thus, physical layout is an important element of an entire production system that embraces operation cards, inventory control, material handling, scheduling, routing, and dispatching. All these elements must be carefully integrated to fulfill the stated objective.

Students sometimes ask, "Is there a type of layout that tends to be the best?" The answer is no. A given layout can be best in one set of conditions and yet be quite poor in a different set of conditions. And since work conditions are seldom static, methods analysts often have the opportunity to make improvements in the layout.

Although it is difficult and costly to make changes in arrangements that already exist, analysts should critically review every portion of every layout. Poor plant layouts result in major costs. Unfortunately, most of these costs are hidden and, consequently, cannot be readily exposed. The indirect labor expense of long moves, backtracking, delays, and work stoppages due to bottlenecks are characteristic of a plant with an antiquated layout.

Layout Types

In general, all plant layouts represent one or a combination of two basic layouts. These are product, or straight-line, layouts and process, or functional, layouts. In the straight-line layout, the machinery is located so that the flow from one operation to the next is minimized for any product class. Thus, in an organization that utilizes this technique, it would not be unusual to see a surface grinder located between a milling machine and a turret lathe, with an assembly bench and plating tanks in the immediate area. This type of layout is quite popular for certain mass-production manufacture because material handling costs are lower than for process grouping.

One of the principal advantages of group technology is that it utilizes a product grouping plant layout. Through group technology, a sufficient volume of work utilizing the same equipment in the same sequence permits this type of layout. Thus, a plant with, say, seven product groups would be laid out with seven lines of flow based on product grouping. The remainder of the plant may be planned on a process type of layout to accommodate all work not falling within any of the seven product groups.

Product grouping has some distinct disadvantages. Since a broad variety of occupations are represented in a small area, employee discontent can escalate. This is especially true when the different opportunities carry a significant money rate differential. Because unlike facilities are grouped together, operator training becomes more cumbersome, since no experienced employee on a given facility may be available in the immediate area to train the new operator. The problem of finding competent supervisors is also exacerbated due to the variety of facilities and jobs that must be supervised. Then, too, this type of layout invariably necessitates greater initial investment because duplicate service lines are required, such as air, water, gas, oil, and power. Another disadvantage of product grouping is the fact that this arrangement of facilities tends to give the casual observer the impression that disorder and chaos prevail. With these conditions, it is often difficult to promote good housekeeping. In general, the disadvantages of product grouping are more than offset by the advantages if production requirements are substantial.

Process, or functional, layout is the grouping of similar facilities. Thus, all turret lathes would be grouped in one section, department, or building. Milling machines, drill presses, and punch presses would also be grouped in their respective sections.

This type of arrangement gives a general appearance of neatness and orderliness, and tends to promote good housekeeping. Another advantage of functional layout is the ease with which a new operator can be trained. Surrounded by experienced employees operating similar machines, the new worker has the opportunity to learn from them. The problem of finding competent supervisors is lessened because the job demands are not as great with this type of grouping. Since these supervisors need be familiar with but one general type or class of facilities, their backgrounds do not have to be as extensive as those of supervisors in shops using product grouping.

Of course, the disadvantage of process grouping is the chance for long moves and backtracking on jobs that require a series of operations on diversified machines. For example, if the operation card of a job specified a sequence of drill, turn, mill, ream, and grind, the movement of the material from one section to the next could prove extremely costly. Another major disadvantage of process grouping is the great volume of paperwork required to issue orders and control production between sections.

Usually, if production quantities of similar products are limited and the plant is "job" or special order in form, then a process layout is more satisfactory.

No two plants have completely identical layouts, even though the nature of their operations is similar. Many times, a combination of process grouping and product grouping is in order. Regardless of what type of grouping is contemplated, the following principal points can improve layout in either a production facility or an office:

1. For straight-line mass production, material laid aside should be in position for the next operation.
2. For diversified production, the layout should permit short moves and deliveries, and material should be convenient to operators.
3. The operator should have an easy line of sight to those portions of the workstations requiring control.
4. Workstation designs should allow operators to change posture regularly during the work period.
5. For multiple-machine operations, the equipment should be grouped around the operator.
6. For efficient stacking, storage should be arranged to minimize searching and rehandling.
7. For better worker efficiency, service centers should be close to production areas. Thus, those service areas needed by several people should be located centrally.
8. Offices should allow a separation distance of at least 4 feet between people.

Volume, Distance, and Travel Charts

Before designing a new layout or correcting an old one, analysts must accumulate all facts that directly and indirectly influence the layout. These include many, if not all, of the following:

1. The present and anticipated sales volume of each product, line, or class.
2. The labor content of each operation on each product.
3. The sitting, standing, and sit/stand requirements of the operations.

4. Identification of workplaces where visual work is extensively employed, such as computer terminals.
5. A complete inventory of existing machines and material handling equipment.
6. The status of existing facilities from the standpoint of condition and book value.
7. Possible changes in product design.
8. Drawings of the existing plant indicating the location of all service facilities, windows, doors, columns, aisles, corriders, reinforced areas, stairs, ramps, and the condition of walkways and floors.
9. The amount of handling taking place between facilities.

Once all of these facts have been gathered, an analyst can construct a flow process chart (see Chapter 3), which in itself gives the general form the layout will take. In the construction of the flowchart, solicit suggestions from operators, inspectors, material handlers, and line supervisors. These employees are closer to the production work than anyone else, and they often provide valuable suggestions.

Other charts that can be helpful in connection with both plant layout and material handling work are the volume, distance, and travel charts. These tools help diagnose problems related to the arrangement of departments and service areas, as well as to the location of equipment within a given sector of the plant. The volume chart presents in matrix form the magnitude of material handling that takes place between two facilities per time period. The unit identifying the amount of handling may be whatever seems most appropriate to the analyst making the study. It can be pounds, tons, frequency of handling, and so on. The volume chart is a useful application mainly in process type layouts. Figure 5–28 illustrates a very elementary volume chart. Note that the volume of the material routed from the No. 4 W. & S. turret lathe to the No. 2 Cinn. Hor. Mill is considerably less than the volume of the material routed from the No. 2 Cinn. Hor. Mill to the No. 4 W. & S. turret lathe.

The reader should understand that when the volume chart is multiplied by the distance chart, we get a travel chart. Thus:

Volume chart (tons)				Distance chart (feet)				Travel chart (foot – tons)		
	A	B	C		A	B	C		A	B	C
A	X	5	7	A	X	100	150	A	X	500	1050
B	1	X	4	× B	100	X	40	= B	100	X	160
C	0	3	X	C	150	40	X	C	0	120	X

Making the Layout

To make the proposed layout, templates of all facilities must be prepared. Templates are usually constructed to a scale of ¼ inch equals 1

FIGURE 5–28

The volume chart is a useful tool in solving material handling and plant layout problems related to process-type layouts

Volume Chart								
From \ **To**	No. 4 W. & S. Turret Lathe	Delta 17" Drill Press	2-Spindle L. & G. Drill	No. 2 Cinn. Hor. Mill	No. 3B. & S. Verticle Mill	Niagara 100Ton Press	No. 2 Cinn. Centerless	No. 3 Excello Thd. Grinder
No. 4 W. & S. Turret Lathe		20	45	80	32	4	6	2
Delta 17" Drill Press			6	8	4	22	2	3
2-Spindle L. & G. Drill				22	14	18	4	4
No. 2 Cinn. Hor. Mill	120				10	5	4	2
No. 3B. & S. Verticle Mill						6	3	1
Niagara 100Ton Press	60	12	2				0	1
No. 2 Cinn. Centerless	15							15
No. 3 Excello Thd. Grinder			15	8				

foot unless the size of the project is quite large, when a scale of ⅛ inch equals 1 foot may be used. If an existing layout is available, make a copy, and cut all facilities out of the print for use as templates. If no layout exists, purchase two-dimensional printed templates as illustrated in Figure 5–29. Of course, you can draw your own templates on a good grade of stiff cardboard and cut them out. The advantages of using stiff cardboard are apparent if the same templates are to be used several times.

Scale models gives the third dimension to plant layouts, and are especially helpful to an analyst endeavoring to sell a contemplated layout to a top executive who has neither the time nor the familiarity to grasp all details of a two-dimensional layout (see Figure 5–30).

Once all the necessary templates have been made, a trial layout can be prepared. Good layouts incorporate the principal points for efficient layout. They also provide adequate output capacity at each workstation without introducing bottlenecks and interrupting the flow of production.

FIGURE 5–29
Standard templates of representative machine tools

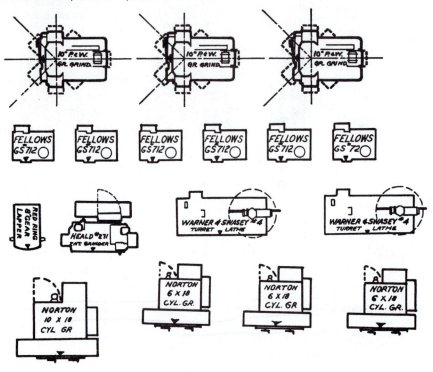

After an ideal layout has been designed, construct a flowchart of the proposed plan to highlight the reduction in the distances traveled, storages, delays, and overall costs. This greatly facilitates final approval of the design. A good technique for testing the layout is to wind colored thread around the map tacks holding down the templates. With the thread follow the flow of the product from its raw material components to its transformation into a finished product. By using a different colored thread for every line of product produced, the flow of all work can be visualized quite rapidly. This pictorial presentation, supplemented with the flowchart, brings to light most of the flaws of the proposed method.

COMPUTER-AIDED LAYOUT. Software can help analysts in developing realistic layout solutions rapidly and inexpensively. The CRAFT (Computerized Relative Allocation Facilities) program (IBM share library No. SDA 3391) is one computer program that has been extensively used. It has the capacity for handling 40 activity centers. These could be departments or work centers within a department. Any one of these activity centers can be identified as being fixed, thus freezing it and allowing freedom of movement in those that can be readily moved. For example, it

FIGURE 5–30
Three-dimensional layout of a portion of an industrial
plant—note detail and accuracy

General Electric Co., Louisville, Kentucky

often is desirable to freeze such activity centers as elevators, restrooms, stairways. Input data includes the number and location of fixed work centers, material handling costs, interactivity center flow, and a representation of a block layout. The governing heuristic algorithm asks: What change in material handling costs would result if work centers were exchanged? The answer is stored, and the computer proceeds in an iterative manner until it converges on a good solution. CRAFT calculates the distance matrix as the rectangular distances from the department centroids.

Another available program is CORELAP (Engineering Management Associates, Boston, Massachusetts). The input requirements for CORELAP are the number of departments, the departmental areas, the departmental relationships, and weights for these relationships. CORELAP then constructs layouts by locating departments, using rectangular areas. The objective is to provide a layout with "high-ranking" departments close together.

ALDEP (IBM Corporation program order no. 360D-15.0.004) constructs plant layouts by randomly selecting a department and locating it in a given layout. The RELATION chart is then scanned, and a department that has a high closeness rating is introduced into the layout. This process continues until it places all departments. ALDEP computes a score for the layout, and the process repeats a specific number of times. It has the ability to provide multifloor layouts.

Computer-aided layouts are principally helpful in process-type plants. Usually manual layouts result in better solutions in less time when designing a plant with a product or straight line layout.

Summary: Plant Layout

A systematic search can uncover improvements in plant layout. Arrange workstations and machines to permit the most efficient processing of a product with a minimum of handling. A layout should not be changed until making a careful study of all the factors involved. Methods analysts should learn to recognize poor layout and to present the facts to the plant engineer. Computer programs can rapidly provide layouts that represent a good first effort in the development of the recommended layout.

Every time a new layout is made or an existing layout is changed, analysts should make recommendations that not only are effective but also reduce the difficulty in making future changes. One example is keeping plant services, such as compressed air and electricity, overhead. Another is maintaining flexibility in connection with material handling equipment and keeping all fixed facilities, such as elevators, in areas that probably will never need to be changed. Storage areas should be located in those sectors where change is contemplated or may likely occur in time, since these areas are the least expensive to alter.

PRINCIPLES OF MOTION ECONOMY

The last of the 10 primary approaches to operation analysis involves improving the arrangement of parts at the workplace and of the motions required to perform the task. Chapter 7 gives considerable detail to the laws of motion economy. In this chapter we focus on the investigation of motion economy as practiced by methods analysts. When studying work performed at a workstation, the analyst should ask: "Are both hands working at the same time and in opposite symmetrical directions?" "Is each hand going through as few motions as possible?" "Is the workplace arranged so that long reaches are avoided?" and "Are both hands being used effectively and not as a holding device?" Negative answers to any of these questions indicate opportunities for improvement of the workstation.

Both Hands Should Work at the Same Time

The left hand, in right-handed people, can be just as effective as the right hand and it should be used. A right handed boxer learns to jab more

FIGURE 5–31
Workstation and fixture designed to allow productive work with both hands

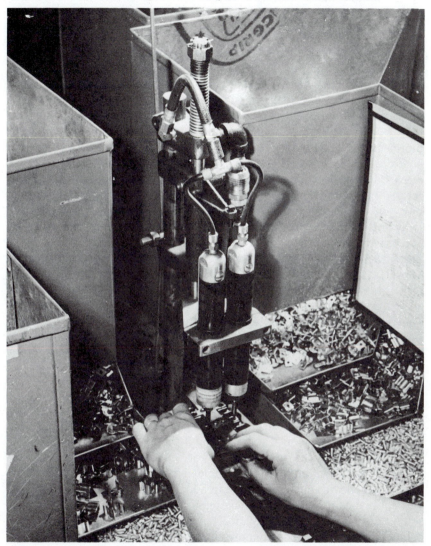

General Electric Co.

effectively with his left hand than with his right hand. A typist is just as proficient with one hand as the other. In a large number of instances, workstations can be designed to do "two at a time." By providing dual fixtures to hold two components, both hands work at the same time, making symmetrical moves in opposite directions (see Figure 5–31).

150					*Motion and Time Study*

For example, production increased 100 percent when analysts developed a fixture that utilized two-handed operation and permitted grinding two motor brushes at a time. The old method involved a one-handed operation, for only one piece was done during each grinding cycle.

Each Hand Should Go Through as Few Motions as Possible

It is just common sense that the more motions the hands make while performing a task, the longer it will take to do the work. All hand motions are a series of reaches, moves, grasps, positions, and releases, and the more of these fundamental motions that can be eliminated or reduced, the more satisfactory the workstation is.

For example, by providing drop delivery and gravity chutes, certain moves and positions are eliminated while reducing release time. Likewise, a belt conveyor that brings material to the workstation and carries away the processed part usually results in reduced "move" time. Position is a time-consuming basic division of accomplishment that in many instances can be minimized through well-designed fixtures. By using tapered channels and pilots, two mating parts can readily be assembled with considerable reduction or the complete elimination of the positioning element.

The Workplace Should Be Arranged to Avoid Long Reaches

The time required to pick up an article depends to a great extent on the distance the hand must move. Likewise, move time is definitely related to distance. If at all possible, the workplace should be arranged so that all parts are within easy reach of the operator. If all components can be reached while both elbows are close to the body, then the work is performed in the "normal" working area. This normal area represents the space within which work can be accomplished in a minimum time.

The work, tools, and parts should not be located beyond the reach of the hand when the arms are fully extended in sitting workplaces. The area encompassed with the arms fully extended in both the horizontal and the vertical planes represents the maximum working area. No workstations should demand work performance beyond this area. Also, those items being handled should not be more than 6 inches above the work surface and should weigh less than 10 pounds. Where heavier parts are involved, either mechanical assists should be utilized or the workplace should be designed for standing.

Figure 5–32A illustrates the conventional assembly workstation for the production of automotive wiring harnesses. Note the hanging harnesses located behind the operator, necessitating her turning around to get the assembly material and then turning back to the conveyor to assemble the harness. The cycle time here averaged 0.85 minutes with the operator having to turn around to acquire material three times per cycle. The

FIGURE 5–32A
Assembly conveyor

Packard Electric, Warren, Ohio

A. Conventional assembly conveyor where automotive wiring harnesses are assembled (Operator must turn around to get the assembly material)

improved method (Figure 5–32B) shows the workstation redesigned to allow a large portion of the assembly material to be placed in front of the operator. The new method minimizes and in some cases eliminates the need to turn around to get material. Furthermore, when working with wire, greater control is required in carrying leads longer distances. The extent of control necessary with leads located closer to the fixture is lessened, thus permitting more effective worker movements.

Avoid Using the Hand as a Holding Device

If either hand is ever used as a holding device during the processing cycle of a part, then it is not performing useful work. Invariably, a fixture can be designed to hold the work satisfactorily, thus allowing both hands to do useful work. Many times, foot-operated mechanisms allow both hands to perform productive work. The therblig "hold" represents an

FIGURE 5–32B

Packard Electric, Warren, Ohio
B. Workstation redesigned to allow a large
portion of the assembly material to be placed in
front of the operator

ineffective basic division of accomplishment and should be eliminated, if possible, from all workstations.

For example, Figure 5–33 shows the old method of hanging automotive harness wire. Note the operator first gets the wire in his left hand, grasping it at the bend (A). Then he slides the wire through his right hand, stopping at the terminal (B). Next he plugs the terminal to a plastic connector (C). The improved method (see Figure 5–34) eliminates the sliding element. The same wire is hung in a comb-type rack being held at the terminal. The operator gets the wire from the rack with his right hand while the left hand gets the connector (A). He then plugs the terminal to the connector (B).

Another example will help clarify the principle of using a fixture as opposed to the hands for holding of work. A company that produced specialty windows in central Pennsylvania was removing a ¾" wide strip of protective paper around all four edges from both sides of Lexan panels. An operator would pick up a single sheet of Lexan and bring it to the work area. Then the operator would pick up a pencil and square and mark the four corners of the Lexan panel. The pencil and square would be laid aside and a template would be picked up and located to the pencil marks.

FIGURE 5–33
Old method of hanging automatic harness wire

A. Operator first gets the wire
with his left hand at the bend

B. Operator slides wire through
his right hand

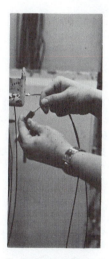

Packard Electric, Warren, Ohio

C. Operator plugs
terminal to plastic
connector.

The operator would then strip the protective paper from around the periphery of the Lexan panels. The standard time developed by MTM–1
was 1.063 min. per piece.

A simple wood fixture was developed to hold three Lexan panels while
each was stripped of ¾″ wide protection paper around the periphery.

FIGURE 5–34
Improved method of hanging automotive harness wire shows the same wire being
hung in a comb rack at the terminal

Packard Electric, Warren, Ohio

A. Operator grasps the wire at the
terminal with his right while his left
hand simultaneously gets a connector

B. Operator plugs terminal to connector

With the fixture method, the worker picked up three Lexan sheets and
located them in the fixture. (See Figure 5–35). The protective paper was
stripped and the sheets were turned 180° and the remaining two sides had
the protective paper removed. The improved method resulted in a standard of 0.46 min. per panel or a savings of 0.603 min. of direct labor per
panel.

FIGURE 5–35
Fixture for stripping ¾" wide protection paper around
periphery of Lexan sheets

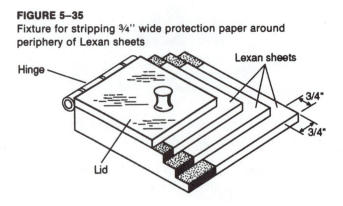

SUMMARY: 10 PRIMARY APPROACHES TO OPERATION ANALYSIS

The 10 primary approaches to operation analysis represent a systematic approach to analyzing the facts presented on the operation and flow process charts. Figure 5–36 depicts how the various steps in methods engineering are integrated into the general flow of the entire analysis. Other tools of analysis used in developing a method include the worker and machine process chart, the operator process chart, and micromotion study. These subjects are discussed in subsequent chapters.

Regardless of the nature of the work, whether continuous or intermittent, process or job shop, soft or hard goods, when systematic operation analysis is applied by competent personnel, real savings result. Remember that these principles are just as applicable in planning new work as in improving work already in production.

Increased output and improved quality are the primary outcomes of operation analysis, but operation analysis also distributes the benefits of improved production to all workers and helps develop better working conditions and methods, so that the worker can do more work at the plant, do a good job, and still have enough energy to enjoy life.

A representative case history, which follows operating analysis, is that of the production of an Auto-Starter, a device for starting AC motors by reducing the voltage through a transformer. A subassembly of the Auto-Starter is the Arc Box. This part sits in the bottom of the Auto-Starter and acts as a barrier between the contacts so that there are no short circuits. The present design consists of the following components:

Asbestos barriers with three drilled holes	6
Spacers of insulating tubing two inches long	15
Steel rods threaded at both ends	3
Pieces of hardware	18
Pieces total	42

In assembling these components, the operator places a washer, lock washer, and nut on one end of each rod. Next, the worker inserts the rods through the three holes in the first barrier. Then the operator places one spacer on each of the rods (three in all), and adds another barrier. Operators repeat this until six barriers are on the rods and separated by the tubing.

It was suggested that the six barriers be made with two slots—one at each end—and that two strips of asbestos for supporting the barriers be made with six slots in each. These would be slipped together and placed in the bottom of the Auto-Starter as needed. The manner of assembly would be the same as that used in putting together the separator in an eggbox.

FIGURE 5–36
The principal steps in a methods engineering program

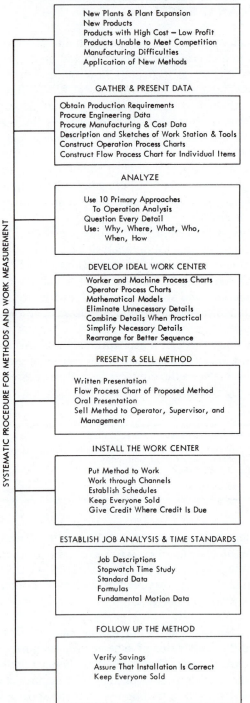

SELECTION OF PROJECT

New Plants & Plant Expansion
New Products
Products with High Cost — Low Profit
Products Unable to Meet Competition
Manufacturing Difficulties
Application of New Methods

GATHER & PRESENT DATA

Obtain Production Requirements
Procure Engineering Data
Procure Manufacturing & Cost Data
Description and Sketches of Work Station & Tools
Construct Operation Process Charts
Construct Flow Process Chart for Individual Items

ANALYZE

Use 10 Primary Approaches
    To Operation Analysis
Question Every Detail
Use: Why, Where, What, Who,
    When, How

DEVELOP IDEAL WORK CENTER

Worker and Machine Process Charts
Operator Process Charts
Mathematical Models
Eliminate Unnecessary Details
Combine Details When Practical
Simplify Necessary Details
Rearrange for Better Sequence

PRESENT & SELL METHOD

Written Presentation
Flow Process Chart of Proposed Method
Oral Presentation
Sell Method to Operator, Supervisor, and
    Management

INSTALL THE WORK CENTER

Put Method to Work
Work through Channels
Establish Schedules
Keep Everyone Sold
Give Credit Where Credit Is Due

ESTABLISH JOB ANALYSIS & TIME STANDARDS

Job Descriptions
Stopwatch Time Study
Standard Data
Formulas
Fundamental Motion Data

FOLLOW UP THE METHOD

Verify Savings
Assure That Installation Is Correct
Keep Everyone Sold

SYSTEMATIC PROCEDURE FOR METHODS AND WORK MEASUREMENT

156

A total of 15 suggested improvements occurred after the analysis was completed, with the following results:

	Old method	New method	Savings
42 parts..............	8	34	
10 workstations.......	1	9	
18 transportations....	7	11	
7,900 feet of travel..	200	7,700	
9 storages...........	4	5	
0.45 hours time.......	0.11 hours time	0.34 hours	
$1.55 costs..........	$0.60	$0.95	

A program of operation analysis resulted in a 17,496-ton annual savings for one Ohio company. By forming a mill section into a ring and welding it, an original rough-forged ring weighing 2,198 pounds was replaced. The new mill section blank weighed only 740 pounds. The saving of 1,458 pounds of high-grade steel, amounting to twice the weight of the finished piece, was brought about by the simple procedure of reducing the excess material that had been cut away in chips.

An analytic laboratory in a New Jersey plant has applied the principles of operation analysis with gratifying results. New workbenches have been laid out in the form of a cross so that each chemist has an L-shaped worktable. This arrangement allows each chemist to reach any part of the workstation by taking only one stride.

The new workbench has consolidated equipment, thus saving space and eliminating the duplication of facilities. One glassware cabinet services two chemists. A large four-place fume head allows multiple activity in this area that was formerly a bottleneck. All utility outlets are relocated for maximum efficiency.

In an effort to streamline its organization, a state government division developed a program of operation analysis that resulted in an estimated annual savings of more than 50,000 hours. This was brought about by combining, eliminating, and redesigning all paperwork activities; improving the plant layout; and developing paths of authority.

Methods improvement is as effective in office procedures as in production operations. One industrial engineering department of a Pennsylvania company was given the problem of simplifying the paperwork necessary for shipping molded parts manufactured in one of its plants to an outlying plant for assembly.[4] A new method was developed that reduced the average daily shipment of 45 orders from 552 sheets of paper forms to 50 sheets. The annual savings in paper alone brought about by this change was significant.

---

[4] From a talk by Lynell Cooper, manufacturing engineer of Westinghouse Electric Corporation, to the central Pennsylvania chapter of the Society for the Advancement of Management.

Methods improvement can and should be a continuing program. For example, in the early 1940s the average savings in the Procter & Gamble Company per member of management was approximately $500. To improve the situation and to maintain an acceptable level of cost reduction, P & G inaugurated a methods program. Manufacturing management attended sessions in the concepts of operation analysis. Also, methods engineers attended a special course; each had one to five years of company experience. After completion of the operation analysis course, these engineers returned to their respective plants to work as methods specialists. The effect of this training was apparent in a very short time. Annual savings increased to approximately $700 per year per member of factory management. Since the program was a success, management trained additional personnel and extended the program to more plants.

By 1950 the position of the industrial engineer had changed from methods specialist to methods coordinator. It had been the practice for the engineer to suggest methods improvements to the line supervisor. Taking the suggested changes as criticism, frequently the supervisor did not participate actively.

Functioning as methods coordinators, engineers spent approximately two thirds of their time helping factory supervisors on their projects and the other third working individually on special factory projects. Each member of plant management was assigned several selected projects where the cost was to be reduced, and each was encouraged to solicit assistance as required from the methods engineer.

Methods coordinators also periodically conducted training courses both for other staff members and for members of the line organization. With the active participation of plant management, the rate of savings had increased to $2,300 per year per member of factory management by 1950.

Subsequently, teams of four to eight employees, made up equally of line supervisors and staff, were formed to work on selected projects having a high potential savings. More enthusiasm on the part of management resulted. Opportunities for the recognition of good work increased. A friendly rivalry developed between teams for first place in plant standing. Display boards were set up in the front office to show team standings. Factory goals for cost reduction were established. By 1954 companywide savings amounted to approximately $4,000 per member of management.

These results convinced top management that the cost reduction or profit improvement program should be plantwide and that all costs should be challenged. Consequently, the program was extended to all plants. To further stimulate the desire of individual plants to make a good showing, P & G circulated a summary sheet comparing the results of the program in each plant.

The director of industrial engineering for Procter & Gamble made the following statement concerning the program:

Early in the program, plants tended to think that they had "skimmed the cream." The easy projects had all been picked off. Next year would be harder, they thought; consequently a lower goal was in order. We had to sell some people on the reasonableness of a higher goal each year.

Higher goals were reasonable because of increased experience at cost reduction. The growth in business was another factor. Getting the less active team members to do more offered real potential for increased savings.

Although we had concluded that the goal setting process should be democratic, some guides were useful. The teams were encouraged to compare themselves with others on a dollars-saved-per-member-of-management basis. Comparisons were also made in terms of goal as a percent of operating expense and as a percent of production value. The desire of plant management groups to show up well in all-plant comparisons was a strong factor in motivating teams to choose goals which required their best efforts for attainment.

Annual savings continued to increase each year, and by 1970 they amounted to $27,000 per member of management. This figure is conservative in that no credit is given for savings which continue beyond the first year.

Since 1970, the program has spread to involve people in all parts of the business at both the nonmanagement and the management levels, making dollars saved per member of management meaningless.

TEXT QUESTIONS

1. In what four ways should an investigation to improve the process of manufacture be made?
2. Explain how rearranging operations can result in savings.
3. What process is usually considered the fastest for forming and sizing operations?
4. How should the analyst investigate the setup and tools to develop better methods?
5. Give some applications of bar coding for the improvement of productivity.
6. When would you recommend the use of energy-efficient motors?
7. How much more efficient is a typical 50 HP energy-efficient motor than a standard 50 HP motor?
8. Why should the methods analyst be responsible for good working conditions?
9. Give the requisites of effective guarding.
10. What are the two general types of plant layout? Explain each in detail.
11. To what scale are templates usually constructed?
12. Explain the fundamental purpose of group technology.
13. What is the best way to test a proposed layout?
14. Which question should the analyst ask when studying work performed at a specific workstation?
15. For what purpose is operations analysis used?
16. Explain the advantages of using a checklist.

17. Why do costs vary little with distance in connection with automated guided vehicles?
18. What are the primary approaches to operation analysis?
19. On what does the extent of tooling depend?
20. Do working conditions appreciably affect output? Explain.
21. How can planning and production control affect set-up time?
22. How can you best handle a material?
23. Why should "terminal" time be minimized?
24. Explain the effect of humidity on the operator.
25. Distinguish between the volume chart, the distance chart, and the travel chart.
26. Why does the travel chart have more application in process grouping than in product grouping?
27. Of the several available computer programs, which one would you prefer to use in connection with a multifloor building?
28. Explain why a sound classification system is the first requisite of a successful group technology program.

GENERAL QUESTIONS

1. Where would you find application for a hydraulic elevating table?
2. What is the difference between a skid and a pallet?
3. Explain the significance of the colored code for stock templates.
4. When would you recommend using three-dimensional models in layout work?
5. What is the general flow of analysis procedure when it is applied to a product that has never been manufactured?
6. In a process like operation analysis, is it necessary to determine the point of diminishing returns? Why or why not?
7. Select five workstations and measure the footcandle intensity at these stations. How do these values compare with recommended practice?
8. Develop a classification system suitable for group technology where we have five product lines utilizing components with 13 geometric configurations and involving three types of thermosetting plastics. Make any assumptions that you feel are appropriate.
9. Derive the equations for the three straight line graphs in Figure 5–7 and calculate algebraically the three break-even points.
10. Check with some local factory where metal removal operations are employed and determine the application of "whisker-reinforced ceramics."
11. Why has Japan taken a leadership role in the use of robots?

PROBLEMS

1. In the Dorben Company, the methods analyst was assigned the task of altering the work methods in the press department to meet OSHA standards

relative to permissible noise exposures. She found the average sound level to be 100 dbA. The standard deviation was 10 dbA.

Supervisors gave the 20 operators in this department earplugs, and reduced the power output from the public-address system from 30 watts to 20 watts. The deadening of the sound level by use of the earplugs was estimated to be 20 percent.

What improvement resulted? Do you feel that this department is now in compliance with the law for 99 percent of its employees? Explain.

$$\text{Hint: Decibel loss} = 10 \log \frac{\text{Power output original}}{\text{Power output planned}}$$

2. In the Dorben Company, the methods analyst designed a workstation where the seeing task was difficult because of the size of the components going into the assembly. She established the desired brightness as being 100 footlamberts on the average, with a standard deviation of 10 footlamberts so as to accommodate 95 percent of the workers. The workstation was painted a medium green having a reflectance of 50 percent.

What would the required illumination in footcandles be at this workstation in order to provide adequate illumination for 95 percent of the workers? Estimate what the required illumination would be if the workstation were repainted with a light cream paint.

3. The analyst in the Dorben Company is considering replacing five 50 HP motors with five 50 HP energy efficient motors. These motors will all be operated seven days per week and three shifts per week at an estimated 85 percent of full load. If the cost of electric power is $0.06 per KWH, how much can the company afford to pay for the five energy-efficient motors?

4. The methods analyst in the Dorben Company is considering the installation of a state-of-the-art solid state electronic energy conservation system that will tune and balance the whole fluorescent system, including lamps, ballast, and power supply in the plant. The system will regulate the voltage and the current on an ongoing basis to establish and hold the system at optimum performance. It will also protect the system with special safety circuits. Based on estimates received from the supplier, the system will save 30 percent of the lighting energy cost, 50 percent of the lamp replacement cost, and all of the present ballast replacement cost. If the cost of the installed system is $15,000, how many months would it take the investment to pay for itself based on the following data:

Lighting Energy Costs
  Number of fixtures—445
  Average KWH per fixture—0.187
  Cost per KWH—$0.085
  Annual operating hours—4,440

Lamp Replacement Cost
  Number of fixtures—445
  Average lamps per fixture—2.25
  Average lamp life—2.5 years
  Cost of lamp installed—$6.50

Ballast Replacement Cost
  Number of fixtures—445
  Average ballasts per fixture—1.1
  Average ballast life—4.5 years
  Cost of ballast installed—$40.00

SELECTED REFERENCES

Bralla, James G.   *Product Design for Manufacturing.* New York: McGraw-Hill, 1986.

Buffa, Elwood S.   *Modern Production Operations Management.* 6th ed. New York: John Wiley & Sons, 1980.

Chang, Ning San.   "Pattern Recognition and Bar Code Technology." In *Handbook of Industrial Engineering,* 2nd ed, edited by Gavriel Salvendy, Chap. 7. New York: John Wiley & Sons, 1992.

Hyer, Nancy Lea., ed.   *Capabilities of Group Technology.* Dearborn, Mich.: Society of Manufacturing Engineers, 1987.

Konz, Stephan.   *Facility Design.* New York: John Wiley & Sons, 1985.

McCormick, E. J.   *Human Factors Engineering.* 4th ed. New York: McGraw-Hill, 1976.

Nof, Shimon Y.   "Industrial Robotics." In *Handbook of Industrial Engineering,* 2nd ed., edited by Gavriel Salvendy, Chap. 16. New York: John Wiley & Sons, 1992.

Sims, Ralph E.   "Material Handling Systems." In *Handbook of Industrial Engineering,* 2nd ed., edited by Gavriel Salvendy, Chap. 68. New York: John Wiley & Sons, 1992.

Smith, Michael J.   "Design for Health and Safety." In *Handbook of Industrial Engineering,* 2nd ed., edited by Gavriel Salvendy, Chap. 41. New York: John Wiley & Sons, 1992.

Society of Manufacturing Engineers.   *Applying Industrial Bar Coding.* Dearborn, Mich.: Society of Manufacturing Engineers, 1985.

Spur, Gunter.   "Numerical Control Machines." In *Handbook of Industrial Engineering,* 2nd ed., edited by Gavriel Salvendy, Chap. 15. New York: John Wiley & Sons, 1992.

Tompkins, James A.   "Facilities Layout." In *Handbook of Industrial Engineering,* 2nd ed., edited by Gavriel Salvendy, Chap. 67. New York: John Wiley & Sons, 1992.

Tompkins, J.   "Plant Layout." In *Handbook of Industrial Engineering,* edited by Gavriel Salvendy. New York: John Wiley & Sons, 1982.

Wemmerlov, Urban and Nancy Lea Hyer.   "Group Technology." In *Handbook of Industrial Engineering,* 2nd ed., edited by Gavriel Salvendy, Chap. 17. New York: John Wiley & Sons, 1992.

SELECTED SOFTWARE

Nicks, J. E.   *Machine Capability Studies and Equipment Justification. Basic Programming Solutions for Manufacturing.* Dearborn, Mich.: Society of Manufacturing Engineers, 1982. Source Code owned by MiCAPP, Inc.

SELECTED VIDEOTAPES

Automated Material Handling. Manufacturing Insights Videotape Series. ½″ VHS: VT251-1368 & ¾″ U-Matic: VT251U-1368. Dearborn, Mich.: Society of Manufacturing Engineers, 1986.

Cutting Tools. Manufacturing Insights Videotape Series. ½″ VHS: VT249-1368 & ¾″ U-Matic: VT249U-1368. Dearborn, Mich.: Society of Manufacturing Engineers, 1986.

Flexible Small Lot Production for Just-In-Time. Manufacturing Insights Videotape Series. ½″ VHS: VT415-1368 & ¾″ U-Matic VT415U-1368. Dearborn, Mich.: Society of Manufacturing Engineers, 1991.

Programmable Controllers. Manufacturing Insights Videotape Series. ½″ VHS: VT254-1368 & ¾″ U-Matic VT254U-1368. Dearborn, Mich.: Society of Dearborn Manufacturing Engineers, 1987.

Setup Reduction for Just-In-Time. Manufacturing Insights Videotape Series. ½″ VHS: VT392-1368 & ¾″ U-Matic VT392U-1368. Dearborn, Mich.: Society Manufacturing Engineers, 1990.

Worker and Machine Relationships

6

Once an operation has been found necessary through analysis of the operation and flow process chart, frequently it may be improved through further analysis. The three process charts discussed in this chapter are helpful tools for the development of the ideal method. While the operation and flow process charts (discussed in Chapter 3) present facts and are used as tools of analysis, the worker and machine process chart, gang process chart, and operator process chart are tools to assist in the development of the ideal work center. That is why these charts are presented at this time rather than in Chapter 3.

WORKER AND MACHINE PROCESS CHARTS

While analysts use the operation and flow process charts primarily to explore a complete process or series of operations, they use the worker and machine process chart to study, analyze, and improve only one workstation at a time. This chart shows the exact relationship in time between the working cycle of the person and the operating cycle of the machine. With these facts clearly presented, possibilities exist for a fuller utilization of both worker and machine time and a better balance of the work cycle.

Today, many of our machine tools are either completely automatic, such as the automatic screw machine, or are partially automatic, such as the turret lathe. In the operation of these types of facilities, the operator is often idle for a portion of the cycle. The utilization of this idle time can increase operator earnings and improve production efficiency.

The practice of having one employee operate more than one machine is known as "machine coupling." Machine coupling is not new. During the Great Depression, one plant was unable to justify the incentive earnings being turned in by the second shift in one of the machining departments.

Investigation revealed that the operators were practicing machine coupling on their own initiative.

Today, some industries have encountered resistance to machine coupling from organized labor. The best way to sell machine coupling is to demonstrate the opportunity for added earnings. Since machine coupling increases the percentage of "effort" time during the operating cycle, the opportunity for greater incentive earnings is enhanced if a company is on an incentive wage payment plan. Also, higher base rates result when machine coupling is practiced, since the operator has greater responsibility and exercises more mental and physical effort under multiple-machine operation.

Constructing the Worker and Machine Process Chart

When constructing the worker and machine process chart, first identify the chart in the usual fashion by indicating at the top of the sheet: Worker and Machine Process Chart. Immediately below this heading, add the following information: the part number, drawing number, description of the operation being charted, present or proposed method, date, and name of the person doing the charting.

Since workers and machine charts are always drawn to scale, select a distance in inches to conform with a unit of time so that the chart can be neatly arranged on the sheet. The longer the cycle time of the operation being charted, the shorter the distance per decimal minute of time. Once exact values have been established for the distance in inches per unit of time, begin charting. On the left side of the paper, list the operations and time for the worker, and to the right show the working time and the idle time of the machine or machines, as the case may be. A solid line drawn vertically represents the employee's working time. A break in the vertical work-time line signifies idle time. Likewise, a solid vertical line under each machine heading indicates machine operating time, and a break in the vertical machine line designates idle machine time. A dotted line under the machine column, indicates loading and unloading machine time during which the machine is neither idle nor productive (see Figure 6–1).

Chart all elements of occupied and idle time for both the worker and the machine until the termination of the cycle. At the foot of the chart show the employee's total working time and total idle time. Likewise, post the total working time and idle time of each machine. The productive time plus the idle time of the worker must equal the productive time plus the idle time of each machine he or she operates.

Accurate elemental time values are necessary before constructing the worker and machine chart. These time values should represent standard times which include an acceptable allowance to take care of fatigue,

FIGURE 6-1
Worker and machine process chart for milling machine operation

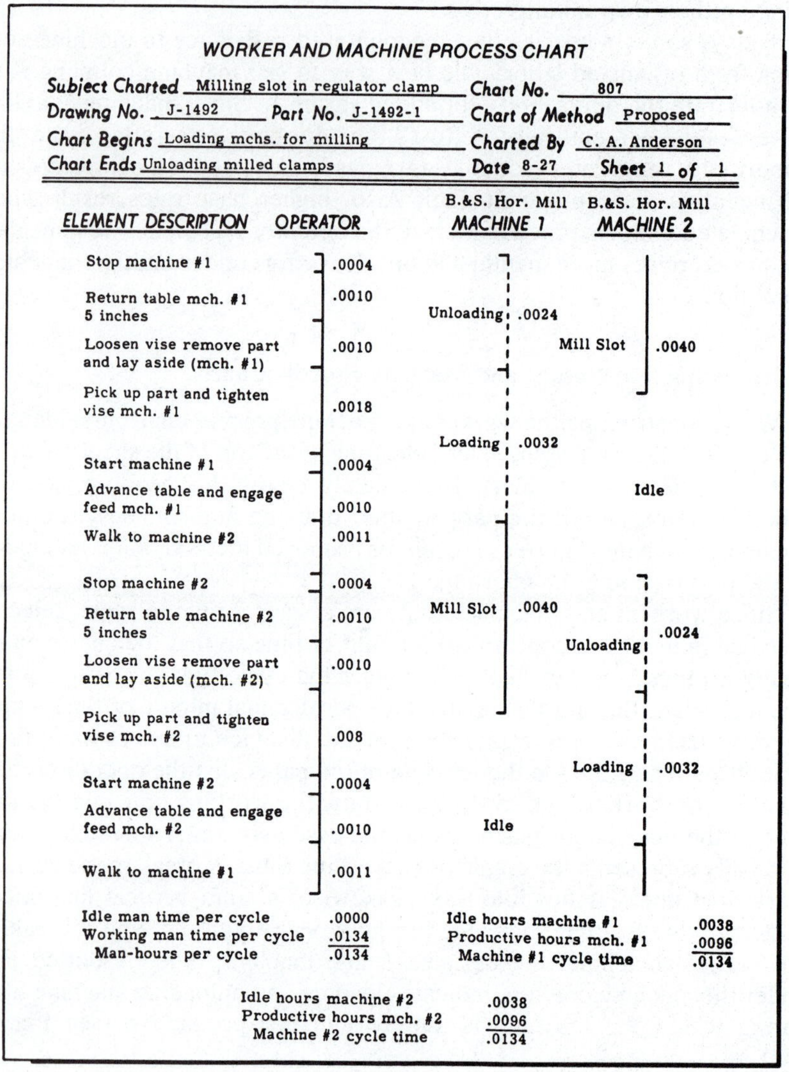

WORKER AND MACHINE PROCESS CHART

Subject Charted Milling slot in regulator clamp   Chart No.   807

Drawing No. J-1492   Part No. J-1492-1   Chart of Method Proposed

Chart Begins Loading mchs. for milling   Charted By C. A. Anderson

Chart Ends Unloading milled clamps   Date 8-27 Sheet 1 of 1

ELEMENT DESCRIPTION	OPERATOR	B.&S. Hor. Mill MACHINE 1	B.&S. Hor. Mill MACHINE 2
Stop machine #1	.0004		
Return table mch. #1 5 inches	.0010	Unloading .0024	
Loosen vise remove part and lay aside (mch. #1)	.0010	Mill Slot	.0040
Pick up part and tighten vise mch. #1	.0018		
Start machine #1	.0004	Loading .0032	
Advance table and engage feed mch. #1	.0010		Idle
Walk to machine #2	.0011		
Stop machine #2	.0004		
Return table machine #2 5 inches	.0010	Mill Slot .0040	.0024
Loosen vise remove part and lay aside (mch. #2)	.0010		Unloading
Pick up part and tighten vise mch. #2	.008		
Start machine #2	.0004		Loading .0032
Advance table and engage feed mch. #2	.0010	Idle	
Walk to machine #1	.0011		
Idle man time per cycle	.0000	Idle hours machine #1	.0038
Working man time per cycle	.0134	Productive hours mch. #1	.0096
Man-hours per cycle	.0134	Machine #1 cycle time	.0134
		Idle hours machine #2	.0038
		Productive hours mch. #2	.0096
		Machine #2 cycle time	.0134

unavoidable delays, and personal delays.[1] In no case should you use overall stopwatch readings in the construction of the chart.

The completed worker and machine process chart clearly shows the areas in which both idle machine time and worker time occur. These areas are generally a good place to start effecting improvements.

---

[1] See Chapter 17.

Do not be deceived by what looks like an appreciable amount of idle work time. In many instances, it is far more economical to have a worker idle for a substantial portion of a cycle rather than chance having an expensive piece of equipment or process be idle for even a small portion of the cycle. To be sure that this proposal is the best solution, analysts must compare the cost of the idle facility with that of the idle worker. It is only when total cost is considered that the analyst can safely recommend one method over another. Figure 6–3 illustrates a worker and machine process chart that represents economies over the one shown in Figure 6–2.

Using the Worker and Machine Process Chart

Construct a worker and machine process chart when your preliminary investigation reveals that the working cycle of the operator is somewhat shorter than the operating cycle of the machine. Once you have completed the chart, the logical place to consider improvement possibilities is during the idle portion of the operator's cycle. While considering the amount of this time, investigate the possibility of assigning the worker the added responsibility of (1) operating a second machine during this idle time, and (2) performing some bench or manual operation, such as filing burrs or gaging parts, during the idle time.

Sometimes more available operator time can be obtained by reducing the speed and feed of the machine. This may permit machine coupling where it would not have been possible otherwise, and thus reduce total costs. As mentioned earlier, machine coupling is not always advisable, as the idle machine time introduced may more than offset the idle operator time saved. The only sure way to make the analysis is on a total cost basis.

Worker and machine charts are effective in determining the extent of coupling justified to assure a "fair day's work for a fair day's pay." They are valuable for determining how idle machine time may be more fully utilized.

GANG PROCESS CHARTS

The gang process chart is, in a sense, an adaptation of the worker and machine chart. After completing a worker and machine process chart, analysts must determine the most economical number of machines one worker can operate. However, several processes and facilities are of such magnitude that it is not a question of how many machines a worker should operate, but of how many workers it should take to operate one machine effectively. The gang process chart shows the exact relationship between the idle and operating cycle of the machine and the idle and operating time per cycle of the workers who service it. This chart shows clearly the

FIGURE 6-2

Worker and machine process chart of thread-grinding operation with one
employee operating one machine

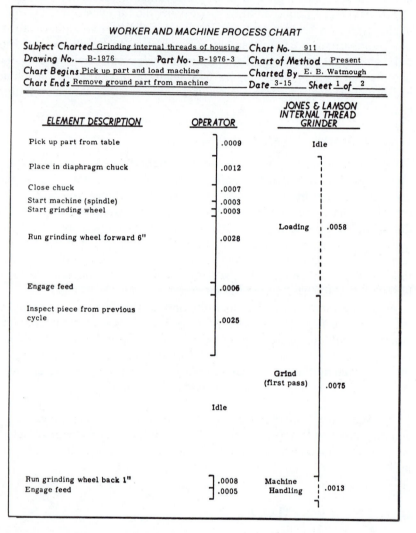

possibilities for improvement by reducing both idle operator time and idle
machine time.

Constructing the Gang Process Chart

After heading the chart "Gang Process Chart" and completely identifying the process being plotted with the part number, drawing number,
description of the operation being charted, present or proposed method,

FIGURE 6–2 (concluded)

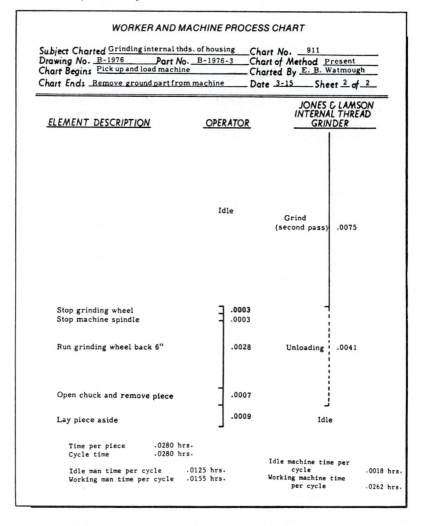

WORKER AND MACHINE PROCESS CHART

Subject Charted <u>Grinding internal thds. of housing</u>   Chart No. <u>911</u>
Drawing No. <u>B-1976</u>   Part No. <u>B-1976-3</u>   Chart of Method <u>Present</u>
Chart Begins <u>Pick up and load machine</u>   Charted By <u>E. B. Watmough</u>
Chart Ends <u>Remove ground part from machine</u>   Date <u>3-15</u>   Sheet <u>2</u> of <u>2</u>

ELEMENT DESCRIPTION	OPERATOR	JONES & LAMSON INTERNAL THREAD GRINDER
	Idle	Grind (second pass) .0075
Stop grinding wheel	.0003	
Stop machine spindle	.0003	
Run grinding wheel back 6"	.0028	Unloading .0041
Open chuck and remove piece	.0007	
Lay piece aside	.0009	Idle

Time per piece   .0280 hrs.
Cycle time   .0280 hrs.

Idle man time per cycle   .0125 hrs.
Working man time per cycle   .0155 hrs.

Idle machine time per cycle   .0018 hrs.
Working machine time per cycle   .0262 hrs.

date, and name of the person doing the charting, select a time scale that will give a neat-appearing chart on the paper being used. Like the worker and machine process chart, the gang process chart is always drawn to scale.

On the left-hand side of the paper list the operations being performed on the machine or process. Immediately to the right of the operation description, enter the loading time, operating time, and idle time of the machine. Further to the right, illustrate the operating time and idle time of each operator performing the process by flow lines in a vertical direction. A solid vertical line indicates that productive work is being done, while a

170   *Motion and Time Study*

FIGURE 6–3
Worker and machine process chart of thread-grinding operation with one
employee operating two machines

WORKER AND MACHINE PROCESS CHART

Subject Charted Grinding internal threads of housing   Chart No. 912
Drawing No. B-1976   Part No. B-1976-3   Chart of Method Proposed
Chart Begins Pick up part and load machine   Charted By E. B. Watmough
Chart Ends Remove ground part from machine   Date 3-17   Sheet 1 of 3

ELEMENT DESCRIPTION	OPERATOR	JONES & LAMSON INTERNAL THREAD GRINDER NO. 1	JONES & LAMSON INTERNAL THREAD GRINDER NO. 2
Stop grinding wheel mch. #1	.0003		
Stop machine spindle machine #1	.0003		
Run grinding wheel back 6" machine #1	.0028	Unloading .0041	
Open chuck and remove part machine #1	.0007		Grind (second pass) .0075
	.0009		
Lay aside part		Idle	
Pick up new part from table	.0009		
Place part in diaphragm chuck machine #1	.0012		
Close diaphragm chuck mch. #1	.0007		
Start mch. (spindle) mch. #1	.0003		
Start grinding wheel mch. #1	.0003		
Run grinding wheel forward 6"	.0028	Loading .0058	Idle
Engage feed	.0005		
Walk to machine #2	.0011		
Stop grinding wheel mch. #2	.0003		
Stop machine spindle mch. #2	.0003		
Run grinding wheel back 6" machine #2	.0028	Grind (first pass) .0075	Unloading .0041

dotted vertical line in connection with the facility shows that either loading or unloading operations are taking place. A void in the vertical flow
line reveals idle time, and the length of the void determines the duration of
the idle time. In the case of the operators, solid vertical lines show that
work is being done, while breaks in the solid lines show idle time. Figure
6–4 illustrates a gang process chart, and it is apparent that a large number
of idle work-hours exist. A better operation of the same process is shown
on the gang process chart in Figure 6–5. The saving of 16 hours per shift
was easily developed through the use of the gang process chart.

FIGURE 6-3 (continued)

WORKER AND MACHINE PROCESS CHART

Subject Charted Grinding internal threads of housing   Chart No. ___912___

Drawing No. _B-1976_   Part No. _B-1976-3_ Chart of Method _Proposed_

Chart Begins _Pick up part and load machine_   Charted By _E. B. Watmough_

Chart Ends _Remove ground part from machine_   Date _3-17_ Sheet _2_ of _3_

ELEMENT DESCRIPTION	OPERATOR	JONES & LAMSON INTERNAL THREAD GRINDER NO. 1	JONES & LAMSON INTERNAL THREAD GRINDER NO. 2
Open chuck and remove part machine #2	.0007		
Lay aside part	.0009		
Pick up new part from table	.0009		Idle
Place part in diaphragm chuck machine #2	.0012		
Close diaphragm chuck Mch. #2	.0007		
Start machine (spindle) mch. #2	.0003		
Start grinding wheel mch. #2	.0003		
			Loading .0058
Run grinding wheel forward 6" machine #2	.0028	Idle	
Engage feed	.0005		
Walk to machine #1	.0011		
Run grinding wheel back 1" machine #1	.0008	Machine Handling .0013	
Engage feed machine #1	.0005		
Inspect pieces from machine #1 and machine #2 previous cycle	.0050		Grind (first pass) .0075
		Grind (second pass) .0075	

Using the Gang Process Chart

The time to construct the gang process chart is after an initial investigation of a given operation indicates that more workers than are necessary are being used to operate a facility or process. The gang process chart is a useful tool for determining the exact number of operators needed to service a machine or process effectively. Once the chart has been constructed, analyze the hours of idle worker time to determine the possibility of utilizing one operator to perform the work elements currently performed by two or more.

FIGURE 6–3 *(concluded)*

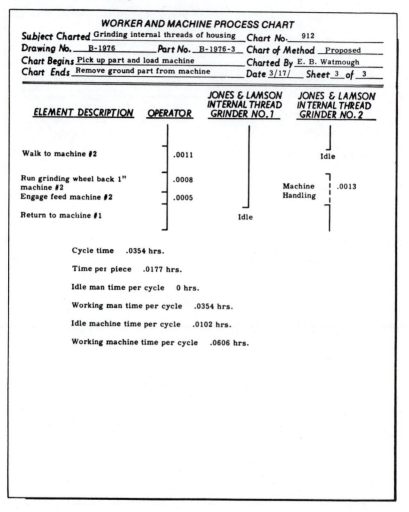

For example, in the gang process chart in Figure 6–4, the company is employing two more operators than are needed. This is apparent because under this process 18.4 idle work-hours are involved in every eight-hour turn.

By relocating some of the controls of the process, it was possible to reassign the elements of work so that four, rather than six, workers effectively operated the extrusion press.

Figure 6–5 illustrates a gang process chart of the proposed method using four operators and saving 16 work-hours per shift. Without the gang process chart, this solution would have been quite difficult.

FIGURE 6–4

GANG PROCESS CHART OF PRESENT METHOD

HYDRAULIC EXTRUSION PRESS   DEPT. 11   BELLEFONTE PA. PLANT

CHARTED BY B.W.N.   4-15-   CHART NO. G-85

MACHINE

OPERATION	TIME
Elevate Billet	.07
Position Billet	.08
Position Dummy	.04
Build Pressure	.05
Extrude	.45
Unlock Die	.06
Loosen & Push Out Shell	.10
Withdraw Ram & Lock Die in Head	.15
WORKING TIME	1.00 MIN.
IDLE TIME	0 "

PRESS OPERATOR

OPERATION	TIME
Elevate Billet	.07
Position Billet	.08
Position Dummy	.04
Build Pressure	.05
Extrude	.45
Unlock Die	.06
Loosen & Push Out Shell	.10
Withdraw Ram & Lock Die in Head	.15
WORKING TIME	1.00 MIN.
IDLE TIME	0 "

ASSISTANT PRESS OPERATOR

OPERATION	TIME
Grease Die & Position Back in Die Head	.12
Idle Time	.68
Run Head & Shell Out	.11
Shear Rod from Shell	.04
Pull Die Off End of Rod	.05
WORKING TIME	.32 MIN.
IDLE TIME	.68 "

FURNACE MAN

OPERATION	TIME
Rearrange Billets in Furnace	.20
Idle Time	.51
Open Furnace Door & Remove Billet	.19
Ram Billet from Furnace & Close Furnace Door	.10
WORKING TIME	.49 MIN.
IDLE TIME	.51 "

DUMMY KNOCKER

OPERATION	TIME
Position Shell on Small Press	.10
Press Dummy Out of Shell	.12
Dispose of Shell	.18
Dispose of Dummy and Lay Aside Tongs	.12
Idle	.43
Grab Tongs & Move to Position	.05
WORKING TIME	.57 MIN.
IDLE TIME	.43 "

ASSISTANT DUMMY KNOCKER

OPERATION	TIME
Move Away from Small Press and Lay Aside Tongs	.12
Idle Time	.68
Guide Shell from Shear to Small Press	.20
WORKING TIME	.32 MIN.
IDLE TIME	.68 "

PULL-OUT MAN

OPERATION	TIME
Pull Rod toward Cooling Rack	.20
Walk Back toward Press	.15
Grab Rod with Tongs and Pull Out	.45
Straighten Rod End with Mallet	.11
Hold Rod while Die Removed at Press	.09
WORKING TIME	1.00 MIN.
IDLE TIME	0 "

IDLE TIME = 2.30 MAN-MINUTES PER CYCLE = 18.4 MAN-HOURS PER EIGHT-HOUR DAY

FIGURE 6–5
Gang process chart of the proposed method of operation of a hydraulic extrusion process

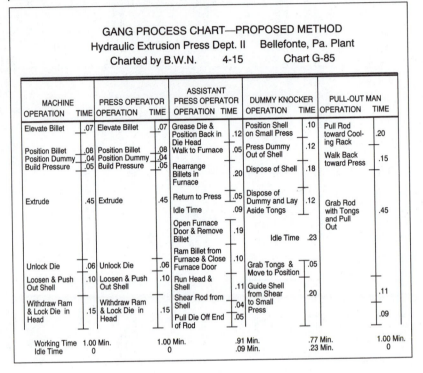

Figures 6–4 and 6–5 show that the gang process chart aids in dividing the available work among the members of the team that operates the equipment, and then clearly determines the job assignments of all involved. Through the construction and use of the gang process chart, equipment can be operated to capacity, labor costs reduced, and employee morale improved as a result of the equitable distribution of work assignments.

QUANTITATIVE TECHNIQUES FOR WORKER AND MACHINE RELATIONSHIPS

Although the worker and machine process chart can determine the number of facilities to be assigned to an operator, this can often be computed in much less time through the development of a mathematical model.

Worker and machine relationships are usually one of three types: (1) synchronous servicing, (2) completely random servicing, and (3) a combination of synchronous and random servicing.

Assigning more than one machine to an operator seldom results in the ideal case where both the worker and the machine are occupied during the whole cycle. Ideal cases such as this are referred to as "synchronous servicing," and the number of machines to be assigned can be computed as:

$$N = \frac{l + m}{l}$$

where

N = Number of machines the operator is assigned
l = Total operator servicing time per machine (loading and unloading)
m = Total machine running time (power feed)

For example, if the total operator servicing time was one minute, while the cycle time of the machine was four minutes, synchronous servicing would result in the assignment of five machines:

$$N = \frac{1 + 4}{1}$$

$$= 5$$

Graphically, this assignment would appear as shown below:

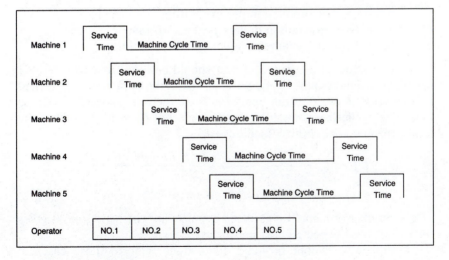

If the number of machines in this example is increased, machine interference takes place, and we have a situation where one or more of the facilities are idle for a portion of the work cycle. If the number of machines is reduced to some figure less than five, then the operator is idle for a portion of the cycle.

In such cases, the minimum total cost per piece usually represents the criterion for optimum operation. In establishing the best method consider the cost of each idle machine and the hourly rate of the operator. Quantitative techniques can determine the best arrangement. The procedure is, first, to estimate the number of facilities that the operator should be assigned by establishing the lowest whole number from the equation:

$$N_1 \leqq \frac{l + m}{l + w}$$

where:

N_1 = Lowest whole number
w = Normal walking time to the next facility

The cycle time with the operator servicing N_1 machines is $l + m$ since in this case the operator is not busy the whole cycle, while the facilities are occupied during the entire cycle.

Using N_1, the total expected cost may be computed as follows:

$$\text{T.E.C.}_{N_1} = \frac{K_1(l + m) + N_1 K_2(l + m)}{N_1}$$

$$= \frac{(l + m)(K_1 + N_1 K_2)}{N_1}$$

where:

T.E.C. = Cost of production per cycle from one machine
K_1 = Operator rate in dollars per unit of time
K_2 = Cost of machine in dollars per unit of time

After this cost is computed, a cost should be calculated with $N_1 + 1$ machines assigned to the operator. In this case, the cycle time is governed by the working cycle of the operator, since there is some idle machine time. The cycle time is now $(N_1 + 1)(l + w)$. Let $N_2 = N_1 + 1$. Then the total expected cost with N_2 facilities is:

$$\text{T.E.C.}_{N_2} = \frac{(K_1)(N_2)(l + w) + (K_2)(N_2)(N_2)(l + w)}{(N_2)}$$

$$= [(l + w)][K_1 + K_2(N_2)]$$

The number of machines assigned depends on whether N_1 or N_2 gives the lowest total expected cost per piece.

"Completely random servicing" situations refer to those cases in which it is not known when a facility needs to be serviced or how long servicing takes. Mean values are usually known or can be determined; with these averages, the laws of probability can provide a useful tool in determining the number of machines to assign the operator.

The successive terms of the binomial expansion give a useful approximation of the probability of 0, 1, 2, 3, . . . n machines down (where n is relatively small), assuming that each machine is down at random times during the day and that the probability of running time is p and the probability of downtime is q.

For example, let us determine the minimum proportion of machine time lost for various numbers of turret lathes assigned to an operator where the average machine operates 60 percent of the time unattended. Operator attention time at irregular intervals is 40 percent on the average. The analyst estimates that three turret lathes should be assigned per operator on this class of work. Under this arrangement, the combinations of machine running (p) or down (q) expressed as probabilities would be:

$$(p + q)^n = (p + q)^3$$
$$= p^3 + 3p^2q + 3pq^2 + q^3$$
$$= (0.60)^3 + (3)(0.60)^2(0.40) + 3(0.60)(0.40)^2 + (0.40)^3$$
$$1.00 = 0.216 + 0.432 + 0.288 + 0.064$$

In tabular form, this would appear as follows:

Machine 1	Machine 2	Machine 3	Probability
	$R = 0.60$	$R = 0.60$	$(0.60)(0.60)(0.60) = 0.216$
		$D = 0.40$	$(0.60)(0.60)(0.40) = 0.144$
$R = 0.60$			
	$D = 0.40$	$R = 0.60$	$(0.60)(0.40)(0.60) = 0.144$
		$D = 0.40$	$(0.60)(0.40)(0.40) = 0.096$
	$R = 0.60$	$R = 0.60$	$(0.40)(0.60)(0.60) = 0.144$
		$D = 0.40$	$(0.40)(0.60)(0.40) = 0.096$
$D = 0.40$			
	$D = 0.40$	$R = 0.60$	$(0.40)(0.40)(0.60) = 0.096$
		$D = 0.40$	$(0.40)(0.40)(0.40) = \underline{0.064}$
			1.000

Thus, the proportion of time that some machines are down may be determined, and the resulting lost time of one operator per three machines may be readily computed. In this example, we have:

No. of machines down	Probability	Machine hours lost per 8-hour day
0	0.216	0
1	0.432	0*
2	0.288	$(0.288)(8) = 2.304$
3	$\underline{0.064}$	$(2)(0.064)(8) = \underline{1.024}$
	1.000	3.328

* Since only one machine is down at a time, the operator can be attending the down machine.

$$\text{Proportion of machine time lost} = \frac{3.328}{24.0} = 13.9 \text{ percent}$$

Make similar computations for more or less machine assignments to determine the assignment resulting in the least machine down-time. The most satisfactory arrangement is usually the arrangement showing the least Total Expected Cost per piece. The Total Expected Cost per piece of a given arrangement is computed by the expression:

$$\text{T.E.C.} = \frac{K_1 + NK_2}{\text{Pieces from } N \text{ machines per hour}}$$

where:

K_1 = Hourly rate of the operator
K_2 = Hourly rate of the machine
N = Number of machines assigned

The pieces per hour from N machines is computed with the mean machine time required per piece, the average machine servicing time per piece, and the expected down or lost time per hour.

For example, under a five-machine assignment to one operator, an analyst determined that the machining time per piece was 0.82 hours, the machine serving time per piece 0.17 hours, and the machine downtime an average of 0.11 hours per machine per hour. Thus, each machine was available for production work only 0.89 hours each hour. The average time required to produce one piece per machine would be $\frac{0.82 + 0.17}{0.89} =$ 1.11. Therefore, the five machines would produce 4.5 pieces per hour. With an operator hourly rate of $12 and a machine hourly rate of $22, we have a total expected cost per piece of:

$$\frac{\$12.00 + 5(\$22.00)}{4.5} = \$27.11$$

Combinations of synchronous and random servicing are perhaps the most common type of worker and machine relationships. Here the servicing time is constant, but the machine downtime is random. Winding, coning, and quilling operations used in the textile industry are characteristic of this type of worker and machine relationship. As in the former examples, algebra and probability can establish the mathematical model that enhances a realistic solution (see queuing theory, Chapter 22).

Using the Digital Computer

As an alternative to the laborious process of constructing the worker and machine process chart in order to arrive at the most favorable division of work elements between internal (work performed while the machine is running) and exernal (work performed while the machine is not running) times, is the use of the digital computer. For example, the analyst will want to know the arrangement of work elements in order to maximize the running time when an operator is assigned two or more machines. The analyst has considerable freedom in determining both the amount of time included in an element and when the element should be performed. For example, a quality control program related to a specific machining sequence may require one part for every five pieces to be inspected. The planner may require that each fifth part be gaged and inspected in a time of 7.5 minutes. Then an inspection time of 1.5 minutes per part would be assigned in connection with the total labor cost. However, a block of 7.5 minutes of time every fifth cycle may not be able to be done internally—especially on short cycle operations. Perhaps the inspection operation can be broken down into several elements so that 1.5 minutes of inspection time takes place on each cycle. Now with only 1.5 minutes of inspection taking place each cycle, it is highly possible this can be scheduled as internal time, thus increasing the machine running time and decreasing the interference time.

Software can be written to solve the problem of machine utilization of each of the machines under the coupling arrangement as well as the machine interference problem. For example, in one company, because of the

TABLE 6–1

Flinchbaugh Engineering, Inc.
Machine simulation
Inspection uniform on each piece
Date 06–26–1991   Time 08:12:32
File name mach. doc.:

Machine #1	Load/unload	1.0
	Cycle time	5.5
	Inspect time	1.5
Machine #2	Load/unload	1.0
	Cycle time	3.5
	Inspect time	1.5

Labor analysis:
Percent labor idle   14.6

	Mach. 1	Mach. 2	
Pcs. loaded	83.0	124.0	
Min. loaded	85.5	127.7	
Pcs. inspected	83.0	123.0	
Min. inspected	120.7	178.9	
Avg. inspect	1.5	1.5	
Std. Hrs.	9.0	9.4	18.4

Machine #1
Machine utilization   89.6

Machine #2
Machine utilization   93.2
Machine interference   8.6

TABLE 6–2

Flinchbaugh Engineering, Inc.
Machine simulation
Inspection 1 in 5 type 1
Date 06–26–1991   Time 08:16:26
File name mach. doc.:

Machine #1	Load/unload	1.0
	Cycle time	3.5
	Inspect time	1.5
Machine #2	Load/unload	1.0
	Cycle time	3.5
	Inspect time	1.5

Labor analysis:
Percent labor idle   19.3

	Mach. 1	Mach. 2	
Pcs. loaded	111.0	110.0	
Min. loaded	114.3	113.4	
Pcs. inspected	111.0	109.0	
Min. inspected	132.4	127.7	
Avg. inspect	1.2	1.2	
Std. Hrs.	8.4	8.3	16.7

Machine #1
Machine utilization   83.4

Machine #2
Machine utilization   82.5
Machine interference   17.1

high cost of its capital equipment, it is emphatic to maximize its running time and minimize machine interference time. Output from the company's software is shown in Tables 6–1 and 6–2. Note that by changing inspection from 7.5 minutes for a complete piece every fifth cycle to 1.5 minutes on every cycle, machine utilization increased on machine 1 from 83.4 percent to 89.6 percent and on machine 2 from 82.5 percent to 93.2 percent. Also, machine interference decreased from 17.1 percent to 8.6 percent. These savings can be the difference between success and failure in today's competitive environment.

Line Balancing

The problem of determining the ideal number of workers to be assigned to a production line is analogous to that of determining the number of workers to be assigned to a production facility; the gang process chart solves both problems. Perhaps the most elementary line balancing situation, yet one that is very often encountered, is a situation in which several operators, each performing consecutive operations, work as a unit. In such a situation, the rate of production is dependent on the slowest operator. For example, we may have a line of five operators assembling bonded rubber mountings prior to the curing process. The specific work assignments might be as follows: Operator 1, 0.52 minutes; Operator 2, 0.48 minutes; Operator 3, 0.65 minutes; Operator 4, 0.41 minutes; Operator 5, 0.55 minutes. Operator 3 establishes the pace, as is evidenced by the following:

Operator	Standard minutes to perform operation	Wait time based on slowest operator	Allowed standard minutes
1	0.52	0.13	0.65
2	0.48	0.17	0.65
3	0.65	—	0.65
4	0.41	0.24	0.65
5	0.55	0.10	0.65
Totals	2.61		3.25

The efficiency of this line can be computed as the ratio of the total standard minutes to the total allowed standard minutes, or:

$$E = \frac{\sum_1^5 S.M.}{\sum_1^5 A.M.} \times 100 = \frac{2.61}{3.25} \times 100 = 80 \text{ percent}$$

where:

E = Efficiency
$S.M.$ = Standard minutes per operation
$A.M.$ = Allowed standard minutes per operation

In a real-life situation similar to this example, the opportunity for significant savings exists. If an analyst can save 0.10 minute on Operator 3, the net savings per cycle is not 0.10 minute but 0.10 × 5 or 0.50 minutes.

Only in the most unusual situations would a line be perfectly balanced; that is, the standard minutes to perform an operation would be identical for each member of the team. The "standard minutes to perform an operation" is really not a standard. It is only a standard in the eyes of the individual who established it. Thus, in our example, where Operator 3 has a standard time of 0.65 minutes to perform the first operation, a different work measurement analyst might have allowed as little as 0.61 minutes or as much as 0.69 minutes. The range of standards established by different work measurement analysts on the same operation might be even greater than the range suggested. The point is that whether the issued standard is 0.61, 0.65, or 0.69, the typical conscientious operator has little difficulty in meeting the standard. In fact, he or she will probably better the standard in view of the performance of the operators on the line with less work content in their assignments. Those operators who have a wait time based on the output of the slowest operator seldom are observed as actually waiting. Instead, they reduce the tempo of their movements to utilize the number of standard minutes established by the slowest operator.

The number of allowed standard minutes to produce one unit of the product is equal to the summation of the standard minutes required times the reciprocal of the efficiency. Thus,

$$\Sigma A.M. = \Sigma S.M. \times \frac{1}{E}$$

The number of operators needed is equal to the required rate of production times the total allowed minutes, where:

N = Number of operators needed in the line
R = Desired rate of production
$N = R \times \Sigma A.M.$

For example, assume that we have a new design for which we are establishing an assembly line. Eight distinct operations are involved. The line must produce 700 units per day, and since it is desirable to minimize storage, we do not want to produce many more than 700 units per day. The eight operations involve the following standard minutes based on existing standard data: Operation 1, 1.25 minutes; Operation 2, 1.38 minutes; Operation 3, 2.58 minutes; Operation 4, 3.84 minutes; Operation 5, 1.27 minutes; Operation 6, 1.29 minutes; Operation 7, 2.48 minutes; and Operation 8, 1.28 minutes. To plan this assembly line for the most economical setup, estimate the number of operators required at 100 percent efficiency as follows:

$$\sum_1^8 S.M. = 15.37 \text{ minutes}$$

$$N = \frac{700}{480} \times \frac{15.37}{E} = \frac{22.4}{E}$$

For 95 percent efficiency, estimate the number of operators to be: $\frac{22.4}{0.95} = 23.6$.

Since it is impossible to have six tenths of an operator, you would endeavor to set up the line utilizing 24 operators.

The next step is to estimate the number of operators to be utilized at each of the eight specific operations. Since 700 units of work are required a day, it will be necessary to produce one unit in about 0.685 minutes $\left(\frac{480}{700}\right)$. Estimate how many operators will be needed on each operation by dividing the number of minutes to have one piece produced into the standard minutes of each operation.

Operation	Standard minutes	Standard minutes / Minutes/unit	No. of operators
Operation 1	1.25	1.83	2
Operation 2	1.38	2.02	2
Operation 3	2.58	3.77	4
Operation 4	3.84	5.62	6
Operation 5	1.27	1.86	2
Operation 6	1.29	1.88	2
Operation 7	2.48	3.62	4
Operation 8	1.28	1.87	2
Total	15.37		24

To determine which is the slowest operation, divide the estimated number of operators into the standard minutes for each of the eight operations.

Operation 1	1.25/2 = 0.625
Operation 2	1.38/2 = 0.690
Operation 3	2.58/4 = 0.645
Operation 4	3.84/6 = 0.640
Operation 5	1.27/2 = 0.635
Operation 6	1.29/2 = 0.645
Operation 7	2.48/4 = 0.620
Operation 8	1.28/2 = 0.640

Thus, Operation 2 determines the output from the line. In this case, it is:

$$\frac{2 \text{ workers} \times 60 \text{ min.}}{1.38 \text{ standard minutes}} = 87 \text{ pieces per hour, or } 696 \text{ pieces per day}$$

If this rate of production is inadequate, increase the rate of production of Operator 2. This can be accomplished by:

1. Working one or both of the operators at the second operation overtime, thus accumulating a small inventory at this workstation.
2. Utilizing the services of a third part-time worker at the workstation of Operation 2.
3. Reassigning some of the work of Operation 2 to Operation 1 or Operation 3. (It would be preferable to assign more work to Operation 1.)
4. Improving the method at Operation 2 to diminish the cycle time of this operation.

FIGURE 6-6
Assembly line involving six workstations

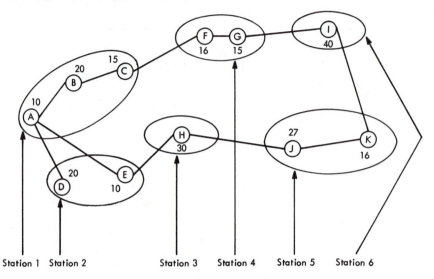

Station 1   Station 2   Station 3   Station 4   Station 5   Station 6

In the preceding example, given a cycle time and operation times, an analyst can determine the number of operators needed for each operation to meet a desired production schedule.

The production line work assignment problem can also be to minimize the number of workstations, given the desired cycle time; or, given the number of workstations, to assign work elements to the workstations, within the restrictions established, to minimize the cycle time.

An important strategy in assembly line balancing is work element sharing. Thus, two or more operators whose work cycle includes some idle time may share the work of another station for the purpose of a more efficient line. For example, Figure 6-6 shows an assembly line involving six workstations. Station 1 has three work elements to accomplish—A, B, and C—for a total of 45 seconds. Note that work elements, B, D, and E cannot begin until A is completed and that there is no precedence among B, D, and E. It may be possible to have element H shared by stations 2 and 4 with only a one-second increase in cycle time (from 45 seconds to 46 seconds) while saving 30 seconds per assembled unit. Element sharing may result in an increase in material handling since parts may have to be delivered to more than one location. In addition, element sharing may necessitate added costs for duplicate tooling.

A second possibility in improving the balance of an assembly line is dividing a work element. Referring again to Figure 6-6, it may be possible to divide element H rather than have half of the number of parts go to station 2 and the other half to station 4.

185

Many times it is not economical to divide an element. An example is
driving home eight machine screws with a power screwdriver. Once the
operator has located the part in a fixture, gained control of the power tool,
and brought it to the work, it would usually be more advantageous to
drive home all eight screws rather than drive only a portion home and
have a different operator drive the rest home. Frequently, however, elements can be divided, and workstations may be better balanced as a result
of the division.

A different sequence of assembly may also produce more favorable
results. Product design generally dictates the assembly sequence. However, often alternatives should not be overlooked. Balanced assembly
lines are not only less costly, but they assist in maintaining worker morale, since in such lines little differential exists in the work content of the
different workers.

The following procedure for solving an assembly line balancing problem is based on General Electric's *Assembly Line Balancing*. General
Electric engineers have prepared a program, written for the GE–225 computer, for assigning work elements to stations on an assembly line. The
method is based on the following:

1. Operators are not able to move from one workstation to another to
   help maintain a uniform work load.
2. The work elements that have been established are of such magnitude
   that further division would substantially decrease the efficiency of
   performing the work element. (Once established, the work elements
   should be identified with a code.)

The first step in the solution of the problem is determining the sequence
of individual work elements. The fewer the restrictions on the order in
which the work elements can be done, the greater the probability of a
favorable balance in the work assignments. To determine the sequence of
the work elements, ask and answer the question: "What other work elements, if any, must be completed before this work element can be
started?"

Complete a precedence chart for the production line under study (see
Figure 6–7). Not only functional design, but available production methods, floor space, and so on, can introduce constraints as far as the work
element sequence is concerned.

A second consideration in the production line work assignment problem is the recognition of zoning restraints. A zone represents a subdivision, which may or may not be physically separated or identified from
other zones in the system. Confining certain work elements to a given
zone may be justified to congregate similar jobs, working conditions, or
pay rates. Or again, it may be desirable to introduce zoning restraints to
identify physically specific stages of a component, such as keeping it in a
certain position while performing work elements. Thus, all work elements

FIGURE 6-7
Partially completed precedence chart. Note that work elements 002 and 003 may be
done in any sequence with respect to any of the other work elements and that 032
cannot be started until 005, 006, 008, and 009 have been completed. Note also that
after 004 has been finished we may start either 033, 017, 021, 007, 008, or 009.

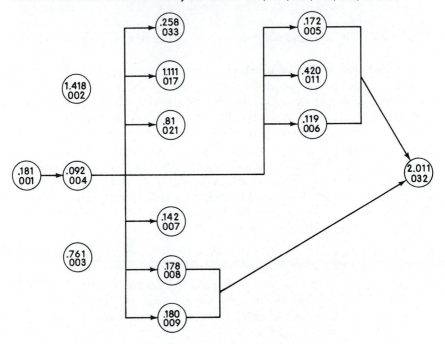

related to one side of a component may be performed in a certain zone
before the component is turned over.

Obviously, the more zoning restraints placed on the system, the fewer
the combinational possibilities open to investigation. Begin by making a
sketch of the system and coding the applicable zones. Within each zone
show the work elements that may be done in this area. The next step is to
estimate the production rate. This is done with the expression:

$$\text{Production per day} = \frac{\text{Working minutes/day}}{\text{Cycle time of system (min./unit)}} \times \text{Rating factor}$$

For example, assuming a 15 percent allowance, we would have 480 − 72,
or 408, working minutes per day. The rating factor would be based on
experience with the type of line under study. It could be either more or
less than standard (100 percent). The cycle time of the system is the cycle
time of the limiting station. Since we know the production requirements
per day, we are able to compute the allowed cycle time of the limiting
station.

This computer-generated assignment does not necessarily represent the optimum. It is a good base from which to work and on which to improve. For example, you may be able to divide an element and thus bring about a more efficient overall arrangement. Or again, you may modify zoning restrictions, or alter precedence relationships.

The computer can select the assignments for each operator, taking into account the cycle time of the system, precedence, and zoning. To illustrate the logic of the computer routine, the following precedence graph is described:

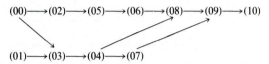

This precedence graph shows that work unit (00) must be completed before (02), (03), (05), (06), (04), (07), (08), (09), and (10); and that work unit (01) must be completed before (03), (04), (07), (08), (09), and (10). Either (00) or (01) can be done first, or they can be done concurrently. Work unit (03) cannot be started until work units (00) and (01) are completed, and so on.

To describe these relationships, the computer uses the precedence matrix illustrated in Figure 6–8. Here the numeral 1 signifies a "must precede" relationship. For example, work unit (00) must precede work units (02), (03), (04), (05), (06), (07), (08), (09), and (10). Also, work unit (09) must precede only work unit (10).

Now a "positional weight" must be computed for each work unit. This is done by computing the summation of each work unit and all the work units that must follow it. Thus, the "positional weight" for work unit (00) would be: Σ 00, 02, 03, 04, 05, 06, 07, 08, 09, 10 = 0.46 + 0.25 + 0.22 + 1.10 + 0.87 + 0.28 + 0.72 + 1.32 + 0.49 + 0.55 = 6.26.

Listing the positional weights in decreasing order of magnitude gives the following:

Unsorted work elements	Sorted work elements	Positional weight	Immediate predecessors
00	00	6.26	—
01	01	4.75	—
02	03	4.40	(00), (01)
03	04	4.18	(03)
04	02	3.76	(00)
05	05	3.56	(02)
06	06	2.64	(05)
07	08	2.36	(04), (06)
08	07	1.76	(04)
09	09	1.04	(07), (08)
10	10	0.55	(09)

FIGURE 6–8
A precedence matrix as used by a digital computer for a line-balancing problem

Estimated work unit time (minutes)	Work unit	Work unit										
		00	01	02	03	04	05	06	07	08	09	10
0.46	00		1	1	1	1	1	1	1	1	1	
0.35	01				1	1			1	1	1	1
0.25	02						1	1		1	1	1
0.22	03					1			1	1	1	1
1.10	04								1	1	1	1
0.87	05							1		1	1	1
0.28	06									1	1	1
0.72	07										1	1
1.32	08										1	1
0.49	09											1
0.55	10											
6.61												

Work elements must now be assigned to various workstations. This is based on the positional weights (those work elements with the highest positional weights are assigned first) and the cycle time of the system. Thus, assign the work element with the highest positional weight to the first workstation. Determine the unassigned time for this workstation by subtracting the sum of the assigned work element times from the estimated cycle time. If there is adequate unassigned time, then assign the work element with the next highest positional weight, provided that the work elements in the "immediate predecessors" column have already been assigned.

This procedure continues until all of the work elements have been assigned.

For example, assume that the required production per day was 300 units and that a 1.10 rating factor was anticipated. Then:

$$\text{Cycle time of system} = \frac{(480 - 72)(1.10)}{300}$$

$$= 1.50 \text{ minutes}$$

| Work | | | | Work | Station time | | |
Sta-tion	Ele-ment	Positional weight	Immediate predecessors	element time	Cumula-tive	Un-assigned	Remarks*
1	00	6.26	—	0.46	.46	1.04	—
1	01	4.75	—	0.35	.81	0.69	—
1	03	4.40	(00), (01)	0.22	1.03	0.47	—
1	~~04~~	~~-4.18-~~	~~(03)~~	~~1.10~~	(2.13)		N.A.
1	02	3.76	(00)	0.25	1.28	0.22	—
1	~~05~~	~~-3.56-~~	~~(02)~~	~~0.87~~	(2.05)		N.A.
2	04	4.18	(03)	1.10	1.10	0.40	—
2	~~05~~	~~-3.56~~	~~(02)~~	~~0.87~~	(1.97)		N.A.
3	05	3.56	(02)	0.87	.87	0.63	—
3	06	2.64	(05)	0.28	1.15	0.35	—
3	~~08~~	~~-2.36~~	~~(04),(06)~~	~~1.32~~	(2.47)		N.A.
4	08	2.36	(04), (06)	1.32	1.32	0.18	—
4	~~07~~	~~-1.76~~	~~(04)~~	~~0.72~~	(2.04)		N.A.
5	07	1.76	(04)	0.72	.72	0.78	—
5	09	1.04	(07), (08)	0.49	1.21	0.29	—
5	~~10~~	~~.55~~	~~(09)~~	~~0.55~~	(1.76)		N.A.
6	10	.55	(09)	0.55	.55	0.95	—

* N.A. means not acceptable.

Under the arrangement illustrated, with six workstations we have a cycle time of 1.32 minutes (workstation 4). This arrangement more than meets the daily requirement of 300. It produces:

$$\frac{(480 - 72)(1.10)}{1.32} = 341 \text{ units}$$

However, with six workstations we have considerable idle time. The idle time per cycle is:

$$\sum_{1}^{6} 0.04 + 0.22 + 0.17 + 0 + 0.11 + 0.77 = 1.31 \text{ minutes}$$

For more favorable balancing, the problem can be solved for cycle times of less than 1.50 minutes. This may result in more operators and in more production per day which may have to be stored. Another possibility includes operation of the line under a more efficient balancing for a limited number of hours per day.

THE OPERATOR PROCESS CHART

The operator process chart, sometimes referred to as a "left- and right-hand process chart" is, in effect, a tool of motion study. This chart shows all movements and delays made by the right and left hands, and the relationship between the relative basic divisions of accomplishment as performed by the hands. The purpose of the operator process chart is to present a given operation in sufficient detail so that the operation can be improved by means of an analysis. Usually it is not practical to make a detailed study through the operator process chart unless a highly repetitive manual operation is involved. Through the motion analysis of the operator chart, inefficient motion patterns become apparent, and violations of the laws of motion economy (see Chapter 7) are readily observed. This chart facilitates changing a method so that a balanced two-handed operation can be achieved and ineffective motions either reduced or eliminated. The result is a smoother, more rhythmic cycle which keeps both delays and operator fatigue to a minimum.

Constructing the Operator Process Chart

Frank and Lillian Gilbreth stated that there are 17 fundamental motions, and that every operation consists of a combination of some of these elements. It is more practical when plotting operator process charts to use only eight basic divisions of accomplishment. These elemental motions with their symbols are:

Reach	RE	Use	U
Grasp	G	Release	RL
Move	M	Delay	D
Position	P	Hold	H

See Chapter 7 for a detailed description of each of these basic divisions of human work.

Although there are several different delay types, such as unavoidable delay, avoidable delay, rest to overcome fatigue, and "balancing delay," in the operator process chart indicate a cessation of work with the identification "delay." Often, delays do not occur to both hands simultaneously. For example, an operator while hand-feeding an engine lathe would be performing "use" with the right hand while the left hand would be delayed. Other examples of delays would be the operator's waiting for a fixture to index so that productive work could begin or sitting down to rest after each cycle of an operation. In both of these examples, the right hand as well as the left hand would be delayed. Delays should be investigated with the thought of eliminating or minimizing them.

As usual, head the chart: Operator Process Chart, and add all necessary identifying information, including the part number, drawing number, operation or process description, present or proposed method, date, and name of the person doing the charting. Immediately below the identifying information sketch the workstation, drawn to scale. The sketch materially aids in presenting the method under study. Figure 6–9 shows a typical operator process chart form, with a section of the sheet laid out in coordinate form to facilitate sketching.

After completely identifying the operation and making a sketch of the workstation showing dimensional relationships, begin constructing the operator process chart. Since this chart is drawn to scale, determine by observation the duration of the cycle. Then you can readily determine the amount of time represented by each ¼ inch of vertical space on the chart. For example, if the operation to be studied has a cycle time of 0.70 minutes and there are 7 vertical inches of available charting space, then each ¼ inch of chart space would equal 0.025 minutes.

It is usually less confusing to chart the activities of one hand completely, and then to chart all the basic divisions of accomplishment performed by the other hand. Although there is no fixed rule on what part of the work cycle should be used as a starting point, it is usually best to start plotting immediately after the "release" of the finished part. If this release is done with the right hand, the next movement that would normally take place would be the first motion shown on the operator process chart. This, for the right hand, would probably be "reach for new part." If you observe that the "reach" element took about 0.025 minutes to perform, indicate this duration by drawing a horizontal line across the right-hand side of the paper ¼ inch from the top. Under the "Symbols" column, write "RE" (for "reach") in black pencil, indicating that an effective motion had been accomplished. Immediately to the right of the symbol enter a brief description of the event, such as "Reach 20 inches" for a ½-inch nut. Immediately below, show the next basic division and so on until completion of the cycle. Then proceed to plot the left hand activities during the cycle. While plotting the left hand's activities, verify that end points of the therbligs actually occur at the same point as indicated—a check for overall plotting errors.

All elements must be large enough to be measured, since it is not possible in most instances to time individual therbligs. For example, in Figure 6–9 the first element performed by the left hand was "Get U-bolt." This element comprised the therbligs "reach" and "grasp." It would not have been possible to time either of these therbligs. Only through use of a motion-picture camera or videotape can time values as short as these be measured.

The observer in Figure 6–9 was using a decimal second watch. By observing an element at a time, the observer broke the elements into one-second periods.

FIGURE 6-9
Operator process chart of assembly of cable clamps

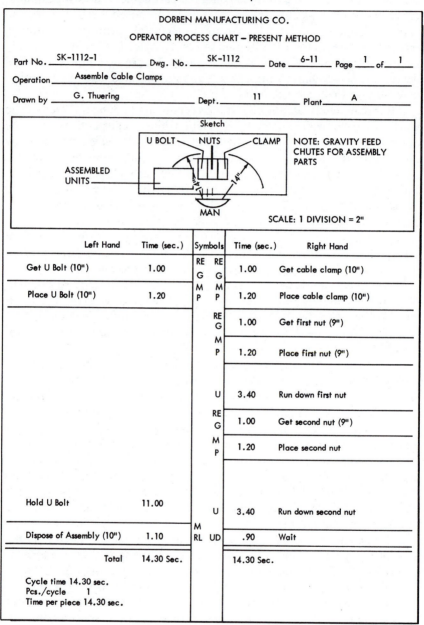

DORBEN MANUFACTURING CO.

OPERATOR PROCESS CHART – PRESENT METHOD

Part No. SK-1112-1   Dwg. No. SK-1112   Date 6-11   Page 1 of 1

Operation Assemble Cable Clamps

Drawn by G. Thuering   Dept. 11   Plant A

Sketch

U BOLT   NUTS   CLAMP

ASSEMBLED UNITS

MAN

NOTE: GRAVITY FEED CHUTES FOR ASSEMBLY PARTS

SCALE: 1 DIVISION = 2"

Left Hand	Time (sec.)	Symbols	Time (sec.)	Right Hand
Get U Bolt (10")	1.00	RE RE / G G	1.00	Get cable clamp (10")
Place U Bolt (10")	1.20	M M / P P	1.20	Place cable clamp (10")
		RE / G	1.00	Get first nut (9")
		M / P	1.20	Place first nut (9")
		U	3.40	Run down first nut
		RE / G	1.00	Get second nut (9")
		M / P	1.20	Place second nut
Hold U Bolt	11.00	U	3.40	Run down second nut
Dispose of Assembly (10")	1.10	M / RL UD	.90	Wait
Total	14.30 Sec.		14.30 Sec.	

Cycle time 14.30 sec.
Pcs./cycle 1
Time per piece 14.30 sec.

After the activities of both the right and the left hand have been charted, create a summary at the bottom of the sheet, indicating the cycle time, pieces per cycle, and time per piece.

Using the Operator Process Chart

Once the operator process chart of an existing method has been completed, see what improvements can be introduced. The "delays" and "holds" are good places to begin. For example, in Figure 6–9 the left hand was a holding device for almost the entire cycle. An analysis of this condition would suggest the development of a fixture to hold the U-bolt. Further consideration of how to get balanced motions of both hands would suggest that, when the fixture holds the U-bolts, then the left hand and the right hand would each assemble a cable clamp completely. Additional study of this chart might result in the introduction of an automatic ejector and gravity chute to eliminate the final cycle element "dispose of assembly."

The best way of doing a job is through systematic analysis of all the detailed elements constituting that job. The operator process chart clearly reveals the work done by each hand in performing an operation and shows the relative time and the relationships of all motions performed by the hands. The operator process chart is an effective tool to:

1. Balance the motions of both hands and reduce fatigue.
2. Eliminate and/or reduce nonproductive motions.
3. Shorten the duration of productive motions.
4. Train new operators in the ideal method.
5. Sell the proposed method.

Making and using operator process charts can bring about improvements.

SUMMARY: WORKER AND MACHINE, GANG, AND OPERATOR PROCESS CHARTS

By understanding the specific functions of the worker and machine, gang, and operator process charts you can select the appropriate one for improving operations. The operator process chart is adapted to all work where the operator goes through a series of manual motions. The worker and machine and gang process charts are used only when machines or facilities are used in conjunction with the operator or operators. Frequently, where an operator is employed in running a machine, it will be worthwhile to make a study both from the standpoint of possible coupling and for motion pattern improvement. In summary, the functions of these three charts are:

1. Worker and machine chart: Used to analyze idle operator time and idle machine time. Ideal for determining the amount of machine

194 *Motion and Time Study*

coupling to be practiced. Used as a training tool to show the relationships of work elements in a multimachine work center.
2. Gang process chart: Used to analyze idle facility time and idle time of
operators servicing a facility or process. Ideal for determining the
labor requirements of a production facility. Used as a training tool to
show the work elements of several operators working with a production facility.
3. Operator process chart: Used to analyze a workstation for proper
layout, proper operator motion patterns, and best sequence of elements. Best chart to use for improving repetitive manual motions.

These three charts along with the operation, flow, flow diagram, and
PERT discussed in Chapter 3 are important tools that methods analysts
should master. Quantitative techniques can determine the optimum arrangement of random servicing work centers. Analysts should be acquainted with sufficient algebra and probability theory to develop a mathematical model that provides the best solution to the problem of machine
or facility assignment. The digital computer can be useful in optimization
solutions of the operator-machine relationship.

TEXT QUESTIONS

1. When is it advisable to construct a worker and machine process chart?
2. Why will higher base rates result when machine coupling is practiced?
3. How does the gang process chart differ from the worker and machine process chart?
4. Explain how you would sell machine coupling to union officials strongly
opposed to the technique.
5. In what way does an operator benefit through machine coupling?
6. What is the difference between synchronous and random servicing?
7. How many machines should be assigned to an operator when:
   a. Loading and unloading time one machine, 1.41 minutes.
   b. Walking time to next facility, 0.08 minutes.
   c. Machine time (power feed), 4.34 minutes.
   d. Operator rate, $13.20 per hour.
   e. Machine rate, $18.00 per hour.
8. What proportion of machine time would be lost in operating four machines
when the machine operates 70 percent of the time unattended and the operator attention time at irregular intervals averages 30 percent? Is this the best
arrangement on the basis of minimizing the proportion of machine time lost?
9. In an assembly process involving six distinct operations, it is necessary to
produce 250 units per eight-hour day. The measured operation times are as
follows:

   a. 7.56 minutes.   *d.* 1.58 minutes.
   b. 4.25 minutes.   *e.* 3.72 minutes.
   c. 12.11 minutes   *f.* 8.44 minutes.

How many operators would be required at 80 percent efficiency?
How many operators will be utilized at each of the six operations?

10. Explain how you can develop software to minimize machine interference under machine coupling arrangements.

GENERAL QUESTIONS

1. In a process plant, which of the following process charts has the greatest application: worker and machine, gang, operation, flow? Why?

2. In the operation of some plant with which you are familiar, outline the "zoning restraints" and discuss how these influence the production line work assignment problem.

PROBLEMS

1. In the Dorben Company's automatic screw machine department, five machines are assigned to each operator. On a given job, the machining time per piece is 0.164 hours, the machine serving time 0.038 hours, and the average machine downtime 0.12 hours per machine per hour. With an operator rate of $12.80 per hour and a machine rate of $14 per hour, calculate the expected cost per unit of output. Exclude material cost.

2. In the Dorben Company, a worker was assigned to operate three like facilities. Each of these facilities is down at random times during the day. A work sampling study indicated that on the average the machines operate 60 percent of the time unattended. Operator attention time at irregular intervals averages 40 percent. This arrangement results in the loss of about 14 percent of the available machine time due to machine interference. If the machine rate is $20 per hour and the operator rate is $12 per hour, what would be the most favorable number of machines (from an economic standpoint) that should be operated by one operator?

3. The analyst in the Dorben Company wishes to assign a number of like facilities to an operator based on minimizing the cost per unit of output. A detailed study of the facilities revealed the following:

Loading machine standard time = 0.34 minutes
Unloading machine standard time = 0.26 minutes
Walk time between two machines = 0.06 minutes
Operator rate = $12.00 per hour
Machine rate (both idle and working) = $18.00 per hour
Power feed time = 1.48 minutes

How many of these machines should be assigned to each operator?

4. A work sampling study revealed that a group of three semiautomatic machines assigned to one operator operate 70 percent of the time unattended. Operator attention time at irregular intervals averaged 30 percent of the time on these three machines. What would be the estimated machine hours lost per eight-hour day because of lack of an operator?

5. Based upon the following data, develop your recommended allocation of work and the number of work stations.

Work unit	Estimated work unit time in minutes
0	0.76
1	1.24
2	0.84
3	2.07
4	1.47
5	2.40
6	0.62
7	2.16
8	4.75
9	0.65
10	1.45

Required production per day is 90 assemblies. In view of just-in-time (JIT) analysis, we do not want to produce more than 105 assemblies a day. A 1.05 rating factor is estimated under an eight-hour working day where a 15 percent allowance is provided. A precedence matrix as follows was developed by the analyst.

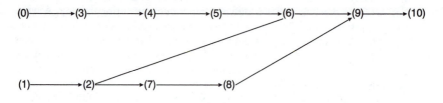

SELECTED REFERENCES

Baker, Kenneth R. *Introduction to Sequencing and Scheduling.* New York: John Wiley & Sons, 1974.

Buffa, E. S., and W. H. Taubert. *Production-Inventory Systems: Planning and Control.* Rev. ed. Homewood, Ill.: Richard D. Irwin, 1972.

Graham, C. F. *Work Measurement and Cost Control.* Oxford, England: Pergamon Press, 1965.

Groff, G. K., and J. F. Muth. *Operations Management: Analysis for Decisions.* Homewood, Ill.: Richard D. Irwin, 1972.

Johnson, Lynwood A., and Douglas C. Montgomery. *Operations Research in Production Planning, Scheduling, and Inventory Control.* New York: John Wiley & Sons, 1974.

Kadota Takeji. "Charting Techniques." In *Handbook of Industrial Engineering,* edited by Gavriel Salvendy. New York: John Wiley & Sons, 1982.

Stecke, K. "Machine Interference." In *Handbook of Industrial Engineering,* edited by Gavriel Salvendy. New York: John Wiley & Sons, 1982.

Motion Study

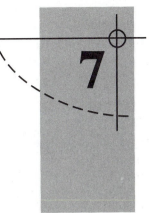

7

Both visual motion study and micromotion study are used to analyze a given method and to help develop an efficient work center. These two techniques are used in conjunction with the principles of operation analysis when there is sufficient volume to justify the additional study and analysis required.

Motion study is the careful analysis of various body motions employed in doing a job. Its purpose is to eliminate or reduce ineffective movements, and to facilitate and speed effective movements. Through motion study, the job is performed more easily and the rate of output is increased. The Gilbreths pioneered the study of manual motion and developed basic laws of motion economy that are still considered fundamental. They were also responsible for the motion-picture technique of making detailed motion studies, known as "micromotion studies," which have proved invaluable in studying highly repetitive manual operations. The reader should understand that all work can be divided into a series of motions and the time to perform a specific motion is the same no matter where that specific motion is employed.

Motion study, in the broad sense, covers two degrees of refinement that have wide industrial application. These are visual motion study and micromotion study.

Visual motion study has considerably broader application, because the activity of the work need not be as great to justify its use economically. This type of study involves a careful observation of the operation and construction of an operator process chart, and a probing analysis of the chart, considering the laws of motion economy.

Chapter 8 describes the micromotion procedure. In view of its much higher cost, micromotion is usually practical only on extremely active jobs whose life and repetitiveness are great. The two types of studies may

197

be compared to viewing a part under a magnifying glass and viewing it under a microscope. The added detail revealed by the microscope finds application only on the most productive jobs.

THE FUNDAMENTAL MOTIONS

The basic division of accomplishment concept, developed by Frank Gilbreth in his early work, applied to all production work performed by the hands of the operator. Gilbreth called these fundamental motions "therbligs" (Gilbreth spelled backward), and concluded that any and all operations are made up of a series of these 17 basic divisions. The 17 fundamental hand motions, modified somewhat from Gilbreth's summary, and their symbols and color designations are shown in Table 7–1. The systems of modern motion and micromotion study and basic motion times (see Chapter 19) had their roots in the Gilbreths' therbligs concepts. These basic divisions will now be described in connection with how they can be improved or eliminated in order to make the work center less fatiguing to the operator and more productive. These will be discussed in a logical sequence for analytical purposes.

Definitions Of Basic Divisions Of Accomplishment

In its "Glossary of Terms Used in Methods, Time Study, and Wage Incentives," the Management Research and Development Division of the Society for the Advancement of Management has provided definitions of the various therbligs. These definitions, in part, are included in the following summary.

1. SEARCH. Search is the basic operation element employed to locate an object. It is that part of the cycle during which the eyes or hands are groping or feeling for the object. It begins the instant the eyes move in an effort to locate an object and ends the instant they are focused on the found object.

Search is a therblig that analysts always endeavor to eliminate. At well-planned workstations work is performed continuously so that it is not necessary for the operator to perform this element. Providing an exact location for all tools and parts is the typical way to eliminate search from a workstation.

A new employee, or one who is not familiar with a job, uses search periodically until skill and proficiency develop. Asking and answering the following questions can eliminate or reduce search time:

1. Are articles properly identified? Perhaps labels or color could be utilized.
2. Can transparent containers be used?
3. Will a better layout of the workstation eliminate searching?
4. Is proper lighting being used?
5. Can tools and parts be pre-positioned?

TABLE 7–1

Therblig Name	Symbol	Color Designation	Symbol
Search	S	Black	
Select	SE	Gray, Light	
Grasp	G	Lake Red	
Reach	RE	Olive Green	
Move	M	Green	
Hold	H	Gold Ocher	
Release	RL	Carmine Red	
Position	P	Blue	9
Pre-position	PP	Sky Blue	
Inspect	I	Burnt Ocher	
Assemble	A	Violet, Heavy	#
Disassemble	DA	Violet, Light	#
Use	U	Purple	
Unavoidable Delay	UD	Yellow Ocher	
Avoidable Delay	AD	Lemon Yellow	
Plan	PL	Brown	
Rest to Overcome Fatigue	R	Orange	

    2. SELECT. Select is the therblig that takes place when the operator chooses one part over two or more analogous parts. This therblig usually follows search, but even with a detailed micromotion procedure it is difficult to determine the exact ending of search and the exact beginning of select. Select does occur without search in selective assembly. In this case, it is usually preceded by inspect. Select can also be classified as an ineffective therblig and should be eliminated from the work cycle by better layout of the workstation and better control of parts.

    To eliminate this therblig, ask:

1.   Are common parts interchangeable?
2.   Can tools be standardized?
3.   Are parts and materials stored in the same bin?
4.   Can parts be pre-positioned in a rack or tray?

3. REACH.   Reach represents the motion of an empty hand, without resistance, toward or away from an object. The basic division reach was known as "transport empty" in Gilbreth's original summary. However, the shorter term is generally accepted by methods analysts today. Reach begins the instant the hand moves toward an object or general location, and it ends the instant hand motion stops on arrival at the object or destination. Reach is usually followed by grasp and preceded by release. Obviously, the time required to perform a reach depends on the distance of the hand movement. The time to perform reach also depends to some extent upon the type of reach. Like grasp, reach is an effective therblig, and usually cannot be eliminated from the work cycle. However, it can be reduced by shortening the distances required for reaching and by providing fixed locations for objects that are reached for during the operator's work cycle. By keeping these elementary principles in mind, workstations can be developed that will keep reach time to a minimum.

4. GRASP.   Grasp is the elemental hand motion of closing the fingers around a part in an operation. Grasp is an effective therblig and usually cannot be eliminated, but in many instances it can be improved. It occurs the instant the fingers of either or both hands begin to close around an object to maintain control of it, and it ends the moment control has been obtained. Grasp is usually preceded by reach and followed by move. Detailed studies have discovered many types of grasp, some taking three times as much time to perform as others. Keep the number of grasps occurring during the work cycle to a minimum, and arrange the parts to be picked up so that the simplest grasp can be used. This is done by having the object by itself in a fixed location, and positioned so that there is no interference from the worktable, bin, or surrounding environment.

These therblig check questions may help improve the grasps performed during a cycle:

1.   Would it be advisable for the operator to grasp more than one part or object at a time?
2.   Can a contact grasp be used rather than a pickup grasp? In other words, can objects be slid instead of carried?
3.   Will a lip on the front of bins simplify grasping small parts?
4.   Can tools or parts be pre-positioned for easy grasp?
5.   Can a vacuum, magnet, rubber fingertip, or other device be used to advantage?
6.   Can a conveyor be used?
7.   Has the jig been designed so that operators may easily grasp the part when removing it?
8.   Can the previous operator pre-position the tool or the work, thus simplifying grasp for the following operator?
9.   Can tools be pre-positioned on a swinging bracket?
10. Can the work table surface be covered with a layer of sponge material so that the fingers can more easily enclose small parts?

5. MOVE. Move is the basic division to signify a hand movement with a load. The load can be in the form of pressure. Move was originally known as "transport loaded." This therblig begins the instant the hand under load moves toward a general location, and it ends the instant motion stops on arrival at the destination. Move is usually preceded by grasp and is usually followed by either release or position.

The time required to perform move depends on the distance, the weight to be moved, and the type of move. Move is a physical effective therblig; it is difficult to eliminate this basic division from the work cycle. Nevertheless, the time to perform move can be diminished by reducing the distances to be moved, lightening the load, and improving the type of move by providing gravity chutes or a conveyor at the terminal point of the move, so that it is not necessary to bring the object being moved to a specific location. Experience has proved that moves to a general location are performed more rapidly than moves to an exact location.

Improve both reach and move therbligs by asking and answering the following questions:

1. Can either of these therbligs be eliminated?
2. Can distances be shortened to advantage?
3. Are the best means (conveyors, tongs, tweezers) being used?
4. Is the correct body member (fingers, wrist, forearm, shoulder) being used?
5. Can a gravity chute be employed?
6. Can transports be effected through mechanization and foot-operated devices?
7. Will time be reduced by transporting in larger units?
8. Is time increased because of the nature of the material being moved or because of a subsequent delicate positioning?
9. Can abrupt changes in direction be eliminated?

6. PRE-POSITION. Pre-position is an element of work that consists of positioning an object in a predetermined place so that it may be grasped in the position in which it is to be held when needed.

Pre-position often occurs in conjunction with other therbligs, one of which is usually move. It is the basic division that arranges a part so that it can be conveniently placed on arrival. The time required to pre-position is difficult to measure since the therblig itself can seldom be isolated. Pre-position takes place if a screwdriver is being aligned while being moved to the screw it is going to drive.

These questions assist in studying pre-position therbligs.

1. Can a holding device at the workstation keep tools in proper positions and the handles in upright positions?
2. Can tools be suspended?
3. Can a guide be used?
4. Can a magazine feed be used?

5. Can a stacking device be used?
6. Can a rotating fixture be used?

7. POSITION. Position is an element of work that consists of locating an object so that it is properly oriented in a specific place.

The therblig position occurs as a hesitation while the hand or hands are endeavoring to place the part so that further work may be more readily performed. Actually, position may be a combination of several very rapid motions. Locating a contoured piece in a die would be a typical example of position. Position is usually preceded by move and followed by release. Position begins the instant the controlling hand or hands begin to agitate, twist, turn, or slide the part to orient it to the correct place, and it ends as soon as the hand begins to move away from the part.

Position frequently can be eliminated or improved through answering these questions:

1. Can such devices as a guide, funnel, bushing, stop, swinging bracket, locating pin, recess, key, pilot, or chamfer be used?
2. Can tolerances be changed?
3. Can the hole be counterbored or countersunk?
4. Can a template be used?
5. Are burrs increasing the problem of positioning?
6. Can the article be pointed to act as a pilot?

8. RELEASE. Release is the basic division that occurs when the aim of the operator is to relinquish control of the object. Release takes the least time of all the therbligs, and little influences the time required for this objective therblig.

Release begins the instant the fingers begin to move away from the part held, and it ends the instant all fingers are clear of the part. This therblig follows a move or position and is usually followed by reach.

Release time may be eliminated or improved by asking:

1. Can the release be made in transit?
2. Can a mechanical ejector be used?
3. Are the bins that contain the part after release of the proper size and design?
4. At the end of the therblig release, are the hands in the most advantageous position for the next therblig?
5. Can multiple units be released?

9. HOLD. Hold is the basic division of accomplishment that occurs when either hand is supporting or maintaining control of an object while the other hand does useful work. Hold is an ineffective therblig and can usually be removed from the work cycle by designing a jig or fixture to hold the work rather than using the hand to do so. The hand is seldom considered an efficient holding device, and hold should not be part of any work assignment.

Hold begins the instant one hand exercises control on the object, and it ends the instant the other hand completes its work on the object. A typical example of hold would occur when the left hand holds a stud while the right hand runs a nut on the stud. During the assembly of the nut to the stud, the left hand would be utilizing the therblig hold.

Hold can often be eliminated by asking and answering these questions:

1. Can a mechanical jig, such as a vise, pin, hook, rack, clip, or vacuum, be used?
2. Can friction be used?
3. Can a magnetic device be used?
4. Should a twin holding fixture be used?

10. ASSEMBLE.   Assemble is the basic division that occurs when two mating parts are brought together. This is another objective therblig; it can more readily be improved than eliminated. Assemble is usually preceded either by position or move, and it is usually followed by release. It begins the instant the two mating parts come in contact with each other, and it ends on completion of the union.

11. DISASSEMBLE.   Disassemble is just the reverse of assemble. It occurs when two mating parts are disunited. The basic division is usually preceded by grasp and is usually followed by either move or release. Disassemble is objective in nature, and improvement possibilities are more likely than is complete elimination of the therblig. Disassemble begins the moment either or both hands have control of the object after grasping it, and ends as soon as the disassembly has been completed, usually evidenced by the beginning of a move or a release.

12. USE.   Use is a completely objective therblig that occurs when either or both hands have control of an object during that part of the cycle when productive work is being performed. When both hands are holding a casting against a grinding wheel, use would be the therblig that indicates the action of the hands. After a screwdriver has been positioned in the slot of the screw, use would occur as the screw is driven home. The duration of this therblig depends on the operation as well as on the performance of the operator. Use is quite easily detected, since this therblig always advances the operation toward the ultimate objective.

While studying the three objective therbligs, assemble, disassemble, and use, ask and answer the following questions:

1. Can a jig or fixture be used?
2. Does the activity justify mechanized or automated equipment?
3. Would it be practical to make the assembly in multiple units?
4. Can a more efficient tool be used?
5. Can stops be used?
6. Is the tool being operated at the most efficient feeds and speeds?
7. Should a power tool be employed?

13. INSPECT.   Inspect is an element included in an operation to assure acceptable quality through a regular check by the employee performing the operation.

Inspect takes place when the predominant purpose is to compare some object with a standard. It is usually not difficult to detect when inspect is encountered, since the eyes are focused upon the object, and a delay between motions is noted while the mind decides to accept or reject the piece in question. The time taken for inspect is determined primarily by the severity of the standard and the amount of deviation of the part in question. Thus, if an operator were sorting out all blue marbles from a bin, little time would be consumed in deciding what to do with a red marble. However, if a purple marble were picked up, there would be a longer hesitation for the mind to evaluate the marble and decide whether it should be accepted or rejected.

These questions may shorten inspect therbligs:

1. Can inspection be eliminated or combined with another operation or therblig?
2. Can multiple gages or tests be used?
3. Will inspection time be reduced by increasing the illumination?
4. Are the articles being inspected at the correct distance from the worker's eyes?
5. Will a shadowgraph facilitate inspection?
6. Does an electric eye have application?
7. Does the volume justify automatic electronic inspection?
8. Would a magnifying glass facilitate the inspection of small parts?
9. Is the best inspection method being used? Has consideration been given to polarized light, template gages, sound tests, performance tests, and so on?

14. PLAN.   The therblig plan is the mental process that occurs when the operator pauses to determine the next action. Plan may take place during any part of the work cycle, and it is usually readily detected as a hesitation after locating all components. This therblig is characteristic of new employees and can usually be removed from the work cycle through proper operator training and experience in the prescribed method.

15. REST TO OVERCOME FATIGUE.   Usually this delay does not appear in every cycle but is evidenced periodically. The duration of rest to overcome fatigue will vary not only with the class of work but also with the individual performing the work.

To reduce the number of occurrences of the therblig rest, consider:

1. Is the best order-of-muscles classification being used?
2. Are temperature, humidity, ventilation, noise, light, and other working conditions satisfactory?
3. Are benches of the proper height?

4. Can the operator alternately sit and stand while performing work?
5. Does the operator have a comfortable chair of the right height?
6. Are mechanical means being used for heavy loads?
7. Is the operator aware of his or her average intake requirements in calories per day?

16. AVOIDABLE DELAY. Any idle time during the cycle for which the operator is solely responsible, either intentionally or unintentionally, is classified as an avoidable delay. Thus, if an operator developed a coughing spell during a work cycle, this delay would be avoidable, for normally it would not appear in the work cycle. A majority of the avoidable delays encountered can be eliminated by the operator without changing the process or method of doing the work.

17. UNAVOIDABLE DELAY. Unavoidable delay is an interruption beyond the control of an operator in the continuity of an operation. It represents idle time in the work cycle experienced by either or both hands because of the nature of the process. Thus, while an operator is hand-feeding a drill with the right hand to a part held in a jig, the left hand would be unavoidably delayed. Since the operator usually has no control over unavoidable delays, the process or tooling has to be changed in some manner to remove them from the cycle.

Therblig Summary

The 17 basic divisions are either effective or ineffective therbligs. Effective therbligs are those that directly advance the progress of the work. These therbligs can frequently be shortened, but it is difficult to eliminate them completely. Ineffective therbligs do not advance the progress of the work and should be eliminated by applying the principles of operation analysis and motion study.

A further classification breaks the therbligs into physical, semimental or mental, objective, and delay groups. Ideally, a work center should comprise only physical and objective therbligs.

A. Effective.
  1. Physical basic divisions.
     a. Reach.
     b. Move.
     c. Grasp.
     d. Release.
     e. Pre-position.
  2. Objective basic division.
     a. Use.
     b. Assemble.
     c. Disassemble.

B. Ineffective.
 1. Mental or semimental basic divisions.
    a. Search.
    b. Select.
    c. Position.
    d. Inspect.
    e. Plan.
 2. Delay.
    a. Unavoidable delay.
    b. Avoidable delay.
    c. Rest to overcome fatigue.
    d. Hold.

PRINCIPLES OF MOTION ECONOMY

Beyond the basic division of accomplishment concept, as first set forth by the Gilbreths, are the principles of motion economy developed by them and added to by others, notably Ralph M. Barnes. These principles are not all applicable to every job, and several are used only in micromotion study. However, those that apply to visual motion study, as well as to the micromotion technique, can be broken down into three basic subdivisions: (1) the use of the human body, (2) the arrangement and conditions of the workplace, and (3) the design of tools and equipment.

The visual principles of motion economy enable analysts to detect inefficiencies in the method by briefly inspecting the workplace and the operation. These basic principles under their respective divisions are as follows:

A. The use of the human body.
 1. Both hands should begin and end their basic divisions of accomplishment simultaneously and should not be idle at the same instant, except during rest periods.
 2. The motions made by the hands should be made symmetrically and simultaneously away from and toward the center of the body.
 3. Momentum should assist workers wherever possible, and should be minimized if it must be overcome by muscular effort.
 4. Continuous curved motions are preferable to straight-line motions involving sudden and sharp changes in direction.
 5. The least number of basic divisions should be used; these should be confined to the lowest practicable classifications. These classifications, summarized in ascending order of the time and fatigue expended in their performance, are:
    a. Finger motions.
    b. Finger and wrist motions.
    c. Finger, wrist, and lower arm motions.

      d. Finger, wrist, lower arm, and upper arm motions.

      e. Finger, wrist, lower arm, upper arm, and body motions.

6. Work done by the feet should be done simultaneously with work done by the hands. However, it is difficult to move the hand and foot simultaneously.

7. The middle finger and the thumb are the strongest working fingers. The index finger, fourth finger, and little finger are not capable of handling heavy loads over extended periods.

8. The feet are not capable of efficiently operating pedals when the operator is in a standing position.

9. Twisting motions should be performed with the elbows bent.

10. To grip tools, workers should use the segments of the fingers closest to the palm of the hand.

B. The arrangement and conditions of the workplace.

1. Fixed locations for all tools and material should be provided to permit the best sequence and to eliminate or reduce the therbligs search and select.

2. Gravity bins and drop delivery should reduce reach and move times; also, wherever possible, ejectors should remove finished parts automatically.

3. All materials and tools should be located within the normal working area in both the vertical and the horizontal planes.

4. A comfortable chair for the operator and the workstation's height should be so arranged that the work can be efficiently performed by the operator alternately standing or sitting.

5. Proper illumination, ventilation, and temperature should be provided.

6. The visual requirements of the workplace should be considered so that eye fixation demands are minimized.

7. Rhythm is essential to the smooth and automatic performance of an operation, and the work should be arranged to permit an easy and natural rhythm wherever possible.

C. The design of tools and equipment.

1. Multiple cuts should be taken whenever possible by combining two or more tools in one, or by arranging simultaneous cuts from both feeding devices, if available (cross slide and hex turret).

2. All levers, handles, wheels, and other control devices should be readily accessible to the operator and designed to give the best possible mechanical advantage and to utilize the strongest available muscle group.

3. Parts should be held in position by fixtures.

4. The use of powered or semiautomatic tools, such as power nut- and screwdrivers and speed wrenches, should always be investigated.

THE LAWS OF MOTION ECONOMY

The Use of the Human Body

Both hands should begin and end their basic divisions of accomplishments simultaneously and should not be idle at the same instant, except during rest periods.

When the right hand is working in the normal area to the right of the body and the left hand is working in the normal area to the left of the body, a feeling of balance tends to induce a rhythm in the operator's performance that leads to maximum productivity. When one hand is working under load and the other hand is idle, the body exerts an effort to put itself in balance. This usually results in greater fatigue than if both hands are doing useful work. This law can readily be demonstrated by reaching out with the right hand about 14 inches, picking up an object weighing about a half pound, and moving it 10 inches toward the body before releasing it. The operation should then be immediately repeated, only this time moving the part away from the body. Repeat the cycle about 200 times and note the discomfort in the body induced by "balance fatigue." Now repeat the operation using both hands simultaneously. Have the left hand reach out radially to the left of the body and the right hand reach out radially to the right of the body, grasp the objects, move both objects toward the body, and release them simultaneously. Repeat the cycle 200 times and note that the body feels less fatigued, even though twice as much load has been handled.

The motions made by the hands should be made symmetrically and simultaneously away from and toward the center of the body.

It is natural for the hands to move in symmetrical patterns: deviations from symmetry in a two-handed workstation result in slow, awkward movements of the operator. The difficulty of patting the stomach with the left hand while rubbing the top of the head with the right hand is familiar to many. Another experiment that can readily illustrate the difficulty of performing nonsymmetrical operations is to draw a circle with the left hand while the right hand is drawing a square. Figure 7–1 illustrates an ideal workstation that allows the operator to assemble two products by going through a series of symmetrical motions made simultaneously away from and toward the center of the body.

Momentum should assist workers wherever possible, and should be minimized if it must be overcome by muscular effort.

As the hands progress through the elements of work constituting the operation, momentum developed during reach and move therbligs is overcome during position and release therbligs. To make full use of the momentum built up, workstations should allow operators to release a finished part in a delivery area while their hands are on their way to get component parts or tools to begin the next work cycle. This allows the

FIGURE 7-1
An ideal workstation that permits the operator to assemble two products by going through a series of symmetrical motions made simultaneously away from and toward the center of the body.

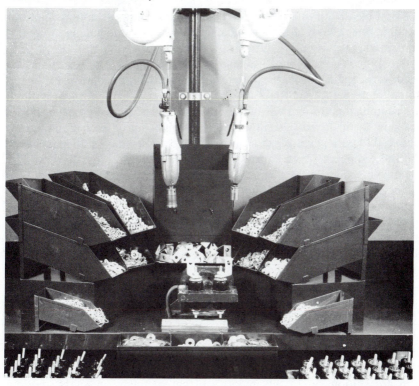

General Electric Co.

hands to perform their reaches with the aid of momentum and makes the therblig easier and faster to perform.

Detailed studies have proved conclusively that both reaches and moves are performed faster if the hand is in motion at the beginning of the therbligs.

Continuous curved motions are preferable to straight-line motions involving sudden and sharp changes in direction.

This law is very easily demonstrated by moving either hand in a rectangular pattern, and then moving it in a circular pattern of about the same magnitude. The greater amount of time required to make the abrupt 90° directional changes is quite apparent. To make a directional change, the hand must decelerate, change direction, and accelerate until it is time to decelerate preparatory to the next directional change. Continuous curved motions do not require deceleration, and consequently are performed faster per unit of distance.

The least number of basic divisions should be used, and these should be confined to the lowest practicable classifications.

Only by identifying the following classifications of motions can you become fully aware of the significance of this fundamental law of motion economy.

1. Finger motions are the fastest of the five motion classes and are readily recognized, for they are made by moving the finger or fingers while the remainder of the arm is kept stationary. Typical finger motions are running a nut down on a stud, depressing the keys of a typewriter, or grasping a small part. Usually there is a significant difference in the time required to perform finger motions with the various fingers. In most cases, the index finger moves considerably faster than the other fingers. This must be considered in designing workstations. Although with practice the fingers of the left hand (in right-handed people) can be trained to move with the same rapidity as the fingers of the right hand, detailed studies have shown that the fingers of the left hand move somewhat slower. R. E. Hoke made a study of the "universal" keyboard used on typewriters and found that 88.9 taps were made with the fingers of the left hand for every 100 made with the right hand.[1]

Finger motions are the weakest of the five motion classes. Continuous finger motions result in fatigue and loss of finger flexibility. Therefore, keep finger forces low; for example, a bar switch is preferable to a trigger switch. Design workstations involving high manual effort to use higher classifications than finger motions.

2. Finger and wrist motions are made while the forearm and upper arm are stationary. In the majority of cases, finger and wrist motions consume more time than do strictly finger motions. Typical finger and wrist motions occur when a part is positioned in a jig or fixture or when two mating parts are assembled. Reach and move therbligs usually cannot be performed by motions of the second class unless the transport distances are very short.

3. Finger, wrist, and lower arm motions are commonly referred to as "forearm motions" and include those movements made by the arm below the elbow while the upper arm is stationary. Since the forearm includes a strong muscle, such motions are usually considered efficient because they are not fatiguing. However, repetitive work with the arms extended and involving force can induce soreness. This usually can be relieved by designing the workstation so that the elbows can be kept at 90° while work is being done. The time required to make forearm motions for a given

[1] R. E. Hoke, *The Improvement of Speed and Accuracy in Typewriting,* The Johns Hopkins University Studies in Education no. 7 (Baltimore: The Johns Hopkins University Press, 1922).

operator depends on the distance moved and the amount of resistance overcome during the movement. Design workstations so that these third-class motions, rather than fourth-class motions, are used to perform the transport therbligs, to minimize cycle times.

4. Finger, wrist, lower arm, and upper arm motions, commonly known as fourth-class or shoulder motions, are probably used more than any other motion class. The fourth-class motion for a given distance takes considerably more time than do the three classes just described. Fourth-class motions are required to perform transport therbligs of parts that cannot be reached without extending the arm. The time required to perform fourth-class motions depends primarily on the distance of the move and the resistance to the move. To reduce static loading of shoulder motions, design tools so that the elbow is not elevated while performing work. For example, by using a socket wrench instead of an open end wrench, the operator can approach the nut to be assembled from an angle without having to lift the elbow.

5. Fifth-class motions include body motions, and those, of course, are the most time-consuming. Body motions include movements of the ankle, knee, and thigh, as well as movements of the trunk.

First-class motions require the least amount of effort and time, while fifth-class motions are considered the least efficient. Therefore, always utilize the lowest practicable motion classification to perform the work properly. This will involve careful consideration of the location of tools and materials so that ideal motion patterns can be arranged.

Work done by the feet should be done simultaneously with work done by the hands.

Since the major part of work cycles is performed by the hands, it is economical to relieve the hands of work that can be done by the feet if this work is performed while the hands are occupied. Since the hands are more skillful than the feet, it would be folly to have the feet perform elements while the hands are idle. Foot pedal devices allowing clamping, the ejection of parts, or feeding can often be arranged, thus freeing the hands for useful work and consequently reducing the cycle time (see Figure 7–2). When the hands are moving, the feet should not be moving, although they can be applying pressure, such as on a foot pedal. Operators cannot efficiently operate pedals while standing. In the development of workstations involving the coordination of the hands and feet, be sure that simultaneous movements of the hands and feet are not required.

The middle finger and the thumb are the strongest working fingers. The index finger, fourth finger, and little finger are not capable of handling heavy loads over extended periods.

Although the index finger is usually the finger that is capable of moving the fastest, it is not the strongest finger. Where a relatively heavy load is involved, it is usually more efficient to use the middle finger or a combination of the middle finger and the index finger.

FIGURE 7–2
This foot-operated press permits the operator's hands to procure parts in preparation for the next cycle while the press is in operation

General Electric Co.

Twisting motions should be performed with the elbows bent.

When the elbow is extended, tendons and muscles in the arm are stretched. If the arm performs twisting motions in this position, it can overstress muscle groups and tendons.

To grip tools, workers should use the segments of the fingers closest to the palm of the hand.

Not only are the segments of the fingers closest to the palm stronger than the other segments, but, being closer to the load held in the hand, they do not induce as great a bending moment as do the more remote segments.

The Arrangement and Conditions of the Workplace

Fixed locations for all tools and materials should permit the best sequence and eliminate or reduce the therbligs search and select.

In driving an automobile, we are all familiar with the shortness of time required to apply the foot brake. The reason is obvious: since the brake pedal is in a fixed location, no time is required to decide where the brake is located. The body responds instinctively and applies pressure to the area where the driver knows the foot pedal is. If the location of the brake foot pedal varied from time to time, considerably more time would be needed to brake the car. Providing fixed locations for all tools and materials at the workstation eliminates, or at least minimizes, the short hesitations required to search and select the various objects needed to do the work (see Figure 7–3).

Gravity bins and drop delivery should reduce reach and move times.

The time required to perform both of the transport therbligs, reach and move, is proportional to the distance that the hands must move in performing these therbligs. By utilizing gravity bins, components can be continuously brought to the normal working area, thus eliminating long reaches to get supplies of parts. Likewise, gravity chutes allow the disposal of parts within the normal area, eliminating the necessity for long moves to dispose of the completed part or parts. Sometimes ejectors can

FIGURE 7–3
Fixed locations for all materials and tools in this workstation minimize search and select hesitations. The operation of subassembly of crankshafts utilizes an indexing table. At the rear of the index table is an induction heating coil used to heat the thrust coolers sitting atop the crankshaft. The heated part drops down onto the top of the cast iron bushing where, on cooling, it shrinks on the shaft and provides the proper clearance required by the pump end bushing.

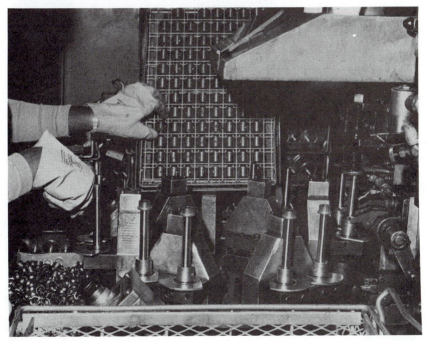

General Electric Co.

FIGURE 7–4
A workstation utilizing gravity bins and a belt conveyor to reduce reach and move times. The conveyor in the background carries other parts past this particular workstation. The operator is feeding the conveyor from under the platform by merely dropping assembled parts onto the feeder belt.

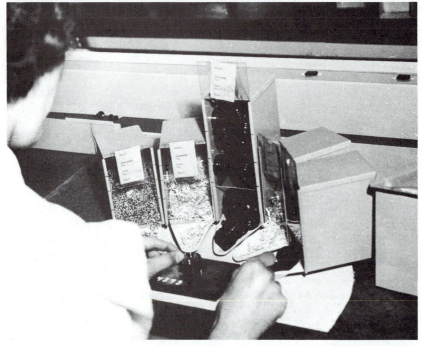

Alden Systems Co.

remove finished products automatically. Gravity chutes make possible a clean workplace area, as finished material is carried away from the work area rather than stacked up all around the workplace (see Figure 7–4).

All materials and tools should be within the normal working area in both the vertical and the horizontal planes.

In every motion there is a distance involved. The greater the distance the larger the muscular effort, control and time. Thus it is important to minimize distances. The normal working area in the horizontal plane of the right hand includes the area generated by the arm below the elbow when moving in an arc pivoted at the elbow. This area represents the most convenient zone within which motions may be made by that hand with a normal expenditure of energy. The normal area of the left hand may be established in a similar manner.

Since movements are made in the third dimension as well as in the horizontal plane, the normal working area also applies to the vertical plane. The normal area relative to height for the right hand includes the area made by the lower arm in an upright position hinged at the elbow moving in an arc. There is a similar normal area in the vertical plane for the left hand (see Figures 7–5 and 7–6).

The maximum working area represents that portion of the workplace within which all tools and materials should be located, where operators can work without excessive fatigue. This area is formed by drawing arcs

FIGURE 7–5
Normal and maximum working areas in the horizontal plane for women (for men, multiply by 1.09)

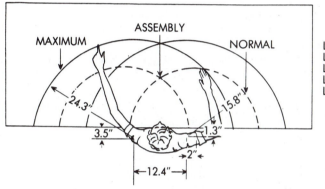

LENGTH OF ARM   28″
LENGTH OF FOREARM   10″
LENGTH OF UPPER ARM   12″
LENGTH OF HAND   6.7′
LENGTH OF END JOINT   .9″
        (2ND FINGER)

FIGURE 7–6
Normal and maximum working areas in the vertical plane for women (for men, multiply by 1.09)

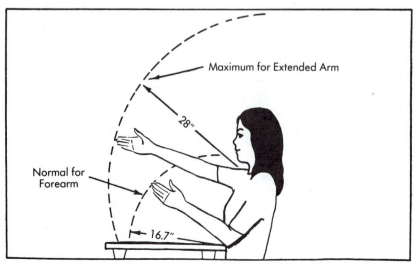

with the arms fully extended and, as with the normal working area, both the horizontal and vertical planes are considered.

Design facilities and workstations by considering such factors as arm reach, leg clearance, and body support, since these human dimensions are important criteria in developing a work environment that is comfortable and efficient.

With the introduction of the video display terminal in the work place, business and industry are finding writer's cramp has been replaced by painful wrists, eyes, necks, and backs. Experience has shown that most of these aches can be relieved or eliminated with a properly adjusted workstation that provides the operator with a comfortable sitting position, flexible enough to reach, use, and observe the screen, keyboard, and documents (see Figure 7–7).

A comfortable chair should be provided for the operator and the work-station height so arranged that the work can be efficiently performed by the operator alternately standing or sitting.

At the workplace, the operator should be seated if possible. Workplaces requiring the operator to stand a significant portion of the day are conducive to high fatigue.

To reduce operator fatigue, selection of the stool or chair used by the operator should receive careful attention. It should accommodate the 1st to 98th percentile of workers. This is equivalent to providing a work-surface height of from 36.5 inches to 44.7 inches. In general, the chair or stool seats should be broad and long enough to support the thighs, yet not so long as to cut into the back of the knees of shorter operators. The contour of the seat should be slightly saddle shaped, and the front end should be chamfered. A good size is 16 inches by 16 inches, with the curved portion beginning about 3 inches back from the front edge. The seat should be lightly padded and aerated. Whenever possible, backs should be provided for all chairs; these should be designed so that they do not interfere with movements of the arms. The backrest should not result in undue pressure against the pelvis or ribs, nor should it cause interference with shoulder blade movement. It should be slightly curved, about 7 inches high and 12 inches wide. Some padding on the back rest eliminates sharp edges. A tilting feature is desirable.

Good chair design should permit several effective working postures. The seats should be adjustable in height within the range of 15 to 21 inches, with half-inch increments of height adjustment. If the operator is working at benches more than 30 inches in height, chair adjustment should permit a range of seat heights of from 18 inches to 27 inches. Chair manufacturers supply industrial chairs that are adjustable in height from the floor to the top edge of the seat. In recent years, both industrial and medical experts have collected data that proves in many instances production costs have been lowered by using chairs and benches of proper height. Figure 7–8 illustrates recommended dimensions for a seated work-place with and without a footrest.

FIGURE 7-7

Arms: When operator's hands are on keyboard, upper arm and forearm should form right angle; hands should be lined up with forearm; if hands are angled up from the wrist, try using attached to front of keyboard; optional arm rests should be adjustable.

Telephone: Cradling telephone receiver between head and shoulder can cause muscle strain; headset allows head, neck to remain straight while keeping hands free.

Document holder: Same height and distance from user as the screen, so eyes can remain focused as they look from one to the other.

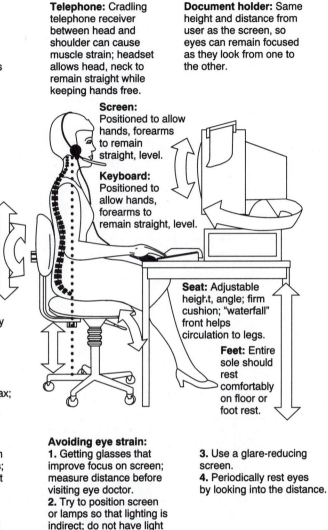

Screen: Positioned to allow hands, forearms to remain straight, level.

Keyboard: Positioned to allow hands, forearms to remain straight, level.

Backrest: Adjustable for occasional variations; shape should match contour of lower back, providing even pressure and support.

Posture: Sit all the way back into chair for proper back support; back, neck should be as comfortably straight ahead; knees should be slightly lower than hips; do not cross legs or shift weight to one side; give joints, muscles a chance to relax; periodically, get up and walk around.

Seat: Adjustable height, angle; firm cushion; "waterfall" front helps circulation to legs.

Feet: Entire sole should rest comfortably on floor or foot rest.

Desk: Thin work surface to allow leg room and posture adjustments; adjustable surface height preferable; table should be large enough for books, files, telephone while permitting different positions of screen, keyboard, mouse pad.

Avoiding eye strain:
1. Getting glasses that improve focus on screen; measure distance before visiting eye doctor.
2. Try to position screen or lamps so that lighting is indirect; do not have light shining directly at screen or into eyes.
3. Use a glare-reducing screen.
4. Periodically rest eyes by looking into the distance.

If the height of the workstation and chair allow the operator to work alternately in a standing or a sitting position, fatigue and job monotony are reduced substantially. Monotony is definitely an important factor in producing fatigue. With the present-day tendency toward specialization and resulting increases in fatigue accidents, reducing monotony is an important priority.

A workstation designed to accommodate a sit/stand operation should not only provide a raising of the seated workplace. The job needs to be

218

Motion and Time Study

FIGURE 7–8
Recommendations for a seated workplace with and without a footrest.

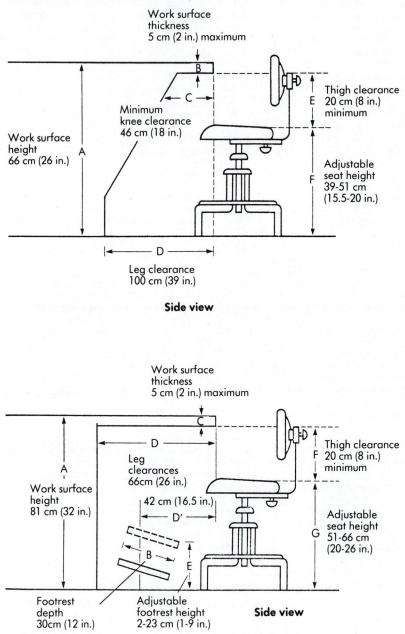

Courtesy Human Factors Section, Eastman Kodak Co.

studied so that the layout takes care of both the standing and sitting situations. A standing work-station that allows work to be done while seated should be approximately 36 inches above the floor with an adjustable footrest. Another alternative is to provide an adjustable platform (4 to 12 inches) on top of the work surface, for operators to use when seated.

If it is not practicable for the operator to alternate between standing and sitting positions, choose a seat that tilts the body slightly forward. Ideally, after the operator is comfortably seated with feet on the floor, position the workbench at the appropriate height to accommodate the operation. Thus, the workstation, too, needs to be adjustable.

In those workplaces where operators must stand while performing such operations as light assembly, clerical, and the like, the working height of most operators' hands is about 42 inches. See Figure 7–9. Where operations are performed that require high downward force, the working height should be lowered to about 36 inches for most operators.

FIGURE 7–9
Recommended standing workplace dimensions.

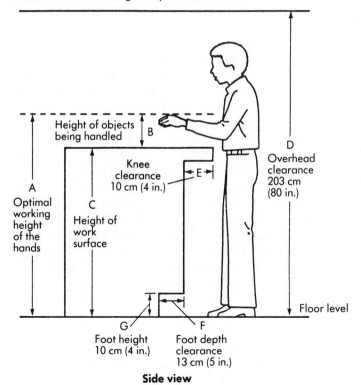

Side view

Courtesy Human Factors Section, Eastman Kodak Co.

The workstation should be designed so that lifting is performed in conjunction with sound biomechanical and metabolic principles. Although the vast majority of heavy work at the workstation and between workstations is handled with mechanical equipment, many work areas require occasional lifting of moderate to heavy loads. Analysts should incorporate the following principles through a program of workstation design and operator training:

1. Some container design providing for stability of load and handles.
2. Floor has adequate friction so that operators do not slip.
3. The load should be compact and no wider than about 30 inches.
4. Operators should not lift anything higher than shoulder height.
5. The load should be kept close to the body and operators should use a smooth two-handed symmetrical lifting motion, avoiding twisting and/or jerking motions.
6. Any load over 100 pounds should be considered hazardous. Operators should be selected and trained for operations requiring regular moving of loads between 30 and 100 pounds.
7. Movement of loads can be made easier by adjusting the terminal point (point where load is released) and access point (where load is picked up) so that both of these points are near the height of the employee's belt level.

Proper illumination, ventilation, and temperature should be provided. Comfortable working conditions are fundamental for peak production and operator satisfaction. Defective illumination is an important cause of operator fatigue, poor quality of product, and low production output. The National Safety Council, in its *Safe Practices Pamphlet No. 50,* mentions that eyes are used in "serious" work about 70 percent of the time. Failure to provide proper lighting increases the consumption of body energy (see Figure 7–10).

Proper ventilation and temperature also help maintain good working conditions by controlling fatigue and reducing the causes of accidents. Laboratory data recorded in diversified industries documents that atmospheric conditions exert an appreciable influence on physical activity.

Members of the medical profession, industrialists, and psychologists have realized that colors are effective in stimulating or depressing. Physical surroundings that relieve eyestrain and create a cheerful atmosphere result in fewer injuries and less absenteeism.

The visual requirements of the workplace should be considered so that eye fixation demands are minimized. Certain visual requirements are characteristic of all work centers. Some equipment or controls may be scanned from nearby or remote points. Other areas require more concentrated attention. Pre-positioning components that require more concentrated observation, such as instruments and dials, lessens eye fixation time as well as eye fatigue.

FIGURE 7–10
Better illumination releases energy for useful work

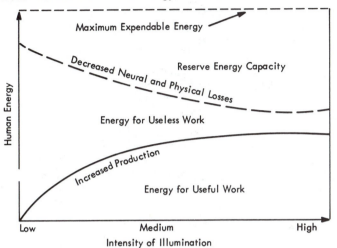

Source: From *National Safety Council Bulletin No. 50*, p. 9.

Almost one half of this country's industrial population wears eye-correction glasses or contact lenses; about 50 percent of this group are bifocal or multifocal users. Consequently, there is considerable variation in the ability of various people to focus on objects at different distances. Figure 7–11 illustrates visual work dimensions for seated workplaces.

Because rhythm is essential to the smooth and automatic performance of an operation, arrange work to permit an easy and natural rhythm wherever possible.

If the basic motions can be sequenced in a regular recurrence of like therbligs, or a regular alternation in the therbligs, the hands instinctively work rhythmically. When work is performed with such a regularity or flow of movement, the operator appears to be working effortlessly; invariably, production is high and operator fatigue is low.

The Design of Tools and Equipment

Multiple cuts should be taken whenever possible by combining two or more tools in one, or by arranging simultaneous cuts from both feeding devices, if available (cross side and hex turret).

Advance production planning for the most efficient manufacture includes taking multiple cuts with combination tools and simultaneous cuts with different tools. Of course, the type of work to be processed and the number of parts to be produced determine the desirability of combining cuts, such as cuts from both the square turret and the hexagon turret.

FIGURE 7–11
Visual work dimensions for seated workplaces

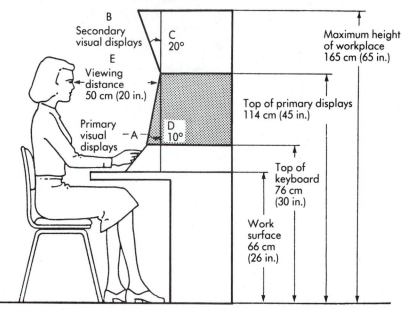

Courtesy Human Factors Section, Eastman Kodak Company

Developed from data in Kennedy and Bates, 1965; and Woodson, Rane, and Conover, 1972.

Figures 7–12 and 7–13 illustrate typical combined and multiple cuts that can be utilized in turret lathe work (see also Figure 7–14).

Figure 7–12 shows combined cuts being made from the cross slide and the multiple turning head on the pentagon turret of a Warner & Swasey 4AC single-spindle automatic chucking machine. Two overhead turning tools and one boring tool are cutting from the multiple turning head while a facing cut is made from the cross slide.

Figure 7–13 shows a forming cut being made from the cross slide while a drilling operation is performed from the turret.

In Figure 7–14, combined cuts are made from the square turret cross slide and the overhead turning tool mounted in a multiple turning head with an overhead pilot bar for extra support. Also, an internal boring operation is made simultaneously.

All levers, handles, wheels, and other control devices should be readily accessible to operators and should be designed to give the best possible mechanical advantage and to utilize the strongest available muscle group.

Many of our machine tools and other devices are mechanically perfect, yet incapable of effective operation because the designer of the facility

FIGURE 7–12
4AC single-spindle automatic chucking machine

Warner & Swasey Co.

overlooked various human factors. Handwheels, cranks, and levers should be of such size and placed in such positions that operators can manipulate them with maximum proficiency and minimum fatigue.

Frequently used controls should be between elbow and shoulder height. Seated operators can apply maximum force to levers located at elbow level; standing operators, to levers located at shoulder height. Handwheel and crank diameters depend on the torque to be expended and the mounting position. Maximum diameters of handgrips depend on the forces to be exerted. For example, for a 10- to 15-pound force, the diameter should be no less than ¼ inch and preferably larger; for 15 to 25 pounds, a minimum of ½ inch should be used; and for 25 or more pounds, a minimum of ¾ inch.

FIGURE 7–13
Forming cut and drilling operation on a single-spindle automatic
chucking machine

Warner & Swasey Co.

Diameters should not exceed 1½ inches, and the grip length should be
at least 3¾ inches to accommodate the breadth of the hand.

Guidelines for crank and handwheel radii are: light loads, radii of 3 to 5
inches; medium to heavy loads, radii of 4 to 7 inches; very heavy loads,
radii of more than 8 inches but not in excess of 20 inches.

Knob diameters of ½ to 2 inches are usually satisfactory. The diameters of knobs should be increased as greater torques are needed.

The hand is seldom an efficient holding device because, when it is
occupied in holding the work, it cannot be free to do useful work. Parts
that have to be held in position while they are being worked on should be
supported by a fixture, freeing the hands for productive motions. Fixtures

FIGURE 7–14
Combined cuts and an internal boring operation being made simultaneously

Warner & Swasey Co.

not only save time in processing parts, but permit better quality, in that the work can be held more accurately and firmly.

The use of powered or semiautomatic tools, such as power nut- and screwdrivers and speed wrenches, should always be investigated.

Power hand tools not only perform work faster than manual tools, but also do the work with considerably less operator fatigue. Greater uniformity of product can be expected when power hand tools are used. For example, a power nut driver can drive nuts consistently to a predetermined tightness in inch-pounds, while a manual nut driver cannot be expected to maintain constant driving pressure in view of operator fatigue.

THE PRACTICAL USE OF MOTION STUDY IN THE PLANNING STAGE

Production personnel in general agree that it is better to concentrate on ideal methods in the planning stage, rather than depending entirely on correcting manufacturing methods after they have been introduced.

Insufficient volume may make it impractical to consider many alternative proposals in the planning stage which may offer substantial saving over existing methods. For example, take an operation that was done on a drill press in which a ½-inch hole was reamed to the tolerance of 0.500 to

0.502 inches. The activity of the job was estimated to be 100,000 pieces. The time study department established a standard of 8.33 hours per thousand to perform the reaming operation, and the reaming fixture cost $2,000. Since a base rate of $7.20 per hour was in effect, the money rate per thousand pieces was $60.

Now assume that a methods analyst suggests broaching the inside diameter because the part can be broached at the rate of five hours per thousand. This would be a saving of 3.33 hours per thousand pieces, or a total saving of 333 hours. At our $7.20 base rate, this would mean a direct labor savings of $2,397.60. However, it would not be practical to go ahead with this idea, since the tool cost for broaching is $2,800. Thus, the change would not be sound unless the labor savings can be increased to $2,800 to offset the cost of the new broaching tools.

Since the labor savings in a new broaching setup would be 3.33 × $7.20 per thousand, 116,800 pieces would have to be ordered before the change in tooling would be justified.

$$\frac{\$2,800 \times 1,000}{\$7.20 \times 3.33} = 116,783 \text{ pieces}$$

However, if the broaching method had been used originally instead of the reaming procedure, it would have paid for itself in

$$\frac{\$2,800 - \$2,000}{\$7.20 \times 3.33/M} = 33,367 \text{ pieces}$$

With production requirements of 100,000 pieces, 3.33 × $7.20 × 66.6 thousand (the difference between 100,000 and 33,400) = $1,596.80 in labor that would have been saved over the present reaming method. Had a motion analysis been made in the planning stage, this saving might have been realized. Figure 7–15 illustrates these relationships with the customary crossover (break-even) chart.

By keeping the principles of motion economy in mind, and breaking the proposed method into its basic divisions, analysts can develop a right-hand, left-hand analysis prior to starting production. Then, by assigning fundamental basic motion time values to the various elements of the operator process chart (see Chapter 19), the practicability of the proposed method can be determined. Let us see how this technique is practiced on a simple manufacturing job. The job involves (1) a light blanking operation, (2) low annual production requirements (200,000 pieces per year), and (3) a competitive price.

With these requirements in mind, a methods analyst proposes performing the job on an arbor press, with an accompanying die to do the trimming. He or she then breaks the job down in the form of a right- and left-hand process chart, as shown in Figure 7–16.

FIGURE 7–15
Crossover chart illustrating the fixed and variable costs of two competing methods

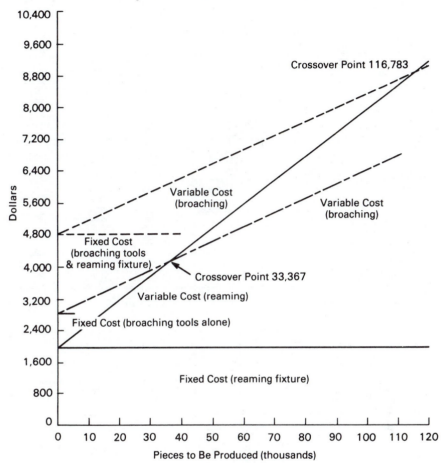

A review of this operator process chart reveals that several of the principles of motion economy have been violated. Motions of the right and left hands are not balanced; unavoidable delays occur in the motion pattern on the left hand; and both hands do not finish their work simultaneously.

Analysis of the poor setup discloses that the first wait of the left hand occurs during the period that the right hand raises the handle of the arbor press. The left hand is then kept idle while the right hand removes the part from the die, and finally, it is unavoidably delayed while the right hand reaches out to gain control of the arbor press handle.

FIGURE 7–16
Right- and left-hand analysis of a proposed method for a blanking operation

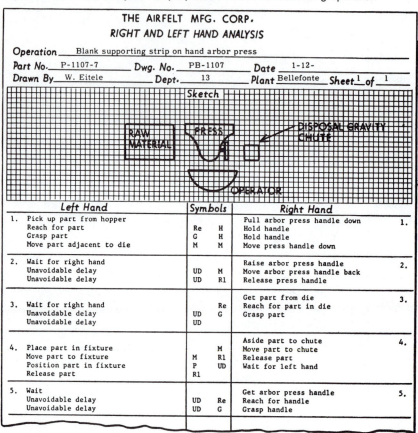

Left Hand	Symbols		Right Hand
1. Pick up part from hopper			Pull arbor press handle down 1.
Reach for part	Re	H	Hold handle
Grasp part	G	H	Hold handle
Move part adjacent to die	M	M	Move press handle down
2. Wait for right hand			Raise arbor press handle 2.
Unavoidable delay	UD	M	Move arbor press handle back
Unavoidable delay	UD	R1	Release press handle
			Get part from die 3.
3. Wait for right hand		Re	Reach for part in die
Unavoidable delay	UD	G	Grasp part
Unavoidable delay	UD		
			Aside part to chute 4.
4. Place part in fixture		M	Move part to chute
Move part to fixture	M	R1	Release part
Position part in fixture	P	UD	Wait for left hand
Release part	R1		
5. Wait			Get arbor press handle 5.
Unavoidable delay	UD	Re	Reach for handle
Unavoidable delay	UD	G	Grasp handle

The right hand must grasp the arbor press handle every cycle as a result of relinquishing it earlier in the cycle to remove the finished part from the die. Therefore, an operator who did not have to remove the finished piece from the die would also be freed of the element "grasp arbor press handle." The improved method in the form of an operator process chart is shown in Figure 7–17.

If this method were put into operation, we would have a balanced setup with considerably shorter cycle time. It is possible to develop a fixture to include leaf springs that lift the part clear of the die on the return stroke of the arbor press. The piece can then be automatically ejected through the back of the press by an air blast actuated from the ram of the press. This motion analysis in the planning stage allows the part to be blanked in the most economical manner.

FIGURE 7–17

Right- and left-hand analysis of a proposed method for a blanking operation in which unavoidable delays have been omitted

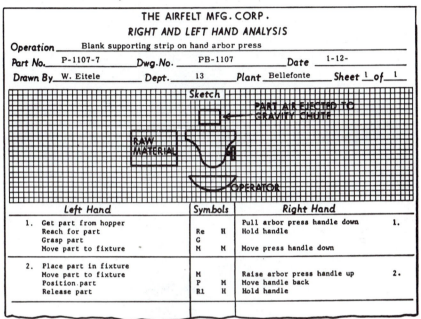

Left Hand	Symbols		Right Hand	
1. Get part from hopper			Pull arbor press handle down	1.
Reach for part	Re	H	Hold handle	
Grasp part	G			
Move part to fixture	M	M	Move press handle down	
2. Place part in fixture				
Move part to fixture	M		Raise arbor press handle up	2.
Position part	P	M	Move handle back	
Release part	R1	H	Hold handle	

Let us take another example and see how motion study in the planning stage helps determine the ideal method. The operation under study is the assembly of the components going into the upper jaw of a pipe vise. The parts involved are the jaw, brace, two lock washers, and two machine screws (see Figure 7–18). The production schedule calls for assembly of 10,000 units per year, priced to meet stiff competition.

Since the parts are all small and can be readily controlled with either hand, a methods analyst may first consider hand assembly at a bench, with the operator seated and all parts fed to the normal work area by gravity bins. This method in the form of a right- and left-hand analysis would appear as shown in Figure 7–19.

A quick review of this method discloses that the left hand is ineffectively employed for the majority of the cycle, since it is occupied by the therblig hold in five distinct areas. To alleviate this condition, the analyst considers making a fixture to hold the part. To eliminate the element "dispose of part," an ejector pin is actuated by a foot pedal (see Figure 7–20). This ejects the part into an accompanying gravity chute. To dispose of the element "lay aside Yankee screwdriver," the tool is suspended overhead. To permit a balanced two-handed motion pattern

FIGURE 7-18

Components of upper jaw of pipe vise—upper panel shows parts in order of
assembly (left to right); lower panel shows assembled parts

throughout the cycle, an additional Yankee screwdriver is suspended
overhead for the left hand. Thus, both screws can be driven home simultaneously with the aid of self-aligning sleeves incorporated into the fixture. Through the use of fundamental motion data, productivity can increase by an estimated 50 percent. A right- and left-hand analysis of this
proposed method is shown in Figure 7-21.

Figure 7-22 shows a workstation for a "doffing yarn" operation. Figure 7-23 illustrates the motion pattern that the operator was following.
The operation in this case was not planned. Stroboscopic photography
reveals that the operator was going through an elaborate body motion
pattern. The planned method shown in Figure 7-24 reduced the number
and difficulty of the motions so that the operator was less fatigued and
production improved greatly.

FIGURE 7–19
Right- and left-hand analysis of the assembly of a pipe-vise jaw

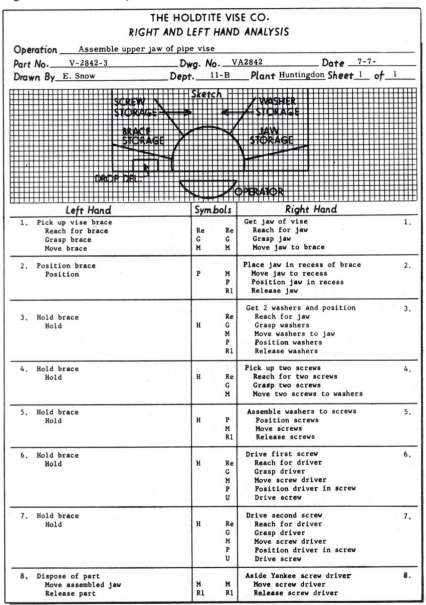

Left Hand	Symbols		Right Hand	
1. Pick up vise brace			Get jaw of vise	1.
Reach for brace	Re	Re	Reach for jaw	
Grasp brace	G	G	Grasp jaw	
Move brace	M	M	Move jaw to brace	
2. Position brace			Place jaw in recess of brace	2.
Position	P	M	Move jaw to recess	
		P	Position jaw in recess	
		Rl	Release jaw	
			Get 2 washers and position	3.
3. Hold brace		Re	Reach for jaw	
Hold	H	G	Grasp washers	
		M	Move washers to jaw	
		P	Position washers	
		Rl	Release washers	
4. Hold brace			Pick up two screws	4.
Hold	H	Re	Reach for two screws	
		G	Grasp two screws	
		M	Move two screws to washers	
5. Hold brace			Assemble washers to screws	5.
Hold	H	P	Position screws	
		M	Move screws	
		Rl	Release screws	
6. Hold brace			Drive first screw	6.
Hold	H	Re	Reach for driver	
		G	Grasp driver	
		M	Move screw driver	
		P	Position driver in screw	
		U	Drive screw	
7. Hold brace			Drive second screw	7.
Hold	H	Re	Reach for driver	
		G	Grasp driver	
		M	Move screw driver	
		P	Position driver in screw	
		U	Drive screw	
8. Dispose of part			Aside Yankee screw driver	8.
Move assembled jaw	M	M	Move screw driver	
Release part	Rl	Rl	Release screw driver	

Motion and Time Study

FIGURE 7–20
Details of guide and ejector pins for a foot-operated mechanism

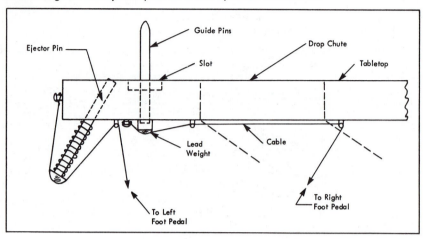

The laws of motion economy and the therblig concept are essential when establishing ideal methods in the planning stage, as well as when improving existing methods. Figure 7–25 illustrates a test station designed to incorporate the principles of motion economy and human factors. (A) shows that access to small components is achieved within the maximum working area. The 15° slope of the bins permits the easy grasp of parts in the front position of the gravity fed bin. Parts used most frequently are within the normal work area on trays mounted to pivot arms (B). The power tool suspended overhead employs a manipulator so it moves about readily (both vertically and horizontally) with very little effort (C). The adjustable fluorescent light provides adequate brightness for seeing the task being performed (D). The work table height is adjustable to accommodate different heights of the workpiece (E). A specially designed tote box brings workpieces and removes them from the workstation (F). The chair assures effective working postures.

TEXT QUESTIONS

1. Explain why a bar switch may be preferable to a trigger switch.

2. Explain why workstation designers should endeavor to have operators perform work elements without having to lift their elbows.

3. What range of height would you recommend for a footrest at a seated workplace?

4. Which seven principles should be employed in connection with workstations where heavy lifting is performed at random intervals?

FIGURE 7–21
Right- and left-hand analysis of a proposed setup for the assembly of a pipe-vise jaw

THE HOLDTITE VISE CO.
RIGHT AND LEFT HAND ANALYSIS

Operation __Assemble upper jaw of pipe vise__
Part No. __V-2842-3__ Dwg. No. __VA-2842__ Date __7-7-58__
Drawn By __E. Snow__ Dept. __11-B__ Plant __Huntingdon__ Sheet __1__ of __1__

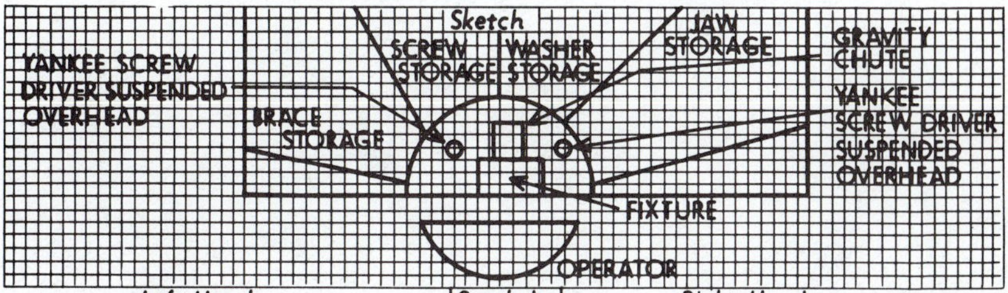

Sketch

Left Hand	Symbols		Right Hand
1. Pick up vise brace			Get jaw of vise 1.
Reach for brace	Re	Re	Reach for jaw
Grasp brace	G	G	Grasp jaw
Move brace	M	M	Move jaw
2. Assemble brace on jaw			Place jaw in fixture 2.
Move brace over jaw	M	M	Move jaw to fixture
Position brace on jaw	P	P	Position jaw in fixture
Release brace	R1	R1	Release jaw
3. Pick up and assemble washer and screw			Pick up and assemble washer and screw 3.
Reach for washer	Re	Re	Reach for washer
Grasp washer	G	G	Grasp washer
Move washer to brace	M	M	Move washer to brace
Position washer	P	P	Position washer
Release washer	R1	R1	Release washer
Reach for screw	Re	Re	Reach for screw
Grasp screw	G	G	Grasp screw
Move screw to washer	M	M	Move screw to washer
Position screw in washer	P	P	Position screw in washer
4. Start screw			Start screw 4.
Move screw	M	M	Move screw
Release screw	R1	R1	Release screw
5. Get driver and drive screw			Get driver and drive screw 5.
Reach for driver	Re	Re	Reach for driver
Grasp driver	G	G	Grasp driver
Move driver	M	M	Move driver
Position screw driver	P	P	Position screw driver
Drive screw	U	U	Drive screw
Release driver	R1	R1	Release driver

5. What viewing distance would you recommend for a seated operator working at a computer terminal?

6. When is visual motion study practical?

7. Define and give examples of the 17 fundamental motions, or therbligs.

8. How may the basic motion "search" be eliminated from the work cycle?

9. What basic motion generally precedes "reach"?

10. What three variables affect the time for the basic motion "move"?

11. How does the analyst or observer determine when the operator is performing the element "inspect"?

FIGURE 7–22
A workstation for doffing yarn

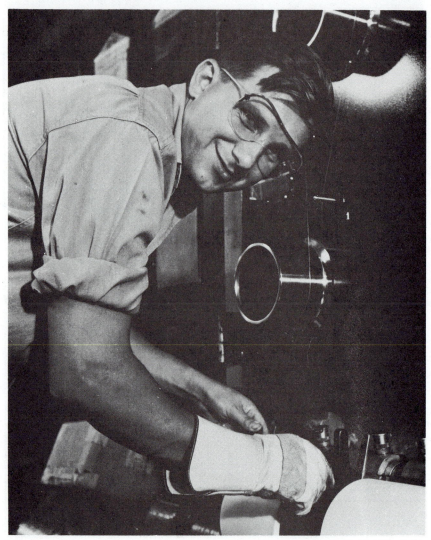

E. I. du Pont de Nemours & Co.

12. Explain the difference between avoidable and unavoidable delays.
13. Which of the 17 therbligs are classed as effective, and usually cannot be removed from the work cycle?
14. Why should fixed locations be provided at the workstation for all tools and materials?
15. Which of the five classes of motions is used most by industrial workers?

FIGURE 7–23
A stroboscopic photograph showing the motion pattern prior to motion planning on a yarn doffing operation

E. I. du Pont de Nemours & Co.

FIGURE 7–24
A stroboscopic photograph showing the motion pattern after motion planning on a yarn doffing operation

E. I. du Pont de Nemours & Co.

FIGURE 7-25
Test station designed to incorporate the principles of motion economy and
human factors

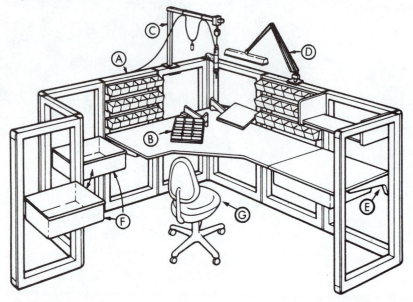

Redrawn from Ray Pukanic and Donald Morelli, "A Systems Approach to Ergonomically Sound
Design of Electronics Assembly/Test Stations," *Industrial Engineering,* July 1985, p. 50.

16. Why is it desirable to have the feet working only when the hands are occupied?
17. Explain the significance of human dimensions.
18. Outline the principal guidelines in the design of operators' handgrips, handwheels, and knobs.
19. What four considerations should be observed to avoid eye strain in connection with a work center involving the use of a video display terminal?
20. Explain where you would locate a document holder for an operator working at a video display terminal.

GENERAL QUESTIONS

1. Take an operation out of your daily routine, such as dressing, analyze it through the operator process chart, and try to improve the efficiency of your movements.
2. Explain why effective visual areas are reduced where concentrated attention will be required.
3. Why does a broach usually cost more than a reamer to perform the same sizing operation?

4. Explain why the adjustable seat height of a workplace with a footrest should be higher than a seated workplace without a footrest.

PROBLEMS

1. A given design has fixed costs of $5,000 and variable costs of $15,000 at 100 percent of company capacity that yields total sales of $30,000. How much more of the plant's capacity must be utilized to break even if the fixed costs rise to $8,000?
2. Compute the range of work-surface heights that would accommodate all workers through the 98th percentile.
3. What would be the effective visual surface area (in square inches) of an average employee whose workstation was centered 26 inches from the center point between the eyes?
4. The methods analyst in the Dorben Company is involved in improving work centers in the assembly of capacitor chips. For a particular series of operations, the company is providing community assistance by employing a group of physically handicapped workers who have a mean height of 61 inches with a standard deviation of 1.8 inches. Based upon the information provided in Figures 7–7 and 7–8, design a work center where the operator is seated and the assembly is performed at a work table that surrounds the operator by 180°.

SELECTED REFERENCES

Barnes, Ralph M. *Motion and Time Study: Design and Measurement of Work.* 7th ed. New York: John Wiley & Sons, 1980.

Clark, Daniel O. and Guy C. Close. "Motion Study." In *Handbook of Industrial Engineering,* ed. Gavriel Salvendy. New York: John Wiley & Sons, 1982.

Eastman Kodak Co., Human Factors Section. *Ergonomic Design for People at Work.* New York: Van Nostrand Reinhold, 1983.

Konz, Stephen. *Work Design.* Columbus, Ohio: Grid, 1979.

Mundel, M. E. *Motion and Time Study.* 5th ed. Englewood Cliffs, N.J.: Prentice Hall, 1978.

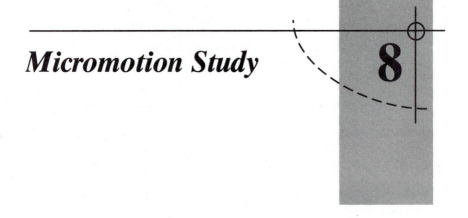

Micromotion Study   8

Micromotion study is the most refined technique for analyzing an existing work center. The cost of conducting a micromotion study is approximately four times as great as the cost of visual motion study. Consequently, it is economically sound to utilize videotape for motion study only when the activity of a specific job or class of work is high. The term *micromotion study* signifies the making of a detailed motion study employing either videotapes or motion pictures. Today videotape is used almost exclusively. On film, the picture in each space, known as a frame, is projected and studied independently, and then collectively with successive frames.

The basic division of motion, or therblig, concept is usually more important in making a micromotion study than in conducting a visual motion study. This is because a work assignment can more readily be broken down into basic elements with the frame-by-frame analysis than with visual motion studies. Since the object of the micromotion procedure is to uncover every possible opportunity for improvement, down to the performance of each therblig, it is fundamental to identify each basic division as performed.

Several important corollaries to the principles of motion economy have application to micromotion study. They include:

1. Establish the best sequences of therbligs.
2. Investigate any substantial variation in the time required for a given therblig and determine the cause.
3. Examine and analyze hesitations to determine and then to eliminate their causes.
4. Aim for cycles and portions of cycles completed in the least amount of time as a goal. Study deviations from these minimum times to determine the cause.

OPERATOR SELECTION FOR MICROMOTION STUDY

In making a micromotion study, observe either the best operator or, preferably, the best two operators. This procedure is quite different from time study, where an average operator is usually selected for study (see Chapter 14). This is not always possible because the operation might be performed by only one person. In such instances, if past performance indicates that the individual is only a fair or a poor operator, then a competent, skilled, cooperative employee should be trained in the operation before the pictures are taken. Only first-class, proficient operators should be selected for filming. This is fundamental for a number of reasons: the proficient employee is usually a dexterous individual who instinctively follows the principles of motion economy related to the use of the human body; this type of operator is usually cooperative and willing to be photographed; extra effort put forth by such an operator shows greater results than similar efforts employed by the mediocre operator.

If the best two operators have been studied, analysis may indicate proficiency in different areas of the cycle for the two operators. This may lead to greater improvement than if only one individual is studied.

The people who are being studied should be given at least a day's notice prior to taking the videotape. This allows them to make any personal preparations they may wish and also permits them to choose wearing apparel that will be conducive to making clear pictures.

To secure the cooperation, advise the supervisor of your plans several days in advance. This allows the boss to adjust personnel to assure that projected production schedules will be maintained. Interruptions brought about by film analysis work may cause a given section to lose several valuable work hours. A supervisor who is not given advance notice of contemplated motion study work can hardly be expected to be cooperative.

MICROMOTION STUDY AS A TRAINING AID

In addition to being used as a methods improvement tool, micromotion study is finding increasing use as a training aid. The world of sports has used this tool for many years, to develop athletes' timing of movement, rhythm, and smoothness of performance. Videotape pictures are made of outstanding performers in a given sport and projected to facilitate detailed analysis of their motions. Less proficient performers are then able to pattern their efforts after the experts.

Industry is finding that it can attain the same results accomplished in the area of athletics. Videotaping highly skilled workers and showing their movements and motion patterns in slow motion makes it possible to train new employees in a minimum of time to follow the ideal motion-method pattern.

One Pennsylvania concern engaged in the manufacture of loose-leaf notebooks, writing paper, envelopes, tablets, and similar paper items continually takes videotapes for training purposes, not only in its own plant but in affiliated plants. By continually exchanging tapes taken at the different plants, it takes advantage of methods improvements introduced throughout the corporation, and trains its employees to follow the new methods in a minimum of time.

Management should take full advantage of industrial videotapes once a micromotion program has been launched. By showing all the pictures taken of the various operations to the operators involved and to their fellow workers, immeasurable goodwill and individual interest are established throughout an organization. Once operators understand the usefulness of micromotion study, their assistance can be enlisted in working out improved methods.

MICROMOTION EQUIPMENT

To be able to do an acceptable job in micromotion study, it is necessary to budget a minimum of $7,000. This figure is only a rough approximation based on current prices and can fluctuate appreciably as market conditions change.

Videotape equipment provides the distinct advantage of instantaneous replay: Immediately after taking pictures on videotape, analysts can observe the operation under study. Videotapes may be used over and over again for different micromotion studies. Figure 8–1 illustrates videotape

FIGURE 8–1
Videotape equipment that permits immediate playback at real-time speeds, in slow motion, or frame by frame, where the time between frames is 1/120 second

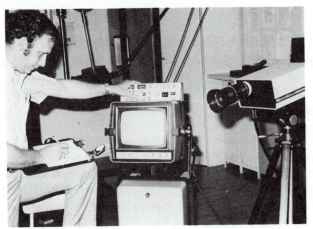

equipment that permits real-time projection speeds, slow motion, and frame-by-frame analysis.

Today, videotape systems are available using an international tape format, 8 mm tape that delivers the same high quality pictures as the ½ inch (1.3 cm) tape used with older models. A single cassette of 8 mm video can give up to two hours playing time, with frequency modulated (FM) recording that produces sound similar to quality audio systems. This compares favorably with 16 mm motion-picture film which permits less than four minutes of exposure using 100-foot rolls of film and exposing at normal speed. Furthermore, the videotape camera-recorder combination is available with superimposed time for studies and analysis on a video player with excellent stop-frame capability. The equipment is light—typically weighing only about four pounds—and conveniently portable. Today's quality equipment includes:

1. A fast f/1.2 lens making the equipment ideal for low light recording— typically as low as 19 lux.
2. An adjustable electric viewfinder allowing analysts a wide range of shooting positions.
3. A motor-driven zoom lens permitting smooth wide angle or tele- photo zooming at the press of a button.
4. A diopter that is built into the viewfinder allowing analysts to adjust the eyepiece to suit their eyesight.
5. A macro setting allowing analysts to take close-ups of subjects 4 mm from the lens.
6. A recorded review button providing analysts a quick review of what was just recorded. By pressing this button, the last five seconds of the scene are played back in the viewfinder.
7. An iris control that automatically adjusts to different lighting levels for optimum exposure. The iris can be adjusted manually to cope with extremely dark or bright lighting situations.
8. Flexible playback modes allow searching or reviewing segments at nine times normal speed at the touch of a button. Other modes permit freezing a specific picture or advancing the film frame by frame.
9. Recorded FM audio signals to give hi-fi sound of good quality.
10. Battery and tape end warning lights, a tape counter, and recording/ playback status are conveniently displayed.

The videotape model illustrated in Figure 8–1 records at 120 frames per second. This speed provides the added advantage of slow-motion playback, which facilitates the study of workstation methods. Of course, recorded tapes may be played back in real time, as well as in a range of slow-motion times. Stop action analysis may also be accomplished with this equipment.

The Ekta Pro 1000 motion analyzer manufactured by Kodak's Motion Analysis Systems Division, can record up to a thousand full-frame images per second. In addition, 2000, 4000, or 6000 can be recorded by splitting frames and recording two, four, or six pictures on each one.

Videotape facilities can incorporate two cameras and a dual camera control unit for the added flexibility of split screen. This feature permits the simultaneous observation and study of two competing methods. It also provides an excellent training device in "performance rating," as is explained in Chapter 15.

Some videotape equipment has scene and frame identification features which preclude the necessity of including a microchronometer or some other frame-identifying device in the scene.

In the equipment illustrated, the scene and frame identification consists of two separate sets of numbers at the top of each picture. The three-digit scene number is followed by a six-digit frame identification. The frame identification number starts at zero count at the beginning of the record mode, and increases by one count for each frame.

Videotape equipment continues to improve, particularly cameras, lenses, and monitors. Also, prices have become attractive. For 8 mm videotape, the equipment cost competes favorably with the equipment cost of 16 mm movie equipment.

Taking Motion Pictures

In chronological sequence, the fundamental steps in making a micro-motion study are:

1. Take videotape or moving pictures of the two best operators.
2. Analyze the film on a frame-by-frame basis, plotting the results on a simo (simultaneous motion) chart.
3. Consider the laws of motion economy and their corollaries and create an improved method.
4. Teach and standardize the new method.
5. Take videotape or moving pictures of the new method.

Check these factors before taking motion pictures on the production floor or in the laboratory. First, be sure that the selected operators have had reasonable advance notice that pictures are to be taken. This is important so that they can be properly groomed and attired. Second, be sure to obtain clearance for the use of the operators from the departmental supervisor. In most plants, it is necessary to obtain clearance from the union before proceeding with the picture taking. Even if this is not required, it is always wise, in the interests of healthy labor relations, to keep the union officers completely informed as to the purposes of all phases of the videotape procedure.

Third, make sure that there is an adequate inventory of material so that it will not be necessary to replenish various supplies during the picture-taking process.

Fourth, on the completion of several practice cycles, conducted to relieve the operator of any nervousness caused by being put under the spotlight, ask the operator to give his or her best performance. At this point begin filming. Several cycles should be taken for analysis; the number depends on their duration. Often, the first cycle is unsatisfactory, as the operator may develop a case of nerves while working under the observation of the camera. This reaction usually is brief, and he or she soon acquires their normal feel for the work. Usually, operators enjoy being photographed, and exhibit excellent cooperation.

Analyzing Tape or Film

Before analyzing the film or tape, run the pictures through several times to determine the cycle that represents the best performance. The shortest cycle is usually the best one to study.

After selecting the cycle to be studied, begin the frame-by-frame analysis. Although the study can be made at any point in the cycle, it is customary to begin the analysis at the frame after "release of finished part" of the previous cycle. Thus, the first basic operation employed in most instances would be the reach made to pick up a part or tool to begin production on the cycle being studied.

Once the starting point of the cycle has been determined, set the frame counter to zero to facilitate recording elapsed elemental times.

After noting the therblig employed by the operator, advance the film a frame at a time until termination of the basic division. Then record on the micromotion data sheet the number of frames exposed from the frame counter.

Reach and move motions are considered as beginning in the frame preceding noticeable movement, and as ending in the frame in which noticeable movement ceases. Where only minor movements occur, such as grasp, release, and position, the beginning point is considered to be the frame in which the preceding motion ended, and the ending point is the frame preceding the one in which the following motion was noticeable.

In reviewing the tape or film, carefully observe the motion class being used to perform the basic division, and record this information on a data sheet, or simo chart. At all times, look for ways to eliminate or improve each basic division.

After briefly describing the basic division in the space provided (such as "move to fixture" in Figure 8–2), draw a horizontal line across the portion of the paper representing the body member being studied. Usually you study only the arms and hands of the operator, although sometimes you must analyze foot and body motions.

FIGURE 8–2

Film analysis of vise clamp assembly

DATE: JANUARY 9   DEPT. I.E. 320   DRAWING: SEE REPORT   PART NO. 114

PART DESCRIPTION: VISE CLAMP

OPERATION: ASSEMBLY   FILM SPEED: 16 FRAMES/SEC.

CLOCK READINGS	LEFT HAND	SYMBOL	MOTION CL.	FINGER	WRIST	FOREARM	SHOULDER	BODY	TIME (IN WINKS)	BODY	SHOULDER	FOREARM	WRIST	FINGER	MOTION CL.	SYMBOL	RIGHT HAND	CLOCK READINGS
2059																		2059
2089	MOVE TO JAW BIN 14"	RE	3						0030							3	RE	MOVE TO BRACE BIN 14"	2089
2105	PICK UP JAW	G	1						0016										2105
2125	MOVE TO FIXTURE 8"	M	3						0020 0036								UD	IDLE	2125
2184	PLACE IN FIXTURE	U	1						0059							1	G	PICK UP BRACE	2184
									0038							3	M	MOVE TO FIXTURE 8"	2222
2451	MOVE TO FIXTURE 12"	M	3						0037							3	M	MOVE TO FIXTURE 12"	2451
2579	SCREW IN SCREWS	U	1						0128							1	U	SCREW IN SCREWS	2579

Record the motion symbol in the space provided, and in combined motions, note the symbols of the elements occurring simultaneously. Thus, a pre-position may occur during a move, such as when the hand aligns a bolt while transporting it. In this case, the element description would be "Move bolt _____ inches," and the symbols recorded would be "P.P. and M."

In the motion class columns, differentiate the productive elements from the nonproductive elements. This is usually done through color; the elements reach, grasp, move, use, and assemble are shown as solid black, and the remainder of the therbligs in red or by cross-hatching if the charts are to be duplicated. Indicate delay times as Class 1 motions in red. Identify all other therbligs on the chart by the motion class utilized in performing the operation.

The time scale selected should provide enough room to identify clearly the shortest basic divisions and create a neat-appearing chart that can be easily studied and diagnosed for improvement. Release is the shortest of all the therbligs, taking but one frame at normal camera speed. Thus, if each division of the time scale is equated to 0.002 of a minute, a neat-appearing chart would result.

After analyzing the first basic division of one hand, and recording the method, advance the film slowly, and analyze the next basic division and so on until completion of the cycle. To avoid confusion, study one hand completely before the study of the other hand is begun. As the second hand is being considered, periodically verify that the motions recorded are occurring in the same relationship with the other hand as is indicated in the analysis sheet. Remember that when you view the projected picture all directions are reversed; the left arm is on the right side of the picture.

After both hands have been completely charted, add a summary at the bottom of the chart showing the cycle time, pieces made per cycle, productive time per cycle, and nonproductive time per cycle. Sometimes a sketch of the workplace is helpful, and this can be included at the bottom of the chart. The sketch is usually not necessary, since the film gives a pictorial presentation of the work area.

Creating an Improved Method

After completing the simo chart, the next step is to use it. The nonproductive sections of the chart represent a good place to start. These sections include the therbligs hold, search, select, position, pre-position, inspect, and plan, and all the delays. The more of these that can be eliminated, the better the proposed method is. However, analysis should not be confined to the red sections of the chart, since possibilities for improvement exist in the productive portions as well. For example, the productive element "reach 24" can be improved by reducing the distance and, consequently, lowering the motion class.

TABLE 8-1
Checklist for the simo chart

---

1. Are both hands beginning their therbligs simultaneously?
2. Are both hands ending their therbligs simultaneously?
3. Can delay or idle times be removed from the cycle?
4. Are the motion patterns of the arms symmetrical?
5. Are motions made radially from the body in opposite directions?
6. Can a fixture be incorporated to eliminate hold?
7. Are all materials and tools within the normal work area in both the horizontal and the vertical planes?
8. Does the chair and workstation allow the operator to get close enough to work?
9. Has the overhead space been utilized to suspend tools and store inventory components?
10. Are motions confined to the lowest possible classifications?
11. Are tools and materials located to permit the best sequence of motions?
12. Can foot-operated mechanisms be introduced advantageously?
13. Have gravity chutes and drop delivery been provided?
14. Are all tools pre-positioned to minimize grasp time and reduce the search and select therbligs?
15. Has an automatic ejector been incorporated into the fixture?
16. Can position times be reduced by providing "pilots" or other devices for mating parts?

A micromotion checklist helps avoid overlooking any possibility of improvement. Table 8-1 illustrates a checklist that covers most of the considerations that apply to the principles of motion economy and their corollaries.

Figure 8-3 illustrates a completed simo chart of the assembly of a variable resistor. Analysts constructed this chart after a new fixture was developed to facilitate assembly. It represents a method considerably improved over the original method of assembly. Yet an analysis of the chart discloses several areas for further improvement. The right- and left-hand movements are not balanced in many parts of the cycle; 10 ineffective motions are performed by the left hand, and 11 ineffective motions by the right hand. Several of the moves and reaches appear to be unduly long, suggesting the shortening of distances at the workstation layout.

The micromotion technique should be used to uncover every inefficiency, regardless of its apparent insignificance. In total, sufficient numbers of minute improvements result in an appreciable annual savings.

Teaching and Standardizing the New Method

To be assured that full benefits are realized from the micromotion study, put the improved method into practice as soon as possible. Ask all operators to follow it in exact detail. Verbal explanations of the motion pattern to be followed are usually inadequate: both time and effort are saved if an instruction chart is used to give specific information on the "how" of the improved method.

FIGURE 8–3
Simo chart assembly of a variable resistor

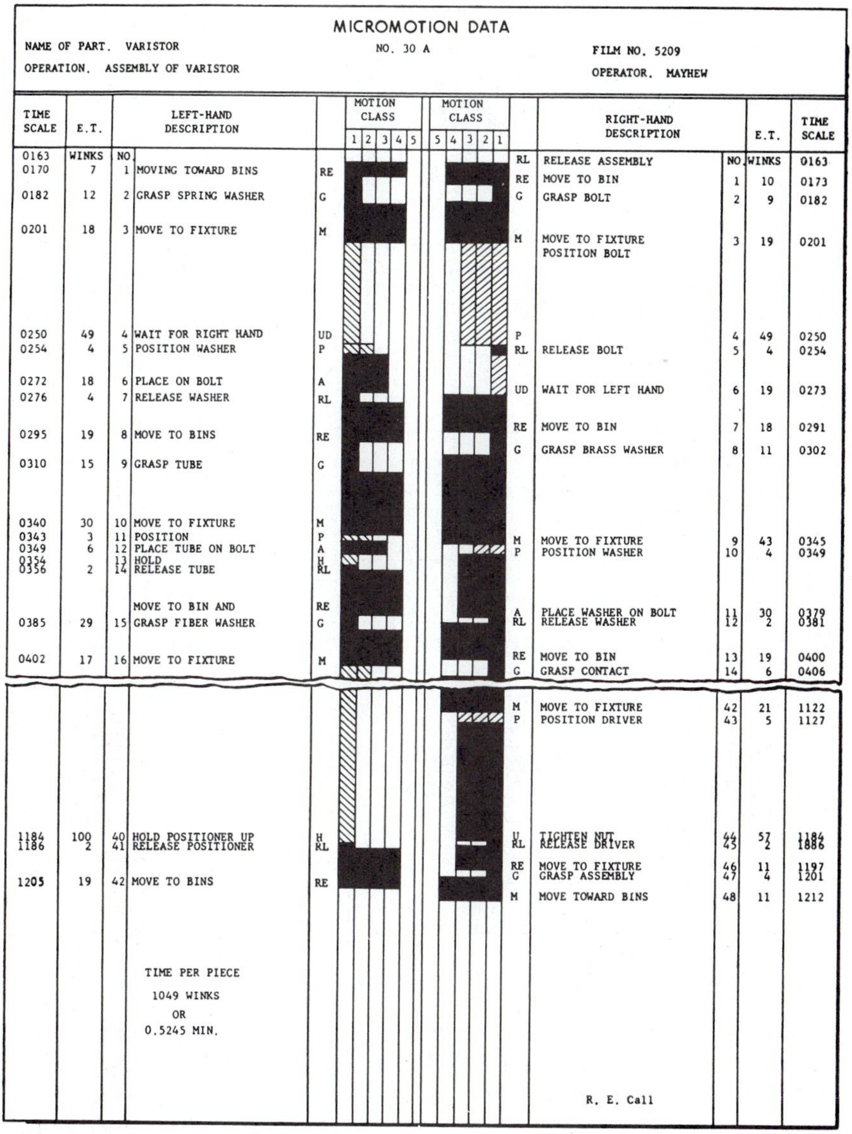

MICROMOTION DATA

NAME OF PART. VARISTOR   NO. 30 A   FILM NO. 5209
OPERATION. ASSEMBLY OF VARISTOR   OPERATOR. MAYHEW

TIME SCALE	E.T.	NO	LEFT-HAND DESCRIPTION	MOTION CLASS		MOTION CLASS	RIGHT-HAND DESCRIPTION	NO	WINKS	TIME SCALE
0163 0170	WINKS 7	1	MOVING TOWARD BINS	RE		RL	RELEASE ASSEMBLY	NO	WINKS	0163
						RE	MOVE TO BIN	1	10	0173
0182	12	2	GRASP SPRING WASHER	G		G	GRASP BOLT	2	9	0182
0201	18	3	MOVE TO FIXTURE	M		M	MOVE TO FIXTURE POSITION BOLT	3	19	0201
0250	49	4	WAIT FOR RIGHT HAND	UD		P		4	49	0250
0254	4	5	POSITION WASHER	P		RL	RELEASE BOLT	5	4	0254
0272	18	6	PLACE ON BOLT	A		UD	WAIT FOR LEFT HAND	6	19	0273
0276	4	7	RELEASE WASHER	RL						
0295	19	8	MOVE TO BINS	RE		RE	MOVE TO BIN	7	18	0291
0310	15	9	GRASP TUBE	G		G	GRASP BRASS WASHER	8	11	0302
0340	30	10	MOVE TO FIXTURE	M						
0343	3	11	POSITION	P		M	MOVE TO FIXTURE	9	43	0345
0349	6	12	PLACE TUBE ON BOLT	A		P	POSITION WASHER	10	4	0349
0354		13	HOLD	H						
0356	2	14	RELEASE TUBE	RL						
			MOVE TO BIN AND	RE		A	PLACE WASHER ON BOLT	11	30	0379
0385	29	15	GRASP FIBER WASHER	G		RL	RELEASE WASHER	12	2	0381
0402	17	16	MOVE TO FIXTURE	M		RE	MOVE TO BIN	13	19	0400
						G	GRASP CONTACT	14	6	0406
						M	MOVE TO FIXTURE	42	21	1122
						P	POSITION DRIVER	43	5	1127
1184	100	40	HOLD POSITIONER UP	H		U	TIGHTEN NUT	44	57	1184
1186	2	41	RELEASE POSITIONER	RL		RL	RELEASE DRIVER	45	2	1186
						RE	MOVE TO FIXTURE	46	11	1197
1205	19	42	MOVE TO BINS	RE		G	GRASP ASSEMBLY	47	4	1201
						M	MOVE TOWARD BINS	48	11	1212

TIME PER PIECE
1049 WINKS
OR
0.5245 MIN.

R. E. Call

In form, the typical micromotion instruction sheet is similar to the operator process chart. A layout of the work area drawn to scale heads the instruction sheet, and the motion sequence of both the right hand and the left hand follows. These elements appear in relationship to one another. Special tools, such as form cutters, gages, jigs, and fixtures, are

FIGURE 8-4
Instruction sheet

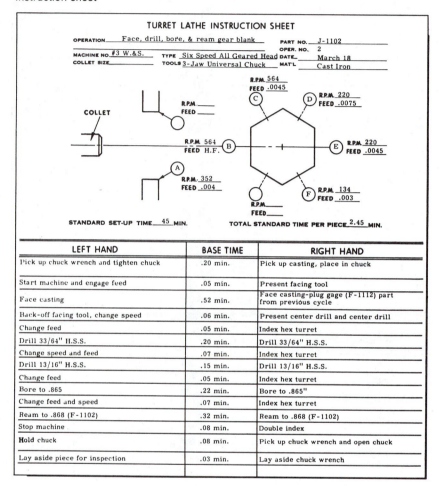

TURRET LATHE INSTRUCTION SHEET

OPERATION __Face, drill, bore, & ream gear blank__   PART NO. __J-1102__

                                                       OPER. NO. 2
MACHINE NO. #3 W.&S.   TYPE Six Speed All Geared Head DATE __March 18__
COLLET SIZE__   TOOLS 3-Jaw Universal Chuck   MAT'L __Cast Iron__

STANDARD SET-UP TIME __45__ MIN.   TOTAL STANDARD TIME PER PIECE __2.45__ MIN.

LEFT HAND	BASE TIME	RIGHT HAND
Pick up chuck wrench and tighten chuck	.20 min.	Pick up casting, place in chuck
Start machine and engage feed	.05 min.	Present facing tool
Face casting	.52 min.	Face casting-plug gage (F-1112) part from previous cycle
Back-off facing tool, change speed	.06 min.	Present center drill and center drill
Change feed	.05 min.	Index hex turret
Drill 33/64" H.S.S.	.20 min.	Drill 33/64" H.S.S.
Change speed and feed	.07 min.	Index hex turret
Drill 13/16" H.S.S.	.15 min.	Drill 13/16" H.S.S.
Change feed	.05 min.	Index hex turret
Bore to .865	.22 min.	Bore to .865"
Change feed and speed	.07 min.	Index hex turret
Ream to .868 (F-1102)	.32 min.	Ream to .868 (F-1102)
Stop machine	.08 min.	Double index
Hold chuck	.08 min.	Pick up chuck wrench and open chuck
Lay aside piece for inspection	.03 min.	Lay aside chuck wrench

identified after the element in which they are used. Speeds, feeds, depth of cut, and other pertinent manufacturing information also appear. Usually a time summary accompanies the instruction sheet, so that the operator can measure performance against standard while being trained in the new method. A typical instruction sheet appears in Figure 8-4.

The line supervisor must be well versed in the proposed method to assist in training the workers. Usually instruction sheets are duplicated so that copies can be filed in the methods section, time study section, and supervisor's office, as well as given to each operator performing the operation. Provide a hanger for the instruction sheet at each operator's workstation so that it can readily be referred to at all times during the operating cycle. If encased in a clear plastic envelope the instruction sheet should remain readable for a long period.

The supervisor, with the methods engineer, should periodically check each operation on the production floor to be sure that the new method is being followed, and to answer any question that may come up on the new procedure. Installation and follow-up are two very important phases in micromotion methods improvement. Often it is difficult for the average operator to justify the new method because the changes may appear so insignificant. Consequently, both the supervisor and the methods engineer must do a real job of selling the improved system.

Unless periodic checks are made for at least several weeks, operators may go back to their old way of doing the job, completely disregarding the new technique. One of the added benefits of the written instruction sheet is that it helps the operator to conform to the established instructions.

As soon as one of the better operators has become skilled in the new method, take motion pictures and construct a simo chart to illustrate the improvements resulting from the study. Use the same operators selected in making pictures of the unimproved method. This will give a better comparison for evaluating the savings than if different operators are used in the final pictures. Once the new pictures have been taken, use them as a training tool for all other operators performing the same operation. Seeing how the operation should be done is much more helpful than reading or hearing how it should be done. The psychological impact of seeing a fellow worker perform the new method is a force in breaking down the inherent human resistance to change.

Comparing the completed simo chart of the improved method to the one drawn of the old method can clearly present the savings. Figure 8–5 illustrates a method proposed after a complete micromotion study. This tool not only shows the amount of savings but vividly explains how these results were made possible.

MEMOMOTION STUDY

One other important motion study technique gives more detail than visual motion study and less than micromotion study. This is the technique referred to as "memomotion study."

Memomotion study is a moving-picture technique developed by Marvin Mundel several years ago to analyze the principal movements in an operation to improve methods, define problem areas, and establish standards.

Memomotion study usually uses a filming speed of 60 frames per minute with an accurate rate time interval between pictures to provide a record for frame-by-frame analysis. However, equipment is available to provide much greater flexibility in the rate of tape or film exposure.

FIGURE 8–5
Simo chart of improved method for the assembly of lace finger in textile machinery

SIMO CHART

OPERATOR: Ken Reisch
DATE: May 21,
OPERATION: Assembly
PART: Lace Finger
METHOD: Proposed
CHART BY: Joseph Riley

TIME SCALE (winks)	ELEMENT TIME	LEFT-HAND DESCRIPTION	SYMBOL	MOTION CLASS	SYMBOL	RIGHT-HAND DESCRIPTION	ELEMENT TIME	TIME SCALE (winks)
4548	12	Reach for finger	RE		RE	Reach for finger	12	4548
4560	19	Grasp finger	G		G	Grasp finger	19	4560
4579	31	Move finger	M		M	Move finger	31	4579
4610	75	Position and release finger	P RL		P RL	Position and release finger	75	4610
4685	15	Reach for clamp	RE		RE	Reach for clamp	15	4685
4700	15	Grasp clamp	G		G	Grasp clamp	15	4700
4715								4715
7541	12	Grasp assembly	G		G	Grasp assembly	12	7541
	18	Move and release assembly	M RL		M RL	Move and release assembly	18	7559
7559								

SUMMARY

%	TIME	LEFT HAND SUMMARY	SYM.	RIGHT HAND SUMMARY	TIME	%
8.56	249	Reach	RE	Reach	245	8.4
7.49	218	Grasp	G	Grasp	221	7.5
12.16	354	Move	M	Move	413	14.2
30.45	887	Position	P	Position	1124	38.7
39.3	1145	Use	U	Use	876	30.1
1.03	30	Idle	I	Idle	0	0.0
.96	28	Release	RL	Release	32	1.1
100.0	3011			TOTALS	3011	100.0

I have standardized on 60 frames per minute, or 1 frame per second. This speed gives a measurement of elapsed elements that is precise enough for establishing standards, and permits significant economies over the regular camera speed of 960 frames per minute for 16 mm silent film. Thus, under memomotion, pictures are exposed just 1/16 as fast as under the normal speed (see Figure 8–6). For example, in the study of a three-minute cycle, the exposure and the analysis of only 4.5 feet of film are required, as compared with 72 feet when the micromotion procedure is used.

$$\frac{3 \text{ min.} \times 60 \text{ frames/min.}}{40 \text{ frames/foot}} = 4.5 \text{ feet}$$

Today, very little memomotion study work is done using moving picture film because of processing and film costs. This method has been replaced by using videotape. One videotape may be "stretched" to approximately 300 hours depending upon the time interval between frames.

Frequently, memomotion study is helpful in the following areas of work:

1. Multiworker activities.   5. Plant layout.
2. Multifacility operations.   6. Recorded picture histories.
3. Irregular cycle studies.   7. Work sampling studies.
4. Long-cycle operations.

Memomotion finds application in the areas of study because:

1. It provides a more accurate record than do visual means for analysis and film appraisal.
2. It records interrelated events more accurately than do visual techniques.
3. It provides a basis for work measurement, since the camera operates at a constant speed.
4. It focuses attention on the major movements taking place at a work center.
5. It reduces the cost of purchasing and analyzing the film to about 6 percent of the cost incurred when filming at normal speed.

Memomotion is particularly helpful in the study of crew activities. Terminal points in the elements of various operators within a group and the relation of terminal points to one another are almost impossible to recognize by visual methods. Frame-by-frame analysis makes it easy to determine the relationships of the various work elements performed by the members of the crew. Consequently, memomotion study lends itself to such uses as crew balance studies and gang process chart work (see Chapter 6).

Multifacility operation studies are quite similar to crew studies from the standpoint of adaptability to memomotion study work. Here it is

FIGURE 8–6

Portion of a memomotion study of four men operating a hydraulic extrusion press (film exposed at rate of one frame per second)—note the distance a given operator moves from one frame to the next)

desirable to follow not only the activities of the operator but also the facilities he or she services. Such factors as interference delays and machine breakdowns are very difficult to study by visual means, and yet they are easily observed through the frame-by-frame procedure. It is quite difficult to study irregular cycles, such as the periodic replenishing of stock supplies, sharpening tools, and moving completed work, by visual means. To begin with, analysts do not know when they are going to take place, and consequently are obliged to record the element and measure its duration while it is taking place. This is most difficult. The memomotion procedure records all the facts pictorially—the method employed and the time consumed—so that an analysis of the facts may be conveniently and leisurely made through frame-by-frame analysis.

Long-cycle operations employing many short elements are usually too costly to study by micromotion techniques and too difficult to study in adequate detail by visual methods. Memomotion provides the ideal alternatively to both of these procedures.

Memomotion studies of a work area highlight the major moves of all personnel coming into that area. Long moves, such as backtracking and trips for tools, raw materials, and supplies, are emphasized, and inefficient areas of the layout are pointed out. Frequently, a work area that appears satisfactory under a conventional micromotion film reveals areas for improvement under memomotion.

Filing a permanent record of a method is frequently helpful. The permanent record simplifies setting up for the same job at a later date; it is also useful for later methods improvement studies, for settling labor disputes, and for operator training.

Another important use of the memomotion camera is in preparing work sampling studies (see Chapter 21). By using memomotion equipment, unbiased work sampling studies may be made. A timer is attached to the typical memomotion camera permiting random observations during the eight-hour day. The "observation times" can be preset by placing small metal clips around the circumference of the time dial. Analysts may use a random numbers table in locating the clips. Also, they can set the timer to take pictures at regular intervals during the day if this is desired. A motor-driven timer operates the synchronous motor that drives the camera.

Another mechanism provides a means for randomly triggering the camera. Here random punchings on a constant-speed tape energize the switching mechanism. The roll of tape can allow operation of the equipment for 160 working hours, based on a minimum of 10 seconds between random observations. A preset timer allows the exposure of 1 to 10 frames at each triggering of the camera. Thus, at random times during the work period, the camera takes observations for 1 to 10 seconds of the work area.

By using memomotion equipment adapted as described previously, the operator can make work sampling studies that give unbiased results.

FIGURE 8–7
Time-lapse TV system shown studying a group activity

Timelapse, Inc.

Work sampling studies made by an observer recording data at random intervals tend to be biased for the following reasons:

1. The mere arrival of the observer at the work scene influences the operator to become productively engaged in work that he or she would not normally be doing at the time of the observation.

2. There is a natural tendency for the observer to record what has just happened or what will be happening at the exact moment of the observation.

The use of memomotion to make work sampling studies saves labor in the collection of the data, since it is only necessary to remove the film at the end of the work period rather than go to the work scene periodically to record data.

For the most part, micromotion study allows the same analysis that memomotion study does. However, memomotion offers the kinds of economies that open the door to the frame-by-frame analysis of activities that normally would not justify the cost of a micromotion study.

Time-lapse TV provides the added feature of instantaneous replay. After taking pictures on videotape, analysts can observe immediately the operation under study. Tapes may be used over and over again for different time-lapse studies. Figure 8–7 illustrates equipment that is available in compression ratios of from 2 to 15 hours of activity to 1 hour of showing.

TEXT QUESTIONS

1. Explain why videotape systems have almost made film systems obsolete.
2. Describe the characteristics of modern videotape systems.
3. How does micromotion study differ from motion study?
4. For micromotion study, what additional considerations should be given to "the use of the human body" and "the arrangement and conditions of the workplace"?
5. Why is the best operator usually selected in the micromotion procedure?
6. What consideration should be given the supervisor relative to taking micromotion pictures in his or her department?
7. How can micromotion study be helpful in training?
8. What are the important steps in making a micromotion study?
9. How does the simo chart differ from the operator process chart?
10. Which motions are classified as ineffective on the simo chart?
11. What class of motions is signified on the simo chart when plotting a delay?
12. What is the purpose of the micromotion instruction sheet?
13. In tape or film analysis, why is it inadvisable to analyze both hands simultaneously?
14. What advantages does memomotion study have over micromotion study?
15. How much film would be required to study a cycle 13.5 minutes in duration using a memomotion speed of 50 frames per minute?
16. What filming speed is generally used in making memomotion studies?
17. What range of speeds is possible in memomotion study?
18. Explain when you would advocate the use of time-lapse TV facilities.

GENERAL QUESTIONS

1. Explain how you would sell the local union steward on the importance of filming only the best two operators when making a micromotion study.
2. Make a sketch of your desk or workstation and show an arrangement and conditions that would agree with the principles of motion economy and their corollaries.
3. Interview at least two factory workers and get their reaction to micromotion study through the film analysis technique.
4. What activity within an industrial organization, other than methods analysis, might include the use of micromotion equipment? How?
5. Discuss the relative merits of the telephoto lens and the wide-angle lens.
6. Why do unions require permission before motion pictures are taken on the production floor?
7. Compare the cost and quality characteristics of the following compact video systems: Canon, GE, RCA, Sharp, Sony.
8. In general, what is the operator's attitude toward being taped?
9. How can industrial relations be improved by micromotion studies?
10. Explain what impact, if any, today's philosophy of throwaway designs has on the use and application of micromotion study.
11. What influence does international competition have on the application of micromotion study?

PROBLEM

1. In the assembly of small tube-type flashlights in the XYZ Company, two size C (2.5 cm diameter by 4.5 cm length) batteries provide the power. A breakdown of the work elements at each of the 10 assembly stations was as follows:
   a. Reach 44 cm with left hand and pick up tube-type housing.
   b. Simultaneously with (*a*), reach 50 cm with right hand and pick up two size C batteries.
   c. Palm one battery and assemble one battery in housing. Place second battery in housing.
   d. Reach 40 cm and pick up assembled head. Screw assembled head onto tube-type housing.
   e. Reach 40 cm and pick up spring-loaded end. Assemble end to tube-type housing.
   f. Place assembled flashlight in cardboard carton with left hand. This move averages 50 cm.

   Design an improved workstation based on the principles of motion economy. Estimate the hourly savings of your improved layout if 10 million flashlights are produced.
2. Review a word processing workstation with which you are familiar. Outline those therbligs that you feel may be improved by making a micromotion study of the operator and the workstation. Develop a design of an improved workstation based on information obtained in Chapters 7 and 8.

SELECTED REFERENCES

Barnes, Ralph M. *Motion and Time Study: Design and Measurement of Work.* 7th ed. New York: John Wiley & Sons, 1980.

Kadota, Takeji. "Charting Techniques." In *Handbook of Industrial Engineering,* ed. Gavriel Salvendy. New York: John Wiley & Sons, 1982.

Knight, James L. and Gavriel Salvendy "Psychomotor Work Capabilities." In *Handbook of Industrial Engineering,* 2nd ed., edited by Gavriel Salvendy, Chap. 37. New York: John Wiley & Sons, 1992.

Mundel, Marvin E. *Motion and Time Study: Principles and Practices.* 5th ed. New York: Prentice Hall, 1978.

Human Factors Considerations

9

For the most part, operation analysis, motion study, and micromotion study have concentrated on the improvement of the workstation. The principal objectives of these techniques are to:

1. Optimize physical work.
2. Minimize the time required to perform tasks.
3. Maximize the product quality per dollar of cost.
4. Maximize the welfare of the employee from the standpoint of earnings, job security, health, and comfort.
5. Maximize the profit of the business or enterprise.

Workstation improvements utilizing these three techniques are based primarily on:

1. Newton's laws of motion.
2. The biomechanics of the human body and the limitations of employees.
3. Optimization methodologies.

A solid grasp of the fundamentals of the human factors and ergonomics approach to work improvement helps analysts to improve existing methods and more thoroughly plan projected work. The areas of study within the human factors and ergonomic approach include the physical environment of the workstation and physiological and psychological factors relating to both the operator and the work force. Chapter 5 covered some study of the physical environment. Before implementing the new or improved method resulting from operation analysis, there should be an understanding of additional important basic principles that relate to a somewhat broader spectrum of the methods engineering program.

MEASUREMENT AND CONTROL OF THE PHYSICAL ENVIRONMENT

The immediate physical environment has a significant impact not only on the operator's and supervisor's performance, but also on the reliability of the process. The principal environmental factors that influence the productivity of the working personnel and the dependability of the process include the visual environment, noise, vibration, humidity, temperature, and atmospheric contaminants.

The Visual Environment

The efficient performance of almost all tasks, whether industrial, office, business, service, or professional, depends to some degree on adequate vision. Adequate lighting is just as important for the dentist who is repairing a molar as it is for the toolmaker who is polishing contours in a mold for producing plastic components.

The principal criteria that govern the visual environment are the amount of lighting, the contrasts between the immediate surroundings and the specific task being carried out, and the presence or absence of glare.

Although experts have conducted much research regarding the amount of light required for a specific job, the exact amounts needed are still controversial. The ability to see increases as the logarithm of the illumination, so that a point is soon reached at which large increases in illumination result in very small increases in the worker's efficiency. In 1932 Lythgoe observed that the relationship between visual acuity and the logarithm of the luminance started to depart from linearity at approximately 10 footlamberts (see Figure 9–1).

The quantity of light needed to perform a task satisfactorily is affected by several independent factors. Notable among these are:

1. The contrast between the object being viewed and the immediate surroundings. Colors also have a significant impact on contrast.
2. The reflectivity of the immediate surroundings.
3. The physical dimensions of the object being viewed.
4. The viewing distance.
5. The time permitted for seeing.

Methods people have a certain amount of control over all of these factors except for the physical dimensions of the object being viewed. Even this is controlled if they plan the work so that it is exposed dimensionally in its most advantageous position from the standpoint of the operator's visual perception.

The contrast between the object being viewed and the immediate surroundings can be thought of as brightness differences and may be expressed in terms of intensity. Contrast may then be expressed as:

FIGURE 9–1

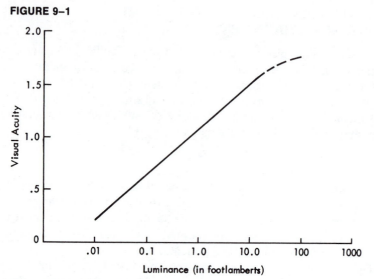

Luminance (in footlamberts)

Source: Redrawn from R. J. Lythgoe, "The Measurement of Visual Acuity." M.R.C. Special Report Series 173. London.

$$C = \frac{B_L - B_D}{B_L} \times 100$$

where:

C = Contrast in percent
B_L = Luminance of the lighter surface[1]
B_D = Luminance of the darker surface

In many work situations, a second form of contrast is evident. This is color contrast. Two or more colors within a work environment may reflect approximately the same amount of incident light, yet be significantly distinguishable because of their difference in hue. Where color is involved, analysts can enhance contrast by selecting colors which are not only widely separated on the color circle, but which also have widely different reflectivities.

The luminance of an object depends on the amount of the incident light which is reflected. Reflectivity is the percentage of the incident flux reflected by a surface. The relationship between reflectivity, luminance, and illumination can be equated as follows:

$$R = \frac{B}{E} \times 100$$

---

[1] The basic unit of luminance is the lambert, which can be defined as the brightness of a perfectly diffusing surface emitting one lumen/cm^2.

TABLE 9–1

Color or finish	Percent of reflected light	Color or finish	Percent of reflected light
White	85	Medium blue	35
Light cream	75	Dark gray	30
Light gray	75	Dark red	13
Light yellow	75	Dark brown	10
Light buff	70	Dark blue	8
Light green	65	Dark green	7
Light blue	55	Maple	42
Medium yellow	65	Satinwood	34
Medium buff	63	Walnut	16
Medium gray	55	Mahogany	12
Medium green	52		

where:

R = Reflectivity (%)

B = Luminance (quantity of reflected light in lumens per square cm)

E = Illumination (amount of light in lumens per square cm which falls on object)

Table 9–1 provides reflectance factors of typical paint and natural wood finishes.

The size of the object has a notable impact on its ability to be seen. The smaller the size, and consequently the less the visual angle, the more difficult it is to see the part. Murrell has classified small work into six categories related to an 18-inch viewing distance, as shown in Table 9–2.[2]

When an object is reasonably large (over 0.030 inch when viewed from 18 inches) and the contrast is high, the object can be seen in a very short period of time. However, if the object is quite small and the contrast is low, the time required to see the object is relatively high. It has been proved experimentally that the time required to see a small object can be increased by a factor of four if the contrast is altered by 20 percent (for example, from 70 percent to 50 percent). This is an important factor in quality control, when an inspector is searching for minute flaws in work passing by rapidly on a conveyor.

Light recommendations in footlamberts appear in Table 9–3. Footcandle requirements for different seeing tasks where the reflectance and contrast are relatively high appear in Table 9–4.

After determining the footlamberts and footcandle requirements for the area under study, analysts select appropriate artificial light sources. Two important parameters related to artificial lighting are efficiency (lumens

[2] K. F. H. Murrell, *Ergonomics* (London, England: Chapman and Hall, 1969), p. 307.

TABLE 9–2

Category	Visual angle (minutes and seconds)	Size (inches)	Example
I	Less than 23 seconds	<0.002	Fine wires (about 0.001 inch in diameter)
II	23–45 seconds	0.002–0.004	Fine scribed line or human hair
III	45 seconds– 1 minute 30 seconds	0.004–0.008	Lines on slide rule or micrometer
IV	1 minute 30 seconds– 3 minutes	0.008–0.015	Size 40 sewing cotton or thickness of 6-point print
V	3–6 minutes	0.015–0.030	Thickness of 12-point print
VI	Over 6 minutes	>0.030	Any large object

TABLE 9–3
Guide brightness for various categories of seeing at specified footcandles and reflectance

Category of seeing task	Guide brightness (footlamberts)	Footcandles for specified reflectance conditions		
		90%	50%	10%
A. Easy	Below 18	Below 20	Below 36	Below 180
B. Ordinary	18–42	20–95	36–84	180–420
C. Difficult	42–100	45–153	84–240	420–1,200
D. Very difficult	120–420	133–455	240–840	1,200–4,200
E. Most difficult	420 up	455	840	4,200

From Ernest J. McCormick, *Human Factors Engineering*, 4th ed. (New York: McGraw-Hill, 1976).

TABLE 9–4
Current light recommendations for different seeing tasks

Tasks	Footcandles in service, i.e., on task or 30 inches above floor
Most difficult seeing tasks . . . . Finest precision work involving: Finest detail; poor contrasts; long periods of time. Extratime assembly; precision grading; extrafine finishing.	200–1,000
Very difficult seeing tasks . . . . Precision work involving: Fine detail; fair contrasts; long periods of time. Fine assembly; high-speed work; fine finishing.	100
Difficult and critical seeing tasks . . . . Prolonged work involving: Fine detail; moderate contrasts; long periods of time. Ordinary benchwork and assembly; machine shop work; finishing of medium-to-fine parts; office work.	50
Ordinary seeing tasks . . . . Involving: Moderately fine detail; normal contrasts; intermittent periods of time. Automatic machine operation; rough grinding; garage work areas; switchboards; continuous processes; conference and file rooms; packing and shipping.	30
Casual seeing tasks . . . . Such as: Stairways; reception rooms; washrooms and other service areas; active storage.	10
Rough seeing tasks . . . . Such as: Hallways; corridors; passageways; inactive storage.	5

Source: International Labour Office, Geneva, Switzerland.

TABLE 9–5
Artificial light sources. The efficiency (column 2), in lumens per watt (lm/W), and color rendering (column 3) of six frequently used light sources (column 1) are indicated. Lamp life and other features are given in column 4. Color rendering is a measure of how colors appear under any of these artificial light sources compared with their color under a standard light source. Higher values for efficiency indicate better energy conservation.

Type	Efficiency (lm/W)	Color rendering	Comments
Incandescent	17–23	Good	Incandescent is a commonly used light source but is the least efficient. Lamp cost is low. Lamp life is typically less than one year.
Fluorescent	50–80	Fair to good	Efficiency and color rendering vary considerably with type of lamp: cool white, warm white, deluxe cool white. Significant energy cost reductions are possible with new energy-saving lamps and ballasts. Lamp life is typically five–eight years.
Mercury	50–55	Very poor to fair	Mercury has a very long lamp life (9–12 years), but its efficiency drops off substantially with age.
Metal halide	80–90	Fair to moderate	Color rendering is adequate for many applications. Lamp life is typically one–three years.
High pressure sodium	85–125	Fair	This lamp is a very efficient light source. Lamp life is three–six years at average burning rates, up to 12 hours per day.
Low pressure sodium	100–180	Poor	This lamp is the most efficient light source. Lamp life is four–five years at average burning rate of 12 hours per day. Mainly used for roadways and warehouse lighting.

Adapted from Lum-i-neering Associates, 1979; Ross and Baruzzini, Inc. 1975; courtesy Human Factors Section, Eastman Kodak Co.

per watt, lm/w) and color rendering. Efficiency is of a particular concern since it is related to cost. Efficiency light sources reduce energy consumption. Color rendering relates to how closely the perceived colors of the object being observed match the perceived colors of the same objects when illuminated by standard light sources. The more efficient light sources (high and low pressure sodium) have only fair to poor color rendering characteristics and consequently may not be suitable for certain inspection operations, where color discrimination is necessary. Table 9–5 provides efficiency and color rendering information for the principal types of artificial light.

THE IMPACT OF COLOR. Both color and texture have psychological effects on people. For example, yellow is the accepted color of butter; therefore, margarine must be made yellow to appeal to the appetite.

Beefsteak is another example. Cooked in 45 seconds on an electronic grill it does not appeal to customers because it lacks a seared, brown, "appetizing" surface. A special attachment had to be designed to sear the beefsteak. In a third example, employees in an air-conditioned midwestern plant complained of feeling cold, although the temperature was maintained at 72° F. When the white walls of the plant were repainted in a warm coral color, all complaints ceased. In the last instance, workers in a factory complained that boxes were too heavy until the plant engineer had the old boxes repainted a light green. The next day several workers said to the supervisor, "Say, those new lightweight boxes sure make a difference."

Perhaps the most important use of color is to improve the environmental conditions of the workers by providing more visual comfort. Analysts use colors to reduce sharp contrasts, to increase reflectance, to highlight hazards, and to call attention to features of the work environment.

Sales are conditioned by colors. People recognize a company's products instantly by the pattern of colors used on packages, trademarks, letterheads, trucks, and buildings. Some research has indicated that color preferences are influenced by nationality, location, and climate. Sales of a product formerly made in one color increased when several colors suited to the differences in customer demands were supplied. Table 9–6 illustrates the typical emotional effects and psychological significance of the principal colors. Figure 9–2 illustrates pairs of colors that give harmonious hues; the figure also shows complementary colors.

In a visual sense, texture refers to the patterns of contrasts in the light reflections that identify a surface. The influence of surface texture on customers is as significant as that of color. Marketing people have been aware for some time of the consuming public's increasing desire for mirror finishes, such as glass-fronted stores and buildings, Koroseal fabrics, and chrome finishes on appliances and automobiles.

Other finishes also create interest and inspire workers with respect for and pride in their environment. For example, hammered finishes simulate the mottled texture of hammered metals, and Rigid-Tex is rigidized metals rolled in any of a number of textured patterns and finished in a variety of colors.

Noise

From the practical analyst's point of view, noise is any unwanted sound. Sound waves originate from the vibration of some object, which in turns sets up a succession of waves of compression and expansion through the transporting medium (air, water, and so on). Thus, sound can be transmitted not only through air and liquids, but also through solids, such as machine tool structures. We know that the velocity of sound waves in air is approximately 340 m/sec (1,100 ft). In viscoelastic mate-

TABLE 9-6
The emotional and psychological significance of the principal colors

Color	Characteristics
Yellow	Has the highest visibility of any color under practically all lighting conditions. It tends to instill a feeling of freshness and dryness. It can give the sensation of wealth and glory, yet can also suggest cowardice and sickness.
Orange	Tends to combine the high visibility of yellow and the vitality and intensity characteristic of red. It attracts more attention than any other color in the spectrum. It gives a feeling of warmth, and frequently has a stimulating or cheering effect.
Red	A high-visibility color having intensity and vitality. It is the physical color associated with blood. It suggests heat, stimulation, and action.
Blue	A low-visibility color. It tends to lead the mind to thoughtfulness and deliberation. It tends to be a soothing color, although it can promote a depressed mood.
Green	A low-visibility color. It imparts a feeling of restfulness, coolness, and stability.
Purple and violet	Low-visibility colors. They are associated with pain, passion, suffering, heroism, and so on. They tend to bring a feeling of fragility, limpness, and dullness.

FIGURE 9-2
Color wheel based on the Munsell color system. Five primary and five secondary colors or hues are indicated. Pairs of adjacent colors on this wheel are known technically as harmonious hues. Colors opposite to each other are complementary.

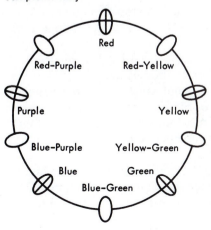

rials, such as lead and putty, sound energy is dissipated rapidly as viscous friction.

Sound can be defined in terms of the frequencies that determine its tone and quality and of the amplitudes that determine its intensity. Frequencies audible to the human ear range from approximately 20 to 20,000 cycles per second. The unit of cycles per second is commonly called Hertz, abbreviated Hz. The fundamental equation of wave propagation is: $c = \lambda f$, where c = sound velocity in m/sec; f = frequency in Hz; and λ = wave length in m. Note that as the wave length increases the frequency (as you would intuitively anticipate) decreases. Figure 9–3 shows the relationship between the frequency and the wavelength of sound in room-temperature air. Methods analysts measure sound intensity with a sound-level meter; the unit of sound intensity is the decibel (db). The greater the amplitude of the sound waves, the greater the sound pressure measured on the decibel scale.

Because of the very large range (about 10^7 to 1) of sound intensities encountered in the normal human environment, the decibel scale has been chosen. In effect, it is the logarithmic ratio of the sound intensity to the sound intensity at the threshold of hearing of a young person. Thus, the sound pressure level L_P in decibels is given by:

$$L_P = 10 \log_{10} \frac{P^2_{\text{rms}}}{P^2_{\text{ref}}} \text{ db}$$

where

P_{rms} = Root-mean-square sound pressure in microbars (or dynes/cm²)
P_{ref} = Sound pressure at the threshold of hearing of a young person at 1,000 Hz (0.0002 microbars)

Since sound-pressure levels are logarithmic quantities, the effect of the coexistence of two sound sources in one location requires that a logarithmic addition be performed. A graphic method is convenient in performing this operation. Figure 9–4 describes the method and gives examples of its application.

The A-weighted sound level used in Figure 9–5 is the most widely accepted measure of environmental noise. The A weighting recognizes that from both the psychological and the physiological points of view, the low frequencies (50–500 Hz) are far less annoying and harmful than are sounds in the critical frequency range of 1,000–4,000 Hz. Above frequencies of 10,000 Hz, hearing acuity (and, therefore, noise effects) again drops off.

The appropriate electronic network is built into sound-level meters to attenuate low and high frequencies, so that the sound-level meter can read in dbA units directly to correspond to the effect on the average human ear.

FIGURE 9–3
Relationship between the frequency and the wavelength of sound in room-temperature air

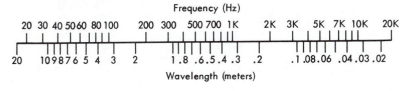

Frequency (Hz)

Wavelength (meters)

FIGURE 9–4
Chart for combining levels of uncorrelated noise signals. To add levels, enter the chart with the "numerical difference between two levels being added." Follow the line corresponding to this value to its intersection with the curved line, then left to read the "numerical difference between total and larger level." Add this value to the larger level to determine the total.

Example: Combine 75 db and 80 db. The difference is 5 db. The 5-db line intersects the curved line at 1.2 db on the vertical scale. Thus, the total value is 80 + 1.2, or 81.2 db.

To subtract levels, enter the chart with the "numerical difference between total and larger level." This value is less than 3 db. Enter the chart with the "numerical difference between total and smaller levels" if this value is between 3 and 14 db. Follow the line corresponding to this value to its intersection with the curved line, then either left or down to read the "numerical difference between total and larger [smaller] levels." Subtract this value from the total level to determine the unknown level.

Example: Subtract 81 db from 90 db. The difference is 9 db. The 9-db vertical line intersects the curved line at 0.6 db on the vertical scale. Thus, the unknown level is 90 − 0.6, or 89.4 db.

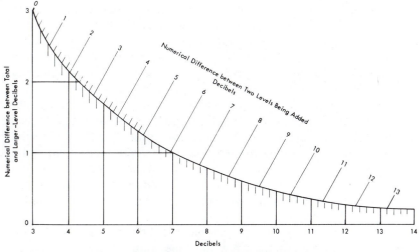

Source: This chart is based on one developed by R. Musa. Reprinted by permission from ASHRAE Handbook of Fundamentals, 1967.

FIGURE 9–5
Decibel values of typical sounds (dbA)

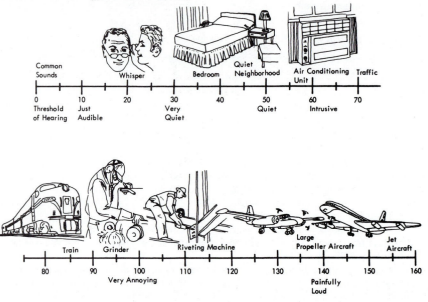

The chances of damage to the ear resulting in "conductive" deafness increase as the frequency approaches the 2,400 to 4,800 Hz range. This loss of hearing is a result of a loss of mechanical flexibility in the middle ear, so that it fails to transmit sound waves to the inner ear. Also, as the exposure time increases, especially where higher intensities are involved, there will eventually be an impairment in hearing. Nerve deafness is the result of damage in the inner ear or in the auditory nerve itself. Both conductive and nerve deafness are most commonly due to excess exposure to noise, and one of their causes is occupational noise. Individuals vary widely in their susceptibility to noise-induced deafness.

In general, we can classify noise as being of two types: broadband noise and meaningful noise. Broadband noise is made up of frequencies covering a significant part of the sound spectrum. This type of noise can be either continuous or intermittent. Meaningful noise represents distracting information that has an impact on the worker's efficiency.

In long-term situations broadband noise can result in deafness, and in day-to-day operations it can result in reduced worker efficiency and ineffective communication.

Continuous broadband noise is typical of such industries as the textile industry and an automatic screw machine shop, where the noise level does not deviate significantly during the entire working day. Intermittent

TABLE 9–7
Permissible noise exposures

Duration per day (hours)	Sound level (dA)
8	90
6	92
4	95
3	97
2	100
1.5	102
1	105
0.5	110
0.25 or less	115

Note: When the daily noise exposure is composed of two or more periods of noise exposure of different levels, their combined effect should be considered rather than the individual effects of each. If the sum of the following fractions $C_1/T_1 + C_2/T_2 \ldots C_n/T_n$ exceeds unity, then the mixed exposure should be considered to exceed the limit value. C_n indicates the total time of exposure at a specified noise level and T_n equals the total time of exposure permitted during the work day.

broadband noise is characteristic of a drop forge plant and a lumber mill. When a person is exposed to noise that exceeds the damage level, the initial effect is likely to be a temporary hearing loss from which there is complete recovery within a few hours after leaving the work environment. If repeated exposure continues over a long time, then irreversible damage to hearing can result. The effects of excessive noise depend on the total energy which the ear has received during the work period. Thus, reducing the time of exposure to excessive noise during the work shift, reduces the probability of permanent hearing impairment.

Both broadband and meaningful noise have proved to be distracting and annoying enough to result in decreased productivity and in increased employee fatigue. However, federal legislation was enacted primarily because of the possibility of permanent hearing damage. The *Federal Register,* vol. 34, no. 96 (May 20, 1969), in reference to the Walsh-Healy Act (the Occupational Safety and Health Act) on occupational noise exposure, provides the data contained in Table 9–7.

The law (50–204.10 occupational noise exposure) states:

Protection against the effects of noise exposure shall be provided when the sound levels exceed those shown in Table I [Table 9–7] of this section when measured on the A scale of a standard sound level meter at slow response. When noise levels are determined by octave band analysis, the equivalent A-weighted sound level may be determined as follows:

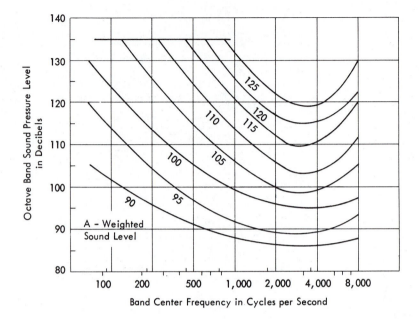

Today OSHA requires a mandatory hearing conservation program, including exposure monitoring, audiometric testing, and training for all employees who have occupational noise exposures equal to or exceeding an eight-hour time-weighted average of 85 db. Although noise levels below 85 db apparently do not cause hearing loss, they contribute to distraction and annoyance resulting in poor worker performance. For example, typical office noises, although not loud, can make it difficult to concentrate resulting in low productivity in design and other creative work. Also, the effectiveness of telephone and face-to-face communications can be considerably distracted by noise levels less than 85 db.

Management can control the noise level in three ways: The best, and usually the most difficult, way is to reduce the noise level at its source. It would be very difficult to redesign such equipment as pneumatic hammers, steam forging presses, board drop hammers, and woodworking planers and jointers so that the efficiency of the equipment would be maintained and yet the noise level would be brought into a tolerable range. In some instances, however, more quietly operating facilities may be substituted for those operating at a high noise level. For example, a hydraulic riveter may be substituted for a pneumatic riveter, electrically operated apparatus for steam-operated apparatus, and an elastomer-lined tumble barrel for an unlined barrel.

If the noise cannot be controlled at its source, then analysts should investigate the opportunity to isolate the equipment responsible for the noise. They can control the noise that emanates from a machine by housing all or a substantial portion of the facility within an insulating enclosure. This has frequently been done in connection with power presses having automatic feeds. Ambient noise can frequently be reduced by isolating the noise source from the remainder of the structure, thus preventing a sounding board effect. This can be done by mounting the facility on a shear-type elastomer, thus damping the telegraphing of noise.

In situations where enclosing the facility would not prevent operation and accessibility, the following steps can assure the most satisfactory enclosure design:

1. Clearly establish the design goals and determine the acoustical performance required of the enclosure. Establish octave band criteria at one meter (three feet) from the major machine surfaces. This distance is consistent with that recommended by NMTBA, GAGI, and so on.
2. Make actual measurements of the octave band noise levels of the equipment to be enclosed at the locations recommended in step 1.
3. Determine the accumulation of noise and then the net noise when multiple facilities are being used (see Figure 9–5).
4. Determine the required spectral attenuation of each enclosure. This is the difference between the design criteria determined in step 1 and the net noise level determined in step 3.
5. Select the acoustical panels and wall configuration for the enclosure. Table 9–8 provides several materials that are popular for relatively small enclosures. A viscolastic damping material should be applied if any of these materials (with the exception of lead) are used. This can provide an additional attenuation of 3 to 5 db.

TABLE 9–8
Octave band noise reduction of single-layer materials commonly used for enclosures

Octave band center frequency	125	250	500	1,000	2,000	4,000
16-gauge steel	15	23	31	31	35	41
7-mm steel	25	38	41	45	41	48
7-mm plywood						
0.32 kg/0.1 square meter	11	15	20	24	29	30
¾-in. plywood						
0.9 kg/0.1 square meter	19	24	27	30	33	35
14-mm gypsum board						
1 kg/0.1 square meter	14	20	30	35	38	37
7-mm fiberglass						
0.23 kg/0.1 square meter	5	15	23	24	32	33
0.2 mm						
0.45 kg/0.1 square meter	19	19	24	28	33	38
0.4-mm lead						
0.9 kg/0.1 square meter	23	24	29	33	40	43

FIGURE 9–6
Spectrum of noise before and after acoustical treatment

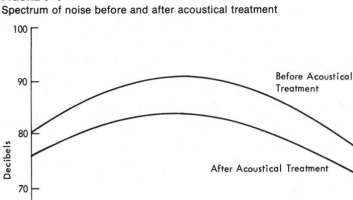

Source: Redrawn from McCormick, 1976.

If the noise cannot be reduced at its source, and if the noisemaking source cannot be acoustically isolated, then acoustic absorption can provide beneficial results. The purpose of installing acoustical materials on walls, ceilings, and floors is to reduce reverberation. Figure 9–6 illustrates the amount of noise reduction typically possible through acoustical treatment.

Lastly, the personnel in the area can wear personal protective equipment, though in most cases OSHA accepts this as only a temporary solution. Personal protective equipment can include various types of earplugs, some of which are able to attenuate noises in all frequencies up to sound-pressure levels of 110 db or more. Also available are earmuffs which attenuate noises to 125 db above 600 Hz and up to 115 db below this frequency.

WANTED SOUND.   Not all sound is undesirable in a work environment. For example, background music has been used in factories for many years to improve the work environment. The majority of production and indirect workers (maintenance, shipping, receiving, etc.) enjoy listening to music while they work. When introducing music to a work environment, first consult the employees to determine the type of music to be played and the schedule for playing. Intervals of 20 to 30 minutes on and 20 to 30 minutes off have worked out well.

Vibration

Vibration can cause detrimental effects on human performance. Vibrations of high amplitude and low frequency have especially undesirable effects on body organs and tissue. The parameters of vibration are frequency, amplitude, velocity, acceleration, and jerk. For sinusoidal vibrations, amplitude and its derivations with respect to time are:

$$\text{Amplitude } (s) = \text{Maximum displacement from static position in inches}$$

$$\text{Maximum velocity } \frac{ds}{dt} = 2\pi(s)(f) \text{ in./sec}$$

$$\text{Maximum acceleration } \frac{d^2s}{dt^2} = 4\pi^2(s)(f^2) \text{ in./sec}^2$$

$$\text{Maximum jerk } \frac{d^3s}{dt^3} = 8\pi^3(s)(f^3) \text{ in./sec}^3$$

where:

f = Frequency
s = Displacement amplitude

Displacement and maximum acceleration are the principal parameters used to characterize the intensity of vibration.

There are three classifications of exposure to vibration:

1. Circumstances in which the whole or a major portion of the body surface is affected; for example, when high-intensity sound in air or water excites vibration.
2. Cases in which vibrations are transmitted to the body through a supporting area; for example, through the buttocks of a woman driving a truck or through the feet of a man standing by a shakeout facility in a foundry.
3. Instances in which vibrations are applied to a localized body area, for example, to the hand when operating a power tool.

Of these three classifications, the second is of the most concern. This category has the greatest effects on working efficiency and on the health, safety, and comfort of the working force.

Research in this area has suggested that the most sensitive frequency ranges are 4 to 8 Hz for vertical vibration and below 2 Hz for horizontal vibration. Various parts of the body resonate at specific frequencies, causing disturbances. Human tolerance of vibration decreases as the exposure time increases. Thus, the tolerable acceleration level increases with decreasing exposure time. Low-frequency, high-amplitude vibrations are the principal cause of the motion sickness that some people experience in sea and air travel. Workers experience fatigue much more rapidly when they are exposed to vibrations in the range from 1 to 250 Hz. Early symptoms of vibration fatigue are headache, loss of appetite, and loss of interest. Vibrations experienced in this range are often characteristic of the trucking industry. The vertical vibrations of many rubber-tired trucks when traveling at typical speeds over ordinary roads range from 3 to about 7 Hz.

Management can protect employees against vibration in several ways: The applied forces responsible for initiating the vibration may be reduced by modifying the speed, feed, or motion, and by properly maintaining the equipment by balancing and/or replacing worn parts. Analysts can place equipment on antivibration mountings (springs, shear type elastomers, compression pads) or alter workers' body positions to lessen the disturbing vibratory forces. Then too, they can reduce the time workers are exposed to the vibration by alternating work assignments within a group of employees. Lastly, they can introduce supports that cushion the body and thus damp higher amplitude vibrations. Seat suspension systems involving hydraulic shock absorbers, coil or leaf springs, rubber shear-type mountings, or torsion bars may be used. In standing operations, a soft elastomer floor mat usually proves helpful.

Thermal Conditions

Although we are able to function within a broad range of thermal conditions, performance deviates substantially if we are subjected to temperatures that vary from those characterized as "normal." When considering the temperature of the work environment, analysts should realize the following:

1. The environmental temperature is the temperature actually experienced by the man or woman in a given environment. This temperature is the sum of the convective heat exchange; conduction from hot or cold floors or tools; the radiation exchange from the walls, floor, and ceiling; and any solar radiation that is transmitted or reflected to the occupant through the transparent areas of the work environment.

2. Effective temperature is an experimentally determined index, which includes the temperature, air movement, and humidity. The normal range is from 18.3° C (65° F) to 22.8° C (73° F), with a relative humidity of 20 to 60 percent (see Figure 9–7).

The normal range of the effective temperature has also been referred to as the thermal comfort zone.[3] Here temperatures of 66° F and 79° F are recommended "as the outer limits for thermostatic regulation in areas where sedentary or light work is done" (see Figure 9–7). Of course, the work load, clothing worn by the worker, and the radiant heat load affect

[3] S. H. Rodgers, *The Thermal Comfort Zone* (Eastman Kodak Company, Human Factors Section; New York: Van Nostrand Reinhold, 1983).

FIGURE 9–7
The thermal comfort zone. The dry bulb temperature and humidity
combinations that are comfortable for most people doing sedentary or
light work are shown as the shaded area on the psychometric chart.
The dry bulb temperature range is from 19° to 26°C (66° to 79°F), and
relative humidities (shown as parallel curves) range from 20 to 85
percent, with 35 to 65 percent being the most common values in the
comfort zone. On this chart ambient dry bulb temperature (A) is plotted
on the horizontal axis and indicated as parallel vertical lines; water
vapor pressure (B) is on the vertical axis. Wet bulb temperatures (C) are
shown as parallel lines with a negative slope; they intersect the dry
bulb temperature lines and relative humidity curves (D) on the chart. In
the definition of the thermal comfort zone, assumptions were made
about the work load, air velocity, radiant heat, and clothing insulation
levels. These assumptions are given in the top left corner of the chart.

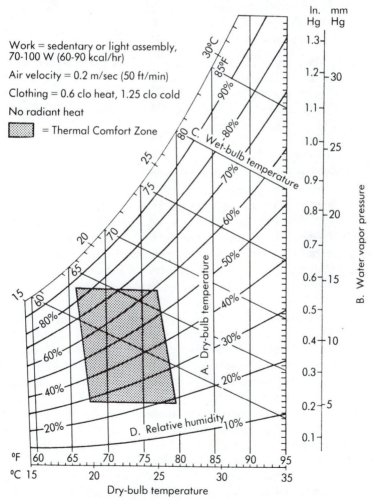

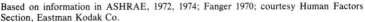

Based on information in ASHRAE, 1972, 1974; Fanger 1970; courtesy Human Factors
Section, Eastman Kodak Co.

that individual's sense of comfort within the normal range of the effective temperature or the comfort zone.

3. The operative temperature is the employee's body temperature which is determined by the cumulative effects of all heat sources and sinks. For an individual to maintain a comfortable skin temperature of approximately 32° C (89.6° F), a convective heat removal commensurate with the operative temperature needs to be effected. This may be expressed as:

$$E = M \pm R \pm C$$

where:

E = Heat loss by evaporation
M = Heat gain from metabolism
R = Heat exchange by radiation
C = Heat exchange by convection

If the environment is of the hot-dry type where the principal source of heat is radiant, then the problem is fundamentally one of heat gain, and it is solved by reducing the heat load. If the environment is hot-moist, the problem centers on heat loss, which is best solved by ventilation and by the dehydration of the ambient air.

The maximum rise in body temperature should be about 1° C or 2° F. Conditions that cause greater change than this can result in heat stress.

To estimate the length of time that a person can be exposed to a particular working environment, estimate or measure the heat load. Measurements can be made on both the environment and the person. When taking measurements of the person, we use one or more of three factors: heart rate, oxygen consumption, and the body temperature.

The heart rate increases with an increase in body temperature resulting from the presence of heat while a particular job is being done. This differential in the heart rate is a measure of the heat load on the employee. As a general rule, the heat load should not cause an increase in the pulse rate of more than 15 beats per minute. A related benchmark is that the combination of the work load and the environment should not cause an increase in the pulse rate of more than 45 beats per minute above the normal rest pulse count. Obviously, conditioning and accommodation to an environment can make a great difference in the effects of the work load and the environment on an employee.

Measures of oxygen consumption alone do not provide a satifactory estimate of the heat load. However, the normal relationship between the heart rate and oxygen consumption cannot be maintained when work is performed in the presence of excess heat, and this variation can be used in an estimation of the heat load.

Body temperature, which is normally approximately 37.0° C (98.6° F), should not be allowed to rise to as much as 38.0° C (100.4° F). Allowing body temperatures to rise to this level can result in the physical collapse of the worker.

In measuring the variables that affect the environment, we are concerned with the dry-bulb temperature, humidity, air velocity, and radiation from the immediate surroundings.

Dry-bulb temperatures are usually measured with an ordinary liquid-in-glass thermometer. Thermocouples, thermistors, and resistance thermometers may also be used. To obtain reliable results, the sensing element must be shielded against radiation from all nearby surfaces that are significantly hotter or colder than the surrounding air.

The instrument most commonly used to determine the humidity of the air is probably the sling psychrometer. This instrument whirls a wet-bulb and a dry-bulb thermometer simultaneously. The whirling is interrupted at intervals to read the thermometers, and is continued until both thermometer readings stabilize. Then, by use of a table, the relative humidity can be established.

If the environment in which the relative humidity is being measured is subject to considerable radiant heat, a properly shielded and aspirated psychrometer should be used instead of the sling psychrometer.

A globe thermometer is used to measure the mean radiant temperature. In this instrument, a thermocouple, thermistor, or thermometer bulb is fixed at the center of a six-inch hollow copper sphere painted on the outside and inside with a matte black finish. Both the air temperature and the air velocity around the globe must be determined to compute the mean radiant temperature, which may be approximated by the following equation:

$$MRT = t_g + cv(t_g - t_a)$$

where:

MRT = Mean radiant temperature (F)
c = Convention coefficient = 0.17 for a 6-in. globe
v = Air velocity at the globe (ft./min.)
t_g = Globe temperature (F)
t_a = Ambient air temperature (F)

A period of from 15 to 25 minutes is required for the air inside the globe to reach equilibrium.

Thermal anemometers of the heated-thermocouple, hot-wire, or heated-thermometer types are best suited for the measurement of the low-velocity, random-direction air movements characteristic of most work centers.

An index has been suggested that combines the effects of temperature, humidity, radiant heat, and air movement. The index can be computed by combining dry-bulb temperatures with wet-bulb temperatures and conventional dry-globe temperature as follows:

$$WBGT_i = 0.7(WBT) + 0.3(GBT)$$
$$WBGT_o = 0.7(WBT) + 0.2(GBT) + 0.1(DBT)$$

where:

$WBGT_i$ = Wet bulb-globe thermometer for indoor use
$WBGT_o$ = Wet bulb-globe thermometer for outdoor use
WBT = Wet-bulb temperature
GBT = Dry-globe temperature
DBT = Dry-bulb temperature

Although this index has not received wide industrial use, it can provide a measure of the thermal conditions related to a work center. When the $WBGT_o$ value reaches a 82° F to 85° F level, the U.S. Marine Corps posts a green flag, thus alerting personnel to the possible curtailment of drill. A similar value at a workstation suggests a need for improvement.

Another index is the effective temperature, which has already been defined. This can be estimated by the use of a nomogram originally developed by a research team within the American Society of Heating, Refrigerating, and Air-Conditioning Engineers, Inc. (see Figure 9–8).

PROTECTION FROM HEAT.   Many industrial activities involve exposure to intense heat from which the worker needs protection. Typical examples include the hot forging of large work, tending a furnace for the production of glass or steel, and tapping a cupola in a foundry. For workers engaged in some of these activities, an air-conditioned enclosure with suitable windows assures protection while permitting them to perform effectively. For example, operators of overhead cranes, mechanical equipment, and the like can be housed in such air-conditioned cubicles.

If the worker needs to get exceptionally close to a source of radiant heat, personal protective equipment is needed. Air-conditioned suits are now available. This type of clothing is usually aluminized on the exterior for further protection. The fabric is light in weight and allows an envelope of circulating cool air. Under less severe thermal exposure, gloves, protective clothing, and a face covering may prove adequate.

LOW-TEMPERATURE EFFECTS ON PERFORMANCE.   Relatively few assignments in modern business and industry require the personnel involved to be exposed to cold environments for long periods of time. The principal jobs that do result in such exposure are outdoor jobs as winter work in the building trades and the police force, and work in cold storage warehouses, such as those for meat and other perishable commodities.

Research has indicated that performance deteriorates as temperature declines. One study indicated a fall of approximately 40 percent in performance as temperature declined from 30° F to −40° F. For the operator to maintain thermal balance under low-temperature conditions, there must be a close relationship between the operator's physical activity (heat production) and the body insulation provided by protective clothing. Fig-

FIGURE 9–8
Chart for determining the effective temperature for fully clothed worker performing light work from measurements of the dry-bulb temperature, wet-bulb temperature, and air velocity. To use the chart, draw a line (for example, A–B) through the measured dry-bulb and wet-bulb temperatures. Read the effective temperature or velocity at the desired intersections.

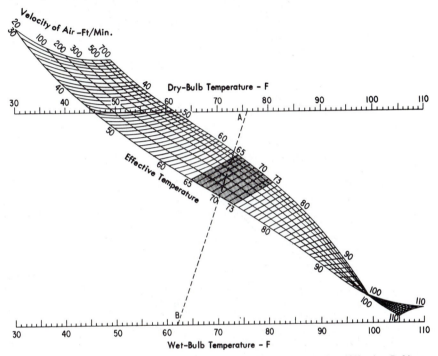

Source: Redrawn from data provided through the courtesy of the American Society of Heating, Refrigerating, and Air-Conditioning Engineers, Inc.

ure 9–9 illustrates this relationship. Here, a clo unit represents the insulation necessary to maintain comfort for a person sitting where the relative humidity is 50 percent, the air movement 20 feet/minute, and the environmental temperature 21° C (70° F).

Radiation

Although all types of ionizing radiation can damage tissue, beta and alpha radiation are so easy to shield that most attention is given today to gamma ray, X ray, and neutron radiation. High-energy electron beams impinging on metal in vacuum equipment can produce very penetrating X rays that may require much more shielding than does the electron beam itself.

FIGURE 9–9

Prediction of the total insulation required as a function of the ambient temperature
(M = Heat production in cal/m²/hr)

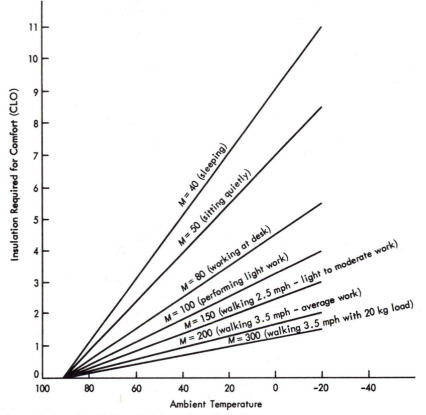

Source: Redrawn from Belding, H. S. Index for evaluating heat stress in terms of physiological strains.
Heat, Pipe, and Air Cond., 27, 129.

Absorbed dose is the amount of energy imparted by ionizing radiation
to a given mass of material. The unit of absorbed dose is the *rad,* which is
equivalent to the absorption of 0.01 joules per kilogram (100 ergs per
gram). The SI unit for absorbed dose is the *gray* (Gy) which is equivalent
to 1 joule/kilogram. Dose equivalent is a way of correcting for the differences in the biological effect on humans of various types of ionizing
radiation. The unit of dose equivalent is called the *rem,* which produces a
biological effect essentially the same as that of one rad of absorbed dose
of X or gamma radiation. The SI unit for dose equivalent is the *Sievert*
(Sv) which is equal to 100 rem. The *roentgen* (R) is a unit of exposure
which measures the amount of ionization produced in air by X or gamma
radiation. Tissue located at a point where the exposure is one roentgen

receive an absorbed dose of approximately one rad. In the international system, exposure is given in terms of coulombs of charge produced by X or gamma radiation per kilogram of air.

Very large doses of ionizing radiation—100 rads or more—received over a short time span by the entire body can cause radiation sickness. An absorbed dose of about 400 rads to the whole body would be fatal to approximately one half of a large group of adults. Small doses received over a longer period of time may increase the probability of contracting various types of cancers or other diseases. The overall risk of a fatal cancer from a radiation dose equivalent of one rem is about 10^{-4}; that is, a person receiving a dose equivalent of one rem has about 1 chance in 10,000 of dying from a cancer produced by the radiation. The risk can also be expressed as the expectation of one fatal cancer in a group of 10,000 persons, if each person receives a dose equivalent of one rem.

Persons working in areas where access is controlled for the purpose of radiation protection are generally limited to a dose equivalent of five rem per year. The limit in uncontrolled areas is usually $\frac{1}{10}$ of that value. Working within these limits should have no significant effect on the health of the individuals involved. All persons are exposed to radiation from naturally occurring radioisotopes in the body, to cosmic radiation, and to radiation emitted from the earth and building materials. The dose equivalent from natural background sources is about 0.1 rem (100 millirem) per year.

FUNDAMENTALS OF WORK PHYSIOLOGY

To design a workstation that will result in high productivity over a period of time during which different workers are engaged, analysts need a background in the fundamentals of work physiology. In many instances, work personnel differ in significant respects, such as age, sex, background, physical and mental characteristics, and health.

Motor Fitness, Reaction Time, and Visual Capacity

The motor fitness elements of strength, endurance, speed of movement, and reaching distance, coupled with visual capacity and the speed and accuracy of response to events, have a significant collective impact on both the rate of productivity and the total productivity over time of most manual operations.

Three factors influence the accuracy of control movements: the number of muscle fibers controlled by each motor nerve ending being utilized, the position of the body members, and neural stimuli. Arms have considerably more motor nerve endings and, consequently, much greater accuracy of control than do the legs. Also, the closer the position of a limb to the body, the more accurately it can be moved. Consequently, controls

FIGURE 9–10
Effective visual areas for the average employee

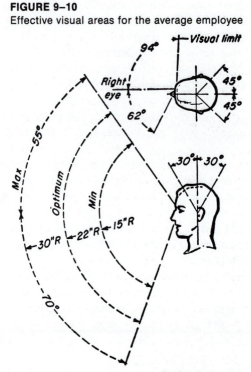

Product Engineering

that are operated by a worker's hands should be located so that it is not necessary for the worker to extend his or her arms to manipulate them.

The average person normally moves the head to the right or left only about 55 degrees. When eye movement is added, the total angle is substantially larger, averaging somewhere between 90 and 100 degrees for the normal person (see Figure 9–10).

The shoulder joint (ball and socket) permits more range of movement than does any other joint in the human body. Also, the shoulder is capable of applying about 30 percent more force than the elbow. The torque which can be applied by right-handed people is about 20 percent greater for the right hand than for the left hand, and is at a maximum when the elbow is at a right angle. Similarly, position influences the maximum weight that can be lifted. For example, if the lifting movement is performed with the palm of the hand facing upward (supine), the typical male adult will lift about 55 pounds. However, if the palm of the hand is palm downward (prone), the load that can be handled will be about 30 percent smaller.

Although there is little difference in the magnitude of the force that the typical operator can exert by either pulling or pushing, the maximum force capable of being exerted decreases as the hands are brought nearer to the body. Also, this maximum push or pull can be achieved over only about three inches of travel.

The usual method of applying force with either leg is by pushing. A worker seated with a satisfactory backrest can produce forces up to about 300 kilograms for thrusts of short duration in the horizontal plane.

Response time is another important ingredient of overall performance. Generally, response time is made up of:

1. The time involved in sensing a signal.
2. The time for the decision process as to the nature of the response.
3. The time to perform the physical movement.

Reaction time, which is a combination of steps 1 and 2 above, is affected by several factors. The size and clarity of the signal affects the time needed to sense a signal; the larger the signal, the faster the reaction time up to some particular point. Also, the reaction time for a visual signal is faster when the signal is observed by the center or near the center of the eye rather than by the periphery of the eye. The type of stimuli—whether visual, auditory, or tactile—also affects reaction time.

The different body members have different response times. The right hand (in right-handed people) has the fastest response time, followed by the left hand, the right foot, and the left foot. The average response time increases with age and is longer for women than for men, though individual differences in these respects are great.

Like response time, reading speed varies greatly for different individuals. An important factor in reading speed is eye fixation. To increase reading speed, one must decrease the number of fixations or the time of each fixation. Typically, a line of print 100 millimeters long requires about six eye fixations if one has had training comparable to that of the university graduate. A youngster in first grade averages about 18 fixations per line.

The number of fixations varies with the number of difficulties encountered. Thus, more eye fixations take place when more unfamiliar words appear in the reading material. Color has little effect on reading speed; however, contrast has a great effect. If the brightness contrast is low, then reading speed is slow.

Memory

The storage capacity of the human memory system is between 10^8 and 10^{15} bits. The memory of humans appears to be of two types, which can be classified as static and dynamic. In the static, or long-term, memory we store relevant information, which we call upon and use from time to time. For example, the value of π, the sine of 30 degrees, and the modulus of elasticity of steel are typical values that most engineers maintain in their static memory bank. In the dynamic, or short-term memory we store information or data needed for immediate use. For example, a telephone number looked up just prior to dialing represents data stored in the dynamic memory. After the data is used, it is no longer retained.

There is considerable variation in the memory or recall capacity of different people. This variation is characteristic of both static and dynamic memory. Memory—in particular, dynamic memory—is influenced by age. As an individual becomes older, short- and long-term memories tend to decline.

Physiological Fatigue

Everyone is familiar with the effects of physiological fatigue. When brush painting a ceiling, the arm held over the head soon tires because of the insufficient flow of blood into the contracted arm muscle. One must stop periodically to relax the muscle and allow the flow of blood through it.

The oxygen used by the body in doing work comes either from the blood or from the chemical compounds within the muscle fibers. If one's ability to supply oxygen to working muscles is sufficient to prevent a buildup of the by-products of metabolism in the body during the course of the workday, the job assignment is referred to as "aerobic." If the job assignment is such that performing it depletes the reserve of oxygen in a muscle or muscles, the assignment is referred to as "anaerobic." The symptoms of this oxygen depletion are muscle pain and an accompanying physiological fatigue or muscle weakness.

It has been estimated that the basal metabolic rate averages 1,700 calories per day. This represents the number of calories needed to maintain the body in an inactive state. Thus, a worker requires additional calories to handle the duties and responsibilities that go with his or her job. Otherwise, a portion of the work is necessarily anaerobic. Some estimates of human energy expenditures are shown in Table 9–9. Anaero-

TABLE 9–9
Energy expenditure and the corresponding oxygen consumption

Level of work	Energy expenditure (calories per minute)	Oxygen consumption (liters per minute)
Unduly heavy (carrying mortar upstairs at 54 feet per minute)	Over 12.5	Over 2.5
Very heavy (shoveling charge into furnace)	Over 10.0	Over 2.0
Heavy (carrying load upstairs at 27 feet per minute)	Over 7.5	Over 1.5
Moderate (wheelbarrowing sand at 220 feet per minute)	Over 5.0	Over 1.0
Light (light assembly work)	Over 2.5	Over 0.5

FIGURE 9–11
Seating depends on reach, vision, and the work load on the operator
(the dimensions shown are based on the small man, which will
accommodate the reach limitations of 95 percent of all men)

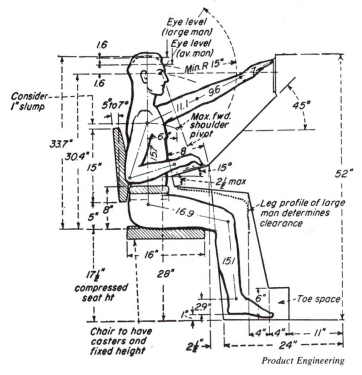

Product Engineering

bic work is related more to the velocity of movement than to the duration of the work. Work of long duration exhausts energy supplies in the muscles (glycogen) rather than causing the operator to suffer from lack of oxygen.

Individual Differences

Human beings are variable in their performance. This variation represents one of the most important considerations in the design of worker and machine systems. Not only is there considerable difference between the performance of different individuals, but also the same individual can vary from moment to moment, from one period of the day to the next, and from day to day. Even when carrying out the simplest functions, the individual varies considerably in his or her performance.

The variations in performance among members of the work force are due in large part to variations in both static and dynamic anthropometry. Figures 9–11 and 9–12 illustrate some of the principal static body dimensions.

FIGURE 9–12

The dimensions of the average adult male based on the 2½ to
97½ percentile range of the subjects measured

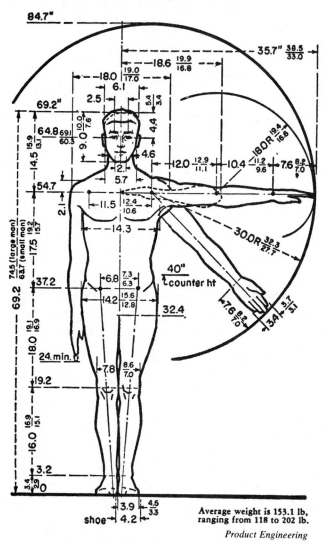

Average weight is 153.1 lb,
ranging from 118 to 202 lb.

Product Engineering

    Then, too, body dimensions vary with age. From birth to about the
middle 20s, most static body dimensions increase. From about the age of
60 on, there is a slight decrease in height and body weight, and often in the
dynamic dimensions, such as functional arm reach. Table 9–10 provides
the mean height and weight, and the standard deviations therefrom, of
white American men and women at various ages. These data demonstrate
the variability of human body dimensions in the work force.

TABLE 9-10

	Male				Female			
	Height (inches)		Weight (pounds)		Height (inches)		Weight (pounds)	
Age	Mean	Standard deviation	Mean	Standard deviation	Mean	Standard deviation	Mean	Standard deviation
3	37.8	1.3	32	3	37.5	1.4	31	4
6	46.1	2.1	47	6	45.7	1.9	45	5
9	52.8	2.4	66	8	52.1	2.3	64	11
12	58.3	2.9	87	12	59.6	2.7	93	18
15	66.3	3.1	128	16	63.4	2.4	117	20
18	68.5	2.6	149	20	64.1	2.3	123	17
22	68.7	2.6	158	23	64.0	2.4	125	19
26	68.7	2.6	163	24	63.7	2.5	127	21
30	68.5	2.6	165	25	63.6	2.4	130	24
34	68.5	2.6	165	25	63.6	2.4	130	24
38	68.4	2.6	166	25	63.4	2.4	136	25
48	68.0	2.6	167	25	63.2	2.4	142	27
58	67.3	2.6	165	25	62.8	2.4	148	28
68	66.8	2.4	162	24	62.2	2.4	146	28
78	66.5	2.2	157	24	61.8	2.2	144	27

Source: Adapted from Stoudt et al., 1960.

Both age and sex have a bearing on the time required for response. Response time differences among people increase as the work becomes more precise, more rushed, or more difficult to perform. Also, as the work environment becomes more adverse, differences in reaction time among the various workers increase.

Women have a slightly longer mean reaction time to both light and sound than do men (see Figure 9–13). Also, the reaction time of both men and women increases with age after they are about 30 years old.

The differences among people as to visual capacity and memory have already been discussed. Similarly, there is considerable variation in the muscle strength and motor skills of the work force. Strength usually reaches a maximum at about age 25. After around age 30, strength declines until, at age 60, it is only about three fourths of what it was at maximum. Not only is there considerable variation in the strength of a given worker due to biological, environmental, and occupational reasons, but also there is considerable variation among workers. For example, the standard deviation of the maximum force that can be exerted by a two-handed vertical pull on a horizontal bar 28 inches above floor level by male workers is about 68 pounds. Among female workers, the standard deviation for this backlift force is about 40 pounds.

Because of the wide variability in the performance of the population as a whole, task forces should be based on the research results obtained from

FIGURE 9–13
Mean reaction time to light and sound for men and
women at various ages

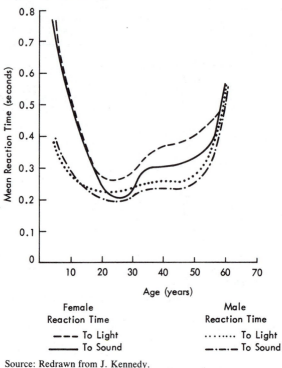

Female
Reaction Time

——— To Light
———— To Sound

Male
Reaction Time

········ To Light
—·—·— To Sound

Source: Redrawn from J. Kennedy.

subjects taken from a population representative of the particular job under study.

Today United States' industrial population includes men and women between the ages of 17 and 70. Needless to say, people in this range have many functional losses from developed disabilities that have a negative effect on their capacity for work. Congenital disorders also can take their toll on expected output based on a workplace design for young healthy men or women.

The Work Regimen

Today, the eight-hour day and the five-day week are generally considered standard in American business and industry. However, some interesting experiments are being made with the 10-hour day and the four-day week. Reports from companies experimenting with this type of schedule are not consistent. Some reports state that 10 hours is too long for an employee to work in one day, and that the added productivity of the last two hours is considerably less, proportionately, than the productivity experienced over the first eight hours. Other objections to the 10-hour day, four-day week, stem from members of management who state that they are obliged to be on the job not only 10 hours for four days but for at least eight hours on the fifth day. Other reports indicate satisfaction with having the plant closed three days every week.

Perhaps the principal proponent of the 10-hour day and four day week are those plants and businesses having relatively long set-up times. For example, heat treating, forging, and melting facilities require a significant amount of time, up to 15 percent of the eight-hour working day or more, to bring the facility and material up to the required temperature before production can begin. By going to a 10-hour day, an additional two hours of production time can be gained with no additional set-up time. Here the economic savings from the longer work day can be significant.

I have also had good experience with some workers who are enthusiastic about working two hours extra for four days in exchange for a three-day weekend.

Experience has proved that the typical worker today responds well, both psychologically and physiologically, to the 40-hour workweek, provided that the environmental conditions are satisfactory and that he or she receives adequate recognition (both monetary and nonmonetary). Thus, the 40-hour, or approximately 40-hour, week continues to be the basis of the work schedule in most U.S. industries as it will many years into the future.

In most businesses, industries, and service organizations, the vast majority of work may be classified as "light" from a physiological standpoint. Of course, some heavy physical work is done by the labor force. Typical heavy work exists in many of our coal mines, in bench molding in the foundry industry, and in connection with "pick and shovel" work related to building and highway construction.

After any activity has continued for a period of time, workers feel the need to take a short break from the routine. When this cessation in work is not taken, there is a pronounced and progressive falloff in productivity, even from highly motivated employees, until there is a forced break. This forced break could be a lunch period or a work stoppage due to equipment difficulty. The falloff in productivity can take place in two ways: first, the cycle time may increase, and second, there may be more rejects or poorer quality due to human errors.

That period of time up to the point at which productivity begins to decline markedly is known as the actile period. At the end of this period, a rest break should take place. The length of the actile period depends on both the work and the worker. Typically, the length of this period in most light work is about one hour. The minimum break that provides satisfactory recovery is usually about five minutes. Most workers take this break periodically (about three times in the morning and three times in the afternoon), whether or not it is scheduled. The results of the periodic break are almost always positive. Workers taking regularly scheduled breaks produce more and have fewer rejects than those forcing themselves to work continuously for four hours in the morning and four hours in the afternoon.

It is general practice to schedule a department or plant shutdown for 10 to 15 minutes in the middle of the morning and a similar work break in the middle of the afternoon. These forced breaks in the workday help assure that employees do not overextend the actile period. Employees do not really know the length of their actile periods, and often they continue to work for a period that exceeds it, to the detriment of their productivity. Conscientious employees, in particular, overextend their actile periods.

In the development of work standards for light work, it is customary to provide at least a 10 percent allowance for personal delays and fatigue. This usually amounts to 24 minutes in the morning and a like amount of time in the afternoon. Alert employees utilize this allowed time periodically, say, eight minutes after every hour's work, to avoid the decline in productivity that takes places at the end of the actile period.

In the performance of heavy work, the actile period is shorter because of muscular exertion. Thus, for optimum productivity, the operator needs to take more breaks during the day. For recovery, the length of the break seldom needs to be more than about five minutes, as is also the case for light work.

This recent case history illustrates the impact of the actile period: Analysts made a 16-day study of capsule sorting involving three different sorting machines. Four operators were randomly assigned to the different machines, so that each operator spent a total of four days on each of the three machines. The information recorded for each sorted can of capsules included: the time used to sort, amount of scag removed, number of capsules per can, time of day, number of good capsules sorted out as scag, capsule size, inspection sample size, and transparency. They computed efficiency scores for cans of work as well as for each total day of work. They based these efficiency scores on the ratio of the actual performance to the expected performance, based on work measurement techniques. The overall average daily efficiency was 95.9 percent. The overall average of the daily efficiencies adjusted for good capsules inadvertently sorted out as scag was 86.3 percent. The overall average number of good capsules sorted out as scag per day was 5,377 per machine. The overall average number of good capsules sorted and passed through inspection per day was 479,498 per machine.

Figure 9–14 illustrates the effect of monotony and boredom on the average performance of all the operators. It also illustrates the immediate increase in efficiency that occurs after rest periods. Figure 9–15 compares the effects of monotony and boredom on the operation of the three different machines.

FIGURE 9–14
A typical relationship between worker efficiency and the time of day under daywork
operation

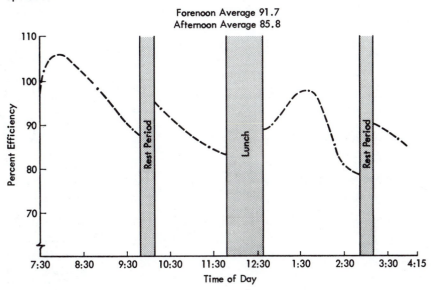

FIGURE 9–15
The effects of monotony and boredom on the operation of three different designs of
machines

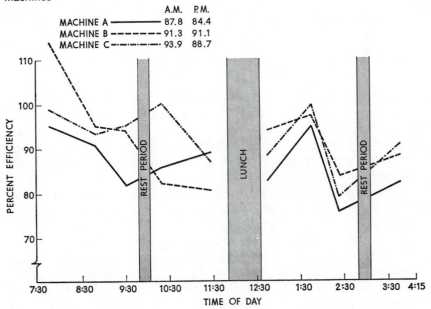

FIGURE 9–16

Operator differences among four typical female employees (the connecting lines indicate significant differences)

	Helen	Joy	Judy	Mary
Overall Efficiency for 16 Days	95.2	90.5	87.1	108.1
Average of Daily Efficiencies	94.6	90.7	87.8	108.9
Average of Daily Efficiencies—Adjusted	92.0	82.9	81.1	93.0
Average Daily Number of Good Capsules in Scag	1,900	4,700	4,700	10,900
Average Daily Number of Capsules Passed	554,700	501,200	445,000	418,100
Years of Sorting Experience	19	14	1	2
Age	48	37	20	21
Wear Eyeglasses	Yes	No	No	Sometimes
Preferred Hand	RH	RH	RH	RH

This study revealed significant differences between machines with fluorescent lighting and those with filament lighting. This study also revealed that boredom or vigilance factors associated with the time of day are apparently a normal (or expected) human fatigue characteristic. The decline in performance, regardless of the machine type used, suggests this conclusion.

Although all four operators were similar in physical capabilities, and all had adequate experience with the work, there were significant differences among them, as is shown in Figure 9–16.

The results of the study indicated that the greatest improvements in work of this type can be made by providing: (1) more frequent rest periods of shorter duration; (2) workstations and work procedures that give the operator maximum opportunity to see imperfections throughout the day; and (3) test procedures that help select people who are more endowed

FIGURE 9–17
The relationship between a well-conditioned operator and a
not-so-well-conditioned operator as to capacity to perform work on a given
task

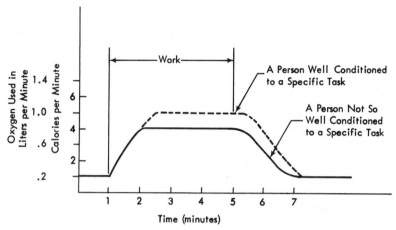

with the visual ability and the mental qualifications that minimize the effect of boredom. Rotating sorting operators to different types of sorting stations, or perhaps regularly rotating these operators with operators having entirely different work requirements, might help reduce the effects of monotony or boredom.

For light and moderately heavy work as well as office work including administration, the current recommendation is a rest break of 10 to 15 minutes both in the morning and the afternoon. Where heavy work is being performed, obligatory rest periods should be distributed throughout the eight hours of the work shift.

When performing heavy work, workers should be so conditioned that they expend about five calories per minute throughout the workday when intermingled with appropriate rest breaks. Employees not capable of expending energy at this moderate rate will need more frequent rest breaks, and overall productivity will be limited. Figure 9–17 illustrates the well-conditioned and not-so-well-conditioned worker from the standpoint of the ability to use up calories, or the capacity to perform work. Conditioning reflects the following differences among individuals:

1. Muscle tone: the degree to which the muscle reflexes are conditioned to a specific task.
2. Endurance: the degree to which fuel is made available, fuel is stored, and oxygen is made abundant, owing to a more adequate circulation of blood through active muscle.
3. Transmission: the facility of transmission of nerve impulses across the motor end plate in the muscle fiber.

4. Anaerobic efficiency: the efficiency (the ratio of the work done, in calories, to the net energy used, in calories) of the body during very heavy work.
5. Aerobic efficiency: the efficiency of the body during moderate work, where oxygen intake and demand are balanced.
6. Body health: the degree to which the physiological processes function normally.
7. Physical fitness: the degree of ability to execute a specific physical task under specific ambient conditions; long-term ability to perform (no aches, pains, strains, excessive tiredness, or hangover).

These factors are interrelated and affect one another. They cannot be thought of as completely separate entities.

Human capacity to perform a specific task is not determined by efficiency alone, but by the combination of all the physiological processes (and some psychological processes) and their interaction with one another. The degree of training and practice influences the conditioning of a worker for a given assignment, and the importance of this should be recognized in defining the production requirements of the job. Thus, operators who continually work on an assignment involving heavy muscular demands become physically conditioned in preparation for this class of work. Their output is considerably greater than that of operators who perform the same operation at infrequent intervals and who, consequently, have not been conditioned for this type of work.

BEHAVIORAL CONCEPTS. The now-famous Hawthorne studies, published in 1939, were among the earliest studies to point out that social behavior is a reality of life that is continually taking place in every work environment. The Hawthorne studies demonstrated that, to a certain extent, the behavior of people in the work environment is conditioned by their social needs. The studies also clearly delineated the effect of informal leaders on the behavior of work groups as distinct from formally assigned leaders.

Other studies have identified a hierarchy of needs that typically apply to all people. These needs, in preferential order, are needs at the lowest level involving the basic necessities of life; then security; then the need for belonging to a group; then the need for status and esteem in some social system; and finally the need for personal self-actualization. The notion is that one tends to progress up this hierarchy of needs. As one's needs are filled in any of the more basic areas, the higher level needs are evoked, and the individual's behavior can be explained less by the satisfaction of lower level needs. Certainly, this idea of a hierarchy of needs seems to apply today. Until one's needs are satisfied at lower, more primitive levels, one does not move up the scale to more sophisticated requirements.

To apply these behavioral concepts in the effective operation of business and industry, work must be designed so that the principles of operation analysis incorporate the fundamental psychological and behavioral concepts of the work force. In the design of the job, there sometimes needs to be an enlargement of the work in certain situations. This enlargement may be characterized by multijob assignments, diversity and flexibility, or worker control. Perhaps as much as 15 to 20 percent of the jobs in industry would benefit through training personnel to perform a broader spectrum of the functions. This broadening provides an overall perception of the whole work process within an organization and some overall picture of that particular organization's goals.

Just as a number of jobs would benefit substantially through broadening, perhaps as many as 80 percent should not be broadened or enlarged. In other words, from the standpoint of the job itself, most of the work force is satisfied. What is needed in the majority of work situations is a clear concept by the entire work force of what the organization's goals are and of how the individual operator's efforts can help fulfill those goals. This communication from management to each and every individual worker can go a long way toward fulfilling the hierarchy of needs of all employees. For long-term success, however, careful placement and advancement procedures must still be practiced.

SAFETY AND HEALTH CONCEPTS

Certainly, one of the objectives of any progressive management team is to provide a safe and healthful workplace for the employees. This requires control over the physical environment of the business or operation. Most injuries are the result of accidents which are caused by an unsafe condition or an unsafe act, or a combination of the two. The unsafe condition is related to the physical environment. This involves the equipment used and all of the physical conditions surrounding the workplace. For example, hazards can stem from lack of guarding or inadequate guarding of the equipment, the location of machines, the condition of storage areas, or the condition of the building.

Some general safety considerations related to the building include adequate floor-loading capacity. This is especially important in storage areas, where overloading has caused many serious accidents every year. The danger signs of overloading include cracks in walls or ceilings, excessive vibration, and the displacement of structural members.

Aisles, stairs, and other walkways should be investigated periodically to assure that they are free of obstacles, are not uneven, and do not have oil or other material that could lead to slips and falls. In many old buildings, stairs should be inspected, since they are the cause of numerous lost time accidents. Stairs should have a slope of 30 to 35 degrees, with tread widths of approximately 9½ inches. Riser heights should not exceed 8 inches. All stairways should be equipped with handrails, have at least 10 footcandles of illumination, and be painted in light colors.

There should be at least two exits on all floors of a building, and the size of the exits should be in accordance with the Life Safety Code of the National Fire Protection Association. This code gives consideration to the occupancy and relative fire hazard that the exit area is servicing. Adequate fire protection should be incorporated, based on both OSHA standards and specific local regulations. Thus, the building should contain adequate fire extinguishers, sprinkler systems, and standpipe and hose.

Aisles should be plainly marked and straight, with well-rounded corners or diagonals at turn points. If aisles are to accommodate vehicle travel, they should be at least three feet wider than twice the width of the broadest vehicle. When traffic is only one way, then two feet wider than the broadest vehicle is adequate. In general, aisles should have at least five footcandles of illumination. The initial installation of sufficient fixtures does not assure adequate illumination. Only a continuing maintenance effort can assure that periodic cleaning of fixtures and replacement of blown lights takes place.

Color should be used throughout to identify hazardous conditions. The color recommendations shown in Table 9–11 are in compliance with OSHA standards.

TABLE 9–11

Color	Used for	Examples
Red	Fire protection equipment, danger, and as a stop signal	Fire alarm boxes, location of fire extinguishers and fire hose, sprinkler piping, safety cans for flammables, danger signs, emergency stop buttons
Orange	Dangerous parts of machines, other hazards	Inside of movable guards, safety starting buttons, edges of exposed parts of moving equipment
Yellow	Designating caution, physical hazards	Construction and material handling equipment, corner markings, edges of platforms, pits, stair treads, projections. Black stripes or checks may be used in conjunction with yellow
Green	Safety	Location of first-aid equipment, gas masks, safety deluge showers
Blue	Designating caution against starting or using equipment	Warning flags at starting point of machines, electrical controls, valves about tanks and boilers
Purple	Radiation hazards	Container for radioactive materials or sources
Black and white	Traffic and housekeeping markings	Location of aisles, direction signs, clear floor areas around emergency equipment

Most machine tools can be satisfactorily guarded so that the probability of being injured while operating the facility is remote. The problem is that in many instances a facility can be guarded but isn't. In these instances take immediate action to see that a guard is provided and that it is workable and routinely used. There are, of course, exceptions, such as a jointer or a circular cutoff saw, where the process does not lend itself to foolproof guarding. In such cases, partial guarding is easily attainable, but complete guarding is excessively expensive or impossible because it interferes with the operator's manipulations. In such cases, consider several alternatives. Sometimes the process may be automated, thus completely freeing the operator from the "nip" point. In other instances, a robot manipulator can be used in place of an operator, or the method can be planned and the operator trained to use manual feeders or devices to keep hands and other portions of the body away from danger points.

In addition to making provision for guarding the facility at the nip point, the operator must have adequate protection from potential accidents resulting from the use of the tool. To control such accidents, management needs to take steps to:

1. Train operators in the correct and safe use of tools.
2. Provide the correct tool for the job.
3. Maintain the tool so that it is always in a safe condition.
4. Insure the use and maintenance of the necessary guards and safety devices or practices.

A system of quality control and maintenance should be incorporated within the tool room and the tool cribs, so that only reliable tools in good working condition are released to workers. Examples of unsafe tools that should not be released to operators include: power tools with broken insulation, electrically driven power tools lacking grounding plugs or wires, poorly sharpened tools, hammers with mushroomed heads, cracked grinding wheels, grinding wheels without guards, and tools with split handles or sprung jaws.

To initiate and maintain a program of safety and employee health, analysts should realize the potential danger of certain materials. A large segment of our business and manufacturing enterprises uses some potentially dangerous chemicals. As a matter of company policy, the composition of every chemical compound used by a concern should be ascertained, its hazards determined, and control measures established to protect employees. The long-term health effect of many materials is still unknown; as hazards are identified by OSHA and other groups, companies are initiating new procedures to deal with these substances.

Materials known to cause health and/or safety problems fall into one of three categories: corrosive materials, toxic or irritant materials, and flammable materials.

Corrosive materials include a variety of acids and caustics that can burn and destroy human tissue upon contact. The chemical action of corrosive materials can take place by direct contact with the skin or through inhalation of their fumes or vapors. To avoid the potential danger resulting from the use of corrosive materials, consider the following measures:

1. Be sure that the methods of material handling are completely foolproof.
2. Assure that the process does not result in any spilling or spattering, especially during initial delivery processes.
3. Be sure that operators exposed to corrosive materials have and are using correctly designed personal protective equipment and waste disposal procedures.
4. Assure that the dispensary or the first-aid area is equipped with the necessary emergency provisions, including deluge showers and eye baths.

Toxic or irritating materials include gases, liquids, or solids, that poison the body or disrupt normal processes by ingestion, absorption through the skin, or inhalation.

To control toxic materials, use the following methods:

1. Completely isolate the process from workers.
2. Provide adequate exhaust ventilation.
3. Provide workers with reliable personal protective equipment.
4. Substitute a nontoxic or nonirritating material.

Flammable materials and strong oxidizing agents present problems of fire and explosion. The spontaneous ignition of combustible materials can take place when there is insufficient ventilation to remove the heat from a process of slow oxidation. To prevent such fires, combustible materials need to be stored in a well-ventilated, cool, dry area. Small quantities should be stored in covered metal containers.

Some combustible dusts, such as sawdust, are not ordinarily known to be explosive. However, explosions can occur when such dusts, flammable vapors, or gases are present in the air in large enough concentrations to ignite. For both gases and dusts, there are limiting concentrations in air below which and above which explosions do not occur. For light dusts, the generally accepted lower explosive limit is 0.015 ounce per cubic foot, and for heavy dusts, 0.5 ounce per cubic foot. Vapors and gases have a wider range over which an explosion is liable to take place. Concentrations in the air of 0.5 percent by volume are frequently listed as lower limits. An increase in temperature depresses the lower limit.

To avoid explosions, prevent ignition by providing adequate ventilation-exhaust systems. Adequate control of the manufacturing processes minimizes the generation of dusts and the liberation of gases and vapors.

Gases and vapors may be removed from gas streams by absorption in liquids or solids, adsorption on solids, condensation, and catalytic combustion and incineration. In absorption, the gas or vapor becomes distributed in the collecting liquid or solid. Equipment for absorption includes absorption towers, such as bubble-cap plate columns, packed towers, spray towers, and wet-cell washers.

For the adsorption of gases and vapors, use a variety of solid adsorbents with an affinity for certain substances. Charcoal, for example, adsorbs many different substances, including benzene, carbon tetrachloride, chloroform, nitrous oxide, and acetaldehyde.

The catalytic combustion process uses a platinum alloy–alumina catalyst to burn hydrocarbons. The minimum catalytic ignition temperature varies from 350° F to 600° F. In catalytic combustion, gases and vapors go through a low-temperature oxidation process and are converted to odor- and color-free gases. The presence of the catalyst merely provides an activated surface on which the reaction proceeds more readily.

JOB FACTORS LEADING TO UNSATISFACTORY PERFORMANCE

In this chapter, we have discussed the principal human factors considerations, including the physical environment, physiological and psychological constraints, and sociological considerations. One additional aspect is task factors that may lead to human error.

The facility, coupled with the worker's task of managing and operating a piece of equipment, may be too demanding for an operator who has difficulty in operating efficiently through a normal shift. Facility outputs depend on the operator's ability to grasp readily the meaning of the output and to respond promptly in the most effective manner.

One of the most common methods of providing information output of facilities is through visual displays. The principal forms of visual displays include: indicator lights, scales, counters, printers, graphic plotters, and cathode-ray tubes. To be effective, a display must be able to communicate information quickly, accurately, and efficiently. By efficiently, we mean that the eye as the sensory organ that gathers the information must be able to do so in a manner that is free from error. Thus, it should be possible to read the display quickly and accurately both from a position of straight on the display and at viewing angles of up to 45 degrees, as the job demands.

Simple indicator lights are typically found in cars, as are legend lights that provide additional information. For example, a legend light in a modern car advises that something is wrong. The legend "alt" informs you that the alternator is not producing sufficient output. Scales display graduated information, such as air speed or the amount of fuel in a tank. Counters are used when a precise quantity or indication is desired in connection with the operation of the facility or process. A counter may be used on a press to inform the operator when to change a die. A printer is usually an electromechanical device for recording information. It may be desirable to incorporate an output printer with a desk calculator. The graphic plotter is similar to a printer, except that the output is in graphic form, which reflects continuous trends. Today people often use plotters and printers with computers to secure permanent records of material processed by computers. Cathode-ray tubes are used to present moving visual images, as in a TV screen. Also, cathode-ray tubes present alphanumeric information. This is done in connection with schedules, such as schedules of airline arrivals and departures, and production schedules in the large job shop.

Indicator Lights

Indicator or warning lights are probably the type of visual displays in greatest use. Several basic requirements should be incorporated into their use.

First, they should be designed so that they immediately get the attention of the worker. They should also indicate to the operator what is wrong and what action he or she should take.

Generally, only one warning light should be used with a given system. Other lights that identify the cause and the action to be taken, and that operate with the single warning light, may be located in less central positions. The warning light should remain on until the condition that caused it to be energized has been remedied.

One flashing light attracts attention quickly, but several lose most of this ability. A flashing light should give four flashes per second. Immediately after the operator takes action, the flashing should stop but the light should remain on until the improper condition has been completely remedied.

The warning light should be red or yellow and of sufficient size and intensity to be noticed immediately. A good rule is to make it twice the size and brightness of other panel indicators, and place it not more than 30 degrees off the operator's expected line of sight.

Display Information

Table 9–12 shows the advantages and disadvantages of using moving pointers, moving scales, and counters.

Operator errors in reading display information increase as the density of information per unit area of the display increases and as the operator time for reading the display and responding decreases. Coding is a method that improves the readability of the display and the operator's viewing efficiency. The best three coding methods are: color, alphanumerics (letters and digits), and shape (geometric figures). These three coding techniques require little space and allow easy identification, though operators may require some training in their interpretation.

TABLE 9–12

	Service rendered			
Indicator	Quantitative reading	Qualitative reading	Setting	Tracking
Moving pointer	Fair	Good (changes are easily detected)	Good (easily discernible relation between setting knob and pointer	Good (pointer position is easily controlled and monitored
Moving scale.	Fair	Poor (may be difficult to identify direction and magnitude)	Fair (may be difficult to identify relation between setting and motion	Fair (may have ambiguous relationship to manual-control motion
Counter.	Good (minimum time to read and results in minimum error)	Poor (position change may not indicate qualitative change)	Good (accurate method to monitor numerical setting)	Poor (not readily monitored)

As shown in Table 9–6, colors have emotional and psychological significance. Universally in Western culture, red indicates a stop situation, as is characterized by traffic control. Red also usually symbolizes danger. On the other hand, green is thought of as a proceed or go-ahead situation.

Yellow is usually thought of as a caution symbol. It is widely used in connection with hunting sportswear, and everyone is aware of its use with traffic lights to convey caution. A recommended coding of simple indicator lights appears in Table 9–13.

Alphanumeric coding provides many more combinations than does color coding. From an efficiency standpoint, it parallels color coding. For highly effective alphanumeric coding, consider the stroke width of the numerals and letters, the width-height ratio, and the type form, or "font."

Based on a viewing distance of up to 28 inches under a range of illuminating conditions, the letter or numeral height should be at least 0.20 inches, and the stroke width at least 0.04 inches, to give a width-height ratio of 1:5. Use a broader stroke with dark letters on a bright background, and a narrower stroke with bright letters on a dark background.

The font refers to the available type styles, such as Gothic, Futura, and Tempo. In general, capital, or uppercase, letters are easier to read for a few words than are lowercase letters. Consequently, use uppercase letters; their width-height ratio should be about 3:5.

Acoustic Signals

In some instances, it is better to use auditory signals than visual presentations. For example, auditory signals are usually more efficient if the worker's job necessitates his or her continual movement about the plant or business, or if the person receiving the signal is in a work area where it

TABLE 9–13
Recommended coding of indicator lights

		Color			
Diameter	*State*	*Red*	*Yellow*	*Green*	*White*
12.5 mm........	Steady	Failure; Stop action; Malfunction	Delay; Inspect	Circuit energized; Go ahead; Ready; Producing	Functional; In position; Normal (on)
25 mm or larger	Steady	System or subsystem in stop action	Caution	System or subsystem in go-ahead state	
25 m or larger	Flashing	Emergency condition			

would be difficult to see a visual signal, such as a dark area or an excessively bright area. Short, simple messages are also usually better handled by auditory means.

The human auditory system is alert continuously. It can detect sources of different signals without orientation of the body, as is usually necessary with visual signals. Since hearing is omnidirectional, and reaction times to sounds are shorter than to visual indications, auditory messages are especially desirable in connection with warning signals. Of course, only acoustic means are satisfactory for speech. There are cases where auditory signals should not be considered as an alternative to visual signals but as an addition to them. In cases where the visual system of the operator may already be overburdened, it may be more efficient to add an auditory system.

Shape and Size Coding

Shape coding, using two- or three-dimensional geometric configurations, permits both tactual and visual identification. It is especially useful where redundant or double-quality identification is desirable, thus helping to minimize errors. Shape coding permits a relatively large number of discriminable shapes. However, if the operator must identify controls without vision, discrimination can be difficult and slow as the number of shapes increase. If the operator is obliged to wear gloves, then shape coding is desirable only for visual discrimination or for the tactual discrimination of only two to four shapes.

Size coding, analogous to shape coding, permits both tactual and visual identification of controls. Size coding is used principally where the controls cannot be seen by operators. Of course, as is the case with shape coding, size coding permits redundant coding, since controls can be discriminated both tactually and visually. In general, try to limit the size categories to three.

Control Size, Displacement, and Resistance Criteria

In both the micro- and macroscopic components of their work assignments, workers continually use various types and designs of controls. The three parameters that have a major impact on performance are the control size, the control resistance when engaged, and the total displacement on activation. A control that is either too small or too large cannot be actuated efficiently. Likewise, the amount of resistance and displacement has an impact on operator performance. The effect of both distance and resistance on performance time will become more apparent after reading Chapter 19 on basic motion times. Tables 9–14, 9–15, and 9–16 provide helpful design information about minimum and maximum dimensions for various control mechanisms.

TABLE 9–14
Control size criteria

| | | | Control size | |
| | | | --- | --- |
Control		Dimension	Minimum (mm)	Maximum (mm)
Pushbutton	Fingertip	Diameter	13	*
	Thumb/palm	Diameter	19	*
	Foot	Diameter	8	*
Toggle switch		Tip diameter	3	25
		Lever arm length	13	50
Rotary selector		Length	25	*
		Width	*	25
		Depth	16	*
Continuous adjustment knob	Finger/thumb	Depth	13	25
		Diameter	10	100
	Hand/palm	Depth	19	*
		Diameter	38	75
Cranks	For rate	Radius	13	113
	For force	Radius	13	500
Handwheel.		Diameter	175	525
		Rim thickness	19	50
Thumbwheel		Diameter	38	*
		Width	*	*
		Protrusion from surface	3	*
Lever handle.	Finger	Diameter	13	75
	Hand	Diameter	38	75
Crank handle		Grasp area	75	*
Pedal		Length	88	†
		Width	25	†
Valve handle.		Diameter	75 inches per inch of valve size	

* No limit set by operator performance.
† Dependent on space available.

TABLE 9–15
Control displacement criteria

		Displacement	
Control	*Condition*	*Minimum*	*Maximum*
Pushbutton	Thumb/fingertip operation	3 mm	25 mm
	Foot Normal	13 mm	–
	Heavy boot	25 mm	–
	Ankle flexion only	–	63 mm
	Leg movement	–	100 mm
Toggle switch	Between adjacent positions	30°	–
	Total	–	120°
Rotary selector	Between adjacent detents: Visual	15°	–
	Nonvisual	30°	–
	For facilitating performance	–	40°
	When special engineering is required	–	90°
Continuous adjustment knob	Determined by desired control/display ratio (mm. of control movement for each mm. of display movement)		
Crank	Determined by desired control/display ratio		
Handwheel	Determined by desired control/display ratio		90°– 120°†
Thumbwheel	Determined by number of positions		
Lever handle.	Fore-aft movement	*	350 mm
	Lateral movement	*	950 mm
Pedals	Normal	13 mm	–
	Heavy boot	25 mm	–
	Ankle flexion (raising)	–	63 mm
	Leg movement	–	175 mm

* None established.
† Provided optimum control/display ratio is not hindered.

HUMAN FACTORS AND THE DESIGN OF THE WORKSTATION

As we have learned, many factors have a significant impact on both the productivity and the well-being of the operator at the workstation. Thus, sound human factor technology applies both to the equipment being used and the general conditions surrounding the work area. From both the equipment point of view and the workstation environment, it is important that adequate flexibility be provided so that variations in employee height, reach, strength, reflex time, and so on, can be accommodated. A work bench that is 32″ high may be just right for a 6′3″ worker but would

TABLE 9–16
Control resistance criteria

		Resistance	
		Minimum	Maximum
Control	Condition	(kg)	(kg)
Push button	Fingertip	0.17	1.14
	Foot: Normally off control	1.82	9.10
	Rested on control	4.55	9.10
Toggle switch	Finger operation	0.17	1.14
Rotary selector	Torque.	1 cm-kg	7 cm-kg
Continuous adjustment knob . .	Torque: Fingertip <1-in. dia	*	0.3 cm-kg
	Fingertip >1-in. dia	*	0.4 cm-kg
Crank	Rapid, steady turning: <3-in. radius	0.91	2.28
	5–8-in. radius	2.28	4.55
	Precise settings	1.14	3.64
Handwheel†	Precision operation: <3-in. radius	*	*
	5–8-in. radius	1.14	3.64
	Resistance at rim: One-hand	2.28	13.64
	Two-hand	2.28	22.73
Thumbwheel.	Torque	1 cm-kg	3 cm-kg
Lever handle	Finger grasp	0.34	1.14
	Hand grasp: One-hand	0.91	–
	Two-hands	1.82	–
	Fore-aft: Along median plane:		
	One-hand−10 in. forward SRP§	–	13.64
	−16–24 in. forward SRP	–	22.73
	Two-hand−10–19 in. forward SRP	–	45
	Lateral:		
	One-hand−10–19 in. forward SRP	–	9.09
	Two-hand−10–19 in. forward SRP	–	22.73
Pedal	Foot: Normally off control	1.82	–
	Rested on control	4.55	–
	Ankle flexion only	–	4.55
	Leg movement	–	80

* Not established.
† For valve handles/wheels: 25 ± cm-kg of torque/cm of valve size
(8 cm-kg of torque/cm of handle diameter).
§ SRP = Seat reference point.

definitely be too high for a 5'6" employee. Adjustable heights of workstations and chairs are desirable to accommodate the full range of workers based upon plus or minus two standard deviations. See Table 9–10.

Just as there is a significant variation in height and size in the work force, as already pointed out, there is equal or greater variation from person to person in visual capacity, ability to hear, ability to feel, manual dexterity, and memory. The better able we are to provide a flexible work center so as to accommodate the total range of our work force, the more satisfactory will be the productivity results and worker satisfaction.

Certainly, the vast majority of workstations can be improved. Applying human factor consideration in connection with methods engineering will lead to more efficient competitive work environments that will improve the well-being of the workers, the quality of the product, the labor turnover of the business, and the prestige of the organization.

TEXT QUESTIONS

1. What are the principal objectives of operation analysis, motion study, and micromotion study?
2. What areas of study relate to the human factors approach toward improvements?
3. Which two factors concern the methods analyst about artificial lighting?
4. What independent factors affect the quantity of light that is needed to perform a task satisfactorily?
5. Explain the color rendering effect of low pressure sodium lamps.
6. What is the relationship between contrast and seeing time?
7. What footcandle intensity would you recommend 30 inches above the floor in the company washroom?
8. Explain how sales may be influenced by colors.
9. Would the combined colors of yellow and blue give a harmonious hue? Explain.
10. What color has the highest visibility?
11. How is sound energy dissipated in viscoelastic materials?
12. A frequency of 2,000 Hz would have approximately what wavelength in meters?
13. What would be the approximate decibel value of a grinder being used to grind a high-carbon steel?
14. Distinguish between broadband noise and meaningful noise.
15. Would you advocate background music at the workstation? What results would you anticipate?
16. According to the present OSHA law, how many continuous hours per day of a 100 dbA sound level would be permissible?
17. What three classifications have been identified from the standpoint of exposure to vibration?
18. In what ways can workers be protected from vibration?
19. What is meant by the environmental temperature? The effective temperature? The operative temperature?
20. Explain what is meant by the "thermal comfort zone"?
21. What is the maximum rise in body temperature that analysts should allow?
22. How would you go about estimating the maximum length of time that a worker should be exposed to a particular heat environment?
23. With a dry-bulb temperature of 80° F, a wet-bulb temperature of 70° F, and an air velocity of 200 feet per minute, what would be the normal effective temperature?
24. Which type of radiation is given the most attention by the safety engineer?

25. What is meant by absorbed dose of radiation? What is the unit of absorbed dose?
26. What is meant by the rem?
27. What insulation would be required for a stenographic pool if the ambient temperature were 40° F?
28. Which three factors influence the accuracy of control movements?
29. What caloric intake would you recommend for an operator doing heavy work? Explain.
30. Explain the preferential order of needs that applies to most workers.
31. What color would you paint a container used for holding radioactive materials?

GENERAL QUESTIONS

1. What steps would you take to increase the amount of light in an assembly department by about 15 percent? The department currently uses fluorescent fixtures, and the walls and ceiling are painted a medium green. The assembly benches are a dark brown.
2. What color combination would you use to attract attention to a new product being displayed?
3. What is the relationship between the heartbeat rate and oxygen consumption?
4. When would you advocate that the company purchase aluminized clothing?
5. Are there possible health hazards in conjunction with electron beam machining? With laser beam machining? Explain.
6. Explain why supine lifting permits heavier lifts than does prone lifting.
7. Explain why the effective visual areas for average employees is greater for the right eye than for the left eye.
8. Is there a satisfactory explanation for the deterioration of dynamic memory with age?
9. Explain the impact of noise levels below 85 dbA on office work.

PROBLEMS

1. A work area has a reflectivity of 60 percent, based on the color combinations of the workstations and the immediate environment. The seeing task of the assembly work could be classified as difficult. What would be your recommended illumination?
2. What would be the level of two uncorrelated noise signals of 86 and 96 decibels?
3. In the XYZ Company, an industrial engineer designed a workstation where the seeing task was difficult because of the size of the components going into the assembly. He established the brightness desired was 100 footlamberts on the average, with a standard deviation of 10 footlamberts so as to accommodate 95 percent of the workers.
   The workstation was painted a medium green having a reflectance of 50 percent. What illumination in footcandles would be required at this

workstation to provide adequate illumination for 95 percent of the workers? Estimate what the required illumination would be if you repainted the workstation with a light cream paint.

4. In the XYZ Company, an industrial engineer was assigned to alter the work methods in the press department to meet OSHA standards relative to permissible noise exposures. He found that the sound level averaged 100 db and that the standard deviation was 10 db. The 20 operators in this department wore earplugs provided by XYZ. Also, the power output from the public-address system was altered from 30 watts to 20 watts. The deadening of the sound level of the earplugs was estimated to be 20 percent effective. What improvement resulted? Do you feel that this department is now in compliance with the law for 99 percent of the employees? Explain.

5. In the XYZ mill room, an all-day study revealed the following noise pollution: 0.5 hrs., 105 dBA; 1 hr., less than 80 dBA; 3.5 hrs., 90 dBA; 2 hrs., 92 dBA; 1 hr., 96 dBA. Is this company in compliance? Explain.

6. Using log-log paper, plot the relationship between the frequency and the wave length of sound in room temperature air and develop an equation for this relationship. See Figure 9–3.

SELECTED REFERENCES

Crocker, N. J., and F. N. Kessler. *Noise and Noise Control,* vol. 2, Boca Raton, Fla., CRC Press, 1981.

Eastman Kodak Co., Human Factors Section. *Ergonomic Design for People at Work.* New York: Van Nostrand Reinhold, 1983.

Eberts, Ray, and Cindelyn Gray Eberts. "Human Information Processing." In *Handbook of Industrial Engineering,* 2nd ed., Chap. 36. Gavriel Salvendy, New York: John Wiley & Sons, 1992.

Granjean, E. *Fitting the Task to the Man.* New York: International Publication Services, 1980.

Harris, Cyril M. *Handbook of Noise Control.* 2nd ed. New York: McGraw-Hill, 1979.

Hutchison, R. Dale. *New Horizons for Human Factors in Design.* New York: McGraw-Hill, 1981.

Karwowski, Waldemar. "Occupational Biomechanics." In *Handbook of Industrial Engineering,* 2nd ed., edited by Gavriel Salvendy, Chap. 39. New York: John Wiley & Sons, 1992.

Konz, Stephen. *Work Design.* Columbus, Ohio: Grid, 1979.

McCormick, Ernest J. *Human Factors Engineering.* 4th ed. New York: McGaw-Hill, 1976.

Murrell, K. F. H. *Ergonomics.* London, England: Chapman and Hall, 1969.

Park, Kyung S. "Human Reliability." In *Handbook of Industrial Engineering,* 2nd ed., edited by Gavriel Salvendy, Chap. 38. New York: John Wiley & Sons, 1992.

VanCott, H. P., and R. G. Kinkade. *Human Engineering Guide to Equipment Design.* Rev. ed. Washington, D.C.: U.S. Government Printing Office, 1970.

Presentation and Installation of the Proposed Method

Presenting and installing the proposed method are the fifth and sixth steps in the systematic development of a work center to produce a product or perform a service. Selling the proposed method is always the first important element in the presentation procedure. This step is as important as any of the preceding steps, since a method not sold usually is not installed. No matter how thorough the data gathering and analysis and the ingenuity of the proposed method, the value of the project is zero unless it is installed.

Humans naturally resent the attempts of others to influence their thinking. When someone approaches us with a new idea, our instinctive reaction is to put up a defense against it. We feel that we must protect our own individuality—preserve the sanctity of our own ego. And all of us are just egotistical enough to convince ourselves that our ideas are better than those of anyone else. It is natural for us to react in this manner even if the new idea is for our own advantage. If the idea has merit, there is a tendency to resent it because we did not think of it first.

Some techniques that analysts use to sell ideas include:

1. · Introducing the idea so that the other person feels it is really his or her idea to a large extent. For example, begin by saying, "You recently gave me the idea that. . . ."
2. Not appearing overly anxious to have an idea accepted. Introduce your thoughts with such statements as: "Do you think this idea has possibilities?" or "Have you considered this?"
3. Presenting objections to the idea. This can get the other person arguing for the support of the idea.

The presentation of the proposed method should emphasize savings. Savings in material (both direct and indirect) and savings in direct and indirect labor should highlight an analyst's report.

The second most important part of the presentation is that dealing with quality and reliability improvement resulting from installation of the improved method. Every progressive manager today recognizes that the key to the continued health of any product or service producer is to be able to supply quality materials in a timely consistent manner.

The third most important part of the presentation is that dealing with the recovery of capital investment.

Once the proposed method has been properly presented and sold, installation can take place. Installation, like presentation, requires sales ability. During installation, continue selling the proposed method to engineers and technicians on their own level, to subordinate executives and supervisors, and to labor and representatives of organized labor. Their job during the presentation and installation of the proposed method involves selling—up, down, and sideways.

THE REPORT ON THE PROPOSED METHOD

Analysts make their presentation in both written and oral form. Even if the company involved does not require a written report, it is good practice to make one for record purposes and for future applications. A well-written report is a major step in selling the proposed method.

The elements of a well-written report are:

1. Title page.   4. Summary.
2. Table of contents.   5. Body.
3. Letter of transmittal.   6. Appendix.

From the standpoint of presentation, the summary is the most important section of the report, since it will be the only part read by busy executives. The summary is based on the body of the report; consequently, it is usually not written until after the body has been completed. However, it is presented early in the report so that those who must pass or disapprove the proposal can obtain the facts quickly to make a decision.

The summary should contain three elements: an abstract explaining briefly the nature of the problem; conclusions outlining the results of the analysis; and recommendations setting forth the proposed method and summarizing estimated savings, quality and reliability improvement, and the recovery of capital expenditures.

The body includes a section on the nature of the problem followed by details relative to the gathering of the data and the methods of analysis. This section contains the operation and flow process charts used to present the facts and the worker and machine, gang, operator process, and micromotion charts used in developing the proposed work center. All sketches, drawings, and specifications showing details of proposed fixtures and tool and machine designs are included. Where robots and/or automation procedures are recommended, details of type, style, and size of the proposed facilities are included. Information as to the anticipated improvement in quality and reliability due to consistency and dependability resulting from the mechanization is emphasized. This section also contains the reasons for the conclusions and recommendations given in the summary.

The entire report should be clear, concise, complete, and accurate. It should be prepared so that it can be easily read and studied.

Reporting the Recovery of Capital Investment

The three most frequently used appraisal techniques for determining the desirability of investing in a proposed method are: (1) the return on sales method, (2) the return on investment or payback method, and (3) the cash flow method.

The following steps may be necessary in making an economic study:

1. Identification of the economic problem requiring study.
2. Determination of the alternatives to be compared and preliminary estimates of the anticipated differences between the alternatives with monetary differences expressed as receipts and disbursements on specific dates.
3. Determination of which alternatives are worthy of consideration for further analysis in a preliminary analysis of the estimates.
4. Determination of the interest rate or minimum attractive rate of return and the calculations to place the money time series on a comparable basis.
5. Determination of a choice between the alternatives giving due consideration to the monetary comparisons and also to those anticipated differences not reduced to money terms.

The return on sales method involves a computation of the ratio of the average yearly profit brought about through using the method, to the average yearly sales or increase in dollar value added to the product, based upon the estimated pessimistic life of the product. Although this ratio provides information on the effectiveness of the method and of the resulting sales efforts, it does not consider the original investment required to get started on the proposed method.

The return on investment method gives the ratio of the average yearly profit brought about through using the method, based upon the estimated pessimistic life of the product, to the original investment. Of two proposed methods that would result in the same sales and profit potential, management would obviously prefer to use the one requiring the investment of the least capital. The reciprocal of the return on investment is often referred to as the "payback" method. This gives the time that it would take to return the original investment.

The cash flow method computes the ratio of the present worth of cash flow, based on a desired percentage return, to the original investment. This method introduces the rate of flow of money in and through the company and the time value of money. The time value of money is important. Because of interest, a dollar today is worth more than a dollar next year or at any later date. For example, at 15 percent compound interest, a dollar today is worth $2.011 five years from now. Expressing it another way, a dollar received five years from now would be worth about 50 cents today. Interest may be thought of as the return obtainable by the productive investment of capital.

The following applies to the present value concept:

Single Payment

Compound amount factor (given P and to find S) $S = P(1 + i)^n$

Present worth factor   (given S and to find P) $P = 1(1 + i)^n$

Uniform Series

Sinking fund factor
(given S and to find R) $= i/(1 + i)^n - 1$

Capital recovery
(given P and to find R) $= i(1 + i)^n/(1 + i)^n - 1$

Compound amount factor
(given R and to find S) $= (1 + i)^n - 1/i$

Present worth factor
(given R and to find P) $= (1 + i)^n - 1/i(1 + i)^n$

where:

i = Interest rate for a given period
n = Number of interest periods
P = Present sum of money (present worth of principal)
S = A sum of money at the end of n periods from the present date. It is equivalent to P with interest i
R = The end-of-period payment or receipt in a uniform series continuing for the coming n periods, the entire series equivalent to P at interest i

An assumed return rate (i) is the basis of the cash flow computation. Then all cash flows following the initial investment for the new method are estimated and adjusted to their present worth, based on the assumed return rate. The total estimated cash flows for the estimated pessimistic life of the product are then summed up as a profit or loss in terms of cash on hand today. This total is then compared with the initial investment.

An illustrative example will clarify the use of these three appraisal methods for prognosticating the potential of a proposed method.

End of year	Increase in sales values due to proposed method	Cost of production with proposed method	Gross profit due to proposed method
1	$ 5,000	$ 2,000	$ 3,000
2	6,000	2,200	3,800
3	7,000	2,400	4,600
4	8,000	2,600	5,400
5	7,000	2,400	4,600
6	6,000	2,200	3,800
7	5,000	2,000	3,000
8	4,000	1,800	2,200
9	3,000	1,600	1,400
10	2,000	1,500	500
Totals	$53,000	$20,700	$32,300
Average	$ 5,300	$ 2,070	$ 3,230

Return on sales = $\frac{3,230}{5,300}$ = 61%   Return on investment = $\frac{3,230}{10,000}$ = 32.3%

Payback = 1/0.323 = 3.09 years

Investment for proposed method: $10,000

Desired return on investment: 10 percent.

Salvage value of jigs, fixtures, and tools: $500.

Estimated life of the product for which the proposed method will be used: 10 years.

Present worth of cash flow:

(3,000)(0.9091) = $2,730   (3,800)(0.5645) = 2,140
(3,800)(0.8264) = 3,140   (3,000)(0.5132) = 1,540
(4,600)(0.7513) = 3,460   (2,200)(0.4665) = 1,025
(5,400)(0.6830) = 3,690   (1,400)(0.4241) = 595
(4,600)(0.6209) = 2,860   (500)(0.3855) = 193
$21,373

Salvage value of tools:

(500)(0.3855) = $193

Total present worth of anticipated gross profit and tool salvage value: $21,566. Ratio of present worth to original investment:

$$\frac{21,566}{10,000} = 2.16$$

The new method has satisfactorily passed all three of these appraisal methods. A 61 percent return on sales and a 32.3 percent return on capital investment certainly represent attractive returns. The return of the $10,000 capital investment will take place in 3.09 years and the cash flow analysis reveals that the original investment will be recovered in four years while earning 10 percent. During the 10-year anticipated life of the product, $11,566 more than the original investment will be earned.

Estimates of the product demand 10 years hence may deviate considerably from reality. Thus, the element of chance is introduced, and the probabilities of success tend to diminish with the increased length of the payoff period. The results of any study are only as valid as the reliability of the input data. Constant follow-up can determine the validity of the assumptions in this study. Do not hesitate to alter decisions when the original data prove to be invalid. Sound financial analysis is intended to facilitate the decision-making process—not to replace good business judgment.

The Oral Presentation

Frequently, analysts are asked to present their method proposals orally. To help assure approval of the proposal, be prepared to present the benefits and advantages accurately and forcefully. Give estimates of the resulting increase in productivity and/or decrease in cost. If quality will be improved or customer service enhanced, this information should be given. This is a good time to bring out that cycle-time reduction results in better products and service delivered to the market faster.

Be sure to plan the presentation in advance. Have data relative to all the advantages of the proposed method, as well as cost information and facts giving the expected savings and the expected recovery of capital investment. List in advance all the information that your bosses may ask for, and then be prepared to supply this information.

Be prepared to answer objections raised to the proposal. These are usually centered on the initial cost, time to adopt the method, and inconvenience while installation is taking place. Point out that these objections have been carefully considered and that plans have been made to cope with them.

The oral presentation of the proposed method requires much study, preparation, and sales ability. A good method will not sell itself; it must be sold.

INSTALLATION

After a proposed method has been approved, the next step is installation. Too frequently, analysts do not stay close enough to the job during installation. Do not assume that installation will take place automatically according to your proposal. A maintenance person, mechanic, or worker can make a slight change or modification without considering the consequences. This may mean that the proposed method does not give the anticipated results.

Stay with the job during installation to assure that all details are carried out in accordance with the proposed plan. During installation, verify that the work center being established is equipped with the facilities proposed, that the planned working conditions are provided, that the tooling is in accordance with recommendations, and that the work is progressing satisfactorily.

During installation, sell the new method to the operator, supervisor, setup man, and so on. Then, by the time the installation is complete, these employees will be readier to give the new method an enthusiastic try.

Once the new work center has been installed, check all aspects to see whether they conform to the specifications established. In particular, verify that the "reach" and "move" distances are the correct length; that the tools are correctly sharpened; that the mechanisms function soundly; that stickiness and sluggishness have been worked out; that safety features are operative; that material is available in the quantities planned; that working conditions associated with the work center are as anticipated; and that all parties have been informed of the new method.

After you are sure that every aspect of the method is ready for operation, have the supervisor assign the operator who will be working with the method. Then stay with the operator as long as is necessary to break him or her in on the new assignment. This period may be a matter of a few minutes or of several hours or even of days, depending on the complexity of the assignment.

Once the operator begins to get a feel for the method and works along systematically, proceed with other work. However, the installation phase should not be considered complete until you have checked back several times during the first few days after installation to assure that the proposed method is working out as planned. Also check with line supervisors to assure that they spot check and monitor the new method.

TEXT QUESTIONS

1. Which communication skills are important in selling a new method?
2. What are the principal elements of a well-written report?
3. What is meant by the "cash flow" appraisal technique?
4. What are the principal concerns of management with regard to a new method that is relatively costly to install?
5. What is meant by the payback method? How is it related to the return on investment method?

GENERAL QUESTIONS

1. What is the relationship between return on capital investment and the risk associated with the anticipated sales of the product for which a new method will be used?

2. When do you feel that the oral report would be more important than the written report in getting approval for a new method?

PROBLEMS

1. How much capital could be invested in a new method if it is estimated that $5,000 would be saved the first year, $10,000 the second year, and $3,000 the third year? Management expects a 30 percent return on invested capital.

2. You have estimated the life of your design to be three years. You expect that a capital investment of $20,000 will be required to get it into production. You also estimate, based on sales forecasts, that the design will result in an after-tax profit of $12,000 the first year and $16,000 the second year, and a $5,000 loss the third year. Management has asked for an 18 percent return on capital investment. Should we go ahead with the investment to produce the new design? Explain.

3. In the Dorben Company a materials handling operation in the warehouse is being done by hand labor. Annual disbursements for this labor and for closely related expenses (social security, accident insurance, and other fringe benefits), are $8,200. The methods analyst is considering a proposal to build certain equipment to reduce this labor cost. The first cost of this equipment will be $15,000. It is estimated that the equipment will reduce annual disbursements for labor and labor extras to $3,300. Annual payments for power, maintenance, and property taxes and insurance are estimated to be $400, $1,100, and $300, respectively. The need for this particular operation is anticipated to continue for 10 years. Because the equipment is specially designed for the particular purpose, it will have no salvage value. It is assumed that the annual disbursements for labor, power, and maintenance will be uniform throughout the 10 years. The minimum rate of return before income taxes is 10 percent. Based on annual cost comparison, should the company proceed with the new material handling equipment?

SELECTED REFERENCES

Emerson, C. Robert, and William R. Taylor. *An Introduction to Engineering Economy*. Davis, Calif.: Cardinal Publishers, 1973.

Fabrycky, W. J., and G. J. Thuesen. *Economic Decision Analysis*. Englewood Cliffs, N.J.: Prentice-Hall, 1974.

Fleischer, G. A. "Economic Risk Analysis." In *Handbook of Industrial Engineering*, 2nd ed., edited by Gavriel Salvendy, Chap. 52. New York: John Wiley & Sons, 1992.

Grant, E.; W. Ireson; and R. S. Leavenworth. *Principles of Engineering Economy*. 6th ed. New York: Ronald Press, 1976.

Lutz, Raymond P. "Discoiunted Cash Flow Techniques." In *Handbook of Industrial Engineering*, 2nd ed., edited by Gavriel Salvendy, Chap. 50. New York: John Wiley & Sons, 1992.

Smith, G. W. *Engineering Economy: Analysis of Capital Expenditures.* 3rd ed. Ames: Iowa State University Press, 1979.

Thuesen, H. G.; W. J. Fabrycky; and G. J. Thuesen. *Engineering Economy.* 5th ed. Englewood Cliffs, N.J.: Prentice-Hall, 1977.

White, J. A.; M. H. Agee; and K. E. Case. *Principles of Engineering Economic Analysis.* New York: John Wiley & Sons, 1977.

Job Analysis and the Job Evaluation

Closely associated with the installation of the ideal method and the standard time required to perform the operation is the job analysis of the work center and the resulting job evaluation. This is the seventh step in the systematic procedure of applying methods engineering. Every time a method is changed, the job description should be altered to reflect the conditions, duties, and responsibilities of the improved method. When a new method is introduced, a job analysis should be made so that a qualified operator may be assigned to the work center and an appropriate base rate provided.

The factory cost of a product includes the cost of direct labor, direct material, and factory expense. Factory expense consists of such items as light, heat, rent, service supplies, and factory supervision. A close review of expense or overhead items shows that these costs are made up of but two components: material and labor. So, in effect, factory cost is made up of just labor and material. Therefore:

Factory cost = Cost of labor + Cost of material

Material cost is easily determined, no matter what the product may be. In every instance, multiply the cost per unit of measure by the number of measures involved. Thus, some yardstick always governs its basis of cost. Steel is bought by the ton, castings and forgings by the piece, water and gas by the cubic foot, electricity by the kilowatt-hour, tubing by the foot, silver by the ounce, cloth by the square yard, oil by the gallon, and so forth. In every case, material cost can be prorated to determine the proportionate share each product item should absorb.

The only other item entering into factory cost is the cost of labor, either direct or indirect. Labor cost, in most commodities, represents the major portion of total costs. To have some conception of the true costs of specific products, standards must be established on labor elements. Hourly monetary rates mean nothing unless they are supplemented with performance standards. If an operator running a No. 2 Brown & Sharpe Universal Mill is paid a base rate of $12.60 per hour while milling a profile on a forging, there is no conception of the unit milling cost until a time standard has been established on the job. If a standard allows six minutes per piece, then a labor cost of $1.26 can be assigned to the operation. However, time standards alone do not give the entire story with regard to labor costs. They form only one side of a rectangle whose area may be thought of as equitable and competitive costs of labor. The other side of the rectangle is sound base rates.

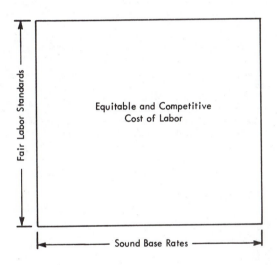

Base rates that are sound must assure money rates commensurate with the local rates for similar work; they must allow adequate differentials for jobs requiring higher skills and responsibilities; and they must be based on techniques that can be explained and justified.

JOB ANALYSIS

Appropriate base rates are a result of job evaluation, a technique for equitably determining the relative worth of the different work assignments within an organization. The basis of job evaluation is job analysis, a procedure for making a careful appraisal of each job and then recording the details of the work so that it can be evaluated fairly by a trained analyst. Figure 11–1 illustrates an analysis of a clerical job for use in a point job evaluation plan. Before a job description is developed, all aspects of the opportunity should be carefully studied to assure that the best

320   *Motion and Time Study*

FIGURE 11–1
Job analysis for a shipping and receiving clerk

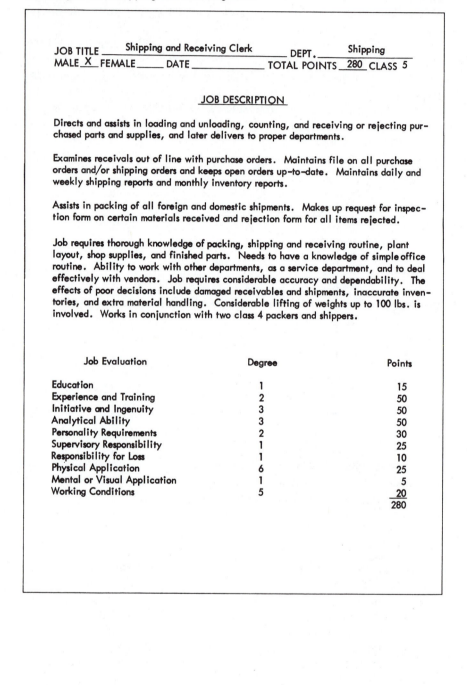

JOB TITLE _____Shipping and Receiving Clerk_____ DEPT. _____Shipping_____
MALE _X_ FEMALE_____ DATE_____ TOTAL POINTS _280_ CLASS 5

JOB DESCRIPTION

Directs and assists in loading and unloading, counting, and receiving or rejecting purchased parts and supplies, and later delivers to proper departments.

Examines receivals out of line with purchase orders. Maintains file on all purchase orders and/or shipping orders and keeps open orders up-to-date. Maintains daily and weekly shipping reports and monthly inventory reports.

Assists in packing of all foreign and domestic shipments. Makes up request for inspection form on certain materials received and rejection form for all items rejected.

Job requires thorough knowledge of packing, shipping and receiving routine, plant layout, shop supplies, and finished parts. Needs to have a knowledge of simple office routine. Ability to work with other departments, as a service department, and to deal effectively with vendors. Job requires considerable accuracy and dependability. The effects of poor decisions include damaged receivables and shipments, inaccurate inventories, and extra material handling. Considerable lifting of weights up to 100 lbs. is involved. Works in conjunction with two class 4 packers and shippers.

Job Evaluation	Degree	Points
Education	1	15
Experience and Training	2	50
Initiative and Ingenuity	3	50
Analytical Ability	3	50
Personality Requirements	2	30
Supervisory Responsibility	1	25
Responsibility for Loss	1	10
Physical Application	6	25
Mental or Visual Application	1	5
Working Conditions	5	_20_
		280

methods are being used and that the operator is thoroughly trained in the prescribed methods.

Typically, various job responsibilities, authorities, and the consequences resulting from poor decisions are items that would be included in a job analysis. Also, the analysis should provide information regarding the machines and tools used in the job and the problem-solving capability required. The physical and social conditions related to the job should be outlined as well.

Job Description

The job description is an essential component of the job analysis (see Figure 11–1). Job descriptions are useful supervisory tools that can aid in the selection, training, and promotion of employees and the assessment of work distribution. The job description should identify the job's specific duties and responsibilities and the minimum requirements of the worker performing the job. The job description should emphasize:

1. Doing the job as described.
2. Doing the right thing to produce at the least cost and the highest quality.

This second characteristic of modern job evaluation is representative of today's nonbureaucratic style in which people will do what is best for the company by focusing on the customer-client relationship.

In writing a job description, first pinpoint an accurate title. The title should unambiguously describe what a worker actually does. Be sure to enlist the worker's help in accurately defining responsibilities. This procedure often can lead to the development of cost effective improvements. A combination of personal interviews and questionnaires along with direct observation results in a concise definition of each job and the duties that each entails. In writing a job description, list the mental and physical functions required to perform the work. Use such definitive words as: direct, examine, plan, measure, and operate. The more accurate the description, the better.

JOB EVALUATION

Job evaluation is a procedure by which an organization ranks its jobs in order of their worth or importance. It was during the World War II years that job-evaluation systems became popular. It was the only way increases in wages could be given, since all wages were frozen by the National War Labor Board, and it had to be shown that inequities existed in a firm if wage adjustments were to be initiated. By introducing a job-evaluation system, it was easy to identify wage and/or salary inequities and obtain permission to provide increases.

The reader should understand that, generally, individuals believe that others in similar positions work less and are paid more than they are.

The main purpose of any job evaluation plan is to determine the proper compensation for the work performed on each job. A well-conceived job evaluation plan includes the following factors:

1. It provides a basis for explaining to employees why one job is worth more or less than another job.
2. It provides a reason to employees whose rate of pay is adjusted because of a change in method.
3. It provides a basis for assigning personnel with specific abilities to certain jobs.
4. It helps determine the criteria for a job when employing new personnel or making promotions.
5. It provides assistance in the training of supervisory personnel.
6. It provides a basis for determining where opportunities for methods improvement exist.

The majority of job evaluation systems in use today are a variation or combination of four principal systems. These are the classification method, the point system, the factor comparison method, and the ranking method.

The classification method, sometimes called the grade description plan, consists of a series of definitions designed to differentiate jobs into wage groups. Once the grade levels have been defined, analysts study each job and assign it to the appropriate level on the basis of the complexity of its duties and responsibilities and its relation to the description of the several levels. The United States Civil Service uses this plan extensively.

To use this method of job evaluation, take the following steps:

1. Prepare a grade description scale for each type of job, for example, machine operations, manual operations, skilled (craft) operations, or inspection.
2. Write the grade descriptions for each grade in each scale, using such factors as:
   a. Type of work and complexity of duties.
   b. Education necessary to perform job.
   c. Experience necessary to perform job.
   d. Responsibilities.
   e. Effort demanded.
3. Prepare job descriptions for each job. Classify each job by "slotting" (placing in a specific category) the job description into the proper grade description.

Both the point system and the factor comparison method are more objective and thorough in their evaluations of the various jobs involved;

both plans study the basic factors common to most jobs that influence their relative worth. Of the two plans, the point system is the more commonly used procedure and is generally considered the more accurate method for occupational rating. In this method, analysts compare all the attributes of a job directly with the attributes in other jobs.

When a point system is installed, the following procedure should be followed:

1. Establish and define the basic factors common to most jobs, indicating the elements of value in all jobs.
2. Specifically define the degrees of each factor.
3. Establish the points to be accredited to each degree of each factor.
4. Prepare a job description of each job.
5. Evaluate each job by determining the degree of each factor contained in it.
6. Sum the points for each factor to get the total points for the job.
7. Convert the job points into a wage rate.

The factor comparison method of job evaluation usually has the following elements:

1. Determining the factors establishing the relative worth of all jobs.
2. Establishing an evaluation scale that is usually similar to a point scale except that the units are in terms of money. For example, a $2,000-per-month benchmark job might attribute $800 to the responsibility factor, $400 to education, $600 to skill, and $200 to experience.
3. Preparing job descriptions.
4. Evaluating key jobs, factor by factor, by ranking each job from the lowest to the highest for each factor.
5. Paying wages on each key job based on various factors. The money allocation automatically fixes the relationships among jobs for each factor, and therefore establishes the ranks of jobs for each factor.
6. Evaluating other jobs factor by factor, on the basis of the monetary values assigned to the various factors in the key jobs.
7. Determining a wage by adding up the money value of the various factors.

Today, the factor comparison system is not popular in business or industry. It is time consuming and costly to introduce. The evaluation is highly subjective and difficult to sell to employees.

The ranking method arranges jobs in order of their importance or according to their relative worth. Here, the entire job is considered; this includes the complexity and degree of difficulty of the duties, the requirements for specific areas of knowledge, skills, amount of experience, and the level of authority and responsibility assigned to the job. This method became popular in the United States during World War II because of its simplicity and ease of installation. At this time, the National War Labor

Board set up the requirement that all companies working on government contracts must have some type of wage classification system. The ranking method satisfied this requirement. Generally speaking, the ranking method is less objective than the other techniques; consequently, it necessitates greater knowledge of all jobs. For this reason, it has not been used extensively in recent years, but has been superseded by the other plans. The following steps are followed when installing the ranking method:

1. Prepare job descriptions.
2. Rank jobs (usually departmentally first) in the order of their relative importance.
3. Determine the class or grade for groups of jobs, using a bracketing process.
4. Establish the wage or wage range for each class or grade.

Selecting Factors

Under the factor comparison method, most companies use five factors. In some point programs, 10 or more factors may be used. However, it is preferable to use a small number of factors. The objective is to use only as many factors as are necessary to provide a clear-cut difference among the jobs of the particular company. The elements of any job may be classified as to:

1. What the job demands that the employee bring in the form of physical and mental factors.
2. What the job takes from the employee in the form of physical and mental fatigue.
3. The responsibilities that the job demands.
4. The conditions under which the job is done.

The selection of factors is usually the first basic task undertaken when introducing job evaluation.

The National Electrical Manufacturers Association (NEMA) states that the relative value of a job depends on the following factors:

1. Education.
2. Experience.
3. Initiative and ingenuity.
4. Physical demand.
5. Mental and/or visual demand.
6. Responsibility for equipment or process.
7. Responsibility for material or product.
8. Responsibility for safety of others.
9. Responsibility for work of others.
10. Working conditions.
11. Hazards.

TABLE 11–1
Points assigned to factors and key to grades

Factors	1st degree	2nd degree	3rd degree	4th degree	5th degree
Skill					
1. Education	14	28	42	56	70
2. Experience	22	44	66	88	110
3. Initiative and ingenuity	14	28	42	56	70
Effort					
4. Physical demand	10	20	30	40	50
5. Mental and/or visual demand	5	10	15	20	25
Responsibility					
6. Equipment or process	5	10	15	20	25
7. Material or product.............	5	10	15	20	25
8. Safety of others................	5	10	15	20	25
9. Work of others.................	5	10	15	20	25
Job conditions					
10. Working conditions.............	10	20	30	40	50
11. Unavoidable hazards	5	10	15	20	25

Source: National Electrical Manufacturers Association.

These factors are present in varying degrees in the various jobs, and any job under consideration falls under some one of the several degrees of each factor. The various factors are not of equal importance. To give recognition to these differences in importance, weights or points are assigned to each degree of each factor, as shown in Table 11–1. Figure 11–2 illustrates a job rating and a substantiating data sheet based on the plan of NEMA.

Each degree of each factor is carefully defined so that it is evident what degree characterizes the work situation under study. For example, education may be defined as appraising the requirements for the use of shop mathematics, drawings, measuring instruments, or trade knowledge. First-degree education may require only the ability to read and write, and to add and subtract whole numbers. Second-degree education could be defined as requiring the use of simple arithmetic, such as addition and subtraction of decimals and fractions, the ability to read simple drawings and use some measuring instruments, such as calipers and scales. It would be characteristic of two years of high school. Third-degree education may require the use of fairly complicated drawings, advanced shop mathematics, handbook formulas, and a variety of precision measuring instruments, plus some trade knowledge in a specialized field or process. It could be equivalent to four years of high school plus short-term trades training. Fourth-degree education could require the use of complicated drawings and specifications, advanced shop mathematics, and a wide variety of precision measuring instruments, plus broad shop knowledge. It may be equivalent to four years of high school plus four years of formal

FIGURE 11-2

Job rating and substantiating form

		JOB RATING - SUBSTANTIATING DATA	
		DORBEN MFG. CO.	
		UNIVERSITY PARK, PA.	

JOB TITLE: Machinist (General) CODE: 176 DATE: Nov. 12

FACTORS	DEG.	POINTS	BASIS OF RATING
Education	3	42	Requires the use of fairly complicated drawings, advanced shop mathematics, variety of precision instruments, shop trade knowledge. Equivalent to four years of high school or two years of high school plus two to three years of trades training.
Experience	4	88	Three to five years installing, repairing, and maintaining machine tools and other production equipment.
Initiative and Ingenuity	3	42	Rebuild, repair, and maintain a wide variety of medium-size standard automatic and hand-operated machine tools. Diagnose trouble, disassemble machine and fit new parts, such as antifriction and plain bearings, spindles, gears, cams, etc. Manufacture replacement parts as necessary. Involves skilled and accurate machining using a variety of machine tools. Judgment required to diagnose and remedy trouble quickly so as to maintain production.
Physical Demand	2	20	Intermittent physical effort required tearing down, assembling, installing, and maintaining machines.
Mental or Visual Demand	4	20	Concentrated mental and visual attention required. Laying out, setup, machining, checking, inspecting, fitting parts on machines.
Responsibility for Equipment or Process	3	15	Damage seldom over $900. Broken parts of machines. Carelessness in handling gears and intricate parts may cause damage.
Responsibility for Material or Product	2	10	Probable loss due to scrapping of materials or work, seldom over $300.
Responsibility for Safety of Others	3	15	Safety precautions are required to prevent injury to others; fastening work properly to face plates, handling fixtures, etc.
Responsibility for Work of Others	2	10	Responsible for directing one or more helpers a great part of time. Depends on type of work.
Working Conditions	3	30	Somewhat disagreeable conditions due to exposure to oil, grease, and dust.
Unavoidable Hazards	3	15	Exposure to accidents, such as crushed hand or foot, loss of fingers, eye injury from flying particles, possible electric shock, or burns.

REMARKS: Total 307 Points--assign to job class 4.

trades training. Fifth-degree education may require a basic technical knowledge sufficient to deal with complicated and involved mechanical, electrical, or other engineering problems. Fifth-degree education could be equivalent to four years of technical university training.

Experience appraises the time that an individual with the specified education usually requires to learn to perform the work satisfactorily from the standpoint of both quality and quantity. Here, first degree could involve up to three months; second degree, three months to one year; third degree, one to three years; fourth degree, three to five years; and fifth degree, over five years.

In a similar manner, each degree of each factor is identified with a clear definition and with specific examples when applicable.

Performing the Evaluation

Considerable judgment is needed to evaluate each job with respect to the degree required of each factor utilized in the plan. Consequently, it is usually desirable to have a committee perform the evaluation. A separate committee should be appointed for each department of the company or business. A typical committee would include a permanent chairman (usually from industrial relations or industrial engineering), a union representative, the department supervisor, the department steward, and a management representative (usually from industrial relations).

When meeting, the committee should evaluate all jobs for the same factor before proceeding to the next factor. For example, all jobs in the department under study should be evaluated for degree of skill before proceeding to other factors, such as effort, responsibility, and job conditions. Using this pattern, the committee measures the job rather than the individual filling the job.

Committee members should assign their degree evaluations independently of the other members. The correlation between and among different evaluators should be reasonably high, such as 0.85 or 0.90 or higher. Then all members should discuss any differences until there is agreement on the level of the factor. Do not end a meeting until the factor under study has been evaluated for all jobs being evaluated in the department.

Classifying the Jobs

After all jobs have been evaluated, tabulate the points assigned to each job. Next, decide the number of labor grades within the plant. This number is a function of the range of points characteristic of the jobs within the plant. Typically, the number of grades runs from 8 (typical of smaller plants and lesser skilled industries) to 15 (typical of larger plants and higher skilled industries). See Figure 11–3. For example, if the point range of all the jobs within a plant ranged from 110 to 365, the following grades could be established:

Grade	Score range (points)	Grade	Score range (points)
12	100–139	6	250–271
11	140–161	5	272–293
10	162–183	4	294–315
9	184–205	3	316–337
8	206–227	2	338–359
7	228–249	1	360 and above

FIGURE 11–3

Evaluation points and base rate range for nine labor grades

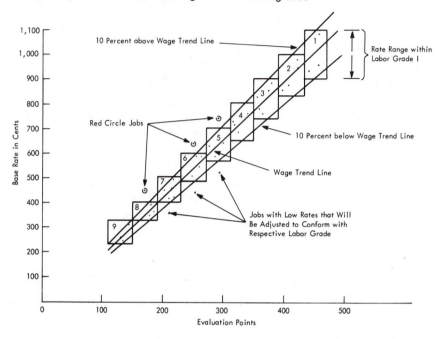

Like ranges are not necessary for the various labor grades. Increasing point ranges might be desirable for more highly compensated jobs.

Now review the jobs falling within the various labor grades in relation to one another to assure fairness and consistency. For example, it would not be appropriate for a Class A machinist to be in the same grade level as a Class B machinist.

The next step is to assign hourly rates to each of the labor grades. These rates are based on area rates for similar work, company policy, and the cost-of-living index. Frequently, analysts establish a rate range for each labor grade. The total performance of each operator determines his or her pay rate within the established range. Total performance refers to quality, quantity, safety, attendance, suggestions, and so on.

Installation of the Job Evaluation Program

After plotting area rates against the point values of the various jobs, develop a rate versus point value trend line. This trend line may or may not take the form of a straight line. Regression techniques are helpful in developing the trend line. After the trend line has been developed, several points will be both above and below the trend line. Points significantly above the trend line represent employees whose present rate is higher than that established by the job evaluation plan, and points significantly below the trend line represent employees whose present rate is less than that prescribed by the plan.

Employees whose rates are less than that called for by the plan should receive immediate increases to the new rate. Employees whose rates are higher than that called for by the plan (such rates are referred to as "red circle rates") are not given a rate decrease. They are, however, not given an increase at the next contract review unless the cost of living adjustment results in a rate higher than their current pay. Of course, any new employee would be paid the new, lower rate as advocated by the job evaluation plan.

Some Negative Considerations

There is no question that a point job evaluation system is probably the most favorable approach to bring both equity and objectivity to a plant or business in connection with the problem of determining how much to compensate individuals. However, the reader should understand that installation of a point plan can initiate several costly and difficult problems.

First, a point plan tends to emphasize exactly how a job should be done rather than how it best be done today in view of current circumstances. Modern management styles point out that all workers should take the approach that their jobs should always be performed in a manner that proves the best for the company and its customers—not necessarily what is the best for the individual. It is for this reason that I emphasize that in preparing the job description, importance should be directed toward doing the right thing so that the company can take full advantage of what an employee can do for the business.

Unless the job description is worded carefully, some employees may refuse to perform important work only because these tasks are not included in the job description. In addition to job duties, the job description should spell out the desirability of employee development, growth, and superior performance.

A point job evaluation plan can also create unnecessary and undesirable power relationships within the company. Job evaluation point scores inform many in the work force all the relative scores of the various jobs, providing an obvious pecking order that can retard cooperation and group decision making, which is so important in modern management.

Another problem that inevitably develops is that individuals soon recognize that a creatively written job description can add enough points so that the job will be assigned the next higher class. Thus, jobs will in time be written to reflect rates higher than they are really worth. This procedure may snowball to where the majority of employees ask for a restudy of their jobs. The organization can end up by paying the majority of its employees too much.

Along this same line, employees will recognize that they can increase their job evaluation points by increasing their responsibility. This often can be achieved by adding unnecessary work or increasing their responsibility by adding another clerical or other employee. These additions may be unnecessary and really add only additional direct and/or overhead cost in addition to the cost increase resulting from the higher paid job resulting from the point additions.

Finally, in some organizations, the points developed by a point job evaluation plan may be inappropriately used for nonpay purposes. For example, they may be used as a basis for privileges such as parking spaces, shift preferences, mailing lists, invitations to meetings, use of company recreation areas, and the like.

Conclusion

Point job evaluation plans provide the most favorable approach to equitable wage payment. However, they also identify comparable worth inequities. For example, a point plan might show that an industrial nurse should be paid $32,000 a year, yet in the community there is no difficulty in employing nurses at $26,000 a year. If the company paid the higher salary, many of the other salaried people would feel they were underpaid by comparison. Comparable worth within a geographic area is a factor to be reckoned with.

Properly designed point plans should result in companies being able to attract and retain qualified and competent employees and provide internal equity.

One of the principal concerns heard in the courtrooms and legislative hearings deals with "comparative worth." It appears that by the end of the 1990s, the principle of "equal pay for equal work" will be amended to "equal pay for comparable work." Certainly, a point job evaluation system is based on the concept of equal pay for comparable work. No longer does a wage gap exist between men and women on comparable work assignments. However, the analyst must realize that there is no inherent worth to any job: it is worth what is provided in the market place. If the analyst deviates from the established point plan to correct an inequity in order to provide for a salary based upon the marketplace alone, the action will almost always create a new inequity. The analyst should also recognize that treating everyone the same is inconsistent with treating individuals equitably in relation to their contribution to the business or industry.

Employees must understand the fairness of the job evaluation plan. It is also important that regular follow-up of the plan be done so that it is adequately maintained. Jobs do change, so it is necessary to review all jobs periodically and to make adjustments when necessary.

George Fry Associates undertook a comprehensive survey of job evaluation practices in over 500 companies. Some of the significant results from this survey appear in the following table.

Unfortunately over half of the companies responding to the survey found their employees' understanding of the plan to be poor or nonexistent. If a job evaluation plan is to succeed over the years, the vast majority of the employees should have at least an average understanding of how the plan works.

Job analysis and job evaluation are important steps after installation of the ideal method. Their principal purpose is to determine the relative worth of the jobs in a company. Job evaluation provides the means for compensating all employees within an organization in proportion to their responsibilities and the difficulty of their work. At the same time, it leads to base pay rates in line with remuneration for similar work in the community. The benefits effected through job evaluation improve personnel relations.

				Percent
Firms using a standard job evaluation plan				66.0
Plan used:				
Ranking				3.5
Grade description				1.0
Factor comparison...................................				10.5
Point system.......................................				85.0
Average of hourly employees covered				65.0
Job evaluation function reports to:				
Industrial relations.................................				69.0
Industrial engineering...............................				16.0
Results used:				
In employment.....................................				87.0
In employee placement..............................				88.0
In wage rate bargaining				68.0
Job evaluation program:				
Is recognized in union contract				83.0
Is an issue during negotiations				59.0
Has gone to arbitration (of these cases,				
management won 74 percent)				27.0

Understanding and acceptance:	*Good*	*Average*	*Poor*	*None*
By employees................	7%	39%	49%	5%
By top-management	50	36	13	1
By middle-management	18	17	5	0
By first-line supervisors	35	52	13	0

Before submitting bids and quotations, the cost of the various direct and indirect materials and of all labor must be precisely determined. Predetermined material costs can easily be calculated. Predetermined labor costs also can be readily computed if good labor standards developed by one of the work measurement techniques prevail. Standards determined by estimates and historical records usually are not sufficiently accurate to meet competitive prices or do not allow manufacture at a necessary profit.

Although labor standards determine the "how long," it is necessary to have sound base rates to measure fairly the "how much" in dollars and cents. Of the various methods of job evaluation used in establishing sound base rates, the point plans tend to give the most reliable results. Point job evaluation systems patterned after the techniques of the National Metal Trades and the National Electrical Manufacturers Association represent a logical approach toward developing the relative worth of the different work assignments within an organization.

TEXT QUESTIONS

1. What two specific subjects should be emphasized in writing the job description?
2. What are the three basic components of factory cost?
3. Is time a common denominator of labor cost? Why or why not?
4. What is job analysis?
5. When were job evaluation systems introduced in large numbers throughout the United States?
6. Which four methods of job evaluation are being practiced in this country today?
7. Why is it that most people feel that others doing work similar to theirs are paid more?
8. Explain in detail how a "point" plan works.
9. Which factors influence the relative worth of a job?
10. Why are estimates unsatisfactory for determining direct labor time standards?
11. What is the weakness of using historical records as a means of establishing standards of performance?
12. Which work measurement techniques give valid results when undertaken by competent trained analysts?
13. What are the principal benefits of a properly installed job evaluation plan?
14. Explain why a range of rates rather than just one rate should be established for every labor grade.
15. Explain what is meant by total operator performance.
16. What are the principal negative considerations that should be understood prior to the installation of a point job evaluation system?

GENERAL QUESTIONS

1. Should cost-of-living increases be given as a percentage of base rates or as a straight hourly increment? Why?
2. Why would a consulting firm such as George Fry and Associates undertake a comprehensive survey of job evaluation practices?
3. Visit a plant with which you are familar and find out if a job evaluation system has been installed. If one has been installed, find out the type of system and the degree of satisfaction by polling a sample of the employees.

PROBLEMS

1. A job evaluation plan based on the point system uses the following factors:
   a. Experience: maximum weight 200 points; five grades.
   b. Education: maximum weight 100 points; four grades.
   c. Effort: maximum weight 100 points; four grades.
   d. Responsibility: maximum weight 100 points; four grades.

   A floor sweeper is rated as 150 points, and this position carries an hourly rate of $6.50. A class 3 milling machine operator is rated as 320 points, which results in a money rate of $10.00 per hour. What grade of experience would be given to a drill press operator with a $8.50 per hour rate and point ratings of grade 2 education, grade 1 effort, and grade 2 responsibility?

2. A job evaluation plan in the Dorben Company provides for five labor grades, of which grade 5 has the highest base rates and grade 1 the lowest. The linear plan involves a range of 50 to 250 points for skill, 15 to 75 points for effort, 20 to 100 points for responsibility, and 15 to 75 points for job conditions. Each of the four factors has five degrees. Each labor grade has three money rates: a "low," a "mean," and a "high" rate.

   If the high money rate of labor grade 1 is $8 per hour and the high money rate of labor grade 5 is $20 per hour, what would be the mean money rate of labor grade 3? What degree of skill is required for a labor grade of 4 if second-degree effort, second-degree responsibility, and first-degree job conditions apply?

3. In the Dorben Company, the analyst has installed a point job evaluation plan covering all indirect employees in the operating divisions of the plant. Ten factors were used in this plan, and each factor was broken up into five degrees. In making the job analysis, the position of shipping and receiving clerk was shown as having second-degree initiative and ingenuity, valued at 30 points. The total point value of this job was 250 points. The minimum number of points attainable in the plan was 100, and the maximum was 500. If 10 job classes prevailed, what degree of initiative and ingenuity would be required to elevate the job of shipping and receiving clerk from job class 4 to job class 5?

   If job class 1 carries a rate of $8 per hour and job class 10 carries a rate of $20 per hour, what rate does job class 7 carry? (Note: Rates are based on the midpoint of job class point ranges.)

SELECTED REFERENCES

Dunn, J. D., and F. M. Rachel. *Wage and Salary Administration: Total Compensation Systems.* New York: McGraw-Hill, 1971.

Ellig, Bruce R. *Compensation Issues of the Eighties.* Amherst, Mass.: Human Resource Development Press, Inc., 1988.

Lawler, Edward E. *What's Wrong with Point-Factor Job Evaluation?* Amherst, Mass.: Human Resource Development Press, Inc., 1988.

Livy, B. *Job Evaluation: A Critical Review.* New York: Halstead, 1973.

McCormick, Ernest J. "Job Evaluation." In *Handbook of Industrial Engineering,* edited by Gavriel Salvendy. New York: John Wiley & Sons, 1982.

Milkovich, George T., Jerry M. Newman, and James T. Brakefield. "Job

Evaluation in Organizations." In *Handbook of Industrial Engineering*, 2nd ed., edited by Gavriel Salvendy, Chap. 34. New York: John Wiley & Sons, 1992.

Otis, Jay, and Richard H. Leukart. *Job Evaluation: A Sound Basis for Wage Administration*. Englewood Cliffs, N.J.: Prentice Hall, 1954.

Risner, Howard. *Job Evaluation: Problems and Prospects*. Amherst, Mass.: Human Resource Development Press, Inc., 1988.

Salvendy, Gavriel, and Douglas W. Seymour. *Prediction and Development of Industrial Work Performance*. New York: John Wiley & Sons, 1973.

Wegener, Elaine. *Current Developments in Job Classification and Salary Systems*. Amherst, Mass.: Human Resource Development Press, Inc., 1988.

Zollitsch, Herbert G., and Adoph Langsner. *Wage and Salary Administration*. 2nd ed. Cincinnati: South-Western Publishing, 1970.

Time Study Requirements

12

The eighth step in the systematic procedure for developing the work center to produce the product in an efficient manner is establishing time standards. These three techniques help determine time standards: estimates, historical records, and work measurement procedures.

Analysts used estimates as a means of establishing standards to a greater extent in years past. With today's increasing competition from foreign producers, there has been an increasing effort to establish standards based on facts rather than judgment. Experience has shown that no individual can establish consistent and fair standards of production by the simple procedure of taking a look at a job and then judging the amount of time required to produce it. Where estimates are used, standards are out of line. Compensating errors sometimes diminish this figure, but experience has shown that over a period of time, estimated values deviate substantially from measured standards. Both historical records and work measurement techniques give much more accurate values than the use of estimates based on judgment alone.

Under the historical method, production standards are based on the records of previously produced similar jobs. In common practice, the worker punches in on a time clock or data collection hardware every time he or she begins a new job, and then punches out after completing the job. This technique tells how long it took to do a job, but never indicates how long it should have taken. Since operators wish to justify their entire working day, some jobs carry personal delay time, unavoidable delay time, and avoidable delay time to a much greater extent than they should, while other jobs do not carry their appropriate share of delay time. I have seen historical records that deviated consistently by as much as 50 percent on the same operation of the same job. As a basis of determining labor standards historical records are better than no records at all. Such records give more reliable results than estimates based on judgment alone, but they do not provide sufficiently valid results to assure equitable and competitive labor costs.

Any of the work measurement techniques—stopwatch (electronic or mechanical) time study, fundamental motion data, standard data, time formulas, or work sampling studies—represent a better way to establish fair production standards. All of these techniques are based on facts. All consider each detail of the work and its relation to the normal time required to perform the entire cycle. Accurately established time standards make it possible to produce more within a given plant, thus increasing the efficiency of the equipment and the operating personnel. Poorly established standards, although better than no standards at all, lead to high costs, labor dissension, and eventually the possible failure of the enterprise.

Successful installation of any of the work measurement techniques requires a wholehearted commitment by management. This commitment involves the allocation of enthusiasm, time, and the necessary financial resources on a continuing basis.

A smoothly operating work measurement program requires considerable planning and effective communication to all members of the enterprise. Prior to the introduction of the program, management should establish clear objectives and policies and hire properly trained and experienced analysts. Good communication is essential during installation and throughout the life of the program. All levels of management, as well as the employees, should be kept informed on the progress of installation and on the mechanics of the program.

As data from the work measurement system become available, they should be used. Sound standards have many applications that can mean the difference between the success or the failure of a business. Companies should use standards for planning purposes, for the comparison of alternative methods, for effective plant layout, for determining capacities, for purchasing new equipment, for balancing the work force with the available work, for production control, for the installation of incentives, and for standard cost and budgetary control.

Time study is a technique for establishing an allowed time standard to perform a given task. This technique is based on measurement of the work content of the prescribed method, with due allowance for fatigue and for personal and unavoidable delays. Frequently, time study is defined as a method of determining a "fair day's work." This concept is discussed before the requirements and responsibilities of those associated with time study are explained. It is necessary to have a clear understanding of what is involved in a fair day's work.

A FAIR DAY'S WORK

Practically everyone connected with industry in any way has often heard the expression "a fair day's work." Yet most of the people who have heard the expression would be perplexed if they were asked to define just what a fair day's work is. The intraplant wage rate inequities agreements of the basic steel industries[1] contain the provision that "the fundamental principle of the work and wage relationship is that the employee is entitled to a fair day's pay in return for which the company is entitled to a fair day's work." A fair day's work is defined in these agreements as the "amount of work that can be produced by a qualified employee when working at a normal pace and effectively utilizing his time where work is not restricted by process limitations." This definition does not make clear what is meant by *qualified employees, normal pace,* and *effective utilization.* Although all of these terms have been defined by the steel industries, a certain amount of flexibility prevails because firm benchmarks cannot be established on such broad terminology. For example, the term *qualified employee* is defined as "a representative average of those employees who are fully trained and able satisfactorily to perform any and all phases of the work involved, in accordance with the requirements of the job under consideration." This definition leaves a doubt as to what is meant by a "representative average employee."

Then the term *normal pace* is defined as "the effective rate of performance of a conscientious, self-paced, qualified employee when working neither fast nor slow and giving due consideration to the physical, mental, or visual requirements of the specific job." The intraplant wage rate inequities agreements specify as an example "a man walking without load, on smooth, level ground at a rate of three (3) miles per hour." Although the three miles an hour concept tends to tie down what is meant by normal pace, still a notable amount of latitude can prevail if we think of normal pace on the thousands of different jobs in American industry.

Again, a feeling of uncertainty arises when one considers the definition of *effective utilization.* This is explained in the agreements as "the maintenance of a normal pace while performing essential elements of the job during all portions of the day except that which is required for reasonable rest and personal needs, under circumstances in which the job is not subject to process, equipment or other operating limitations."

In general, a fair day's work is one that is fair to both the company and the employee. This means that the employee should give a full day's work for the time that he or she gets paid, with reasonable allowances for personal delays, unavoidable delays, and fatigue. He or she is expected to operate in the prescribed method at a pace that is neither fast nor slow,

---

[1] With United Steelworkers of America.

but one that may be considered representative of all-day performance by the experienced, cooperative employee.

TIME STUDY REQUIREMENTS

Certain fundamental requirements need to be realized before the time study is taken. If the standard is required on a new job, or if it is required on an old job on which the method or part of the method has been altered, the operator should be thoroughly acquainted with the new technique before the operation is studied. The method must be standardized at all points where it is to be used before the study begins. Unless all details of the method and working conditions have been standardized, the time standards have little value and become a continual source of mistrust, grievances, and internal friction.

Analysts should tell the union steward, the departmental supervisor, and the operator that the job is to be studied. Each of these parties can then make specific plans in advance and take the steps necessary to allow a smooth, coordinated study. The operator should verify that he or she is performing the correct method and should become acquainted with all details of the operation. The supervisor should check the method to make sure that feeds, speeds, cutting tools, lubricants, and so forth, conform to standard practice as established by the methods department. Also, the supervisor must investigate the amount of available material so that no shortage takes place during the study. If several operators are available for the study, the supervisor should determine which operator will give the most satisfactory results. The union steward should then make sure that only trained, competent operators are selected for time study observation. He or she should explain to the operators why the study is being taken and should answer pertinent questions raised by the operator from time to time.

Responsibilities of the Time Study Analyst

All work involves varying degrees of skill and physical and mental effort. In addition to such variations in job content, there are differences in the aptitude, physical application, and dexterity of the workers. It is an easy matter for the analyst to observe an employee at work and to measure the actual time taken to perform a task. It is a considerably more difficult matter to evaluate all variables and determine the time required for the "normal" operator to perform the job.

Because of the many human interests and reactions associated with the time study technique, it is essential that there be full understanding on the part of the supervisor, the employee, the union steward, and the time study analyst. In general, time study analysts have these responsibilities:

1. To probe, question, and examine the present method to assure that it is correct in all respects before the standard is established.
2. To discuss the equipment, method, and operator's ability with the supervisor before studying the operation.
3. To answer questions relating to time study practice or to a specific time study that may be asked by the union steward, the operator, or the supervisor.
4. To cooperate with the supervisor and the operator at all times to obtain maximum help from both.
5. To refrain from any discussion with the operator under study or other operators that might be construed as criticism of the individual being studied.
6. To show on each time study complete and accurate information specifically identifying the method under study.
7. To record accurately the times taken to perform the individual elements of the operation being studied.
8. To evaluate the performance of the operator honestly and fairly.
9. To behave properly at all times to obtain and keep the respect and confidence of the representatives of both labor and management.

The qualifications of time study analysts are similar to those required for success in any field in which the major efforts are directed toward establishing ideal human relations.

Good time study analysts must have the mental ability to analyze diversified situations and make rapid, sound decisions. They should have inquisitive, probing, and open minds that seek to improve, and ask not only "why" but also "how."

To supplement a keen mind, it is essential that time study analysts have practical shop training in the areas in which they establish standards. If they are to be associated with the metal trades, they should have backgrounds as journeyman machinists or the equivalent knowledge of the correct use and application of machines, hand tools, jigs, fixtures, and gages. This would include specific knowledge of cutting feeds, speeds, and depths of cuts to get the maximum results consistent with the desired quality of the product and ultimate tool life.

Since time study analysts directly affect the pocketbooks of workers and the profit and loss statement of companies, it is essential that their work be completely dependable and accurate. Inaccuracy and poor judgment will not only affect the operator and the company financially, but may also result in complete loss of confidence by the operator and the union, which may undo harmonious labor relations that have taken management years to build up.

To achieve and maintain good human relations, the following personal requirements are essential for successful time study analysts:

1. Honesty.
2. Tact, human understanding.
3. Resourcefulness.
4. Self-confidence.
5. Good judgment, analytic ability.
6. Pleasing, persuasive personality, supplemented with optimism.
7. Patience, self-control.
8. Bountiful energy tempered with a cooperative attitude.
9. Well-groomed, neat appearance.
10. Enthusiasm for the job.

All of these qualifications might not be required of the top executives of a concern. However, the magnitude of the labor relations problems today demands only the most competent people enter the field of time study. No other one individual within a company comes in contact with as many personnel from different levels of the organization as does the time study analyst. Therefore, it is imperative that this person have the best qualifications.

Time study analysts should learn to recognize the human qualities in various employees and then be guided by a realization of the limitations of human nature. Thus, in order to receive cooperation, analysts must determine and follow through with the best method of approach to workers. This calls for analysis of each employee's attitudes toward the job, other workers, the company, and the time study analyst.

The Supervisor's Responsibility

Any and all supervisors are management's representatives throughout the plant. Next to the operator, the supervisor is closer to specific jobs than is any other person in the plant. In view of this, he or she must accept certain responsibilities in connection with the establishment of time standards.

In the interest of harmonious labor relations within the department a supervisor must maintain equitable time standards. Both "tight" and "loose" standards are the direct cause of endless personnel problems, and the more of these that can be avoided, the easier and pleasanter is the job. Of course, if all standards were loose, supervisory responsibilities would be relatively easy. However, this situation could not exist practically, since competition would not be met if all standards were loose.

The supervisor should notify the operator in advance that his or her work assignment is to be studied. This clears the way for both the time study analyst and the operator. The operator has the assurance that the supervisor knows a rate is to be established on the job. Therefore, the operator can bring out specific difficulties that he or she feels should be corrected before a standard is set. In addition, the time study analyst is more at ease whenever his or her presence is anticipated.

The supervisor should see that the proper method established by the methods department is being utilized and that the operator selected is competent and has adequate experience on the job. Although the time study analyst is required to have a practical background in the area of work being studied, analysts can hardly be expected to be infallible in specifications of all methods and processes. Thus, the supervisor becomes an ally in verifying that the cutting tools are properly ground, that the correct lubricant is being used, and that a proper selection of feeds, speeds, and depths of cuts is being made.

If it is questionable whether a fair time study can be taken for any reason, the supervisor should immediately tell the time study analyst. In general, the supervisor is responsible for assisting and cooperating with the time study analyst in any way that aids in defining or clarifying an operation. The supervisor should carefully consider any suggestions for improvement brought out by the time study analyst. Prior to stopwatch study, the supervisor should have established an ideal method in conjunction with the methods department.

The supervisor should make certain that operators use the prescribed method by conscientiously assisting and training all employees in perfecting this method. A supervisor should freely answer any questions asked by the operator regarding the operation.

Anytime a methods change takes place within a department, the supervisor should notify the time study department immediately so that the standard is adjusted appropriately. This procedure should be followed regardless of the degree of the change. Methods changes include such things as changes in material handling to and from the workstation, in inspection procedure, in feeds and speeds, in workstation layout, and in processes.

When a time study has been completed, the supervisor should sign the original study, thus indicating compliance with the study taken. Supervisors who accept and carry out their responsibilities toward time study practice can be assured of operating harmonious departments that are looked on with favor by management, the union, and the employees themselves. Those supervisors who fall short of these responsibilities contribute to the establishment of inequitable rates that result in numerous labor grievances, pressure from management, and considerable dissatisfaction from the union.

The Union's Responsibility

Most unions are opposed to work measurement and would prefer to see all standards established by arbitration. However, unions recognize that standards are necessary for the profitable operation of a business and that management continues to develop them by means of the principal work measurement techniques.

Furthermore, every union steward knows that poor time standards cause labor just as many problems as they cause management. In the interest of operating a healthy union within a profitable business, the union should accept certain responsibilities toward time study.

Through training programs, the union should educate all of its members in the principles, theories, and economic necessity of time study practice. All of us tend to fear anything that we know little about. Operators can hardly be expected to be enthusiastic about time study if they know nothing about it. This is especially true in view of its background (see Chapter 2). Therefore, the union should accept the responsibility of helping to clarify and explain this important tool of management.

The union representative should make certain that the time study includes a complete record of the job conditions as to work method and workstation layout. Also, the representative should ascertain that the current job description is accurate and complete.

The union representative should see that the elemental breakdown has been made with clearly defined end points, thus helping to assure the consistency of elemental times. The union's leaders should assure that the study has been taken over a long enough time to accurately reflect all of the variations that normally take place in performing the operation, as well as the typical unavoidable delays. Time study is a sampling technique, and samples of insufficient size can lead to erroneous results.

The union should urge its members to cooperate with the time study analyst and to refrain from practices that would place their performances at the low end of the rating scale. In the final analysis, encouraging operators to deceive the time study analyst adds up to but one thing: a poor standards structure that includes both loose and tight rates.

The union should see that updated standards are put into effect whenever a methods change is made. When methods are revised, the union should be sure that the time study department is notified through specified lines of authority.

Unions that train their members in the elements of time study, encourage cooperativeness, and stay abreast of management's program benefit by more cooperation at the bargaining table, fewer work stoppages, and better satisfied members. Unions that encourage distrust of time study, and keep operators uninformed face a multitude of grievances from their members; a balky management negotiating team; and over a period of time, sufficient work stoppages to create hardships for all the parties.

The Operator's Responsibility

Every employee should be sufficiently interested in the welfare of the company to give wholehearted support to every practice and procedure inaugurated by management. Although this situation is seldom realized, it certainly can be approached if a company's management demonstrates its desire to operate with fair standards, fair base rates, good working conditions, and adequate employee fringe benefits in the form of insurance and retirement programs. Once management has taken the initiative in these areas, every employee can be expected to cooperate in all operations and production control techniques.

Individual operators should give new methods a fair trial. They should wholeheartedly cooperate in helping to work out the "bugs" characteristic of practically every innovation. Making suggestions for further improvement of the methods should be an accepted part of each operator's responsibilities. The operator is closer to the job than anyone else, and can make a real contribution to the company by helping to establish ideal methods.

The operator should assist the time study analyst in breaking down the job into elements, thus assuring that all details of the job are specifically covered. Also, the operator should work at a steady, normal pace while the study is being taken and introduce as few foreign elements and extra movements as possible. The worker should use the exact prescribed method and make no effort to deceive the time study analyst by introducing an artificial method to lengthen the cycle time in hopes of receiving a more liberal standard.

CONCLUSION

To many practicing time study analysts, the responsibilities we have assigned to the operator, the union, and the supervisor may be considered a utopian goal which can never be realized. However, as stated, if management takes the initiative, these conditions can be approximated, and the result will be a competitive business that is profitable for all parties.

Time study represents one of the most important and exacting forms of work in any industrial, commercial, or governmental enterprise. When intelligently used and fully understood by all parties, it offers marked benefits to workers, management, and the general public.

TEXT QUESTIONS

1. What type of reliability can we expect from estimates?
2. How is a fair day's pay determined by the intraplant wage rate inequities agreements of the basic steel industries?
3. What benchmark for normal pace is given under these agreements?
4. Why should the supervisor sign the time study?
5. Why should the time study analyst have excellent personal qualities?
6. Explain how poor time standards increase the difficulties of the union steward.
7. How can management increase the cooperation of the union steward, the supervisor, and the operator in their dealings with the time study analyst?

GENERAL QUESTIONS

1. Is it customary for the union to cooperate with the time study department to the extent recommended in this text? Is this cooperation likely to occur?
2. If the requirements for a time study analyst are so high, why is it that industry does not pay more for time study work?
3. How does the walking pace of three miles per hour agree with your concept of a normal performance?
4. Why is it that universities spend so little time today in the teaching of work measurement if it is so important to the practicing industrial engineer?

PROBLEM

1. The typical plant that has not had the benefit of a methods, standards, and wage payment system is operating at an estimated 50 percent of standard. A good methods program should increase productivity to 80 percent of standard. A standards program that is well conceived, operated, and maintained should increase productivity to 95 percent of standard. And, a well-designed and conscientiously implemented incentive system can further increase productivity to 120 percent of standard.

   The cost of installing a complete methods, standards, and wage payment system is approximately $30,000 per year for every 100 people coming under the plan.

   The XYZ Company employs 650 people. The total value added by production averages 80 percent of the payroll dollar. The average cost of labor (including fringe benefits) is $11.25 per hour.

   What return on investment can be expected if the XYZ Company installs a complete methods, standards, and wage payment system? Show all your calculations.
2. Take a simple operation that you perform regularly such as brushing your teeth, shaving, or combing your hair and estimate the time it takes you to perform the operation. Now measure the time it takes while working at a normal pace. Is your estimate within plus or minus 20 percent of the estimated time?

SELECTED REFERENCES

Barnes, Ralph. *Motion and Time Study,* 7th ed. New York: John Wiley & Sons, 1980.

Gomberg, William. *A Trade Union Analysis of Time Study.* 2nd ed. Englewood Cliffs, N.J.: Prentice Hall, 1955.

Gomberg, William. "Labor Relations: The Special Problems of the Industrial Engineer," In *Handbook of Industrial Engineering,* ed. Gavriel Salvendy. New York: John Wiley & Sons, 1982.

Griepentrog, Carl W., and Gilbert Jewell. *Work Measurement: A Guide for Local Union Bargaining Committees and Stewards*. Milwaukee: International Union of Allied Industrial Workers of America, AFL–CIO, 1970.

Krick, Edward V. *Methods Engineering*. New York: John Wiley & Sons, 1962.

Smith, George, L. *Work Measurement—A Systems Approach*. Columbus, Ohio: Grid, 1978.

United Auto Workers. *Time Study—Engineering and Education Departments. Is Time Study Scientific?* Publication No. 325, Detroit: Solidarity House, 1972.

Time Study Equipment

The minimum equipment required to carry on a time study program includes a stopwatch, a time study board, time study forms, and a pocket calculator.

In addition, time-recording devices are being used successfully; time-recording machines, motion-picture cameras, and videotape equipment offer some advantages over the stopwatch.

The equipment needed for time study or work measurement is not nearly as elaborate or costly as that required for micromotion study. In general, the ability and personality of the time study analyst is more important than the equipment he or she chooses to use.

THE STOPWATCH

Of the several types of stopwatches in use today, the majority fall into the following classifications:

1. Decimal minute watch (0.01 minute).
2. Decimal minute watch (0.001 minute).
3. Decimal hour watch (0.0001 hour).
4. Electronic stopwatch.

The decimal minute watch in Figure 13–1 has 100 divisions on its face, and each division is equal to 0.01 minute. Thus, a complete sweep of the long hand requires one minute. The small dial on the watch face has 30 divisions, and each of these divisions is equal to one minute. For every revolution of the sweep hand, the small hand moves one division, or one minute.

This watch is started by moving the side slide toward the crown. Moving the side slide away from the crown stops the watch with the hands in

FIGURE 13–1
Decimal minute watch

Meylan Stopwatch Co.

their respective positions. To continue operation of the watch from the point where the hands stopped, move the slide toward the crown. Depressing the crown moves both the sweep hand and the small hand back to zero. Releasing the crown puts the watch back into operation unless the side slide is moved away from the crown.

The decimal minute watch tends to be a favorite with time study analysts because of its ease of reading and recording. The sweep hand moves only 60 percent as fast as the long hand of the decimal hour watch, and thus terminal points are more discernible. In recording the time values, the analyst's task is simplified by the fact that the elemental readings are in hundredths of a minute, eliminating the need for the ciphers that are registered when using the decimal hour watch, which is read in ten thousandths of an hour.

The decimal minute watch (0.001 minute) is similar to the decimal minute watch (0.01 minute). In the former, each division of the larger hand is equal to one thousandth of a minute. Thus, it takes 0.10 minute for the sweep hand to circle the dial rather than one minute, as with the 0.01 decimal minute watch. This watch is used primarily for timing short elements for standard data purposes (see Chapter 18). In general, the 0.001 minute watch has no starting slide on its side. It is started, stopped, and returned to zero by successive depressions of the crown.

A special adaption of the decimal minute watch that many time study people find convenient to use is illustrated in Figure 13–2. The two long

FIGURE 13–2
Decimal minute split (double-action)
watch

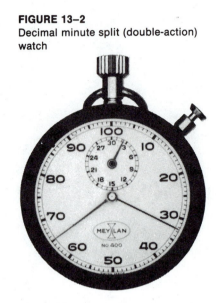

hands indicate decimal minutes and complete one turn of the dial in one minute. The small dial is graduated in minutes, and a complete sweep of its hand represents 30 minutes.

To start this watch, the crown is depressed, and both sweep hands simultaneously start from zero. Depressing the side pin at the termination of the first element stops the lower sweep hand only. The time study analyst can then observe the elapsed time for the element without the difficulty of reading a moving hand. Depressing the side pin causes the lower hand to rejoin the upper hand, which has been moving uninterruptedly. At the termination of the second element, depressing the side pin repeats the procedure.

The decimal hour stop watch has 100 divisions on its face also, but each division on this watch represents 1 ten thousandth (0.0001) of an hour. Thus, a complete sweep of the large hand on this watch represents one hundredth (0.01) hour, or 0.6 minute. The small hand registers each turn of the long hand; a complete sweep of the small hand takes 18 minutes, or 0.30 hour (see Figure 13–3). The decimal hour watch can be started and stopped, and its hands are returned to zero, in the same manner as is the decimal minute (0.01) stopwatch.

Since the hour represents a universal unit of time in measuring output, the decimal hour watch is a practical, widely used timer. Somewhat more skill is required to read this watch while timing short elements on account of the speed of the sweep hand. For this reason, some time study analysts prefer the decimal minute watch with its slower moving hand.

FIGURE 13–3
Decimal hour watch

By mounting four watches on a board, with linkage between them, during the course of a study the analyst can always read one watch which has stopped hands and maintain a cumulative record of the overall elapsed time. Figure 13–4 illustrates such an arrangement. These four crown action watches are actuated by the lever on the right. Pressure on the lever starts watch 1 (far left), cocks watch 2, stops watch 3, and starts watch 4. At the end of the first element, a clutch activating watch 4 is disengaged, and the lever is again pressed. This stops watch 1, starts watch 2, and resets watch 3 to zero, while watch 4 continues to run, as it will measure the overall time as a check. Watch 1 is now waiting to be read, while the next element is being timed by watch 2.

A more common practice is to use only one watch attached to the observation board, as illustrated in Figure 13–13 on page 360.

Most stopwatches record times with accuracies of plus or minus 0.025 minutes over 60 minutes of operation. Government specifications of stopwatch equipment allow a deviation of 0.005 minutes per 30-second interval. All stopwatches should be checked periodically to verify that they are not providing out-of-tolerance readings. To help assure continued accuracy of reading, stopwatches must be maintained properly. They should be protected from moisture, dust, and abrupt temperature changes. Once a year they should be cleaned and lubricated. If the watches are not used regularly, they should be wound and allowed to run down periodically.

Electronic stopwatches cost approximately $150. These watches provide resolution to one-hundredth second and accuracy to ±0.002 percent. They weigh about 0.20 kilogram and are about 10 centimeters long by 8

FIGURE 13–4
Time study board with four watches and form mounted, placed in proper position

centimeters wide and 4 centimeters deep. They permit timing any number of individual elements while also counting the total elapsed time. Thus, they can provide all of the advantages of snapback stopwatch study, with none of the disadvantages. (When the instrument is in the snapback mode, depressing the read button registers the time for that event and automatically resets to zero and begins accumulating time for the next event whose time is displayed by pushing the read button at the event's termination.) Electronic stopwatches operate on rechargeable batteries. Typically, the batteries must be recharged after about 14 hours of continuous service. (See Figure 13–5.) Professional electronic stopwatches have built-in battery indicators to avoid an untimely interruption of a study due to battery failure.

The only disadvantage, other than cost, is that some difficulty may be encountered in reading the display of the LED electronic stopwatch in studies made in the bright sunlight. The "sunlight" timer is an LCD electronic stopwatch designed for time studies in the bright sunlight. (LED is for "light emitting diode"; LCD is for liquid cryst display"). (See Figure 13–7.)

The electronic timer in Figure 13–6 provides cumulative and snapback studies; in both cases a stopped digital read-out can be recorded. When in

FIGURE 13–5
Digital electronic time-study board

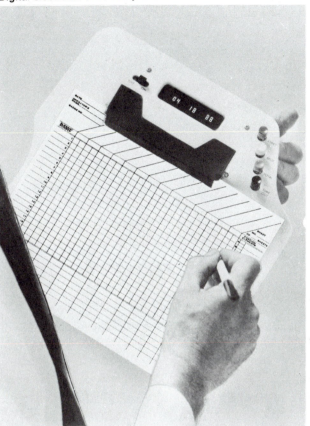

Meylan Stopwatch Co.

the cumulative mode, the timer accumulates the time and shows the total elapsed time from the start of the first event. At the termination of each element, pressing the read button provides a numerical read-out while the instrument continues to accumulate time. At the termination of the next element, pressing the read button again gives a stopped reading of the total accumulated time up to this point.

COMPUTER ASSISTED ELECTRONIC STOPWATCHES

The DataMyte 1000 all-solid-state battery operated data collector is a practical alternative to either a mechanical or electronic stopwatch. This instrument was first developed by the Electro/General Corporation (now DataMyte Corporation) in 1971 and is in wide use throughout the world today. It allows the keying in of observational data and recording the data

FIGURE 13-6
Electronic timer with start-
stop or continuous timing
options

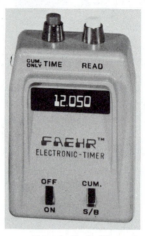

*Courtesy Faehr Electronic
Timers, Inc.*

FIGURE 13-7
Electronic stopwatches can include either L.E.D. display (for general office and
industrial indoor use) or L.C.D. display (for use in bright sunlight)

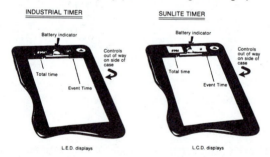

Courtesy of Faehr Electronic Timers, Inc.

in a solid-state memory in computer language. Elapsed time readings are recorded automatically. All input data and elapsed time data may be transmitted directly from the DataMyte to a computer terminal through an output cable. The computer prepares printed summaries, eliminating the laborious task of manually computing normal and allowed elemental times and operation standards.

This instrument is self-contained and can be carried throughout a plant or organization. The rechargeable battery power provides about 12 hours of continuous operation. Figure 13–8 shows an analyst using the Data-Myte. Time studies taken with the DataMyte and a computer take an estimated 50 to 60 percent of the time taken with a stopwatch and hand calculator.

Recently DataMyte Corporation developed Model 1010. It contains a DataMyte and a high-speed printer; thus the printed elemental and operation standards can be developed without interfacing with a computer. The 1010 is also compatible with most mini and microcomputers. Figure 13–9 illustrates those portions of a stopwatch time study handled by the Data-Myte and the interfacing computer.

Another computer-assisted electronic stopwatch that has been developed recently by Faehr Electronic Timers, Inc. is the COMPU–RATE (see Figure 13–10). Here manual entries are required only for the element column and the top four lines of the form, allowing the analyst to concentrate on observing the work and operator performance. Studies can be in thousandths of a minute or one hundred thousandths of an hour. The LCD displays can be contrast adjusted to fit the lighting conditions at the workstation. This instrument uses three "C" cell rechargeable NiCad batteries providing about 120 hours of running time before the batteries need to be recharged.

The COMPU–RATE software system will compute all the typical time study calculations including mean and median element values, adjustment of mean times to normal time after inputting performance rating and allowed times after inputting appropriate allowances. Results are summarized to provide standard times in minutes and/or hours per piece and pieces per hour. An edit function has been provided to correct mistakes.

The GageTalker Corporation (formerly Observational Systems) is marketing the OS–3 Plus Event Recorder. This versatile work measurement recorder has application not only for setting and updating standards but for use in making machine down times studies and for work sampling (see Chapter 21). Figure 13–11 illustrates this equipment, which permits the analyst to select the time units most appropriate for the study being made—0.001 min., 0.0001 hr. or 0.1 sec. The OS-3, after collecting and summarizing input data interfaces with printers for development of hard copy reports. Here the total time, frequency, mean times, performance ratings, normal times, allowances, standard times, and pieces per hour

FIGURE 13–8
Operator using a DataMyte 1000 Data Collector in conjunction with the DataMyte
1010 Time Study System

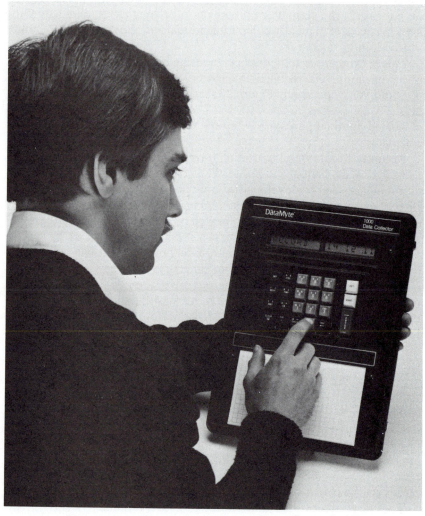

Courtesy DataMyte Corporation

FIGURE 13-9

Application of DataMyte and a computer to the taking and calculation of a time study

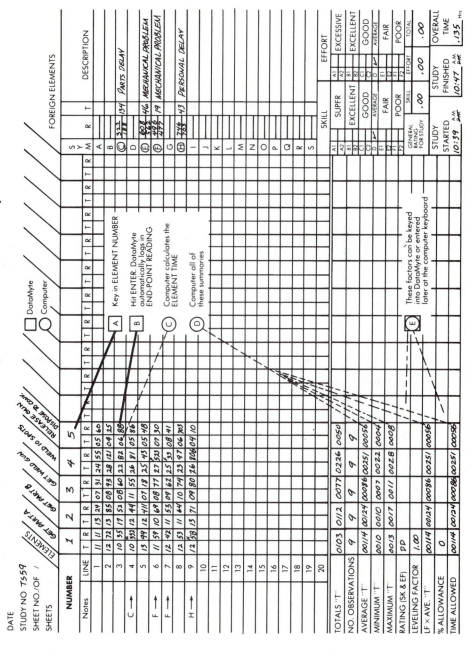

FIGURE 13–10
Computer-assisted electronic stopwatch

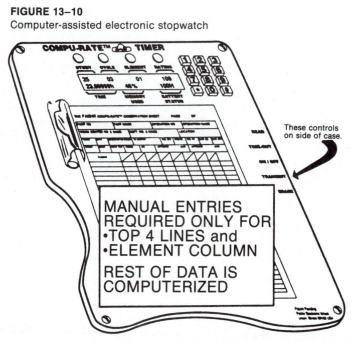

MANUAL ENTRIES
REQUIRED ONLY FOR
•TOP 4 LINES and
•ELEMENT COLUMN
REST OF DATA IS
COMPUTERIZED

These controls
on side of case.

Courtesy of Faehr Electronic Timers, Inc.

are printed. In addition, the standard deviation, the maximum and minimum element values, and frequencies are available.

TIME-RECORDING MACHINES

Several versatile time study machines facilitating the accurate measurements of time intervals are on the market today. In the absence of
time study analysts, these machines measure the time that a facility is
productive. For example, Figure 13–12 illustrates an eight-channel recorder where any two terminals may be connected to a normally open
sensor which closes only when the machine or activity is productive. On
the chart paper, a stylus records the state of the facility continuously. The
model illustrated records the activity of eight separate facilities. Chart
speeds range from 6 inches per hour to 480 inches per hour, depending on
the precision of measurement desired.

An adaptation of this equipment is its use with push button control,
where each channel can measure a specific work element. This adaptation
is especially useful in work sampling studies in which a professional
wishes to self-evaluate the distribution of his or her time. For example,
the eight channels may be assigned as follows:

FIGURE 13–11
The OS-3 Event Recorder

POWER
The OS-3 is powered by internal rechargeable batteries or by an AC-charger connected to this port. Batteries recharge in six hours and provide forty-eight hours of continuous data collection.

ON/OFF SWITCH
On enables data collection. OFF switches to low power mode. Data will remain intact for thirty days.

DISPLAY
The 80 character, dot matrix liquid crystal display combined with a full upper/lower case alphanumeric character set (96 ASCII characters) allows descriptive prompting and labeling.

CHARGING INDICATOR

NUMERIC PAD
The embedded ten Key numeric pad is designed for rapid data entry.

Left hand Keys.

KEYBOARD
The unique over/under typewriter configuration of the OS-3 keyboard allows rapid touch type entry of alphabetic information.

Right hand Keys.

RS-232C SERIAL INTERFACE
Direct printing of formatted reports and transmission of data to a host computer for detailed analysis utilizes this industry standard protocol.

Also allows access to the Audio capabilities. Data and Setup information can be conveniently and reliably stored on standard cassette tapes.

16 EXTERNAL CHANNELS
The OS-3 can automatically record switch closures from external devices.

INTELLIGENT CURSOR
The intelligent cursor indicates current character set as well as display position.

U upper case
L lower case
numeric
S special
!!! acknowledge

FUNCTION SWITCHES
Ten Function Switches serve as an extension of the Keyboard or as a ten channel event recorder.

END
The END key provides a convenient way of exiting from any program segment.

SHIFT
Allows entry of a character from outside the base character set.

lower case
special
upper case
numeric

Courtesy of GageTalker Corp.

Channel 1. Creative development.
Channel 2. Conference.
Channel 3. Dictation.
Channel 4. Incoming phone calls.
Channel 5. Outgoing phone calls.
Channel 6. Supervision and assigning work.

FIGURE 13–12
Eight-channel time study machine

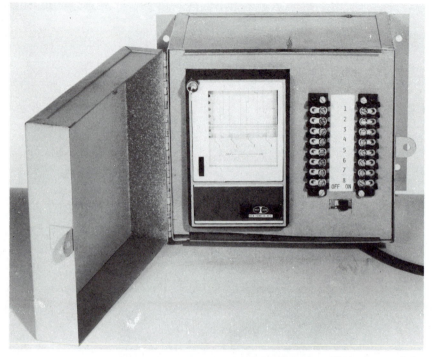

Meylan Stopwatch Co.

Channel 7. Reading mail.
Channel 8. Personal and interruptions.

By pushing a button related to the appropriate channel, the professional continuously accounts for the time expended during the workday.

VIDEOTAPE AND MOTION PICTURE CAMERAS

Videotape and motion picture cameras are ideal for recording operators' methods and elapsed time. The cost of film and the delay necessitated by sending films out to be developed prohibits the use of motion picture cameras in most instances today. The use of videotape is increasing dramatically. With the reduction in cost of quality equipment and the refinement of fundamental motion data systems, the use of videotape equipment will continue to grow.

By taking pictures of the operation and then studying them a frame at a time, analysts can record exact details of the method used and assign normal time values. They can also establish standards by projecting films at the same speed that the pictures were taken and then performance rating the operator. Because all the facts are there, observing the video-tape is a fair and accurate way to rate performance. Then, too, potential methods improvements can be revealed through the camera eye that would seldom be uncovered with a stopwatch procedure.

These advantages, supplemented with the memomotion procedure that allows longer cycle filming with minimum exposure of film or videotape, have increased the popularity of the camera and videotape with time study analysts (see Chapter 8). Figure 8–6 on page 252 illustrates a por-tion of a memomotion film taken at the rate of one frame per second.

THE TIME STUDY BOARD

When the stopwatch is being used, it is convenient to have a suitable board to hold the time study form and the stopwatch. The board should be light, so as not to tire the arm, and yet strong and hard enough to provide a suitable backing for the time study form. One-quarter-inch plywood or a smooth plastic make suitable materials. The board should have arm and body contacts (see Figure 13–13) for comfortable fit and ease of writing while it is being held.

The watch is mounted in the upper right-hand corner of the board (for right-handed people). To the left a spring clip holds the time study form. By standing in the proper position, a time study analyst can look over the top of the watch to the workstation and follow the operator's movements while keeping both the watch and the time study form in the immediate field of vision.

While performance rating operations under study, analysts may attach an alignment chart, developed by the author, to the board (see Figure 13–13). This chart permits analysts to determine synthetically the allowed time for several of the effort elements comprised by the study. The ratio of the synthetic value for a particular element to the mean value actually taken by the operator serves as a guide for determining the performance factor. This chart establishes standards for the therbligs reach, grasp, move, position, and release when these are performed collectively.

TIME STUDY FORMS

All the details of the study are recorded on the time study form. To date, little standardization in the design of forms used by various indus-tries has taken place. The form must provide space to record all pertinent information concerning the method being studied. This is often done by constructing an operator process chart (see Chapter 6) on one side of the form. In addition to making a permanent record of the relative location of the tools and materials in the work area, analysts can list such methods data as feeds, depths of cuts, speeds, and inspection specifications. Of

FIGURE 13-13
Time study board for a right-handed observer (note the alignment chart on the board to assist the observer in determining the performance rating factor; see page 427 for an explanation of this alignment chart)

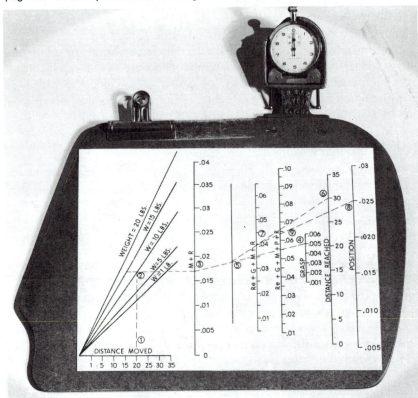

course, analysts should identify the operation being studied by including such information as the operator's name and number, operation description and number, machine name and number, special tools used and their respective numbers, department where the operation is performed, and prevailing working conditions. Providing too much is better than too little information concerning the job being studied.

The time study form should also include space for the signature of the supervisor, indicating approval of the method under observation. Likewise, the inspector should sign every study taken, acknowledging his or her acceptance of the quality of the parts being produced during the time study.

Analysts should design the form so that it contains watch readings, foreign elements (see Chapter 14), and rating factors (see Chapter 15), and

FIGURE 13-14
Front of time study form designed for either elemental or overall rating

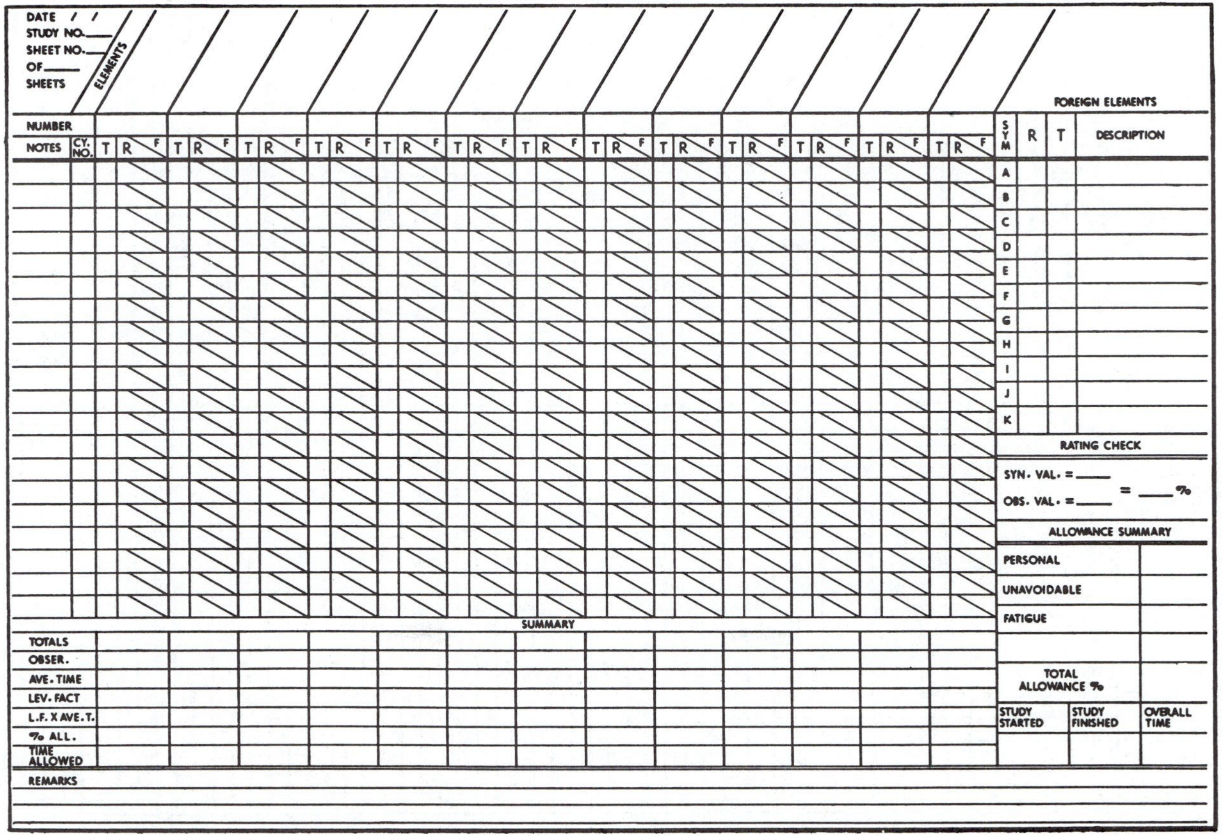

FIGURE 13–15
Back of time study form. Note operator chart (act breakdown) on which all details of
the method under study may be recorded.

SKETCH			STUDY NO._____DATE_____

(form fields)

STUDY NO._____DATE_____
OPERATION_____
DEPT._____OPERATOR_____NO._____
EQUIPMENT_____
_____MCH. NO._____
SPECIAL TOOLS, JIGS, FIXTURES, GAGES_____
CONDITIONS_____
MATERIAL_____
PART NO._____DWG. NO._____
PART DESCRIPTION_____

ACT BREAKDOWN		ELEM. NO.	SMALL TOOL NUMBERS, FEEDS, SPEEDS, DEPTH OF CUT, ETC.	ELEMENTAL TIME	OCC. PER CYCLE	TOTAL TIME ALLOWED
LEFT HAND	RIGHT HAND					

EACH PIECE_____   TOTAL_____
SET-UP_____   HRS. PER C_____
FOREMAN_____   INSPECTOR_____
OBSERVER_____   APPROVED BY_____

still has space to calculate the allowed time. Figures 13–14 and 13–15 illustrate a time study form that has been developed by the author. It is sufficiently flexible to study practically any type of operation.

On this form, analysts can record the various elements of the operation horizontally across the top of the sheet, as well as recording the cycles studied vertically row by row.

The "R" column is divided into two sections. The large area is for recording the watch reading, and in the small section marked "F" is shown the element performance factor if elemental rating is used. The "T" column is provided for elemental elapsed values (see Chapter 14) if leveling of the entire study is performed. The "T" column is used for normal elemental times if elemental rating is done. In the diecasting operation (see Figure 14–6, page 381), an analyst performed an elemental rating so the "T" values shown are the product of the R (reading) value and the F (rating factor) value.

AUXILIARY EQUIPMENT

Time study analysts need additional equipment to facilitate calculating studies rapidly and accurately. The foremost of these is an electronic calculator to solve problems involving multiplication, division, and proportion rapidly.

Although, in most cases, modern self-powered machine tools show their speeds in an obvious place, sometimes the running speed is not evident. Then, too, the speeds indicated by the manufacturer are based on pulley diameters which may have been altered during the setup or changed during the maintenance or overhaul of the machine.

To determine the speed being used, time study analysts can use a speed indicator. This instrument has few parts, is simple in operation, and gives relatively accurate revolutions of shafting, wheels, and spindles in either direction.

Training Equipment

Two inexpensive pieces of equipment assist in the training of time study analysts: the first of these, a random elapsed time describer, is illustrated in Figure 13–16. This device can be programmed (by means of a specially contoured cam) so that successive elements, in this case one through nine, are completed and so that each is accomplished in a known period. The trainee records the durations of the elements as they take place. The trainee is signaled by the lighting of a bulb and by a buzzer at the end of each element. Either the buzzer or the light may be made inoperative if desired.

This exercise provides practice in reading the watch at terminal points and in recording the elapsed time. After becoming proficient with data generated through the given cam, an operator can be challenged by a new

FIGURE 13–16
Random elapsed time describer for training time study analysts

cam capable of generating shorter elemental cycles. This device teaches the trainee the technique of using both sound and sight to identify terminal points. It also provides practice in handling two or more short elements that occur successively and are followed by a relatively long element.

A second helpful training tool is the metronome used by students of music. This device can be set to provide a predetermined number of beats per minute. For example, the metronome can be set to provide 104 beats per minute. This is also the number of cards dealt per minute when dealing at a normal pace. By synchronizing the delivery of a card at a four-hand bridge table so that a card is delivered with each beat of the metronome, we are able to demonstrate normal pace. This speed of movement involving a series of reaches, grasps, moves, and releases can be easily identified with practice. To illustrate 80 percent performance, the instructor need only set the metronome to 83 beats per minute and then synchronize the card dealing accordingly. Instructors find this device very helpful in demonstrating various levels of performance in card dealing. Once a new time study trainee becomes proficient in accurately rating hand movements involving the dealing of cards, he or she can make the transition to evaluating shop operations more easily.

TEXT QUESTIONS

1. What equipment is needed by time study analysts to carry on a program of time study?
2. What features of the decimal minute watch make it attractive to time study analysts?
3. Where does the decimal minute watch (0.001 minute) have application?
4. How does the duration of one division on the decimal hour watch compare to one division on the decimal minute watch?
5. Describe in detail the operation of the DataMyte work measurement facility. What advantage does it have over the stopwatch? The electronic timer? What disadvantages?
6. Describe the COMPU-RATE electronic stopwatch.
7. What is the principal advantage of the LCD electronic stopwatch?
8. What does LED stand for in connection with electronic stopwatches?
9. What is the principal advantage of the time-recording machine?
10. What is the memomotion procedure? What are its advantages and limitations?
11. Why is it desirable to provide space for an operator process chart on the time study form?
12. Describe the speed indicator and how it is used.
13. How many feet of film would be exposed on a five-minute cycle while using the memomotion technique?
14. How do government specifications of stopwatch equipment compare with the accuracy of most stopwatches?
15. Explain how a "random elapsed time describer" can be used to train new work measurement analysts.
16. How can the metronome be used as a training tool for performance rating?

GENERAL QUESTIONS

1. Explain the statement, "The criteria for success are the ability and personality of the time study analyst rather than the equipment he or she chooses to use."
2. Why hasn't industry adopted a standardized time study form?
3. Do you feel that the typical shop worker would favor the DataMyte facility over the stopwatch? Why or why not?
4. What would be the advantages of using videotape where the features of playback sound, as well as operation recorded end points of progressive elements, are incorporated?

PROBLEMS

1. To demonstrate various levels of performance to a group of union stewards, the time study supervisor of the XYZ Company is using the metronome while dealing bridge hands. How many times per minute should the metronome beat

to demonstrate the following levels of performance: 60 percent, 75 percent, 100 percent, 125 percent?

2. What would be the synthetic time value to reach 20 inches for a 15-pound casting, move it 15 inches, and place it in a fixture requiring 0.020 minutes position time and a grasp involving 0.004 minutes? See Chapter 19.

3. If government regulations permit an average deviation of 0.005 minutes per 30-second interval and a standard deviation of 0.001, what is the maximum deviation in seconds per minute that would be permitted at the 99 percent confidence level?

SELECTED REFERENCES

Barnes, Ralph M. *Motion and Time Study: Design and Measurement of Work.* 7th ed. New York: John Wiley & Sons, 1980.

Lowry, Stewart M.; Harold B. Maynard; and G. J. Stegemerten. *Motion and Time Study.* 3rd ed. New York: McGraw-Hill, 1940.

Mundel, M. E. *Motion and Time Study: Improving Productivity.* 5th ed. Englewood Cliffs, N.J.: Prentice Hall, 1978.

Nadler, Gerald. *Work Design: A Systems Concept.* Rev. ed. Homewood, Ill.: Richard D. Irwin, 1970.

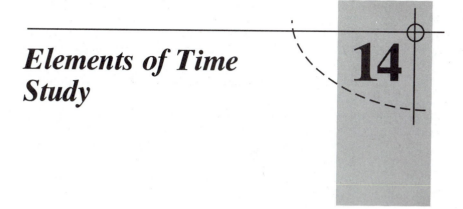

Elements of Time Study

14

The actual taking of a time study is both an art and a science. To be sure of success in this field, analysts must be able to inspire confidence in, exercise judgment with, and develop a personable approach to everyone with whom they come in contact. In addition, their backgrounds and training should prepare them to thoroughly understand and perform the various functions related to the study. These elements include selecting the operator, analyzing the job and breaking it down into its elements, recording the elapsed elemental values, performance rating the operator, assigning appropriate allowances, and working up the study.

CHOOSING THE OPERATOR

The first approach to beginning a time study is made through the departmental or line supervisor. After reviewing the job in operation, both the supervisor and the time study analyst should agree that the job is ready to be studied. If more than one operator is performing the work assignment for which the standard is to be established, consider several things when selecting just which operator to use in taking the study. In general, an operator who is average or somewhat above average in performance gives a more satisfactory study than a low-skilled or highly superior operator. The average operator usually performs the work consistently and systematically. That operator's pace will tend to be in the approximate range of normal (see Chapter 15), thereby making it easier for the time study analyst to apply a correct performance factor.

Of course, the operator should be completely trained in the method, should like the work, and demonstrate an interest in doing a good job. The operator should be familiar with time study procedures and practice, and have confidence in time study methods, as well as in the time study analyst. The operator should also be cooperative enough to follow through willingly with suggestions made by both the supervisor and the time study analyst.

At times, the analyst has no choice of operators because the operation is performed by only one worker. In cases of this nature, the analyst must be very careful when establishing the performance rating because the operator may be performing at either of the extreme ends of the rating scale. In one-worker jobs, the method used must be correct and the analyst must approach the operator tactfully.

Approach to the Operator

The analyst's approach to the selected operator may determine the cooperation received. Approach the operator in a friendly manner and show that you understand the operation to be studied. Give the operator an opportunity to ask any questions about the timing technique, method of rating, and application of allowances. In some instances, the operator may have never been studied before. It is worthwhile to answer all questions frankly and patiently. Encourage the operator to offer suggestions and, when the operator does so, receive them willingly, thus showing that you respect the skill and knowledge of the operator.

Show interest in the worker's job, and at all times be fair and straightforward toward the worker. This approach wins over the worker's confidence in your ability. The resulting respect and goodwill not only help in establishing a fair standard, but also make future work assignments on the production floor pleasanter.

ANALYZING MATERIALS AND METHODS

Perhaps the most common mistake made by time study analysts is neglecting to make sufficient analysis and records of the method being studied. The time study form illustrated in Chapter 13 has space for a sketch or photograph of the work area. All sketches should be drawn to scale and show each detail that affects the method. Sketches should clearly show the location of the raw material and finished parts bins with respect to the work area. Clearly record distances that operators must move or walk. Indicate the location of all tools involved in the operation, thus illustrating the motion pattern utilized in performing successive elements.

Immediately below the pictorial presentation of the method, allow space for an operator process chart (see Chapter 6) of the method being studied. On high-activity work, complete this chart before beginning the actual timing of the operation. By completing this right- and left-hand chart, analysts identify the method under study and can observe opportunities for methods improvement. This facilitates the elemental breakdown of the study to be taken, and helps analysts to acquire a better conception of the skill being executed.

The value of completely identifying the method under study cannot be overemphasized. Since management usually guarantees a standard as long as the method studied is in effect, it is mandatory that the method studied be completely known. For example, after machining a casting, the operator may have been studied while placing the casting in a tote box on a pallet four feet from the workstation. The element "lay aside finished casting" may have consumed 0.05 minute. Assume that some time after the standard was established management changed the location of the skid so that a chute and drop delivery could remove the completed part. This could reduce the "lay aside finished casting" element to 0.01 minute. The savings of 0.04 minute per cycle would be substantial if the job could be completed in 0.40 minute or less. This small methods change could result in a loose rate that would be troublesome to the union, the company, and the worker's fellow employees. Unless a complete record of the method originally used to "lay aside finished casting" had been made, the time study department would have no authority to restudy the job.

More pronounced methods changes are often made without informing the time study department, such as changing the job to a different machine, increasing or decreasing feeds and speeds, or using different cutting tools. Changes of this nature, of course, seriously affect the validity of the original standard. Often the first time they are brought to the time study department's attention is when a grievance has been submitted that a rate is too tight, or the cost department complains about a loose standard. Investigation frequently reveals that a change in method has been the cause of the inequitable standard. To know what part or parts of the job should be restudied, analysts must have information as to the method used when the job was initially studied. If this information is not available and the rate appears too loose, the only recourse analysts have is to let the loose rate prevail for the life of the job—a situation that management resents—or to change the method again and then immediately restudy the job—a program that the operator and the union will criticize.

Analysts should record information on the type of material being processed and on the material used in cutting tools (ceramic, high-speed, carbide). A design change calling for a part to be made from "60–40 brass" rather than "70–30 brass" could have a pronounced effect on cycle time. Likewise, a change from high-speed tools to carbide tools could decrease machining time by more than 50 percent.

No job should be time studied until it is *ready* to be studied, that is, until the method employed is correct. However, methods must be continually improved in order to progress. A plant that does not continually emphasize methods improvement falls into the doldrums and eventually is unable to operate profitably. Since methods changes are continually taking place, a complete analysis of the existing materials and methods should be made and recorded prior to beginning the actual watch readings.

RECORDING SIGNIFICANT INFORMATION

Analysts need to make note of the machines, hand tools, jigs or fixtures, working conditions, materials used, operation performed, name and clock number of the operator, department, date of the study, and time study observer's name. Perhaps all this detail may seem unimportant to the novice, but an experienced analyst realizes that the more pertinent information he or she records, the more useful the study becomes over the years. The time study becomes a source of establishing standard data (see Chapter 18) and the development of formulas (see Chapter 20). It will also be useful in methods improvement, operator evaluation, tool evaluation, and machine performance.

When machine tools are used, specify the name, size, style, capacity, and serial or inventory number. For example, "No. 3 Warner & Swasey ram-type turret lathe, Serial No. 111408, equipped with 2-jaw S.P. air chuck" would identify the facility under use. Cutting tools should be described completely, such as "½-inch, two-flute, high-speed, 18–4–1 gun drill." Likewise, the descriptions of general facilities should be complete and well defined, such as "2-pound ball peen hammer," "1-inch micrometer," "combination square," "12-inch bastard file."

Identify dies, jigs, gages, and fixtures by their numbers and short descriptions. For example, "flange trimming die F–1156," "progressive trimming, piercing, forming die F–1202," "Go No-Go Thread Plug Gage F–1101."

Note the working conditions for several reasons: First, the prevailing conditions have a definite relationship to the "allowance" added to the leveled or normal time. If conditions improve in the future, the allowance for personal time, as well as fatigue, may be diminished. Conversely, if for some reason the working conditions become poorer than when the time study was first taken, the allowance factor should be raised.

Second, if the working conditions prevailing during the study are different from the normal conditions that exist on that job, then they would have an effect on the usual performance of the operator. For example, in a drop forge shop, if a study were taken on an extremely hot summer day, it can be readily understood that the working conditions would be poorer than they usually are, and operator performance would reflect the effect of the intense heat. The following examples illustrate the description that should be included when recording the job conditions: "Normal for job, wet, hot (90° F), operator standing," or "poor, temperature 85° F, operator seated, clean, noisy."

Customarily analysts indicate first the working conditions as they compare with the average conditions for the job. Following this description is
a short account of the exact conditions observed.

Identify raw materials completely, giving such information as heat
number, size and shape, weight, quality, and previous treatments.

Describe the operation performed specifically. For example, "broach
⅜ inch × ⅜ inch keyway in 1-inch bore" is considerably more explicit
than the description "broach keyway." There could be several inside
diameters on the part, each having a different keyway, and unless the hole
that is being broached is specified and the keyway size is shown, misinter-
pretation may result.

Identify the operator being studied by name and clock number. There
could easily be two John Smiths in one company. Then, too, the clock
number does not completely identify an employee, as labor turnover
results in the assignment of the same clock number to more than one
employee over a period of years.

The Observer's Position

After making the correct approach to the operator and recording all the
significant information, the time study analyst is ready to record the time
consumed by each element.

Take a position a few feet to the rear of the operator so as not to
distract or interfere. Stand while taking the study. Trying to take studies
while seated will be subject to criticism from the operators and will soon
cause analysts to lose their respect. Then, too, standing observers are
better able to move about and follow the movements of the operator's
hands as the operator goes through the work cycle. Figure 14–1 illustrates
the correct position of the observer with respect to the operator while
studying the set-up elements of a robotic welding operation. The observer
is using a DataMyte instrument. The observer has a simultaneous view of
his instrument and the hands of the operator.

During the course of the study, avoid any conversation with the operator, as this would tend to upset your routine and the worker's.

Dividing the Operation Into Elements

For ease of measurement, the operation is divided into groups of
therbligs known as "elements." To divide the operation into its individual
elements, watch the operator for several cycles. However, if the cycle
time is over 30 minutes, write the description of the elements while taking
the study. If possible, determine the elements into which the operation is
to be divided before the start of the study. Elements should be broken
down into divisions that are as fine as possible and yet not so small that
accuracy of reading is sacrificed. Elemental divisions of around 0.04 minute are about as fine as can be read consistently by an experienced time

FIGURE 14–1
Time study observer using a DataMyte collector to study the set-up elements of a
robotic welder.

study analyst. However, if the preceding and succeeding elements are
relatively long, an element as short as 0.02 minute can be readily timed.

To identify end points completely and develop consistency in reading
the watch from one cycle to the next, consider sound as well as sight in
the elemental breakdown. Thus, the terminal point of elements can be
associated with the sounds made, such as when a finished piece hits the
container, when a facing tool bites into a casting, when a drill breaks
through the part being drilled, and when a pair of micrometers are laid on
a bench.

Record each element in its proper sequence and include a basic division of work terminated by a distinctive sound or motion. Thus, the
element "up part to manual chuck and tighten" would include the basic
divisions reach for part, grasp part, move part, position part, reach for
chuck wrench, grasp chuck wrench, move chuck wrench, position chuck
wrench, turn chuck wrench, and release chuck wrench. The point of

termination of this element would be evidenced by the sound of the chuck wrench being dropped on the head of the lathe. The element "start machine" could include reach for lever, grasp lever, move lever, and release lever. The rotation of the machine with the accompanying sound would identify the point of termination so that readings could be made at exactly the same point in each cycle.

Frequently, different time study analysts within a company adopt a standard elemental breakdown for given classes of facilities to assure uniformity in establishing terminal points. For example, all single-spindle bench-type drill press work may be broken down into standard elements, and all lathe work may be composed of a series of predetermined elements. Having standard elements as a basis for operation breakdown is especially important in the establishment of standard data (see Chapter 18).

The principal rules for accomplishing the elemental breakdown are:

1. Ascertain that all the elements being performed are necessary. If some are unnecessary, the time study should be discontinued and a methods study should be made to develop the proper method.
2. Keep manual and machine time separate.
3. Do not combine constants with variables.
4. Select elements so that a characteristic sound identifies terminal points.
5. Select elements so that they can be readily and accurately timed.

When dividing the job into elements, keep machine or cutting time separate from effort or handling time. Likewise, constant elements (those elements for which the time does not deviate within a specified range of work) should be kept separate from variable elements (those elements for which the time does vary within a specified range of work).

After making a proper breakdown of all the elements constituting the operation, describe each element accurately. The termination of one element is automatically the beginning of the succeeding element and is referred to as the "breaking point." Observers should readily recognize the description of this terminal point or breaking point. This is especially important when the element does not include sound at its ending. On cutting elements, record the feed, speed, depth, and length of the cut immediately after the element description. Typical element descriptions are: "Pk. up pt. from table & pos. in vise" and "Dr. ½" D. 0.005" F 1200 R.P.M." In the interest of brevity, analysts abbreviate and use symbols to a great extent. This system of notation is acceptable only if the element is completely described in terminology and the symbols are meaningful to all who may come in contact with the study. Some companies use standard symbols throughout their plants, and everyone connected with a plant is familiar with the terminology.

When an element is repeated, a second description is not necessary. Instead, indicate in the space provided for a description of the element, the identifying number that was used when the element first occurred.

The time study form offers the flexibility required by diversified studies. For example, occasionally, recording an element that successively repeats itself is impossible due to the space limitations of the time study form. This can be handled by recording the watch readings of the repeating elements in the same column where the first occurrence of the element appears. Figure 14–2 illustrates this method of recording.

When there are more than 15 elements in the study, use a second sheet to record those elements in excess of 15. If more than 20 cycles are to be observed and the study comprises seven or fewer elements, then use the right-hand half of the form by repeating the elements and continuing the study in the open spaces. However, if more than seven elements occur and more than 20 cycles are to be recorded, use a second sheet. Figure 14–3 illustrates a 60-cycle study comprising four elements.

Taking the Study

There are two techniques of recording the elemental time while taking the study. The continuous method, as the name implies, allows the stopwatch to run for the entire duration of the study. In this method, the analyst reads a watch at the breaking point of each element while the hands of the watch are moving. A stopped hand is read under the continuous method when using a double-action watch, as shown in Figure 13–2 on page 348. Also, an electronic time study instrument can provide a stopped numerical value. See Figure 13–6 on page 352.

In the snapback technique, after reading the watch at the termination point of each element, the analyst then snaps the hands back to zero. As the next element takes place, the hands move from zero. Once the elapsed time is read directly from the watch at the end of this element, the hands are again returned to zero. This is the procedure throughout the entire study.

The time study analyst should advise the operator as to the start of the study as well as the exact time of day at which the study is being taken so that the operator will be able to verify the overall time. Record the time of day that the study is started on the time study form (see Figure 14–4), just before starting the stopwatch. Figures 14–4 and 14–5 illustrate a completed time study using the continuous method.

THE SNAPBACK METHOD. The snapback method has certain advantages and disadvantages when compared with the continuous technique. These should be considered before standardizing on one way of recording the values. Some time study analysts use both methods, believing that studies made up predominantly of long elements are adapted to snapback

FIGURE 14-2

Method of recording an element that successively repeats itself

DATE / / STUDY NO.__ SHEET NO.__ OF SHEETS															

NUMBER	1	2	3	4	5	6	7	8	9	10	11	12	13	14	15
NOTES CY.NO	T R F	T R F	T R F	T R F	T R F	T R F	T R F	T R F	T R F	T R F	T R F	T R F	T R F	T R F	T R F
1	5	12	16	26	35										
2			30	46	57										
3			62	71	80	101	9	19	40	57	62	68	201	11	32
4	37	44	40	60	71										
5			75	63	92										
6			87	307	17	36	43	52	81	90	83	400	35	44	63
7	60	77	82	91	501										
8			6	16	24										
9			27	38	47	66	78	88	606	17	23	29	50	80	86
10	90	701	5	14	24										
11			9	20	40										
12			44	52	64	84	93	686	35	46	52	58	86	94	820
13	25	33	37	47	86										
14			60	60	70										
15			83	95	1006	25	33	43	70	81	88	91	125	36	65
16	60	60	73	82	91										
17			96	1206	15										
18			19	20	30	50	67	77	1302	11	17	23	64	64	84
19	60	87	1401	11	22										
20			26	37	47										

SUMMARY															
TOTALS															
OBSER.															
AVE. T.															
LEV. FACT															
L.F. CAV. T.															
% ALL.															
TIME ALLOWED															

FOREIGN ELEMENTS			GENERAL RATING CHECK	SYN. VAL. ——= % OBS. VAL.	REMARKS:
	R	T			
A				STUDY STARTED / STUDY FINISHED / OVERALL TIME	
B				A.M. P.M. / A.M. P.M. / MIN.	
C			F		
D			G		
E			H		

readings, while short-cycle studies are better suited to the continuous procedure.

Since elapsed element values are read directly in the snapback method, it requires no clerical time for making successive subtractions, as does the continuous method. Then, too, elements that are performed out of order

FIGURE 14–3
Time study form showing the recording of a four-element, 60-cycle study

DATE / /
STUDY NO.__
SHEET NO.__ OF __ SHEETS

NOTES	CY. NO.	1 T	1 R	2 T	2 R	3 T	3 R	4 T	4 R	T	R	1 T	1 R	2 T	2 R	3 T	3 R	4 T	4 R	T	R	1 T	1 R	2 T	2 R	3 T	3 R	4 T	4 R	T	R
	1		4		14		20		23				64		73		80		83				53		65		72		75		
	2		27		36		42		46				87		97		503		6				80		91		97		1001		
	3		51		62		69		74				10		20		25		28				5		15		22		25		
	4		78		88		94		98				33		45		52		55				29		39		46		49		
	5		101		13		19		23				59		69		75		78				53		61		67		71		
	6		26		35		41		44				84		95		602		5				74		84		90		93		
	7		48		56		62		65				9		19		26		29				97		1106		13		16		
	8		69		79		85		88				34		43		49		52				20		30		37		40		
	9		91		200		5		9				55		63		68		72				44		54		60		63		
	10		13		22		28		31				76		91		98		702				67		76		82		85		
	11		35		45		51		54				5		15		21		24				89		99		1202		5		
	12		58		66		71		74				29		39		46		49				9		19		25		28		
	13		79		89		97		301				53		63		70		73				32		42		48		51		
	14		5		15		22		26				77		86		93		96				55		66		72		74		
	15		30		41		46		49				800		10		18		24				78		88		94		98		
	16		52		60		66		69				29		37		45		48				1302		12		18		21		
	17		73		83		89		92				52		62		69		72				25		35		41		44		
	18		96		406		12		15				76		85		91		94				49		61		68		71		
	19		19		29		35		38				98		908		15		18				75		85		91		95		
	20		42		51		57		60				23		35		45		48				1400		12		18		21		

SUMMARY

TOTALS															
OBSER.															
AVE. T.															
LEV. FACT															
L.F. × AV. T.															
% ALL.															
TIME ALLOWED															

FOREIGN ELEMENTS				GENERAL RATING CHECK			SYN. VAL. / OBS. VAL. —— = %	REMARKS:
	R	T						
A				STUDY STARTED	STUDY FINISHED	OVERALL TIME		
B				____ A.M. P.M.	____ A.M. P.M.	____ MIN.		
C				F				
D				G				
E				H				

by the operator can be readily recorded without special notation. Proponents of the snapback method state that delays are not recorded. Also, since elemental values can be compared from one cycle to the next, a decision can be made as to the number of cycles to study. Actually, it is erroneous to use observations of the past few cycles to determine how

FIGURE 14–4
Completed time study illustrating the continuous method of watch reading and overall performance rating (front of form)

DATE 3/11
STUDY NO. 108
SHEET NO. 1 OF 1 SHEETS

NUMBER		1		2		3		4		5		6		7		8	
Element		Pick up part from table & bring to vise & file edge		Position part in vise & tighten vise by hand. Position table		Start machine & engage feed		Mill 3/4" slot		Return table eight inches, stop machine		Open vise & remove part		Brush out chips		Reposition part for second cut	
NOTES	CY. NO	T	R	T	R	T	R	T	R	T	R	T	R	T	R	T	R
	1	14	14	16	30	4	34	68	102	6	8	19	27	14	41	11	52
	2	12	12	15	27	5	32	73	405	7	12	16	28	15	43	10	53
	3	A 22	36	17	53	4	57	64	1021	5	26	21	47	18	65	10	75
Blow Nose	4	12	34	B 16	32	(10)	42	70	1512	8	20	21	41	14	55	11	66
	5	12	30	12	42	4	46	70	1816	7	23	22	45	16	61	9	70
	6	13	23	17	40	4	44	66	2110	5	15	20	35	15	50	10	68
Drop Part	7	16	81	16	97	5	2402	73	75	5	80	(34)	2514	14	28	9	37
Incomp. Brushing	8	15	92	17	2709	4	13	D (50)	64	6	70	18	88	(7)	95	10	3005
	9	14	83	17	3800	4	4	71	75	6	81	21	3902	14	16	11	27
	10	17	88	12	4100	–	M	–	80	5	85	21	4206	14	20	11	31
	11	12	88	15	4403	5	8	70	78	6	84	20	4504	14	4528 / 4514	10	4514 / 4504
	12	15	85	18	4703	4	7	68	4775	6	81	17	98	15	4813	9	22
	13	11	84	17	5001	6	7	71	78	7	85	18	5103	16	19	10	29
	14	13	92	17	5309	5	14	73	87	6	93	16	5409	15	24	9	33
	15	20	5600	18	18	4	22	71	93	5	98	19	5717	13	30	9	39
	16	13	5906	14	20	4	24	66	90	6	96	18	6014	14	28	11	39
	17	11	94	19	6213	5	18	70	88	6	94	20	6314	12	26	10	36
	18	12	98	20	6518	6	24	71	95	5	6600	20	20	15	35	9	44
	19	14	6809	16	25	5	30	75	6905	4	9	15	24	17	41	11	52
	20	12	7117	18	35	6	41	72	7213	5	18	19	37	15	52	9	61

NUMBER		9		10		11		12		13		14		15	
Element		Tighten vise & return table		Start machine		Run table forward 4" & engage feed		Mill second 3/4" slot		Return table eight inches		Stop machine & open vise		Brush chips & aside part	
NOTES	CY. NO	T	R	T	R	T	R	T	R	T	R	T	R	T	R
	1	-21	73	3	76	8	84	77	261	9	70	12	82	18	300
	2	23	76	5	81	10	91	80	571	9	80	6	86	20	606
	3	21	96	3	99	11	1110	75	85	11	96	10	1206	16	22
Blow Nose	4	19	85	4	89	9	98	82	1680	11	91	8	99	19	1718
	5	22	92	2	94	9	1903	71	74	9	83	8	91	19	2010
	6	20	80	5	85	10	95	80	2275	C 10	36	9	45	20	65
Drop Part	7	21	58	3	61	8	69	76	2645	(4)	49	10	59	18	77
Incomp. Brushing	8	20	25	5	30	11	41	E 80	3732	9	41	12	53	16	69
	9	21	48	5	53	8	61	71	4032	10	42	14	56	15	71
	10	22	53	–	M	–	M	–	4340	10	50	8	58	18	76
	11	20	4548 / 4528	4	52	10	62	75	4637	9	46	7	53	17	78
	12	21	43	4	47	10	57	78	4935	10	45	8	53	20	73
	13	20	49	4	53	9	62	80	5242	11	53	8	61	18	79
	14	19	52	5	57	8	65	77	5542	9	51	9	60	20	80
	15	19	58	6	64	10	74	80	5854	11	65	9	74	19	93
	16	20	59	5	64	12	76	74	6150	9	59	7	63	17	83
	17	19	55	5	60	11	71	79	6450	8	58	8	66	20	86
	18	20	64	6	70	9	79	80	6759	8	67	10	77	18	95
	19	23	75	5	80	10	90	81	7071	9	80	8	88	17	7105
	20	21	82	6	88	11	99	80	7379	9	88	10	98	22	7420

SUMMARY

	1	2	3	4	5	6	7	8	9	10	11	12	13	14	15
TOTALS	2.80	3.27	.94	12.62	1.16	3.61	2.80	1.99	4.12	.86	1.84	14.76	1.81	1.81	3.67
OBSER.	20	20	18	18	20	19	18	20	20	19	19	19	19	20	20
AVE. T.	.140	.163	.047	.701	.058	.190	.147	.100	.206	.045	.097	.777	.095	.090	.184
LEV. FACT	1.10	1.10	1.10	1.00	1.10	1.10	1.10	1.10	1.10	1.10	1.10	1.00	1.10	1.10	1.10
L.F. & AV. T.	.154	.179	.052	.701	.0[illegible]	.208	.162	.110	.226	.050	.107	.777	.104	.099	.202
% ALL.	15	15	15	10	15	15	15	15	15	15	15	10	15	15	15
TIME ALLOWED	.177	.206	.060	.806	.711	.239	.186	.126	.260	.058	.123	.855	.120	.114	.232

FOREIGN ELEMENTS	R	T		GENERAL RATING CHECK	SYN. VAL. / OBS. VAL.		REMARKS:
					#2 .164/.150 = 109%		
				STUDY STARTED	STUDY FINISHED	OVERALL TIME	
A	914 / 606	308	Get drink	10:32 A.M.	11:46 A.M.	74 MIN.	
B	1416 / 1234	182	Interrupted by foreman				
C	2326 / 2275	51	Gage part	F			
D	2914 / 2713	201	Chip in eye	G			
E	3652 / 3041	611	Replace milling cutter	H			

many additional cycles should be studied. This practice can lead to studying entirely too small a sample.

W. O. Lichtner points out one recognized disadvantage of the snapback method: Individual elements should not be removed from the operation and studied independently because elemental times are dependent on

FIGURE 14–5
Completed time study form (back of form)

SKETCH

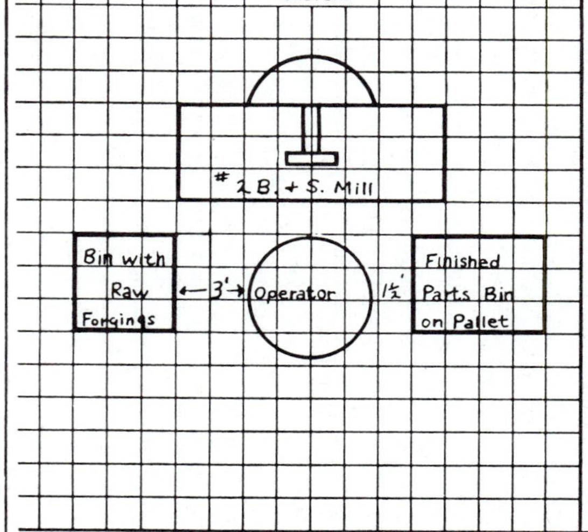

2 B. + S. Mill

Bin with Raw Forgings — 3' → Operator — 1½' → Finished Parts Bin on Pallet

STUDY NO. 132   DATE 9-15

OPERATION Mill diagonal slots in regulator clamp

DEPT. 11-5   OPERATOR B.V. Harvey   NO. 14

EQUIPMENT #2 B. + S. Horizontal Mill

MCH. NO. 3-17694

SPECIAL TOOLS, JIGS, FIXTURES, GAGES Milling fixture
#F-1941, 18-4-1 high speed cutters

CONDITIONS Normal

MATERIAL X-4130 steel forging

PART NO. S-11691-1   DWG. NO. SA-11694

PART DESCRIPTION 2" clamp for series 411

ELEM. NO.	SMALL TOOL NUMBERS, FEEDS, SPEEDS, DEPTH OF CUT, ETC.	ELEMENTAL TIME	OCC. PER CYCLE	TOTAL TIME ALLOWED
1		.177	1	.177
2		.206	1	.206
3		.060	1	.060
4	95 R.P.M. 5⅜"/min.	.806	1	.806
5		.074	1	.074
6		.239	1	.239
7		.186	1	.186
8		.126	1	.126

ACT BREAKDOWN

LEFT HAND	RIGHT HAND
Reach 12"; grasp forging on mill bed, move 12" hold.	Reach 8" grasp file,
	Move file to forging, file edge (3 strokes)
Position forging in vise	Move file 8", release.
	Reach 12" for vise handle,
	grasp handle, turn handle, 90 to 110 degrees
Idle	Reach for start button

No.	Description		Value	Value
9	14", apply pressure, reach for feed lever, grasp lever, Move lever.		.260	.260
10	Get new piece from bin. Idle during mill cut		.058	.058
11	Reach for stop button, apply pressure. Reach 8", grasp control wheel, Move 45° release wheel.		.123	.123
12	Reach 15" to part, grasp part, Move part 8" and release. Reach 14" for vise handle, grasp handle, turn handle 90-180°.	95 R.P.M. 5⅜"/ min.	.855	.855
13	Idle — Reach 12" grasp brush, move brush to vise 12" and brush chips.		.120	.120
14	Reach 8" to part, grasp, Move 8" to vise, position — Move brush to table 1/2", release brush, reach for vise handle.		.114	.114
15	Grasp vise handle, turn handle 90 to 180 degrees.		.232	.232
	Idle — reach 8", grasp control wheel move 45 degrees and release			
	Idle — Reach 14" for start button, apply pressure.			
	Reach 14" for control wheel, grasp wheel, turn wheel			
	Reach 18" grasp feed lever move lever — 15 degrees, release.			
	Idle during mill cut — Lay aside finished piece.			
	Reach for control wheel			
	Idle — 14", grasp wheel, turn 60 degrees.			
	Reach 14" for stop button, apply pressure. — Release control wheel, reach 12" and grasp vise handle, move handle 90°.			
	Reach for part 12", grasp, Move part to table — Reach 12" grasp brush, move brush to vise 12" brush			
	release. — chips, lay brush aside.			

EACH PIECE 3.636 MIN.   TOTAL 3.636

HRS. PER C 6.06

SET-UP

FOREMAN W. H. Armstrong   INSPECTOR C. D. Anderson

OBSERVER George Thuring   APPROVED BY B.L.M.

the preceding and succeeding elements.[1] Consequently, by omitting such factors as delays, foreign elements, and transposed elements, erroneous values prevail in the readings accepted.

Another objection to the snapback method that has received considerable attention from labor groups, is the time lost while snapping the hand back to zero. Lowry, Maynard, and Stegemerten state: "It has been found that the hand of the watch remains stationary from 0.00003 to 0.000097 hour at the time of snapback, depending upon the speed with which the button of the watch is pressed and released."[2] This would mean an average loss of time of 0.0038 minute per element, which would be a 3.8 percent error in an element 0.10 of a minute in duration. Of course, the shorter the element, the greater the percent of error introduced, and the longer the element, the smaller the error. Although experienced time study people tend in reading the watch to allow for the "snapback time" by reading to the next higher digit, a sizable cumulative error can develop with the snapback method. The new electronic watches negate this disadvantage since no time is lost in moving a hand back to zero. Figures 14–6 and 14–7 illustrate a time study of a die casting operation using the snapback method.

In summary, the snapback procedure has the following disadvantages:

1. Time is lost in snapping back; therefore, a cumulative error is introduced into the study. This can be avoided by using electronic watches.
2. Short elements (0.06 minute and less) are difficult to time.
3. A record of the complete study is not always given in that delays and foreign elements may not be recorded.
4. There is no verification of the overall time with the sum of the elemental watch readings.

THE CONTINUOUS METHOD.   The continuous method of recording elemental values is superior for several reasons. Probably the most significant reason is that this type of study presents a complete record of the entire observation period and, as a result, appeals to the operator and the union. The operator is able to see that no time has been left out of the study and that all delays and foreign elements have been recorded. With all the facts clearly presented, it is easier to explain and sell this technique of recording times.

The continuous method is also better adapted to measuring and recording very short elements. Since no time is lost in snapping the hand back to zero, accurate values can be obtained on successive elements of 0.04

---

[1] W. O. Lichtner, *Time Study and Job Analysis* (New York: Ronald Press 1921), p. 168.

[2] S. M. Lowry, H. B. Maynard, and G. J. Stegemerten, *Time and Motion Study and Formulas for Wage Incentives*, 3rd ed. (New York: McGraw-Hill, 1940), p. 191.

FIGURE 14–6
Snapback study of a die casting operation (front of form)

DATE **3** /1/81
STUDY NO. **2-23**
SHEET NO. **1**
OF **1**
SHEETS

ELEMENTS
1. Remove part, keep die, lubricate die
2. Place part in fixture, ram in
3. Get 2 lead pigs, move 20', place in pot

CY. NO.	1 T	1 R	2 T	2 R	3 T	3 R
1	27	30⁹⁰	21	63¹⁰⁰	16	120⁸⁰
2	22	27¹⁰⁰	21	21		
3	28	31⁹⁰	21	23		
4	30	35⁹⁵	20	20¹⁰⁰		
5	28	28¹¹⁰	20	20¹⁰⁰		
6	25	25¹⁰⁰	20	18¹¹⁰		
7	28	31⁹⁰	22	24⁹⁵		
8	28	28¹⁰⁰	29	24⁹⁵		
9	32	32⁹⁰	21	23⁹⁰		
10	26	26⁹⁰	20	19¹⁰⁵		
11	30	35¹¹⁰	20	24⁹⁰		
12	30	35¹⁰⁰	21	21¹⁰⁰		
13	27	27¹⁰⁰	30	30¹⁰⁰		
14	29	40¹⁰⁵	13	15⁹⁵		
15	24	27¹⁰⁵	19	19⁹⁵		
16	28	28⁹⁵	24	24⁹⁰		
17	32	32⁹⁰	13	15⁹⁰		
18	27	22¹⁰⁰	22	22⁹⁰		
19	28	31⁹⁰	21	22⁹⁵	19	98 120 75
20	26	29¹¹⁰	21	19¹¹⁰		

	1	2	3
TOTALS	5.61	4.18	1.94
OBSER.	20	20	2
AVE.-TIME	0.8906	0.209	0.97
LEV. FACT	0.2805	0.209	0.97
% ALL	17	17	17
TIME ALLOWED	0.328	0.245	1.135

SUMMARY

REMARKS: Machine cycle time 0.35 minutes
Needs to get lead (element 3) once every 25 cycles

FOREIGN ELEMENTS

SYM	R	T	DESCRIPTION
A			
B			
C			
D			
E			
F			
G			
H			
I			
J			
K			

RATING CHECK

SYN. VAL. = _____ = _____ %
OBS. VAL. = _____

ALLOWANCE SUMMARY

PERSONAL	5
UNAVOIDABLE	8
FATIGUE	4
TOTAL ALLOWANCE %	17

STUDY STARTED	STUDY FINISHED	OVERALL TIME
3:42 PM	3:55	min. 13

FIGURE 14–7

Snapback study of die casting operation (back of form)

ACT BREAKDOWN		ELEM. NO.	SMALL TOOL NUMBERS, FEEDS, SPEEDS, DEPTH OF CUT, ETC.	ELEMENTAL TIME	OCC. PER CYCLE	TOTAL TIME ALLOWED
LEFT HAND	RIGHT HAND					

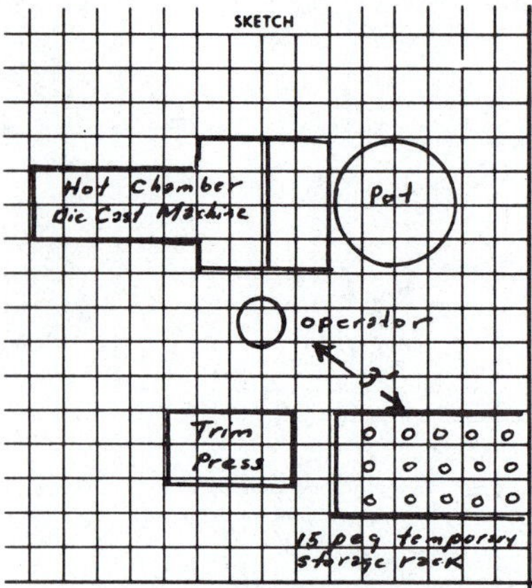

STUDY NO. *2-85*   DATE *3/18/*

OPERATION *Die cast negative plate*

DEPT. *41-11*   OPERATOR *Dinh*   NO. *114*

EQUIPMENT *Hot chamber die casting machine (Danly)*

MCH. NO. *86-4112*

SPECIAL TOOLS, JIGS, FIXTURES, GAGES *Negative plate single cavity die*

CONDITIONS *Hot, operator standing, wears gloves*

MATERIAL *Pb die cast alloy*

PART NO. *J-1476*   DWG. NO. *JC-1476-1*

PART DESCRIPTION *Negative plate for B-1 battery*

SKETCH — *Hot Chamber Die Cast Machine, Pot, operator, Trim Press, 15 peg temporary storage rack*

LEFT HAND	RIGHT HAND	ELEM. NO.	SMALL TOOL etc.	ELEMENTAL TIME	OCC. PER CYCLE	TOTAL TIME ALLOWED
Idle	*As press opens, reach 14" grasp part & remove from die*	1		0.328	1	0.328
	Transfers part to	2	*Trim fx · JB-1476-8*	0.245	1	0.245
Hold part	*left hand, gets lubrication can, lubricate die*	3	*Get 2 pys - 60" total*	1.135	0.04	0.045
Hold part	*Inspect die for voids, completeness of fill & finish*					
Lock fixture	*Place part in fixture*					
Idle	*Engage feed Trim part*					
Open fixture	*Get part and move to rack with post (3')*					

EACH PIECE *0.618*   TOTAL *0.618*

SET-UP   HRS. PER C *1.03*

FOREMAN *A. B. Jones*   INSPECTOR *Dick Guill*

OBSERVER *John Darben*   APPROVED BY *Ken Knott*

minute and on elements of 0.02 minute when followed by a relatively long element. With practice, a good time study analyst using the continuous method can catch accurately three successive short elements (less than 0.04 minute) if they are followed by an element of about 0.15 minute or longer. This is possible by remembering the watch readings of the terminal points of the three short elements and then recording their respective values while the fourth, longer element is taking place.

Of course, more clerical work is involved in calculating the study when the continuous method is used. Since the watch is read at the breaking point of each element while the hands of the watch continue their movements, it is necessary to make successive subtractions of the consecutive readings to determine elapsed elemental times. For example, the following readings might represent the terminal points of a 10-element study: 4, 14, 19, 121, 25, 52, 61, 76, 211, 16, and the elemental values of this cycle would be 4, 10, 5, 102, 4, 27, 9, 15, 35, and 5.

Recording the Time Consumed by Each Element

When recording the watch readings, note only the necessary digits and omit the decimal point, thus giving as much time as possible to observe the performance of the operator. If using a decimal minute watch, and the terminal point of the first element occurs at 0.08 minute, record only the digit 8 in the "R" (reading) column of the time study form. If using the decimal hour watch, and the end point of the first element is 0.0052, the recorded reading would be 52. Table 14–1 illustrates the procedure of using a decimal minute watch.

The small hand on the watch indicates the number of elapsed minutes so that the observer can refer to it periodically to verify the correct first digit to record after the large hand sweeps past the zero. For example, after 22 minutes have passed in taking a given study, the observer may not

TABLE 14–1

Consecutive reading of watch in decimal minutes	Recorded reading
0.08.	8
0.25.	25
1.32.	132
1.35.	35
1.41.	41
2.01.	201
2.10.	10
2.15.	15
2.71.	71
3.05.	305
3.17.	17
3.25.	25

recall whether the value to record after the termination of the current element should be prefixed by "22" or "21." Glancing at the small hand of the watch tells the analyst that it has moved past 22, indicating that 22 is the correct prefix of the reading to be recorded.

Record all watch readings in consecutive order in the "R" column until the cycle is completed. Study subsequent cycles in a similar manner, recording their elemental values.

USING THE DATAMYTE. A typical time study using the DataMyte 1000 Data Collector would be similar to the following: Assume the work measurement analyst takes a 20-cycle study lasting about 45 minutes for the spot welding of a special SAE 1112 steel hinge to a cold rolled steel tank. The analyst, after observing the method, breaks the job down into the following elements:

Element 1: Get and bring tank (on conveyor) to work area.

Element 2: Pick up hinge and place in fixture.

Element 3: Position overhead spot welder and spot weld 6 spots.

Element 4: Open fixture and move assembled tank 10 feet on conveyor.

During the course of the study four foreign elements took place:

Element A: Interrupted by supervisor.

Element B: Leaves workstation to get drink of water.

Element C: Delays getting tank on conveyor.

Element D: Sorts out and discards defective hinge.

In taking the study, the analyst should:

1. Turn on the instrument and set to input mode 1, data-plus-time.
2. Key in a study identification code of up to 12 characters.
3. Key in a 6-character date.
4. Key in a 4-character begin time.
5. Key in 1 (element 1) key "enter" at end of element.
6. Key in 2 (element 2) key "enter" at end of element.
7. Key in 3 (element 3) key "enter" at end of element.
8. Key in 4 (element 4) key "enter" at end of element.

The preceding cycle would continue for the 20 cycles. The foreign elements are keyed in as they occur and "enter" is stroked at their termination. Elapsed time is logged automatically every time "enter" is stroked. If the analyst is assigning an overall performance rating factor, it would be entered at the end of the study. Performance ratings can, however, be entered at any time—at the end of each element, at the end of each cycle, and so forth.

The DataMyte is limited to approximately 500 observations per 4K memory. For example, a 16K DataMyte accommodates approximately 2,000 observations.

At the conclusion of the data collections, the instrument can be connected to a computer terminal where the data is dumped. The data is saved by assigning a file name, for example MS84 (machine shop 84th study). The analyst can now call up the work measurement program on the computer terminal to develop the standard desired. For example, the TSS1 program provides all normal elemental times. The operation standard that includes applying appropriate allowances and summarizing the allowed elemental times is handled by adding the appropriate statements in BASIC to the TSS1 program.

Figure 14–8 shows an analyst using a DataMyte 1000 Data Collector for the purpose of developing a standard for loading food items from roller bins onto a conveyor belt. Figure 14–9 shows the DataMyte downloading the time study to a personal computer for data review and standard and report generation. Figure 14–10 illustrates the output from the study with the normal elemental times computed in minutes for 10 work elements. Also, the idle time has been tabulated.

FIGURE 14–8
An analyst is using a DataMyte 1000 Data Collector to develop a standard for loading food items onto a conveyor belt

Courtesy Nabisco Brands, Inc.

FIGURE 14–9
An analyst is downloading a time study from a DataMyte 1000 Data Collector to a
personal computer for data viewing and report generation

Courtesy Nabisco Brands, Inc.

Difficulties Encountered

During the time study, analysts may observe variations from the sequence of elements originally established. Occasionally analysts may
miss specific terminal points. These difficulties complicate the study; the
less often they occur, the easier it is to calculate the study.

When missing a reading, analysts should immediately indicate an "M"
in the "R" column of the time study form. In no case should they make an
approximation and endeavor to record the missed value, because this
practice can destroy the validity of the standard established for the specific element. If the element were used as a source of standard data,
appreciable discrepancies in future standards might result. Occasionally,
the operator omits an element; this is handled by drawing a horizontal line
through the space in the "R" column. This should happen infrequently,
since it is a sign of an inexperienced operator or an indication of a lack of
standardization of method. Of course, the operator can inadvertently omit
an element—for example, by forgetting to "vent cope" in making a bench
mold or by failing to "apply parting dust." If elements are omitted repeatedly, analysts should stop the study and investigate the necessity of performing the omitted elements at all. This should be done in cooperation

FIGURE 14–10
Output from study of loading food items onto a conveyor

```
                Time Study programs by DataMyte Corporation
                      Model 1031-04   version 07

                         Time Study Summary
```

Description *TIME STUDY OF SALES BRANCH ORDER PICKER*

```
Study Id: PICKER11
Date: 06/23/86
Begin Time: 1230 (12:30 PM)

********** Deletions - time not used in summary ********************************

                    Time Deleted:    0.00


********** Summary ************************************************************
   obs      raw     rated    min    max     ave    rate                 nrm/occ
```

1 *PICK ONE CASE FROM BIN*   0.0771
```
   211   16.2700   16.2700   0.02   0.34   0.0771  100.0
```

2 *PICK TWO CASES FROM BIN*   0.1158
```
    19    2.2000    2.2000   0.02   0.34   0.1158  100.0
```

3 *PICK ONE CASE FROM PALLET*   0.0805
```
   147   11.8300   11.8300   0.02   0.40   0.0805  100.0
```

4 *PICK TWO CASES FROM PALLET*   0.0778
```
     9    0.7000    0.7000   0.03   0.21   0.0778  100.0
```

5 *PICK ONE CASE FROM END OF LINE*   0.1588
```
    39    6.8800    6.1920   0.02   0.62   0.1764   90.0
```

6 *PICIC THREE CASES FROM BIN OR PALLET*   0.0900
```
     7    0.6300    0.6300   0.05   0.14   0.0900  100.0
```

7 *PICIC FOUR CASES FROM BIN OR PALLET*   0.0975
```
     4    0.3900    0.3900   0.04   0.22   0.0975  100.0
```

8 *PREPARE DISPLAY MATERIAL*   1.3200
```
     1    1.3200    1.3200   1.32   1.32   1.3200  100.0
```

9 *DISCARD DAMAGED CASES*   0.2050
```
     4    0.8200    0.8200   0.11   0.25   0.2050  100.0
```

10 *PAPERWORK / COMMUNICATIONS*   0.8475
```
     4    3.3900    3.3900   0.15   2.84   0.8475  100.0
```

88 *IDLE*   3.0067
```
     3    9.0200    9.0200   0.28   4.52   3.0067  100.0
```

```
Totals                                       98.7

Time used:    53.45

Total time:   53.45
```

Courtesy Nabisco Brands, Inc.

with the supervisor and the operator so that the best method can be established. The observer is expected to be constantly on the alert for possibilities of better ways to perform the elements; as ideas come to mind, the observer should jot them down in the "note" section of the time study form for future study and possible development.

Also the observer may see the performance of elements out of sequence. This happens fairly frequently when a new or inexperienced employee is studied on a long-cycle job made up of many elements. Avoiding as many of these disturbances as possible is one of the prime reasons for studying a competent, fully trained employee. However, when elements are performed out of order, immediately go to the element that is being performed and draw a horizontal line through the middle of its "R" space. Directly below this line write the time the operator began the element, and above it record the completion time. This procedure is repeated for each element performed out of order and for the first element that is performed back in the normal sequence. Figure 14–4 illustrates a typical study, and elements 7, 8, and 9 on Cycle 11 show how the analyst handled the elements performed out of order.

During a time study the operator may encounter unavoidable delays, such as an interruption by a clerk or a supervisor or tool breakage. Furthermore, the operator may intentionally cause a change in the order of work by going for a drink of water or stopping to rest. Interruptions of this nature occurring in the work cycle are referred to as "foreign elements."

Foreign elements can occur either at the breaking point or during the course of an element. The majority of foreign elements, particularly if they are controlled by the operator, occur at the termination of one of the elements constituting the study. When a foreign element occurs during an element, signify the event by an alphabetical designation in the "T" column of this element. If the foreign element occurs at the breaking point, record it in the "T" column of the work element which follows the interruption. The letter *A* is used to signify the first foreign element, the letter *B* to signify the second, and so on.

As soon as the foreign element has been properly designated with an alphabetical symbol, write a short description of it in the space provided immediately after the corresponding reference letter. Indicate the time that the foreign element begins in the lower part of the "R" block of the foreign element section, and add the watch reading at the termination of the foreign element in the upper part. These values can then be subtracted at the time the study is calculated, to determine the exact duration of the foreign element. Write this value in the "T" column of the foreign element section. Figure 14–4 illustrates the correct handling of several foreign elements.

Sometimes, investigation reveals that foreign elements have a definite relation to the job being studied. In such cases, consider the elements irregular and level the elapsed time, adding the proper allowance, and properly prorate the result to the cycle time to achieve a fair standard.

Occasionally, a foreign element is of such short duration that it is impossible to record it in the fashion outlined. Typical examples of this would be dropping a wrench on the floor and quickly picking it up, wiping one's brow with a handkerchief, or turning to speak briefly to the supervisor. In cases similar to these examples, where the foreign element may be 0.06 minute or less, the most satisfactory method of handling the interrup-

TABLE 14–2

Cycle time in minutes	Recommended number of cycles
0.10	200
0.25	100
0.50	60
0.75	40
1.00	30
2.00	20
2.00– 5.00	15
5.00–10.00	10
10.00–20.00	8
20.00–40.00	5
40.00–above	3

Source: Information taken from the Time Study Manual of the Erie Works of the General Electric Company, developed under the guidance of Albert E. Shaw, manager of wage administration.

tion is to allow it to accumulate in the element where it occurs and immediately circle the reading, indicating that a "wild" value has been encountered. Add a short comment in the "note" section of the time study form across from the element where the interruption occurred, justifying the circling procedure.

THE NUMBER OF CYCLES TO STUDY. The question of how many cycles must be studied to arrive at an equitable standard is a subject that has caused considerable discussion among time study analysts, as well as union representatives. Since the activity of the job, as well as its cycle time, directly influences the number of cycles that can be studied from an economic standpoint, one cannot be completely governed by the sound statistical practice that demands a certain size sample based on the dispersion of the individual element readings.

The General Electric Company has established Table 14–2 as a guide to the number of cycles to be observed. The Westinghouse Electric Corporation has considered activity as well as cycle time and has evolved the values shown in Table 14–3 as a guide for its time study personnel.

Since time study is a sampling procedure, the mean of the sample of the observations should be reasonably close to the mean of the population. Thus, analysts should take sufficient readings so that when their values are recorded, a distribution of values with a dispersion characteristic of the population dispersion is obtained.

Some training programs for time study analysts recommend that an observer take readings and plot the values to obtain a frequency distribution.[3] Although there is no assurance that the population of elemental times is normally distributed, still experience has shown that the

---

[3] For example, the Kurt Salmon Company.

TABLE 14–3

When the time per piece of cycle is over (hours)	Minimum number of cycles to study (activity)		
	Over 10,000 per year	1,000 to 10,000	Under 1,000
8.000	2	1	1
3.000	3	2	1
2.000	4	2	1
1.000	5	3	2
0.800	6	3	2
0.500	8	4	3
0.300	10	5	4
0.200	12	6	5
0.120	15	8	6
0.080	20	10	8
0.050	25	12	10
0.035	30	15	12
0.020	40	20	15
0.012	50	25	20
0.008	60	30	25
0.005	80	40	30
0.003	100	50	40
0.002	120	60	50
Under 0.002	140	80	60

Source: Westinghouse Electric Company.

FIGURE 14–11
The normal frequency distribution

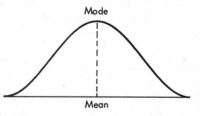

Mode

Mean

variations in the performance of a given operator approximate the normal bell-shaped curve (see Figure 14–11).

The number of cycles that should be studied to assure a reliable sample may be determined mathematically, and this value, tempered with sound judgment, gives analysts a helpful guide in deciding the length of observation.

Statistical methods may be used in determining the number of cycles to study. It is known that averages of samples ($\bar{x}$) drawn from a normal distribution of observations are distributed normally about the population mean μ. The variance of $\bar{x}$ about the population mean μ equals $\dfrac{\sigma^2}{n}$, where n equals the sample size and σ^2 equals the population variance.

Normal curve theory leads to the following confidence interval equation:

$$\bar{x} \pm z\frac{\sigma}{\sqrt{n}}$$

The preceding equation assumes that the population standard deviation is known. This, in general, is not true, but the population standard deviation may be estimated by the sample standard deviation s, where:

$$s = \sqrt{\frac{\displaystyle\sum_{i=1}^{i=n}(x_i - \bar{x})^2}{n - 1}}$$

or for computational purposes:

$$s = \sqrt{\frac{\Sigma x_i^2}{n - 1} - \frac{(\Sigma x_i)^2}{n(n - 1)}}$$

When estimating σ in this way, we are dealing with the quantity

$$\frac{\bar{x} - \mu}{\dfrac{s}{\sqrt{n}}}$$

which is not normally distributed except for large samples ($n > 30$). Its distribution is the "t" distribution, which should be used in the formulas that follow. The confidence interval equation then is:

$$\bar{x} \pm t\frac{s}{\sqrt{n}} \tag{1}$$

For example, if 25 readings for a given element showed that $\bar{x} = 0.30$ and $s = 0.09$, there would be 95 percent confidence that μ would be contained in the interval 0.337 to 0.263 or that $\bar{x}$ is within ± 12.3 percent of μ. (See Table A3–3, Appendix 3, for values of t.)

$$\bar{x} \pm t\frac{s}{\sqrt{n}}$$

$$0.30 + 2.06\frac{0.09}{5} = 0.337$$

$$0.30 - 2.06\frac{0.09}{5} = 0.263$$

$$\frac{0.037}{30} = 12.3 \text{ percent}$$

If the accuracy is deemed unsatisfactory when Equation 1 is utilized in the above manner, it is possible to solve for N, the required number of readings for a given accuracy, by equating $\dfrac{ts}{\sqrt{N}}$ to a percentage of $\bar{x}$:

$$\frac{ts}{\sqrt{N}} = k\bar{x}$$

$$N = \left(\frac{st}{k\bar{x}}\right)^{2}$$

where k = An acceptable percentage of $\bar{x}$.

In the above example, if $\bar{x}$ is required to be within ± 5 percent of μ with 95 percent confidence:

$$N = \left(\frac{st}{k\bar{x}}\right)^{2}$$

$$= \left(\frac{(0.09)(2.06)}{(0.05)(0.30)}\right)^{2}$$

$$= 152 \text{ observations}$$

If n readings have been taken on a time study, the selection of the most appropriate element for computing the desired number of readings may be a problem. Select the element having the greatest coefficient of variation $\dfrac{s}{\bar{x}}$ for this purpose. It is also possible to solve for N before taking the time study by interpreting historical data of similar elements or by actually estimating $\bar{x}$ and s from several snapback readings of certain elements to be studied.

One concern has equated the number of observations in terms of the range of the cycle time and the average cycle time.[4] This has been expressed as:

$$N = 205\frac{R}{\bar{x}} - 42$$

where:

N = Number of observations required
R = Range of cycle time
$\bar{x}$ = Mean cycle time

This has been derived as follows:

$$R = Ks$$

---

[4] A. C. Spark Plug Division, General Motors Corporation.

TABLE 14–4

N	K*	A†	$\bar{x} = 0.10$ minute	$\bar{x} = 0.20$ minute
40	4.3216	0.3222	0.0402	0.0805
50	4.4982	0.2859	0.0472	0.0944
75	4.8060	0.2310	0.0624	0.1248
100	5.0152	0.1990	0.0756	0.1512
125	5.1727	0.1774	0.0875	0.1749
150	5.2985	0.1617	0.0983	0.1967
175	5.4029	0.1494	0.1085	0.2169
200	5.4921	0.1396	0.1180	0.2360

* K values obtained from *Biometrika* by L. H. C. Tippett.
† A values obtained from *Statistical Methods for Research Workers* by R. A. Fisher.

where:

K = Average ratio between the range and the standard deviation in a succession of samples
s = Standard deviation of the sample taken

To conform to the desired limits of 95 percent probability and 3 percent permissible error, the possible error, As, and the permissible error, $.03\bar{x}$, are assumed to be equal.
Therefore:

$$As = 0.03\bar{x}$$

$$s = \frac{0.03\bar{x}}{A}$$

where:

A = Probability factor at 95 percent

Substituting for s in the range formula:

$$R = \frac{(K)(0.03\bar{x})}{A}$$

The values of K and A vary with the number of observations (N) taken. Substituting in the formula

$$R = \frac{(K)(0.03\bar{x})}{A}$$

the representative values in Table 14–4 are shown.

On graph paper, plotting R (range) against $\bar{x}$ (average time) for various samples (N), we find a straight-line relationship from the formula $R = M\bar{x} + B$. Therefore, for sample sizes of 40, 150, 200, we get slope M:

$$N = \;\; 40 \qquad M = 0.402 \qquad B = 0$$
$$N = 150 \qquad M = 0.983 \qquad B = 0$$
$$N = 200 \qquad M = 1.180 \qquad B = 0$$

To introduce N into the above formula, N was plotted against M for various sample sizes, and the resulting straight-line formula is:

$$N = M_1 M + B$$
$$N = 205M - 42 \text{ (approximate)}$$

From the formula $R = M\bar{x} + B$, since $B = 0$ we get M in terms of R and $\bar{x}$ as:

$$M = \frac{R}{\bar{x}}$$

Substituting this expression from M in the formula $N = M_1 M + B$, we get

$$N = 205\frac{R}{\bar{x}} - 42$$

Given an equation involving N (number of observations), R (range of cycle time), and $\bar{x}$ (average cycle time), an alignment chart may be designed to facilitate the solution of N. The N alignment chart illustrated in Figure 14–12 is evolved as follows:

Rearranging the equation to conform to the general N chart equation:

$$\frac{205R}{\bar{x}} - N = 42$$

N varies from 0 to 240
R varies from 0 to 0.30
$\bar{x}$ varies from 0 to 0.75

Comparing with the general alignment chart equation

$$Af(u) + Bf(v) = Cf(w) + CK$$

Then:

$$R = f(u)$$
$$N = f(v)$$
$$O = f(w)$$
$$A = \frac{205}{\bar{x}}$$
$$B = -1$$
$$C = 1$$
$$K = 42$$

Select chart width 8 inches, chart height 12 inches.

FIGURE 14–12
N-type alignment chart for determining the number of time study observations required (above 40) for 95 percent probability and 3 percent error

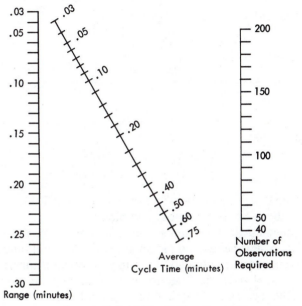

A. C. Spark Plug Division, General Motors Corp.

Develop scale modulus:

$$M_x = \frac{\text{Length of scale}}{\text{Range of } f(u)} = \frac{12}{0.30} = -40$$

(when *B* is negative, M_x is negative)

$$M_y = \frac{\text{Length of scale}}{\text{Range of } f(v)} = \frac{12}{240} = 0.05$$

Calculate *A*:

$$A = \frac{BDM_x}{AM_y + BM_x} = \frac{(-1)(8)(-40)}{\dfrac{205}{\bar{x}}(0.05) + (-1)(-40)} = \frac{0.320}{\dfrac{10.25}{\bar{x}} + 40}$$

Calculate *E*:

$$E = \frac{CM_x M_y(f(w) + K)}{AM_y + BM_x} = \frac{(1)(-40)(0.05)(42)}{\dfrac{205}{\bar{x}}(0.05) + (-1)(-40)}$$

$$E = \frac{-84}{\dfrac{10.25}{\bar{x}} + 40}$$

TABLE 14–5
Sample

R	$x = -40R$	N	$y = 0.05N$	$\bar{x}$	$A = \dfrac{\dfrac{320}{10.25}}{\bar{x}} + 40$	$E = \dfrac{\dfrac{-84}{10.25}}{\bar{x}} + 40$
0..........	0	0	0	0	—	—
0.05........	−2	40	2	.05	1.306	−0.343
0.20........	−8	150	7.5	.40	4.876	−1.280
0.30........	−12	240	12.0	.75	5.963	−1.565

The scale equations:

$$X = M_x f(u) = -40R$$
$$Y = M_y f(v) = 0.05N$$

Values are then developed in tabular form as shown in Table 14–5. Using these values, the N chart is laid out and plotted.

These mathematical derivations represent only a guide and do not replace common sense and good judgment, which are still fundamental if a time study analyst is to do a good job in a practical manner.

Analysts must also decide when to observe the recommended number of cycles. If 30 cycles are to be taken, should they be taken as one successive group, as two groups of 15, or as three groups of 10? The mean of three groups of 10 taken at random times during the day probably gives a better estimate of the population mean than one group of 30. The taking of a time study is a sampling procedure, and the average of several small samples usually provides more reliable estimates of parent parameters than does one sample of a size equivalent to the total of small samples. For example, if we had a barrel of 10,000 bolts, we would be more confident of the quality of the entire 10,000 if we averaged the results from

TABLE 14–6

Average at the end of the	Average	Average at the end of the	Average
1st cycle	0.261	11th cycle	0.281
2d cycle.........	0.280	12th cycle	0.280
3d cycle.........	0.248	13th cycle	0.281
4th cycle	0.251	14th cycle	0.281
5th cycle	0.274	15th cycle	0.280
6th cycle	0.261	16th cycle	0.281
7th cycle	0.264	17th cycle	0.281
8th cycle	0.283	18th cycle	0.281
9th cycle	0.279	19th cycle	0.281
10th cycle	0.280	20th cycle	0.281

three independent samples. Thus, we might take one sample of 40 from one end, another sample of 40 from the middle, and a third sample of 40 from the other end. This would be a more reliable procedure than would taking a sample of 120 from one end.

Individual samples should be large enough so that their mean is somewhat stabilized. When stabilization occurs, the addition of one more cycle does not substantially alter the mean. For example, we might have the following averages given in Table 14–6.

In Table 14–6 the average was stabilizing at the end of the 10th cycle and nothing really practical was accomplished by the analyst in observing more than 11 cycles.

RATING THE OPERATOR'S PERFORMANCE

Before leaving the workstation, analysts should give a fair performance rating to the study. On short-cycle, repetitive work, it is customary to apply one rating to the entire study. However, where the elements are of long duration and entail diversified manual movements, it is more practical to evaluate the performance of each element as it occurs during the study. The time study form illustrated in Figures 14–4 and 14–5 includes provision for both the overall rating and the individual element rating. The study made of die casting (Figure 14–6 and 14–7) used elemental rating where the elements were over 0.20 minutes in duration. The study of the milling operation (Figures 14–4 and 14–5) where elements were as short as 0.04 and 0.05 minutes used the overall rating procedure.

Since the actual time required to perform each element of the study was dependent to a high degree on the skill and effort of the operator, it is necessary to adjust the time of the good operator up to normal and the time of the poor operator down to normal.

In the performance rating or leveling system, the observer evaluates the operator's effectiveness in terms of a "normal" operator performing the same element. This effectiveness is given a value expressed as a decimal or percentage and assigned to the element observed. A "normal" operator is defined as a qualified, thoroughly experienced operator working under conditions as they customarily prevail at the workstation, at a pace neither too fast nor too slow but representative of average. The normal operator exists only in the mind of the time study analyst, whose concept has developed as a result of thorough and exacting training and experience in the technique of measuring wide varieties of work.

The basic principle of performance rating is to adjust the actual observed mean time for each acceptable element performed during the study to the time that would be required by the normal operator to perform the same work. To do a fair job of rating, the time study analyst must be able to disregard personalities and other varying factors, and consider only the amount of work being done per unit of time as compared to the amount of work that the normal operator would produce. Chapter 15 explains more fully the performance rating or leveling techniques in common use.

APPLYING ALLOWANCES

No operator can maintain an average pace for every minute of the working day, just as it would be impossible for a football game to entail 60 minutes of actual, continuous play time. Three classes of interruptions take place occasionally for which extra time must be provided. The first of these is personal interruptions, such as trips to the rest room and drinking fountain; the second is fatigue, which, as we all know, affects even the strongest individual on the lightest work. Lastly, unavoidable delays for which some allowances must be made, such as tool breakage, interruptions by the supervisor, slight tool trouble, and material variation.

Since the time study is taken over a relatively short period, and since foreign elements have been removed in determining the normal time, an allowance must be added to the leveled or base time to arrive at a fair standard that can be achieved by the normal operator when exerting average effort. Chapter 16 outlines in detail the means for arriving at realistic allowance values.

CALCULATING THE STUDY

After properly recording all the necessary information on the time study form, observing an adequate number of cycles, and performance rating the operator, the analyst should thank the operator and proceed to the next step—computation of the study. In some plants, clerical help computes the study, but in most cases the analyst does the work. Analysts usually prefer to analyze their own studies because they are vitally interested in the resulting standard and do not care to let other individuals tamper with their calculations. Then, too, when analysts compute their own studies, they have no difficulty in interpreting their own notes and reading recorded values, so there is less chance of error.

The initial step in computing the study is to verify the final stopwatch reading with the overall elapsed clock reading. These two values should check within plus or minus a half minute, and if a sizable discrepancy is noted, check the study stopwatch readings for error.

When using the continuous method, subtract the watch reading from the preceding reading for the elapsed time; record this in ink or red pencil. Follow this procedure throughout the study, subtracting each reading from the succeeding one. Analysts must be especially accurate in this phase because carelessness at this point can completely destroy the validity of the study. If elemental performance rating has been used, multiply the elapsed elemental times by the leveling factor; record in ink or red pencil in the "T" spaces of the time study form. The hand-held calculator permits this step to be performed quite rapidly.

Elements that have been missed by the observer are signified by placing an "M" in the "R" column. When computing the study, disregard both the missed element and the succeeding one, since the subtracted value in the study includes the time for performing both elements.

Elements missed by the operator may be disregarded since they have no effect on the preceding or succeeding values. Thus, if the operator happened to omit element 7 of cycle 4 in a 30-cycle study, the analyst should have but 29 values of element 7 with which to calculate the mean observed time.

To determine the elapsed elemental time on out-of-order elements, it is merely necessary to subtract the value appearing in the lower half of the "R" block from the value in the upper half.

For foreign elements, deduct the time required for the foreign element from the cycle time of the element in which it occurred. Obtain the time taken by the foreign element by subtracting the lower reading in the "R" space of the foreign element section of the time study form from the upper reading.

After all elapsed times have been calculated and recorded, study them carefully for abnormality. There is no set rule for determining the degree of variation permitted to keep the value for calculation. If a broad variation on a certain element can be attributed to some influence that was too brief to be handled as a foreign element, yet long enough to affect substantially the time of the element, such as dropping a tool or blowing the nose, or if the variation may be attributed to errors in reading the stopwatch, then these values should be immediately circled and excluded from further consideration in working up the study. However, if wide variations are due to the nature of the work, then it would not be wise to discard any of these values.

Elements that are "paced" by the facility or process have little variation from cycle to cycle, while considerably wider variation in "operator-controlled" elements would be expected. When unexplainable variations in time occur, be quite careful before circling any such values. Be aware that this is not a performance rating procedure, and by arbitrarily discarding high values or low values, analysts may end up with an incorrect standard. A good rule is, "When in doubt, do not discard the value."

If elemental rating is used, then after the elemental elapsed time values have been computed, determine the normal elemental time by multiplying each elemental value by its respective performance factor. Then record this normal time for each element in the "T" columns (see Figure 14–6). Next, determine the mean elemental normal value by dividing the number of observations into the total of the times in the "T" columns.

After determining all elemental elapsed times, check to assure that no arithmetic errors have been made. This can best be done by adding the "total" column for each element and the foreign and abnormal element times. The total of these values must be equal to the last reading of the stopwatch. If this total does not equal the last watch reading, the study should be checked for arithmetic errors before proceeding with further computation. This check applies only when using the continuous method.

If the overall study is leveled, then compute the means of the elapsed elemental times, and apply the performance factor to these values in the space provided to determine the various normal elemental time values.

After the normal elemental times have been evolved, add the percentage allowance to each element to determine the allowed time. The nature of the job determines the amount of allowance to be applied, as shown in Chapter 16. Suffice to say at this point that 15 percent is an average allowance for effort elements, and that 10 percent is representative of the allowance applied to process-controlled elements.

Upon determining the allowed time for each element, summarize these values in the space provided on the reverse side of the time study form to obtain the allowed time for the entire job. This is commonly referred to as the "standard" or the "rate" for the job.

Summary: Computing the Study

To summarize, the steps taken in computing a typical study with continuous watch readings and overall performance rating are as follows:

1. Make subtractions of consecutive readings to obtain elemental elapsed times, and record in red pencil.
2. Circle and discard all abnormal or "wild" values where an assignable cause is evident.
3. Summarize the remaining elemental values.
4. Determine the mean of the observed values of each element.
5. Determine the elemental normal time by multiplying the performance factor by the mean elapsed time.
6. Add the appropriate allowance to the elemental normal values to obtain the elemental allowed times.
7. Summarize the elemental allowed times on the reverse side of the time study form to obtain the standard time. (Elements occurring more than once per cycle need be shown only once with the number of occurrences and the resulting product.)

If elemental leveling has been used on a continuous study, then the steps taken in computing the study would be:

1. Make subtractions of consecutive readings to obtain elemental elapsed times.
2. Determine the normal times of each individual element by multiplying the performance factor by the elapsed time value, and record them in red pencil in the "T" column.
3. Circle and discard all abnormal or "wild" values where an assignable cause is evident.

4. Determine the mean of the elemental normal times.
5. Add the appropriate allowance to the elemental normal values to obtain the elemental allowed times.
6. Summarize the elemental allowed times on the reverse side of the time study form to obtain the standard time. (Elements occurring more than once per cycle need be shown only once with the number of occurrences and the resulting product.)

TEXT QUESTIONS

1. What considerations should be given to the choice of the operator to be studied?
2. Of what use is the operator process chart on the time study form?
3. Why is it essential to record complete information on tools and the facility on the time study form?
4. Why are working conditions important in identifying the method being observed?
5. Why would a time study analyst who is hard of hearing have difficulty in recording stopwatch readings of terminal points?
6. Differentiate between constant and variable elements. Why should they be kept separate when dividing the job into elements?
7. Explain how repeating elements may be recorded on the time study form.
8. What advantages does the continuous method of watch recording offer over the snapback method?
9. Explain why the electronic time study instrument has broadened the use of the snapback procedure.
10. Why is the time of day recorded on the time study form?
11. What variations in sequence will the observer occasionally encounter during the course of the time study?
12. Explain what a foreign element is and how foreign elements are handled under the continuous method.
13. What factors enter into the determination of the number of cycles to observe?
14. Why is it necessary to performance rate the operator?
15. When should individual elements of each cycle be rated?
16. Define a "normal" operator.
17. For what reasons are allowances applied to the normal time?
18. What is the significance of a "circled" elapsed time?
19. What steps are taken in the computation of a time study conducted by the continuous overall performance rating procedure?
20. Based upon the Westinghouse guide sheet, how many observations should be taken on an operation whose annual activity is 750 pieces and whose cycle time is estimated at 15 minutes? What would be the number of observations needed according to the General Electric guide sheet?

21. You anticipate using the DataMyte for developing some standard data on some short cycle elements involved in the use of 100-ton punch presses. You will need to take approximately 1,000 observations. How many K of memory are required?

GENERAL QUESTIONS

1. What do you feel the difference is between (*a*) a normal operator, (*b*) an average operator, and (*c*) a standard operator?
2. Why are time study procedures the only techniques for supplying reliable information on standard times?
3. To what positions of importance does time study work lead?
4. In what way can "loose" time standards result in poor labor relations?
5. Why are poor time standards a "headache" to union officials?
6. How would you approach a belligerent operator if you were the time study analyst?
7. You were using the General Electric guide sheet to determine the number of observations to study. It developed that 10 cycles were required, and after taking the study you used the standard error of the mean statistic to estimate the number of observations needed for a given confidence level. The resulting calculation indicated that 20 cycles should be studied. What would be your procedure? Why?
8. Based on the data provided in Figures 14–6 and 14–7, would it be a good idea to have the operator load and unload a second die casting machine? Explain.

PROBLEMS

1. The time study analyst at the Dorben Company developed the following snapback stopwatch readings where elemental performance rating was used. The allowance for this element was assigned a value of 16 percent. What would be the allowed time for this element?

Snapback Reading	Performance Factor
28	100
24	115
29	100
32	90
30	95
27	100
38	80
28	100
27	100
26	105

2. What would be the required number of readings if the analyst wanted to be 87 percent confident that the mean observed time was within ±5 percent of the true mean and he established the following values for an element after observing 20 cycles: 0.09, 0.08, 0.10, 0.12, 0.09, 0.08, 0.09, 0.12, 0.11, 0.12, 0.09, 0.10, 0.12, 0.10, 0.08, 0.09, 0.10, 0.12, 0.09?
3. The following data resulted from a time study taken on a horizontal milling machine:

Pieces produced per cycle: 8.

Mean measured cycle time: 8.36 minutes.

Mean measured effort time per cycle: 4.62 minutes.

Mean rapid traverse time: 0.08 minutes.

Mean cutting time (power feed): 3.66 minutes.

Performance factor: +15. 115%

Machine allowance (power feed): 10 percent.

Effort allowance: 15 percent.

An operator works on the job a full eight-hour day and produces 380 pieces.

a. How many standard hours does the operator earn?

b. What is his efficiency for the day?

c. If his base rate is $14 per hour, compute his earnings for the day if he is paid in direct proportion to this output.

4. A work measurement analyst in the Dorben Company took 10 observations of a high-production job. He performance rated each cycle and then computed the mean normal time for each element. The element with the greatest dispersion had a mean of 0.30 minutes and a standard deviation of 0.03 minutes. If it is desirable to have sampled data within ±5 percent of the true data, how many observations should this time study analyst take of this operaion?

5. In the Dorben Company, the work measurement analyst took a detailed time study of the making of shell molds. The third element of this study had the greatest variation in time. After studying nine cycles, the analyst computed the mean and standard deviation of this element, with the following results:

$$\bar{x} = 0.42 \qquad S = 0.08$$

If the analyst wanted to be 90 percent confident that the mean time of her sample was within ±10 percent of the mean of the population, how many total observations should she have taken? Within what percent of the average of the total population is $\bar{x}$ at the 95 percent confidence level under the measured observations?

SELECTED REFERENCES

Barnes, Ralph M. *Motion and Time Study: Design and Measurement of Work.* 7th ed. New York: John Wiley & Sons, 1980.

Mundel, Marvin E. *Motion and Time Study: Improving Productivity.* 5th ed. Englewood Cliffs, N.J.: Prentice Hall, 1978.

Nadler, Gerald. *Work Design: A Systems Concept.* Rev. ed. Homewood, Ill.: Richard D. Irwin, 1970.

Performance Rating

While taking the study, time study observers carefully observe the performance of the operator during the entire course of the study. Seldom does the performance being executed conform to the exact definition of "normal," often referred to as "standard." Thus, some adjustment must be made to the mean observed time to derive the time required for the normal operator to do the job when working at an average pace. Time study analysts must increase the actual time taken by above-standard operators to that required by the normal worker, and decrease the time taken by below-standard operators to the value representative of normal performance. Only in this manner can they establish a true standard for normal operators.

Performance rating is probably the most important step in the entire work measurement procedure. Certainly, it is the step most subject to criticism, since it is based entirely on the experience, training, and judgment of the work measurement analyst. Because of its importance, this chapter not only describes the most accepted techniques used today but also reviews some of the more important historical methods, such as the Westinghouse system. An understanding of the historical philosophies and the resulting methods is helpful in developing an appreciation for this very important step in the systematic procedure for methods and work measurement.

Performance rating is a technique for equitably determining the time required to perform a task by the normal operator after the observed values of the operation under study have been recorded. Chapter 14 defined a "normal" operator as a qualified, thoroughly experienced operator working under conditions as they customarily prevail at the workstation, at a pace that is neither too fast nor too slow, but representative of average.

There is no one universally accepted method of performance rating, although the majority of the techniques used are based primarily on the judgment of time study analysts. For this reason, the analysts must have the high personal characteristics discussed earlier in this text. No other component of the time study procedure is subject to as much controversy and criticism as the performance rating phase. Regardless of whether the rating factor is based on the speed or tempo of the output or on the performance of the operator when compared to that of the normal worker, judgment is still the criterion for the determination of the rating factor. Some attempts have been made to accumulate fundamental motion data and to compare these values to the mean observed values to arrive at a factor that can be used for the evaluation of the entire study. This technique, known as "synthetic rating," is discussed later in this chapter.[1]

THE CONCEPT OF NORMAL PERFORMANCE

Just as there is no universal method of performance rating, so there is no universal concept of normal performance. In general, a company engaged in the manufacture of low-cost, highly competitive products has a "tighter" conception of normal performance than a company producing a line of products protected by patents.

In defining "normal" performance, it is helpful to enumerate benchmark examples that are familiar to all. When this is done, method and work requirements must be carefully defined. Thus, if the benchmark of dealing 52 cards in 0.50 minute is established, a complete and specific description should be given of the distance of the four hands dealt, with respect to the dealer, and of the technique of grasping, moving, and disposing of the cards. Likewise, if the benchmark of 0.38 minute is established for walking 100 feet (3 miles per hour), it should specify whether this is walking on level ground or not, whether this is walking with or without load, and if with load, how heavy the load. Supplementing the benchmark examples should be a clear description of the characteristics of an employee carrying out a normal performance. A representative description of such an employee might be:

> A worker who is adapted to the work, and has attained sufficient experience to perform the job in an efficient manner with little or no supervision. The worker possesses coordinated mental and physical qualities enabling him or her to proceed from one element to another without hesitation or delay, in accordance with the principles of motion economy. The worker maintains a good level of efficiency through knowledge and proper use of all tools and equipment related to the job. He or she cooperates and performs at a pace best suited for continuous performance.

---

[1] Robert Lee Morrow, *Time Study and Motion Economy* (New York: Ronald Press, 1946), pp. 129–30.

The more clear-cut and specific the definition of normal, the better the definition is. The definition should clearly describe the skill and effort involved in the performance, so that all the workers have a comprehensive understanding of the concept of normal that prevails in the plant.

Uses and Effects of the Concept of Normal Performance

Even though personnel departments endeavor to provide only "normal" or "above normal" employees for each position within the company, individual differences still exist. These differences among a given group of employees within the plant can become more pronounced as time goes on. Differences in inherent knowledge, physical capacity, health, trade knowledge, physical dexterity, and training cause one operator to outperform another consistently and progressively. The degree of variation in the performance of different individuals approximates the ratio of 1 to 2.25.[2] Thus, in a random selection of 1,000 employees, the frequency distribution of the output would approximate the normal curve, with less than three cases on the average falling outside the three-sigma limits (99.73 percent of the time). If 100 percent should be taken as normal, then based on a ratio of the slowest operator to the fastest as 2.25 to 1, $\bar{x} - 3_\sigma$ could equal 0.61, and $\bar{x} + 3_\sigma$ would then equal 1.39. This would mean that 68.26 percent of the people would be within a limit of plus or minus one sigma, or between performance rating values of 0.87 and 1.13. Graphically, the expected total distribution of the 1,000 people would appear as shown in Figure 15–1.

FIGURE 15–1
Normal distribution curve of the output of 1,000 people selected at random

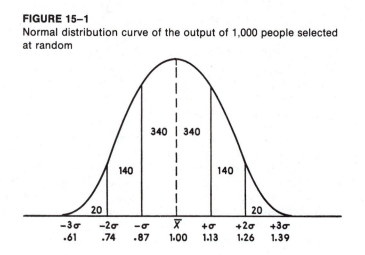

---

[2] Ralph W. Presgrave, *The Dynamics of Time Study,* 4th ed. (Toronto, Can.: The Ryerson Press, 1957), pp. 75–76.

FIGURE 15–2
Typical productivity increase graph

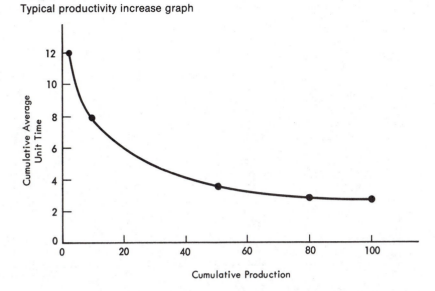

Cumulative Production

Since employees are usually carefully screened by personnel departments before being assigned to specific jobs, the mean of any selected
group may exceed the 100 percent figure representative of a sample taken
arbitrarily from the population, and the dispersion of their output may be
considerably less than the 2.25-to-1 ratio. In fact, some concerns believe
that careful testing programs to select the right person for the job, and
intensively training that employee in the correct method of performance,
results in similar output within close limits by different operators assigned
to the same job.

In the majority of cases, significant differences in output prevail among
those assigned to a given class of work. Therefore, analysts must adjust
the performance of the operator being studied to a predetermined concept
of normal.

The Learning Curve

Industrial engineers, human engineers, and other professionals interested in the study of human behavior recognize that learning is time
dependent. Even the simplest of operations may take hours to master.
Complicated work may take days and even weeks of effort before the
operator achieves coordinated mental and physical qualities enabling him
or her to proceed from one element to another without hesitation or delay.
A typical learning curve appears in Figure 15–2.

Once the operator reaches the flattening section of the learning curve,
the problem of performance rating is simplified. However, it is not always

FIGURE 15–3
Estimated unit times based on a 20 percent reduction each time the quantity doubles

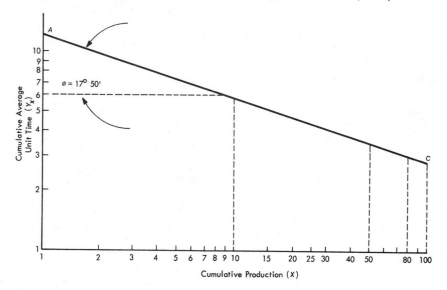

convenient to wait this long in the development of a standard. Analysts may be obliged to establish the standard at the point in the learning curve where the slope is greatest. In such cases analysts must have acute powers of observation and be able to execute mature judgment based on thorough training so that an equitable normal time is computed.

It is helpful to have learning curves representative of the various classes of work being performed in the company. This information can be useful, not only in determining at what production stage it would be desirable to establish the standard, but also by providing a guide as to the expected level of productivity that the average operator with a known degree of familiarity of the operation achieves after producing a fixed number of parts.

By plotting learning curve data on logarithmic paper, analysts may linearize the data and thus have it in a form that is easier to use. For example, plotting both the dependent variable (cumulative average unit time) and the independent variable (cumulative production) shown in Figure 15–2 on log-log paper results in a straight line, as seen in Figure 15–3.

A new learning curve situation does not necessarily result every time a new design is put into production. Former designs similar to new designs have a pronounced effect on the point at which the learning curve begins to flatten. Thus, if a company introduces a completely new design of a complex electronic panel, the assembly of this panel would involve a

TABLE 15-1

Cumulative production	Cumulative average hours per unit	Ratio to previous cumulative average
1	100.0	—
2	90.0	90
4	81.0	90
8	72.9	90
16	65.6	90
32	59.0	90
64	53.1	90
128	47.8	90

much different learning curve than the introduction of a panel similar to one that had been in production for the past five years.

The theory of the learning curve proposes that as the total quantity of units produced doubles, the time per unit declines at some constant percentage. For example, if analysts expect a 90 percent rate of improvement, then as production doubles, the average time per unit declines 10 percent.

Table 15-1 illustrates the decline in the cumulative average hours per unit of production with successive doubling of the production quantity where a 90 percent rate of improvement exists.

The smaller the percent rate of improvement, the greater the progressive improvement with production output. Typical rates of learning are as follows: large or fine assembly work (such as aircraft), 70-80 percent; welding, 80-90 percent; machining, 90-95 percent.

Thus, the percent rate of improvement (learning) equals 100 times the cumulative average hours per unit at a given total production divided by the cumulative average hours per unit when the total production was 50 percent of the present quantity. For example, from Table 15-1:

$$\text{Percent learning} = \frac{81}{90} \times 100 = 90 \text{ percent or}$$

$$= \frac{72.9}{81} \times 100 = 90 \text{ percent}$$

When linear graph paper is used, the learning curve is a hyperbola of the form $Y_x = KX^N$. On log-log paper, the curve is represented by:

$$\log Y_x = \log K + N \log X$$

where

$Y_x =$ Cumulative average value of X units
$K =$ Value in time of the first unit

TABLE 15–2

Learning curve percentage	Slope
70	0.514
75	0.415
80	0.322
85	0.234
90	0.152
95	0.074

X = Number of units produced

N = Exponent representing the slope (tan ϕ in Figure 15–3)

By definition, the learning in percent is then equal to:

$$\frac{K(2X)^N}{K(X)^N} = 2^N$$

And taking the log of both sides:

$$N = \frac{\log \text{ of learning percent}}{\log 2}$$

For 80 percent learning:

$$N = \frac{\log \text{ of } 0.80}{\log \text{ of } 2} = \frac{9.9031 - 10}{0.3010}$$

$$= -0.322$$

$$\text{arc tan } 0.322 = 17°50'$$

Table 15–2 presents the slopes of the common learning curve percentages.

An example will help clarify the above relationships. Assume that 50 units are produced at a cumulative average of 20 work-hours per unit. What is the learning curve percentage when 100 units are produced at a cumulative average of 15 work-hours per unit?

X_1 = Cumulative production to point one (50 units)

X_2 = Cumulative production to point two (100 units)

$\dfrac{X_2}{X_1}$ = 2 (when the production quantity is doubled)

and

$$\frac{\text{Tc.a. 2}}{\text{Tc.a. 1}} = \frac{K(X_2)^N}{K(X_1)^N} = 2^N$$

where:

Tc.a. = Cumulative average hours per unit for any number of units in logarithmic form

$$\log Tc.a. = \log K + N \log X$$
$$\log 20 = \log K + N \log 50$$
$$\log 15 = \log K + N \log 100$$
$$\log 20 - \log 15 = N (\log 50 - \log 100)$$
$$N = \frac{\log 20 - \log 15}{\log 50 - \log 100}$$
$$= \frac{1.30103 - 1.17609}{1.69897 - 2.00000}$$
$$= \frac{0.12494}{-0.30103}$$
$$N = -0.415$$

and

$$2^{-0.415} = 75 \text{ percent (learning curve percentage)}$$

The above derivation is based on the "cumulative average learning curve" since the relationships refer to the average hours to produce a unit based on all units produced up to a particular point.

When working with the work-hours required to produce a specific unit, we are dealing with the "unit learning curve," which refers to the hours required to produce a specific unit. The log plot of the cumulative average is asymptotically parallel to the log plot of the unit curve. The cumulative average line is straight, while the individual line curves downward from unit 1 until it becomes parallel to the cumulative average line. For practical estimating purposes, the individual and cumulative curves become about parallel somewhere between the 10th and 20th unit.

To plot the unit time versus the quantity, calculate two points where the quantity is above 20 and plot them on log-log paper. To calculate the unit time value of the selected points, multiply the cumulative average time of these points by a conversion factor. The conversion factor used for making the unit plot is $1 + N$. This is obtained as follows:

$$Y_x = KX^N = \text{Cumulative mean for } X \text{ planes}$$
$$T = XY_x = KX^{N+1} = \text{Total time for } X \text{ planes}$$

Since the time for each individual plane is a function of X

$$f(X) = \text{Time for each plane}$$

Then:

$$T = XY_X = \int_0^x f(X)dx = KX^{N+1}$$

FIGURE 15–4
Unit and cumulative learning curves for 85 percent learning. The two curves
are shown to be parallel after the tenth unit is produced.

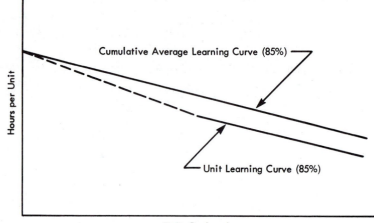

Differentiating:

$$\frac{dT}{dX} = f(x) = (N + 1)\ KX^N$$

$$= (N + 1)\ Y_x = \text{Time for } X\text{th plane}$$

Thus the conversion factor for an 85 percent learning curve would be:

$$1 + N = 1 + (-0.234) = 0.766$$

Multiplying this conversion factor by the cumulative average time for
20 units and the cumulative average time for 40 units would enable us to
plot the unit time curve for the portion that is asymptotically parallel to
the cumulative average plotting (see Figure 15–4).

Software for microcomputer programs can determine expected staffing
requirements and estimate the costs of projected production or scheduling, by utilizing learning curve theory. For example, the program written
by Gary Whitehouse, Yasser Hosni, and Farid Guediri of the University
of Central Florida computes the rate of improvement (learning efficiency)
on input of data on observed times. It also computes the cumulative
average work-hours/unit for the nth unit as well as the work-hours required to produce the nth unit.

Being able to estimate the time for the first unit produced and the time
for successive units can be extremely helpful in connection with estimates
of relatively low quantities if the analyst has available standard data and
learning curve information. Since standard data is usually based upon

worker performance when the quantity is such that we are at the flat portion of the learning curve, it needs to be adjusted upward to assure adequate time (cost) is allowed per unit under low quantity conditions. For example, let us assume the analyst wants to know the time to produce the first unit of a complex assembly. His standard data analysis suggests a time of 1.47 hours (which is the average accumulated time for the nth unit). This nth unit, in this case is estimated to be 300 assemblies. It is at that point where the learning curve begins to flatten. Furthermore, the analyst expects a 95 percent learning rate. Here, the exponent N representing the slope would be $- 0.074$. Then K, the value in time for the first unit would be:

$$K = \frac{1}{(300)^{- 0.074}} \times 1.47 = 2.24 \text{ hours}$$

Thus, the analyst's costs would be based upon 2.24 hours of time to produce one assembly—not the 1.47 hours developed from his standard data.

Computations involving use of the learning curve may be computed rapidly using software. For example, appendix A3–16 provides software written in Basic that provides the reader the opportunity to compute:

1. The exponent for a learning curve.
2. The time to produce the first unit of a given learning curve.
3. The time to produce the nth unit.
4. The computation of launching costs. These costs are the difference between standard costs and actual costs up to the point that standard times are achieved.

CHARACTERISTICS OF A SOUND RATING SYSTEM

The first and most important characteristic of any rating system is accuracy. Since the majority of rating techniques rely on the judgment of the time study observer, perfect consistency in rating is impossible.

However, rating procedures that readily permit time study analysts within a given organization to study different operators employing the same method to arrive at standards that are not more than 5 percent in deviation from the average of the standards established by the group are considered adequate. The rating plan that carries variations in standards greater than the ±5 percent tolerance should be improved or replaced.

Other factors being similar, the rating plan that gives the most consistent results is the most useful. Inconsistency in rating does more than any other one thing to destroy the operator's confidence in the time study procedure. For example, assume that through stopwatch study an analyst developed a standard that allowed 7.88 hours to mill 100 castings. At some later date the analyst studied a similar casting entailing a slightly longer cut and established a standard of 7.22 hours. Considerable complaint would be registered by the operator, even though both studies were within the ±5 percent criterion of accuracy.

A rating plan that achieves consistency when used by the various time study analysts within a given plant, and yet is outside the accepted definition of normal accuracy, can be corrected. A rating plan that gives inconsistent results when used by the different analysts is sure to end in failure. Time study people who find difficulty in rating consistently after proper and thorough training would be better off seeking some other means of livelihood. It is not difficult to correct the rating habits of an analyst who rates consistently high or consistently low, but it is most difficult to instill rating ability into the mind of an analyst who rates too high today and too low tomorrow.

A rating system that is simple, concise, easily explained, and keyed to well-established benchmarks is more successful than a complex rating system requiring involved adjusting factors and computational techniques that confuse the average shop employee.

In view of these accuracy limitations, every company accumulates a number of standards termed "tight" or "loose" by the production floor. If these inequities are within the 5 percent tolerance range, few grievances develop. However, if an operator can earn as much as a 50 percent premium on one job, and not even make standard on another, then employees will display general dissatisfaction for the whole rate program. Inequities in different rates are not necessarily entirely due to poor performance rating. Loose rates frequently are due to methods improvements inaugurated over a period of time without restudy of the job from the time study point of view.

Rating at the Workstation

Performance rating should be done only during the observation of elemental times. As the operator progresses from one element to the next, analysts carefully evaluate speed, dexterity, freedom from false moves, rhythm, coordination, effectiveness, and all the other factors influencing output by the prescribed method. At this time the performance of the operator in comparison to normal performance is most clearly evident. Once the performance has been judged and recorded, it should not be changed. This does not imply that faulty judgment on the part of the observer is not possible. In case the leveling is questioned, the job or operation should be restudied to prove or disprove the recorded evaluation.

How often should ratings take place? Most studies of 30 minutes or less can be satisfactorily rated by applying one factor to all of the effort elements. In longer studies, the operator's performance may fluctuate and more reliable results can be obtained by periodic rating—perhaps every cycle. Where elements are rather long and some require a high degree of skill or dexterity, rating of each element of each cycle may provide the best results.

Immediately after completing the study and recording the final performance factor (if overall rating was used), the observer should advise the operator of the performance rating. Even if elemental rating is used, the analyst can approximate the operator's performance. This practice gives the operator an opportunity to express his or her opinion about the fairness of the performance factor directly to the person responsible for its development. Thus, they can achieve an understanding prior to computation of the standard. Where this procedure is practiced, time study analysts find that they receive greater respect from the operators and that they tend to be more conscientious in their rating of the operators' performance. Also, considerably fewer rate grievances are submitted, because most rates have been agreed on, or at least satisfactorily explained and sold, before they are issued.

Rating Elements Versus Overall Study

The question of how frequently the performance should be evaluated during the course of a study often comes up. Although no set rule can be established as to the interval limit that permits concise rating, the more frequently the study is rated, the more accurate the evaluation of the operator's demonstrated performance.

On short-cycle repetitive operations, little deviation in operator performance is realized during the course of the average length study (15 to 30 minutes). In cases like this, it is perfectly satisfactory to evaluate the performance of the entire study and to record the rating factor for each element. Of course, power-fed or machine-controlled elements are rated normal, or 1.00, as their speed cannot be changed or modified at will by operators. In short-cycle studies, an observer who endeavors to performance rate each successive element of the entire study, will be so busy recording values that he or she will be unable to do an effective job of observing, analyzing, and evaluating the operator's performance.

When the study to be taken is relatively long (over 30 minutes) or is made up of several long elements, operator performance may vary during the course of the study. In such studies, analysts should periodically evaluate and rate performance. They can consistently and accurately rate elements longer than 0.10 minute as they occur. However, if a study is made of a series of elements shorter than 0.10 minute, then no effort should be made to evaluate each element of each cycle of the study, as time does not permit this operation. It is satisfactory to rate the overall time of each cycle or perhaps of each group of cycles.

METHODS OF RATING

The Westinghouse System

One of the oldest and most widely used systems of rating was developed by the Westinghouse Electric Corporation. It is outlined in detail by Lowry, Maynard, and Stegemerten.[3] This method considers four factors in evaluating the performance of the operator. These are skill, effort, conditions, and consistency.

They define skill as "proficiency at following a given method," and further relate it to expertise, demonstrated by the proper coordination of mind and hands.

The skill of an operator results from experience and inherent aptitudes, such as natural coordination and rhythm. Practice develops skill, but it cannot entirely compensate for deficiencies in natural aptitude. All the practice in the world could not make major league baseball pitchers out of most athletes.

A person's skill in a given operation increases over a period of time because increased familiarity with the work brings speed, smoothness of motions, and freedom from hesitations and false moves. A decrease in skill is usually caused by impairment of ability brought about by physical or psychological factors, such as failing eyesight, failing reflexes, and the loss of muscular strength or coordination. Therefore, a person's skill can vary from job to job and even from operation to operation on a given job.

The Westinghouse system of leveling or rating lists these six degrees or classes of skill that represent an acceptable proficiency for evaluation: poor, fair, average, good, excellent, and super. The observer evaluates the skill displayed by the operator and places it in one of these six classes. Table 15–3 illustrates the characteristics of the various degrees of skill with their equivalent numerical values. The skill rating is then translated into its equivalent percentage value, which ranges from plus 15 percent for superskill to minus 22 percent for poor skill. This percentage is then combined algebraically with the ratings for effort, conditions, and consistency to arrive at the final leveling, or performance rating factor.

This rating method defines effort as a "demonstration of the will to work effectively." It is representative of the speed with which skill is applied, and can be controlled to a high degree by the operator. When evaluating the effort given, the observer must rate only the effective effort demonstrated. Many times an operator applies misdirected effort with high tempo to increase the cycle time of the study and yet retain a liberal rating factor. The six classes of effort for rating purposes are poor, fair, average, good, excellent, and excessive. Excessive effort has been assigned a value of plus 13 percent, and poor effort, a value of minus 17

[3] S. M. Lowry, H. B. Maynard, and G. J. Stegemerten, *Time and Motion Study and Formulas for Wage Incentives,* 3rd ed. (New York: McGraw-Hill, 1940), pp. 207–50.

TABLE 15–3
Skill

+0.15.............	A1	Superskill
+0.13.............	A2	Superskill
+0.11.............	B1	Excellent
+0.08.............	B2	Excellent
+0.06.............	C1	Good
+0.03.............	C2	Good
0.00.............	D	Average
−0.05.............	E1	Fair
−0.10.............	E2	Fair
−0.16.............	F1	Poor
−0.22.............	F2	Poor

Source: S. M. Lowry, H. B. Maynard, and
G. J. Stegemerten, *Time and Motion Study
and Formulas for Wage Incentives,* 3rd ed.
(New York: McGraw-Hill, 1940), p. 233.

percent. Table 15–4 gives the numerical values for the different degrees of effort and also outlines the characteristics of the various categories.

The conditions referred to in this performance rating procedure affect the operator and not the operation. Time study people rate conditions normal or average in more than a majority of instances, as conditions are evaluated in comparison with the way in which they are customarily found at the workstation. Elements affecting working conditions include temperature, ventilation, light, and noise. Thus, if the temperature at a given work station was 60° F whereas it was customarily maintained at 68° to 74° F, the conditions would be rated as lower than normal. Conditions that affect the operation, such as poor condition of tools or materials,

TABLE 15–4
Effort

+0.13.............	A1	Excessive
+0.12.............	A2	Excessive
+0.10.............	B1	Excellent
+0.08.............	B2	Excellent
+0.05.............	C1	Good
+0.02.............	C2	Good
0.00.............	D	Average
−0.04.............	E1	Fair
−0.08.............	E2	Fair
−0.12.............	F1	Poor
−0.17.............	F2	Poor

Source: S. M. Lowry, H. B. Maynard, and
G. J. Stegemerten, *Time and Motion Study
and Formulas for Wage Incentives,* 3rd ed.
(New York: McGraw-Hill, 1940), p. 233.

TABLE 15–5
Conditions

+0.06	A	Ideal
+0.04	B	Excellent
+0.02	C	Good
0.00	D	Average
−0.03	E	Fair
−0.07	F	Poor

Source: S. M. Lowry, H. B. Maynard, and
G. J. Stegemerten, *Time and Motion Study
and Formulas for Wage Incentives,* 3rd ed.
(New York: McGraw-Hill, 1940), p. 233.

would not be considered when applying the performance factor for working conditions. Six general classes of conditions have been enumerated, with values ranging from +6 percent to −7 percent. These "general state" conditions are listed as ideal, excellent, good, average, fair, and poor. Table 15–5 gives the respective values for these conditions.

The last of the four factors that influence the performance rating is the consistency of the operator. Unless the analyst uses the snapback method, or makes and records successive subtractions as the study progresses, the consistency of the operator must be evaluated as the study is worked up. Elemental time values that repeat constantly would, of course, have a perfect consistency. This situation occurs very infrequently, as there always tends to be dispersion due to the many variables, such as material hardness, the tool cutting edge, the lubricant, the skill and effort of the operator, erroneous watch readings, and the presence of foreign elements. Elements that are mechanically controlled would, of course, have values of near perfect consistency, but such elements are not rated. There are six classes of consistency: perfect, excellent, good, average, fair, and poor. Perfect consistency is rated +4 percent, and poor consistency is rated −4 percent, while the other classes fall in between these values. Table 15–6 summarizes these values.

TABLE 15–6
Consistency

+0.04	A	Perfect
+0.03	B	Excellent
+0.01	C	Good
0.00	D	Average
−0.02	E	Fair
−0.04	F	Poor

Source: S. M. Lowry, H. B. Maynard, and
G. J. Stegemerten, *Time and Motion Study
and Formulas for Wage Incentives,* 3rd ed.
(New York: McGraw-Hill, 1940), p. 233.

No fixed rule can be cited when rating consistency. Some operations of short duration are free of delicate positioning manipulations and give relatively consistent results from one cycle to the next. Such operations would have a more exacting requirement of average consistency than a job of long duration demanding high skill in its positioning, engaging, and aligning elements. The time study analyst's knowledge of the work determines the justified range of variation for a particular operation to a large extent.

Some operators consistently perform poorly in an effort to deceive observers. This is easily accomplished by counting to oneself, thereby setting a pace that can be accurately followed. Operators familiar with this performance rating procedure sometimes perform at a pace that is consistent and yet below the effort rating curve. In other words, they may be performing at a pace that is poorer than poor. In cases like this, the operator cannot be leveled. The study must be stopped and the situation brought to the attention of the operator or the supervisor or both.

Once the skill, effort, conditions, and consistency of the operation have been assigned, and their equivalent numerical values established, analysts determine the performance factor by algebraically combining the four values and adding their sum to unity. For example, if a given job is rated C2 on skill, C1 on effort, D on conditions, and E on consistency, the performance factor would be evolved as follows:

Skill	C2	+.03
Effort	C1	+.05
Conditions	D	+.00
Consistency	E	−.02
Algebraic sum		+.06
Performance factor		1.06

The performance factor is applied only to the effort, or manually performed, elements; all machine-controlled elements are rated 1.00.

Many companies have modified the Westinghouse system to include only skill and effort factors entering into the determination of the rating factor. They contend that consistency is very closely allied to skill, and that conditions are rated average in most instances. If conditions deviate substantially from normal, the study could be postponed, or the effect of the unusual conditions could be taken into consideration in the application of the allowance (see Chapter 16).

In 1949 Westinghouse Electric Corporation developed a new rating method, the "performance rating plan."

In the performance rating plan, in addition to using the physical attributes displayed by the operator, the company made an attempt to

TABLE 15–7

	+ Above		0 Expected	- Below	
DEXTERITY:					
1. Displayed ability in use of equipment and tools, and in assembly of parts.	6	3	0	2	4
2. Certainty of movement.	6	3	0	2	4
3. Coordination and rhythm.		2	0	2	
EFFECTIVENESS:					
1. Displayed ability to continually replace and retrieve tools and parts with automaticity and accuracy.	6	3	0	2	4
2. Displayed ability to facilitate, eliminate, combine, or shorten motions.	6	3	0	4	8
3. Displayed ability to use both hands with equal ease.	6	3	0	4	8
4. Displayed ability to confine efforts to necessary work.			0	4	8
PHYSICAL APPLICATION:					
1. Work pace.	6	3	0	4	8
2. Attentiveness.			0	2	4

evaluate the relationship between those physical attributes and the basic divisions of work. The characteristics and attributes that the Westinghouse performance rating technique considers are: (1) dexterity, (2) effectiveness, and (3) physical application.

These three major classifications do not in themselves carry any numerical weight, but have been assigned attributes that do carry numerical weight. Table 15–7 gives the numerical values of the nine attributes evaluated under this system.

The first major classification, dexterity, has three attributes, the first of which is *Displayed ability in use of equipment and tools, and in assembly of parts.*

When considering this attribute, the primary concern is with the "do" portion of the work cycle after the "get" operations (reach, grasp, move) have taken place.

The second attribute under dexterity is *Certainty of movement.*

In evaluating this attribute, the number and degree of hesitations, pauses, or roundabout moves is important. The basic divisions of accomplishment that give the operator a low rating for this attribute are change direction, plan, and avoidable delay. All of these affect certainty of movement.

The last attribute considered under dexterity is *Coordination and rhythm*. This attribute is evidenced by the degree of the displayed performance, by smoothness of motions, and by freedom from spasmodic spurts and lags.

The second major classification, effectiveness, is efficient, orderly procedure. The classification has four individual attributes. The first of these is *Displayed ability to continually replace and retrieve tools and parts with automaticity and accuracy.*

Here concern centers on the worker's ability to repeatedly place tools, materials, and parts in specified locations and positions, and to retrieve them automatically and accurately by eliminating such ineffective basic divisions of work as searching and selecting.

The second of the individual attributes in effectiveness is *Displayed ability to facilitate, eliminate, combine, or shorten motions*. Scrutiny falls on the proficiency of the basic divisions position, pre-position, release, and inspect. The transport therbligs are usually predetermined by the established method. However, a skilled worker can eliminate or shorten the elements of pre-position, position, and inspect through manipulative ability.

The third attribute under effectiveness is *Displayed ability to use both hands with equal ease*. The analyst rates the degree of effective utilization of both hands.

The fourth and last attribute under effectiveness is *Displayed ability to confine efforts to necessary work*. This attribute rates the presence of unnecessary work that could not be removed when taking the study. It carries only a negative weight, for no percentage is added when the work is confined to the necessary work because this condition is expected.

The third major classification, physical application, is the demonstrated rate of performance and has two attributes. The first of these is *Work pace*. Its rates compare the speed of movement to preconceived standards for the particular work under consideration.

The second attribute for physical application is *Attentiveness*. It is rated as the degree of displayed concentration.

Both of the Westinghouse rating techniques demand considerable training to differentiate the levels of each of the attributes. Their training entails a 30-hour course during which approximately 25 hours are spent rating videotapes or films and discussing the attributes and the degree to which each is displayed. The procedure generally followed is:

1. A film is shown and the operation explained.
2. The film or tape is reshown and rated.
3. The individual ratings are compared and discussed.
4. The film or tape is reshown, and the attributes pointed out and explained.

5. Step 4 is repeated as often as is necessary to reach understanding and agreement.

Elemental rating is not practical using either of the Westinghouse systems. Except in the case of very long elements, analysts would not have time to evaluate the dexterity, effectiveness, and physical application of each element of the study. Westinghouse rating procedures are appropriate for either cycle rating or overall study rating.

Synthetic Rating

In an effort to develop a method of rating that would not rely on the judgment of a time study observer and would give consistent results, R. L. Morrow established a procedure known as "synthetic leveling."[4]

The synthetic leveling procedure determines a performance factor for representative effort elements of the work cycle by comparing actual elemental observed times to times developed through fundamental motion data (see Chapter 19). Thus, the performance factor may be expressed algebraically as:

$$P = \frac{F_t}{O}$$

where:

P = Performance or leveling factor
F_t = Fundamental motion time
O = Observed mean elemental time for the elements used in F_t

This factor would then be applied to the remainder of the manually controlled elements comprised by the study. Of course, machine-controlled elements are not rated, as is the case in all rating techniques.

A typical illustration of synthetic rating appears in Table 15–8.
For element 1,

$$P = \frac{0.096}{0.08} = 120 \text{ percent}$$

and for element 4,

$$P = \frac{0.278}{0.22} = 126 \text{ percent}$$

The mean of these is 123 percent, and this is the factor used for rating all of the effort elements. Obviously, synthetic performance rating is a sampling technique.

---

[4] Morrow, *Time Study*, p. 241.

TABLE 15-8

Element no.	Observed average time in minutes	Element type	Fundamental motion time in minutes	Performance factor
1	0.08	Manual	0.096	123
2	0.15	Manual	–	123
3	0.05	Manual	–	123
4	0.22	Manual	0.278	123
5	1.41	Power fed	–	100
6	0.07	Manual	–	123
7	0.11	Manual	–	123
8	0.38	Power fed	–	100
9	0.14	Manual	–	123
10	0.06	Manual	–	123
11	0.20	Manual	–	123
12	0.06	Manual	–	123

More than one element should be used in establishing a synthetic rating factor because research has proved that operator performance varies significantly from element to element, especially in complex work.

Actually, all experienced time study analysts unconsciously follow the synthetic rating procedure to some extent. Past experience crams time study analysts' minds full of benchmarks established on similar work. Consequently, they know that the normal performance of advancing the drill of a 17-inch single-spindle Delta drill is 0.03 minute, and of indexing the hex turret of a No. 4 Warner & Swasey turret lathe is 0.06 minute, and of blowing out a vise or fixture with an air hose and laying the finished part aside is 0.08 minute. These benchmarks and many others, when compared to actual performance, certainly influence and even determine the rating factor given the operator.

A major objection to the synthetic leveling procedure is the time required to construct a left- and right-hand chart of the elements selected for the establishment of basic motion times. Establishing a standard for the entire job synthetically eliminates the laborious task of recording elemental times, making subtractions, determining the mean elapsed time, and determining the normal time synthetically for several elements so as to arrive at a performance factor, and applying the performance factor. Many standards are established in this manner using "standard data" or "fundamental motion data."

Speed Rating

Speed rating is a performance evaluation method which considers only the rate of accomplishment of the work per unit time. In this method, observers measure the effectiveness of the operator against the conception of a normal operator doing the same work, and then assign a percentage to indicate the ratio of the observed performance to normal performance. They particularly emphasize the observer's having complete knowledge of the job before taking the study. For example, the pace of machine workers in a plant producing aircraft engine parts would appear considerably slower than the pace of machine workers producing farm machinery components. The greater precision of aircraft work requires such care that the movements of the various operators appears unduly slow to one not completely familiar with the work performed.

In speed rating, 100 percent is usually considered normal. Thus, a rating of 110 percent indicates the operator was performing at a speed 10 percent greater than normal, and a rating of 90 percent would mean that he or she was performing at a speed 90 percent of normal. Some companies using the speed rating technique call 60 percent standard or normal. This is based on the standard hour approach, that is, producing 60 minutes of work every hour. On this basis, a rating of 80 would mean that the operator was working at a speed of 80/60, which equals 133 percent, or 33 percent above normal. A rating of 50 would indicate a speed of 50/60, or 83⅓ percent of normal or standard.

In the speed rating method, analysts first appraise the performance to determine whether it is above or below normal. Then they try to place the performance in the precise position on the rating scale that correctly evaluates the numerical difference between the standard and the performance demonstrated.

A form of speed rating referred to as "pace rating" has received considerable attention from the basic steel industry. In effect, pace rating is speed rating. However, in an effort to identify completely a normal pace on different jobs, benchmarks have been provided on a broad range of work. Thus, in addition to card dealing, such effort operations as shoveling sand, coremaking, brick handling, and walking have been clearly identified as to method, and quantified as to normal rate of production. Once time study analysts become familiar with a series of benchmarks closely allied to the work under study, they are much better equipped to evaluate the speed performed.

Time study people use speed rating for elemental, cycle, or overall rating. Figure 14–6 shows a completed study where analysts performance rated snapback readings, using speed rating.

Objective Rating

The objective rating method developed by M. E. Mundel eliminates the difficulty of establishing a normal speed criterion for every type of work.[5] This procedure establishes a single work assignment to which the pace of all other jobs is compared. After the judgment of pace, a secondary factor

---

[5] M. E. Mundel, *Motion and Time Study*, 4th ed. (Englewood Cliffs, N.J.: Prentice Hall, 1960), pp. 319–45.

assigned to the job indicates its relative difficulty. Factors influencing the difficulty adjustment are: (1) amount of body used, (2) foot pedals, (3) bimanualness, (4) eye-hand coordination, (5) handling or sensory requirements, and (6) weight handled or resistance encountered.

Numerical values, resulting from experiments, have been assigned for a range of degrees of each factor. The sum of the numerical values for each of the six factors comprises the secondary adjustment. By this method, the normal time can be expressed as follows:

$$T_n = (P_2)(S)(O)$$

where:

T_n = Computed established normal time
P_2 = Pace rating factor
S = Job difficulty adjustment factor
O = Observed mean elemental time

This performance rating procedure gives consistent results. Comparing the pace of the operation under study to an operation completely familiar to the observer is easier than judging simultaneously all the attributes of an operation to a concept of normal for that specific job. The secondary factor does not affect inconsistency since this factor merely adjusts the rated time by the application of a percentage. A percentage value table gives values for the effects of various difficulties in the operation performed.

OPERATOR SELECTION

To eliminate the performance rating step entirely in the calculation of the standard, some companies select the operators and then consider the average observed time as the normal time. Using this method an analyst studies more than one operator and observes enough cycles to calculate a reliable average time (within ±5 percent of the population average). Of course, the success of this method depends on the selection of the employees and their performance during the study. If the performances of the operators are slower than normal, then too liberal a standard results; conversely, if the observed operators produce at a pace more rapid than normal, then the standard is unduly tight. There is always the possibility of having but one or two available operators and the chance that they may differ from normal. If, in an effort to avoid delay in establishing a standard, the observer makes the study, the result is a poor time standard.

ANALYSIS OF RATING

As is true of all procedures requiring the exercise of judgment, the simpler and more concise the plan, the easier it is to use and, in general, the more valid the results.

The performance rating plan that is easiest to apply, easiest to explain, and gives the most valid results is straight speed or pace rating augmented by synthetic benchmarks. As has been explained, 100 is considered normal in this procedure, and performance greater than normal is indicated by values directly proportional to 100. Thus, a rating of 120 indicates a performance 20 percent higher than normal. A rating of 60 indicates that the operator performed at a pace of only 0.60 of normal. The speed rating scale usually covers a range of from 0.50 to 1.50. Operators performing outside this productivity range of 3 to 1 may be studied, but this is not recommended. The closer the performance is to normal, the better the chance of achieving a fair normal time.

Four criteria determine whether or not time study analysts using speed rating can consistently establish values not more than 5 percent above or below the rating average calculated by a group of trained analysts. These are:

1. Experience in the class of work performed.
2. Use of synthetic benchmarks on at least two of the elements performed.
3. Selection of an operator who gives performances somewhere between 115 and 85 percent of normal.
4. Use of the mean value of three or more independent studies.

Certainly, the most important of these four criteria is experience in the class of work performed. This does not necessarily mean that analysts must at one time have been actual operators in the work being studied, although this would be desirable. From past personal experience, either by observation or operation, analysts should be sufficiently familiar with the work to understand every detail of the method being used. Thus, on a job being performed on a turret lathe, they should recognize the tooling, have knowledge of the correct speeds and feeds, the correct rake and clearance angles, the lubricant, horsepower requirements, method of holding the work, and so on. On an assembly job taking place in a fixture, the observers should be familiar with the difficulty in positioning the components in the fixture, and should know the class of fit between all mating parts and have a clear understanding of the relationship between time and class of fit. They should know the proper sequence of events and the weights of all the parts handled.

Experience is the key to accurate performance rating. An analyst with 10 years' experience in the metal trades would have considerable difficulty in establishing standards in a women's shoe factory, and of course the reverse would be true.

Although operator performance varies from element to element, the average performance of two or more elements of the study gives a reasonably good estimate of the cycle performance. Preestablishing synthetically the normal time required to perform several elements involved in the study gives an indication of the overall performance.

The alignment chart shown on the time study board (see Figure 13–13 on page 360) permits analysts to establish synthetically the normal time for the elements "pick up pc. & position in fixture, jig, or die," "pick up pc. & bring to work station," "pick up pc. & lay aside."[6] Since very few work assignments can be performed without utilizing at least one of these elements, time study analysts have a useful guide for establishing a "correct" performance factor. For example, if a 10-pound casting is picked up and placed in a three-jaw turret lathe chuck, and the element entailed a reach of 30 inches, a move of 20 inches, a class 4 grasp (see Chapter 19 on MTM), and a P3SE position, the synthetic time could be determined graphically. By going from 20 inches on scale 1 vertically to 10 pounds on scale 2 and then moving horizontally to scale 3, a time of 0.017 minute for the move and release is obtained. This point connected to the G4 grasp of 0.005 minute (scale 4) gives a turning point on scale 5. This turning point on scale 5 connected with a reach of 30 inches on scale 6 gives a time of 0.04 minute for the reach, grasp, move, and release. This point connected with a P3SE position entailing 0.025 minute on scale 8 gives the time to perform the complete element, 0.065 minute on scale 9.

The synthetic value cannot establish a rating for every element of each cycle of the study. Through experience and training in performance rating analysts can evaluate operators quite precisely; the alignment chart and the resulting synthetic standards provide an additional check and guide bolstering analysts' confidence in their ability to establish fair normal times.

Whenever more than one operator is available to be studied, select the one who is thoroughly experienced on the job, who has a reputation of being receptive to time study practice, and who consistently performs at a pace near standard or slightly better than standard. The closer the operator performs to a normal pace, the easier it will be to level him or her. Sizable errors in judgment during rating invariably result from improper evaluation of an operator performing at either extremity of the rating scale. For example, if 0.50 minute is normal for dealing a deck of cards into four bridge hands, performance within ±15 percent of this conception of normal is fairly easy to identify. However, a performance 50 percent faster or slower than normal, causes considerable difficulty in establishing an accurate rating factor for the performance demonstrated.

---

[6] The author developed part of the data used in the design of this alignment chart after analyzing over 600 feet of film. The remainder of the data used was taken from Methods-Time Measurement tables.

Take three or more studies before arriving at the standard. The total
error, due to rating and the determination of a mean elemental elapsed
time that deviates from the population mean, becomes minimal when the
averages of several independent studies are used in computing the time
standard. The independent studies can be made on the same operator at
different times of the day, or on different operators. The point is that, as
the number of studies increases, compensating errors diminish the overall
error. A recent example clarifies this point. The author, along with two
other trained industrial engineers, reviewed performance rating training
films involving 15 different operations. The results are tabulated in Table
15–9. The standard ratings were not disclosed until all 15 operations had
been rated. The average deviation of all three engineers for the 15 different operations was only 0.87 of a point on a speed rating scale with a range
of from 50 points to 150 points. Yet Engineer C was 30.0 high on operation
7, and Engineer B was 20.0 low on operation 2. In only one case (operation 3), when the known ratings were within the 70 to 130 range, did the
average rating of the three analysts exceed ±5 points of the known rating.

The normal time for the operation should be determined by averaging
the normal times of the independent studies. This procedure not only
reduces the error inherent in the performance rating process, but also
results in the determination of a mean measured time with less deviation
from the population mean measured time.

TABLE 15–9
Performance ratings of three different engineers observing 15 different operations

Operation	Standard rating	Engineer A Rating	Engineer A Deviation	Engineer B Rating	Engineer B Deviation	Engineer C Rating	Engineer C Deviation	Average of Engineers A, B, C Rating	Average of Engineers A, B, C Deviation
1	110	110	0	115	+5	100	−10	108	−2
2	150	140	−10	130	−20	125	−25	132	−18
3	90	110	+20	100	+10	105	+15	105	+15
4	100	100	0	100	0	100	0	100	0
5	130	120	−10	130	0	115	−15	122	−8
6	120	140	+20	120	0	105	−15	122	+2
7	65	70	+5	70	+5	95	+30	78	+13
8	105	100	−5	110	+5	100	−5	103	−2
9	140	160	+20	145	+5	145	+5	150	+10
10	115	125	+10	125	+10	110	−5	120	+5
11	115	110	−5	120	+10	115	0	115	0
12	125	125	0	125	0	115	−10	122	−3
13	100	100	0	85	−15	110	+10	98	−2
14	65	55	−10	70	+5	90	+25	72	+7
15	150	160	+10	140	−10	140	−10	147	−3
Average of 15 operations	112	115.0		112.3		111.3		112.9	
Average deviation	0	+3.0		+0.3		−0.7		+0.9	

Performance rating, like any other form of work involving judgment, must be accomplished by competent, well-trained individuals. A person with a background of experience, access to reliable synthetic time values, and sound judgment in operator selection, calculates reasonably accurate results.

Training for Rating

To be successful, analysts must develop track records for setting accurate standards that are accepted by both labor and management. Furthermore, rates must be consistent to maintain the respect of all parties. In general, analysts should regularly establish standards within ±5 percent of the true rate, when studying operators performing somewhere in the range of 0.70 to 1.30 of normal. Thus, if several operators are performing the same job, and different analysts, each studying a different operator, establish time standards on the job, then the resulting standard from each individual study should be within ±5 percent of the mean of the group of studies.

To assure consistency in rating, both with their own rates and with the rates established by the others, analysts should continually participate in organized training programs. Of course, the training in performance rating should be more intense for the neophyte time study analyst.

One of the most widely used methods for training analysts in performance rating is the observation of videotape or motion-picture films illustrating diversified operations performed at different productivity levels. Each film or tape has a known level of performance. After it is shown, the correct rating is compared with the values established independently by the various trainees. If analysts' values deviate substantially from the correct value, then specific information should justify the rating. For example, the observer may have been misled by high performance in the handling of the material to and from the workstation, while poor performance prevailed during the cycle at the workstation. Then, too, the analyst may have underrated the operator due to an apparently effortless sequence of motions, whereas the operator's smooth, rhythmic blending of movements is really an indication of high dexterity and manipulative ability.

The operations selected should be simple, yet should contain a number of fast motions. The observation of short elements cultivates both speed and concentration in the observer and trains the "second sight."

As successive operations are reviewed, analysts should plot their ratings against the known values (see Figure 15–5). A straight line indicates perfection, whereas high irregularities on both sides of the line indicate inconsistency, as well as inability to evaluate performance.

In the example, the analyst rated the first film 75, whereas the correct rating was 55. He rated the second 80, while the proper rating was 70. In

FIGURE 15–5
Chart showing a record of seven studies with the analyst tending to rate a
little high on studies 1, 2, 4, and 6, and a little low on studies 3 and 7. Only
study 1 was rated outside the range of desired accuracy.

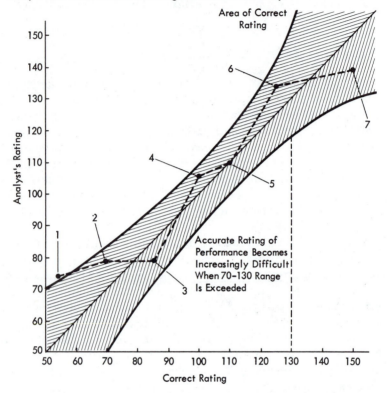

all but the first case, the analyst was within the company's established
area of correct rating. It is interesting to note that this concern indicates
that only at the 100 percent, or normal, level of performance is the ±5
percent criterion of accuracy valid. When performance is below 70 percent of normal or above 130 percent of normal, an experienced time study
analyst expects an error much larger than 5 percent.

It is also helpful to plot successive ratings on the *X*-axis and to indicate
the positive or negative magnitude of deviation from the known normal on
the *Y*-axis (see Figure 15–6). The closer the time study analyst's rating
comes to the *X*-axis, the more nearly correct he or she is.

To quantitatively determine an analyst's ability to performance rate,
compute the percentage of the analyst's rating contained within specified
limits of the known ratings. This can be done as follows:

1. Compute the mean difference ($\bar{x}_d$) between the rater's rating and the
    actual rating for *n* tests (*n* should be at least 15 observations).

FIGURE 15-6
Record of an analyst's rating factors on 15 studies

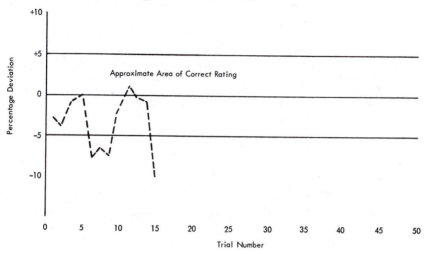

2. Compute the standard deviation s_d of the differences in rating.
3. Compute the normal deviate z_1, where:

$$z_1 = \frac{+5 \text{ (or some other figure of accuracy)} - \bar{x}_d}{s_d}$$

4. Compute the normal deviate z_2, where:

$$z_2 = \frac{-5 \text{ (or some other figure of accuracy)} - \bar{x}_d}{s_d}$$

5. Compute the area under the normal distribution between ± 5 (or some other figure of accuracy) centered at $\bar{x}_d$ which is assumed equal to μ_d and s_d is assumed equal to σ_d.

For example, suppose that, after reviewing 15 rating tapes, a given analyst had the achievement shown in Table 15-10.
The following computations would be made:

$$\bar{x}_d \text{ (mean difference)} = \frac{\Sigma d}{n} = \frac{50}{15} = 3.33$$

$$s_d \text{ (standard deviation)} = \sqrt{\frac{\Sigma d^2 - \frac{(\Sigma d)^2}{n}}{n-1}} = 7.7$$

$$z_1 = \frac{5.00 - 3.33}{7.7} = 0.217$$

TABLE 15-10

Tape no.	Correct rating	Analyst's rating	Difference (d)	Difference squared
1.............	115	105	−10	100
2.............	125	120	−5	25
3.............	85	95	+10	100
4.............	105	105	0	0
5.............	100	105	+5	25
6.............	95	110	+15	225
7.............	120	125	+5	25
8.............	140	150	+10	100
9.............	100	105	+5	25
10.............	60	75	+15	225
11.............	100	100	0	0
12.............	110	105	−5	25
13.............	90	90	0	0
14.............	130	125	−5	25
15.............	80	90	+10	100

and

$$P(z_1) = \int_0^{z_1} \frac{1}{\sqrt{2\pi}} e^{-\frac{z^2}{2}} \, dz = 0.0859$$

$$z_2 = \frac{-5 - 3.33}{7.7} = -1.08$$

and

$$P(z_2) = \int_0^{z_2} \frac{1}{\sqrt{2\pi}} e^{-\frac{z^2}{2}} \, dz = 0.3597$$

$$P(z_1) + P(z_2) = 0.4456$$

In this example, the analyst would receive a rating of 0.4456. This represents the portion of ratings that lie within ±5 rating points of the ideal ratings. This area of correct rating, together with the distribution of the difference between the analyst's rating and the correct rating, is shown in Figure 15–7. If the number of films observed is less than 15, use the student's "t" distribution.

A recent statistical study involving the performance rating of 6,720 individual operations by a group of 19 analysts over a period of approximately two years confirmed several facts that had been accepted by most industrial engineers. This study concluded that the "level of performance" is a factor that significantly affects errors in rating performance. Analysts overrated low levels of performance and underrated high levels of performance.

This study also concluded that the operation studied has an effect on errors in rating performance. Thus, complex operations tend to be more difficult to performance rate than do simpler operations—even for experienced analysts. At low levels of performance, overrating is greater for difficult operations than for simple operations, while at high levels of performance, underrating is larger for the easy-to-perform operations.

Today, no known tests can accurately evaluate the ability of a person to rate performance. Experience has shown, however, that only analysts showing a tendency toward consistency and proficiency after a brief training in rating can do an acceptable job in rating.

A survey made by one large manufacturer disclosed that an experienced industrial engineering employee did not level any more accurately than a less experienced analyst. This survey also indicated that those in the higher work classifications did not level better than those in the lower classifications, and that those in machining areas did no better than those in assembly areas.

A group of undergraduate engineers was given 15 minutes' instruction in rating and then requested to performance rate a film illustrating a series of heavy labor operations. The students were given no orientation in the class of work, and the operations observed differed substantially from the operations used in giving the students a concept of proper performance. These students were extremely successful in applying the concepts learned on light operations to heavy labor operations.

In another instance, 34 undergraduate engineering students performance rated five industrial films involving operations ranging from mold making to bench assembly. The students found four films were representative of normal performance, and one was 1.15, or 15 percent above normal. The results are shown in Table 15–11. Most students did not deviate greatly from the known values.

On request, one division of the General Electric Company checked the validity of the performance rating technique on three different occasions.

FIGURE 15–7
Graphic representation of the analyst's correct ratings (±5) and the distribution of all his ratings

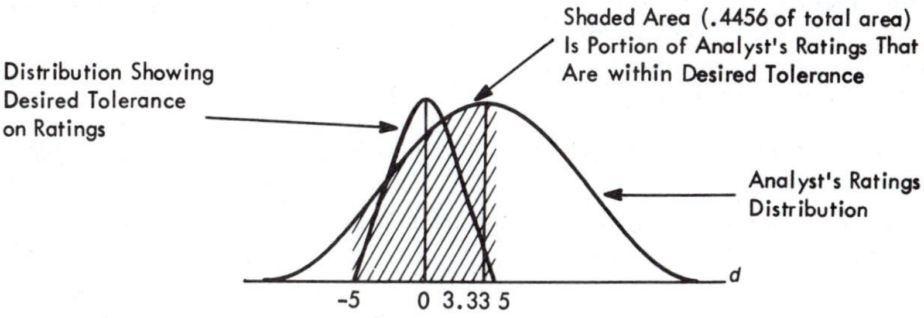

TABLE 15–11

Known rating	Number of students in each rating bracket					
	85–90	*91–95*	*96–100*	*101–105*	*106–110*	*111–125*
1.15	0	3	4	12	6	9
1.00	4	5	19	2	3	1
1.00	2	3	7	12	5	5
1.00	7	5	14	8	0	0
1.00	1	2	26	4	1	0

After brief training in the rating methods, a union steward and the line supervisor rated a given operation with the time study supervisor. At the completion of the study, the time study analyst rated the study without advising the other three men on the values he used, so that they would not be influenced by his opinion. These three men then went into their respective offices so that they would not be influenced by one another's opinions, and independently and secretly rated the operation. The results were as follows: time study analyst, 100; planning and wage payment supervisor, 99; line supervisor, 103; union steward, 103.

At a later date, a similar test case involved an entirely different class of work. The results were: time study analyst, 100; planning and wage payment supervisor, 99; line supervisor, 103; union steward, 103.

On the third occasion, an entirely new set of conditions prevailed, and in this instance all four men rated identically—109!

The foregoing studies bring out (1) that the concept of normal performance can be taught quickly, and (2) that the concept is transferable to some degree to dissimilar operations.

By actual observation of different work assignments throughout a plant under the guidance of a time study supervisor, it is possible to achieve excellent training in rating. The supervisor explains in detail the "why" of values after the trainees have already written their ratings independently. The independent values should be recorded to determine the consistency of the group and the necessity for additional, and perhaps more intensive, training. Companies with extensive and well-designed programs of training in rating have been successful in eliminating the tendency to overrate and underrate.

TEXT QUESTIONS

1. Why has industry been unable to develop a universal conception of "normal performance"?
2. Which factors enter into large variances in operator performance?
3. What are the characteristics of a sound rating system?

4. When should the performance rating procedure be accomplished? Why is this important?
5. What governs the frequency of performance rating during a given study?
6. Explain the Westinghouse system of leveling.
7. How does the Westinghouse performance rating method differ from the leveling method?
8. Under the Westinghouse rating system, why are "conditions" evaluated?
9. What is synthetic rating? What is its principal weakness?
10. What is the basis of speed rating, and how does this method differ from the Westinghouse system?
11. What is the purpose of the "secondary adjustment" in the objective rating technique? What factors are considered in the "secondary adjustment"?
12. Which four criteria are fundamental for doing a good job in speed rating?
13. Why is training in performance rating a continuous process?
14. Using the data in Figure 15–5, determine the average percentage of correct rating within ±5 rating points.
15. Why is performance rating the most important step in the entire work measurement procedure?
16. Why should more than one element be used in the establishment of a synthetic rating factor?
17. What is meant by the term "launching costs"?

GENERAL QUESTIONS

1. Would there be any objection to studying an operator who was performing at an excessive pace? Why or why not?
2. In what ways can an operator give the impression of high effort and yet produce at a mediocre or poor level of performance?
3. If an operator strongly objected to the performance rating factor on completion of the study, what would your next step be if you were the time study analyst?
4. How would you go about maintaining a uniform concept of normal performance in a multiplant enterprise whose various plants are located in different sections of the country?

PROBLEMS

1. The Dorben Company produced 200 units of a new design. The total time required to produce these 200 units was maintained by having all the operators involved punch in and punch out on this line of work. The total recorded time was 25,412 hours. After 400 units of the new product were produced, 42,808 hours had been utilized. From the data recorded, what is the learning curve associated with this new product?
2. An analyst is studying a complete assembly operation that takes place in his plant in conjunction with a new product line. He anticipates that an 85 percent rate of learning will take place. To develop helpful learning curve information

for future planning, he wishes to compute the rate of learning that takes place on this assembly work.

After 50 units were produced, the analyst noted a total assembly time charge of 1,000 work-hours. Only 500 more work-hours were needed to produce an additional 50 units. What learning curve percentage was taking place?

3. The work measurement analyst in charge of training time study analysts decided to have all trainees review 20 film loops, where the rate of each loop was known. Each trainee then computed her own record, which was based on the proportion of ratings that fell within plus or minus five points of the known ratings.

   One analyst computed her average difference in rating as −4.08 points on 20 films. The standard deviation was 6.4. What percentage of this analyst's ratings was contained within the desired rating?

   (Note: Assume that the sample values are the population values.)

4. The Dorben Company is using synthetic leveling on its low-skill highly repetitive operations. A time study analyst for the company finds that the mean time required by a given operator for element 2 averages 0.05 minutes. An MTM analysis of this element involves the following:

   One class A 20-inch reach; one class |C| grasp; one class C 24-inch move; one position involving a semisymmetrical assembly with light pressure and a relatively easy-to-handle part; one normal release.

   What performance factor will be assigned to the effort elements of this study? What would the allowed time for element 2 be if a P.D. & F. allowance of 20 percent were utilized?

5. The Dorben foundry is producing an order of 20 large castings. For the first 10 castings, the average time per casting is 40 work-hours. What would the learning curve percentage be if the average time per casting were 35 hours on completion of the order?

6. In the Dorben Company's machine shop, a 95 percent learning curve is in effect. On a new job, the time for the first unit was measured and found to be 1.5 hrs.
   a. What would the accumulated average time be for the 416th unit?
   b. What would be the time for the 416th unit only?
   c. What would be the launching costs based upon a machine rate of $30.00 per hour?

SELECTED REFERENCES

Barnes, Ralph M. *Motion and Time Study: Design and Measurement of Work.* 7th ed. New York: John Wiley & Sons, 1980.

Hancock, Walton M. "The Learning Curve." In *Handbook of Industrial Engineering,* 2nd ed., edited by Gavriel Salvendy, Chap. 6. New York: John Wiley & Sons, 1992.

Mundel, Marvin E. *Motion and Time Study: Improving Productivity.* 5th ed. Englewood Cliffs, N.J.: Prentice-Hall, 1978.

Nadler, Gerald. *Work Design: A Systems Concept.* Rev. ed. Homewood, Ill.: Richard D. Irwin, 1970.

SELECTED SOFTWARE

Nicks, J. E., Learning Curves and Launching Curves and Launching Costs, p. 81, Basic Programming Solutions for Manufacturing. Dearborn, Mich.: Society of Manufacturing Engineers, 1982.

Whitehouse, Gary E., Yasser A. Hosni, and Farid Guediri, Learning Curve IIE Microsoftware, Industrial Engineering and Management Press. Norcross, Ga: Institute of Industrial Engineers, 1985.

Allowances

After the calculation of the normal time, sometimes referred to as the "rated" time, one additional step must be performed to arrive at a fair standard. This last step is the addition of allowance to take care of the many interruptions, delays, and slowdowns brought on by fatigue in every work assignment. For example, in planning a motorcar trip of 1,000 miles, we know that the trip cannot be made in 20 hours when driving at a speed of 50 miles per hour. An allowance must be added to take care of periodic stops for personal needs, for driving fatigue, for unavoidable stops brought on by traffic congestion and stoplights, for possible detours and the resulting rough roads, for car trouble, and so forth. Thus, we may estimate that we will make the trip in 25 hours, since we feel that 5 additional hours would be necessary to take care of all delays. Similarly, analysts must provide an allowance for workers if the resulting standard is to be fair and readily maintainable by an average worker performing at a steady, normal pace.

The watch readings of any time study are taken over a relatively short period of time. To determine the average or selected time, remove abnormal readings, unavoidable delays, and time for personal needs from the study. Therefore, normal time does not include unavoidable delays and other legitimate lost time; consequently, analysts must make some adjustment to compensate for such losses.

In general, allowances are made to cover three broad areas: personal delays, fatigue, and unavoidable delays. The application of allowances is considerably broader in some concerns than in others. For example, Table 16–1 reveals which items 42 firms included in their allowances.

Allowances are frequently applied carelessly because they have not been established on sound time study data. This is especially true of allowance for fatigue, where it is difficult if not impossible to establish

438

TABLE 16–1

Allowance factor	No. of firms	Percent
1. Fatigue	39	93
A. General	19	45
B. Rest periods	13	31
Did not specify A or B	7	17
2. Time required to learn	3	7
3. Unavoidable delay	35	83
A. Man	1	2
B. Machine	7	17
C. Both, man and machine	21	50
Did not specify A, B, or C	6	14
4. Personal needs	32	76
5. Setup or preparation operations	24	57
6. Irregular or unusual operations	16	38

Source: J. O. P. Hummel, "Motion and Time Study in Leading American Industrial Establishments" (Master's thesis, Pennsylvania State University).

values based on rational theory. Many unions, well aware of this situation, have bargained for additional fatigue allowance as if it were a "fringe" issue. (Fringe benefits are company expenses unrelated to employee output, such as insurance and pensions.) Allowance must be as accurate and correct as possible; otherwise, all the care and precision put into the study up to this point can be completely nullified.

Allowances are applied to three categories of the study: allowances applicable (1) to the total cycle time, (2) to machine time only, and (3) to effort time only.

Express allowances applicable to the total cycle time as a percentage of the cycle time and include such delays as personal needs, cleaning the workstation, and oiling the machine. Machine time allowances include time for tool maintenance and power variance, while representative delays covered by effort allowances are fatigue and certain unavoidable delays.

There are two frequently used methods of developing standard allowance data. One is the production study, which requires observers to study two or perhaps three operations over a long period of time. Observers record the duration of and reason for each idle interval (see Figure 16–1). After establishing a reasonably representative sample, observers summarize their findings to determine the percent allowance for each applicable characteristic. Data obtained in this fashion, like that for any time study, must be adjusted to the level of normal performance.

Since observers must spend a long time directly observing one or more operations, this method is exceptionally tedious, not only to analysts, but also to the operator or operators. Another disadvantage is the tendency to take too small a sample, which may result in biased results.

FIGURE 16-1
Lost time analysis chart

```
┌─────────────────────────────────────────────────────────────────────────────┐
│                    LOST TIME ANALYSIS OF TIME STUDY                            │
│                                                                               │
│   Dwg._____ Part _____ Date _____   │
│   Operation_____  │
│   Symbol_____   │
│   A.  Personal_____   │
│   B.  Start work late_____   │
│   C.  Stop work early_____   │
│   D.  Talk with foreman or instructor_____   │
│   E.  Talk with other persons_____   │
│   F.  Search for tools_____   │
│   G.  Search for drawings_____   │
│   H.  Rework fault of operator_____   │
│   I.  Rework fault of another operator_____   │
│   J.  Rework fault of machine or fixtures_____   │
│   K.  Idle-wait for crane (excess over allowed)_____   │
│   L.  Idle-wait for inspector (excess over allowed)_____   │
│   M.  Wait in line at tool crib (excess over allowed)_____   │
│   N.  Wait in line at dispatch office (excess over allowed)_____   │
│   O.  Wait in line at B/P station _____   │
│   P.  Tool maintenance_____   │
│   Q.  Oil machine_____   │
│   R.  Clean work station_____   │
│   S.  Circled readings (circled reading minus ave. for ele.)_____   │
│   T.  Miscellaneous minor delays_____   │
│   U.  Lost time developing methods during study_____    │
│   V.  _____   │
│   W.  _____   │
│   X.  _____   │
│   Y.  _____   │
│   Z.  _____   │
│                                                           Total_____     │
│                                                                               │
│   1.  Gross over-all_____Mins._____Hrs.                           │
│   2.  Total lost _____  "  _____  "                            │
│   3.  % lost time compared with net actual (2 ÷ 4)              _____      │
│   4.  Net actual or productive_____Mins._____Hrs.             │
│   5.  Allowed time        _____  "  _____  "                 │
│                                                                               │
│   _____    │
│   _____    │
│                                                                               │
│     Note--Place lost time symbol alongside description of lost time on study   │
│             and staple this card to study.                                     │
└─────────────────────────────────────────────────────────────────────────────┘
```

The second technique of establishing the allowance percentage is through work sampling studies (see Chapter 21). This method involves taking a large number of random observations, thus requiring only part-time, or at least intermittent, services of the observer. In this method, no stopwatch is used, as observers merely walk through the area under study at random times and note briefly what each operator is doing.

The number of delays recorded, divided by the total number of observations during which the operator is engaged in productive work, approximate the allowance required by the operator to accommodate the normal delays encountered.

In using work sampling studies for the determination of allowances, observers must practice several precautionary measures. First, observers must not anticipate observations, but record only the actual happenings.

Second, a given study should not cover dissimilar work, but should be confined to similar operations on the same general type of equipment. Third, the larger the number of observations and the longer the period of time over which the data is taken, the more valid are the results. Analysts should take daily observations over a span of at least two weeks.

PERSONAL DELAYS

Personal delays include those cessations in work necessary for maintaining the general well-being of the employee, for example, trips to the drinking fountain and the rest room. The general working conditions and class of work influence the time necessary for personal delays. Thus, working conditions involving heavy work performed at high temperatures, such as that done in the pressroom of a rubber-molding department or in a hot-forge shop, would require greater allowance for personal needs than would light work performed in comfortable temperature areas. Detailed production checks have demonstrated that a 5 percent allowance for personal time, or approximately 24 minutes in eight hours, is appropriate for typical shop working conditions. The amount of time needed for personal delays varies to some extent with the person as well as the class of work. The 5 percent figure cited appears to be adequate for the majority of male and female workers.

FATIGUE

Closely associated with the allowance for personal needs is the allowance for fatigue, although this allowance is usually applied only to the effort portions of the study. Fatigue allowances have not reached the state where their qualifications are based on sound, rational theories, and they probably never will. Consequently, next to performance rating, the fatigue allowance is the least defensible and the most open to argument of all the factors making up a time standard. For example, many proponents of the MTM system (see Chapter 19) feel that no fatigue allowance should be used in the development of the majority of standards since MTM values are based on a work rate that can be sustained for an eight-hour working day by the average healthy employee. However, fair fatigue allowances for different classes of work can be approximated by empirical means. Fatigue is not homogeneous in any respect; it ranges from strictly physical to purely psychological and includes combinations of the physical and psychological. It has a marked influence on some people, but apparently has little or no effect on others.

Whether the fatigue that sets in is physical or mental, the results are similar: there is a lessening in the will to work. The major factors that affect fatigue are well known and have been clearly established. Some of these are:

1. Working conditions.
   a. Light.
   b. Temperature.
   c. Humidity.
   d. Air freshness.
   e. Color of room and environment.
   f. Noise
2. Nature of work.
   a. Concentration necessary to perform task.
   b. Monotony of similar body movements.
   c. Position employee must assume to perform the operation.
   d. Muscular tiredness due to stressing some muscles.
3. General health of the worker, physical and mental.
   a. Physical stature.
   b. Diet.
   c. Rest.
   d. Emotional stability.
   e. Home conditions.

Although fatigue can be reduced, it can never be eliminated. In general, heavy work is diminishing in industry because of the marked progress in the mechanization of both material handling and processing elements. As industry becomes more automated, there will be less muscular tiredness due to the stressing of muscles. Thus, we have made real progress toward decreasing physical fatigue. The major problem of fatigue is not physical but psychological, and industry through its scientific selection programs is substantially reducing this factor by putting the right people on the right jobs. A person who has an unfavorable reaction to monotony should not be placed on a monotonous job. Since it is not customary to provide fatigue allowance for the general health factors that influence the degree of fatigue, such conditions as emotional stability, rest, diet, and physical stature are usually considered in employee selection.

Because fatigue cannot be eliminated, proper allowance must be made for the working conditions and repetitiveness of work. These influence the degree to which fatigue sets in. Experiments have shown that fatigue must be plotted as a curve, not as a straight line. Figure 16–2 illustrates a typical work curve showing the relationship between the load in pounds and the time for handling each load.

Figure 16–3 shows the effect of fatigue in an assembly operation. After 80 blocks of work were completed, an experiment on exhaustion was performed in which the work load was increased by allowing only one five-minute break after completion of a block of work. Many industrial studies have shown a drop in production toward the end of the working period attributable to fatigue alone. Usually the rate of production increases during the early part of the day and then falls after the third hour. There is a short period of increased production after the lunch period, but

FIGURE 16–2
Typical work curve

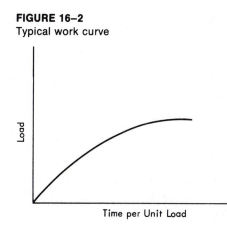

Time per Unit Load

FIGURE 16–3
Learning curve for assembly work showing the impact of
financial incentive and the effect of fatigue after the flattening
of the learning curve

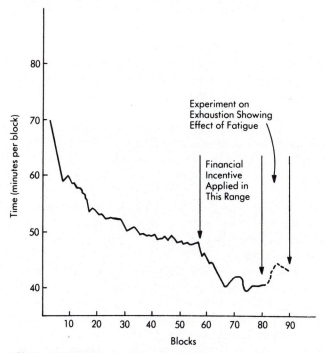

1 Block = 1,296 cycles.
Source: Redrawn from W. Rohmert and K. Schlaich, "Learning of Complex
Manual Tasks," *International Journal of Production Research* 5, no. 2 (1966),
p. 137.

this soon begins to taper off, and output usually continues to decline for the balance of the working day.

Perhaps the most widely used method of determining the fatigue allowance is measuring the decline in production throughout the working period. Thus, the production rate for every quarter of an hour may be measured during the course of the working day. Any decline of production that cannot be attributed to methods changes or personal or unavoidable delays may be attributed to fatigue and expressed as a percentage. However, many outside factors, such as the state of health or outside interference, can influence the fatigue factor. Thus, many studies should be made to obtain a reasonable sample before deciding on the final fatigue allowance for a given facility. Eugene Brey[1] has expressed the coefficient of fatigue as follows:

$$F = \frac{(T - t)\ 100}{T}$$

where:

F = Coefficient of fatigue

T = Time required to perform the operation at the end of continuous work

t = Time required to perform the operation at the beginning of continuous work

Many attempts have been made to measure fatigue, none of which have been completely successful. The tests of fatigue may be classified into three groups: (1) physical, (2) chemical, and (3) physiological.

Physical tests include the various dynamometer tests of changes in the rate of working, such as the hand dynamometer, the mercury dynamometer, the water dynamometer, and the Martin spring balance for registering the force exerted by six different sets of large body muscles.

Chemical tests include various techniques for analysis of the blood and the body secretions, such as saliva, to note changes resulting from fatigue.

The physiological tests of fatigue include the pulse count, blood pressure, the respiratory rate, oxygen consumption, and the production of carbon dioxide. Table 16–2 shows the rate of change in physiological reactions due to fatigue.

In recent years, considerable attention has been directed to the physiological requirements of various work assignments. This branch of the scientific study of the worker and environment is recognized as "occupational physiology." Here the effort per unit of time plus the physiological recovery time is measured for the different work assignments within a plant. As these studies progress, more quantitative data for estimating

---

[1] Eugene E. Brey, "Fatigue Research in Its Relation to Time Study Practice," *Proceedings, Time Study Conference, Society of Industrial Engineers,* Chicago, February 14, 1928.

TABLE 16–2

Factor	Percent decrease	Percent increase
1. Pulse count	—	113.5
2. Pulse pressure	—	77.0
3. Respiratory rate	—	60.5
4. CO_2 combining power	41.0	—
5. Metabolism	—	43.0
6. White blood cells	—	57.0
7. Red blood cells	—	6.0
8. Blood pressure, diastolic	28.0	—
9. Blood pressure, systolic	—	20.5
10. Blood sugar	—	12.0

Source: Moss and Roe, *Physiological Reactions due to Fatigue;* and Joseph W. Siphron, "Fatigue Allowances in Time Study," Master's thesis, The Pennsylvania State University, 1935.

fatigue allowances will inevitably result. Laws of physiological economy will someday develop that will supplement and may even supersede some of the principles of motion economy.

In the development of equitable fatigue allowances, one of the foremost problems of time study is determining at which portion of the fatigue curve a time study was made. The next step is to derive a fatigue allowance that can be used as a constant for work performed at the given workstation in the future.

For most industrial operations, fatigue allowances are arbitrarily broken into three elements, each of which has a spread of influence on the total fatigue allowance. These are: operations involving strenuous work, operations involving repetitive work, and operations performed under disagreeable working conditions. Of course, more than one of these conditions can apply in any specific operation.

Taking controlled production studies from an adequate sample of work can result in fatigue allowance values that prove equitable for the various degrees of each of the preceding factors. The apparent adequacy of fatigue allowances determined by measuring the decline in production through all-day production studies is due to the fact that the fatigue allowance for a given job is not a critical value, but may be safely established within a rather broad range.

The International Labour Office has tabulated the effect of working conditions to arrive at an allowance factor for personal delays and fatigue. This tabulation appears in Table 16–3. The factors considered include: standing while working, abnormal positions demanded, use of force, illumination, atmospheric conditions, job attention required, noise level, mental strain, monotony, and tediousness.

In using this table, determine an allowance factor for each element of the study. For example, element 3 of a given study may involve the application of a 40-pound force. Because of this use of force, analysts

TABLE 16–3

A. Constant allowances:
   1. Personal allowance 5
   2. Basic fatigue allowance 4
B. Variable allowances:
   1. Standing allowance 2
   2. Abnormal position allowance:
      a. Slightly awkward 0
      b. Awkward (bending) 2
      c. Very awkward (lying, stretching) 7
   3. Use of force, or muscular energy (lifting, pulling, or pushing):
      Weight lifted, pounds:
         5 0
        10 1
        15 2
        20 3
        25 4
        30 5
        35 7
        40 9
        45 11
        50 13
        60 17
        70 22
   4. Bad light:
      a. Slightly below recommended 0
      b. Well below 2
      c. Quite inadequate 5
   5. Atmospheric conditions (heat and humidity)—variable 0–10
   6. Close attention:
      a. Fairly fine work 0
      b. Fine or exacting 2
      c. Very fine or very exacting 5
   7. Noise level:
      a. Continuous 0
      b. Intermittent—loud 2
      c. Intermittent—very loud 5
      d. High-pitched—loud 5
   8. Mental strain:
      a. Fairly complex process 1
      b. Complex or wide span of attention 4
      c. Very complex 8
   9. Monotony:
      a. Low 0
      b. Medium 1
      c. High 4
  10. Tediousness:
      a. Rather tedious 0
      b. Tedious 2
      c. Very tedious 5

would use an allowance of 9 percent in structuring the allowance computation for this element. This tabulation provides a basic 9 percent allowance for all effort elements, for personal delays, and for fatigue. Analysts add applicable variable allowances as enumerated in the table to this basic 9 percent allowance.

FATIGUE ALLOWANCES UNDER HEAVY LOADS AND SHORT CYCLES

On occasion the work measurement analyst will encounter operations, such as in heavy press work, where the cycle time is short (less than 45 sec.) and the weights handled are relatively large (over 20 lbs.). In such cases, the percentage provided in Table 16–3 are insufficient in providing an adequate fatigue allowance. In these cases, the author has successfully used the following analysis:[2]

E_W = energy expenditure for work
T = time observed (min.)
N = number of lifts per time T
H_2 = upper lifting height (in.)
H_1 = lower lifting height (in.)
L_u = load lifted up in lbs.
L_d = load lifted down in lbs.

$$E_W = \frac{1.824 \times T + N \times E_{lift}}{T}$$

$$E_{lift\ low} = \frac{142(32 - H_1) + (2.08L_u + 0.8L_d)(H_2 - H_1)}{10,000}$$

$$E_{lift\ high} = \frac{22.8(H_2 - 32) + (3.22L_u + 1.03L_d)(H_2 - H_1)}{10,000}$$

$$E_W = E_{lift\ low} + E_{lift\ high}$$

VRA (variable relaxation allowance) = 31.5 (E_W − 4.67)

For example, let us assume an operator with a cycle time of 0.5 minutes was loading and unloading a 40 pound forging into a trimming press. Then:

T = 0.5 min.
N = 2
L_u = 40 lbs.
L_d = 40 lbs.
H_1 = 6 in. (to chute to remove forging from area)
H_2 = 38 in. (from temporary storage to die)

$$E_{lift\ low} = \frac{142(32 - 6) + (2.08)(40)(38 - 6) = 0.635\ \text{Kcal/lift}}{10,000}$$

$$E_{lift\ high} = \frac{22.8(38 - 32) + (3.22)(40)(38 - 6) = 0.425\ \text{Kcal/lift}}{10,000}$$

---

[2] Andris Freivalds and Joseph H. Goldberg, "Specifications of Bases for Variable Relaxation Allowances: Standing, Abnormal Positions, and Use of Force or Muscular Energy," *The Journal of Methods-Time Measurement,* Volume XIV. pp. 2–8.

Note the $0.8L_d$ and the $1.03L_d$ dropped out of the above equations since we are loading in only one direction.

$$E_{lift} = 0.635 + 0.425 = 1.06 \text{ Kcal/lift}$$

$$E_W = \frac{1.824 \times 0.5 + 2 \times 1.06}{0.5} = 6.07 \text{ Kcal/lift}$$

$$VRA = 31.5\,(6.067 - 4.67) = 44\%$$

Thus, this 44 percent allowance would be used instead of the 9 percent shown in Table 16–3 for this loading and unloading element. Figure 16–4 shows the relationship between cycle time and percentage allowance for 40 and 60 pound loads.

FIGURE 16–4
Allowance for fatigue in handling 60 and 40 pound loads during short cycle operations

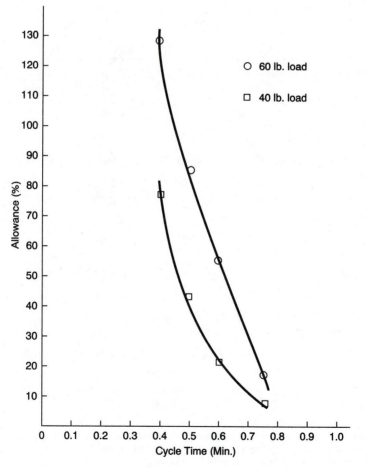

There are two ways of applying the fatigue allowance. It can be a percentage added to the normal time, as has been explained. In this method, the allowance is based on a percentage of the productive time only. As an alternative technique, companies establish periodic rest periods to compensate for the fatigue allowance.

The former way is preferred because it allows physically stronger employees to participate in greater earnings if the company is using incentive wage payment. Compulsory rest periods penalize strong employees who are not as subject to fatigue by restricting their output. However, rest periods definitely reduce fatigue. If 10-minute rest periods are introduced into a plant, as they frequently are, the fatigue allowance that formerly prevailed should be proportionately modified. For example, if a bench assembly operation earned a fatigue allowance of 8 percent, and at some later date, through negotiation, a 10-minute rest period was provided in the morning and another 10-minute rest period in the afternoon, the fatigue allowance on this class of work would be diminished:

$$\frac{20}{\text{Normal productive time}} = \underline{\hspace{2cm}} \text{ percent}$$

The normal daily productive time on this class of work may be 400 minutes. The fatigue allowance accounted for by the 20-minute rest period would then be 20/400, or 5 percent. Thus, future standards in this area carry a fatigue allowance of 8 percent minus 5 percent, or 3 percent.

UNAVOIDABLE DELAYS

This class of delays applies to effort elements and includes interruptions from the supervisor, dispatcher, time study analyst, and others; material irregularities; difficulty in maintaining tolerances and specifications; and interference delays where multiple machine assignments are made.

As can be expected, every operator has numerous interruptions during the course of the working day. These can be due to a wide range of reasons. The supervisor or group leader may interrupt the operator to give instructions or to clarify certain written information. Then the inspector may interrupt to point out the reasons for some defective work that passed through the operator's workstation. Interruptions also frequently occur from planners, expediters, fellow workers, production personnel, and others.

Unavoidable delays are frequently a result of material irregularities. For example, the material may be in the wrong location, or it may be running slightly too soft or too hard. Again, it may be too short or too long, or it may have excessive stock on it, as in forgings when the dies begin to wash out, or on castings due to incomplete removal of risers. When material deviates substantially from standard specifications, it becomes necessary to restudy the job and establish allowed time for the extra elements introduced by the irregular material, as the customary unavoidable delay allowance may prove inadequate.

Machine Interference

If more than one facility is assigned to an operator during the working day, one facility or more must wait until the operator completes work on another facility he or she is servicing. As more facilities are assigned to the operator, the "interference" time delay increases. In practice, machine interference has been found to "occur predominantly from 10 to 30 percent of the total working time, with extremes of from 0 to 50 percent."[3] The amount of machine interference depends on the number of facilities assigned, the randomness of the required servicing time, the proportion of service time to the running time, the length of the running time, and the mean length of the service time.

Although many expressions, tables, and charts have been developed to determine the magnitude of machine interference, the expression developed by William R. Wright (*Mechanical Engineering,* vol. 58) is relatively simple and has proved to be satisfactory when the number of machines assigned is seven or more. When from two to six facilities are assigned, Wright recommends the use of empirical curves, as illustrated in Figure 16–5. For seven or more facilities, Wright developed:

$$I = 50[\sqrt{[(1 + X - N)^2 + 2N]} - (1 + X - N)]$$

where:

I = Interference expressed as a percentage of the mean attention time
X = Ratio of mean machine running time to mean machine attention time
N = Number of machine units assigned to one operator

Wright's formula for interference was developed from the solution of a problem in the congestion of telephone lines made by Thornton C. Fry of the Bell Telephone Laboratories.[4] The conditions applying to Dr. Fry's work included:

1. A telephone call requiring the use of a given trunk line is delayed whenever the trunk line is used by a call on a different line. The second call continues to be delayed until the first call is completed.
2. All telephone calls are of equal duration.
3. Calls are assigned individually to groups of channels and collectively at random.

---

[3] H. B. Maynard, *Industrial Engineering Handbook* (New York: McGraw-Hill, 1956), pp. 3–78.

[4] Thornton C. Fry, *Probability and Its Engineering Uses* (Princeton, N.J.: D. Van Nostrand, 1928).

FIGURE 16-5
Interference in the percentage of attention time when the
number of facilities assigned to one operator is six or
less

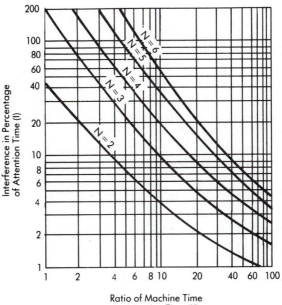

Ratio of Machine Time
to Attention Time (X)

Assumptions 2 and 3 do not always apply in multiple machine activity. However, as the number of machines increases, the random nature of servicing becomes more pronounced. Wright found his equation applicable in four entirely different industries where the number of machines assigned was greater than six.

For example, in the development of a standard for quilling production, an operator has been assigned 60 spindles. The mean running time per package (unit of output), determined by stopwatch study, is 150 minutes. The standard mean attention time per package, developed by time study, is three minutes.

The computation of the machine interference expressed as a percentage of the mean operator attention time would be:

$$I = 50[\sqrt{[(1 + X - N)^2 + 2N]} - (1 + X - N)]$$

$$I = 50 \left[\sqrt{\left(1 + \frac{150}{3.00} - 60\right)^2 + 120} - \left(1 + \frac{150}{3.00} - 60\right) \right]$$

$$I = 50[\sqrt{(1 + 50 - 60)^2 + 120} - (1 + 50 - 60)]$$

$$I = 1,160 \text{ percent}$$

Thus, in this example we would have:

Machine running time..	150.00 min.
Attention time, including the personal, fatigue, and unavoidable delay allowance	3.00 min.
Interference time (11.6)(3.00)	34.80 min.
Standard time for 60 packages	187.80 min.

Standard time per package $\dfrac{187.80}{60} = 3.13$ min.

The amount of interference that takes place is related to the performance of the operator. Thus, the operator demonstrating a low level of effort experiences more machine interference than would the operator who through higher effort reduces the time spent in attending the stopped machine. The analyst determines the normal interference time which, when added to (1) the machine runing time required to produce one unit and (2) the normal time spent by the operator in servicing the stopped machine, equals the cycle time. The cycle time divided into the running time of each machine multiplied by the number of machines assigned to the operator gives the average machine running hours per hour. Thus, we have the expression:

$$O = \frac{NT_1}{C}$$

where:

O = Machine running hours per hour
N = Number of machines assigned to the operator
T_1 = Running time (hours) to produce one piece
C = Cycle time to produce one piece

and

$$C = T_1 + T_2 + T_3$$

where:

T_2 = Time (hours) spent by normal operator attending the stopped facility
T_3 = Time lost by normal operator working at normal pace because of interference

Using waiting line theory, analysts have developed tables where the interval between service times is exponential and where the service time is either constant or exponential. Table A3–14, Appendix 3, gives these values for various "k," which is the ratio of service time to facility running time $k = \dfrac{T_2}{T_1}$.

With reference to the previous example,

$$k = \frac{3.00}{150.00} = 0.02$$
$$N = 60$$

From Table A3–14, Appendix 3, with exponential service time and $k = 0.02$ and $N = 60$, we have a waiting time (interference delay) of 16.8 percent of the cycle time. Denoting the interference time by T_3, we have $T_3 = 0.168C$, where C is the cycle time to produce one unit per spindle. Then:

$$150 + 3.00 + 0.168C = C$$
$$0.832C = 153$$
$$C = 184 \text{ minutes}$$

and

$$T_3 = 0.168C = 30.9 \text{ minutes}$$

The interference time computed by Wright's equation closely agrees with that developed by the queuing model. However, as N (the number of machines assigned) becomes smaller, the proportion of difference between the two techniques increases.

AVOIDABLE DELAYS

It is not customary to provide any allowance for avoidable delays, which include visits with other operators for social reasons, uncalled-for work stoppages, and idleness other than rest to overcome fatigue. Of course, operators may take these delays at the expense of output, but no allowance is provided for these cessations of work in the development of the standard.

EXTRA ALLOWANCES

In typical metal trade and related operations, the allowance for personal, unavoidable, and fatigue delays usually approximates 15 percent. However, in certain cases, an extra allowance may be needed to arrive at a fair standard. Thus, due to a substandard lot of raw material, analysts may add an extra allowance to take care of an unduly high generation of rejects caused by the poor material. Again, a situation may arise in which, because of the breakage of a jib crane, the operator is obliged to place a 50-pound casting in the chuck of the machine. Therefore, an extra allowance takes care of the additional fatigue brought on by the manual handling of the work.

Whenever practical, establish allowed time for the additional work of any operation by breaking it down into elements and then including these times in the specific operation. If this is not practical, then provide an extra allowance.

A form of extra allowance frequently used, especially in the steel industry, is a percentage added to a portion or all of the cycle time when the operator must observe the process to maintain the efficient progress of the operation. This allowance is frequently referred to as "attention time" allowance, which covers these situations:

1. The inspector in the electrolytic tinning operation while observing the tin plate coming off the line.
2. The first helper in an open hearth while observing the condition of the performance or molten bath.
3. The first helper in an open hearth while reporting to or receiving instructions from the melter.
4. The crane operator in a shipping department while receiving directions or signals from the crane hooker.
5. The hi-mill roller in the seamless hot mill while required, during the rolling process, to watch for any change in the length of pipe to signal the operator to return the pipe if too short and adjust the rolls as required.[5]

The U.S.X. Company adds a 35 percent allowance to the actual required time for required attention. Thus, an operator can earn a 35 percent premium on a process-paced operation if the process runs efficiently during the entire working day.

Using an extra allowance on operations where a large portion of the cycle time is based on the machine or process cycle is quite popular in plants practicing incentive wage payment. Management adds the allowance so that operators may make earnings equivalent to those of workers assigned to operations that are not machine paced. A typical practice is to allow 30 percent extra allowance on the machine-controlled portion of the cycle time. This extra allowance provides the incentive for the operator to keep the facility productively employed during the entire working period.

Without this extra allowance, operators would find it impossible to make the same earnings as fellow employees. For example, if the machine-controlled portion of a cycle is two minutes and the operator-controlled portion is one normal minute, operators would have to work at a pace 25 percent above normal to realize a 7 percent increase in productivity.

$$\frac{3 \text{ min. (normal cycle time)}}{2 \text{ min. (mach. control)} + \dfrac{1 \text{ min. (normal effort time)}}{1.25 \text{ (operator performance)}}} = 1.07$$

[5] From United States Steel Company Time Study Manual.

1.07 − 1.00 = 0.07 increase in productivity when the operator is working at 25 percent above normal during the effort part of the cycle

Now, if management adds an extra allowance of 25 percent to the machine-controlled portion of the cycle, operators would be able to achieve 25 percent incentive earnings if they worked at a pace 25 percent above normal and did not utilize more allowance than was provided for personal delays, unavoidable delays, and fatigue.

Clean Workstation and Oil Machine

The time required to clean the workstation and lubricate the operator's machine may be classified as an unavoidable delay. Often, analysts include this time, when performed by the operator, as a total cycle time allowance. The type and size of equipment, and the material being fabricated, have considerable effect on the time required to do these tasks. When these elements are the responsibilities of the operator, management must provide an applicable allowance. One concern has established the accompanying table of allowances to cover these items. See Tables 16–4, 16–5, and 16–6.

TABLE 16–4
Clean machine allowance chart

	Percent per machine		
Item	*Large*	*Medium*	*Small*
1. Clean machine when lubricant is used....................	1	¾	½
2. Clean machine when lubricant is not used	¾	½	¼
3. Clean and put away large amounts of tools or equipment...	½	½	½
4. Clean and put away small amounts of tools or equipment...	¼	¼	¼
5. Shut machine down for cleaning (this percentage is for machines equipped with chip pans, which are stopped at intervals to permit sweeper to clean away large chips)	1	¾	½

TABLE 16–5
Machine classification

Large machine	*Medium machine*
1. Turret lathe (20-in. chuck or over)	1. Turret lathe (10-in. to 20-in. chuck)
2. Boring mill (60 in. and over)	2. Boring mill (under 60 in.)
3. Punch press (100T and over)	3. Punch press (40T to 100T)
4. Planer (over 48 in.)	

TABLE 16–6
Oil machine allowance chart

Item	Percent per machine		
	Large	*Medium*	*Small*
1. Machine oiled or greased by hand........	1½	1	½
2. Machine oiled automatically.............	½	½	½

Frequently, supervisors give operators 10 or 15 minutes at the end of the day to perform the elements "clean workstation" and "oil machine." When this is done, the standards established would not include any allowance for cleaning and oiling the machine.

Power Feed Machine Time Allowance

The allowance required for power feed elements usually differs from that required for effort elements. Consider two factors in applying allowances for power feed: shutdowns and tool maintenance.

Allowances are made for shutdowns because of minor repairs. In case a major repair to the facility is needed, then an extra allowance would be provided. This extra allowance would not be applied within the standard, but would be an independent standard covering machine repair.

Tool maintenance allowance provides time for the operator to maintain tools after the original setup. In the setup time, the operator is expected to provide first-class tools properly ground. Generally, little tool maintenance takes place during the course of the average production run. In long runs, tools have to be sharpened periodically. The percentage allowance for tool maintenance varies directly with the number of perishable tools in the setup. For example, one manufacturer's tool maintenance allowance table is as follows:

Allowance for tool maintenance	Percent
1. One or more tools ground in tool crib	1
2. One tool sharpened by the operator	3
3. Two or more tools cutting at one time sharpened by the operator........	6

APPLICATION OF ALLOWANCES

The fundamental purpose of allowances is to add enough time to normal production time to enable the average worker to meet the standard when performing at a normal pace. It is customary to express the allowance as a multiplier so that the normal time, consisting of productive work elements, can be adjusted readily to the time allowed. Thus, if a 15 percent allowance were provided on a given operation, the multiplier would be 1.15.

Allowances must be carefully included with the time study standard. The allowance is based on a percentage of the daily production time and not on the overall workday. For example, if a study revealed that in an eight-hour working day 50 minutes of delay time are to be allowed during 400 minutes of normal production time, then the percentage allowance applicable would be 50/400, or 12.5 percent. However in the development of allowance tables, the percentage computed is often based on the work day (usually 480 minutes) since the production time is not known. In applying the allowance the total allowance is converted to an allowance factor. For example, the computation of a total allowance may be:

Personal	5.0 percent
Fatigue	6.5
Unavoidable delay	4.0
Total	15.5 percent

$$\text{Allowance Factor} = \frac{100 \text{ percent}}{100 \text{ percent} - 15.5 \text{ percent}} = 1.183$$

Thus, normal time would be multiplied by 1.183 to determine the allowed time. Allowance is based on the normal production time, since the percentage is applied to this value on subsequent studies.

Typical Allowance Values

In an industrial survey comprising 42 different plants, the smallest average total allowance in use was 10 percent. This was used in a plant producing household electrical appliances. The greatest average allowance was 35 percent, in effect in two different steel plants. The average allowance of all the plants was 17.7 percent.

Typical allowances established in a certain plant for standard operations appear in Table 16–7. These values may or may not apply in other plants.

SUMMARY

In establishing time study allowance values, exercise the same care that prevails in taking individual studies. It would be folly to divide a job carefully into elements, precisely measure the duration of each element in hundredths of a minute, accurately evaluate the performance of the operator, and then arbitrarily assign an allowance picked at random. Practices of this nature lead to inaccurate standards. If the allowances are too high, obviously manufacturing costs are unduly inflated, and if the allowances are too low, tight standards result, which cause poor labor relations and the eventual failure of the system.

TABLE 16–7
Time study allowances for standard operations

Symbol	Facility or operation	Method	Total applied to effort time	Total applied to machine time	Personal	Clean work station	Oil machine	Shutdown	Tool mainte-nance	Unavoid-able delays and fatigue
21	Anneal	Oven	10	–	5	½	–	–	–	4½
22	Assembly	Bench	13	–	5	–	–	–	–	8
23	Assembly	Floor	14.5	–	5	1	–	–	–	9½
24	Blacksmith	Drop forge	21	–	7	½	–	–	–	13
25	Brake	Power	15	–	5	½	–	–	–	9½
26	Braze	Electric	15	–	5	½	½	–	–	9½
27	Drill	Hand feed	15	–	5	½	½	½	2	6½
28	Drill	Power feed	15	12	5	½	½	m–2	m–4	–
29	Engrave	Pantograph	18	–	5	½	½	e–½	e–2	e–6½
30	Lathe, English	Over 36 inches	15	15	5	2	1	m–2	m–5	e–7
31	Lathe, engine		15	15	5	2	1	m–2	m–5	e–7
32	Lathe, turret		17	15	5	2	1	m–2	m–5	e–9
33	Milling		16	15	5	2	1	m–2	m–5	e–8
34	Grinding	Blanchard	15	15	5	2	1	m–2	m–5	e–7
35	Grinding	Thread	17	15	5	2	1	m–2	m–5	e–9
36	Grinding	External and internal	16	15	5	2	1	m–2	m–5	e–8
37	Punch press	Up to 100 tons	14		5	½	1	½	–	7
38	Saw	Circular	14		5	½	½	½	1	6½
39	Saw	Do-All	15		5	½	½	1½	2	5½
40	Shear	Square	15		5	½	½	1	–	8
41	Welder	Spot	17		5	½	–	2	3	6½
42	Paint	Spray	17		5	2	–	1	1	8

m: applied to machine time only.
e: applied to effort time only.

458

TEXT QUESTIONS

1. Compare the interference delay allowance using the queuing model and Wright's equation, where $N = 20$, mean facility running time is 120 minutes, and attention time is 3 minutes.
2. Which three broad areas are allowances intended to cover?
3. To which three categories of the time study are allowances provided? Give several examples of each.
4. What are the two methods used in developing allowance standard data? Briefly explain the application of each technique.
5. Give several examples of personal delays. Which percentage allowance seems adequate for personal delays under typical shop conditions?
6. What are some of the major factors that affect fatigue?
7. Under which groups have the tests of fatigue been classified?
8. Why do many proponents of the MTM fundamental motion data system advocate that no allowance be provided for fatigue?
9. Which operator interruptions would be covered by the unavoidable delays allowance?
10. Which percentage allowance is usually provided for avoidable delays?
11. When are "extra allowances" provided?
12. What fatigue allowance should be given to a job if it developed that it took 1.542 minutes to perform the operation at the end of continuous work and but 1.480 minutes at the beginning of continuous work?
13. Why are allowances based on a percentage of the productive time?
14. What is meant by occupational physiology?
15. Define "attention time." Why is it necessary to apply an allowance to attention time?
16. Based on the International Labour Office's tabulation, what would be the allowance factor on a work element involving a 42-pound pulling force in inadequate light where exacting work was required?
17. Explain what relation, if any, there is between financial incentives and the effect of fatigue.
18. What are the advantages of having operators oil and clean their own machines?
19. When would you find it desirable to add to the allowances shown in Table 16–3?

GENERAL QUESTIONS

1. Do you believe that the operation or the operator determines the extent to which personal delay time is utilized? Why or why not?
2. Why is fatigue allowance frequently applied only to the effort areas of the work cycle?
3. Should fatigue allowances vary with the different shifts for a given class of work? Why or why not?

460 *Motion and Time Study*

4. What are the objections to determining fatigue allowances by measuring the decline of production not attributable to methods changes or personal or unavoidable delays?
5. Give several reasons for not applying an "extra" allowance to operations when the major part of the cycle is machine controlled and the internal time is small compared to the cycle time.
6. With the extension of automation and the tendency of many companies to go to the four-day week, do you feel that personal allowances should be increased? Why or why not?
7. With reference to Figure 16–3, why did the time for assembly decrease after 85 blocks of work were completed?

PROBLEMS

1. The work measurement analyst is planning to develop a table of allowances for a given class of work in the maintenance department, using the work sampling technique. The areas for which she wants to establish allowances and the variation she expects to find the actual allowance 95 percent of the time are as follows:

> Personal allowance......... 3 to 7 percent
> Crane wait .... 2 to 6 percent
> Grind tools............... 5 to 9 percent
> Avoidable delays........... 1 to 4 percent
> Unavoidable delays .... 10 to 20 percent

How many random observations should she take? Over what period of time should she take these observations? Explain how she will determine when to take each day's observations.
2. Based on Table 16–3, develop an allowance factor for an assembly element where the operator is standing in a slightly awkward position, regularly lifts a weight of 15 pounds, and has good light and atmospheric conditions. The attention required is fine, the noise level is continuous at 70 dbA, and the mental strain is low, as is the monotony and the tediousness of the work.
3. Calculate the fatigue allowance for an operation where the cycle time is 0.4 minutes and the operator loads and unloads a 50-pound gray iron casting each cycle. An investigation of the work station indicates $H_1 = 12$ in. and $H_2 = 40$ in.

SELECTED REFERENCES

Davis, H. L.; T. W. Faulkner; and C. L. Miller. "Work Physiology." *Human Factors*, vol. 2 (1969), pp. 157–66.
Department of Defense 5010.15.1-M Basic Volume, Appendix II.
International Labour Office. *Introduction to Work Study*. Geneva, Switzerland: Atar, 1964.
Konz, S. *Work Design*. Columbus, Ohio: Grid, 1979.

Moodie, Colin L. "Assembly Line Balancing." In *Handbook of Industrial Engineering,* 2nd ed. edited by Gavriel Salvendy, Chap. 56. New York: John Wiley & Sons, 1992.

MTM Association. *Work Measurement Allowance and Survey,* Fair Lawn, N.J.: 1976.

Stecke, Kathryn E. "Machine Interference: Assignment of Machines to Operators." In *Handbook of Industrial Engineering,* 2nd ed., edited by Gavriel Salvendy, Chap. 57. New York: John Wiley & Sons, 1992.

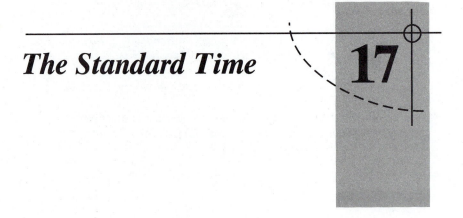

The Standard Time

The standard time for a given operation is the time required for an average, fully qualified, trained operator working at a normal pace, to perform the operation. Standard time is the sum of the allowed time for all of the individual elements comprised by the time study.

To determine elemental allowed times, multiply the mean elapsed elemental time by a conversion factor. Thus, the expression:

$$T_a = (M_t)(C)$$

where:

T_a = Allowed elemental time
M_t = Mean elapsed elemental time
C = Conversion factor found by multiplying the performance rating factor by one plus the applicable allowance

For example, if the mean elapsed elemental time of element 1 of a given time study was 0.14 minute, the performance factor was 0.90, and an allowance of 18 percent was applicable, the allowed elemental time would be:

$$T_a = (0.14)(0.90)(1.18) = 0.148$$

Allowed elemental times are rounded off to three places to the right of the decimal point. Thus, in the preceding example, 0.1483 minute is recorded as 0.148 minute. If the result were 0.1485 minute, then the allowed time would be 0.149 minute.

The allowed elemetal time is nothing more than the normal time plus an allowance to provide for personal and unavoidable delays and fatigue. In Figure 17–1, the allowed time for element 1 results from adding 12 percent

FIGURE 17–1: Completed time study using speed rating on all elements. This operator was rated 10 percent faster than normal. This rating is confirmed by element 1 where 255 TMU are assigned to this element and the operator's mean time was 230 TMU, indicating the operator was working at 111 percent of normal. A 12 percent allowance was applied to all elements. This results in a conversion factor of 1.232, which when multiplied by the average element time, provides elemental allowed times.

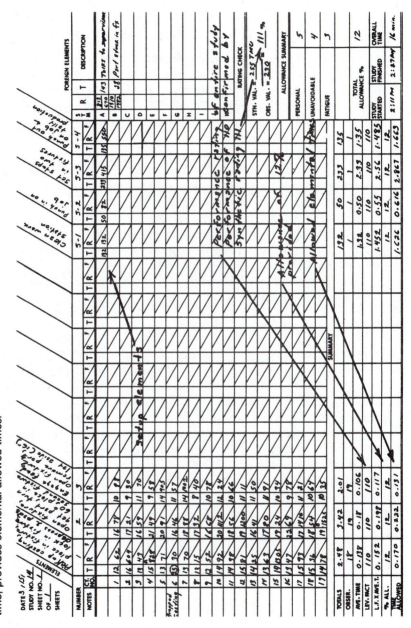

to the normal time (Leveling factor × Average time) of 0.152 minute or 0.170 minute.

THE ELECTRONIC HAND-HELD CALCULATOR

With an electronic hand-held calculator, work standards may be computed accurately and rapidly. An advanced, professional calculator permits easy calculations with a constant, such as a performance rating factor, allowance, or conversion factor. The constant key stores a number and an operation for use in repetitive calculations. Typically users can perform the calculations of $+$, $-$, $\div$, y^x, $\sqrt[x]{y}$, and $\Delta\%$.

The procedure is to first enter the operation, which would be "multiply" in the case of converting averaged observed elemental times to allowed elemental times. Then the repetitive number, the conversion factor, would be entered into storage.

After the constant (conversion factor) is stored, additional calculations for determining the allowed elemental times are completed by entering the variable (mean elemental time) and pressing the "equal" key. By using an electronic calculator, a time study analyst can establish four time studies (each representing about 30 minutes of observation time) during an eight-hour day. See Figure 17–2.

EXPRESSING THE STANDARD TIME

The sum of the elemental allowed times gives the standard in minutes per piece using a decimal minute watch, or hours per piece using a decimal hour watch. The majority of industrial operations have relatively short cycles (less than five minutes); consequently, it is usually more convenient to express standards in hours per hundred pieces. For example, the standard on a press operation might be 0.085 hour per hundred pieces. This is a more satisfactory method of expressing the standard than is 0.00085 hour per piece or 0.051 minute per piece. Thus, an operator producing 10,000 pieces during the working day earns 8.5 hours of production, and performs at an efficiency of 106 percent. This is expressed as:

$$E = \left(\frac{He}{Hc}\right)(100)$$

where:

E = Percent efficiency
He = Standard hours earned
Hc = Clock hours on the job

In another instance, the standard time may have resulted in 11.46 minutes per piece. This would be converted into decimal hours per hundred pieces as follows:

FIGURE 17–2
Back of the time study form shown in Figure 17–1. The study indicates a piece time of 0.523 minute (0.8717 hours/hundred) and a setup time of 6.772 minutes (0.1129 hour).

SKETCH

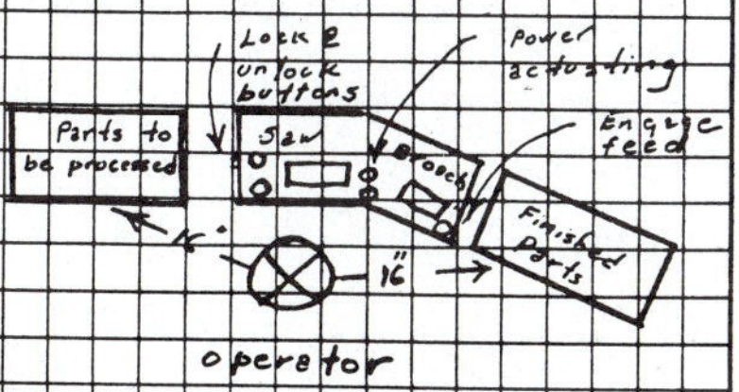

STUDY NO. 14 DATE 3/15/

OPERATION Lug saw & center cut die casting

DEPT. F-114 OPERATOR Rainbow NO. 127

EQUIPMENT Power air saw + piercing broach

MCH. NO. F 114-146

SPECIAL TOOLS, JIGS, FIXTURES, GAGES J-1117-9 & J-1117-10

CONDITIONS Operator standing, clean, average noise, adequate light

MATERIAL Aluminum 380

PART NO. A-1117 DWG. NO. C-1117

PART DESCRIPTION Bracket for mounting assembly

ACT BREAKDOWN

LEFT HAND	RIGHT HAND
Idle	Get casting (16") Move to fixture, place in fixture,
Press pneumatic locking actuator	Idle
Idle	Press actuating button for power saw
Press unlocking actuator	Idle
Idle	Get part from fixture.
Turn body 60°	Turn body 60° Place part in piercing fixture
Lock fixture	
Idle	Reach to control Engage feed
Broach 10	Broach 10.
Move to control to open fixture. Engage	Idle
Idle	Get completed part, move 16" to tote box, place in tote box

ELEM. NO.	SMALL TOOL NUMBERS, FEEDS, SPEEDS, DEPTH OF CUT, ETC.	ELEMENTAL TIME	OCC. PER CYCLE	TOTAL TIME ALLOWED
1	J-1117-9, 3600 r.p.m	0.170	1	0.170
2	J-1117-10, 30'/min	0.222	1	0.222
3		0.131	1	0.131
	Total (each piece)			0.523
S-1		1.626		1.626
S-2		0.616		0.616
S-3		2.867		2.867
S-4		1.663		1.663
	Total (setup)			6.772

EACH PIECE 0.523 min. TOTAL

SET-UP 0.1129 hrs HRS. PER C 0.8717

FOREMAN Eugene Reiter INSPECTOR Jerome Gates

OBSERVER Paul Roche APPROVED BY Dick Henshaw

$$S_h = 1.667 S_m$$

where:

S_h = Standard expressed in hours per hundred pieces
S_m = Standard expressed in minutes per piece
1.667 = Constant developed by converting minutes to decimal hours and multiplying by 100

Thus:

$$S_h = (1.667)(11.46)$$
$$= 19.104 \text{ hrs.}/C$$

If an operator produced 53 pieces in a given working day, the standard hours produced would be:

$$H_e = (0.01)(P_a)(S_h)$$

where:

H_e = Standard hours earned
P_a = Actual production in pieces
S_h = Standard expressed in hours per hundred

In this example:

$$H_e = (0.53)(19.104) = 10.125 \text{ hours}$$

Once the allowed time has been computed, the standard is given to the operator in the form of an operation card. The card can be computer generated, or run off on a ditto machine or any copier. The operation card serves as the basis for routing, scheduling, instruction, payroll, operator performance, cost, budgeting, and other necessary controls for the effective operation of a business. Figure 17–3 illustrates a typical production operation card.

Temporary Standards

Employees require time to become proficient in any operation that is new or somewhat different. Frequently, time study analysts establish a standard on an operation that is relatively new and on which there is insufficient volume for the operator to reach top efficiency. That is, he or she has not reached the flat portion of the learning curve. If analysts base their grading of operators on the usual conception of output, the resulting standard may seem unduly tight, and the operator will in all probability be unable to make any incentive earnings. On the other hand, if analysts consider that the job is new and that the volume is low, and establish a liberal standard, trouble may occur if the size of the order is increased, or if a new order for the same job is received.

FIGURE 17-3
Typical production operation card (the "F" numbers refer to the fixtures used with the involved operation)

PRODUCTION OPERATION CARD

DESCRIPTION Shower head face   DWG. NO. JB-1102   PART NO. J-1102-1

MADE FROM 2½" diam. 70-30 extruded brass rod

Routing 9-11-12--14-12-18   DATE 9-15

OP. NO.	OPERATION	DEPT.	MACHINE AND SPECIAL TOOLS	SET-UP MINUTES	EACH PC. MINUTES
1	Saw slug	9	J. & L. Air Saw	15 min	.077
2	Forge	11	150 Ton Maxi F-1102	70 min	.234
3	Blank	12	Bliss 72 F-1103	30 min	.061
4	Pickle	14	HCL. Tank	5 min	.007
5	Pierce 6 holes	12	Bliss 74 F-1104	30 min	.075
6	Rough ream and chamfer	12	Delta 17" D.P. F-1105	15 min	.334
7	Drill 13/64" holes	12	Avey D.P. F-1106	15 min	.152
8	Machine stem and face	12	#3 W. & S.	45 min	.648
9	Broach 6 holes	12	Bliss 74½	30 min	.167
10	Inspect	18	F-1109,F-1110,F-1112		Daywork

Perhaps the most satisfactory method of handling situations like this is through the issuance of temporary standards. By doing this, time study analysts establish the standard giving consideration to the difficulty of the work assignment and the number of pieces to be produced. Then, by using a learning curve for the work being studied and existing standard data, analysts can develop an equitable standard for the work at hand. The resulting standard is considerably more liberal than if the job involved a large volume. The standard, when released to the production floor, is clearly marked as a "temporary" standard and includes the maximum quantity for which it applies. Temporary standards issued on vouchers of a different color than permanent standard vouchers indicate clearly that the rate is temporary and subject to restudy in the event of additional orders or an increase in the volume of the present order.

Releasing too many temporary standards can result in a lowering of the approved conception of normal. Also, operators may strongly object to the changing of temporary standards to permanent standards, since the tighter permanent standard appears to them as a rate or wage-cutting procedure. Only new work that is definitely foreign to operators and that involves limited quantities of production justifies issuing temporary standards. When temporary standards are released, they should be in effect for the duration of the contract or for 60 days, whichever is shorter. Upon expiration, they should be replaced by permanent standards. Several union contracts specifically state that temporary standards lasting longer than 60 days must become permanent.

Setup Standards

The elements of work commonly included in setup standards involve all events that took place between completion of the previous job and the start of the present job. The setup standard also has "teardown" or "put-away" elements; these include punching in on the job, getting tools from the tool crib, getting drawings from the dispatcher, setting up the machine, punching out on the job, removing tools from the machine, returning tools to the crib, and tallying production. Figures 17–1 and 17–2 illustrate a study that involved four setup elements (S-1, S-2, S-3, S-4).

In establishing setup times, analysts use the identical procedure followed in establishing standards for production. First, ascertain that the best setup methods and a standardized procedure are in effect. Then carefully break down the work into elements, accurately time them, performance rate them, and calculate the appropriate allowance. The importance of valid setup times cannot be overemphasized, especially in job shops, where setup time represents a high proportion of the overall time.

Be especially alert when timing setup elements because there is no opportunity to get a series of elemental values for determining the mean times. Also, analysts cannot observe the operator perform the elements in advance and, consequently, they are obliged to divide the setup into elements while the study is taking place. Since setup elements for the most part are long in duration, there is a reasonable amount of time to break the job down, record the time, and evaluate the performance as the operator proceeds from one work element to the next.

There are two ways of handling setup times: first they can be distributed over a specific manufacturing quantity, such as 1,000 or 10,000 pieces. This method is satisfactory only when the magnitude of the production order is standard. For example, industries that ship from stock and reorder on a basis of minimum-maximum inventories are able to control their production orders to conform to economical lot sizes. In cases like this, the setup time can be equitably prorated over the lot size. Suppose that the economical lot size of a given item was 1,000 pieces and that reordering was always done on the basis of 1,000 units. If the standard setup time in a given operation was 1.50 hours, then the allowed operation time could be increased by 0.15 hour per 100 pieces to take care of the makeready and put-away elements.

This method would not be at all practical if the size of the order were not controlled. In a plant that requisitions on a job-order basis, that is, releases production orders specifying quantities in accordance with customer requirements, it would be impossible to standardize on the size of the work orders issued to the plant. Thus, this week an order for 100 units may be issued, and next month an order for 5,000 units of the same part. In the example cited above, the operator would be allowed but 0.15 hour to set up the machine for the 100-unit order, which would be inadequate. On the 5,000-unit order, the operator would be given 7.50 hours, which would be considerably too much time.

It is more practical to establish setup standards as separate allowed times (see Figures 17-1 and 17-2). Then, regardless of the quantity of parts to be produced, a fair standard prevails. In some concerns, the setup is performed by a person other than the operator who does the job. The advantages of having separate setup personnel are quite obvious. Lower skilled workers can be utilized as operators when they do not have to set up their own facilities. Setups are more readily standardized and methods changes more easily introduced when the responsibility for setup rests with but one individual. Also, when sufficient facilities are available, production can be continuous if the next work assignment is set up while the operator is working on the present job.

Partial Setups

Frequently, it is not necessary to set up a facility completely to perform a given operation, because some of the tools of the previous operation are required in the job that is being set up. For example, in hand screw machine or turret lathe setups, careful scheduling of similar work to the same machine allows partial setups from one job to the next. Instead of having to change six tools in the hex turret, it may be necessary to change only two or three. This savings in setup time is one of the principal benefits of a well-formulated group technology program.

Since the sequence of work scheduled to a given machine seldom remains the same, it is difficult to establish partial setup times to cover all the possible variations. The standard for a complete setup for a given No. 4 Warner & Swasey turret lathe might be 0.80 hour. However, if this setup is performed after job X, it might take only 0.45 hour; after job Y, it may require 0.57 hour; while following job Z, 0.70 hour may be necessary. The possible variations in partial setup time are so broad that the only practical way to accurately establish their values is on the basis of standard data (see Chapter 18) for each job in question.

Maintenance of Standard Times

Both labor and management have emphasized the necessity of establishing standard times that are fair. Once fair standards are introduced, it is equally important that they be maintained.

The standard time directly depends on the method used by the operator during the course of the time study. Method, in the broad sense, refers not only to the tools and facilities being used, but also to such details as operator motion pattern, workstation layout, material conditions, and working conditions. Since method controls the time standard, it is essential that methods changes and alterations be controlled if equitable standards are to be maintained. If methods changes are not controlled, inequities soon develop in the standards established and much of the work spent on the development of consistent time standards is undone.

Just as accountants periodically audit the financial records of a company, so should analysts check all established time standards at regular intervals to see whether they are in line with the method being used. An audit of time standards principally involves investigating the method being used by the operator. Frequently, minor changes have been made by the operator, the supervisor, or even the methods department, and no record of these changes has been given to the time study analyst. Some workers conceal methods changes for which they are responsible, so that they can increase their earnings or diminish their effort while achieving the same production. Of course, changes in method that increase the time required to perform the task may develop. These changes may be initiated by the supervisor or inspector and may be of insufficient consequence, in their opinion, to adjust the standard.

Usually the performance of workers approximates the normal curve as described on page 406. However, several common variations from the normal curve are symptomatic of restrictions in output and indicate the desirability of conducting an audit. Figure 17–4 illustrates that the

FIGURE 17-4
Distribution of performance in plant where standards are loose compared with
expected distribution of average performance of 115 percent of normal

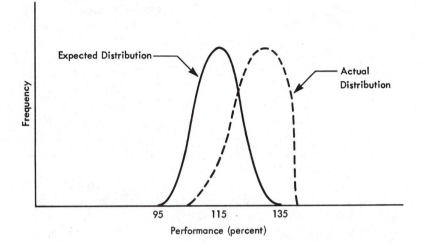

Performance (percent)

FIGURE 17-5
Distribution of performance in plant where methods and/or material have not
been standardized

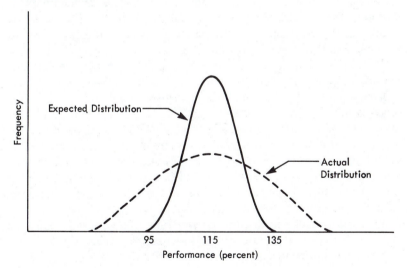

Performance (percent)

standards are loose and that workers are holding back so they do not earn
at a rate above 140 percent, feeling that if they perform beyond this point,
the time standards will be adjusted downward.

Figure 17-5 illustrates the output in an environment where the method
has not been standardized. Variation in material is another cause of this

flat distribution. In both of these instances, an audit can assure that the best method is used. Thus, the developed standard reflects the time required for average experienced operators working with good skill and effort to perform an operation at a pace that can be maintained for eight hours, giving due allowance for personal and unavoidable delays and fatigue.

The operation being time-studied must be analyzed for possible methods improvements prior to the establishment of the standard. Operation analysis, work simplification, motion study, and standardization of the method and conditions always precede work measurements. A standard does not get out of line if the method that was time-studied is maintained by operators. If methods study has developed the ideal method, and if this method is standardized and followed by operators, then there is less need to maintain time standards.

Frequently, however, methods changes are introduced—both favorable and unfavorable. If these changes are extensive, they are brought to the attention of management. Tight standards are brought to management's attention by operators. Standards that become very loose are brought to management's attention through the payroll department where excessive earnings by given workers are reported. However, minor accruing methods changes frequently take place unnoticed and weaken the entire standards structure. To maintain standards properly, the time study department should periodically verify that the method being used is the method that was studied when the standard was established. This can readily be done by referring to the original time study that includes a complete description of the method. A time study analyst should enlist the cooperation of the supervisor to verify standards that may be out of line. Being close to the operators, a supervisor should be aware of standards that may be either loose or tight. He or she can advise the time study analyst about the sequence in the audit of existing standards. Observe the operation as performed to determine the correctness of the description, sequences, frequencies, conditions, and standard time allowances. It is usually satisfactory to check the accuracy of the time standard by measuring several cycles of the overall time, performance rating the data, and adding an appropriate allowance.

If the cycle overall time and the existing time standards vary more than 5 percent, take a detailed time study to identify the cause of the discrepancy. In the majority of cases where the audit reveals a discrepancy of more than 5 percent, a change in method is the cause. If the operator is using an inferior method, then he or she should immediately be instructed in the ideal method. If the operator developed a better method than existed when the standard was developed, then the new method should become the accepted standard. If the operator developed the improvement, he or she should be justly rewarded.

Such audits should be carried out by a representative of the standards staff. Only completely qualified analysts should audit time standards. The audit frequency is determined when the standard is developed. This is based on an estimate of the hours of application of the standard in one year. For example, one large company uses the following data to determine the frequency of the audit of methods and standards:[1]

Hours of application per standard per year	Frequency of audit
0–10	Once per three years
10–50	Once per two years
50–600	Once per year
Over 600	Twice per year

When properly done, auditing methods and standards takes time and, consequently, is expensive. However, this maintenance assures the success of the program.

CONCLUSION

It is common practice to issue standards for setup operations, independent of "piece" standards, and to specify the allowed time in decimal hours or decimal minutes. Piece standards are expressed in hours per hundred pieces for ease in payroll computation, scheduling, and control. On extremely short-cycle operations, such as punch press work, die casting, or forging, standards are expressed in hours per thousand pieces, since during the course of an eight-hour shift, several thousand pieces can be produced.

Schedule work to take advantage of partial setups to improve delivery dates, decrease total cost, and allow greater remuneration for operators.

Maintain time standards to assure a satisfactory rate structure. This calls for a continuing analysis of methods. Periodically check all standards to verify that the methods employed are identical with those in use at the time the standards were established.

TEXT QUESTIONS

1. Define the term *standard time.*
2. How is the conversion factor determined?
3. What is the conversion factor in Figure 17–1?
4. Why is it usually more convenient to express standards in time per hundred pieces rather than time per piece?

---

[1] By H. B. Brandt, former associate director, Industrial Engineering Division, Procter & Gamble Co.

Motion and Time Study

5. For which reasons are temporary standards established?
6. Express the standard of 5.761 minutes in hours per hundred pieces. What would the operator's efficiency be if 92 pieces were completed during a working day? What would his efficiency be if he set up his machine (standard for setup = 0.45 hours) and produced 80 pieces during the eight-hour workday?
7. Which elements of work are included in the setup standard?
8. What is the preferred method of handling setup time standards?
9. How is the operator compensated on partial setups on hand screw machines?
10. Determine the conversion factor and the allowed time for a job that had an average time of 5.24 minutes and carried a performance factor of 1.15 and an allowance of 12 percent.
11. Explain why it is necessary that time standards be properly maintained.
12. Explain how you would use learning curves to establish temporary standards.

GENERAL QUESTIONS

1. For what reasons might it be advantageous to express allowed times in minutes per piece?
2. How can the excessive use of temporary standards cause poor labor relations?
3. How does a company effectively maintain standards and avoid the reputation of being a "rate-cutter"?
4. If an audit revealed that a standard as originally established was 20 percent loose, explain in detail the methodology to be followed to rectify the rate.

PROBLEMS

1. Based on the data provided in Figure 17–1, what would be the efficiency of an operator who set up the machine and produced an order of 5,000 pieces in a 40-hour workweek?
2. Establish the money rate per hundred pieces from the following data:
   Cycle time (averaged measured time): 1.23 minutes
   Base rate: $8.60 per hour.
   Pieces per cycle: 4.
   Machine time (power feed): 0.52 minutes per cycle.
   Allowance: 17 percent on effort time; 12 percent on power feed time.
   Element 1 average time: 0.09 minutes.
   MTM time for element 1: 132 TMU. (One TMU = 0.00001 hr.)
   Plant uses synthetic performance rating.
3. The following data resulted from a study taken on a horizontal milling machine:
   Pieces produced per cycle: 8.
   Average measured cycle time: 8.36 minutes.
   Average measured effort time per cycle: 4.62 minutes.

Average rapid traverse time: 0.08 minutes.
Average cutting time power feed: 3.66 minutes.
Performance factor: +15 percent.
Allowance (machine time): 10 percent.
Allowance (effort time): 15 percent.
The operator works on the job a full eight-hour day and produces 380 pieces. How many standard hours does the operator earn? What is her efficiency for the eight-hour day?

4. A time standard is established allowing the operator 11.28 min. per piece. The sales department expects to sell at least 2,000 of these parts in the next year. How many audits of this standard would you recommend be scheduled during the next 12 months?

SELECTED REFERENCES

Graham, C. F. *Work Measurement and Cost Control.* London, England: Pergamon Press, 1965.

Panico, Joseph A. "Work Standards: Establishment, Documentation, Usage and Maintenance." In *Handbook of Industrial Engineering,* 2nd ed., edited by Gavriel Salvendy. New York: John Wiley & Sons, 1992.

Rotroff, Virgil H. *Work Measurement.* New York: Reinhold Publishing, 1959.

Smith, George L. *Work Measurement—A Systems Approach.* Columbus, Ohio: Grid, 1978.

SELECTED SOFTWARE

Whitehouse, Gary E. Stopwatch Time Study Analysis, Work Measurement IIE Microsoftware, Industrial Engineering and Management Press, Norcross, Ga: Institute of Industrial Engineers, 1985

Standard Data

Standard time data are elemental time standards taken from time studies; these standards have proved to be satisfactory. Analysts classify and file elemental standards so that they can readily be abstracted when needed. Just as a cook refers to a recipe to determine how many minutes to cream butter and sugar, how long to beat the batter, and thus how much time is required to bake the cake, so an analyst refers to standard data. Thus, an analyst determines how long it should take the normal operator to pick up a small casting and place it in a jig, close the jig and lock the part with a quick-acting clamp, advance the spindle of the drill press, and perform the remainder of the elements required to produce the part.

Standard data can be of several levels of refinement: motion, element, and task. The more refined the standard data element, the broader its range of usage. Thus motion standard data has the greatest application, but it takes longer to develop a standard than when using either element or task standard data. Element standard data has wide application and allows the development of a standard faster than using motion data. This chapter is devoted to element standard data and Chapter 19 to motion standard data.

The application of standard time data is fundamentally an extension of the same process used to arrive at allowed times through stopwatch time study. The principle of standard data application is not new; many years ago, Frederick W. Taylor proposed that each established elemental time be properly indexed so that it could be used to establish time standards for future work. When we speak of standard data today, we refer to all the tabulated elemental standards, curves, alignment charts, and tables that are compiled to allow the measurement of a specific job without the necessity of a timing device, such as the stopwatch.

TABLE 18-1

Operation	Number of time studies taken for development of standard data	Number of standards set in one year from standard data developed	Percent of standards set that would be covered by stopwatch studies
Coremaking......................	60	7,500	0.8
Snag grinding	40	656	6.1
Visual inspection	53	422	12.6 ·
Turret lathe operation.............	100	600	16.7

Source: Phil Carroll, Jr., "Notes on Standard Elemental Data," *Modern Machine Shop*, April 1950, p. 176.

Work standards calculated from standard data are relatively consistent in that the tabulated elements comprised by the data resulted from many proven stopwatch time studies. Since the values are tabulated, and it is only necessary to accumulate the required elements in establishing a standard, the various time study personnel within a given company arrive at identical standards of performance for a given method. Therefore, consistency is assured for standards established by the different analysts within a plant as well as for the various standards computed by a given time study observer.

Standards on new work can usually be computed more rapidly through standard data than by means of stopwatch time study. This rapidity allows the establishment of standards on indirect labor operations; usually this is impractical if done by stopwatch methods. Typically, one work measurement analyst establishes five rates per day using stopwatch methods. This compares to 25 rates a day using the standard data technique. Using standard data permits the establishment of time standards over a wide range of work. Table 18-1 illustrates the coverage possible when standard data elements are determined.

DEVELOPMENT OF STANDARD TIME DATA

To develop standard time data, analysts must distinguish constant elements from variable elements. A constant element is one for which the allowed time remains approximately the same for any part within a specific range. A variable element is one for which the allowed time varies within a specified range of work. Thus, the element "start machine" would be a constant, and the element "drill ⅜-inch diameter hole" would vary with the depth of the hole, the feed, and speed of the drill.

Index and file standard data as they are developed. Keep setup elements separate from those elements incorporated in each piece time, and constant elements separate from the variable elements. Typical standard data would be tabulated as follows:

A. Machine or operation.
  1. Setup.
     a. Constants.
     b. Variables.
  2. Each piece.
     a. Constants.
     b. Variables.

Compile standard data from different elements on time studies of a given process over a period of time. Include only those studies proved valid through use. In tabulating standard data, be careful to define end points clearly. Otherwise, there may be an overlapping of time in the recorded data. For example, in the element "out stock to stop" done on a bar feed No. 3 Warner & Swasey turret lathe, the element could include reaching for the feed lever, grasping the lever, feeding the bar stock through the collet to a stock stop located in the hex turret, closing the collet, and reaching for the turret handle. Then, again, this element may involve only the feeding of bar stock through the collet to a stock stop. Since standard data elements are compiled from a great number of studies taken by different time study observers, carefully define the limits or end points of each element. Figure 18–1 illustrates a form to summarize data taken from an individual time study to develop standard data on die-casting machines.

To fill a specific need in a standard data tabulation, analysts may resort to work measurement of the particular element in question. This is handled quite accurately by using the electronic stopwatch or the mechanical "fast" watch (see Chapter 13), both of which record elapsed times in 0.001 of a minute. In this analysis, the snapback method records the elapsed elemental time, as analysts are usually interested in determining the allowed time for only a few of the elements comprised by the study. Upon completion of the observations, summarize the elemental elapsed times and determine the mean, as in the case of a typical time study. Next, performance rate the average values and add an allowance to arrive at fair allowed times.

Sometimes, because of the brevity of individual elements, measuring their duration separately is impossible. You can, however, determine their individual values by timing groups collectively and using simultaneous equations to solve for the individual elements.

For example, element *a* might be "pick up small casting," element *b* might be "place in leaf jig," *c* might be "close cover of jig," *d* "position jig," *e* "advance spindle," and so on. These elements could be timed in groups as follows:

FIGURE 18-1

DIE-CASTING MACHINE

Mach. No.

Part No. ____ & Type _____ Operator _____ Date _____
              No. of Parts   Method of Placing   Total Wt. of Flsh,
Of _____ in Tote Pan _____ Parts in Tote Pan _____ Parts, Gate & Sprue ____
        No. of Parts   Liquid Metal _____
_____ per Shot _____   Plastic Metal _____ Chill ____ Skim ____ Drain ____
Capacity in Lbs.   Describe
Holding Pot   _____   Greasing _____

Describe
Loosening of Part _____

Describe
Location _____

ELEMENTS	TIME	END POINTS
Get metal in holding pot	_____	All waiting time while metal is being poured in pot.
Chill metal	_____	From time operator starts adding cold metal to liquid metal in pot until operator stops adding cold metal to liquid metal in pot.
Skim metal	_____	From time operator starts skimming until all scum has been removed.
Get ladleful of metal	_____	From time ladle starts to dip down into metal until ladleful of metal reaches edge of machine or until ladle starts to tip for draining.
Drain metal	_____	From time ladle starts to tip for draining until ladleful reaches edge of machine.
Pour ladle of metal in machine	_____	From time ladleful of metal reaches edge of machine until foot starts to trip press.
Trip press	_____	From time foot starts moving toward pedal until press starts downward.
Press time	_____	Complete turnover of press.
Hold plunger down	_____	From time plunger stops downward motion until plunger starts moving upward.
Press button and raise slug	_____	From time plunger stops moving until slug is raised out of cavity.
Remove slug--drop slug--lift	_____	From time slug is raised out of cavity until slug is pushed into tote pan or pot.
Trip pedal to open dies	_____	From time foot starts moving to pedal until dies start to open.
Wait for dies to open	_____	From time dies start to open until die stops moving.
Remove part from die	_____	From time die stops moving until part is free of die cavity.
Place part in tote pan	_____	From time part is free of die cavity until part is placed in tote pan.

$a + b + c$ equals element 1 equals 0.070 min. $= A$ (1)
$b + c + d$ equals element 3 equals 0.067 min. $= B$ (2)
$c + d + e$ equals element 5 equals 0.073 min. $= C$ (3)
$d + e + a$ equals element 2 equals 0.061 min. $= D$ (4)
$e + a + b$ equals element 4 equals 0.068 min. $= E$ (5)

By adding these five equations:

$$3a + 3b + 3c + 3d + 3e = A + B + C + D + E$$

Then let

$$A + B + C + D + E = T$$
$$3a + 3b + 3c + 3d + 3e = T = 0.339 \text{ min.}$$

and

$$a + b + c + d + e = \frac{0.339}{3} = 0.113 \text{ min.}$$

Therefore:

$$A + d + e = 0.113 \text{ min.}$$

Then:

$$d + e = 0.113 \text{ min.} - 0.07 \text{ min.} = 0.043 \text{ min.}$$

since

$$c + d + e = 0.073 \text{ min.}$$
$$c = 0.073 \text{ min.} - 0.043 = 0.03 \text{ min.}$$

Likewise:

$$d + e + a = 0.061$$

and

$$a = 0.061 - 0.043 = 0.018 \text{ min.}$$

Substituting in Equation 1

$$b = 0.070 - (0.03 + 0.018) = 0.022$$

Substituting in Equation 2

$$d = 0.067 - (0.022 + 0.03) = 0.015 \text{ min.}$$

Substituting in Equation 3

$$e = 0.073 - (0.015 + 0.03) = 0.028 \text{ min.}$$

In determining standard data elements by simultaneous equations, be consistent when reading the watch at the terminal points of the established elements. Inconsistency in establishing the terminal points results in erroneous standard data elements.

TABLE 18-2
Typical operating conditions for producing holes in various materials with twist drills

Material	Tool Material	Speed (sfm)	Feed Rate				
			1/8"D	1/4"D	1/2"D	3/4"D	1"D
Aluminum and its alloys	High speed steel	350	0.003	0.006	0.010	0.0155	0.0190
Bakelite	Tungsten carbide	60–70	0.0015	0.003	0.0035	0.0045	0.005
Copper and its alloys							
(high machineability)	High speed steel	200	0.003	0.006	0.010	0.0155	0.0190
(low machineability)	High speed steel	70	0.003	0.006	0.010	0.0155	0.0190
Fiberglass-epoxy	Tungsten carbide	650	0.0025	0.0030	0.0035	0.0042	0.0050
Glass	Tungsten carbide	15–25	light	light	light	light	light
High temperature alloys							
cobalt based	High speed steel with cobalt	20	0.0015	0.003	0.0035	0.0045	0.005
Iron based	High speed steel with cobalt	25	0.002	0.0035	0.006	0.0085	0.0105
Iron							
cast (soft)	High speed steel	140–150	0.003	0.006	0.010	0.0155	0.019
	Tungsten carbide	90–165	0.003	0.005	0.008	0.0105	0.0125
cast (medium hard)	High speed steel	80–110	0.003	0.005	0.008	0.0105	0.0125
hard chilled	Tungsten carbide	30	0.002	0.0035	0.006	0.0085	0.0105
malleable	High speed steel	90–120	0.003	0.005	0.008	0.0105	0.0125
	Tungsten carbide	100–150	0.002	0.0035	0.006	0.0085	0.0105
ductile	High speed steel	60	0.003	0.005	0.008	0.0105	0.0125
	Tungsten carbide	80–100	0.002	0.0035	0.006	0.0085	0.0105
Magnesium and its alloys	High speed steel	150–400	0.003	0.006	0.010	0.0155	0.0190
Plastics	High speed steel	100	0.003	0.005	0.008	0.0105	0.0125
	Tungsten carbide	100–200	0.002	0.0035	0.006	0.0085	0.0105
Rubber (hard)	High speed steel	100–300	0.002	0.0035	0.006	0.0085	0.0105
Steel							
plain to 0.25 C	High speed steel	80	0.003	0.005	0.008	0.0105	0.0125
plain 0.25 to 0.50 C	High speed steel	65	0.003	0.005	0.008	0.0105	0.0125
plain 0.5 to 0.9 C	High speed steel	55	0.003	0.005	0.008	0.0105	0.0125
alloy, low C	High speed steel	70	0.003	0.006	0.010	0.0155	0.019
alloy, med C	High speed steel	50–60	0.002	0.0035	0.006	0.0085	0.0105
maraging	High speed steel	55	0.003	0.005	0.008	0.0105	0.0125
stainless (austenitic)	High speed steel with cobalt	55	0.002	0.0035	0.006	0.0085	0.0105
stainless (ferritic)	High speed steel	65	0.002	0.0035	0.006	0.0085	0.0105
stainless (martensitic)	High speed steel with cobalt	65	0.003	0.006	0.010	0.0155	0.019
Zinc alloys	High speed steel	250	0.003	0.006	0.010	0.0155	0.0190

Source: *Tool and Manufacturing Engineers Handbook*, 4th ed. (Dearborn, Mich.: Society of Manufacturing Engineers, 1983), Vol. 1, *Machining*, Table 9–20, pp. 9-90–9-91.

Calculation of Cutting Times

By learning the feeds and speeds for different types of material, analysts can calculate and tabulate the cutting times for different machining operations. Table 18–2 gives recommended speeds and feeds for high-speed and tungsten carbide drills used on various kinds of material. Various technical handbooks include this information that cutting tool manufacturers can also provide.

DRILL PRESS WORK. A drill is a fluted end-cutting tool used to originate or enlarge a hole in solid material. In drilling operations on a flat surface, the axis of the drill is at 90° to the surface being drilled. When a hole is drilled completely through a part, add the lead of the drill to the length of the hole to determine the entire distance the drill must travel to make the hole. When a blind hole is drilled, do not add the lead of the drill to the hole depth, because the distance from the surface being drilled to the farthest penetration of the drill is the distance that the drill must travel (see Figure 18–2).

Since the commercial standard for the included angle of drill points is 118°, the lead of the drill may readily be found through the expression:

$$l = \frac{r}{\tan A}$$

where:

l = Lead of drill
r = Radius of drill
$\tan A$ = Tangent of ½ the included angle of the drill

To illustrate, calculate the lead of a general purpose drill 1 inch in diameter:

$$l = \frac{0.5}{\tan 59°}$$

$$l = \frac{0.5}{1.6643}$$

$$l = 0.3 \text{ inches lead}$$

After determining the total length that the drill must move, divide the feed of the drill in inches per minute into this distance to find the drill cutting time in minutes.

Express drill speed in feet per minute, and feed in thousandths of an inch per revolution. To change the feed into inches per minute when the feed per revolution and the speed in feet per minute are known, substitute in the following equation:

$$F_m = \frac{3.82(f)(Sf)}{d}$$

where:

F_m = Feed in inches per minute
f = Feed in inches per revolution
Sf = Surface feet per minute
d = Diameter of drill in inches

FIGURE 18–2
Distance *L* indicates the distance the drill must travel when drilling through (illustration at left) and when drilling blind holes (illustration at right) (lead of drill is shown by distance *l*)

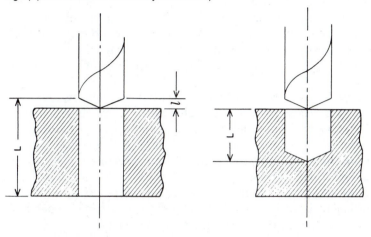

For example, to determine the feed in inches per minute of a 1-inch drill running at a surface speed of 100 feet per minute and a feed of 0.013 inch per revolution, figure

$$F_m = \frac{(3.82)(0.013)(100)}{1} = 4.97 \text{ inches per minute}$$

To determine how long it would take for this 1-inch drill running at the same speed and feed to drill through 2 inches of a malleable iron casting, substitute in the equation:

$$T = \frac{L}{F_m}$$

where:

T = Cutting time in minutes
L = Total length drill must move
F_m = Feed in inches per minute

and we should have

$$T = \frac{2 \text{ (thickness of casting)} + 0.3 \text{ (lead of drill)}}{4.97}$$

$$= 0.464 \text{ minutes cutting time}$$

The cutting time so calculated does not include any allowance, which must be added to determine the allowed time. The allowance includes time for variations in material thickness and tolerance in the setting of stops, both of which affect the cycle cutting time to some extent. Also, add personal and unavoidable delay allowances to arrive at an equitable allowed elemental time.

All speeds may not be available on the machine being used. For example, the recommended spindle speed for a given job might be 1,550 rpm; however, the fastest speed that the machine is capable of running may be 1,200 rpm. In that case, use 1,200 rpm as the basis of computing allowed times.

LATHE WORK. Many variations of machine tools are classified in the lathe group. These include the engine lathe, turret lathe, and automatic lathe (automatic screw machine). Operators use all of the lathe group machine tools primarily with stationary tools or with tools that translate over the surface to remove material from the revolving work, which may be in the form of forgings, castings, or bar stock. In some cases, the tool revolves while the work is stationary, as on certain stations of automatic screw machine work. Thus, a slot in a screwhead can be machined in the slotting attachment on the automatic screw machine.

Many factors alter speeds and feeds, such as the condition and design of the machine tool, material being cut, condition and design of the cutting tool, coolant used for cutting, method of holding the work, and method of mounting the cutting tool. Table 18–3 outlines the approximate cuts, feeds, and speeds for certain metallic and nonmetallic turning.

As in drill press work, express feeds in thousandths of an inch per revolution, and speeds in surface feet per minute.

To determine the cutting time for L inches of cut, divide the length of cut in inches by the feed in inches per minute or, expressed algebraically:

$$T = \frac{L}{F_m}$$

where:

T = Cutting time in minutes
L = Total length of cut
F_m = Feed in inches per minute

and

$$F_m = \frac{3.82(S_f)(f)}{d}$$

where:

f = Feed in inches per revolution
S_f = Speed in surface feet per minute
d = Diameter of work in inches

MILLING MACHINE WORK. Milling refers to the removal of material with a rotating multiple-toothed cutter. While the cutter rotates, the work is fed past the cutter; thus, a milling machine differs from a drill press, where the work usually is stationary. In addition to machining plane and irregular surfaces, operators use a milling machine for cutting threads, slotting, and cutting gears.

In milling work, as in drill press and lathe work, the speed of the cutter is expressed in surface feet per minute. Feeds or table travel are usually represented in terms of thousandths of inch per tooth.

To determine the cutter speed in revolutions per minute from surface feet per minute and the diameter of the cutter, use the following expression:

$$N_r = \frac{3.82 S_f}{d}$$

where:

N_r = Cutter speed in revolutions per minute
S_f = Cutter speed in feet per minute
d = Outside diameter of cutter in inches

To determine the feed of the work in inches per minute into the cutter, use the expression

$$F_m = f n_t N_r$$

where:

F_m = Feed of the work in inches per minute into the cutter
f = Feed in inches per tooth of cutter
n_t = Number of cutter teeth
N_r = Cutter speed in revolutions per minute

The number of cutter teeth suitable for a particular application may be expressed as:

$$n_t = \frac{F_m}{F_t \times N_r}$$

where:

F_t = Chip thickness

Table 18–4 gives suggested feeds and speeds for milling under average conditions.

In computing cutting time on milling operations, take into consideration the lead of the milling cutter when figuring the total length of cut under power feed. You can determine this by triangulation as illustrated in the example of slab-milling a pad shown in Figure 18–3.

TABLE 18–3

Suggested operating parameters for machining various materials with carbide tools

Work Material	Cutting Speed, sfm (m/min)		Feed Rate, ipr (mm/rev)		Depth of Cut, in. (mm)	
	Roughing	Finishing	Roughing	Finishing	Roughing	Finishing
Free-machining carbon steels: AISI 1100 and 1200 series, 140–190 Bhn	250–1,100 (76–335)	1,000–2,000 (305–610)	0.010–0.085 (0.25–2.16)	0.005–0.015 (0.13–0.38)	0.125–0.675 & up (3.18–17.15)	Up to 0.180 (4.57)
Plain carbon steels: AISI 1000 series, 185–240 Bhn	200–800 (61–244)	700–1,600 (213–488)	0.010–0.085 (0.25–2.16)	0.005–0.015 (0.13–0.38)	0.125–0.675 & up (3.18–17.15)	Up to 0.180 (4.57)
Alloy steels: AISI 1300, 4000, 5000, 8000, and 9000 series, 190–240 Bhn	175–600 (53–183)	550–1,200 (168–366)	0.010–0.085 (0.25–2.16)	0.005–0.015 (0.13–0.38)	0.125–0.675 & up (3.18–17.15)	Up to 0.180 (4.57)
Cast irons: gray, nodular, and malleable, 150–210 Bhn	200–1,200 (61–366)	200–750 (61–229)	0.010–0.055 (0.25–1.40)	0.005–0.0015 (0.13–0.38)	0.125–0.675 & up (3.18–17.15)	Up to 0.180 (4.57)
Martensitic stainless steels: wrought 400 and 500 series, and PH types, 175–210 Bhn	175–450 (53–137)	450–850 (137–259)	0.010–0.040 (0.25–1.02)	0.005–0.015 (0.13–0.38)	0.125–0.500 (3.18–12.70)	Up to 0.180 (4.57)

Material						
Austenitic stainless steels: wrought 200 and 300 series, 140–190 Bhn	125–425 (38–130)	425–650 (130–198)	0.010–0.040 (0.25–1.02)	0.005–0.015 (0.13–0.38)	0.125–0.500 (3.18–12.70)	Up to 0.180 (4.57)
Superalloys: iron, nickel, titanium, and cobalt-based alloys, 240–300 Bhn	30–150 (9–46)	150–400 (46–122)	0.010–0.025 (0.25–1.02)	0.005–0.015 (0.13–0.38)	0.100–0.300 (2.54–7.62)	Up to 0.100 (4.57)
Tool steels: wrought high-speed, shock resistant, and hot and cold work, 210–240 Bhn	100–300 (30–91)	275–750 (84–229)	0.010–0.065 (0.25–1.65)	0.005–0.015 (0.13–0.38)	0.125–0.675 & up (3.18–17.15)	Up to 0.180 (4.57)
Nonferrous free-matching alloys: aluminum, copper, zinc, and brass alloys, 80–120 Bhn	400–1,200 (122–366)	1,000–2,000 (305–610)	0.010–0.085 (0.25–2.16)	0.005–0.015 (0.13–0.38)	0.125–0.675 & up (3.18–17.15)	Up to 0.180 (4.57)
Nonmetallics: nylons, acrylics, and phenolic resins	350–800 (107–244)	800–1,500 (244–457)	0.010–0.040 (0.25–1.02)	0.005–0.015 (0.13–0.38)	0.125–0.500 (3.18–12.70)	Up to 0.180 (4.57)

Source: *Tool and Manufacturing Engineers Handbook*, 4th ed., (Dearborn, Mich.: Society of Manufacturing Engineers, 1983), Vol. 1, *Machining*, Table 3–12, p. 3–25, Kennametal, Inc.

TABLE 18–4
Typical speeds and feeds for milling various materials with representative tool materials (feed per tooth—in.; cutting speed—sfm)

Material Milled	Tool Material	Face Mills	Slab Mills	Form Mills
Aluminum alloys	High speed steel	0.010–0.025 300–1,200	0.015–0.025 300–1,200	0.010–0.020 300–1,200
	Uncoated carbide	0.010–0.020 2,000–4,000	0.010–0.020 2,000–4,000	0.008–0.015 2,000–4,000
	Coated carbide	0.010–0.020 4,000–6,000	0.010–0.020 4,000–6,000	0.008–0.015 4,000–6,000
Brass	High speed steel	0.010–0.025 150–300	0.008–0.020 100–300	0.008–0.015 100–300
	Uncoated carbide	0.010–0.020 500–1,500	0.010–0.020 500–1,500	0.008–0.015 500–1,500
	Coated carbide	0.010–0.020 1,500–3,000	0.008–0.020 1,500–3,000	0.008–0.015 1,500–3,000
Bronze	High speed steel	0.010–0.025 50–225	0.008–0.020 50–200	0.008–0.015 50–200
	Uncoated carbide	0.010–0.020 300–1,500	0.010–0.020 300–1,400	0.008–0.015 200–1,400
	Coated carbide	0.010–0.020 1,500–2,700	0.010–0.020 1,400–2,500	0.008–0.015 1,400–2,500
Cast irons 150–180 Brinell	High speed steel	0.010–0.025 80–120	0.010–0.025 70–110	0.010–0.015 60–80
	Uncoated carbide	0.010–0.016 275–800	0.010–0.020 275–900	0.008–0.015 250–800
	Coated carbide	0.010–0.016 800–1,100	0.010–0.020 900–1,200	0.008–0.015 800–1,100
Cast irons 180–225 Brinell	High speed steel	0.010–0.020 60–80	0.008–0.015 50–70	0.008–0.012 50–60
	Uncoated carbide	0.008–0.015 250–500	0.008–0.015 225–500	0.006–0.012 225–500
	Coated carbide	0.008–0.015 500–800	0.008–0.015 500–750	0.006–0.012 500–750
Cast irons	High speed steel	0.005–0.012 40–60	0.005–0.01 35–50	0.005–0.01 35–50
	Uncoated carbide	0.005–0.010 200–400	0.005–0.010 200–400	0.005–0.010 200–400
	Coated carbide	0.005–0.010 400–600	0.005–0.010 400–600	0.005–0.010 400–600
Steels 100–150 Brinell	High speed steel	0.015–0.020 80–130	0.008–0.015 80–130	0.008–0.010 70–100
	Uncoated carbide	0.010–0.018 400–900	0.008–0.015 350–800	0.004–0.010 350–800
	Coated carbide	0.010–0.018 900–1,500	0.008–0.015 800–1,300	0.004–0.010 800–1,300
Steels 150–250 Brinell	High speed steel	0.010–0.020 50–70	0.008–0.015 50–70	0.006–0.010 50–70
	Uncoated carbide	0.010–0.015 300–700	0.008–0.015 300–700	0.004–0.010 300–700
	Coated carbide	0.010–0.015 700–1,200	0.008–0.015 700–1,200	0.004–0.010 700–1,200
Steels 250–350 Brinell	High speed steel	0.005–0.010 35–60	0.005–0.010 35–50	0.005–0.010 35–50
	Uncoated carbide	0.008–0.015 225–600	0.007–0.012 200–600	0.003–0.008 200–600

TABLE 18–4 (concluded)

Material Milled	Tool Material	Face Mills	Slab Mills	Form Mills
	Coated carbide	0.008–0.015 600–1,000	0.007–0.015 600–1,000	0.003–0.008 600–1,000
Steels 350–450 Brinell	High speed steel	0.003–0.008 20–35	0.005–0.008 20–35	0.003–0.008 20–35
	Uncoated carbide	0.005–0.012 180–400	0.007–0.012 150–400	0.003–0.008 150–400
	Coated carbide	0.005–0.012 400–600	0.007–0.012 400–600	0.003–0.008 400–600

Source: *Tool and Manufacturing Engineers Handbook,* 4th ed. (Dearborn, Mich.: Society of Manufacturing Engineers. 1983), Vol. 1, *Machining,* Table 10–5, p. 10–52, Valenite Div., Valeron Corp.

In this case, to arrive at the total length that must be fed past the cutter add the lead *BC* to the length of the work (8 inches). Clearance for removal of the work after the machining cut is handled as a separate element because greater feed under rapid table traverse is used. By knowing the diameter of the cutter, you can determine *AC* as being the cutter radius, and calculate the height of the right triangle *ABC* by subtracting the depth of cut *BE* from the cutter radius *AE*:

$$BC = \sqrt{AC^2 - AB^2}$$

In the preceding example, assume that the cutter diameter is 4 inches, and that it has 22 teeth. The feed per tooth is 0.008 inch, and the cutting

FIGURE 18–3
Slab-milling a casting 8 inches in length

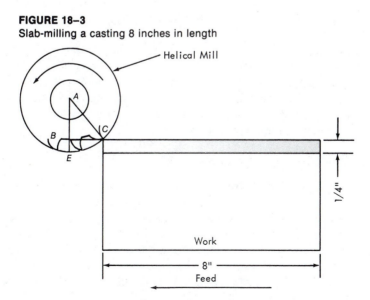

491

speed is 60 feet per minute. Compute the cutting time by using the equation

$$T = \frac{L}{F_m}$$

where:

T = Cutting time in minutes
L = Total length of cut under power feed
F_m = Feed in inches per minute

Then L would be equal to 8 inches + BC and

$$BC = \sqrt{4 - 3.06} = 0.975$$

Therefore:

$$L = 8.975$$
$$F_m = fn_tN_r$$
$$F_m = (0.008)(22)N_r$$

or

$$N_r = \frac{3.82S_f}{d} = \frac{(3.82)(60)}{4} = 57.3 \text{ revolutions per minute}$$

Then

$$F_m = (0.008)(22)(57.3) = 10.1 \text{ inches per minute}$$

and

$$T = \frac{8.975}{10.1} = 0.888 \text{ minutes cutting time}$$

Through a knowledge of feeds and speeds, analysts can determine the required cutting or processing time for various work performed in their plants. The illustrations cited in drill press, lathe, and milling work are representative of the techniques used in establishing raw cutting times. The necessary applicable allowances must be added to these values to create fair elemental allowed values.

Determining Horsepower Requirements

When developing standard data times for machine elements, tabulate horsepower requirements for various materials in relationship to depths of cut, cutting speeds, and feeds. Frequently you can use standard data for planning new work. So that existing equipment is not overloaded, it is important that the analyst have information as to the work load being assigned to each machine for the conditions under which the material is being removed. For example, in the machining of high-alloy steel forgings on a lathe capable of a developed horsepower of 10, it would not be feasible to take a ⅜-inch depth of cut while operating at a feed of 0.011 inch per revolution and a speed of 200 surface feet per minute. Table 18–5 indicates a horsepower requirement of 10.6 for these conditions. Consequently, the work would need to be planned for a feed of 0.009 inch at a speed of 200 surface feet, since in this case the required horsepower would be but 8.7. (See Table A3–12, Appendix 3, for horsepower requirements.)

Plotting Curves

Because of space limitations, tabularizing values for variable elements is not always convenient. By plotting a curve or a system of curves in the form of an alignment chart, you can express considerable standard data graphically on one page.

Figure 18–4 illustrates a nomogram used for determining turning and facing time. For example, if the problem is to determine the production in pieces per hour to turn 5 inches of a 4-inch diameter shaft of medium carbon steel on a machine utilizing 0.015-inch feed per revolution and having a cutting time of 55 percent of the cycle time, the answer could be readily determined graphically. Connecting a recommended cutting speed of 150 feet per minute for medium carbon steel shown on scale 1 to the 4-inch diameter of the work shown on scale 2 results in a speed of 143 rpm shown on scale 3. The 143-rpm point is connected with the 0.015-inch feed per revolution shown on scale 4. This line extended to scale 5 shows a feed of 2.15 inches per minute. This feed point connected with the length of cut shown on scale 6 (5 inches) gives the required cutting time on scale 7. This cutting time of 2.35 minutes, when connected with the percentage of cutting time shown on scale 8 (in this case, 55 percent), gives the production in pieces per hour on scale 9 (in this case, 16).

Using curves has some distinct disadvantages. First, it is easy to introduce an error in reading from the curve because of the amount of interpolation usually required. Second, there is the chance of outright error through incorrect reading or misalignment of intersections of the various scales.

When we refer to the plotting of curves to show the relationship between time and those variables that affect time, our solution may take the form of a single straight line, a curved line, a system of straight lines as in the ray chart, or a special arrangement of lines characteristic of an alignment chart or nomogram. In the plotting of simple, one-line curves, observe certain standard procedures. First, plot time on the ordinate of the charting paper, and the independent variable on the abscissa. Second, if practical, all scales should begin at zero to show their true proportions. Last, the scale selected for the independent variable should have a sufficient range to fully utilize the paper on which the curve is plotted. Figure

TABLE 18–5

Horsepower requirements for turning high-alloy steel forgings for cuts ³⁄₈ inch and ½ inch deep at varying speeds and feeds

Surface feet	³⁄₈-in. depth cut (feeds, in./rev.)						½-in. depth cut (feeds, in./rev.)					
	0.009	0.011	0.015	0.018	0.020	0.022	0.009	0.011	0.015	0.018	0.020	0.022
150	6.5	8.0	10.9	13.0	14.5	16.0	8.7	10.6	14.5	17.3	19.3	21.3
175	8.0	9.3	12.7	15.2	16.9	18.6	10.1	12.4	16.9	20.2	22.5	24.8
200	8.7	10.6	14.5	17.4	19.3	21.3	11.6	14.1	19.3	23.1	25.7	28.4
225	9.8	11.9	16.3	19.6	21.7	23.9	13.0	15.9	21.7	26.1	28.9	31.8
250	10.9	13.2	18.1	21.8	24.1	26.6	14.5	17.7	24.1	29.0	32.1	35.4
275	12.0	14.6	19.9	23.9	26.5	29.3	15.9	19.4	26.5	31.8	35.3	39.0
300	13.0	16.0	21.8	26.1	29.0	31.9	17.4	21.2	29.0	34.7	38.6	42.5
400	17.4	21.4	29.1	34.8	38.7	42.5	23.2	28.2	38.7	46.3	51.5	56.7

FIGURE 18–4

Nomogram for determining facing and turning time

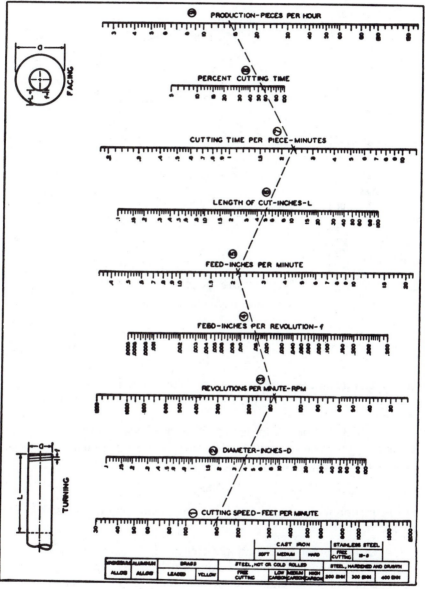

Crobalt, Inc.

FIGURE 18–5

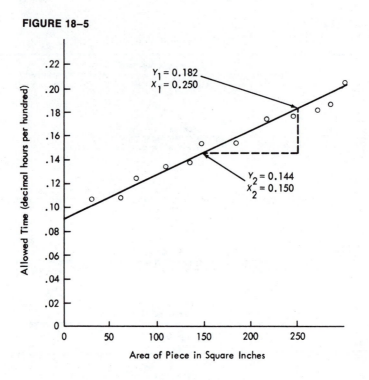

Area of Piece in Square Inches

18–5 illustrates a chart expressing "forming" time in hours per hundred pieces for a certain gage of stock over a range of sizes expressed in square inches.

Each of the 12 points in this chart represents a separate time study. Looking at the plotted points you can see a straight-line relationship between the various studies. The equation for a straight line is:

$$Y = mx + b$$

(The equation for a straight line is also frequently expressed as: $Y = a + bx$.)

Y = Ordinate (hours per hundred pieces)

x = Abscissa (area of piece in square inches)

m = Slope of the straight line, or the proportionate change of time on the Y-axis for each unit change on the X-axis

b = Intercept of the straight line with the Y-axis when $x = 0$

The curve drawn by inspection shows an intercept value of 0.088 on the Y-axis; the slope is calculated by the equation:

$$m = \frac{y_1 - y_2}{x_1 - x_2}, \text{ where } x_1 y_1 \text{ and } x_2 y_2 \text{ are specific points on the curve}$$

$$= \frac{0.182 - 0.144}{250 - 150} = 0.00038$$

The Least Squares Method

It is also possible to solve for m and b using the method of least squares. In this technique, the resulting slope and y-axis intercept will give a straight line where the sum of the squares of the vertical deviations of observations from this line is smaller than the corresponding sum of the squares of deviations from any other line. The two equations that are solved simultaneously are:[1]

$$\Sigma y = Nb + m\Sigma x \tag{1}$$
$$\Sigma xy = b\Sigma x + m\Sigma x^2 \tag{2}$$

The derivations of these equations are available in most mathematical texts.

In the example cited, solve for m and b using the least squares method as shown in Table 18–6. Substituting in Equation 1 and 2:

$$12b = 1.834 \ - 2{,}073m \tag{1}$$
$$2{,}073b = 348.82 - 453{,}801m \tag{2}$$

Multiplying Equation 1 by 2,073 and Equation 2 by 12:

TABLE 18–6

Study	x or area	y or time	xy	x^2
1........	25	0.104	2.60	625
2........	65	0.109	7.09	4,225
3........	77	0.126	9.70	5,929
4........	112	0.134	15.01	12,544
5........	135	0.138	18.63	18,225
6........	147	0.150	22.05	21,609
7........	185	0.153	28.31	34,225
8........	220	0.174	38.28	48,400
9........	245	0.176	43.12	60,025
10........	275	0.182	50.05	75,625
11........	287	0.186	53.38	82,369
12........	300	0.202	60.60	90,000
	2,073	1.834	348.82	453,801

[1] These equations are frequently expressed: $\Sigma y = na + b\Sigma x$ and $\Sigma xy = a\Sigma x + b\Sigma x^2$.

$$24{,}876b = 3{,}801.882 - 4{,}297{,}329m$$
$$\underline{24{,}876b = 4{,}185.840 - 5{,}445{,}612m}$$
$$0 = -383.958 + 1{,}148{,}283m$$

or

$$m = \frac{383.958}{1{,}148{,}283} = 0.000334$$

and substituting in Equation 1:

$$12b = 1.834 - 0.692$$
$$b = \frac{1.142}{12} = 0.095$$

Solving by Regression Line Equations

As shown, the least squares method calls for solving simultaneous equations. These equations can get unwieldy. To solve these equations for the slope m and the y intercept b, use direct substitution. As in the least squares solutions use identical totals, which include:

$$\Sigma x, \ \Sigma x^2, \ \Sigma y, \ \Sigma xy$$

and N, the number of data. The regression line equation to solve for the constant b is:

$$b = \frac{(\Sigma x^2)(\Sigma y) - (\Sigma x)(\Sigma xy)}{(N)(\Sigma x^2) - (\Sigma x)^2}$$

and the slope m is computed as follows:

$$m = \frac{(N)(\Sigma xy) - (\Sigma x)(\Sigma y)}{(N)(\Sigma x^2) - (\Sigma x)^2}$$

Solving for b and m in the foregoing example:

$$b = \frac{(453{,}801)(1.834) - (2{,}073)(348.82)}{(12)(453{,}801) - (2{,}073)^2}$$
$$= \frac{109{,}168}{1{,}148{,}283} = 0.095$$

and

$$m = \frac{(12)(348.82) - (2{,}073)(1.834)}{(12)(453{,}801) - (2{,}073)^2}$$
$$= \frac{383.96}{1{,}148{,}283} = 0.000334.$$

FIGURE 18-6

CLASSIFICATION

KIND___Fillet_____
TYPE___Flat Position_____
ELECTRODE___E-6020 D.H._____

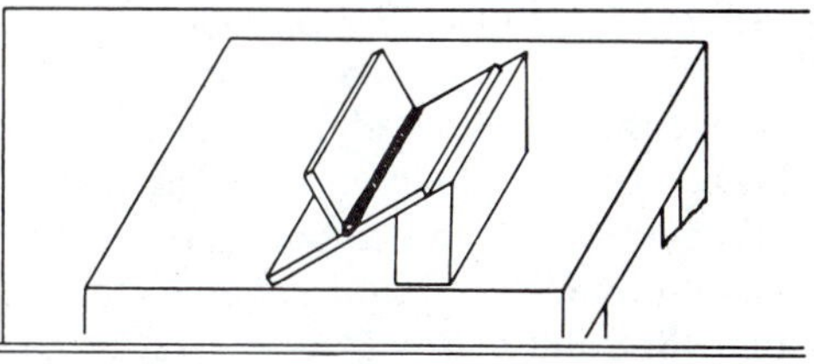

WELDING PROCEDURE GENERAL
PROCESS OF WELDING___Shielded Metallic Arc_____ POSITION ___Flat_____
MATERIAL___Mild Steel to Mild Steel_____ P.D.S.__1550, 1555___S.A.E.___1010___
ELECTRODE___E-6020_____ __D.H.__ ___Convex___ ___Heavy___
　　　　　A.W.S. CLASS　　TYPE　SHAPE OF WELD　　COATING
POWER SOURCE (A.C. OR D.C. – AND POLARITY IF D.C.)_____D.C. Straight_____
BACKING___None_____PEENING___None_____CHIPPING___None_____
PREHEAT_____None_____
STRESS RELIEVING___None_____

WELDING PROCEDURE–DETAILS

SIZE OF WELD	SIZE OF ELECTRODE	THICKNESS OF PLATE	NUMBER OF PASSES	WELDING CURRENT (AMPERES)	WELDING VOLTAGE (@ ARC)	*MAN HOURS PER INCH WELD	*WELDING SPEED FT./HR.
1/8	1/8	1/8	1	160-190	26-28	.0025	33.3
3/16	5/32	3/16	1	160-190	26-28	.0028	29.8
1/4	3/16	1/4	1	180-230	32-36	.0033	25.3
3/8	1/4	3/4	1	280-330	32-36	.0050	16.7
1/2	1/4	3/4	2	280-330	32-36	.0078	10.7
5/8	1/4	1"	2	280-330	32-36	.0123	6.8
3/4	1/4	1 1/2	4	280-330	32-36	.0196	4.3
1	1/4	1 1/2	6	280-330	32-36	.0318	2.6

*NOTE: INCLUDES CHANGE ELECTRODE TIME, ARC TIME, WELD CLEANING TIME AND WELDING TIME.

USING STANDARD DATA

For ease of reference, tabularize constant standard data elements and file them under the machine or the process. Variable data can be tabularized or may be expressed in term of a curve or an equation, and are also filed under the facility or operation class.

In some instances, it may be desirable to combine constants with variables and tabularize the summary, where standard data are broken down to cover a given machine and class of operation. This quick reference data expresses the allowed time to perform a given operation completely. Figure 18–6 illustrates welding data in which the constants "change electrode" and "arc" are combined with the variables "weld cleaning" and "welding." The result is the work-hours required to weld 1 inch for various sizes of welds.

Table 18–7 illustrates standard data for a given facility and operation class where elements have been combined. By identifying the job as to distance that the strip of sheet stock is moved per piece, you can find the allowed time for the complete operation.

Frequently setup elements are combined or tabularized in combinations to diminish the time required for summarizing a series of elements. Table 18–8 illustrates standard setup data for No. 5 Warner & Swasey turret lathes applicable to a specific plant. To determine the setup time with this data, visualize the tooling in the square and hex turret and refer to the table. For example, if a certain job required a chamfering tool, turning took, and facing tool in the square turret, and needed two boring tools, one reamer, and a collapsible tap in the hex turret, the setup standard time would be 69.70 minutes plus 25.89 minutes, or 95.59 minutes. Find the value of the relevant tooling under the "square turret" column (line 8) and the most time-consuming applicable tooling in the "hex turret" section, in this case, tapping. This gives a value of 69.7 minutes. Since three additional tools are in the hex turret (1st bore, 2nd bore, and ream), multiply 8.63 by 3 and get 25.89 minutes. Adding the 25.89 minutes to the 69.70 minutes gives the required setup time.

More frequently, standard data are not combined but are left in their elemental form, thus giving greater flexibility in the development of time standards. Representative standard data applicable to a given plant would appear as shown in Table 18–9.

The data include the applicable personal delay and fatigue allowance. This standard time data is used to establish allowed setup and piece times to drill the two hold-down bolt holes in the cast-iron housing casting illustrated in Figure 18–7.

The allowed setup time would equal

$$B + C + D + F + G + H + I = 8.70 \text{ minutes}$$

Element A is not included in the setup time because of the simplicity of the job. The drawing is not issued to the operator because the operation card shows the drill jig number, drill size, plug gage number, and feed and spindle speed. Element E is not included in the setup since in this case first-piece inspection is not performed by the inspector. According to the operation card issued, the operator periodically inspects every 10th piece with the go no-go gage provided. Since the casting would weigh less than four pounds, and two parts can easily be handled in one hand, the data outlined would be applicable.

After investigating the design of the drill jig to be used and making an analytic motion study of the job to determine the elements required to perform the operation, the analyst makes an elemental summary. This appears in Table 18–10.

TABLE 18–7
Standard data for blanking and piercing strip stock hand feed with piece automatically removed on Toledo 76 punch press

L (distance in inches)	T (time in hours per hundred hits)
1	0.075
2	0.082
3	0.088
4	0.095
5	0.103
6	0.110
7	0.117
8	0.123
9	0.130
10	0.137

TABLE 18–8
Standard setup data for No. 5 turret lathes

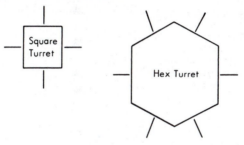

Basic tooling

No.	Square turret	Partial	Hex turret Cham-fer	Bore or turn	Drill	S. tap or ream	C. tap	C. die
1.	Partial	31.5	39.6	44.5	48.0	47.6	50.5	58.5
2.	Chamfer	38.2	39.6	46.8	49.5	50.5	53.0	61.2
3.	Face or cut off	36.0	44.2	48.6	51.3	52.2	55.0	63.0
4.	Tn bo grv rad	40.5	49.5	50.5	53.0	54.0	55.8	63.9
5.	Face and chf.	37.8	45.9	51.3	54.0	54.5	56.6	64.8
6.	Fa and cut off.	39.6	48.6	53.0	55.0	56.0	58.5	66.6
7.	Fa and tn or tn and cut off	45.0	53.1	55.0	56.7	57.6	60.5	68.4
8.	Fa, tn, and chf	47.7	55.7	57.6	59.5	60.5	69.7	78.4
9.	Fa, tn, and cut off	48.6	57.6	57.5	60.0	62.2	71.5	80.1
10.	Fa, tn, and grv.	49.5	58.0	59.5	61.5	64.0	73.5	81.6

11. Circled basic tooling from above . . . . _____
12. Each additional tool in square. . . . . 4.20x_____ = _____
13. Each additional tool in hex . . . . 8.63x_____ = _____
14. Remove and set-up three jaws . . . . 5.9 _____
15. Set up subassembly or fixture . . . . 18.7 _____
16. Set up between centers. . . . . 11.0 _____
17. Change lead screw . . . . 6.6 _____

Total setup _____ min.

TABLE 18–9
Standard data

Application: Allen 17-inch vertical single-spindle drill.

Work size: Small work—up to four pounds in weight and such that two or more parts can be handled in each hand.

Setup elements: *Minutes*

  A. Study drawing 1.25
  B. Get material and tools and return and place ready for work 3.75
  C. Adjust height of table 1.31
  D. Start and stop machine 0.09
  E. First-piece inspection (includes normal wait time for inspector) 5.25
  F. Tally production and post on voucher 1.50
  G. Clean off table and jig 1.75
  H. Insert drill in spindle 0.16
  I. Remove drill from spindle 0.14

Each piece elements:

  1. Grind drill (prorate) 0.78
  2. Insert drill in spindle 0.16
  3. Insert drill in spindle (quick-change chuck) 0.05
  4. Set spindle 0.42
  5. Change spindle speed 0.72
  6. Remove tool from spindle 0.14
  7. Remove tool from spindle (quick-change chuck) 0.035
  8. Pick up part and place in jig
     a. Quick-acting clamp 0.070
     b. Thumbscrew 0.080
  9. Remove part from jig
     a. Quick-acting clamp 0.050
     b. Thumbscrew 0.060
 10. Position part and advance drill 0.042
 11. Advance drill 0.035
 12. Clear drill 0.023
 13. Clear drill, reposition part, and advance drill (same spindle) 0.048
 14. Clear drill, reposition part, and advance drill (adjacent spindle) 0.090
 15. Insert drill bushing 0.046
 16. Remove drill bushing 0.035
 17. Lay part aside 0.022
 18. Blow out jig and part and lay part aside 0.081
 19. Plug gage part . . . . 0.12 per
                                                                                       hole

---

To this 0.278-minute time must be added the actual drilling time to drill the two holes. This can readily be determined as previously outlined. For a ½-inch diameter drill used for drilling cast iron, use a surface speed of 100 feet per minute and a feed of 0.008 inch per revolution. One hundred surface feet per minute would equal 764 revolutions per minute.

$$\text{rpm} = \frac{12S_f}{\pi d}$$

FIGURE 18–7
Cast-iron housing

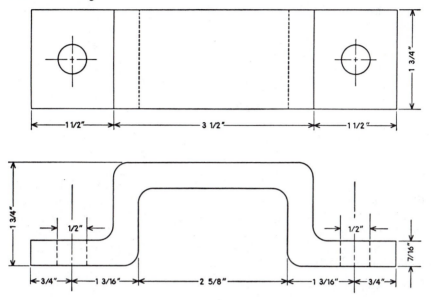

where:

S_f = Surface feet per minute
π = 3.14
d = Diameter of drill in inches

However, investigation of the drill press for which this work has been routed reveals that 600 rpm or 900 rpm are the closest speeds that are

TABLE 18–10

Element number	Element	Allowed time (minutes)
8	Pk. pt. pl. jig (quick-acting clamp)	0.070
9	Rem. pt. jig (quick-acting clamp)	0.050
10	Pos. pt. & adv. dr.	0.042
13	Cl. dr. rep. pt. & adv. dr.	0.048
12	Cl. dr.	0.023
17	Lay pt. aside	0.022
19	Plug gage pt. (10%)	0.012
1	Gr. dr. (once per 100 pcs.)	0.008
2	Insert dr. in sp. (once per 100 pcs.)	0.002
6	Rem. tool from sp. (once per 100 pcs.)	0.001
		0.278 min

available to the recommended 764 rpm. The analyst proposes to use the slower speed due to the condition of the machine, and determines the drilling time as follows:

$$T = \frac{L}{F_m} \times 2$$

$$L = 0.437 + \frac{0.25}{\tan 59°} = 0.588 \text{ inches}$$

$$F_m = (600)(0.008)$$
$$= 0.48 \text{ inches per minute}$$

$$T = \left(\frac{0.588}{4.8}\right) 2 = 0.244 \text{ minute}$$

An appropriate allowance must be added to this 0.244 minute cutting time. If 10 percent allowance on the actual machining time is used, the cutting time is 0.268 minute and handling time, 0.278 minute, for a total time of 0.546 minute to drill one casting on the single-spindle press available. This standard, supplemented with the setup time of 8.70 minutes, would be released as:

Setup: 0.145 hours
Each piece: 0.91 hours per hundred

Thus, an operator who set up this job and ran 1,000 pieces in an eight-hour working day would be performing at an efficiency of 116 percent:

$$\frac{(0.91)(10) + 0.145}{8} = 116 \text{ percent}$$

Computerized Standard Data

Today computers store standard data and retrieve, accumulate, and develop standards for new work in advance of production. Standard data, whether stored in the motion, elemental, or task form, is easily retrieved, accumulated, and adjusted for applicable allowances using a computer. Several software systems have a fundamental motion database. For example MOST, WOCOM, 4M, and Univation use motion data as a base for standards development. Other systems such as CSD do not have a built-in database, but allow any company to introduce its own data.

Store data in both the motion and element forms for maximum efficiency and for the development of the best standards. Software can select, retrieve, and modify the appropriate motions and/or elements for generating a work standard.

For the processing elements such as metal cutting, or metal joining, logic modules can serve as simple process time calculators, or they can optimize the process parameters with various degrees of sophistication. For example, using it as a time calculator, analysts enter the speed, feed, number of cuts, and length of cuts into a computer that performs the necessary arithmetic operations to calculate the time for the cut. In the advanced machining logic modules, the input includes the machine scheduled to perform the operation, work piece material, and the characteristics of the cut. The computer selects the optimum speed, feed, and number of cuts consistent with the horsepower available and condition of the machine. The system may also select the tooling and the proper sequence of cuts.

Assembly logic modules may be used with standard data stored in the computer. In such cases the system can analyze the several parameters relating to an assembly operation and can determine which work elements should be applied. The applicable work elements are summed, allowances applied and the final job standard and operator instructions are created. The work measurement analyst interacts with the computer by responding to its queries. The computer then selects the appropriate standard data based on the input responses of the analyst.

CONCLUSION

When properly applied, standard data permit the establishment of accurate time standards before the job is performed. This feature makes their use especially attractive when estimating the cost of new work for quotation and subcontracting purposes.

Time standards can be calculated much more rapidly by using standard data, and the consistency of the standards established can be assured. Thus, this technique allows the economical development of indirect labor standards. For example, the standard data elements in Table 18–11 allow the rapid development of standards related to the maintenance and repair or replacement of railroad car wheels. Standards developed from standard data tend to be completely fair to both the worker and management in that they are the result of already proven standards. The elemental values used in arriving at the standards have proved satisfactory as components of established and acceptable standards in use throughout the plant.

The use of standard data simplifies many managerial and administrative problems in plants having unions that operate as bargaining agents. Union contracts contain many clauses pertaining to such matters as the type of study to be taken (continuous or snapback), the number of cycles to be studied, who shall be studied, and who shall observe the study. These restrictions frequently make it difficult for analysts to arrive at a standard that is equitable to both the company and the operator. By using the standard data technique, analysts may avoid restrictive details. Thus, not only is the determination of a standard simplified, but also the sources of tension between labor and management are alleviated.

TABLE 18–11
Standard data for railroad car wheels maintenance

Activity code	Valid preceding codes	Activity	TTL direct labor (minutes)	Valid following codes
6000.00	0	Move car into shop and position, ft	7.00	6001.00
6001.00	6000.00	Place chocks under wheel	0.58	6002.00
6002.00	6001.00	Disconnect air brake rigging (top rod)	2.69	6003.00
6003.00	6002.00	Raise floor jacks to contact body	3.54	6004.00
6004.00	6003.00	Raise car to clear center pin, inches	1.16	6005.00
6005.00	6004.00	Roll truck out to clear center plate	0.74	6006.00
6006.00	6005.00	Place safety stands	0.85	6007.00
6007.00	6006.00	Remove center pin	0.67	6008.00
6008.00	6007.00	Roll truck to electric hoist	0.45	6009.00
6009.00	6008.00	Place chocks under wheel	0.38	6010.00
6010.00	6001.00–6009.00	Burn off pedestal key bolt	0.87	6011.00
6011.00	6010.00	Lift truck side frame off wheel set	1.21	6012.00
6012.00	6011.00	Get wheel and axle readings	1.06	6013.00
6013.00	6012.00–6014.00 6015.00–6016.00	Write on axle	0.65	6014.00–6015.00
6014.00	6012.00–6013.00	Write on wheel	0.43	6013.00–6015.00 6016.00
6015.00	6011.00–6012.00 6013.00–6014.00	Gage adapters	2.02	6011.00–6014.00
6016.00	6011.00–6012.00 6013.00–6014.00 6015.00	Roll out bad wheel	0.40	6017.00
6017.00	6013.00–6014.00 6015.00–6016.00	Fork lift bad wheel to stub track	9.59	6018.00
6018.00	6017.00	Run to new wheel track, ft	2.08	6019.00
6019.00	6018.00	Fork lift new wheel to electric hoist	10.01	6020.00
6020.00	6019.00–6021.00	Roll wheel under truck frame	0.35	6021.00–6022.00
6021.00	6020.00–6022.00	Place lube pad in center plate	0.76	6020.00–6022.00
6022.00	6020–6021.00	Place adapter on bearing	0.49	6022.00–6023.00
6023.00	6022.00	Lower truck side frame on wheel set	1.44	6024.00
6024.00	6023.00	Install pedestal key and bolt	0.55	6025.00
6025.00	6024.00	Remove chocks from wheel	0.38	6026.00

In general, the more refined the elemental times, the greater the coverage possible for the data. Consequently, in job shop practice, it is practicable to have individual elemental values as well as grouped or combined values, so that the data for a given facility has enough flexibility to allow the setting of rates for all types of work scheduled to the machine.

The use of fundamental motion data for establishing standards, in particular on short-cycle jobs, is becoming more widespread. This data is so basic that it permits the predetermination of standards on practically any class of manual elements. Of course, the "objective" or "do" basic divisions, such as "use," must be handled as variables, and tabularized data, curves, or algebraic expressions established.

As the examples indicate, the application of standard data is an exacting technique. Careful and thorough training in methods and shop practice are fundamental before analysts can accurately establish standards using standard data. Analysts must know and recognize the need for each element in the class of work. Supplementing this background, the men or women working with standard data should be analytic, accurate, thorough, conscientious, and completely dependable.

TEXT QUESTIONS

1. What do we mean by "standard data"?
2. What is the approximate ratio in the time required for setting standards by stopwatch methods and by standard data methods?
3. What advantages are there to establishing time standards by using standard data rather than by taking individual studies?
4. What would the time for the element "mill slot" depend on?
5. Of what use is the "fast" watch in the compilation of standard data?
6. Compute the times for elements a, b, c, d, and e when elements $a + b + c$ were timed at 0.057 minute and $b + c + d$ were found to equal 0.078 minute; $c + d + e$, 0.097 minute; $d + e + a$, 0.095 minute; and $e + a + b$, 0.069 minute.
7. What would be the lead of a ¾-inch diameter drill with an included angle of 118°?
8. What would be the feed in inches per minute of a ¾-inch drill running at a surface speed of 80 feet per minute and a feed of 0.008 inch per revolution?
9. How long would it take the preceding drill to drill through a casting 2¼ inches thick?
10. How are feeds usually expressed in lathe work?
11. How long would it take to turn 6 inches of 1-inch bar stock on a No. 3 W. & S. turret lathe running at 300 feet per minute and feeding at the rate of 0.005 inch per revolution?
12. A plain milling cutter 3 inches in diameter with a width of face of 2 inches is being used to mill a piece of cold-rolled steel 1½ inches wide and 4 inches long. The depth of cut is 3/16 inch. How long would it take to make the cut if the feed per tooth is 0.010 inch and a 16-tooth cutter running at a surface speed of 120 feet per minute is used?
13. What are some of the disadvantages of using curves to tabulate standard data?
14. Which standard procedures should be followed in the plotting of simple curves?
15. What would be the horsepower requirements of turning a mild steel shaft 3 inches in diameter if a cut of ¼ inch with a feed of 0.022 inch per revolution at a spindle speed of 250 rpm were established?

GENERAL QUESTIONS

1. What do you think is the attitude of labor toward standards established through the use of standard data?
2. Using the drill press standard data shown in the text, preprice a drilling job with which you are familiar. How does this standard compare with the present rate on the job? (Be sure that identical methods are compared.)
3. Does complete standard data supplant the stopwatch? Explain.
4. If you were president of your local union, would you advocate that standards be established by standard data? Why or why not?

PROBLEMS

1. The analyst in the Dorben Company made 10 independent time studies in the hand paint spraying section of the finishing department. The product line under study revealed a direct relation between spraying time and product surface area. The following data were collected:

Study no.	Leveling factor	Product surface area	Standard time
1	0.95	170	0.32
2	1.00	12	0.11
3	1.05	150	0.31
4	0.80	41	0.14
5	1.20	130	0.27
6	1.00	50	0.18
7	0.85	120	0.24
8	0.90	70	0.23
9	1.00	105	0.25
10	1.10	95	0.22

Compute the slope and intercept constant, using regression line equations. How much spray time would you allow for a new part with a surface area of 250 square inches?

2. The work measurement analyst in the Dorben Company wants to develop an accurate equation for estimating the cutting of various configurations in sheet metal with a band saw. The data from eight time studies for the actual cutting element provided the following information:

No.	Lineal inches	Standard time
1	10	0.40
2	42	0.80
3	13	0.54
4	35	0.71
5	20	0.55
6	32	0.66
7	22	0.60
8	27	0.61

What would be the relation between the length of cut and the standard time, using the least squares technique?

3. The work measurement analyst in the XYZ Company wishes to develop standard data involving fast, repetitive manual motions for use in a light assembly department. Because of the shortness of the desired standard data elements, the analyst is obliged to measure them in groups as they are performed on the factory floor. On a certain study, this analyst is endeavoring to develop standard data for five elements which will be denoted as *A, B, C, D,* and *E*. Using a fast (0.001) decimal minute watch, the analyst studied a variety of assembly operations and arrived at the following data:

$$A + B + C = 0.131 \text{ minutes}$$
$$B + C + D = 0.114 \text{ minutes}$$
$$C + D + E = 0.074 \text{ minutes}$$
$$D + E + A = 0.085 \text{ minutes}$$
$$E + A + B = 0.118 \text{ minutes}$$

Compute the standard data values for each of the elements *A, B, C, D,* and *E*.

4. The work measurement analyst in the Dorben Company is developing standard data for prepricing work in the drill press department. Based on the following recommended speeds and feeds, compute the power feed cutting time of ½-inch high-speed drills with an 118° included angle to drill through material that is 1-inch thick. Include a 10 percent allowance for P.D. & F.

Material	Recommended speed (ft./min.)	Feed (in./rev.)
Al (copper alloy)	300	0.006
Cast iron	125	0.005
Monel (R)	50	0.004
Steel (1112)	150	0.005

SELECTED REFERENCES

Brisley, C. L., and R. J. Dossett. "Computer Use and Non-Direct Labor Measurement Will Transform Profession in the Next Decade." *Industrial Engineering*. August 1980, pp. 34–43.

Cywar, Adam W. "Development and Use of Standard Data." In *Handbook of Industrial Engineering,* ed. Gavriel Salvendy. New York: John Wiley & Sons, 1982.

Fein, Mitchell. "Establishing Time Standards by Parameters." *Proceedings of the Spring Conference of the American Institute of Industrial Engineers,* 1978.

Metcut Research Associates. *Machining Data Handbook.* Cincinnati: Metcut Research Associates, 1966.

Ostwald, Philip F. *Cost Estimating for Engineering and Management.* Englewood Cliffs, N.J.: Prentice-Hall, 1974.

Pappas, Frank G., and Robert A. Dimberg. *Practical Work Standards.* New York: McGraw-Hill, 1962.

Rotroff, Virgil H. *Work Measurement.* New York: Reinhold Publishing, 1959.

Basic Motion Times

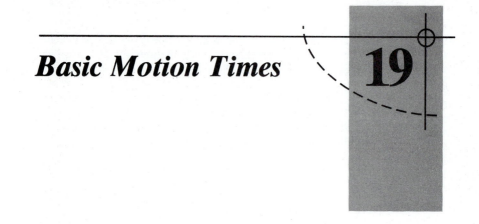

Since the time of Frederick W. Taylor, management has realized the desirability of assigning standard times to the various basic divisions of accomplishment. It has never been a question of need, but a question of how time values for various basic divisions can be assigned practically. Recent years have seen considerable progress in the assignment of reliable time values to basic elements of work. These time values are referred to as basic motion times. They are also known as synthetic and predetermined times.

Basic motion times are a collection of valid standard times assigned to fundamental motions and groups of motions that cannot be precisely evaluated with ordinary stopwatch time study procedures. They are the result of studying a large sample of diversified operations with a timing device, such as a motion-picture camera or videotape machine capable of measuring very short elements. The time values are synthetic in that they are often the result of logical combinations of therbligs. For example, analysts have established a series of time values for different categories of grasp. The therbligs search, select, and grasp can be part of the grasp time. The time values are basic in that further refinement is not only difficult but impractical. Thus, basic motion times are frequently referred to as *synthetic basic motion times*.

Since 1945 there has been a growing interest in the use of basic motion times as a modern method of establishing rates quickly and accurately without using the stopwatch or other time recording devices. In view of this, you may want to reread Chapter 7 after completing Chapter 18 on standard data. One by-product of synthesized time standards has probably been as useful as, if not more useful than, the time standards themselves. That is the development of methods consciousness to a refined degree in all parties associated with the establishment of standards using synthetic values. For this reason, the entire subject of synthetic basic motion times should be meshed or integrated with the motion and micro-motion techniques, even though it is also closely allied to the work measurement phase.

Even with the laws of motion economy and the concept of basic division of accomplishment clearly established, the methods analyst without reliable time values for the basic divisions has only a portion of the facts necessary to engineer a method prior to beginning actual production. Without time values for the basic divisions, how can analysts be sure that the proposed method is the best? With reliable motion times, analysts would be able to evaluate their proposed methods for the average or normal worker, who would eventually be the recipient of these ideas. The method used determines the time required to do a task. Whenever time values for all forms of activity are available, the most favorable methods can be determined in advance.

Today, practicing methods analysts can obtain information from approximately 50 different systems of established synthetic values. Essentially, these predetermined time systems are sets of motion–time tables with explanatory rules and instructions on the use of the motion–time values. Considerable specialized training is essential prior to the practical application of any of the techniques. In fact, most companies require certification before analysts are permitted to establish standards using the Work-Factor, MTM, MOST, or MODAPTS systems.

If a trained analyst establishes a standard on a given method using two of the existing basic motion time systems, he or she will probably arrive at two different answers. The reason for this is that different concepts of normal performance may have been utilized in the original development of the standard data. For example, in the early 1940s, the Westinghouse Corporation sponsored a series of studies involving sensitive drill presses. These studies were used by Messrs. Maynard, Stegemerten, and Schwab of Methods Engineering Council to develop MTM (Methods Time Measurement). Now anyone familiar with metal trade operations recognizes that normal performance here is not at all difficult to achieve if the individual performing the work has been adequately trained and possesses average skill and gives average effort. Here performances of 125 percent or more can be attained readily if the operator is willing to demonstrate high effort, has good skill qualifications, has been trained thoroughly, and is well advanced on the learning curve.

On the other hand, during the Great Depression (early 1930s) when practically everyone felt it was not only desirable but absolutely necessary to work very hard to hold a job, concepts of normal performance were tighter than what was considered normal 15 years later. This was especially true in industries such as the garment industry and those involving light assembly work such as fabrication of radios, washing machines, refrigerators, and similar white ware.

It was at this time, the Philco Corporation sponsored a series of studies in selected factory operations as a direct result of union reactions against stopwatch time studies. Philco engineers (Quick, Shea, and Koehler) developed these studies into a system originally called "Motion-Time Standards". Later, additional research was done by these engineers and others working for the Radio Corporation of America where the system became known as "Work-Factor."

Now one can appreciate that normal performance based upon the Work-Factor system allowed less time than that allowed by the MTM system for the same operation and using identical methods. Clifford N. Sellie, chief executive officer of Standards International Inc., identifies two different concepts of normal as (1) "normal" or "daywork" or "low task," (2) "incentive pace" or "required time" or "high task." He also indicates the difference in times of the two systems as being approximately 25 percent.

About 1940, A. E. Shaw of the Erie, Pennsylvania, works of the General Electric Company used Work-Factor in connection with the assembly of refrigerators which were at that time being produced in Erie. This application led to the development of the Engstrom tables, which allowed for extra controls as required. Engstrom's work was based primarily on studies from punch-press and stamping operations and in values for "get" and "place." Shortly thereafter, H. S. Geppinger at the General Electric Company developed "Dimensional-Motion-Times." These basic values are called MTS or DMT and have been used for many years at General Electric and other companies.

Other contributions to the development of predetermined time systems include the following:

MTA. A. B. Segur, based upon research begun in 1919 and assistance from the philosophies of Frank and Lillian Gilbreth, developed values known as "Motion-Time-Analysis" or MTA. Segur's work is known for the fact that consideration is given to both acceleration and decelleration since body motions are slower as a stop is approached and are faster after a start is initiated.

M-H and WTM. Minneapolis-Honeywell Regulator Company and General Motors Corporation developed similar systems based in part on early work in the area. The Minneapolis-Honeywell Regulator Company's work was known as the "M-H" system and the General Motor's system was called "Work Time Measurement."

ETS. The earliest industrial pioneer of basic time systems that mainly was based upon the work of Segur was Western Electric Company's "Elemental Time Standards," developed in the early 1920s. It has been widely used in various Western Electric plants.

BMT and SMS. Basic Motion Times (BMT) was developed by J. D. Woods and Gordon and Simplified Motion System (SMS) by Walter Dill. Both systems were designed to simplify the adding for stops and starts caused by motion difficulties.

BMT was based largely upon the RCA (Work-Factor) and the Westinghouse (MTM) systems while the SMS system was influenced mainly by the RCA and Minneapolis Honeywell (M-H) systems.

Clifford Sellie has classified all predetermined time systems as falling into one of three groups. These are:

1. *Acceleration-Deceleration Systems.* These systems recognize that body motions move at different velocities during their movement. Values determined using this approach suggest that 40 percent of the total time is used during the acceleration period, 20 percent for constant velocity, and 40 percent for deceleration.

2. *Average-Motion Systems.* Here recognition is given to representative or average motion difficulties that are usually encountered in industrial operations.

3. *Additive Systems.* Under these systems, basic time values are used. To these basic values, additions for motion difficulties encountered are assigned. These additives range from 10 to 50 percent.

To give the reader a good background in the field of basic motion times, we will review in some detail Work-Factor, which according to the above classification would be an additive system, and MTM, which would be a pioneer in the average-motion system classification. Today, acceleration-deceleration systems are not widely used for standards setting. However, they are excellent tools for detailed methods analysis. Systems characteristic of this classification will not be discussed.

In order to give the reader an appreciation and understanding of three of the popular systems in use today, MOST (Maynard Operation Sequence Technique), MODAPTS (Modular Arrangement of Pts), and MicroMotion and MacroMotion Analyses systems will be discussed briefly. The reader should understand that MOST is derived from MTM so would be representative of average-motion systems. Also, this is true in connection with MODAPTS. On the other hand, MicroMotion Analyses is based primarily on Work-Factor while MacroMotion analyses drives input from both MicroMotion Analyses and MTM.

Figure 19–1 illustrates the derivation of all of these basic motion time systems.

WORK-FACTOR®¹

SMC Wofac², developer of the Work-Factor system, is one of the pioneer concerns in establishing standards synthetically with motion-time values. Work-Factor data became available in 1938, after four years of

---

¹ Work-Factor is a registered service mark (trademark) of Science Management Corporation, identifying its services as consultants to industry and its system of predetermined fundamental motion-time standards for the motion times themselves and the techniques used to apply them to methods determination and work measurement.

² Originally known as the Work-Factor Company, Inc.

FIGURE 19–1
Family tree of predetermined times

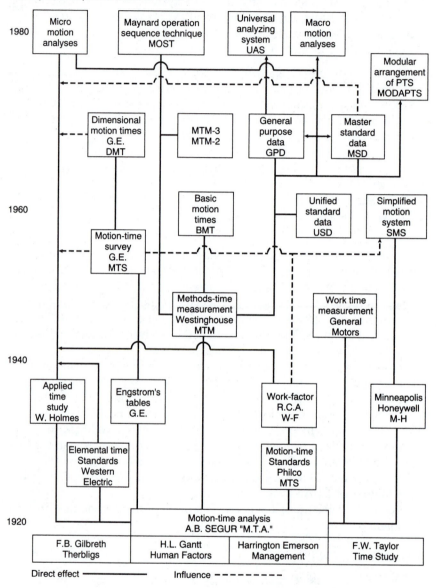

Courtesy of Standards International, Chicago, Illinois

gathering values by the micromotion technique, stopwatch procedures, and the use of a specially constructed photoelectric time machine.

The Work-Factor system has achieved flexibility by developing three different procedures of application, depending on the objectives of the analysis and the accuracy required. These procedures are the Detailed®,

Ready®, and Brief® techniques. Each technique is self-sufficient, requiring no dependence on a higher or lower level system. However, the systems are completely compatible and can be mixed. In addition, a fourth technique, Mento-Factor®, provides precise standards for mental activity.

Detailed Work-Factor contains precise time standards for either day-work measurement or incentive pay plans. Providing a precise tool for methods analyses, it is primarily used for short-cycle operations and repetitive work. Also, it is commonly used for developing standard data.

Detailed Work-Factor has eight elemental descriptions. Its Motion Time Table has 764 time values, and it is the most detailed of all modern predetermined motion-time systems.

Ready Work-Factor is suited to operations that do not require as precise an analysis as Detailed Work-Factor. Generally this is used for medium quantity production. Analysts can obtain ready time standards in about one third the time required for Detailed; the loss in accuracy normally does not exceed +5 percent. Ready Work-Factor is also useful for training supervisors and employees in work simplification and work-time concepts because many of its times and rules are readily memorized. The Ready Work-Factor System has nine elemental descriptions and the Motion Time Table has 154 time values.

Brief Work-Factor offers the simplest Motion Time Table, combining the various standard elements into work segments. It is applied to work situations requiring much less detailed measurement, such as short-run production, the manual portion of operations that are principally machine time, and nonrepetitive operations with long cycle times occurring in job shop maintenance, clerical, and many other indirect labor functions. Brief Work-Factor analyses take about one tenth the time required for a Detailed analysis, and vary from Detailed by ±10 percent. Often operation times are established as rapidly as operations are performed, and times are based on observations of only one or two cycles. Brief Work-Factor has five elemental descriptions, and its Motion Time Table has only 32 time values. A subset of Brief Work-Factor, called Abridged Brief, has only five time values, yet has an accuracy similar to regular Brief.

All of the Work-Factor systems contain sufficiently accurate time values for the small amount of mental work associated with most productive work. However, when mental work represents much of the task, the Detailed Mento-Factor System may be used. The Mento-Factor System measures mental activity; Detailed or Ready Work-Factor measures the manual portion of the operation.

Detailed Mento-Factor provides elemental times for all identifiable mental processes required in useful work. It can be used when precise measurement is required for human mental functions occurring in operations in inspection (audio, visual, kinesthetic), reading, proofreading, calculating, using a computer, color-matching, and similar operations. Its time tables cover 14 fundamental mental processes and have 710 time values.

Detailed Work-Factor System

Work-Factor recognizes the following variables influence the time required to perform a task:

1. The body member making the motion, such as arm, forearm, finger-hand, foot.
2. The distance moved (measured in inches).[3]
3. The weight carried (measured in pounds, converted into Work Factors).
4. The manual control required (care, directional control or steering to a target, changing direction, stopping at a definite location; measured in Work Factors).

Through analysis of films, Work-Factor's developers concluded, as the Gilbreths did many years earlier, that finger motions can be performed more rapidly than arm motions, and that arm motions are made in less time than body motions. Work-Factor motion times have been compiled for these body members:

1. *Finger-hand.* This includes the movements of the fingers and thumb and movement of the hand about the wrist joint.
2. *Arm.* This includes the movements of the lower arm around the elbow when used as hinge joint and all movements of the whole arm hinged at the shoulder except for swivels. Movement of the hand, fingers, and lower arm may occur simultaneously.
3. *Forearm swivel.* Here the lower arm rotates about the axis of the forearm, such as when turning a screwdriver, or the full arm rotates about the entire axis and the swivel takes place at the shoulders.
4. *Trunk.* This includes a forward, backward, sidewise, or swiveling motion of the trunk of the body.
5. *Foot.* This includes motions of the foot hinged at the ankle while the upper and lower leg remain in a fixed location.
6. *Leg.* This includes movements of the leg hinged at the waist, motions in which the legs move the torso as in bending, and movements of the knee to the side.

All proponents of fundamental motion data techniques have recognized that distance affects all motions. The longer the distance, the more time is required. Work-Factor has tabularized values for finger and hand movements from 1 to 4 inches, and for arm movements from 1 to 40

---

[3] All Work-Factor Systems have been metricated. The metric version measures distance in centimeters and weight in kilograms.

inches. Work-Factor measures distance as a straight line between the starting and stopping points of the motion arc. The actual motion path is measured only when a change in direction is involved.

Following is a list of the points at which distance should be measured for the various body members:

Body member	Point of measurement
Finger or hand............	Fingertip
Arm	Knuckles (knuckle having greatest travel should be used)
Forearm swivel	Knuckles
Trunk...................	Shoulder
Foot....................	Toe
Leg.....................	Ankle, knee, or hip
Head turn	Nose

Weight or resistance influence time according to the body member being used and the sex of the operator. It is measured in pounds for all body members except for "forearm swivel" motions, where pound-inches of torque represent the unit of measurement.

Manual control is the most difficult variable to quantify. However, the Work-Factor system concluded that in most instances, work motions involve one or more of the following four types of control:

1. *Definite stop work-factor.* Here some manual control is required to stop a motion within a fixed interval. Definite stop does not exist when the motion is terminated by some physical barrier or made to an indefinite location. The motion must be terminated by the muscular coordination of the operator.

2. *Directional control or steer work-factor.* Here manual control is necessary to bring or guide a part to a specific location or to make a motion through an area with limited clearance.

3. *Care or precaution work-factor.* Here manual control prevents spilling or damage, such as moving a full vessel of acid or handling a thin piece of glass. Manual control also maintains directional control of a motion, as in drawing a straight line free hand.

4. *Change of direction work-factor.* Here manual control is exercised when the motion involves a direction change to get around some obstacle. For example, moving a nut in back of a panel would require a change in direction once the moving hand reached the front of the panel.

A Work-Factor has been defined as the index of the additional time required over and above the basic time. It is a unit for identifying the effect of the variables of manual control and weight. The other two variables that affect the time to perform manual motions—body member used and distance—do not employ Work-Factors as a measure of magnitude. Here member used and inches represent the quantitative yardstick. The simplest or basic motions of a body member involve no Work-Factors. As complexities are introduced into a manual motion through the addition of weight or control, Work-Factors are added. And, of course, each Work-Factor that is added represents an additional increment of time.

Table 19–1 illustrates the Work-Factor Transport table. There are separate tables for body members; each table contains values for distances. In the forearm swivel and head turn, degrees indicate motion distance while all other motions are shown in inches. At the bottom of each table are weight or resistance limits for men and women. The forearm swivel motion resistance section is given in pounds-inches of torque, while all others are listed in pounds. The table gives the maximum resistance encountered that can be considered basic and also the top values for a specific number of Work-Factors.

The Work-Factor system divides all tasks into eight "Standard Elements of Work." These are:

1. TRANSPORT.   The element transport connects the other standard elements. It is broken into two classifications:

> *Reach:* When a body member is relocated to go to a destination, location, or object.
>
> *Move:* When a body member is used to relocate an object.

2. GRASP.   The element grasp is the act of obtaining manual control of an object; it begins after the hand has moved directly to the object and ends when manual control has been obtained and a move can occur. The Work-Factor system recognizes four types of grasps:

a. *Simple grasp:* Used for isolated easy-to-grasp objects and requires only one single motion.

b. *Manipulative grasp:* Includes all grasps of isolated or orderly stacked objects that require more than one motion of the fingers to gain control. There may be arm motions, several finger motions, or combinations of both.

c. *Complex grasp:* Defined as the grasp of an object from a random (jumbled) pile. The system provides a complete table for complex grasps. Complex grasps involve more than one motion and sometimes include arm movements.

d. *Special grasps:* Includes transferring an object from one hand to the other and grasps of more than one part.

Objects to be grasped are classified as:

a. *Cylinders and regular cross-sectioned solids:* Defined as all objects having cylindrical or regular (all sides and angles equal) cross sections such as square, hexagons, and so on.

TABLE 19–1
Work-factor motion time table for detailed analysis

TRANSPORT

ARM (A): Measured at Knuckles

MOTION DISTANCE in Inches	BASIC	1	2	3	4
1	18	26	34	40	46
2	20	29	37	44	50
3	22	32	41	50	57
4	26	38	48	58	66
5	29	43	55	65	75
6	32	47	60	72	83
7	35	51	65	78	90
8	38	54	70	84	96
9	40	58	74	89	102
10	42	61	78	93	107
11	44	63	81	98	112
12	46	65	85	102	117
13	47	67	88	105	121
14	49	69	90	109	125
15	51	71	92	113	129
16	52	73	94	115	133
17	54	75	96	118	137
18	55	76	98	120	140
19	56	78	100	122	142
20	58	80	102	124	144
22	61	83	106	128	148
24	63	86	109	131	152
26	66	90	113	135	156
28	68	93	116	139	159
30	70	96	119	142	163
35	76	103	128	151	171
40	81	109	135	159	179

Weight, lb.
	BASIC	1	2	3	4
Male	-2	7	13	20	>20
Female	-1	3½	6½	10	>10

LEG (L): Measured at Ankle

MOTION DISTANCE in Inches	BASIC	1	2	3	4
1	21	30	39	46	53
2	23	33	42	51	58
3	26	37	48	5?	65
4	30	43	55	66	76
5	34	49	63	75	86
6	37	54	69	8?	?5
7	40	59	75	?0	103
8	43	63	80	?6	110
9	46	66	85	10?	11?
10	48	70	89	107	1?3
11	50	72	94	112	1?9
12	52	75	97	117	134
13	54	77	101	121	139
14	56	80	103	125	144
15	58	82	106	130	149
16	60	84	108	133	153
17	62	86	111	13?	158
18	63	88	113	137	161
19	65	90	115	140	164
20	67	92	117	142	166
22	70	96	121	147	171
24	73	99	126	151	175
26	75	103	130	155	179
28	78	107	134	159	183
30	81	110	137	163	187
35	87	118	147	173	197
40	93	126	155	182	206

Weight, lb.
	BASIC	1	2
Male	8	-42	>42
Female	4	-21	>21

TRUNK (T): Measured at Shoulder

MOTION DISTANCE	BASIC	1	2	3	4
1	26	38	49	58	67
2	29	42	53	64	73
3	32	47	60	72	82
4	38	55	70	84	96
5	43	62	79	95	109
6	47	68	87	105	120
7	51	74	95	114	130
8	54	79	101	121	139
9	58	84	107	128	147
10	61	88	113	135	155
12	66	94	123	147	169
14	71	100	130	158	182
16	75	105	136	167	193
18	80	111	142	173	203
20	84	116	148	179	209
22	88	121	153	185	215
24	92	125	158	190	220
26	95	130	163	196	226
28	99	134	168	201	231
30	102	139	173	206	236

Weight, lb.
	BASIC	1	2
Male	-11	58	>58
Female	-5½	29	>29

FINGER–HAND (F, H): Measured at Finger Tip

MOTION DISTANCE in Inches	BASIC	1	2	3	4
1	16	23	29	35	4?
2	17	25	32	38	4?
3	19	28	36	43	4?
4	23	33	42	50	58

Weight, lb
	BASIC	1	2	3
Male	2/3	2½	4	>4
Female	1/3	1¼	2	>2

FOOT (Ft): Measured at Toe

	BASIC	1	2	3	4
1	20	29	37	44	51
2	22	32	40	48	55
3	24	35	45	55	63
4	29	41	53	64	73

Weight, lb.
	BASIC	1	2
Male	-5	-22	>22
Female	-2½	-11	>11

FOREARM SWIVEL (FS): Measured at Knuckle

	BASIC	1	2	3	4
45°	17	22	28	32	?7
90°	23	30	37	43	49
135°	28	36	44	52	58
180°	31	40	49	57	65

Torque, lb.-in.
	BASIC	1	2
Male	-3	-13	>13
Female	-1½	-6½	>6½

WALK

TYPE		30-inch PACES — 2	OVER 2
General	Analyze from Table	260	120 + 80/Pace
Restricted		300	120 + 100/Pace

Add 100 for >120° - 180° Turn at Start or Finish of Walk

Up Steps (8-inch rise, 10-inch flat)	126/step
Down Steps	100/step

HEAD TURN (HT): Measured at Nose Tip

Degrees Turn	Distance in Inches	Number of Work-Factors			
		Basic or 1	2	3	4
>22½-45	>2-4	40	51	58	66
-90	-8	60	76	86	99

TABLE 19–1 *(continued)*

COMPLEX GRASP * FROM RANDOM PILES																													
SIZE (Major Dimension or Length) (inches)		SOLIDS AND BRACKETS Thickness (inches)				THIN FLAT OBJECTS Thickness (inches)								CYLINDERS AND REGULAR CROSS-SECTIONED SOLIDS Diameter (inches)														Add for Entangled, Nested, or Slippery Objects†	
		>.047 (>3/64)				–.016 (–1/64)				–.047 (–3/64)				–.063 (–1/16)		–.125 (–1/8)		–.188 (–3/16)		–.500 (–1/2)				>.500 (>1/2)					
		Blind		Visual		Blind		Visual		Blind		Visual		Blind		Blind		Blind		Blind		Visual		Blind		Visual			
		n	s	n	s	n	s	n	s	n	s	n	s	n	s	n	s	n	s	n	s	n	s	n	s	n	s	n	s
–.063	–1/16	120	172	B	B	–	–	–	–	131	189	B	B	S	S	S	S	S	S	S	S	S	S	S	S	S	S	17	26
–.125	–1/8	79	111	B	B	108	154	B	B	85	120	B	B	85	120	S	S	S	S	S	S	S	S	S	S	S	S	12	18
–.188	–3/16	64	88	B	B	102	145	B	B	74	103	B	B	79	111	74	103	S	S	S	S	S	S	S	S	S	S	12	18
–.250	–1/4	48	64	B	B	72	100	B	B	56	76	B	B	79	111	68	94	64	88	S	S	S	S	S	S	S	S	8	12
–.500	–1/2	40	52	B	B	64	88	B	B	48	64	B	B	62	85	56	76	56	76	44	58	B	B	S	S	S	S	8	12
–1.000	–1	40	52	32	40	64	88	60	82	48	64	44	58	62	85	56	76	48	64	48	64	44	58	40	52	32	40	8	12
–4.000	–4	37	48	20	22	53	72	36	46	45	60	28	34	56	76	48	64	40	52	40	52	36	46	37	48	20	22	8	12
>4.000	>4	46	61	20	22	70	97	44	58	62	85	36	46	56	76	48	64	40	52	40	52	36	46	37	48	20	22	9	14

* Special Grasp conditions should be analyzed in detail
n = Non-simo; s = Simo
B = Use Blind column, since Visual Grasp offers no advantage
S = Use Solid Table

† Add the indicated amounts when objects: (1) are entangled (not requiring two hands to separate); (2) are nested together because of shape or film; (3) are slippery (as from oil or polished surface). When objects both entangle and are slippery, or both nest and are slippery, use double the value in the table.

TABLE 19–1 *(continued)*

ASSEMBLE

AVERAGE NUMBER OF ALIGNS (A1S MOTIONS)

TARGET DIMENSION (Inches)	CLOSED TARGETS — RATIO OF PLUG DIAMETER TO TARGET DIMENSION					
	–.225	–.290	–.415	–.900	–.935 †	>.935 ‡
>.875	(D*) 18	(D*) 18	(D*) 18	(¼) 25	(¼) 51	(¼) 59
–.875	(D*) 18	(D*) 18	(SD*) 18	(¼) 25	(¼) 51	(¼) 59
–.625	(SD*) 18	(SD*) 18	(¼) 25	(½) 31	(½) 57	(½) 65
–.375	(½) 31	(1) 44	(1) 44	(1½) 57	(1½) 83	(1½) 91
–.225	(1) 44	(1) 44	(1) 44	(1½) 57	(1½) 83	(1½) 91
–.175	(1) 44	(1¼) 51	(1½) 57	(1½) 57	(1½) 83	(1½) 91
–.124	(2½) 83	(2½) 83	(2½) 83	(2½) 83	(2½) 109	(2½) 117
>.025–.074	(3) 96	(3) 96	(3) 96	(3) 96	(3) 122	(3) 130

TARGET DIMENSION (Inches)	OPEN TARGETS — RATIO OF PLUG DIAMETER TO TARGET DIMENSION					
	–.225	–.290	–.415	–.900	–.935 †	>.935 ‡
>.875	(D*) 18	(D*) 18	(D*) 18	(D*) 18	(¼) 51	(¼) 59
–.875	(D*) 18	(D*) 18	(D*) 18	(SD*) 18	(¼) 51	(¼) 59
–.625	(SD*) 18	(SD*) 18	(SD*) 18	(½) 31	(½) 57	(½) 65
–.375	(¼) 25	(½) 31	(½) 31	(¾) 38	(¾) 64	(¾) 72
–.225	(½) 31	(½) 31	(½) 31	(¾) 38	(¾) 64	(¾) 72
–.175	(¾) 38	(1) 44	(1) 44	(1) 44	(1) 70	(1) 78
–.124	(1¼) 51	(1¼) 51	(1¼) 51	(1¼) 51	(1¼) 77	(1¼) 85
>.025–.074	(1½) 57	(1½) 57	(1½) 57	(1½) 57	(1½) 83	(1½) 91

*Letters indicate Work-Factors in Move preceding Assemble.

†Requires A(X)S Upright for all ratios >.900 (Table value includes A1S Upright and A1 Insert).

‡ Requires A(Y)S Upright and A(Z)P Insert for all ratios >.935 (Table value includes A1S Upright and A1P Insert).

DISTANCE BETWEEN TARGETS

Distance Between Targets (Inches)	Percent Addition to Aligns	Method of Align
– 1	Neg.	Simo
– 2	10	Simo
– 3	30	Simo
– 5	50	Simo
– 7	70	Simo
–15	Align and Insert first Plug, then Assemble* second end.	
>15	Align and Insert first Plug, turn head toward second Plug, React (60 Time Units), Assemble* second Plug.	

*If connected, treat 2nd Assemble as Open Target with no Upright. Index may be required up to 7 inch distances.

GRIPPING DISTANCE

Distance from Grip Point to Align Point (Inches)	Percent Addition to Aligns	Length of Upright Motion (Inches)
– 2	Neg.	1
– 3	10	1
– 5	20	2
– 7	30	2
–10	40	3
–15	60	5
–20	80	6
>20	100	7 and up

BLIND TARGETS

Blind Distance (Inches)	Percent Addition to Aligns	
	Permanent (Blind at all times)	Temporary (Blind during Assemble)
– ½	20	0
– 1	30	10
– 2	40	20
– 3	70	30
– 5	130	50
– 7	250	70
–10	380	120

GENERAL RULES FOR ASSEMBLE

1. When required, add W and P Work-Factors to all Assemble Motions according to Transport Rules.
2. Reduce number of Aligns by 50 percent when hand is rigidly supported.
3. Where Gripping Distance, two Targets, and Blind Targets are involved, add each percentage to original Align. Do not pyramid percentages.
4. Aligns for Surface Assemble are taken from –.225 column and are A1SD Motions.
5. Index is F1S, A1S, or FS45 ° S.

TABLE 19–1 (concluded)

POST-DISENGAGE TRAVEL

DISENGAGE RESISTANCE (Pounds)	POST-DISENGAGE TRAVEL (Inches)
– 2	Negligible
– 7	3
–13	6
–20	10

MENTAL PROCESS (MP)—Simple

Focus (Fo)	20
React (Rn)	20
Impact (I)	30
Mento (Mt)	10

PRE-POSITION

SHAPE, SIZE (INCHES), AND WEIGHT (POUNDS) OF OBJECT[1]		ONE HAND			TWO HANDS	
		VERY SMALL V5F1	OPTIMUM V3F1	MEDIUM V4F1	MEDIUM 2T(f)+V4dD	LARGE 2T(f)(V)+V4dD
1. Cylinders – Regular Cross-section[2]	Diameter	–.375	>0 – 1.25	>1.25 – 2.50	>1.25 – 4.50	..
	Major Dimension	–.375	>.375 – 4.00	>4.00 – 16.00	>1.25 – 30.00	..
2. Solids, Thin Flats, etc.[3]	Width	–.375	>0 – 1.25	>1.25 – 2.50	>2.50 – 10.00	>10.00 – 16.00
	Thickness	–.375	>0 – 1.25	>1.25 – 16.00	>0 – 4.50	>0 – 4.50
	Major Dimension	–.375	>.375 – 4.00	>1.25 – 16.00	>2.50 – 16.00	>10.00 – 16.00
3. Weight Limits – All Objects (Pounds)	Male	–.667	–.667	–.667	–1.00	–3.50
	Female	–.333	–.333	–.333	–0.50	–1.75

WORK-FACTOR TIME UNITS[4]

NO. OF POSITIONS SATISFACTORY FOR USE	% PP REQUIRED	PP–V ●	PP–V s	PP–O ●	PP–O s	PP–M ●	PP–M s	PP–M₂ ●	PP–M₂ s	PP–L
1. From Stack (all wrong way)	100%	80	120	48	72	64	96	70	100	100
2. One Specific Face Up 1 Edge Only	75%	60	90	36	54	48	72	53	75	75
2 Adjacent Edges	63%	50	75	30	45	40	60	44	63	63
2 or more Opposite Edges	50%	40	60	24	36	32	48	35	50	50
3. Two or More Faces Up 1 Edge Only	50%	40	60	24	36	32	48	35	50	50
2 Adjacent Edges	25%	20	30	12	18	16	24	18	25	25
2 or more Opposite Edges	0%									

Notes:
[1] Pre-positions requiring one Finger Motion or Wrist Turn can be done Simo with the Move. Other Pre-positions are taken from this Table, or analyzed.
[2] Cylinders with a diameter > 4.50 inches or a Major Dimension > 30.00 inches require Special Analyses.
[3] Solids and Thin Flats with a width or Major Dimension > 16.00 inches or a thickness > 4.50 inches require Special Analyses.
[4] Objects weighing more than weight limits on this Table require Special Analyses.
● = Non-Simo; s = Simo.

b. *Thin, flat objects:* Flat objects with an effective dimension of ³⁄₆₄ inch or less in thickness.

c. *Solids and brackets:* Defined as objects over ³⁄₆₄ inch thick not otherwise classified as cylinders or thin, flat objects.

3. PRE-POSITION. Pre-position occurs whenever it is necessary to turn and orient an object so that it will be in the correct position for a subsequent element. Pre-position frequently occurs on a percentage basis since the object is in a usable position sometimes but must be oriented at other times. For example, a nail (0.100 inch × ¾ inch), in 50 percent of the cases, is grasped in a usable position; in the remaining 50 percent of the cases, it must be pre-positioned. Using the Work-Factor pre-position table (Table 19–1), the analysis would be: PP-0-50 percent = 24 units.

4. ASSEMBLE. This element occurs whenever two or more objects are joined together, usually through mating or nesting. Assemble also occurs when one object is placed in an exact position relative to another. The system provides a complete assemble table. Assemble time depends on:

a. *Size of target:* The target is the part of an assembly that accepts the plug.

b. *Plug dimensions:* A plug is the part of an assembly that fits into the target.

c. *Plug-target ratio:* Assembly difficulty, and therefore assemble time, increases as the dimension of the plug approaches the dimension of the target. Therefore, assembly time is a function of the plug-target ratio. (Applies only to mating two parts)

$$\frac{\text{Plug dimension}}{\text{Target dimension}} = \text{Plug-target ratio}$$

d. Type (shape) of target: There are two targets in Work Factor terminology—closed and open. Closed targets are closed about the entire perimeter so that align motions are required along two axes. Open targets require align motions along one axis only.

When these facts are known, it is a simple matter to determine assemble time from the table: Add allowances for increased difficulty due to the distance between targets (two at a time), gripping distance (the distance from the hand to the end of the plug), and blind target (when the target cannot be seen during assembly).

5. USE. This element usually refers to machine time, special process time, and time involving tool use. Use may involve manual motions, as in tightening a nut with a wrench or threading a pipe; in such cases, the motions are to be analyzed and evaluated according to all rules and time values obtained from the transport tables. Pure machine and process times are calculated, developed from formulas, or timed.

6. DISASSEMBLE. As the name implies, disassemble is the opposite

of assemble. Usually it is a single motion. Take time values from the transport table.

7. MENTAL PROCESS. This applies to all mental activities and processes. It is a time interval in which reactions and nerve impulses take place.

Mental processes that can be measured include:

Eye motions	Inspections	Computations
Focus	Quality	Read
Shift	Quantity	Action
Reactions	Identify	Concept

8. RELEASE. Release, the opposite of grasp, is the act of relinquishing control of objects. There are three types:

a. *Contact release:* Involves no motion. It is merely lifting the hand from an object.
b. *Gravity release:* Occurs when objects drop from the grasp as contact is broken and before the releasing finger motions are complete.
c. *Unwrap release:* Involves unwrapping the fingers from around the grasped object and is not complete until the unwrapping motions are complete.

All time values on the Work-Factor motion-time table are in 0.0001 minute. These values are in Work-Factor time, which is "that time required for the average, experienced operator, working with good skill and good effort (commensurate with good health and physical and mental well-being) and under standard working conditions to perform an operation on one unit or piece." To determine the standard time, analysts must add an allowance to the Work-Factor values, since the time includes no allowance for personal needs, fatigue, unavoidable delays, or incentive potential.

A typical breakdown study of a draw operation on a Bllis double-action 240-ton press is shown in Figure 19–2. The symbols used in this analysis carry the following meanings:

W—Weight or resistance
S—Steering or directional control
P—Precaution or care
U—Change direction
D—Definite stop
A—Arm
L—Leg
F—Finger

R—Reach
Gr—Grasp
Re Gr—Regrasp
M—Move
Rl—Release
Ru—React
BD—Balancing delay
RP—Relax pressure
WA—Work area
LH—Left hand

FIGURE 19–2

No	Elemental Description	Analysis	Units	Cum. Time	Cum. Time	Units	Analysis
	LEFT HAND					RIGHT HAND	*491 A*
1	R for blank	A20D	80	80			
2	G blank - 4 lbs.	F1W	23	103			
3	M blank to die	A40WSD	159	262			
4	R1 blank, clear						
5	fingers	F3W	28	290	290	290	E11-5LH	Aid LH
6	R to blank on die	F3D	28	318	399	109	A40D	R to trip lever
7	Re Gr blank	Gr-C	D	318			Gr lever -10 lb.
8	Push blank a sumstpins	AZP	29	347	428	29	F1W²	resist
9	withdraw hand	A10	42	389			Pull lever to
10	wait	BD	117	506	506	78	A1DW²	trip press
11	R to stack of blanks	A30D	96	602	529	23	F1W	R1 lever
12	Gr stack	Gr-C	o	602	625	96	A30D	R to oil rag
13	Press down to hold	A1W	26	628	642	17	F2	Gr rag
14							M rag to oil pan
15				727	85	A124D	and dip into it
16				759	32	A6	Raise rag from pan
17				785	26	A4	Shake off excess oil
18				861	76	A18D	M rag to stack of blanks
19	Hold blank	BD	342	970	970	109	A4DU	Apply oil
20	RP on blank	½A1W	13	983			Strike blank with rag
21	R1 blank	R1-C	D	983	1054	84	2A1D	and hand to free
22	R to edge of blank	A16+A3D	84	1067	1105	51	A15	M rag to side
23	Gr blank	F1W	23	1090			
24	Turn blank over, R1 simo	2A14W	138	1228			
25	R to center of blank	A13D	67	1295			
26	Gr blank - 4 lbs.	Gr-C	o	1295			
27	Press down to hold	A1W	26	1321	1321	216	BD	Hold rag; M near blank
28	Hold blank	BD	109	1430	1430	109	A4DU	Apply oil to blank
29	RP on blank	½A1W	13	1443	1481	51	A15	Toss rag near pan
30	R1 blank	R1-C	o	1443	1561	80	A2DD	R to trip lever
31					1590	29	F1W2	Gr lever-10 lb resist
32							Wait for completion
33	Wait and R near						of machine cycle;
34	piece on punch	BD	443	1886	1886	296	BD+A15W2	pull lever to stop
35	Catch piece on palm	Rn	20	1906	1909	23	F1W	R1 handle
36	M piece to stack						
37	body turn simo	A40WPD	159	2065			
38	Toss piece into chute	A5W	43	2108			
39	Return to WA	L13W2	101	2209	2209	300	BD	Wait & Aid LH

Notes:

1. The time for Element 34 RH—Wait for completion of machine cycle is calculated as follows:

> Wait time = Machine time + time elapsed when press is tripped—
> time elapsed when operator is ready to stop machine
> = 1380 + 506 − 1590
> = 296

2. Note the value of double cumulative time column in finding areas of potential methods improvement. The BDs in elements 19 and 34 LH and 27 and 39 RH total 1301 WF'Us (.1301 minutes) or 59 percent of the cycle. This indicates that the hands are not productively occupied during the entire cycle and that the possibility of revising the method to reduce the cycle time should be investigated before finalizing the standard.

In making a Work-Factor study, the analyst first lists all necessary motions made by both hands to perform the task; then he or she identifies each motion in terms of the distance moved, body member used, and work factors involved. In recording distances, fractional inches are not used. Thus, motions of 1 inch or less are recorded as the nearest integral number. The analyst then selects from the table of values the appropriate figure for each of the basic motions and summarizes these to obtain the total time required by the normal operator to perform the task.

To this total time, add percentage allowances for personal delays, fatigue, and unavoidable delays to determine the allowed time.

Ready Work-Factor System

Ready Work-Factor measures work where cycle times are .06 minutes or longer and great precision is not required. Times in the Ready tables are averages and can be traced to the Detailed tables. Ready time values are generally 0 to 5 percent higher than the Detailed values.

Essentially, the Detailed rules apply to Ready with a few minor exceptions. The Ready Work-Factor table includes the following:

1. A table for transport motions, applied to all body members. Time values are classified in ranges.
2. A grasp table listing Work-Factor units for five grasp types, with variations for visual, simo, entangled, nested, and slippery conditions. These values cover the many values for simple, manipulative, complex, and special grasps in Detailed Work-Factor. Five of the seven values apply to complex grasps, classed according to the type and size of the object being grasped. A parts classification table is provided as an objective guide.
3. A pickup table providing time values which combine the standard elements transport and grasp into values for picking up objects (i.e., the motion sequence reach to object, grasp object, and move object to new location).
4. A pre-position table similar to the detailed table.
5. An assemble table in condensed form.
6. A visual inspection table.
7. A release table with values for two different release situations.

The Ready time tables are simple numbers arranged in easily remembered sequences, such as 4, 5, 6, 7, 8 or 3, 5, 7, 9, 11 (see Table 19–2).

Figure 19–3 illustrates a Ready Work-Factor technique, the analysis for the operation "Assemble screws to plate." The analysis formats are somewhat simpler than their Detailed counterparts; the work segment pickup reduces the analysis effort.

Brief Work-Factor System

Brief Work-Factor is a rapidly applied technique for determining the approximate time required to perform the manual portion of any job. Equally applicable to jobs in the home, field, office, or factory, it is particularly advantageous for estimating labor costs in advance of actual

TABLE 19–2
Ready Work-factor (time table)

Ready Work-Factor ·

TIME TABLE

© 1976, 1973, 1971, 1966, 1965, 1963 Science Management Corporation, Formerly Wofac Corporation. All rights reserved.   * Service Mark   Printed in U.S.A.

All times are expressed as Ready Work-Factor Time Units. One Ready Work-Factor Time Unit = 0.0010 minute.
SYMBOLS: — indicates "up to and including". > indicates "greater than".

		Work-Factors						ASSEMBLE		Mechanical [6]					
		0	1	2	3	4				Open (I)			Closed (X)		
		Very Easy	Easy	Aver-age	Diffi-cult	Very Difficult				Ratios					
TRANSPORT		Weight Limits, lbs.						Target, in.		−.4	−.9	>.9	−.4	¬9	>.9
Finger, Hand		−1	−2	−3	−5	>5			>⅜	2	3¹	7¹	2	3¹	7¹
Arm		−2	−4	−6	−10	>10			−⅜	3¹	4²	8²	5³	6⁴	10⁴
Foot		−3	−8	>8	---	---			−⅛	6⁴	6⁴	10⁴	9⁷	9⁷	13⁷
Leg		−5	−16	>16	---	---		Tolerance, in.		**Surface** [6]					
Trunk		−7	−32	>32	---	---			>⅜	3			3		
Motion Distance,[1] in.									−⅜	5²			6³		
Very Short	−4 A[2]	2	3	4	5	6			−⅛	8⁵			12⁹		
Short	−10 B	4	5	6	7	8				% Add-ons for Assemble					
Medium	−20 C	5	7	9	11	13		Distance, in.		−1	−2	−3	−5	−7	−15	>15
Long	−30 D	7	9	11	13	15		Gripping Distance		---	---	10	20	30	50	70
Very Long	−40 E	9	11	13	15	17		Distance Between		---	20	30	50	70	2Asy	2Asy
GRASP	Simple	Pinch[3]	Wrap and Transfer[3]		Assemble Motions			Temporary Blind							150	+5 ---
	Manipulative	---	---	−2	−3	−4		Permanent Blind		30	50	70	150	250	500	---
	Complex[7]	Major Dimension, in.			>¼	−¼				Miscellaneous					
		Diameter, in.		>¼	−¼	ALL		Index:	Mechanical 3 Surface 4						
		Thickness, in.		>³⁄₆₄	−³⁄₆₄			Simo:	50% of (Aligns + Add-ons)						
	Visual			3	5	8		Seat:	Turn; Use Transport Rules						
	Blind	1	2	4	6				Dislodge; Use Assemble Rules						
PICK-UP [4]	Distance, in.							Weight, lbs.		---	−2	−4	−6	−10	>10	
	−4 A	8	9	10	12	15			% Add-on to Total Asy Time	0	30	50	70	100	
	−10 B	12	13	14	16	19		**MENTAL PROCESS**	Focus 2, Inspect 3, React 2					
	−20 C	17	18	19	21	24		**BODY TURN**	Head Turns: −22½° O, −45° 4, −90° 6					
	−30 D	21	22	23	25	28			Body Turns: −90° (1 Foot) 10					
	−40 E	25	26	27	29	32			−90° (2 Feet) 20, −180° 26					
	Distance, in.	Add-on for Weight						**WALK**	General 12 + 8 per Pace					
	−10	---	1	2	3	4			Restricted 12 + 10 per Pace					
	>10	---	2	4	6	8		(in terms of	Up and Down Steps: General 10 per Step					
	Add-on for:							30-inch Paces)	Restricted 13 per Step					
									Stand up 13 Sit Down 9					
	Blind	---		1		---		**RELAX PRESSURE**	1	**APPLY PRESSURE**	See Note 2				
	Simo	---		2⁵				**CIRCULAR** Diameter,		No. Work-Factors					
PRE-POSITION		One Hand			Two Hands			**MOTIONS** in.		Single Circle		Multiple Circles			
		Major Dimension, in.							−1			1			
	% Occurrence	>⅜−4	−10	−⅜	−10	>10			0						
	25		1			2			>1			0			
	50	2		3		4									
	75	3	4	5	5	6									
	100	4	5	6	7	8									
	Simo	Add 50% to time			---	---									
RELEASE		Gravity	Unwrap		---										
		1	2												

[1] For Trunk Motion multiply Distance by 2.
[2] Use Transport class A for Apply Pressure and for Forearm Swivel.
[3] For weight over 3 pounds multiply time by 2.
[4] Times based on Visual Grasp. Work-Factors in Pick-up are considered to be identical to those in Grasp. Times include Reach, Grasp and Move, not Release.
[5] Add blind value also.
[6] Small numbers are Align times, large numbers are total Assemble times.
[7] For Complex Grasp: add 2 Units for simo and 1 Unit each for Entangled, Nested and Slippery.

Motion and Time Study

FIGURE 19–3
Example of a Ready Work-Factor analysis

No.	Elemental Description	Hand L	Hand R	Analysis	Time Units Elemental	Time Units Cumulative	No.
	Operation Name & Description — Asy 4 screws to plate , tighten with airdriver				**Company**		
1	PU screws from pile , M to plate	X	X	PU 10-3B5	19	19	1
2	PP screws	X	X	A-5-50%	3	22	2
3	Asy screws to holes in plate	X	X	CT-³/₈-.9	4	26	3
4	DB .3"	X	X	4x.3	1	27	4
5	SF	X	X	.5 (4+1)	.3	30	5
6	Start threads	X	X	Start	10	40	6
7	Turn down V 2 turns	X	X	V 2 x 3	6	46	7
8	Asy 2 more screws to plate	X	X	EL 1-7	46	92	8
9	PU suspended airdriver		X	PU 20-1	18	110	9
10	Asy driver to 1ˢᵗ screw		X	CT-³/₈ 7.9	10	120	10
11	GD 5"		X	4x.2	1	121	11
12	Ind		X	Ind m	.3	124	12
13	Drive screw		X	Std Data	11	135	13
14	M driver to 2ᴺᴰ screw		X	4-2	4	139	14
15	Drive 2ⁿᵈ screw		X	EL 10-13	25	164	15
16	Drive 2 more screws		X	2x EL 14,15	58	222	16
17	Aside driver		X	20-1	7	229	17
18	Re driver		X	1-	2	231	18
19							19
20							20

wofac — Date 4/81 — Analyst EaB — Total 231 — Time in Minutes .231 — Multiplier — 13.5/1 (69/1)

The symbols mean: PU—Pickup; PP—Preposition; Asy—Assemble; DB—Distance between; SF—Simo-factor; V—Average; GD—Gripping distance; Ind—Index; IndM—Mechanical index; Rel—Release; B—Blind; s—simo; CT—Closed target; El—Element.

production. It can be used for establishing incentive rates on small quantity, nonrepetitive, or long-cycle operations.

Brief Work-Factor is convenient for studying operations of many minutes or hours. In some circumstances the motion descriptions and time values are applied as rapidly as the operator performs his or her task. Hence, when one cycle of the operation is completed, the time standard is essentially completed.

As with Ready Work-Factor, Brief Work-Factor time values can be traced back to the Detailed System. Brief times vary from Detailed by no more than 10 percent.

Brief Work-Factor depends for its speed of application on a simple time table (see Table 19–3) and the use of Work-Segments. Six such segments are included:

Pickup: Reach plus grasp plus pre-position plus move plus release.

Assemble: Unite two parts physically or place one part in an exact location.

TABLE 19–3
Brief Work-Factor Time Table

Distance in Inches	Short	– 4″	15
	Average	–20″	20
	Long	>20″	25

Tolerance in Inches	Loose	–2″	0
	Average	–1/2″	5
	Close	–1/16″	10

One Motion Grasp – Subtract	5
Weight >6 lbs.	+5
Object Major Dimension – 1/4″	+5
Pre–position Required	+5

Blind	+5
Weight >6 lbs.	+5
Index	+5
Seat	+5

3. MOVE ASIDE (MA) & MOVE MOTIONS (M)
– 4″
> 4″

4. OTHER
WORK SEGMENTS

Distance in Inches	DEGREE OF DIFFICULTY		
	Easy	Ave.	Difficult
Short –4″	5		
Average –20″	5		10
		10	
Long >20″	10		15
Short, Easy Repetitive Motions =5 /Pair			

Start Threads (ST)	10
Turn Down (TD) – Per Twist	
By Hand	5
Using Hand Tool –1	10
Using Hand Tool –2	25
Sit or Stand (S)	10
Bend (B) – Up or Down	10
Turn (T)	10
Walk (W) – Per 30″ Pace	10
Read – Per Word	5
Inspect – Per Character	5
Write – Per Character	10
Hammer – Per Blow	10

All time values are expressed as Brief Work—Factor Time Units.
One Brief Work–Factor Time Unit = 0.001 Minute.
Time values outside the boxes are used when estimating with Brief Work–Factor System. Time values within the boxes apply when measuring with Brief Work–Factor System.
Symbols: – Indicates "Up to and including". > Indicates "Greater than."
+ Indicates Add-on to base values. Subtract: Reduce time value according
© 1976, 1975 by Science Management Corporation. All rights reserved.

Move aside: Move plus release.

Move motion: Transports that connect other segments without being part of them.

Use: Machine and process time.

Other work segments: Mental processes and other activities, such as writing, walking, using hammers, and so on.

Figure 19–4 is for prepricing the operation "Copy four-page report." Note the simplicity of the analysis structure.

FIGURE 19–4

Description		Time units
1. TEN	up	10
2. P20	master from bench	20
3. 12W10	to copier	120
4. MA5	master to work area	5
5. P20	1st page to copier with right hand; open cover with left hand	20
6. A10	page to mark on copier	10
7. M5	close cover	5
8. P20-5	start button	15
9. Copy 45	1st page	45
10. M5	open cover	5
11. P20	1st page to work area	20
12. 3E1 5-11	copy 3 more pages of report	360
13. Wait 60	for last copy	60
14. P20	report from catch tray to work area	20
15. 6/2 M5	job report	15
16. M10	report to stapler	10
17. A5	report to stapler	5
18. 2M5	staple report	10
19. P20-5	pencil to report	15
20. A10	pencil to corner of report	10
21. V8 Write 10	name of recipient	80
22. MA5	pencil to bench	5
23. P20-5	report from left hand to tray on machine	15
24. P20	masters from work area	20
25. 12W10	return to desk	120
26. MA5	masters to desk	5
27. S10	down	10
		1035
		Total = 1.035 min.

The symbols used carry the following meanings: P—Pickup; A—Assemble; M—Move motion, MA—Move aside; S—Stand; W—Walk; V—Average.

Mento-Factor System

The Mento-Factor System is used when very accurate standards must be established for operations principally mental in content. Such operations include proofreading, visual inspection, and problem solving. The Mento-Factor System also develops and improves methods for mental work. Thirteen fundamental mental processes are the basis of this system.

1. Eye motions: Move the eyes from one fixation point to the next.
2. See: Register visual stimuli.
3. Conduct: Transmit impulses along the nervous system.
4. Discriminate: Distinguish between similar characteristics.
5. Span: Mentally organize groups of characters.
6. Identify: Determine what has been seen.

FIGURE 19–5
Inspect small nuts in hand to count

The operator is required to inspect a random group of unplated steel nuts ³⁄₈ by ³⁄₈ by ¹⁄₈ inch grasped from a pile prior to insertion in an envelope for packing. The correct number of nuts to be put into the envelope is four. If inspect reveals fewer, additional ones must be grasped. If inspect reveals more, the excess must be discarded. The operator grasps two, three, four, or five nuts, with occurrences of 10, 35, 50, and 5 percent, respectively.

The Mento-Factor analysis for the mental work involved is:

1.	See nuts in hand.	See 1–4	8
2.	Conduct impulse to brain.	Con 1	1
3.	Span Count: 2 at 10% occurrence.	Sp 1–1–2—10%	2
	3 at 35% occurrence.	Sp 1–1–3—35%	12
	4 at 50% occurrence.	Sp 1–1–4—50%	23
	5 at 5% occurrence.	Sp 1–1–5—5%	3
4.	Identify number of nuts in hand.	IdA-4	24
5.	Decide whether too few, enough, too many.	DeA-3	19
6.	Conduct impulse to hand.	Con 3	2
		Total	94

---

7. Decide: Determine the appropriate action to be taken based on the preceding Identity.

8. Convert: Translate one set of symbols into a corresponding set.

9. Memorize: Store information in the brain temporarily.

10. Recall: Recollect information previously memorized.

11. Compute: Combine digits arithmetically.

12. Sustain: A time effect relating to Compute.

13. Transfer attention: Disengage mental concentration from one mental task and engage it in another.

These fundamental mental processes are combined into Mento Intervals which in turn are combined into Mento Operation cycles. A detailed discussion of the Mento-Factor System is obviously beyond the scope of this textbook. However, the analysis of an operation (see Figure 19–5) illustrates this technique.

METHODS–TIME MEASUREMENT

MTM-1

In 1948 *Methods-Time Measurement* was published, giving time values for the fundamental motions reach, move, turn, grasp, position, disengage, and release.[4] The authors defined MTM as "a procedure which analyzes any manual operation or method into the basic motions required

---

[4] H. B. Maynard, G. J. Stegemerten, and J. L. Schwab, *Methods-Time Measurement* (New York: McGraw-Hill, 1948), p. 12.

to perform it, and assigns to each motion a pre-determined time standard which is determined by the nature of the motion and the conditions under which it is made."[5]

MTM-1 data, like those of Work-Factor, are the result of frame-by-frame analysis of motion-picture films involving diversified areas of work. The data taken from the various films were "leveled" (adjusted to the time required by the normal operator) by the Westinghouse technique. This method of performance rating is described in Chapter 15. Then they tabulated and analyzed the data to determine the degree of difficulty caused by variable characteristics. For example, not only the distance but also the type of reach affects reach time. Further analysis categorized five distinct cases of reach, each requiring different time allotments to perform for a given distance. These are:

1. Reach to object in fixed location, or to object in other hand, or on which other hand rests.
2. Reach to single object in location which may vary slightly from cycle to cycle.
3. Reach to object jumbled with other objects so that search and select occur.
4. Reach to a very small object or where accurate grasp is required.
5. Reach to indefinite location to get hand in position for body balance, or next motion, or out of way.[6]

In addition, they found move time was influenced not only by the distance and the weight of the object being moved, but also by the specific type of move. The three cases of move are:

1. Move object to other hand or against stop.
2. Move object to approximate or indefinite location.
3. Move object to exact location.[7]

Table 19–4 summarizes MTM-1 values. The time values of the therblig grasp vary from 2.0 TMU to 12.9 TMU (1 TMU equals 0.00001 hour), depending on the classification of the grasp. Two cases of release and 18 cases of position also influence time.

The steps followed in applying MTM–1 technique are similar to those followed for Work-Factor. First, the analyst summarizes all left-hand and right-hand motions required to perform the job properly. Then he determines from the methods-time data tables the leveled time in TMU for each motion. The nonlimiting motion values should be circled or deleted, as only the limiting motions will be summarized, provided that it is "easy" to perform the two motions simultaneously (see section X of

---

[5] Ibid.
[6] Ibid.
[7] Ibid.

TABLE 19-4

TABLE I—REACH—R

Distance Moved Inches	Time TMU				Hand in Motion		CASE AND DESCRIPTION
	A	B	C or D	E	A	B	
¾ or less	2.0	2.0	2.0	2.0	1.6	1.6	A Reach to object in fixed loca- tion, or to object in other hand or on which other hand rests.
1	2.5	2.5	3.6	2.4	2.3	2.3	
2	4.0	4.0	5.9	3.8	3.5	2.7	
3	5.3	5.3	7.3	5.3	4.5	3.6	B Reach to single object in location which may vary slightly from cycle to cycle.
4	6.1	6.4	8.4	6.8	4.9	4.3	
5	6.5	7.8	9.4	7.4	5.3	5.0	
6	7.0	8.6	10.1	8.0	5.7	5.7	
7	7.4	9.3	10.8	8.7	6.1	6.5	
8	7.9	10.1	11.5	9.3	6.5	7.2	C Reach to object jumbled with other objects in a group so that search and select occur.
9	8.3	10.8	12.2	9.9	6.9	7.9	
10	8.7	11.5	12.9	10.5	7.3	8.6	
12	9.6	12.9	14.2	11.8	8.1	10.1	
14	10.5	14.4	15.6	13.0	8.9	11.5	D Reach to a very small object or where accurate grasp is required.
16	11.4	15.8	17.0	14.2	9.7	12.9	
18	12.3	17.2	18.4	15.5	10.5	14.4	
20	13.1	18.6	19.8	16.7	11.3	15.8	
22	14.0	20.1	21.2	18.0	12.1	17.3	E Reach to indefinite location to get hand in position for body balance or next motion or out of way.
24	14.9	21.5	22.5	19.2	12.9	18.8	
26	15.8	22.9	23.9	20.4	13.7	20.2	
28	16.7	24.4	25.3	21.7	14.5	21.7	
30	17.5	25.8	26.7	22.9	15.3	23.2	

TABLE II—MOVE—M

Distance Moved Inches	Time TMU			Hand in Motion B	Wt. Allowance			CASE AND DESCRIPTION
	A	B	C		Wt. (lb.) Up to	Fac- tor	Con- stant TMU	
¾ or less	2.0	2.0	2.0	1.7	2.5	0	0	
1	2.5	2.9	3.4	2.3				A Move object to other hand or against stop.
2	3.6	4.6	5.2	2.9	7.5	1.06	2.2	
3	4.9	5.7	6.7	3.6				
4	6.1	6.9	8.0	4.3				
5	7.3	8.0	9.2	5.0	12.5	1.11	3.9	
6	8.1	8.9	10.3	5.7				
7	8.9	9.7	11.1	6.5	17.5	1.17	5.6	
8	9.7	10.6	11.8	7.2				
9	10.5	11.5	12.7	7.9	22.5	1.22	7.4	B Move object to approximate or in- definite location.
10	11.3	12.2	13.5	8.6				
12	12.9	13.4	15.2	10.0	27.5	1.28	9.1	
14	14.4	14.6	16.9	11.4				
16	16.0	15.8	18.7	12.8	32.5	1.33	10.8	
18	17.6	17.0	20.4	14.2				
20	19.2	18.2	22.1	15.6				
22	20.8	19.4	23.8	17.0	37.5	1.39	12.5	
24	22.4	20.6	25.5	18.4				
26	24.0	21.8	27.3	19.8	42.5	1.44	14.3	C Move object to ex- act location.
28	25.5	23.1	29.0	21.2				
30	27.1	24.3	30.7	22.7	47.5	1.50	16.0	

TABLE III—TURN AND APPLY PRESSURE—T AND AP

Weight	Time TMU for Degrees Turned											
	30°	45°	60°	75°	90°	105°	120°	135°	150°	165°	180°	
Small— 0 to 2 Pounds	2.8	3.5	4.1	4.8	5.4	6.1	6.8	7.4	8.1	8.7	9.4	
Medium—2.1 to 10 Pounds	4.4	5.5	6.5	7.5	8.5	9.6	10.6	11.6	12.7	13.7	14.8	
Large— 10.1 to 35 Pounds	8.4	10.5	12.3	14.4	16.2	18.3	20.4	22.2	24.3	26.1	28.2	

APPLY PRESSURE CASE A—10.6 TMU.   APPLY PRESSURE CASE B—16.2 TMU.

TABLE 19–4 *(continued)*

TABLE IV—GRASP—G

Case	Time TMU	DESCRIPTION
1A	2.0	Pick Up Grasp—Small, medium or large object by itself, easily grasped.
1B	3.5	Very small object or object lying close against a flat surface.
1C1	7.3	Interference with grasp on bottom and one side of nearly cylindrical object. Diameter larger than ½″.
1C2	8.7	Interference with grasp on bottom and one side of nearly cylindrical object. Diameter ¼″ to ½″.
1C3	10.8	Interference with grasp on bottom and one side of nearly cylindrical object. Diameter less than ¼″.
2	5.6	Regrasp.
3	5.6	Transfer Grasp.
4A	7.3	Object jumbled with other objects so search and select occur. Larger than 1″ x 1″ x 1″.
4B	9.1	Object jumbled with other objects so search and select occur. ¼″ x ¼″ x ⅛″ to 1″ x 1″ x 1″.
4C	12.9	Object jumbled with other objects so search and select occur. Smaller than ¼″ x ¼″ x ⅛″.
5	0	Contact, sliding or hook grasp.

TABLE V—POSITION*—P

CLASS OF FIT		Symmetry	Easy To Handle	Difficult To Handle
1—Loose	No pressure required	S	5.6	11.2
		SS	9.1	14.7
		NS	10.4	16.0
2—Close	Light pressure required	S	16.2	21.8
		SS	19.7	25.3
		NS	21.0	26.6
3—Exact	Heavy pressure required.	S	43.0	48.6
		SS	46.5	52.1
		NS	47.8	53.4

*Distance moved to engage—1″ or less.

TABLE VI—RELEASE—RL

Case	Time TMU	DESCRIPTION
1	2.0	Normal release performed by opening fingers as independent motion.
2	0	Contact Release.

TABLE VII—DISENGAGE—D

CLASS OF FIT	Easy to Handle	Difficult to Handle
1—Loose—Very slight effort, blends with subsequent move.	4.0	5.7
2—Close — Normal effort, slight recoil.	7.5	11.8
3—Tight — Considerable effort, hand recoils markedly.	22.9	34.7

TABLE VIII—EYE TRAVEL TIME AND EYE FOCUS—ET AND EF

Eye Travel Time $= 15.2 \times \dfrac{T}{D}$ TMU, with a maximum value of 20 TMU.

where T = the distance between points from and to which the eye travels.
D = the perpendicular distance from the eye to the line of travel T.

Eye Focus Time = 7.3 TMU.

TABLE 19-4 *(concluded)*

TABLE IX—BODY, LEG, AND FOOT MOTIONS

DESCRIPTION	SYMBOL	DISTANCE	TIME TMU
Foot Motion—Hinged at Ankle.	FM	Up to 4″	8.5
With heavy pressure.	FMP		19.1
Leg or Foreleg Motion.	LM —	Up to 6″	7.1
		Each add'l. inch	1.2
Sidestep—Case 1—Complete when lead-ing leg contacts floor.	SS-C1	Less than 12″	Use REACH or MOVE Time
		12″	17.0
		Each add'l. inch	.6
Case 2—Lagging leg must contact floor before next motion can be made.	SS-C2	12″	34.1
		Each add'l. inch	1.1
Bend, Stoop, or Kneel on One Knee.	B,S,KOK		29.0
Arise.	AB,AS,AKOK		31.9
Kneel on Floor—Both Knees.	KBK		69.4
Arise.	AKBK		76.7
Sit.	SIT		34.7
Stand from Sitting Position.	STD		43.4
Turn Body 45 to 90 degrees— Case 1—Complete when leading leg contacts floor.	TBC1		18.6
Case 2—Lagging leg must contact floor before next motion can be made.	TBC2		37.2
Walk.	W-FT.	Per Foot	5.3
Walk.	W-P	Per Pace	15.0

TABLE X—SIMULTANEOUS MOTIONS

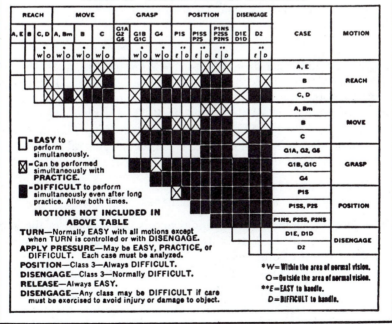

= EASY to perform simultaneously.

= Can be performed simultaneously with PRACTICE.

= DIFFICULT to perform simultaneously even after long practice. Allow both times.

MOTIONS NOT INCLUDED IN ABOVE TABLE

TURN—Normally EASY with all motions except when TURN is controlled or with DISENGAGE.

APPLY PRESSURE—May be EASY, PRACTICE, or DIFFICULT. Each case must be analyzed.

POSITION—Class 3—Always DIFFICULT.

DISENGAGE—Class 3—Normally DIFFICULT.

RELEASE—Always EASY.

DISENGAGE—Any class may be DIFFICULT if care must be exercised to avoid injury or damage to object.

*W = Within the area of normal vision.

O = Outside the area of normal vision.

**E = EASY to handle.

D = DIFFICULT to handle.

MTM Association for Standards and Research, Fair Lawn, New Jersey

Table 19–4), to determine the time required for a normal performance of the task. For example, if the right hand reached 20 inches to pick up a nut, the classification would be R20C and the time value would be 19.8 TMU. If, at the same time, the left hand reached 10 inches to pick up a cap screw, a designation of R10C with a TMU value of 12.9 would be in effect. The right hand would be limiting, and the 12.9 value of the left hand would not be used in calculating the normal time.

The tabulated values do not carry any allowance for personal delays, fatigue, or unavoidable delays; when analysts use these values to establish time standards, they must add appropriate allowances to the summary of the synthetic basic motion times. Proponents of MTM-1 state that "no fatigue allowance is needed in the vast majority of MTM-1 applications. The MTM-1 values are based on a work rate that can be sustained for eight hours, five days per week, for the working life of the employee providing he or she stays healthy."[8]

Today, MTM has received worldwide recognition. In the United States, it is administered, advanced, and controlled by the MTM Association for Standards and Research. This nonprofit association is one of 12 associations comprised by the International MTM Directorate. Much of MTM systems' success is the result of an active committee structure made up of members of the association. The MTM family of systems continues to grow. In addition to MTM-1, it has introduced MTM-2, MTM-3, MTM-V, MTM-C, MTM-M, Adam, 4M Computerized Work Measurement, MTM-MEK, and MTM-UAS.

MTM-2

In an effort to further the application of MTM to work areas where the detail of MTM-1 would economically preclude its use, the International MTM Directorate initiated a research project to develop less refined data suitable for the majority of motion sequences. The result of this effort was MTM-2. Defined by the MTM Association of the United Kingdom, it is a system of synthesized MTM data and is the second general level of MTM data. It is based exclusively on MTM and consists of:

1. Single basic MTM motions.
2. Combinations of basic MTM motions.

The data is adapted to the operator and is independent of the workplace or equipment used. It is not possible to replace any element in MTM-2 by means of other elements in MTM-2.

In general, MTM-2 should find application in work assignments where:

1. The effort portion of the work cycle is more than one minute.

---

[8] Delmar W. Karger and Walton M. Hancock, *Advanced Work Measurement* (New York: Industrial Press, 1982), p. 56.

FIGURE 19–6
Percentage variation of MTM-1 when compared with MTM-2 at increasing cycle lengths

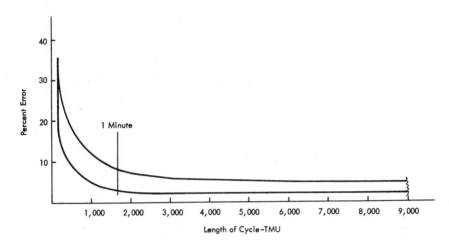

2. The cycle is not highly repetitive.
3. The manual portion of the work cycle does not involve a large number of either complex or simultaneous hand motions.

The variability between MTM-1 and MTM-2 depends to a large extent on the length of the cycle. This is reflected in Figure 19–6, which shows the range of percentage deviation of MTM-2 from MTM-1. This range of "error" is considered to be the expected range 95 percent of the time.

MTM-2 recognizes 11 classes of actions, which are referred to as "categories." These 11 categories and their symbols are:

GET	G
PUT	P
GET WEIGHT	GW
PUT WEIGHT	PW
REGRASP	R
APPLY PRESSURE	A
EYE ACTION	E
FOOT ACTION	F
STEP	S
BEND & ARISE	B
CRANK	C

In using MTM-2, analysts estimate distances by classes; these distances affect the times of the GET and PUT categories. As in MTM-1,

FIGURE 19–7

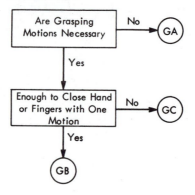

they base the distance moved on the length of the path traveled by the knuckle at the base of the index finger in hand motions and the path traveled by the fingertips if only the fingers move.

The codes for the five tabulated distance classes are:

Inches	Code
0–2	2
Over 2–6	6
Over 6–12	12
Over 12–18	18
Over 18	32

Usually they consider the categories GET and PUT simultaneously. Three variables affect the time required to perform both of these categories. These variables are the case involved, the distance traveled, and the weight handled. GET can be considered a composite of the therbligs reach, grasp, and release, while PUT is a combination of the therbligs move and position.

The three cases of GET are A, B, and C. Case A implies a simple contact grasp, such as when the fingers push an ashtray across the desk. If an object such as a pencil is picked up by simply enclosing the fingers with a single movement, it is a case B grasp. If the type of grasp is neither an A nor a B, then it is a case C GET.

Analysts can resort to the decision diagram (Figure 19–7) to assist in the determination of the correct case of GET. Tabular values in TMU of the three cases of GET applied to each of the five coded distances appear in Table 19–5.

TABLE 19–5
Tabular values for GET

		Distance code				
Case	Description	2	6	12	18	32
GA............	No grasp necessary	3	6	9	13	17
GB	Simple closing of fingers or hand	7	10	14	18	23
GC	Any other grasp	14	19	23	27	32

PUT involves moving an object to a destination with the hand or fingers. It starts with an object grasped and under control at the initial place and includes all transporting and correcting motions necessary to place the object. PUT ends with the object still under control at the intended place.

PUT is selected after considering three variables:

1. PUT is distinguished by the correcting motions employed.
2. The distance moved.
3. Weight of the object or its resistance to motion.

Just as there are three cases of GET, there are three cases of PUT. The case of PUT depends on the number of correcting motions required. A correction is an unintentional stop, hesitation, or change in the direction of the motion at the terminal point.

1. PA: No correction—this is evident as a smooth motion from start to finish and is the action employed in laying aside an object, or placing it against a stop or in an approximate location. This is the most common PUT.
2. PB: One correction—this PUT occurs most often in positioning easy to handle objects where a loose fit occurs. It is difficult to recognize. The decision diagram (Figure 19–8) has been designed to identify it by exception.
3. PC: More than one correction—multiple corrections or several very short unintentional motions are usually obvious. These unintentional motions are caused generally by difficulty of handling, close fits, lack of symmetry of engaging parts, or uncomfortable working positions.

Analysts identify cases of PUT by the decision model shown as Figure 19–8. In cases of doubt when using this model, analysts assign the higher class. An explanation of the three cases of PUT, as well as tabular values for each class applied to the five coded distances, is given in Table 19–6.

PUT is accomplished in one of two ways: insertion or alignment. An insertion involves placing one object into another, such as a shaft into a

FIGURE 19–8

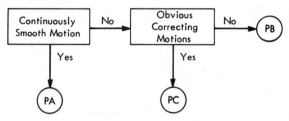

sleeve. When there is an insertion, the terminal point for a correction is the point of insertion. An alignment involves orienting a part on a surface, such as bringing a rule up to a line.

The variable distances are similar to GET. When an engagement of parts follows a correction, an additional PUT is allowed when the engagement distance exceeds 1 inch.

In MTM-2 the authors determine weight similarly to MTM-1. The time value addition for GET WEIGHT is 1 TMU per effective kilogram. Thus, if a load of 6 kilograms is handled by both hands, the time addition due to weight would be 3 TMU since the effective weight per hand is 3 kilograms.

For PUT WEIGHT, additions are 1 TMU per 5 kilograms of effective weight up to a maximum of 20 kilograms.

The category REGRASP is also similar to MTM-1. Here, however, a time of 6 TMU is assigned. The authors of MTM-2 point out that for a REGRASP to be in effect, the hand must retain control.

APPLY PRESSURE has a time of 14 TMU. The authors point out that this category can be applied by any member of the body and that the maximum permissible movement for an apply pressure is ¼ inch.

EYE ACTION is allowed under either of the following cases:

1. When eyes must move to see various aspects of the operation involving more than one specific section of the work area.
2. When eyes must concentrate on an object to recognize a distinguishable characteristic.

TABLE 19–6
Tabular values for PUT

		Distance code				
Case	*Description*	*2*	*6*	*12*	*18*	*32*
PA	Smooth continuous motion	3	6	11	15	20
PB	Some irregularity in motion pattern	10	15	19	24	30
PC	Correcting motions obvious	21	26	30	36	41

FIGURE 19–9

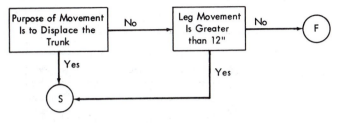

The estimated value of EYE ACTION is 7 TMU. The value is only allowed when EYE ACTION is independent of hand or body motions.

CRANK occurs when the hands or fingers move an object in a circular path of more than ½ revolution. A PUT is a CRANK of less than ½ revolution. Under MTM-2 the category CRANK has only two variables: the number of revolutions and the weight or resistance. A time of 15 TMU is allotted for each complete revolution. Where weight or resistance is significant, PUT WEIGHT is applied to each revolution.

FOOT movements are 9 TMU; and STEP movements, 18 TMU. The time for STEP movement is based on a 34-inch pace. The decision diagram (Figure 19–9) can be helpful in ascertaining whether a given movement is a step or a foot movement.

The category BEND & ARISE occurs when the body changes its vertical position. Typical movements characteristic of BEND & ARISE include sitting down, standing up, and kneeling. A time value of 61 TMU is assigned to this category. The authors indicate that when an operator kneels on both knees the movement should be classed as 2 B.

Table 19–7 summarizes MTM-2 values. Motions performed simultaneously with both hands cannot always be performed in the same time as motions performed by one hand only. Figure 19–10 reflects motion

TABLE 19–7
A summary of MTM-2 data (all time values are in TMU)

Code	GA	GB	GC	PA	PB	PC
2.............	3	7	14	3	10	21
6.............	6	10	19	6	15	26
12............	9	14	23	11	19	30
18............	13	18	27	15	24	36
13............	17	23	32	20	30	41

GW – 1/KG   PW – 1/5 KG

A	R	E	C	S	F	B
14	6	7	15	18	9	61

FIGURE 19-10

SIMULTANEOUS MOTIONS

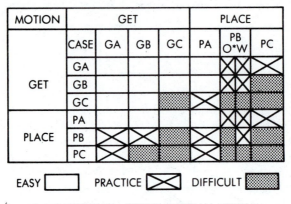

MOTION		GET				PLACE		
	CASE	GA	GB	GC	PA	PB O*W	PC
GET	GA							
	GB							
	GC							
PLACE	PA							
	PB							
	PC							

EASY ☐   PRACTICE ☒   DIFFICULT ▨

*O = OUTSIDE; W = WITHIN NORMAL VISION

FIGURE 19-11
MTM-2 simultaneous hand motion
allowances

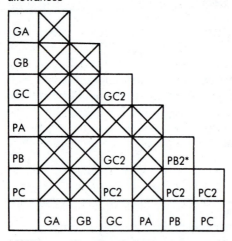

* If PB _____ is performed simultaneously with PB
_____ , an addition of PB2 is made only if the actions
are outside the normal area of vision.

patterns where the time required for simultaneous motions is the same as
that required for easy motions performed by one hand. In these instances,
an open rectangle appears. An X in the rectangle indicates that with
practice simultaneous motions can be made. A darkened rectangle indicates that it is difficult, even with practice, to perform the motions simul-

taneously. Figure 19–11 shows how much additional time difficult simultaneous motions require.

As with all fundamental motion data systems, the novice should not try to apply the data until he has been properly trained in their use and application.

MTM-3

The third level of Methods-Time Measurement is MTM-3. This level was developed to supplement MTM-1 and MTM-2. MTM-3 is helpful in work situations where an interest in saving time at the expense of some accuracy makes it the best alternative.

Analysts use MTM-3 to study and improve methods, evaluate alternative methods, develop standard data and formulas, and establish standards of performance. MTM-3 is not used for operations that require either eye focus or eye travel time since the data do not consider these motions.

The accuracy of MTM-3 is within ±5 percent with a 95 percent confidence level when compared with MTM-1 analysis in cycles of approximately four minutes, exclusive of limiting process time, and in operations not utilizing eye focus and eye travel times. It has been estimated that MTM-3 can be applied in about one seventh the time of MTM-1.

The MTM-3 system consists of only these four categories of manual motions:

1. Handle: A motion sequence with the purpose of getting control of an object with the hand or fingers and placing the object in a new location.
2. Transport: A motion with the purpose of placing an object in a new location with the hand or fingers.
3. Step and Foot Motions: These are the same as defined in MTM-2.
4. Bend and Arise: These, too, are the same as defined in MTM-2.

Table 19–8 presents MTM-3 data. Ten time standards ranging from 7 to 61 TMU form the basis for the development of any standard subject to the limitations earlier noted.

TABLE 19–8
Maynard Research Council, Inc.: MTM-3

		Handle		Transport	
Inches	*Code*	*HA*	*HB*	*TA*	*TB*
6	6	18	34	7	21
6	32	34	48	16	29
		SF 18		B 61	

MTM-V

MTM-V was developed by Svenska MTM Gruppen, the Swedish MTM Association for use in metal cutting operations. It is of particular use in machine shops with short runs. MTM-V provides for work elements involved in: (1) bringing the work to the jig, fixture, or chuck; removing the work from the machine; and placing it aside; (2) operating the machine; (3) checking the work to assure quality of output; and (4) cleaning the nip point area of the machine to maintain facility output and product quality.

MTM-V does not cover process time involving feeds and speeds. Analysts use this system to establish set-up times for all typical machine tools. Therefore, such elements as setting up and dismounting fixtures, jigs, stops, cutting tools, and indicators can be prepriced.

All manual cycle times of 24 minutes (40,000 TMU) or more established by MTM-V are within ±5 percent of that produced by MTM-I at the 95 percent confidence level. MTM-V is about 23 times faster than MTM-1.

MTM-V has 12 groups of elements making up its standard data system. These elements fall into two categories: simple and complex. These elements and their symbols follow:

Elements	Symbol
Simple	
Handle object	HO
Handle tool	HH
Get or return	HL
Rotate	SK
Inspect	GR
Operate	MA
Complex	
Fasten/loosen	FL
Measure	MT
Process	BE
Couple object	KP
Gage	KO
Mark	MR

The authors have subdivided these 12 groups to provide 488 time values. All of these values are traceable to MTM-1 data. The five-place alpha numeric coding system developed for MTM-V is shown in Figure 19–12. Table 19–9 illustrates an MTM-V analysis for a mark and drill operation. The process time (PT) of 845 TMU was determined by actual measurement or developed from the feed and speed of the drill. Analysts use MTM-V in the same way as other MTM systems; that is, they can use it with data developed from other sources. MTM-V is especially useful in developing standard data for specific machine tools.

FIGURE 19–12
Example of the coding structure used with MTM-V

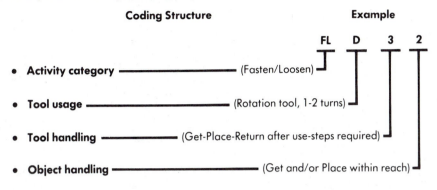

MTM-C

MTM-C is "a two level standard data system used to establish time standards for clerical-related work tasks." Typical clerical areas of application for MTM-C include key punching, filing, data entry, and typing. The system is widely used in the banking and insurance industries. Both levels of MTM-C are traceable to MTM-1 data.

TABLE 19–9
A sample MTM-V analysis mark and drill operation

Description	F	Motion	TMU
Mark work piece		MRA 30	190
Drill into spindle		HO2	40
Put part into vise		HO2	40
Crank down drill		MAC2	50
Set part to drill point		MAF0	30
Tighten vise by hand		SKC0	100
Pin to handle—strike—tighten		FLA20	110
Additional strikes		FLA10	70
Start machine		MAA2	20
Lower drill <2 revolutions		MAC2	50
Process time		PT	—
Stop machine		MAA2	20
Pin to handle—strike—loosen		FLA20	110
Open vise and remove piece		SKA2	70
Remove drill—prying tool		FLA22	140
Brush chips away		BED30	280
Lift wedge		HO2	40
		Total TMU:	1360

= 0.82 Minutes

Level 1 categories and their symbols are:

Level 1 elements	Symbol
Get Place	11 X X X X
Open Close	21 X X X X
Fasten Unfasten	31 X X X X
Organize File	4 X X X X X
Read Write	5 X X X X X
Typing	6 X X X X X
Handling	7 X X X X X
Walk Body Motions	8 X X X X X
Machines	9 X X X X X

The system provides three distinct ranges for reach and move (Get Place). A six-place numeric coding system (similar to MTM-V) provides a detailed description of the operation being studied. MTM-C develops standards in the same way as the other MTM systems. Analysts can combine it with existing proven standard data or standard data developed from other sources and/or techniques. MTM-C is available in manual or automated forms; for the latter an MTM-C Data Set can be incorporated into 4 M or ADAM.

A brief description of the nine level 1 categories used in MTM-C follows:

1. Get Place: This category includes those basic divisions of accomplishment involved to get an object, move it aside without relinquishing control, and release it. For example, the coding and description of an element in this category might be: 112210—Get small stack with medium move.

2. Open Close: Such typical operations as opening books, doors, drawers, binder rings, zippered objects, envelopes, and files are characteristic of this category. An example of the coding for a representative operation would be: 212100—Open hinge cover, medium.

3. Fasten Unfasten: This category includes the attaching and removal of clips, clamps, bands, and staples used to join materials. A representative coding for this familiar work element is: 312130—Fasten with a large paper clip.

4. Organize File: This category includes the basic elements involved with filing activities and some of the organizational handling of work directly or indirectly related to filing. An example of the coding and description of this category is: 410400—Arrange a pile into a stack.

5. Read Write: This category includes a prose reading speed of 330 words per minute. Writing time values have been developed for letters, numerals, and symbols. Values are a weighted average based on frequency of occurrence of each type of character in normal prose. An example of coding and a representative description would be: 510600—Read average prose, per word.

6. Typing: This category includes all of the actions related to preparation for typing, the manual typing functions, and related process times of the machine. An example of the coding and the description of this category follows: 613530—Insert single object in typewriter, long distance.

7. Handling: This category includes all of the clerical activities not covered in the other categories. An example of the coding and the description of an element in this category might be: 760600—Adhere envelope flap.

8. Walk Body Motions: This category includes walking values based on "per pace." Body motions include sitting, standing, and horizontal and vertical body movements while in chair. An example of the coding and description of an element in this category follows: 860002—Move while in swivel chair.

9. Machines: The machine data are representative of a group of similar types of equipment. Data for keyboard calculators and key punch machines are typical examples of this category.

Level 2 elements and their symbols are:

Level 2 elements	Symbol
Aside	A
Body Motions	B
Close	C
Fasten	F
Get	G
Handling	H
Identify	I
Locate File	L
Open	O
Place	P
Read	R
Typing	T
Unfasten	U
Write	W

Level 2 data is directly traceable to Level 1 and to MTM-1. A brief description of each element in Level 2 follows:

1. Get/Place/Aside: These elements are applied collectively or separately. The coding and element example of this category with collective basic divisions would be: G5PA2—Get a pencil for use and aside it later.

2. Open/Close: Getting the object being opened or closed is included in these data. These data are applied individually or in combination as follows: C65—Close string, tie envelope or OC4—Open and close binder rings.

3. Fasten/Unfasten: In the case of fasten (F), the element is made up of the getting of the objects involved and the actual fastening action. In the case of unfasten (U), the element includes the getting of the objects involved and the unfastening action.

4. Identify: The data for this element includes eye travel time values along with eye focuses required to identify (I) single or multiple words and sets of numbers.

5. Locate File: The data for this element is for typical filing activities. The first position of the coding is L. The second position is also a letter that corresponds to the activity of filing such as LI (insert), LR (remove), LT (tilt and replace).

6. Read/Write: The reading data includes the reading of words and single numbers and/or characters. It also contains detailed Read-and-Compare, as well as Read-and-Transcribe data. The writing data includes common clerical items, such as address, date, initials, and names. The coding and element description for two representative elements would be RW20—Read 20 words; RCN25—Read and compare 25 numbers.

7. Handling: This element includes the actual paper handling activities from Level 1, "organize" data and "handling" data. In the majority of elements, objects have been obtained with a "get" as well as the action handling elements. In the coding for handling elements, H is the first coding position. The second position is the initial letter of the element activity. An example of coding for folding a sheet with 2 folds would be: HF12.

8. Body Motions: These elements include walking, sitting and standing, bending, and arising, and the horizontal body motions alone or in a chair.

9. Typing: These elements include three major sections of data known as: Handling, Keystroke, and Correction. Examples of the coding and descriptions follow: THI32—Put three sheets and two carbons in machine and remove; TKE17E—Type one 7-inch line with an electric machine, elite; TCL41—Correct single error on four sheets with correction fluid.

MTM-C Level 1 can be calculated faster than MTM-2. Also, the speed of MTM-C Level 2 is faster than MTM-3.

Assume that a standard for replacing a page in a three-ring binder is developed first using MTM-1 (see Table 19–10), then using MTM-C Level 1 (see Table 19–11), and finally using MTM-C Level 2 (see Table 19–12). Note how closely the three standards agree:

Techniques	Number of elements	Standard
MTM-1	57	577.8
MTM-C Level 1	21	577
MTM-C Level 2	11	575

TABLE 19–10

MTM-1 ANALYSIS OF MTM-C

VALIDATION

MTM ASSOCIATION FOR STANDARDS AND RESEARCH				ANALYST:		
ELEMENT TITLE:	Replace page in 3-ring binder			DATE:		
STARTS:	Get binder from shelf at left					
INCLUDES:	Get binder, open cover, locate correct page, open rings, replace old sheet,					
ENDS:	close rings, Aside binder to shelf					

LEFT HAND DESCRIPTION	F	LH MOTION	TMU	RH MOTION	F	RIGHT HAND DESCRIPTION
1. GET BINDER—OPEN COVER						
Reach to binder		R30B	25.8			
Grasp binder		G1A	2.0			
Move to desk		M30B	24.3			
Release		RL1	2.0			
Reach to cover		R7B	9.3			
Grasp edge		G1A	2.0			
Open cover		M16B	15.8			
Release		RL1	2.0			
			83.2			
2. LOCATE CORRECT PAGE						
Reach to edge	3	R3D	14.6	EF	2	Read first page data
Grasp	3	G1B	21.9			
Move up	3	M4B	10.5			
Regrasp		G2	20.7			
			—			
Move pages back		M8B	43.8	EF	2×3	Identify pages
Release		RL1	10.6			
Reach to hold		R8B	2.0			
Grasp		G5	10.1	R4B (circled)		To edge of page
			0.0	G5		Contact
Contact	3	G5 (circled)	8.0	MfB	4	Slide back up
			0.0	RL2	4	Release
Move	3	MfB (circled)	7.5	R1B	3	To corner
			0.0	G5		Contact
Regrasp pages		G2	5.6			
			87.6	EF	4×3	Identify pages
Move pages back		M8B	10.6			
Release		RL1	2.0			
			255.5			

TABLE 19–10 (concluded)

MTM-1 ANALYSIS OF MTM-C

MTM ASSOCIATION FOR STANDARDS AND RESEARCH	ELEMENT TITLE: Replace page in 3-ring binder STARTS: INCLUDES: Continues ENDS:	VALIDATION ANALYST: DATE:

3. REPLACE PAGE

LEFT HAND DESCRIPTION	F	LH MOTION	TMU	RH MOTION	F	RIGHT HAND DESCRIPTION
To ring		R7A	7.4	R7A		To ring
Grasp		G1A	2.0	G1A		Grasp
Pull open		APB	16.2	APB		Pull open
Open		MfA	2.0	MfA		Open
Release		RL1	2.0	RL1		Release
To edge of paper		R6D	10.1			
Grasp		G1B	3.5			
To basket		M30B	24.3			
Release		RL1	2.0			
To center ring		(R4B)	10.1	(R6D) [R-E]		To new sheet
Grasp		G1A	3.5	G1B		Grasp
Press to close		APB	15.2	M12C		To rings
Close		MfA	16.2	MfC		Align to ring
Release		RL1	2.0	P2SE		To ring
			16.2	P2SE		Align
			2.0	MfA		Down on rings
			8.6	RL1		Release
			2.0	R6B		To center ring
			16.2	G1A		Grasp
			2.0	APB		Press to close
				MfA		Close
			167.5	RL1		Release

4. CLOSE COVER AND ASIDE BINDER

LEFT HAND DESCRIPTION	F	LH MOTION	TMU	RH MOTION	F	RIGHT HAND DESCRIPTION
Reach to cover		R7B	9.3			
Grasp edge		G1A	2.0			
Close cover		M16B	15.8			
Release		RL1	2.0			
Reach to binder		R6B	8.6			
Grasp		G1A	2.0			
Regrasp		G2	5.6			
Move to shelf		M30B	24.3			
Release		RL1	2.0			
			71.6			

ELEMENT SUMMARY

	TMU
1. Get Binder-Open Cover	83.2
2. Locate Correct Page	255.5
3. Replace Page	167.6
4. Close Cover and Aside Binder	71.6
TOTAL	577.8

TABLE 19–11

	MTM-C OPERATION ANALYSIS		VALIDATION	
			Sheet of	

MTM ASSOCIATION FOR STANDARDS AND RESEARCH

MTM-C LEVEL 1

Replace page in 3-ring binder

DEPARTMENT: Clerical	ANALYST: CNR	DATE: 11/77

No.	Description	Reference	Element TMU	Occurrence per Cycle	TMU per Cycle
1.	OPEN BINDER				
	Get binder from shelf	113 520	21	1	21
	Aside to desk	123 002	22	1	22
	Get cover	112 520	14	1	14
	Open cover	212 100	15	1	15
2.	LOCATE CORRECT PAGE				
	Read on first page	510 000	7	2	14
	Locate approximate	451 120	16	3	48
	Identify page number	440 630	22	3	66
	Locate correct page	450 130	18	4	72
	Identify pages	440 630	22	3	66
3.	REPLACE PAGES				
	Get binder rings	112 520	14	1	14
	Open rings	210 400	21	1	21
	Get old sheet	111 100	10	1	10
	Aside sheet to basket	123 002	22	1	22
	Get new sheet	111 100	10	1	10
	Insert sheet in binder	462 104	64	1	64
	Get rings	112 520	14	1	14
	Close rings	222 400	21	1	21
4.	CLOSE COVER AND ASIDE BINDER				
	Get cover	111 520	8	1	8
	Close cover	222 100	13	1	13
	Get binder	112 520	14	1	14
	Aside binder to shelf	123 002	22	1	22
		TOTAL TMU PER CYCLE			571
		ALLOWANCES ____ %			
		STANDARD HOURS PER ____ UNIT			
		UNITS PER HOUR			

TABLE 19–12

MTM-C OPERATION ANALYSIS				VALIDATION
				Sheet of

| MTM ASSOCIATION FOR STANDARDS AND RESEARCH | MTM-C LEVEL 2 Replace page in 3-ring binder | | | |

| DEPARTMENT: Clerical | | ANALYST: CNR | | DATE: 2/77 |

No.	Description	Reference	Element TMU	Occurrence per Cycle	TMU per Cycle
	Get and aside binder	G5A2	29	1	29
	Open cover	O1	29	1	29
	Read first page	RN2	14	1	14
	Locate pages	LC12	129	1	129
	Identify pages	130	22	6	132
	Open rings	O4	35	1	35
	Remove sheet	G1A2	32	1	32
	New sheet on rings	HI14	84	1	84
	Close rings	C4	35	1	35
	Close cover	C1	27	1	27
	Aside binder	G5A2	29	1	29
	TOTAL TMU PER CYCLE				575

MTM-M

MTM-M has been defined by the US/Canada MTM Association as "a system of objective methods—time standard data, based on regression analysis of empirical data, for evaluation of operator work using a stereoscopic microscope." Karger and Hancock have defined MTM-M in a more practical manner stating that "MTM-M is a specialized methods—time functionally oriented standard data system that is not a higher-level MTM-1 based system, yet it is designed to produce time standards that are compatible with MTM-1 standards for work performed partially or totally under a binocular stereoscopic microscope having a magnification range not exceeding thirty power."[9]

In the development of MTM-M, basic times from MTM-1 were not used although the beginning and end point definitions of motion elements were compatible with MTM-1. The data used was original data developed through the efforts of the US/Canada MTM Association.

This system has four major tables and one subtable. All these tables are related to direction of movement. The five directions of movements and their symbols follow:

[9] Ibid., p. 185.

Symbol	Movement
II	Inside to inside
IO	Inside to outside
OO	Outside to outside
OI	Outside to inside
IF	Infield to final target

Analysts consider four variables in the selection of the appropriate data: (1) type of tool; (2) condition of tool; (3) terminating characteristic of the motion; and (4) distance/tolerance ratio.

Factors other than the direction of movement and these four variables have an impact on motion performance time. They include:

1. Tool load state, empty or loaded.
2. Microscope power.
3. Distance moved.
4. Positioning tolerance.
5. Purpose of the motion as determined by the manipulations involved at the motion termination. For example, workers may use tweezers to contact grasp an object, or to pick up an object.
6. Simultaneous motions.

MTM-M is a higher level system, similar to MTM-2. For example, to move a loaded tweezer from outside the microscope field to inside the microscope field, an analyst would code the element:

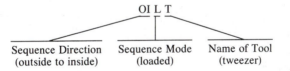

OI L T

Sequence Direction	Sequence Mode	Name of Tool
(outside to inside)	(loaded)	(tweezer)

Table 19–13 contains a portion of MTM-M data for the inside to inside direction of movement table. To illustrate use of this table, consider the element where an empty tweezer is used to contact grasp an object with a motion starting and ending within the microscope field at 15X. The required distance is 0.1 inch, the tolerance is 0.01 inch, and the motion is simultaneous. The initial coding would be IIET signifying an inside to inside direction of movement with an empty tweezer. Since a contact grasp is used, the next code element is A. The coding IIET-A signifies a specific row of the table for this element. The next step is to compute the distance tolerance relationships by dividing the distance moved of 0.1 inch by the target tolerance 0.01 inch. This value is 10. Looking across the "range" row, "5" is the assigned code range for distance divided by field size of 6 to 12. Reading down range number 5 to row IIET-A, leads to a TMU value of 11.0. This value is the basic time value to perform the motion. Since the motion was simultaneous, add 2.1 TMU (see Simo

TABLE 19–13

MTM-M

I.I. (Inside field to inside field)

Code Range (D/T)			1	2	3	4	5	6	7	8	9	10	11	
Range Distance + Tolerance			0.50 to 0.75	0.75 to 1.5	1.5 to 3.0	3.0 to 6.0	6.0 to 12	12 to 25	25 to 50	50 to 100	100 to 200	200 to 350	350 to 725	
Tool and Cond.	Code	Characteristic												Simo Additive
Grasping ET	A*	GTC (Tool contact GR)	4.3	5.7	7.5	9.2	11.0	12.8	14.6	16.3	18.1	19.6	21.3	2.1
	B	GT (Tool grasp)	3.7	7.6	12.2	16.7	21.3	26.0	30.7	35.2	39.8	43.8	48.2	12.8
LT	A*	No release - scrub	3.7	4.2	4.7	5.2	5.7	6.2	6.8	7.3	7.8	8.2	8.7	+
	B	RLTC – 2 DIM move	3.0	3.5	4.2	4.8	5.4	6.1	6.7	7.3	8.0	8.5	9.1	3.9
	C	RLTC – 3 DIM move	5.6	6.1	6.8	7.4	8.0	8.7	9.3	9.9	10.6	11.1	11.7	3.9
	D	RLT (Tool release)	4.5	6.4	8.6	10.7	12.9	15.2	17.4	19.5	21.7	23.6	25.7	9.6
EF	A	All conditions	10.6	12.6	15.0	17.5	19.9	22.4	24.9	27.3	29.7	31.8	34.2	0.0
LF	A	No release	3.3	5.5	8.0	10.5	13.0	15.7	18.3	20.8	23.3	25.5	28.0	15.7
	B*	RLM or RLMC	4.1	7.9	12.5	17.0	21.5	26.3	30.7	35.4	40.0	43.9	48.3	5...
ED	A	All conditions	5.0	8.6	12.7	16.9	21.1	25.4	29.7	33.9	38.1	41.7		
Probing ES	A	All conditions	8.9	11.8	15.3	18.9	21.7							
LS	A	All conditions	2.0	3.1	6.7	10.3								
EP	A	All conditions	4.6	6.4	8.6	12.4								
LP	A	All conditions	2.8	6.5	11.7									
Cutting EC	A	All conditions												
LC	A	All conditions												
Stripping EZ	A	Thermal - All conditions												

	Sequence direction	Tool load condition Name of tool

"Light faced type" - TMU values are e〳
+ No simultaneous motions occurre〳
*Power factor - magnification pow〳
Chart value is adjusted as follows〳

     ETA - When mag po〳
         0.68 (30 - 20〳
     LTA - When mag p〳
         subtract 0.0〳
     LFB - When mag〳
         0.94 (20 -〳

II   - Inside to inside
IO   - Inside to outside
OO - Outside to outside
OI   - Outside to inside field
IF   - Infield to target

E - Empty
L - Loaded

T - Tweezers
F - Fingers
D - Diagonal cutter
S - Soldering iron
P - Probe
C - Razor blade
Z - Thermal stripper

Additive column). The normal time then to perform this motion would be 13.1 TMU.

The total coding for this motion would be:

II—Inside to inside motion

E—Empty condition

T—Tweezer

A—Contact grasp at terminal point

5—D/T range number 5

S—Simultaneous motion

With the growing amount of microminiature manufacturing, the application of fundamental data similar to MTM-M will expand. Such data allows analysts to establish equitable standards that would be difficult to

FIGURE 19–13
Operator performing microscopic assembly of a delayed equilizer using a Bausch & Lomb stereo 1 fixed power pod. The operator performs two different silver soldering operations using two soldering irons (large and small tips) with various size land areas.

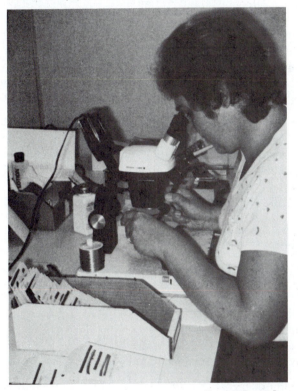

Courtesy Locus, Inc.

establish by stopwatch procedures. For example, Figure 19–13 illustrates the assembly of a delayed equilizer under a Bausch & Lomb stereo 1 fixed power pod. The operator performs two different silver soldering operations and uses tweezers, snips, fine pliers, swab, and two soldering irons to perform the operation. Establishing reliable elemental standards by direct observation is impossible. Only by using standard data similar to MTM-M or by micromotion procedures is it possible to establish sound elemental and operation standards for microscope work.

Other Specialized MTM Systems

Three other specialized MTM systems are mentioned briefly here. The first of these, MTM-TE, was developed for electronic tests. This system has two levels of data that were developed from MTM-1 for basic test application. Level 1 includes the elements: get, move, body motions, identify, adjust, and miscellaneous data. Level 2 includes: get and place, read and identify, adjust, body motions, and writing. A third level of data is also available in the form of synthesized data of Level 1. MTM-TE data does not cover "trouble shoot" relative to electronic test operations. It does, however, provide guidelines for investigation and recommendations for work measurement of this activity.

The second specialized system, MTM–MEK, was designed to measure one of a kind and small lot production. This two-level system developed from MTM-1 can analyze all manual activities as long as the following requirements are met:

1. The operation is not highly repetitive or organized, although it may contain similar elements that require different methods. The method used to perform a given operation typically varies from cycle to cycle.
2. The workplace, tools, and equipment used are universal in character.
3. The task is complex and necessitates employee training; yet the lack of a specific method requires a high degree of versatility by the operator.

The objectives of MTM–MEK are to:

1. Provide accurate measurement of an activity connected with one of a kind or small lot production.
2. Provide an easily definable description of unorganized work, thus generally identifying a procedure.
3. Provide fast application.
4. Provide accuracy relative to MTM-1.
5. Require minimum training and application practice.

The data in MTM–MEK consists of 51 time values in these eight categories: Get and place, handle tool, place, operate, motion cycles, fasten or loosen, body motions, and visual control.

In addition, the authors developed standard data blocks containing data for a wide range of assembly tasks in one of a kind small lot production. This data consists of 290 time values in the following categories: fasten, clamp and unclamp, clean and/or apply lubricant/adhesive, assemble standard parts, inspect and measure, mark and transport.

The third specialized system, MTM–UAS, is a third level system. The authors developed it to provide a process description, as well as to determine the allowed times in any activity related to batch production. MTM–UAS is applicable to various activities as long as the following characteristics of batch production are present:

1. Similar tasks.
2. Workplace specifically designed for the task.

3. Good levels of work organization.
4. Detailed instructions.
5. Well-trained operators.

The MTM–UAS analyzing system consists of 77 time values in seven of the eight categories used in MTM–MEK. These are: get and place, place, handle tool, operate, motion cycles, body motions, and visual control. MTM–UAS is about eight times faster than MTM-1. At cycle times of 4.6 minutes or more, the standard produced by MTM–UAS is within ±5 percent of that produced by MTM-1 with a 95 percent confidence level. Figure 19–14 illustrates the total absolute accuracy at 90 percent confidence level of the MTM systems.

COMPUTERIZED MTM WORK MEASUREMENT SYSTEMS

It has been estimated that the use of a computer base system, at the basic level, should result in application speeds 5 to 10 times faster than manual application speeds.[10] Computer-developed standards are more error free since the computer does not accept an input that is not logical. Software (programs) have been developed for the majority of the fundamental motion data systems. This software incorporates tables of simultaneous motions (for example Table X of MTM-1), handles frequencies of occurrence, and calculates elemental times, as well as cumulative times of successive elements. For example, if an analyst using MTM-1 inputs PISE for the left hand and PISE for the right hand, the program will query "within or without the area of normal vision?" A 32K memory microcomputer can accommodate the development of standards using any of the fundamental motion data systems described in this chapter.

MTM Association has developed two programs for computerized work measurement based on the original MTM-1 system. These two programs are 4M(MTM-1) and ADAM. 4M Data MOD II is recognized by the International MTM Directorate as a technically sound development with substantiated claims of performance. The MTM-1 motions, which comprise 90 percent of the manual work performed in the usual industrial situation, are reach, grasp, move, position, and release. Combining these motions creates the motion aggregates of get and place. Special get and place notations convey to the computer program, a basis for determining the basic MTM-1 motions along with the proper time values. The 4M Data System analyzes and calculates data the analyst places in four files. These are the element master file, analysis detail file, part-operation master file, and the operation element sequence file. The system then performs these functions: (1) element calculation; (2) standard calculations; and (3) mass update.

---

[10] Chester R. Dobrowski, Director–Systems and Services, MTM Association for Standards and Research.

FIGURE 19-14

Total absolute accuracy at 90 percent confidence level of the various MTM systems

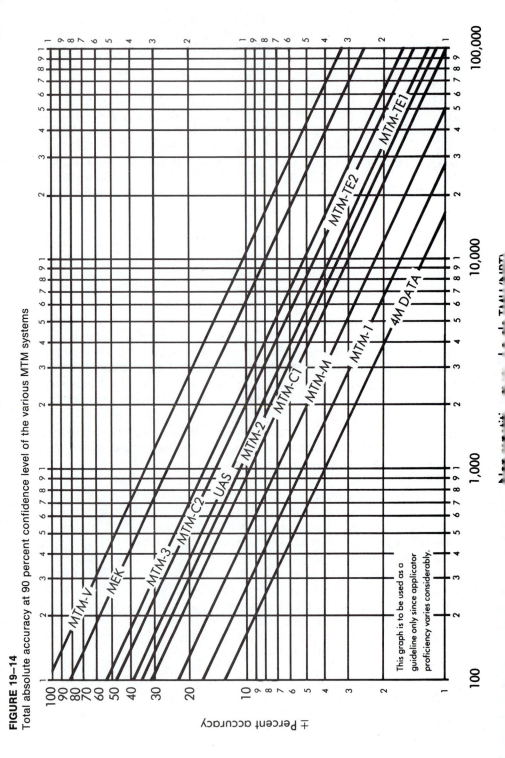

In making the element calculation, 4M Data applies MTM-1 data. To make operation standard calculations, the analyst lists in sequence the code number of the elements required to perform the operation. The 4M Data System retrieves the elements from the element file; applies allowances; calculates the standard; and prints or displays it showing the time per cycle, standard hours per unit, standard minutes per unit, and standard units per hour.

The mass update function allows the changing or deletion of any element with the resulting alteration being incorporated in all operation standards using that particular element. The program can produce six reports:

1. Operation standard report. This report lists the standard time to perform each element, the standard for the operation in various measurement units, and five method improvement indexes for the operation. The first of these is an index that shows the percentage of hand utilization during the manual content of the job. Fifty percent would indicate the use of only one hand, while 100 percent is an ideal usually not attained. The other four indexes suggest additional ways methods can be improved. One of these shows the percentage of reach, move, and body motion times within the analysis which would be affected by reducing the distances at the workstation. Another percentage is the amount of position time within the total. A third percentage is listed for the rest of the motion categories, principally the grasp and release times. The last index provides the amount of process time during which the operator does not perform manual work.

2. Operation analysis report. This report contains the same information as the operation standard report. In addition it includes all the 4M notations for both hands along with left hand, right hand, and net times to perform each operation.

3. Operation analysis report (abbreviated). This report is the same operation analysis report except it is printed in 8½″ × 11″ format.

4. Operator instruction report. This report provides the detail in the operation standard report but omits the time values.

5. Summary operation instructions. This report is used for long operations having many elements where a summary instruction report frequently suffices.

6. MTM-1 analysis. This report is a detailed methods analysis.

ADAM 2.0

Automated Data Application and Maintenance is a complete system for creating and maintaining standard data, as well as labor standards. The user is free to combine basic elements to form intermediate levels of elements and/or use basic elements developed from time study or other means. ADAM uses MTM-2 or MTM-UAS as its generic database. Some of the special features of ADAM 2.0 are:

1. Outputs to either a CRT monitor or to a printer.
2. Built-in editing program allows rapid revisions or corrections.
3. Accepts and applies compound frequencies of occurrence.
4. Accepts data entered in free format.
5. Accepts, stores, and evaluates formulas entered in free format.
6. Stores and retrieves standard data elements.
7. Produces "where used" report and makes mass changes.
8. Uses defined header information and summary calculations.
9. Provides an activity log.

Both 4M and ADAM can develop standards and operation instructions using the MTM functional systems (MTM-V, MTM-C, MTM-M, MTM–MEK, and MTM–UAS). For example, Figure 19–15 illustrates an analysis of replacing a page in an account book using ADAM/MTM-C, and Figure 19–16 shows an analysis of the same element with 4M/MTM-C. Both analyses provide identical times of 575 TMU.

MOST

An outgrowth of MTM called MOST (Maynard Operation Sequence Technique) is a simplified system developed by Kjell B. Zandin and originally applied at Saab-Scania in Sweden in 1967. H. B. Maynard and Company is currently marketing MOST. The company states that analysts can establish MOST standards at least five times faster than MTM-1 standards, with little if any sacrifice in accuracy.

MOST utilizes larger blocks of fundamental motions than MTM-2 and, consequently, analysis of the work content of an operation can be made faster. In contrast to MTM-2, which is built around 37 time values (see Table 19–4) for describing manual work, MOST utilizes only 16 time fragments. MOST identifies three basic sequence models: general move, controlled move, and tool use.

The general move sequence identifies the spatial free movement of an object through the air, while the controlled move sequence describes the movement of an object when it remains in contact with a surface or is attached to another object during the movement. The tool use sequence has been developed for the use of common hand tools.[11]

To identify the exact way a general move is performed, analysts consider four subactivities: action distance, which is primarily horizontal distance; body motion, which is mainly vertical; gain control; and place. Analysts assign time-related index numbers to the applicable subactivity. MOST uses index numbers 0, 1, 3, 6, 10, and 16. It is relatively easy to memorize these values and their application to the four subactivities of general move.

---

[11] Kjell B. Zandin, *MOST Work Measurement Systems* (New York: Marcel Dekker, 1980), pp. 68–72.

FIGURE 19–15

An example using ADAM/MTC-C analysis for replacing a page in an account book

ADAM/MTM-C ANALYSIS

MTM ASSOCIATION FOR STANDARDS AND RESEARCH

Code: AB010

Name: replace page in account book

LINE DESCRIPTION	ELEMENT	TIME	FREQ	TOTAL
1. GET AND ASIDE ACCOUNT BOOK, OPEN COVER				
Get, aside-appr- account book	G5A2	29	1	29
Open account book	O1	29	1	29
Read page numbers	RN	7	2	14
2. LOCATE PAGES, REMOVE AND REPLACE				
Locate sheet(s)	LC12	129	1	129
Identify (no et) 3 words/to 9 dgts	130	22	6	132
Open binder rings	O4	35	1	35
Get, aside-appr- old page	G1A2	32	1	32
Insert-bndr rngs- sheet(s)	HI14	84	1	84
3. CLOSE RINGS AND COVER, ASIDE BBOK				
Close binder rings	C4	35	1	35
Close account book	C1	27	1	27
Get, aside-appr- account book	G5A2	29	1	29
Manual time				575
Machine time				0
Stnd. minutes				0.397
Units/stnd. hr.				151.23

FIGURE 19-16
An example using 4M/MTM-C analysis for replacing a page in an account book

4M/MTM-C ANALYSIS

| 4M DATA SYSTEM | R0715 | 4M ELEMENT ANALYSIS | | | | Requested by DJR | Date 07/19/ | Page 1 |

Element AB010   Replace page in account book   Learning level 100
Practice opportunity _____ Updated _____   Total mu—MANUAL   5750
                                                             —PROCESS   0

		LH Motions	RH or Body Motions	Freq.	Process Time	LH	RH	Net Manual
010	G5A2	Get and aside single object to approx location			0			290
020	O1	Open book, cover or folder			0			290
030	RN	Read number or word (1–3 digits or 1–10 words per word)		2	0			140
040	LC12	Locate exact sheet in folder of 11–30 sheets & identify			0			1290
050	I30	Identify 3 words or 7–9 digits— no eye travel		6	0			1320
060	O4	Open binder rings			0			350
070	G1A2	Get and aside sheet, card or pad to approximate location			0			320
080	HI14	Insert sheet(s) onto binder rings			0			840
090	C4	Close binder rings			0			350
100	C1	Close book			0			270
110	G5A2	Get and aside single object to approximate location			0			290
							TOTAL MU	5750

| MAI 0 Percent | RMB 0 Percent | GRA 0 Percent | POS 0 Percent | PROC 0 Percent |

About 50 percent of manual work occurs as general move. A typical general move may include the subactivities of walking to a location, bending to pick up an object, reaching and gaining control of the object, arising after bending, and placing the object.

The controlled move sequence covers such manual operations as cranking, pulling a starting lever, turning a steering wheel, or engaging a starting switch. In performing controlled move sequences, the following subactivities may prevail: action distance, body motion, gain control, move controlled, process time, and align.

The final sequence in MOST is tool use/equipment use. Cutting, gaging, fastening, and writing with tools are all covered by this sequence. The tool use/equipment use model embraces a combination of general move and controlled move activities. Other subactivities unique to this activity include: fasten, loosen, cut, surface treat, record, think, and measure.

MOST Work Measurement Systems have two adaptations: Mini and Maxi MOST. Mini MOST measures identical, short-cycle operations, and Maxi MOST measures long-cycle operations with significant variation in actual method from cycle to cycle.

All MOST Work Measurement Systems are available in both manual and computerized versions. The computerized version permits the retrieval of suboperation data and the arithmetic operations involved to develop a standard of performance for the input characteristics of the method under study.

MOST is another predetermined motion data system that can prove beneficial to work measurement analysts. Using MOST, analysts can establish standards more rapidly than with the more detailed analysis of MTM-1 and MTM-2. However, the more detailed analysis should establish more reliable standards, especially when the cycle time is short and/ or there is little variation in the operator's method in successive cycles and the number of steps required in each cycle is minimal. When cycle time is relatively long and where a deviation in the method is repeated in cycles, MOST establishes a valid standard more economically.

Micro and Macro Motion Analyses

Standards, International, a work management consulting firm headquartered in Chicago, Illinois, developed two specialized predetermined time systems; MICRO Motion Analyses for precise methods specifications and time standards and MACRO Motion Analyses for general purpose data. These systems were developed to provide improvement over MTM and Work-Factor, which the Standards, International, firm had been using for several years in conjunction with their consulting work. Specifically, the existing systems did not appear to be adequate for some special types of motions, and describing these motions and assigning appropriate time values for these motions were subject to the individual user's judgment. Also, some individual analysts found some of the tables difficult to use since interpolation was often required.

The resulting tables were developed by Standards, International, with much input from several of their clients.

These basic tables follow. See Tables 19–14A through 19–14F. The author urges the reader to review these tables in detail. Having carefully read the sections on Work-Factor and MTM, you should have no difficulty in understanding the meaning and applied method of the elements and their respective time values. As in MOST, these values have been proven valid in thousands of applications.

MODAPTS

The original database for MODAPTS was developed by G. C. Heyde, currently a member of the board of directors of the International MODAPTS Board. Mr. Heyde had been using Master Standard Data (MSD), founded by Dick Crossan and Harold Nance in 1962. See Figure 19–1. However, he was desirous to use a method that would allow the development of sound standards more easily and rapidly. He became familiar with MTM-2 in the early 1960s, and using it as a base developed a system that contained only whole integer time values and could be memorized easily. Then in 1966, MODAPTS was introduced. The name is an acronym for MODular Arrangement of Predetermined Time Standards. Today, the MODAPTS database has 44 elements. The system is based on the idea that all body movements can be expressed in terms of multiples of a single unit of time called a MODE. A MODE has been defined as the normal time required to complete a simple finger move and has been assigned a value of 0.129 seconds or 0.00215 minutes. Every motion is identified by a two-part code, the first being an alphabetic component to identify the body part involved and the second a multiplier of one MODE to give the time to complete the activity.

For example, a finger move is coded M1 and its normal time would be 0.00215 minutes. A hand move, coded M2 would be given 0.0043 minutes, and a forearm move is coded M3. Body motions are coded similarly; a foot move is F3, eye use E2, and a bend and arise is coded B17. All MODAPTS values were derived empirically from testing in the field involving more than 20 industrial organizations, a wide range of jobs, and many factory people.

Studies have indicated that when compared to MTM-1, MODAPTS standards usually fall within 2 to 3 percent greater. Thus, MODAPTS standards will establish standards that are 2 to 3 percent on the loose side when compared to MTM-1.

MODAPTS elements are presented in three groups: movement elements, terminal elements, and supporting elements. There are elements for both small/light objects and for large/heavy objects.

TABLE 19–14

Multivariable chart with two points of entry and with reference included for element: up spindle, swing and traverse head, and down spindle (Western radial drill)

Traverse of Head in Inches / Maximum Swing of Head in Inches

Staircase of Maximum Swing values (in inches) listed under each Traverse column:

Traverse of Head (in)	Maximum Swing of Head (in)
0-1	0-1; 0-1, 2; 3, 4, 6; 8, 10
2	0-1; 4, 8, 10
3	0-1; 8, 10
4	0-1, 2, 3; 4, 6, 10
5	0-1; 2, 3, 6; 10
6	0-2, 3; 4, 8, 10
7	0-1; 2, 4, 6; 10
8	0-1, 2, 3; 6, 10
9	0-1; 3, 6, 10
10	0-1; 3, 6, 10

Depth of Previous Hole in Inches

Reference Line	0-1	2	3	4	5	6	7
	.010	.013	.015	.017	.019	.022	.024
.038	.048	.051	.053	.055	.057	.060	.062
.040	.050	.053	.055	.057	.059	.062	.064
.042	.052	.055	.057	.059	.061	.064	.066
.044	.054	.057	.059	.061	.063	.066	.068
.046	.056	.059	.061	.063	.065	.068	.070
.048	.058	.061	.063	.065	.067	.070	.072
.050	.060	.063	.065	.067	.069	.072	.074
.052	.062	.065	.067	.069	.071	.074	.076
.055	.065	.068	.070	.072	.074	.077	.079
.058	.068	.071	.073	.075	.077	.080	.082
.060	.070	.073	.075	.077	.079	.082	.084
.063	.073	.076	.078	.080	.082	.085	.087
.066	.076	.079	.081	.083	.085	.088	.090
.069	.079	.082	.084	.086	.088	.091	.093
.072	.082	.085	.087	.089	.091	.094	.096
.076	.086	.089	.091	.093	.095	.098	.100
.079	.089	.092	.094	.096	.098	.101	.103
.083	.093	.096	.098	.100	.102	.105	.107
.088	.098	.101	.103	.105	.107	.110	.112

TABLE 19–14A

VISUAL & MENTAL PROCESSES	CODE	TIME
READ — Per Word	ER	50
INSPECT — Per Criteria — Per Occurrence	EI	50
RECALL, DECIDE, REACT or CALCULATE — Per Criteria — Per Occurrence	MA	50
MOVE HEAD TO MICROSCOPE & SEE	HM	130
WRITE — Per Character	WC	90

MOTION-PATTERNS

TOOL HANDLING					
TYPE OF TOOL	MOTION	DISTANCE			
	PATTERN	9"	18"	30"	36"
Screw Driver, Mallet Spin-Tites, Knives File (w/Rd. Handle)	OX-RH-E OX-RH-G	210 180	270 230	340 290	360 310
Pliers, Wire Strippers Scissors, Cutters	OX-DH-E OX-DH-G	240 200	300 260	360 310	380 330
Pencils, Brushes	OX-TH-E OX-TH-G	210 170	270 220	330 280	350 300
Soldering Iron: — in Holster	OX-SI-E OX-SI-G	310 280	380 340	440 390	470 410
Air Tool: — on Bench	OX-ATB-E OX-ATB-G	260 240	340 300	400 360	430 380
Air Tool: — in Holster	OX-ATH-E OX-ATH-G	370 340	450 410	520 470	540 490
Air Tool: — Suspended	OX-ATS-E OX-ATS-G	170 140	220 170	260 210	270 220

Motion Pattern E: Pick up tool; move directly to work; place tool aside.
Motion Pattern G: Pick up tool; move to general area (move to assemble, etc., must be added); place tool aside.

RELATED ELEMENTS	CODE	TIME
TWEEZERS GRASP	GT	140
PALM & UNPALM TOOL	PXT	30
START THREADS	STT	90
THREAD ON OR OFF — Per Twist	TO	30
USE: SCREWDRIVER — Per Twist	SDT	70
NUT RUNNER — Per Twist	NRT	70
BOX OR END WRENCH — Per Turn	BEW	180
RATCHET — Per Turn	RW	60
HAMMER — Per Hit	HB	90
TIGHTEN OR LOOSEN w/HAND TOOL	ASYX	100

TABLE 19–14A (concluded)

STANDARDS, INTERNATIONAL®
INCORPORATED
WORK MANAGEMENT SPECIALISTS

MACRO™
MOTION ANALYSES

Do not attempt to use these tables to determine standards unless you understand the proper application of the data.

This note of caution is presented to prevent the difficulties that may result from misapplication of the time values. Values shown are in decimal minutes (to 4 places) and at required time.

Compatible selections: MTS, W-F, DMT, ETS or MICRO. For interchange with MTM, UAS, MODAPTS, or MSD, add 25% allowance and convert to decimal hours.

NOTE: These are condensed tables; more detail and fuller tables are provided in the manuals.

STANDARDS, INTERNATIONAL INC.
RESEARCH ENGINEERS & MANAGEMENT CONSULTANTS
Chicago, Illinois

Source: © 1989 All rights reserved. Permission courtesy of Clifford N. Sellie, CEO, Standards, International.

TABLE 19–14B

M<small>ACRO</small>™ MOTION ANALYSES

Frequently used General Purpose Data Elements recommended for:

1. Evaluation of manual motions (and costs) by methods personnel and production supervisors.

2. Development of standards for short run, long cycle operations by methods/standards personnel.

OBTAIN AND PLACE				DIST. RANGE IN INCHES		
WT.	CONDITIONS OF OBTAIN	PLACE ACCURACY	CODE	6″	18″	30″
2 LBS. OR LESS	EASY GRASP	Approximate	OEA	90	150	190
		Close	OEC	110	170	210
		Tight	OET	130	190	230
	DIFFICULT GRASP	Approximate	ODA	100	160	200
		Close	ODC	120	180	220
		Tight	ODT	140	200	240
	GRASP HANDFUL	Approximate	OHA	150	190	230
OVER 2 LBS. THRU 18 LBS.		Approximate	OWHA	160	230	270
		Close	OWHC	170	250	290
		Tight	OWHT	180	270	310
OVER 18 LBS. THRU 48 LBS.		Approximate	OW2HA	190	290	320
		Close	OW2HC	200	310	340
		Tight	OW2HT	210	330	360

PLACE ONLY	CODE	6″	18″	30″
Approximate	PA	40	60	80
Close	PC	60	80	100
Tight	PT	70	100	130
Add for: Weights: OVER 2 LBS THRU 24 LBS. 40; OVER 24 LBS. 60				

TABLE 19–14B (concluded)

ASSEMBLE NON-ROUND OBJECT	CODE	UP TO & INCL. 18 LBS.	OVER 18 LBS. THRU 48 LBS.
UP TO & INCL. 1/2" CLEARANCE	ANT	90	120
OVER 1/2" CLEARANCE	ANC	50	70

ASSEMBLE	CODE	PLUG		
		UP TO & INCL. 1/4"	OVER 1/4" THRU 1/2"	OVER 1/2" THRU 1"
Loose Fit	ASYL	60	50	40
Normal Fit	ASYN	80	70	60
Add for: Simo (S) 15 Temporary Blind (TB) 15		Apply Pressure (AP) 40 Regrasp & Apply Pressure (RAP) 60		

CIRCULAR MOTIONS	CODE	DIAMETER		
		3" & UNDER	OVER 3" THRU 12"	OVER 12"
Revolution without Deceleration	CR	50	80	90
Revolution to General Location	CRG	60	110	120
Revolution to Exact Location	CRE	80	140	150

MOTION-PATTERNS

WALKING					(Measured at Toe or Heel)	
NUMBER OF STEPS	OPEN/BASIC (WO)		CONFINED (WC)		RESTRICTED (WR)	
	DIST. FT.	TIME	DIST. FT.	TIME	DIST. FT.	TIME
1 (1 Leg)	2.5'	120	2.5'	130	2'	140
1 (Both Legs)	2.5'	200	2.5'	220	2'	240
2	5'	260	5'	280	4'	300
5	13.5'	500	13.5'	550	11'	600
10	26'	900	26'	1000	21'	1100
Each Add'l Pace — Add	2.5'	80	2.5'	90	2'	100
Add for Turn Over 120°	—	100	—	100	—	100
Stairs/Up (SU) 130 /Down (SD) 100	Bend (B) 160 Arise (AB) 130		Sit (SI) 230 Stand (ST) 280			

TABLE 19–14C

Note: Index is required for assembly of non-round plugs and targets when clearance is less than 25% of plug diameter. Analyses are for: clearances down to and including dimensions specified; plug sizes up to and including dimensions specified.
Analyses: F1S=23; A1S=26; FS45°S=22

Clearance In.	PLUG DIMENSION								
	3/64"	3/32"	5/32"	7/32"	5/16"	11/16"	1"(a)	2"(a)	Over 2"(a)
.450								26	26
.325								26	26
.185							26	26	26
.110						23	26	26	26
.080						23	26	26	26
.050				23	23	23	26	26	26
.030			23	23	23	23	26	26	26
.020		23	23	23	23	23	26	26	26
.012		23	23	23	23	23	26	26	26
.007	23	23	23	23	23	23	26	26	26
.004	23	23	23	23	23	23	26	26	26

(a) Index can also be FS45°S, usually when index is perpendicular to arm.
Where weight or care is required, add to above analysis.
OPEN TARGETS - Use 50% occurrence.

Turning Diameter	Circumference	One revolution, or first of series			Additional Revolution	
		Motion arrested by rigid object	To general location	To specific location	Less Decel.	Spinning (1)
1"	4"	AEXU 33	AEXUD 43	AEXUDS 53	AEXU 33	
2"	7"	41	55	68	41	
3"	10"	47	64	79	47	
4"	13"	AX 47	AXD 67	AXDS 88	AX 47	(AX+1)–A1 31
5"	16"	52	73	94	52	36
6"	19"	56	78	100	56	40
7"	22"	61	83	106	61	45
8"	26"	66	90	113	66	50

(1) The Spinning Motion applies only when operator retains control of the crank.
Free spinning cranks would be considered as process controlled.

Type of Tool	Motion Pattern	Distance						
		6"	9"	12"	18"	24"	30"	36"
Round Handle: Screw Driver, Mallet	OX-RH-E	180	210	240	270	310	340	360
Spin-Tites, Knives, File	OX-RH-G	150	180	200	230	260	290	310
Double Handle: Pliers, Wire Strippers	OX-DH-E	200	240	260	300	330	360	380
Scissors, Cutters	OX-DH-G	170	200	220	260	280	310	330
Thin Handle: Pencils, Brushes	OX-TH-E	170	210	230	270	300	330	350
	OX-TH-G	140	170	190	220	250	280	300
Soldering Iron: — In Holster	OX-SI-E	270	310	340	380	410	440	470
	OX-SI-G	250	280	300	340	360	390	410
Air Tool: — On Bench	OX-ATB-E	230	260	300	340	370	400	430
	OX-ATB-G	200	240	260	300	330	360	380
Air Tool: — In Holster	OX-ATH-E	330	370	410	450	480	520	540
	OX-ATH-G	300	340	370	410	440	470	490
Air Tool: — Suspended	OX-ATS-E	150	170	190	220	240	260	270
	OX-ATS-G	120	140	150	170	190	210	220

Motion Pattern E: Pick up tool; move directly to work; place tool aside.
Motion Pattern G: Pick up tool; move to general area (move to assemble, etc., must be added); place tool aside.

TABLE 19–14C (concluded)

MICRO™
MOTION ANALYSES

*Designed for quick and accurate calculations of time
values for motion elements by computer or manually.*

Do not attempt to use these tables to determine standards unless
you understand the proper application of the data. Values shown
are in decimal minutes (to 4 places) and at required time. Compatible selections: MTS; WF; DMT; MTA. (For interchange with MTM
or exact derivatives of MTM, add 25% allowance and convert to
decimal hours.)

SELECTION SOURCES

Finger-Hand, Arm and Forearm Tables were selected from simplified combinations of Motion-Time-Analysis (A.B. Segur) and the motion difficulty categories used in Motion-Time-Standards (Philco)
& Work-Factor (RCA).

Arm-Elbow Table is based on a simplified version of Motion-Time-Analysis (A.B. Segur).

Jumbled Grasp Tables were selected from Dimensional Motion-Times (General Electric) & Elemental Time Standards (Western
Electric). The Additional Grasp Difficulties are based on Elemental Time Standards, Work-Factor (RCA) & Dimensional Motion
Times.

Assembly Table was selected from Elemental Time Standards
(Western Electric), Work-Factor (RCA) & Dimensional Motion Times
(General Electric).

Surface Assembly Table, including Flexibility Allowance, was
selected from Elemental Time Standards (Western Electric).

Foot, Leg and Trunk Motions were selected from Work-Factor (RCA)
& Methods-Time Measurement (Westinghouse) time values.

Mental Activities time values are based on Work-Factor (RCA) &
Methods-Time Measurement (Westinghouse).

Alpha-Mnemonic Coding

A -Actuate	J -Job	R -Rotate
B -Brush	Preparation	S -Set-up
C-Clamp & Unclamp	L -Locate	T -Tie
D-Direct	M-Machine	U -Use
E-Enter	Manipulate	V -Variable Use
F -Fasten	O -Obtain	W-Walk
G-Gauge	P -Place	X -Reverse Action
H-Handle	Q -Quantity Count	Z -Special
I -Inspect		

TABLE 19–14D

MOTION DIFFICULTIES	U	C	S	D	X
Basic · no motion difficulties					X
Deceleration · control to stop motion				D	
Steer-control to guide into small area of termination			S		
Care · control to avoid injury or damage		C			
Change Direction · out of normal arc	U				

MOTION EXAMPLES	U	C	S	D	X
Arrested by rigid object					X
Toss · object released in transit					X
To indefinite location					X
Return to relaxed position					X
Transport Loaded Stopped in specified area with Assembly Clearance of 2″ and over* { over 4″ move				D	
{ 4″ or less					X
Stopped in specified area with Assembly Clearance under 2″* { over 4″ move			S	D	
{ 4″ or less			S		
Transport Empty Reach to Graps { motion not arrested by object				D	
{ when short motion rule applies					X
{ object arrests motion (generally CT, GR.)					X
Place object aside {				D	
{ when short motion rule applies					X
Move to inspect				D	
Passes near object that could cause injury or damage { over 4″ move, within 2″ of object		C			
{ 4″ or less, within 1″ of object		C			
Travels between 2 real or imaginary lines { over 4″ move, lines 2″ or less apart		C			
{ 4″ or less, lines 1″ or less apart		C			
Move { Fluid within 1″ of opening of average dia. thru 6″		C			
Fluid { Fluid within 2″ of opening of average dia. over 6″		C			
To by-pass obstruction; obstruction ratio over 1 not over 2	U				

Weight or resistance are applied according to body member used;
see Motion-Time Tables.
*See Assembly and Grasp rules for exceptions.

VISUAL PROCESSES

(D) Distance Eye to line of Travel (In.)	(ET) Eye Travel Time (T) Travel between points							
	3″	6″	9″	12″	15″	18″	24″	30″
6″	37	73						
12″	18	37	54	73	91			
18″	12	24	37	48	61	73	96	
24″	9	18	27	37	46	55	73	91
30″	7	15	22	29	37	44	59	73
36″	6	12	18	24	30	37	49	61

(HT) HEAD TURNS   45° 40   90° 60

(EF) FOCUS   Each 3″ x 3″ Area; or once for Scan ·Larger Area.   20/occurrence
(EI) INSPECT   For Presence or Absence   30/point
(ER) READ   30/word up to six letters, plus 30/each additional three letters.
30/number up to three digits, plus 30/each additional three digits.

MENTAL PROCESSES

(MD) DECIDE Where possible choices exceed two.   20/occurrence
(MM) MEMORIZE Words thru six letters and numbers through three digits. Each additional three letters or three digits requires another Memorize.   20/occurrence
(MR) RECALL Phrase or number.   20/occurrence
(MC) CALCULATE Simple mathematics.   20/occurrence

TABLE 19–14D (concluded)

Inches Moved	MOTION DIFFICULTIES						
	Basic	1	2	3	4	5	6
(F) FINGER-HAND					Measured at Finger Tip		
1	16	23	29	35	40	45	50
2	17	25	32	38	44	50	56
3	19	28	36	43	49	55	61
4	23	33	42	50	58	66	74
½ F1	8	12	15	18	20	23	25
Inching	10	14	18	22			
Wrap Gr. lbs.	2	7	13	20	40	UP	—
Pinch Gr. lbs.	1	3½	7	10	20	UP	—
Apply Pr. lbs.	⅔	2½	4	UP	—	—	—
(A) ARM					Measured at Knuckles		
1	18	26	34	40	46	52	58
2	20	29	37	44	50	56	62
3	22	32	41	50	57	64	71
4	26	38	48	58	66	74	82
5	29	43	55	65	75	85	95
6	32	47	60	72	83	94	105
7	35	51	65	78	90	102	114
8	38	54	70	84	96	108	120
9	40	58	74	89	102	115	128
10	42	61	78	93	107	121	135
11	44	63	81	98	112	126	140
12	46	65	85	102	117	132	147
13	47	67	88	105	121	137	153
14	49	69	90	109	125	141	157
15	51	71	92	113	129	145	161
16	52	73	94	115	133	151	169
17	54	75	96	118	137	156	175
18	55	76	98	120	140	160	180
19	56	78	100	122	142	162	182
20	58	80	102	124	144	164	184
22	61	83	106	128	148	168	188
24	63	86	109	131	152	173	194
26	66	90	113	135	156	177	198
28	68	93	116	139	159	179	199
30	70	96	119	142	163	184	205
36	77	104	130	152	173	192	212
42	83	111	138	162	182	202	222
Lbs.	2	7	13	20	40	UP	—
(FS) FOREARM SWIVEL					Measured at Knuckles		
45°	17	22	28	32	37	42	47
90°	23	30	37	43	49	55	61
135°	28	36	44	52	58	64	70
180°	31	40	49	57	65	73	81
Torque	3	13	UP	—	—	—	—

ADVANCED WORK MEASUREMENT TABLES

(AE) ARM-ELBOW
Measured at Knuckles

Use only for highly repetitive motions, on short cycle operations performed within the Ideal Work Area for arm motions.

Inches Moved	MOTION DIFFICULTIES			
	Basic	1	2	3
1	18	26	34	40
2	19	28	36	43
3	20	30	39	48
4	21	33	43	53
5	23	36	48	59
6	24	39	52	64
7	25	41	55	68
8	26	43	58	72
9	27	45	61	76
10	28	47	64	79
11	29	48	67	83
12	30	49	69	86
Lbs.	2	7	13	20

When writing analyses:

Identify methods and parts dimensions, and equipment settings, in elemental description column.

Identify motion-time table dimensions in analysis column.

571

TABLE 19-14E

(ASY) ASSEMBLY TABLE   Plug Dimensions   Regular Assembly

Clear.	Assembly Motions	3/64 .047	3/32 .094	5/32 .156	7/32 .219	5/16 .313	11/16 .687	1 1.00	2 2.00
.450	Pos.	10	8	8	7	7	6	5	3
	Up-Ins	5	7	7	7	7	8	9	11
(.400)	Tot.	15	15	15	14	14	14	14	14
	SIMO	20	19	19	18	18	17	17	16
.325	Pos.	16	11	9	9	9	7	6	4
	Up-Ins	6	10	10	10	10	12	13	15
(.250)	Tot.	22	21	19	19	19	19	19	19
	SIMO	30	27	24	24	24	23	22	21
.185	Pos.	27	19	18	16	15	13	11	8
	Up-Ins	11	15	15	15	15	17	19	22
(.125)	Tot.	38	34	33	31	30	30	30	30
	SIMO	52	44	42	39	38	37	36	34
.110	Pos.	36	26	24	23	22	18	16	11
	Up-Ins	14	19	19	19	19	21	22	26
(.100)	Tot.	50	45	43	42	41	39	38	37
	SIMO	68	58	55	54	52	48	46	43
.080	Pos.	41	31	27	26	25	22	20	14
	Up-Ins	16	20	20	20	20	22	24	28
(.062)	Tot.	57	51	47	46	45	44	44	42
	SIMO	78	67	61	59	58	55	54	49
.050	Pos.	49	38	33	30	29	26	24	19
	Up-Ins	18	22	22	22	22	24	26	31
(.040)	Tot.	67	60	55	52	51	50	50	50
	SIMO	92	79	72	67	66	63	62	60
.030	Pos.	57	47	41	38	35	31	29	22
	Up-Ins	21	24	24	24	24	26	28	34
(.025)	Tot.	78	71	65'	62	59	57	57	56
	SIMO	107	95	86	81	77	73	72	67
.020	Pos.	70	59	52	47	44	40	37	24
	Up-Ins	23	26	26	26	26	29	31	36
(.015)	Tot.	93	85	78	73	70	69	68	60
	SIMO	128	115	104	97	92	89	87	72
.012	Pos.	82	72	63	55	52	48	45	26
	Up-Ins	25	27	27	27	27	30	33	37
(.009)	Tot.	107	99	90	82	79	78	78	63
	SIMO	148	135	122	110	105	102	101	76
.007	Pos.	94	83	74	63	59	55	52	28
	Up-Ins	26	28	28	28	28	31	34	38
(.005)	Tot.	120	111	102	91	87	86	86	66
	SIMO	167	153	139	123	117	114	112	80
.004	Pos.	104	94	84	70	65	62	59	32
	Up-Ins	27	29	29	29	29	32	34	38
(.003)	Tot.	131	123	113	99	94	94	93	70
	SIMO	183	170	155	134	127	125	123	86

Values show: Position Time, Upright and Insert Time, Total Time, and Simo Time.
Table does not include Indexing: add as required.

For Chamfered Assemblies: select Position Time based on Surface Clearance, Upright and Insert Time based on Internal Clearance. For Open Targets use ½ Position Time.

ADDITIONAL ASSEMBLY DIFFICULTIES

Inches	Blind Temp. %	over ¼ Perm. %	Distance Between %	Gripping Add %	Distance Upright Inches
1	10	25	—	—	1
2	20	50	—	—	1
3	30	75	—	10	1
5	50	125	25	20	2
7	75	250	50	30	3
10	125	400	75	50	4
15			(a)	75	5
20			(b)	100	6
25			(b)	125	7

◄━━━━━ Apply to Position values. ◄━━━━━

(a) Treat assemblies as separate.
(b) Treat assemblies as separate, including provision for focus and inspect.

TABLE 19–14E (concluded)

(SA) SURFACE ASSEMBLY TABLE Grams Flexibility FLAT SURFACE

Clearance	Basic	450	200	100	50	15	5
.450	21	30	50	53	58	70	71
.325	27	39	60	64	67	80	82
.185	40	57	72	79	85	100	102
.110	53	70	85	89	98	107	120
.080	59	78	95	98	108	128	135
.050	70	93	108	113	125	146	155
.030	80	109	125	130	147	170	182
.020	93	128	144	152	170	195	210
.012	105	141	160	170	190	220	245
.007	117	152	173	183	203	240	280

(AL) ALIGNMENT TABLE Grams Flexibility ALIGN ALONG STOP

Clearance	Basic	450	200	100	50	15
.450	13	23	36	38	43	58
.325	20	32	44	48	52	66
.185	29	40	55	58	63	79
.110	40	52	66	70	75	90
.080	48	60	75	78	84	99
.050	58	72	87	90	96	113
.030	66	82	96	100	108	123
.020	73	95	104	113	120	136
.012	85	103	113	122	134	153
.007	93	113	125	135	150	175
.004	100	125	140	146	165	195
.002	105	134	150	160	180	210

APPLICATION OF PRESSURE: Use lbs. shown in Finger-Hand table

RELAX PRESSURE: Use only when following Maintained Pressure. Same weight limits as in "Apply Pressure" for Basic. One wt. difficulty maximum.

PRE-POSITION TABLE Average Required Motion-Times

		Major Dimensions			Major Sides	Example for an 8" x 12" plate; over 2 lbs.
		½" L	4" L	8" L		
CYLINDRICAL PARTS		½"	½"x2" x2"	1"x3" x3"	8"x8"	
IRREGULAR SHAPES						

DESCR.	(a) % ALLOWED	ONE HAND P.P.			TWO HAND P.P.	
		5F1	3F1	4F1	F1+A4D+F1	F1W+A8D+F1W
End for End	50	40	24	32	35	50
Either Side (b)						
— Any edge	0	—	—	—	—	—
— 1 of 3 Edges	0	—	—	—	—	—
— 1 of 2 Edges	25	20	12	16	18	25
— 1 Edge	50	40	24	32	35	50
One Side (b)						
— Any Edge	50	40	24	32	35	50
— 1 of 3 Edges	50	40	24	32	35	50
— 1 of 2 Edges	62½	50	30	40	44	63
— 1 Edge	75	60	36	48	53	75
From Stack						
— All Wrong	100	80	48	64	70	100
Alternate Stacking	50	40	24	32	35	50

The above table shows average pre-position values for dimensions up to & incl. 8". Can be used as a guide for dimensions over 8".
(a) The mathematical occurrence of Pre-Positioning is higher than the percentage shown for Sides and Edges. However, some of the required Pre-Positioning could be accomplished by a single Wrist-turn, Finger or Arm motion which usually can be performed simo with the Move motion. The above values are combinations calculated on a practical basis for these occurrences.
(b) For flat objects, interpret word "Side" to mean "Surface."

MISCELLANEOUS

Sit	Sit	230
Stand up	Std.	280

MEASUREMENTS RULES

Measure weights, lengths, and distances up to and including dimensions shown.
Measure clearances down to and including dimensions shown.

TABLE 19–14F

Inches Moved	MOTION DIFFICULTIES						
	Basic	1	2	3	4	5	6
(FT) FOOT				Measured at Toe or Heel			
1	20	29	37	44	51	57	64
2	22	32	41	48	55	61	68
3	24	35	45	55	63	70	78
4	30	42	53	64	73	81	90
Lbs. 5	22	UP	—	—	—	—	—
(L) LEG			Measured at Ankle (At Hip for Sit or Arise)				
1	21	30	39	46	53	60	67
2	23	33	43	51	58	64	71
3	26	37	47	58	66	74	82
4	30	44	55	67	76	85	94
5	33	49	63	75	86	98	109
6	37	54	69	83	95	108	121
7	40	59	75	90	104	117	131
8	44	62	81	97	110	124	138
9	46	67	85	102	117	132	147
12	53	75	98	117	135	152	169
15	59	82	106	130	148	167	185
18	63	87	113	138	161	184	207
24	72	100	125	151	175	199	223
30	81	110	137	163	187	212	236
36	89	120	149	176	199	222	247
40	93	125	155	183	206	229	252
Lbs. 8	42	UP	—	—	—	—	—
(T) TRUNK				Measured at Shoulder			
3	32	46	60	73	83	93	103
6	46	68	87	104	120	136	152
9	58	84	107	129	148	167	186
12	67	94	123	148	170	190	213
15	74	103	133	164	187	210	233
18	80	110	142	174	203	232	260
24	90	125	158	190	220	250	280
Lbs. 11	58	UP	—	—	—	—	—

(W) WALKING — Measured at Toe or Heel

NUMBER OF STEPS	(O) OPEN (or Basic)		(C) CONFINED		(R) RESTRICTED	
	DIST. (ft)	TIME	DIST. (ft)	TIME	DIST. (ft)	TIME
1 (1 Leg)	1¼	100	1¼	110	1¼	120
1 (1 Leg)	2½	120	2½	130	2	140
1 (Both Legs)	2½	200	2½	220	2	240
2	5	260	5	280	4	300
3	8	340	8	370	6½	400
4	11	420	11	460	8½	500
5	13½	500	13½	550	11	600
Each Add'l Pace—Add	2½	80	2½	90	2	100
Turn over 120°—Add	—	100	—	100	—	100
(WS) Stairs—Up	130		Stairs—Down			100

(KN) KNEE MOTIONS — Pivot at Ankle and Hip

Motion	To 2"	37	Motion	To 4"	41

SIDE-STEP OR TURN

Side-Step (Aver. 12") 2 Leg Movements (Most 1 leg sidesteps simo with other work)	SS-2	150
Turn 45° to 90° - 1 Leg Movement Turn to supply or aside	90°T-1	120
Turn 45° to 90° - 2 Leg Movements Turn to supply or aside Turn to next work station (angled)	90°T-2	190
Turn 90° to 180° - both legs to position Turn to side or rear; to bench or container	180°T-2	250

TABLE 19-14F (concluded)

BEND, STOOP, STRAIGHTEN UP

	Lbs.	11	58	UP
Bend (into container, etc., arm reaching below knee level) - average bend	B	110	140	170
- full 90° bend	90°B	130	160	190
Straighten up from bend (Arise from bend) - average bend	AB	80	110	140
- full 90° bend	90°AB	90	130	160
Stoop (or squat down towards floor)	ST	150	160	170
Straighten up from stoop (Arise from stoop)	AST	180	190	210

SIMPLE GRASP AND/OR RELEASE

Type	Grasp	Release
Contact	0	0
Pinch	½F1 thru 1# then ½ F1W*	½F1 thru 1# then ½F1W
Wrap	F1 thru 2# then F1W*	F1 thru 2# then F1W

(G) GRASP TABLES — Small parts, jumbled

Cross Section	LENGTH (Major Dimensions) 1/16	1/8	1/4	1/2	SIMO 1/16	1/8	1/4	1/2
1/64x1/64	*	*	*	*	*	*	*	*
x1/32	*	108	95	72	*	154	135	100
x length	*	92	84	65	*	130	118	89
1/32x1/32	*	100	85	71	*	142	120	99
x1/16	131	94	81	68	189	133	114	94
x length		85	71	61		120	99	84
1/16x1/16	120	91	79	65	172	129	111	90
x1/8		84	75	63		118	105	86
x length			70	55			97	75
1/8 x1/8		81	71	59		114	99	81
x1/4			66	56			91	76
x length				51				69
1/4 x1/4			65	55			90	75
x length				51				69

*Too small for Finger Grasp; use Tweezer Grasp

(G) GRASP TABLES — Miscellaneous sizes, jumbled

	1″	4″	8″	1″	4″	8″
1/8 x1/8	56	48	48	76	64	64
x1/4	53	45	45	72	60	60
x2	45	41	41	60	54	54
1/4 x1/4	50	42	40	67	55	52
x1/2	47	41	39	63	54	51
x2	41	36	35	54	46	45
1/2 x1/2	45	35	34	60	45	43
x2	42	31	31	55	39	39
1 x1	32	20	19	40	22	21
x2		19	18		21	19

(GB) BLIND GRASP

	1″	4″	8″	1″	4″	8″
1/8 x1/8	56	48	48	76	64	64
x1/4	53	45	45	72	60	60
x2	45	41	41	60	54	54
1/4 x1/4	54	46	44	73	61	58
x1/2	51	45	43	69	60	57
x2	45	40	39	60	52	51
1/2 x1/2	49	39	38	66	51	49
x2	46	35	35	61	45	45
1 x1	40	37	36	52	48	46
x2		36	35		46	45

ADDITIONAL GRASP DIFFICULTIES

Slippery	Add 7		Simo
Nested	Add 18		Values
Entangled			add
Simple Hooking or Interference	35 x _____ %		50%
Shake to Clear Part	104 x _____ %		to
Shake and Twist	150 x _____ %		these
Locked — Two Hands	252 x _____ %		figures

A representative motion sequence could be coded M3G3M4P0. This represents a move with the arm to pick up a flat object (M3G3). Following this, the object is moved to a general location such as to the other hand (M4P0). The normal time for this sequence would be 10 MODS or 0.0215 minutes. Figure 19–17 illustrates the application of the three groups of MODAPTS elements.

Without going into further detail relative to the identification of each of the 44 MODAPTS database elements, the International MODAPTS Association, Inc. has brought out the following advantages of the system:

1. The system is simple to understand and easy to learn and use.
2. The activities in a job are identified easily in terms of MODAPTS base elements.
3. The data can be memorized readily by the analyst as a picture and, consequently, reading data from the card will be necessary only in the early stages of its use by the professional analyst.
4. There are fewer calculations than with other systems.

APPLYING BASIC MOTION TIMES

As a means of improving methods, analysts' general knowledge of the fundamentals of basic motion-time standards can prove invaluable. For example, methods analysts with a grounding in MTM design workstations to utilize the "G1A" grasps requiring but 2 TMU rather than the more difficult grasps which require as much as 12.9 (G4C) TMU. Likewise, these analysts' design for utilizing contact releases rather than normal releases and symmetrical positioning rather than semisymmetrical or nonsymmetrical positioning. One of the important uses for any of the synthetic basic motion-time techniques lies in the area of methods. Analysts who appreciate these techniques look more critically at each and every workstation, thinking about how improvements may be made. Reaches and moves of 20 inches appear unduly long, and these analysts immediately consider the savings possible through shortened motion patterns. Positioning elements that involve heavy pressure are automatically signs of a need for improvement. Operations that require eye travel and eye focus time can usually be improved.

Of course, analysts can establish work standards through basic motion times. The use of data for this purpose requires a much greater knowledge of the techniques of application. In no case should untrained, inexperienced analysts establish time standards to be used for rate purposes. Thus, analysts must know whether the distance moved is the linear distance taken by the hand or the circumferential distance taken by the arc that the hand makes. They must know whether the distance is measured from the center of the hand, the knuckles, or the fingertips. They must know when application of pressure prevails and when it does not. They

FIGURE 19–17

The MODAPTS Plus card consists of 4 main sections:

- MOD Unit.
- Movement Classes.
- Terminal Classes.
- Auxiliary Values.

The MOD Unit
_____

for quick planning
= 1/7th second
standard times
(includes 10 3/4% allowances)
_____

Normal times
for building job
specifications
= 0.0000358 hours
= 0.00215 minutes
= 0.129 seconds
= 1/8 seconds
(no allowances included)

The section on the MOD Unit provides information about the MOD (short for Module), which is a value used to express a unit of physical work. This unit of measurement is chosen for its particular convenience in a given situation. As you can see from the key, a MOD is equal to 1/8th of a second, 0.0000358 hours, 0.00215 minutes, or 0.129 seconds.

For example, if a movement or activity has a MOD value of 3, then the predetermined time for that activity is 3 × 0.129 seconds, or 0.387 seconds.

As you read the following definitions of the Movement Classes, Terminal Classes and Auxiliary Values, notice that each has one of thirteen MOD values assigned to it (0, 1, 2, 3, 4, 5, 6, 8, 10, 12, 17, 18, and 30).

The *Movement Classes* refer to movements through space done by the finger-hand-arm-shoulder system. Normally a movement is the activity that is required to position a part of the arm to perform a terminal activity.

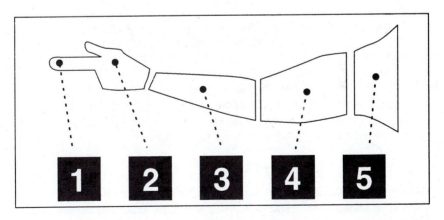

(continued)

FIGURE 19–17 *(continued)*

The *Terminal Classes* are activities that are done at the end of a movement and are in close proximity to the things being worked on. This section includes two main types of activities:

- GET (G0, G1, G3 and, in special cases, G4, G8, G12) activities that involve "obtaining control" of things and
- PUT (P0, P2, P5 and, in special cases, P10) activities that involve putting "things to destinations."

The activities in the Terminal Classes are further categorized as requiring either:

- Low Conscious Control or
- High Conscious Control.

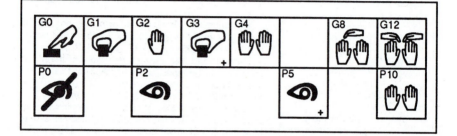

must clearly understand how the elements of alignment and orientation affect positioning time. These and many other considerations must be mastered before analysts can expect to establish consistent and accurate time standards with this tool.

Only after thorough training can a group of analysts arrive at consistent time standards when using any of the synthetic motion-time techniques. In no case should any analyst endeavor to establish time standards with only a superficial knowledge of these techniques.

The Development of Standard Data

One of the most important uses of basic motion times is in the development of standard data elements. With standard data, operations can be prepriced much faster than by the laborious procedure of summarizing long columns of fundamental motion times. In addition to saving time, applying standard data reduces clerical errors, since less arithmetic is involved.

With sound standard data, it is economically feasible to establish standards on indirect work such as maintenance, material handling, clerical and office, inspection, and similar expense operations. Thus, analysts can preprice elements involving long cycles and consisting of many short

FIGURE 19–17 (continued)

The *Auxiliary Values* refer to 17 other activities that are not done with the finger-hand-arm shoulder system, like walking, bending, deciding, etc.

Each of these four sections described above will be discussed further in subsequent chapters.

Since MODAPTS Plus has only 28 different movement/activity categories that have only 13 different MOD values, MODAPTS Plus is easy to learn and simple to use. Once the pictures and their corresponding alphanumeric codes begin to have meaning, they can be easily memorized. Then MODAPTS Plus can be applied without any further reference to the card.

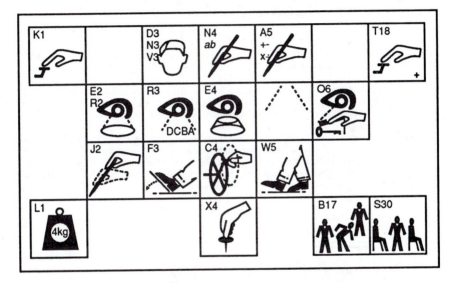

duration elements economically with standard data. For example, one company developed standard data applicable to radial drill operations in its toolroom. Time study people developed standard data for the elements required to move the tool from one hole to the next and to present and back off the drill. Then they combined these standard data elements in one multivariable chart so that they could be summarized rapidly (see Table 19–14).

Figure 19–18 illustrates another application of basic motion times in the development of standard data. The work sheet shown gives all the standard elements for trimming blanks on punch presses. In only minutes you can predetermine a standard for trimming blanks with weights of up to 40 pounds.

A third example illustrating the flexibility of basic motion times is the development of a formula for prepricing a clerical operation. This formula for sorting time slips includes the following elements:

FIGURE 19–17 *(concluded)*

Low and High Conscious Control

Some activities are done with Low Conscious Control, and some activities are done with High Conscious Control. *The key to understanding MODAPTS Plus is to be able to distinguish between these two types of control.* This concept also helps to explain the logic behind MODAPTS Plus and makes its application simple.

Control carried out by the nervous system is at three quite different levels:

- Nonconscious Controls—things that do not occupy the conscious mind at all (e.g., body temperature regulation, secretion of digestive juices, etc.).
- Low Conscious Controls—things that occupy the conscious mind and need a small amount of attention (e.g., tossing a pencil aside anywhere on a desk).
- High Conscious Controls—things that occupy the conscious mind and need a great deal of attention (e.g., threading a needle).

As you can see from these examples, the last two levels of conscious control are quite different. Actions needing a high level of conscious control require a lot of information to be fed back into the brain. This information almost always occurs when workers must use their eyes throughout the entire activity.

Once you understand the difference between high conscious control actions and low conscious control actions, you will have mastered the key to understanding MODAPTS Plus.

Recall that Terminal Class activities have both "Get" and "Put" elements. One group (containing both Gets and Puts) is for high conscious control elements, and the other is for low. *Determining the degree of conscious control in Get and Put elements is the critical factor in the application of MODAPTS Plus.*

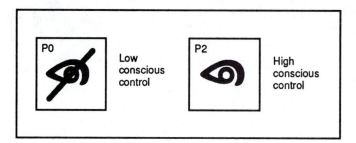

1. Pick up pack of departmental time slips and remove rubber band.
2. Sort time slips into direct labor (incentive), indirect labor, and day-work.
3. Record the total number of time slips.
4. Get pile of time slips, put rubber band around pack, and put aside.
5. Get pile of time slips and bunch.
6. Sort incentive time slips into "parts" time slips.
7. Count piles of incentive "parts" time slips.
8. Record the number of "parts" time slips and the number of incentive time slips.
9. Sort "parts" time slips into numerical sequence.
10. Bunch piles of numerical time slips and place in one pile on desk.

FIGURE 19–17 *(concluded)*

Consider the following pairs of high and low conscious control motions:

Example 1:
  A. Pick up a pin from the desk.
  B. Pick up a pencil from an uncluttered desk.

Example 2:
  A. Take a match out of a half-open matchbox.
  B. Reach out and grasp a box of matches on the table.

Example 3:
  A. Place your pencil exactly on the period of the last sentence.
  B. Place your pencil anywhere on this page.

In Example 1, picking up a pin from the desk requires high conscious control because it is relatively difficult to get control of such a small object when it is lying flat. Picking up a pin takes a lot longer than picking up a pencil. The latter motion requires only low conscious control.

Similarly, in Example 2, reaching out and grasping a box of matches on the table is a simple, easy motion; but picking a match out of a half-open matchbox requires high conscious control because of the physical difficulty of getting at the matches in a confined space and selecting one from a jumble of matches.

In Example 3, placing a pencil anywhere on this page requires almost no conscious control, but placing it exactly on the period at the end of the last sentence requires careful control and takes a lot longer.

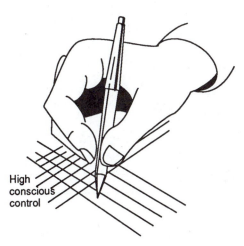

High
conscious
control

Methods people analyze each element from the fundamental motion standpoint. After assigning basic values and determining variables the resulting algebraic equation allows the rapid prepricing of the clerical operation.

Frequently, a stopwatch can be a helpful complement in developing standard data elements. Some portions of an element may be more readily determined by basic motion times, and other portions may be better adapted to stopwatch measurement.

FIGURE 19–18

Standard data for prepricing press operations

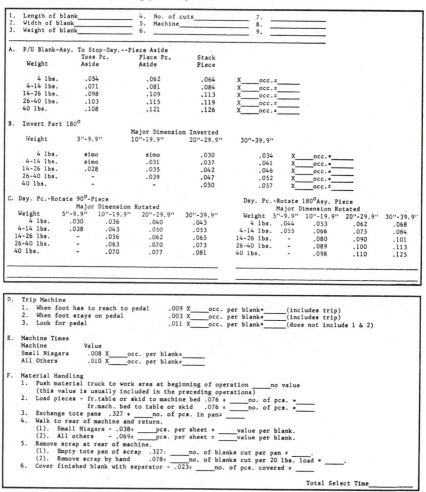

Verify standard data elements with a decimal minute (0.001) watch as an added check on their validity. Predetermined basic motion times converted to standard data elements that have been verified establish fair standards that are more consistent than those established through stopwatch procedure.

SUMMARY

This chapter discusses several of the more popular basic motion time systems. There are many others, including several proprietary systems developed by industry. Many years ago, Frederick W. Taylor visualized the development of standards for basic divisions of work similar to those in use. In his paper on "Scientific Management," he predicted that the time would come when a sufficient volume of basic standards would be developed making further stopwatch studies unnecessary. We have just about reached this state today in connection with effort elements. Certainly today, the vast majority of standards are developed by using either standard data and/or basic motion times.

Basic motion-time values are becoming more accurate as additional studies are made. However, there is still a need for further research, testing, and refinement. For example, there is a question as to the validity of adding basic motion times for the purpose of determining elemental times since therblig times may vary once the sequence is changed. Thus, the time for the basic element "reach 20 inches" may be affected by the preceding and succeeding elements and is not dependent entirely on the class of reach and distance. In general, all fundamental motion techniques that have proved successful have made an effort to take care of the additivity of the elements of the motion in one way or another. Methods-Time Measurement, for instance, has provided three motions for the movement of the hand, and several cases according to the nature of a move or reach. Work-Factor recognizes five elements of difficulty. Similar observations might be made about other fundamental motion techniques.

What has been done already has proved applicable in most instances. However, some gaps remain to be filled. For example, more research is needed in the development of data for combined motions. Reaches that involve a move during the first part of the motion are a typical illustration. Likewise, moves that involve palming a component while en route are typical of combination motions that require more research.

When analyzing motion patterns with existing data, consider not only the main purpose of the motion pattern, but also its complexity and characteristics. For example, if the hand is empty while moving toward an object, the motion is classified as a reach. If the hand holds an object while moving toward another object, consider not only the main purpose of the motion but also what the hand does with the object during the motion. If the hand is palming an object while reaching for something else, the motion cannot be classified as a basic reach. The time necessary to perform the combined motion depends on factors other than distance. Do not neglect the physical characteristics of the motion. When an object is palmed while the hand is moving, in addition to the move, a simultaneous operation takes place. The result might be a reduction in the average speed. This allows the hand to establish control of the object during the distance moved. The longer the distance, the more time the hand has to palm the object in that particular motion. Thus, the longer the combined motion, the more the motion approaches the time required for a simple reach of the same distance.

Additional study is necessary for operations that involve simultaneous motions. Intuitively, there should be a relationship between the degree of bimanualness that the normal operator is able to perform and the motion pattern employed.

Further study increases the accuracy and the resulting application of basic motion times. Today, this technique is probably the most important tool of work measurement analysts.

One of the paramount values of basic motion times is that analysts must carefully examine all factors affecting the motion patterns of jobs. This minute scrutiny inevitably reveals opportunities for the refinement of work methods.

In one company, $40,000 was allocated for advanced tooling to increase the rate of production on a brazing operation. Prior to retooling, analysts made a work measurement study of the existing method. Using fundamental motion data, they discovered that by providing a simple fixture and rearranging the loading and unloading area, production could be increased from 750 to 1,000 pieces per hour. The total cost of the synthetic basic motion-time study was $40. As a result of the study, that company did not embark on the $40,000 retooling program.

A manager of one of our leading electrical appliance companies has emphasized that basic motion times permit predetermining whether or not expenditures for new facilities and tools are warranted. They can accurately forecast the cost of the reductions achieved by such expenditures.

The combining of basic motion times, in the form of standard data, permits the application of basic motion times to a wide variety of work. Standard data thus developed are finding application on indirect and expense operations as well as on direct work. In this form basic motion times will have their broadest application in the future.

Basic motion-time systems have a most important place in the field of methods and work measurement; however, they are no better than the people using them. That is, they should not be installed without professional help or a complete understanding of their application. Specialized training leading to certification is available from such organizations as SMC Wofac, MTM Association for Standards and Research, H. B. Maynard & Company, Inc., Standards, International, Inc., and International MODAPTS Association, Inc.

TEXT QUESTIONS

1. What are the advantages of using basic motion times?
2. Which variables are considered by the Work-Factor technique?
3. How did Work-Factor develop its values?
4. What is the time value of one TMU?
5. Who pioneered the MTM system?
6. Which other two terms are used frequently to identify basic motion times?
7. Explain why most companies today require certification before they utilize basic motion times for the establishment of standards.

8. Would it be easy or difficult to perform a GB get with the left hand while simultaneously performing a PC place with the right hand? Explain.

9. Who was originally responsible for thinking in terms of developing standards for basic work divisions? What was his contribution?

10. Calculate the equivalent in TMUs of 0.0075 hours per piece, of 0.248 minutes per piece, of 0.0622 hours per hundred, of 0.421 seconds per piece, of 10 pieces per minute.

11. How is MTM related to the analysis of method?

12. For which reasons was MTM-2 developed? Where does MTM-2 have special application?

13. Which classes of action are recognized by MTM-2?

14. Explain the relationship of basic motion times to standard data.

15. If you have finished drilling a hole 3 inches deep on a Western radial drill, how long would it take to present the drill and drill a second hole in a steel forging ½ inch in diameter and 3 inches deep? Traverse: 6 inches; swing of head: 8 inches; 0.007-inch feed; 50 feet per minute surface speed.

16. How has Work-Factor endeavored to take care of the additivity of the elements of a motion pattern?

17. What are the four main sections identified on the MODAPTS Plus card?

18. On what basic principle is MODAPTS based?

19. When using the MODAPTS system what activities are represented by "Terminal Classes"? Differentiate between low and high conscious control.

20. For what reasons were Micro and Macro Motion Analyses developed?

21. Explain the background of both Micro and Macro Motion Analyses.

GENERAL QUESTIONS

1. What does the future hold for synthetic basic time values?

2. Which of the several basic motion time techniques outlined is the easiest to apply?

3. Which basic time method, in your opinion, will give the most reliable results? Why?

4. Give several objections to the application of Work-Factor that you might receive from a worker, and explain how you would overcome them?

5. Describe as vividly as you can how you would explain MODAPTS to a worker in your forge shop who knows nothing about it. Tell what it is and how it is applied.

6. Some companies have been experiencing a tendency for their time study people to become more liberal in their performance rating over a period of years. How do fundamental motion data offset this tendency toward creeping loose standards?

7. Is there consistency between MTM-1 and MTM-2 in the handling of simultaneous motions?

8. If MTM-3 were used to study an operation of approximately three minutes' duration, what could you say about the accuracy of the standard?

PROBLEMS

1. Determine the time for the dynamic component of M20 B20.
2. A 30-pound bucket of sand having a coefficient of friction of 0.40 is pushed 15 inches away from the operator with both hands. What would be the normal time for the move?
3. A ¾-inch diameter coin is placed within a 1-inch diameter circle. What would be the normal time for the position element?
4. Give the MTM breakdown required to grasp the cotter pin with the pliers shown in the following sketch.

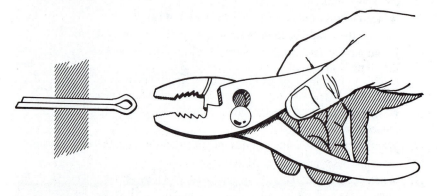

SELECTED REFERENCES

Antis, William; John M. Honeycutt, Jr.; and Edward N. Koch. *The Basic Motions of MTM*. 3rd ed. Pittsburgh: The Maynard Foundation, 1971.

Baily, Gerald B., and Ralph Presgrave. *Basic Motion Time-Study*. New York: McGraw-Hill, 1958.

Birn, Serge A.; Richard M. Crossan; and Ralph W. Eastwood. *Measurement and Control of Office Costs*. New York: McGraw-Hill, 1961.

Geppinger, H. C. *Dimensional Motion Times*. New York: John Wiley & Sons, 1955.

Karger, Delmar W., and Franklin H. Bayha. *Engineered Work Measurement*. New York: Industrial Press, 1957.

Karger, Delmar W., and Walton M. Handcock. *Advanced Work Measurement*. New York: Industrial Press, 1982.

Maynard, Harold B.; G. J. Stegemorten; and John L. Schwab. *Methods Time Measurement*. New York: McGraw-Hill, 1948.

Quick, Joseph H.; James H. Duncan; and James A. Malcolm. *Work-Factor Time Standards*. New York: McGraw-Hill, 1962.

Sellie, Clifford N. "Predetermined Motion-Time Systems and the Development and Use of Standard Data." In *Handbook of Industrial Engineering*, 2nd ed., edited by Gavriel Salvendy, Chap. 63. New York: John Wiley & Sons, 1992.

Zandin, Kjell B. *MOST Work Measurement Systems*. New York: Marcel Dekker, 1980.

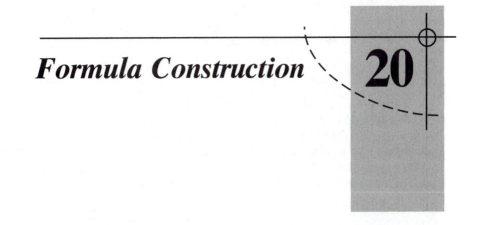

Formula Construction 20

As applied to time study, formula construction involves the design of an algebraic expression or a system of curves that establishes a time standard in advance of production by substitution of known values peculiar to the job for the variable elements. A time study formula represents a simplification of standard data. It has particular application in nonrepetitive work where it is impractical to establish standards by an individual time study for each job.

APPLICATION OF FORMULAS

Time formulas are applicable to practically all work. They have been successfully used in office operations, foundry work, maintenance work, painting, machine work, forging, coil winding, grass cutting, window washing, floor sweeping, welding, and many other areas. If sufficient time studies are collected involving standardized elements to give a reliable sampling of data, it is possible to design a formula for a given range of work in any type of job.

Analysts should apply the formula only to jobs that fall within the limits of the data used in developing it. If they exceed the boundaries of the formula without the supporting proof of individual time studies, erroneous standards, with all the dangers brought about by inequitable rates, may result.

Once analysts build a formula to cover a given operation, it should immediately apply to all pertinent jobs within the range for which it was designed that do not already carry time standards.

Advantages and Disadvantages of Formulas

The advantages of using formulas rather than individual time studies for setting standards parallel those of using standard data (see Chapter 18). They may be summarized as follows:

1. More consistent time standards result.
2. Elimination of duplicated time study effort on similar operations.
3. Standards are established much more rapidly.
4. Less experienced, less trained persons can calculate time standards.
5. Accurate, rapid estimates for labor costs may be made before production is begun.

Probably the most significant advantage of the formula over the standard data method is that a less costly person can work with formulas than with standard data. Any high school graduate proficient in algebra can work out a time study formula and solve for the allowed time required to perform the task. Then, too, standards are established more rapidly by using formulas than by accumulating standard data elements. Since columns of figures must be added in the standard data method, there is a greater chance for omission or arithmetic error in setting a standard than in using a formula.

In the development of formulas, caution must be exercised in the treatment of constants. There is a natural tendency to treat more elements as constants than should be. Thus, errors can creep into the formula design. It is true that a formula will give consistent results; and so it is also true that if it is not accurate, it will give standards that are consistently wrong.

Another disadvantage of the formula lies in its application. Sometimes, to arrive at a standard at the earliest possible time, work measurement analysts use formulas in instances where the variables are beyond the range of the data used in developing the formula. Thus, the formula may be used where it has no application, and the resulting value is far from valid.

Characteristics of the Usable Formula

A time study formula must be both completely reliable and practical if it is to be used with confidence. For the expression to be reliable, it must always give accurate results. If the developed formula is accurate, it should verify the individual standards used in its development within ±5 percent. The greater the number of studies that can be used in developing it, the better the chances for ending up with a reliable formula. At least 10 independent studies should be available for a given class of work before analysts attempt to design a formula. If tabularized standard data are available, use these standards for derivation of the formula. Frequently, when a formula is built, standard data, existing individual time studies, and new individual time studies constitute the basic data used in its construction. Of course, all mathematical calculation must be free from error before complete confidence can be vested in a formula.

The reader should understand that the formula is a compilation of standard elements that are used in a range of work characteristic of a given work order or facility. Thus, when gathering former time studies or making new studies, selection of previous studies should be based on those where the elements have been standardized.

In the case of making new studies, be sure that your breakdown of elements is based on like elements of studies taken earlier.

The formula gives as accurate results as the data used to construct it. These studies and the elements they comprise must be consistent in their end points and methodology, if the formula is to give consistently valid results.

A practical formula is clear, concise, and simple as possible. The simplest formulas are best understood and most easily applied. Cumbersome expressions involving the taking of terms to powers should be avoided. Symbols of unknowns should not be repeated throughout the formula but should appear in only one place, together with their applicable suffixes, prefixes, and coefficients. The area of work that each symbol represents should be specifically identified. Liberal substantiating data should be included in the formula report so that any qualified, interested party can clearly identify the derivation of the formula. The limitations of the formula must be noted by describing in detail its applicable range. Analysts using formulas so constructed apply them rapidly and accurately, with little difficulty in obtaining the required information.

STEPS TO FOLLOW IN FORMULA CONSTRUCTION

The basic step in formula construction is to determine which class of work is involved and which range of work is to be measured. For example, a formula might be developed for curing bonded rubber parts between two and eight ounces in weight. The class of work covered by the formula would be "curing molded parts," and the range of work involved would be two to eight ounces. After this overall analysis is finished and analyzed, the next step is collecting the formula data. This step involves gathering former studies with standardized work elements and standard data elements that prove satisfactory, as well as taking new studies, to obtain a sufficiently large sample to cover the range of work for which the formula is needed. It is important that like elements (standardized elements) in the different studies be consistent in their end points. This is essential in determining the variables that influence time, as well as in arriving at an accurate value for constant elements.

Next, post the elemental time study data on a work sheet for analysis of the constants and variables. Combine the constants and analyze the

FIGURE 20-1

SHEET NO. 1			MASTER TABLE OF			
OF 1 SHEETS						

				S-1	S-2	S-3	S-4
FORMULA 73	JOB CHARACTERISTICS	STUDY					
DATE June 10,		OPERATOR		Petrecca	Winters	Ekey	Kumpf
		Performance Factor		110	110	90	110
		Diameter of Core		1 7/8	2 1/4	1 7/8	2 3/4
		Length of Core		7 7/8	6 1/2	9 1/4	4 1/4
PART Cylindrical Core		Number of Clamps		1	1	2	1
OPERATION Made core from oil sand mix		L/D Ratio		4.2	2.89	4.93	1.55
PERFORMED ON Bench		Area		2.76	3.97	2.78	5.93
		Volume		21.7	25.8	25.6	25.2
		Cl + D1 + Fl		1.242	1.244	1.499	.890
COMPILED BY J. Bodesky							

SYMBOL	OPERATION DESCRIPTION	NORMAL TIME MINUTES	REFERENCE	Operation Class				
A-1	Close core box	.046	Average	C	.049	.041	.050	.053
B-1	Clamp core box (C- clamps)	.112 N	Time vs.No.Cl.	V	.110	.111	.243	.120
C-1	Fill partly full of sand	Y x CT	Time v.LD v.Vol.	V	.085	.094	.093	.119
D-1	Ram	Y x CT	Time v.LD v.Vol.	V	.225	.255	.242	.248
E-1	Place rod and wire	.0153L+.03	Time vs.Length	V	.150	.103	.168	.088
F-1	Fill and ram	Y x CT	Time v.LD v.Vol.	V	.35	.668	.900	.280
G-1	Strike off with slick	.0157A+.07	Time vs. Area	V	.115	.135	.129	.152
H-1	Remove vent wire	.047	Average	C	.048	.050	.048	.056
J-1	Rap box	.043	Average	C	.039	.042	.045	.050
K-1	Remove clamps	.061N	Time vs.No.Cl.	V	.057	.062	.116	.048
M-1	Open box	.046	Average	C	.046	.040	.038	.052
N-1	Roll out core	.0057L+.045	Time vs.Length	V	.098	.082	.102	.075
P-1	Clean box	$\sqrt{.0067+.000016V^2}$	Time vs.Volume	V	.107	.112	.120	.110

variables so that the factors influencing time can be expressed either algebraically or graphically.

Once the constant values have been selected and the variable elements equated, then simplify the expression by combining constants and unknowns where possible. The next procedure is to develop the synthesis in which the derivation of the formula is fully explained, so that persons using it, and any other interested party, understand its application and development.

Before putting the formula to use, check it thoroughly for accuracy, consistency, and ease of application. After writing the formula report describing the method used, working conditions, and limitations of application, the formula is ready for installation.

In collecting the individual time studies for the formula, it is perfectly acceptable for the observer to use existing time studies if they have been satisfactory, if the elements are standardized, if the constant and variable elements in the studies have been properly separated, and if the studies were taken under prevailing conditions and methods. When taking new studies for formula work, use the same exacting care, principles, and procedures as when taking a study for an individual standard. However, when taking studies for a time formula, break the various studies into like elements, with end points terminating at identical places in the work cycle, thus standardizing the elements. Also, make studies of different

DETAIL TIME STUDIES

S-5	S-6	S-7	S-8	S-9	S-10	S-11	S-12	S-13	S-14	S-15	S-16	S-17
Plasan	Markley	Noyes	Kinachan	Winters	Kumpf	Petrccca	Judd	Judd	Geiger	Plasan	Ekey	Noyes
105	105	105	98	120	100	108	95	102	117	85	.95	100
1 3/4	2	1 7/8	1 1/8	1	1 3/4	1 1/4	1 1/4	1 1/4	7/8	1 1/8	2	1 7/8
9 1/2	8	4 1/4	10	13 1/8	5 1/8	6 3/4	7 5/8	4 1/8	8 1/4	6 1/4	10	12 5/8
2	1	1	2	1	1	1	1	1	1	1	2	2
5.43	4	2.27	8.89	13.13	2.73	5.40	6.10	3.30	9.43	5.56	5.0	6.31
2.40	3.14	2.78	.99	.78	2.40	1.23	1.23	1.23	.60	.99	3.14	2.78
22.8	25.1	11.8	9.90	10.3	12.3	8.27	9.33	5.05	4.96	6.19	31.4	35.1
1.557	1.316	.733	1.489	2.138	.804	1.159	1.181	.797	1.277	1.032	1.740	2.299
.045	.042	.047	.050	041	.043	.042	.047	.048	.049	.046	.043	.048
.250	109	.087	.232	.218	.098	.114	.097	.128	091	.099	.229	.240
.121	.141	.082	.129	.206	.073	.102	.079	.062	.133	.100	.107	.195
.272	.245	.160	.344	.436	.194	.281	.252	.090	.342	.162	.285	.480
.171	.142	.093	.180	.222	.110	.125	.140	.090	.142	.120	.182	.210
.932	.725	.258	.889	1.290	.392	.665	.850	.432	.802	.626	1.089	1.390
.109	120	104	.078	.077	.198	.089	.084	.086	.074	.085	.119	106
.047	.042	.041	.052	.051	.043	.042	.042	.044	.040	.051	.052	.045
.035	.041	.045	.057	.048	.040	.039	.042	.041	.050	.038	.047	.041
.120	.039	.062	.116	.120	.071	.059	.056	.070	.058	.058	.140	.145
.038	.041	.045	.054	.046	.066	.038	.065	.045	.040	.035	.038	.052
.109	.091	.068	.108	.120	.080	.082	.090	.065	.090	.071	.110	.118
.115	.118	.090	.089	088	.090	.087	.085	.085	.085	.088	.138	.150

operators to get as large a cross section as possible and select jobs from the entire range of the proposed formula.

The number of studies needed to construct a formula is influenced by the range of work for which the formula is to be used, the relative consistency of like constant elements in the various studies, and the number of factors that influence the time required to perform the variable elements.

At least 10 studies should be available before a formula is constructed. If fewer than 10 are used, the accuracy of the formula may be impaired through incorrect curve construction and data not representative of typical performance. Of course, the more studies used, the more data available, and the more normal conditions will be reflected.

Analyze the Elements

After a sufficient number of time studies have been gathered, summarize the data on one work sheet for analysis purposes. Figure 20–1 illustrates a "Master Table of Detail Time Studies" form designed for this purpose. In addition to the information called for on the form, post any specific information affecting the variable elements, such as surface area, volume, length, diameter, hardness, radius, and weight, under its corresponding study number in the "Job Characteristics" section of the form.

Write the name of the operator studied, the rating factor, and the part number of the job in the separate column for all of this pertinent information for each time study. In the left-hand column headed "Symbol," place an identifying term for each element. Frequently analysts use the letters of the alphabet followed by a suffix number: A–1, B–1, or C–1. When more than 26 elements are involved, then repeat the alphabet and use the suffix 2. These symbols link elements from the Master Table to elements grouped in the synthesis.

Under the "Operation Description," record every element that has occurred on the individual time studies. Express the element description clearly, so that any interested person in the future can visualize exactly the work content of the element. A "C" or a "V" in the column headed "Operation Class" indicates whether an element is a constant or a variable.

Next, enter the normal elemental times from individual studies in the appropriate spaces. After these data have been posted, compare the elemental time values for each element and determine the reasons for variance. As can be expected, even in the constant elements, a certain amount of variation prevails because of inconsistency in operator method and in performance rating. In general, the constant elements should not deviate substantially. Determine the allowed time for each constant by averaging the values of the different studies. Post this average time in the "Normal Time" column, and write in the "Reference" column the word *average*.

Variable elements tend to vary in proportion to some characteristic or characteristics of the work, such as size, shape, or hardness. Study these elements carefully to determine which factors influence the time, and to what extent. By plotting a curve of time versus the independent variable, you may deduce an algebraic expression of the time representing the element. This procedure is explained on page 491. If an equation can be computed, then post it in the "Normal Time" column, adding notes about the curve or curves used in its deduction in the adjacent "Reference" column.

Compute Expressions for Variables

If your analysis of elemental data reveals that one variable characteristic governs the elemental time, construct a graphic relationship following the procedure outlined in Chapter 18 (see Figure 18–5 on page 494). Thus, time is the dependent variable in the plotting, since it is the value to be predicted. In those cases where time is estimated in terms of a single independent variable, y is the random variable, time, whose distribution depends on the independent variable, x. In most relations, x is not random; it is fixed for all practicality, and you are concerned with the mean of the corresponding distribution of time y (the successive elements observed through stopwatch analysis) for given x (see Figure 20–9 on page 602).

Plotted data may take a number of forms—the straight line, parabola, hyperbola, ellipse, exponential forms, or no regular geometric form whatsoever. Graphic procedures may be quite useful for obtaining predicting equations.

If the data plots as a straight line on uniform graph paper, then equate time as a function of the slope times the variable v and the y intercept, as discussed on page 494.

Another frequent linear relationship is the reciprocal function expressed by the equation:

$$y = \frac{1}{b + mv}$$

The reciprocal function involves a linear relationship for values of $\frac{1}{y}$ and v since, by taking the reciprocal of both sides of the equation, we get:

$$\frac{1}{y} = b + mv$$

The hyperbolic curve takes the form illustrated in Figure 20–2 and is expressed by the equation:

$$\frac{y^2}{a^2} - \frac{x^2}{k^2} = 1$$

If the plotted data take the form of a segment of the hyperbola, compute k graphically by drawing a tangent to the curve through the origin

FIGURE 20–2
Segment of hyperbolic curve

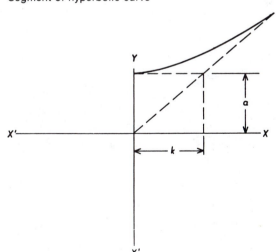

FIGURE 20–3
Segment of ellipse

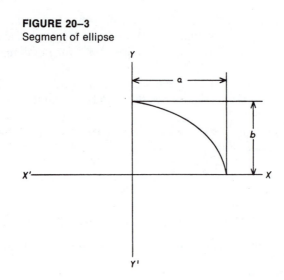

and use the resulting line as the diagonal of the rectangle whose height is *a*, the distance from the origin to the *y* intercept. Substituting time and the variable characteristic for *x* and *y* would give:

$$T = \sqrt{a^2 + \frac{(v^2)(a^2)}{k^2}}$$

The ellipse segment shown in Figure 20–3 is expressed by the equation:

$$\frac{x^2}{a^2} + \frac{y^2}{b^2} = 1$$

When the plotted time values take the form of an ellipse, compute *a* and *b* graphically and substitute their values in the expression:

$$T = \sqrt{b^2 - \frac{b^2 v^2}{a^2}}$$

In other instances, the data may take the form of a parabola, as illustrated in Figure 20–4. This curve is expressed by the equation:

$$x^2 = \frac{a^2 y}{b}$$

and after substituting graphic solutions for *a* and *b*, we have the equation:

$$T = \frac{b v^2}{a^2}$$

FIGURE 20–4
Segment of parabola

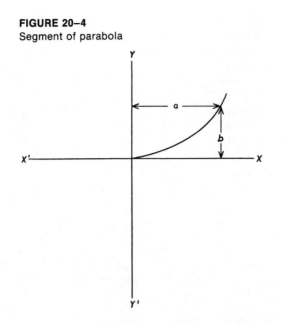

As in previous illustrations, *T* is equal to time and *v* is equal to the variable characteristic governing the elemental time.

One additional relationship that has much application in formula development is the power function:

$$y = bm^v$$

When the data plotted on uniform graph paper do not follow a straight line, parabola, hyperbola, or circle, try using semilogarithmic paper to determine whether the points fall close to a straight line on a transformed scale. If the paired data give a straight line when plotted on semilog paper, then the curve of *y* on *x* is exponential. Therefore, for any given *x*, the mean of the distribution of the *y*'s (time in our work) is given by bm^x.

In the element "strike arc and weld," analysts obtained the following data from 10 detailed studies:

Study no.	Size of weld	Minutes per inch of weld
1	⅛	0.12
2	³⁄₁₆	0.13
3	¼	0.15
4	⅜	0.24
5	½	0.37
6	⅝	0.59
7	¹¹⁄₁₆	0.80
8	¾	0.93
9	⅞	1.14
10	1	1.52

FIGURE 20–5
Plotted curve on regular coordinate paper takes exponential form

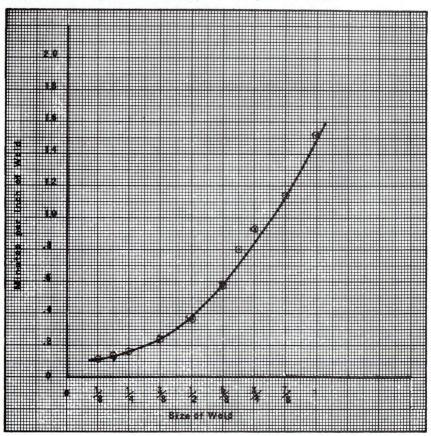

Plotting the data on rectangular coordinate paper resulted in a smooth curve (see Figure 20–5). Plotting the data on semilogarithmic paper created the straight line as shown in Figure 20–6.

The following derivation states this plotting in equation form:
Select any two points on the straight line:

Point 1 $X = 1$, $Y = \log 1.52$
Point 2 $X = 3/8$, $Y = \log 0.24$

Solving for the slope m:

$$m = \frac{Y_1 - Y_2}{X_1 - X_2}$$

$$= \frac{\log 1.52 - \log 0.24}{1 - 3/8}$$

$$= 1.28$$

FIGURE 20–6
Data plotted on semilogarithmic paper take the form of a straight line with the equation $Y = AB^x$

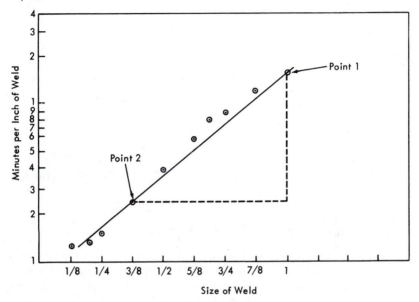

To determine the equation of the line from the equation $Y - Y_1 = m(X - X^1)$, we get:

$$\log Y - \log 1.52 = 1.28(X - 1)$$
$$\log Y - 0.18184 = 1.28X - 1.28$$
$$\log Y = 1.28X - 1.10.$$

From the exponential form

$$Y = AB^x$$
$$\log Y = \log A + X \log B$$

and

$$\log A = -1.10$$
$$A = 0.07944$$
$$\log B = 1.28$$
$$B = 19.05$$

and

$$Y = (0.07944)(19.05)^x$$

Rounding off:

$$Y = (0.08)(19)^x$$

This equation would then be a component of the formula expression and would be less cumbersome than referring to the curve shown in Figure 20–6. It can be checked as follows:

With ½-inch weld:

$$\text{Time} = (0.08)(19)^{0.5}$$
$$= 0.35 \text{ minutes}$$

This checks quite closely with the time study value of 0.37 minutes.

Sometimes, plotting the data on logarithmic paper results in a straight line. For example, the equation $Y = AX^m$ written in logarithmic form becomes:

$$\log Y = \log A + m \log X$$

Here log Y is linear with X; a straight line results from plotting the data on logarithmic paper.

Graphic Solutions for More Than One Variable

When more than one variable affects time, use graphic techniques even if the relationships are nonlinear. For example, solve for time with two or more variables, when one or more is nonlinear, by constructing a chart for each variable. The first chart plots the relationship between time and one of the variables in selected studies in which the values of the other variables tend to remain constant. The second chart shows the relationship between the second variable and time that has been adjusted to remove the influence of the first variable. If there is an additional variable affecting the time required to perform the element, extend the procedure with a third chart.

This example illustrates the application of this procedure. Analysts want to construct a formula for rolling various widths and lengths of 0.072-inch thick, cold-rolled sheet metal on a bench roll. Element 1, shown on the master table of detail time studies, is "pk. up pc. & pos.," and element 3 is "lay aside pc. & pos." Since both of these elements involve the handling time of the same amount of stock, their values can be combined to simplify the final algebraic expression. These values appear in Table 20–1.

First inspection seemed to suggest that the time for element 1 plus element 3 would vary with the area of the part handled. However, plotting indicated something else was influencing the time required to perform these elements. Further analysis revealed that the long and narrow pieces required considerably more time than the nearly square parts, even though their respective areas were about the same. Analysts decided that two variables were affecting time: these were the area of the part and the increased difficulty of handling the longer pieces. The latter may be expressed as a ratio of length to width.

TABLE 20–1

Study no.	Width	Length	Element 1 + element 3
1...............	2	16	0.18
2...............	3	18	0.13
3...............	8	1	0.09
4...............	10½	32	0.14
5...............	7	84	0.55
6...............	8½	69	0.38
7...............	20	30	0.22
8...............	11	55	0.26
9...............	10	75	0.68
10...............	20½	41	0.66
11...............	10½	90	1.80
12...............	16	64	1.44

FIGURE 20–7
Time plotted against the ratio of length to width for
constant areas of about 600 square inches

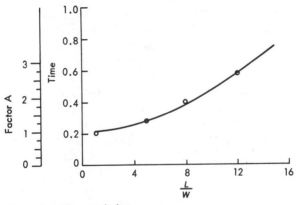

Area = C = 600 square inches.

Four of the 12 studies on the master work sheet (studies 5, 6, 7, and 8) involved stock having areas of about 600 square inches. These four studies had a relatively wide spread of "length divided by width" ratios which was apparently responsible for the range in allowed elemental times.

Since these four studies had about the same areas, plotting a simple curve showed the effect of L/W. (See Figure 20–7.) Now, to see what the effect of area is on the elemental time, it will be necessary to adjust all the tabularized time study values so as to take into account the influence of the variable L/W. To get a factor that can be used to divide the time study values so as to arrive at an adjusted time that varies with area, we construct a scale parallel to the Y axis and call it "Factor A" (adjustment

FIGURE 20–8
Adjusted time plotted against area

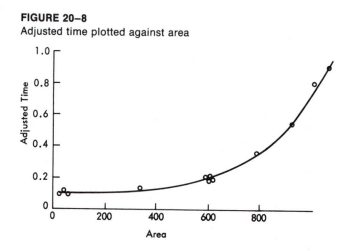

factor). By extending the lowest point on our L/W curve horizontally to the adjustment axis, we get a distance from the origin that can be evaluated as unity on the adjustment axis. Proportional values can then be constructed so as to get a scale on the adjustment axis.

Now by taking the L/W value of all the studies appearing on the master table, we can determine adjustment factors graphically for each study by moving horizontally to the Factor A axis from the specific point on the L/W versus Time curve. Once the adjustment values have been determined for each study, then the adjusted time can be computed by dividing the allowed elemental time study values by the adjustment factor. The resulting adjusted times can then be plotted against area (see Figure 20–8).

The system of curves so developed represents the graphic solution and can be used for establishing standards within their range in the following manner: First, the area and the length over width ratio of the sheet is calculated. Using the L/W ratio, we can refer to the first curve to obtain an adjustment factor. Referring to the second curve, we can determine an adjusted time. The product of the adjusted time and the adjustment factor will give the allowed time to perform elements 1 and 3. By extending this procedure, analysts can solve graphically elemental time values when more than two variable characteristics govern the elemental time.

Sometimes variables remain relatively constant within a specific group. However, once the limits of the specific group are extended, the variable shows a pronounced change in value. For example, the handling of boards of 1-inch white pine to a planer may be classified:

Group	Allowed time
Small (up to 300 square inches)...............	0.070 min.
Medium (300 to 750 square inches)............	0.095 min.
Large (750 to 1,800 square inches)	0.144 min.

This method of grouping gives erroneous values at the extremities of each group. Thus, in the above example, a board of 295 square inches would be allowed 0.070 minute handling time while one of 305 square inches would be given 0.095 minute. The chances are that in the former case, 0.070 minute would represent a tight standard, and that in the latter case, 0.095 minute would be somewhat loose.

When elements are so grouped, the grouping must be clearly and specifically defined so there are no questions as to the category in which an element belongs.

Least Squares and Regression Techniques

Although graphical procedures may be quite useful in establishing a predicting equation for one dependent variable as a function of one independent variable, one may want to resort to more sophisticated methods of curve fitting, such as least squares and regression techniques. This is indeed even more true when one dependent variable y is viewed as a function of more than one independent variable. However, least squares and regression techniques are often mistakenly employed by the novice, with resultant blunders.

As has already been pointed out, we are interested in determining the mean of the distribution of times for a given variable x. There is a distribution of the random variable time for a given value of the independent variable x. Figure 20–9 illustrates such a relationship with the true straight line of the equation $y = B_0 + B_1 x$ passing through the mean of each of the four distributions of y. This relationship is referred to as the regression curve of y on x.

When the analyst is gathering data, he or she is only able to approximate the mean of the time distributions. Any observed time may differ from the true mean by some finite amount. To determine the value of the estimates b_0 and b_1 of the parameters B_1 and B_0 for a given set of data (n paired observations $x_i y_i$), where it is assumed that the regression of y on x is linear, we want to determine the equation of the line that best fits the data. By best fit, we mean the equation of the line $y' = b_0 + b_1 x$, where the sum of the squared vertical distances ($y - y'$) for the points comprised by the data is a minimum.

FIGURE 20–9

Four distributions of time for four values of x and the resulting true curve passing through the means of these distributions

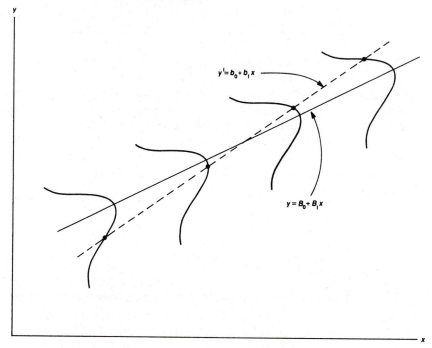

The normal equations involving the parameters b_1 and b_0 whose values for the line of best fit of the data are:

$$\sum_{i=1}^{n} y_i = b_0 n + b_1 \sum_{i=1}^{n} x_i \tag{1}$$

$$\sum_{i=1}^{n} x_i y_i = b_0 \sum_{i=1}^{n} x_i + b_1 \sum_{i=1}^{n} x_i^2 \tag{2}$$

These equations can readily be solved simultaneously to provide the numerical values of the slope b_1 and the y intercept b_0. For example, referring to page 496, we can solve for b_1 and b_0 using the above equations from the data provided in Table 20–2.

Substituting in Equation 1 and 2:

$$12b_0 = 1.834 - 2{,}073b_1 \tag{1}$$
$$2{,}073b_0 = 348.82 - 453{,}801b_1 \tag{2}$$

Multiplying Equation 1 by 2,073 and Equation 2 by 12:

TABLE 20–2

Study	x or area	y or time	xy	x^2
1..........	25	0.104	2.60	625
2..........	65	0.109	7.09	4,225
3..........	77	0.126	9.70	5,929
4..........	112	0.134	15.01	12,544
5..........	135	0.138	18.63	18,225
6..........	147	0.150	22.05	21,609
7..........	185	0.153	28.31	34,225
8..........	220	0.174	38.28	48,400
9..........	245	0.176	43.12	60,025
10..........	275	0.182	50.05	75,625
11..........	287	0.186	53.38	82,369
12..........	300	0.202	60.60	90,000
	2,073	1.834	348.82	453,801

$$24,876b_0 = 3,801.882 - 4,297,329b_1$$
$$24,876b_0 = 4,185.840 - 5,445,612b_1$$
$$0 = -383.958 + 1,148,283b_1$$

or

$$b_1 = \frac{383.958}{1,148,283} = 0.000334$$

Substituting in Equation 1:

$$12b_0 = 1.834 - 0.692$$
$$b_0 = \frac{1.142}{12} = 0.095$$

Transform the preceding equations to solve directly for the slope b_1 and the y intercept b_0 by substituting in the following relationships:

$$b_0 = \frac{(\Sigma x^2)(\Sigma y) - (\Sigma x)(\Sigma xy)}{(n)(\Sigma x^2) - (\Sigma x)^2}$$

$$b_1 = \frac{(n)(\Sigma xy) - (\Sigma x)(\Sigma y)}{(n)(\Sigma x^2) - (\Sigma x)^2}$$

Use the least squares technique also in plotting curves and exponential relationships as straight lines on semilog and log-log paper.

For example, in the case of the exponential relationship involving semilog paper where:

$$\log y = \log b_0 + x \log b_1$$

solve for the y intercept b_0 from the equation:

$$\log b_0 = \frac{(\Sigma x^2)(\Sigma \log y) - (\Sigma x)(\Sigma x \log y)}{(n)(\Sigma x^2) - (\Sigma x)^2}$$

and for the slope b_1 from the equation:

$$\log b_1 = \frac{(n)(\Sigma x \log y) - (\Sigma x)(\Sigma \log y)}{(n)(\Sigma x^2) - (\Sigma x)^2}$$

In those cases where the dependent variable y is a nonlinear function of one independent variable x, and a simple graphic procedure is not indicated, fit the data by the polynomial equation:

$$y' = b_0 + b_1 x + b_2 x^2 + b_3 x^3 \ldots , b_q x^q$$

For example, assume that we have a set of n points of data $(y_i x_i)$. To estimate the coefficients b_0, b_1, b_2, b_3 . . . , b_q we minimize:

$$\sum_{i=1}^{n} \left[y_i - (b_0 + b_1 x_i + b_2 x_i^2 + b_3 x_i^3 \ldots , b_q x_i^q) \right]^2$$

Here we are merely applying the least squares technique by minimizing the sum of the squares of the vertical distances from the points to the curve. By differentiating partially with respect to b_0, b_1, b_2 . . . , b_q, equating these partial derivatives to zero, and rearranging some of the terms, we obtain $q + 1$ normal equations. These are:

$$\Sigma y = nb_0 + b_1 \Sigma x + \ldots , b_q \Sigma x^q$$
$$\Sigma xy = b_0 \Sigma x + b_1 \Sigma x^2 + \ldots , b_q \Sigma x^{q+1}$$
$$\Sigma x^q y = b_0 \Sigma x^q + b_1 \Sigma x^{q+1} \ldots , b_q \Sigma x^{q+q}$$

MULTIPLE REGRESSION.   At times analysts recognize that more than one independent variable is influencing the dependent variable (time). If two independent variables are involved in a linear relationship, then we are fitting a plane to a set of n points to minimize the sum of the squares of the vertical distances from the points to the plane. Thus, we are minimizing

$$\sum_{i=1}^{n} \left[y_i - (b_0 + b_1 x_i + c_1 z_i) \right]^2$$

(See Figure 20–10.)

The resulting normal equations are:

$$\Sigma y = nb_0 + b_1 \Sigma x + c_1 \Sigma z$$
$$\Sigma xy = b_0 \Sigma x + b_1 \Sigma x^2 + c_1 \Sigma xz$$
$$\Sigma zy = b_0 \Sigma z + b_1 \Sigma xz + c_1 \Sigma z^2$$

This example demonstrates the multiple regression technique. Analysts decide to construct a formula for bagging hardware in burlap bags of various sizes. They find that time is not only dependent on the number of

FIGURE 20–10
Two independent variables characterize the
best fit of a plane in a geometric cube

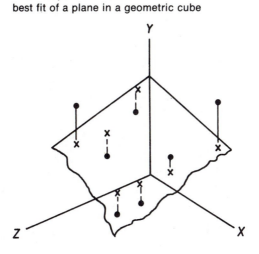

components placed in each bag, but also on the total weight of the components placed in each bag. A master table of detailed time studies (see page
591) contained the information in Table 20–3. These data refer to the
second element, which is "place hardware components in bag."

Substituting

$$\Sigma x = 51.91, \ \Sigma z = 93, \ \Sigma x^2 = 312.75$$
$$\Sigma xz = 515.27, \ \Sigma z^2 = 1{,}003, \ \Sigma y = 2.252$$
$$\Sigma xy = 12.772, \ \Sigma zy = 23.429$$

TABLE 20–3

Study no.	y Time	x Total weight of components	z Number of components
1	0.264	6.62	11
2	0.130	1.15	6
3	0.202	5.61	8
4	0.126	4.01	4
5	0.220	6.14	9
6	0.332	7.25	14
7	0.222	7.02	6
8	0.155	1.75	8
9	0.345	5.45	17
10	0.256	6.91	10
	2.252	51.91	93

into the normal equations gives us:

$$2.252 = 10b_0 + 51.91b_1 + 93c_1$$
$$12.772 = 51.91b_0 + 312.75b_1 + 515.27c_1$$
$$23.429 = 93b_0 + 515.27b_1 + 1,003c_1$$

The solution of this system of equations provides the following results: $b_0 = 0.016$, $b_1 = 0.0139$, $c_1 = 0.0147$. Now estimate the time y to bag parts having the total weight x and the number of components z by direct substitution in the equation:

$$y^1 = 0.016 + 0.0139x + 0.0147z$$

Thus, multiple linear regression predicts the dependent variable of time by taking into account all of the independent variables. After collecting the data, using stopwatch or DataMyte time studies, analysts identify all independent variables. They use this multiple linear regression equation:

$$y = a + b_1x_1 + b_2x_2 + \ldots b_nx_n$$

where

$y =$ Time to perform the operation
$a =$ Intercept
$b =$ Regression coefficients from 1 to n
$x =$ Independent variables from 1 to n

Several different regression programs can solve the cumbersome matrix mathematics involved in multiple linear regression. Among these, the IBM System 360 Scientific Subroutine Package is practical for work measurement analysts.

SOME PRECAUTIONS. Regression methods may present difficulties of interpretation and analysis to the uninformed. These difficulties occur primarily with the statistical and engineering analyses used in the selection of the independent variables and with the significance that each variable plays in the predicting equation. The form of the equation selected (that is, linear or nonlinear) and the inclusion of cross products of the variables in the equation when interaction effects are present may further complicate the analysis. Under these circumstances, consult an analyst competent in regression techniques. However, in the interest of ease of application, it is a good general practice to make predicting equations as simple as possible. From a statistical point of view, complicated predicting equations may result in time estimates that are little, if any, more reliable than equations of a simpler form.

Develop Synthesis

The purpose of the synthesis in the formula report is to completely explain the derivation of the various components in the formula to facilitate its use. Furthermore, a clearly developed synthesis helps explain and sell the formula in the event that someone questions its suitability at some later date.

In order that the final expression may be in its simplest form, combine all constants and symbols wherever possible, with due respect to the accuracy and flexibility of the formula. Classify elemental operations under specific headings, such as preparations, handling by hand, handling by jib crane, and operation. Constant values appear under each elemental class. Combine these values to simplify the final expression of the specific formulas. Details of the combinations are explained in the synthesis. A typical illustration of a synthesis for a constant element would be:

$$A1 + C1 + \text{Constant from equation } 4 + H1 = 0.11 + 0.17$$
$$+ 0.09 + 0.05 = 0.42 \text{ minute}$$

The treatment of variable elements requires a more detailed explanation due to the added complexities associated with this class of elements. For example, to combine elements $B1$ and $D1$, in that they have been influenced by the ratio of L/D, the two equations may be:

$$B1 = 0.1 \frac{L}{D} + 0.08$$

$$D1 = 0.05 \frac{L}{D}$$

and the combination of these would give

$$B1 + D1 = 0.15 \frac{L}{D} + 0.08$$

Thus, the synthesis would clearly show where the 0.15 L/D came from.

Compute Expression

The final expression may not be entirely in algebraic form. It may be more convenient to express some of the variables as systems of curves, nomographs, or single curves. Then, too, some of the variable data may be made up into tables referred to by a single symbol in the formula. A typical example of a formula would be:

$$\text{Normal time} = 0.07 + \text{Chart 1} + \text{Curve 1} + \frac{0.555N}{C}$$

where:

N = Number of cavities per mold
C = Cure time in minutes

In applying this formula, analysts would refer to Chart 1 for the applicable tabularized value. They would also refer to Curve 1 to obtain the numerical value of another variable entering into the formula. Knowing the number of cavities in the mold and the cure time in minutes, they would substitute in the latter part of the expression and summarize the values of the three variables with the constant time of 0.07 minute to get the normal time.

Check for Accuracy

Upon completion of the formula, verify it before releasing it for use. The easiest and fastest way to check the formula is to use it to check existing time studies. This can best be done by tabularizing the results under the following headings: "Part Number," "Time Study Value," "New Formula Value," "Difference," and "Percentage Difference."

Investigate any marked differences between the formula value and the time study value and determine the cause. The formula should show an average difference of less than 5 percent from the time study values of those studies used in its derivation. If, at this point, the formula does not have the expected validity, accumulate additional data by taking more stopwatch and/or standard data studies. After testing the formula over a range of work, explain it to those who are affected directly and indirectly by it so they will accept it. At this point, go over the formula with the supervisor of the department where it will be used. Understanding its derivation and application and having verified its reliability, the supervisor will back the standards resulting from the formula's use.

Write Formula Report

Consolidate all data, calculations, derivations, and applications of the formula in the form of a complete report prior to putting the formula into use. This will make all the facts relative to the process employed, operating conditions, and scope of formula available for future reference.

The Westinghouse Electric Corporation includes 14 separate sections in its formula reports. These are:

1. Formula number, division, data, and sheet number.
2. Part.
3. Operation.
4. Workstation.
5. Normal time.
6. Application.
7. Analysis.
8. Procedure.
9. Time studies.
10. Table of detail elements.
11. Synthesis.
12. Inspection.
13. Wage payment.
14. Signatures of constructor and approver.

FORMULA NUMBER. For ease of reference, identify all paperwork used to develop the formula by assigning a number to it. This formula number should have a prefix which identifies the department or division in which the formula will be used, and a number which gives the chronological arrangement of the formula with reference to other formulas being used in the department. Thus, the formula 11N–56 would be the 56th formula designed for use in department 11N. Include the date that the formula was put into use to connect working conditions in effect at the time it was designed with standards established through its application.

PART. List the part number or numbers together with drawing numbers with a concise description of the work, clearly defining the products for which the formula has application. For example, "Parts J–1101, J–1146, J–1172, J–1496, side plates ranging from 12 × 24 inches to 48 × 96 inches" would identify the parts for which the formula may be used.

OPERATION AND WORKSTATION. Clearly describe the operation covered by the formula, such as "roll radius" or "fabricate inner and outer members" or "broach keyways" or "assemble farm tank." In addition, describe the workstation, with information as to equipment, jigs, fixtures, and gages as well as their size, condition, and serial number. A photograph often helps to define the method in effect at the time of compilation of the formula.

NORMAL TIME. When expressing the formula for the normal time, use separate equations for the "setup" time and the "each piece" time. Immediately following the formula add a key that outlines the meaning of the symbols used in each equation. Include all tables, nomograms, and systems of curves at this point, as well as a sample solution, so that the person using the formula clearly understands it.

APPLICATION. After concisely stating the expressions for the allowed time, give a clear explanation of the application of the formula. This should describe in detail the nature of the work on which the formula may be used, and specifically state the limits within which it may be applied. For example, the application of a formula for curing bonded rubber parts may be written: "This formula applies to all curing operations done in 24 × 28-inch platen presses when the number of cavities range between 8 and 100 and the cubic inches of rubber per cavity range between 0.25 and 3."

ANALYSIS. Under the analysis, give a detailed account of the entire method employed, including tools, fixtures, jigs, and gages and their application; workplace layout; methods of material handling; method of obtaining supply materials; and nature of setup, detailing how the operator is assigned the job, the distance to the tool crib, and other pertinent data.

In the analysis section of the report include the breakdown of the allowances used in the formula. Clearly state the reason for any special or extra allowance. Break out personal time, fatigue, and unavoidable delay allowances independently, so that if any question comes up about the inclusion of an allowance in the formula, you can clearly show which allowances were included and why.

PROCEDURE. After completing the analysis of the job, write up the procedure used by the operator in performing the work. The best way of doing this is to include all the elements appearing on the master work sheet in their correct chronological sequence. While abbreviations may be liberally used on the time studies and on the master table, avoid their use in writing up the operator's procedure. Write individual elements in exact detail. Then everyone will understand which work elements fall within the scope of the formula.

TIME STUDIES. The formula report need not include the actual time studies used in the compilation of the formula, as they are usually available in their own independent files. However, you should refer to the time studies used. This can be done in tabular form as follows:

Time study no.	Part	Drawing	Plant	Date taken	Taken by
S–112	J–1102	JB–1102	A	9–15–	J. B. Smith
S–147	J–1476	JA–1476	B	10–24–	A. B. Jones
S–92	J–1105	JB–1105	B	6–11–	J. B. Smith

TABLE OF DETAIL ELEMENTS. This table is a reference source for information about the normal element time and its derivation. The information for composing this table comes from the master table of detailed elements. Record it in chart form as follows:

Symbol	Element description	Normal time	Reference
Al	Close core box	0.046 min.	Average
Cl	Fill partly full of sand	A × C.T.	All studies
Jl	Rap box	0.043 min.	Average

SYNTHESIS. After recording the table of detailed elements, include a synthesis of the report. The synthesis explains the manner in which the allowed time was derived.

INSPECTION, PAYMENT, AND SIGNATURES. Upon completion of the synthesis you, as the constructor, should record the following information in order: the inspection requirements appearing on the drawings of the jobs covered by the formula; the type of wage payment plans where the formula will be used, such as daywork, piecework, and group incentive; and finally your signature and that of your supervisor.

REPRESENTATIVE FORMULA REPORT. The following formula report clarifies the procedures discussed in this chapter:

FORMULA NO.: M–11–No. 15
DATE: September 15, 19–
SHEET: 1 of 15

PART: Cylindrical oil sand mix cores ⅞″ in diameter to 2¾″ diameter and 4″ long to 13″ long.
OPERATION: Make core complete in wood core box.

WORKSTATION: 30″ × 52″ Bench 36″ high.
NORMAL TIME: Piece time in decimal minutes equals:

$$0.173N + 0.0210L + \sqrt{0.0067 + 0.000016V^2} + Y \times CT + 0.0157A + 0.327$$

Where

N = Number of "C" clamps
L = Length of core in inches
V = Volume of core in cubic inches
Y = Adjustment factor (Curve 1)
CT = Adjusted time (Curve 2)
A = Cross-sectional area of core

EXAMPLE:

Calculate normal time to make a green sand core 2″ in diameter and 8″ long. Volume would be 25.1 cubic inches and with 8″ length, only one clamp would be required. L/D would be equal to 8/2 or 4.

THEN:

$$\text{Normal time} = (0.173)(1) + (0.0210)(8) + \sqrt{0.0067 + (0.000016)(25.1^2)}$$
$$+ (1.5)(.76) + (0.0157)(3.14) + 0.327 = 1.987 \text{ minutes.}$$

APPLICATION

This formula applies to all cylindrical cores made of oil sand mix between diameters ranging from 7/8 inch to 2¾ inches and lengths ranging from 4 inches to 13 inches. This work is performed manually in wooden split core boxes.

ANALYSIS

The following hand tools are provided the operator for making cores covered by this formula: C-clamps, slick, and rammer. The work is done at a bench with the operator alternately sitting and standing. Oil sand mix for producing the cores is piled on the floor by the move worker approximately 4 feet from the operator. Periodically, the operator replenishes a supply of sand on the bench from the inventory on the floor, through the use of a short-handled shovel. Rods and reinforcing wires are requisitioned from the storeroom by the operator who usually acquires a full day's supply with each requisition.

The work voucher placed above the operator's workstation by the departmental supervisor indicates the sequence of jobs to be performed and is the authority for the operator to begin a specific job. Operation cards and drawings must be obtained by the operator from the tool crib.

The standard allowance which must be added when applying this formula is 17 percent. This involves 5 percent for personal delays, 6 percent for unavoidable delays, and 6 percent for fatigue.

Procedure

The working procedure followed in making oil sand mix cores covered by this formula includes 13 elements exclusive of setup and put-away elements. These are:

1. Pick up two sections of core box and close together.
2. Clamp core box shut using one clamp for cores 9 inches or less and two clamps for cores greater than 9 inches in length.
3. Fill core box partly full of sand (about one third depending on core).
4. Ram the sand solid in the core box.
5. Place the vent rod and reinforcing wire.
6. Fill the remainder of the core box with sand and ram solid.
7. With slick, strike off sand on both ends of core box so sand core is flush with box.
8. Remove the vent wire from the core and lay aside.
9. Rap box lightly.
10. Remove clamps and lay aside.
11. Open core box.
12. Roll out core on rack beside workstation.
13. Clean core box with kerosene rag.

Time Studies

The following summary includes the time studies used in developing this formula:

Time study no.	Part number		Plant	Date taken	Taken by
S-13	P-1472	PB-1472	A	6-15	Black
S-111	P-1106	PB-1106	A	12-11	Black
S-45	P-1901	PB-1901	A	7-13	Hirsch
S-46	P-1907	PB-1907	B	7-14	Black
S-47	P-1908	PA-1908	B	7-14	Black
S-32	P-1219	PA-1219	A	8-10	Black
S-76	P-1711	PA-1711	A	11-12	Obe
S-70	P-1701	PB-1701	B	11-9	Obe
S-17	P-1311	PB-1311	B	6-16	Black
S-18	P-1312	PB-1312	B	6-16	Black
S-59	P-1506	PB-1506	A	7-26	Hirsch
S-60	P-1507	PB-1507	A	7-26	Hirsch
S-50	P-1497	PB-1497	B	7-19	Obe
S-51	P-1498	PA-1498	B	7-20	Obe
S-52	P-1499	PA-1499	B	7-21	Obe
S-53	P-1500	PA-1500	B	7-21	Obe
S-54	P-1501	PA-1501	B	7-22	Obe

TABLE OF DETAIL ELEMENTS

Symbol	Element description	Decimal minute normal time	Reference
A-1	Close core box	0.046	Average
B-1.........	Clamp core box	$0.112N$	Time vs. no clamps
C-1.........	Fill partly full of sand	$Y \times CT$	Time vs. L/D & vol.
			Time vs. vol.
D-1	Ram	$Y \times CT$	Time vs. L/D & vol.
E-1.........	Place rod and wire	$0.0153L + 0.03$	Time vs. length
F-1.........	Fill and ram	$Y \times CT$	Time vs. L/D & vol.
			Vs. vol.
G-1	Strike off	$0.0157A + 0.07$	Time vs. area
H-1	Remove vent wire	0.047	Average
J-1	Rap box	0.043	Average
K-1	Remove clamps	$0.061N$	Time vs. no clamps
M-1	Open box	0.046	Average
N-1	Roll out core	$0.057L + 0.045$	Time vs. length
P-1.........	Clean box	$\sqrt{0.0067 + 0.000016V^2}$	Time vs. volume

SYNTHESIS

Standard time = Sum of allowed elemental time
= A–1 + B–1 + C–1 + D–1 + E–1 + F–1 + G–1
+ H–1 + J–1 + K–1 + M–1 + N–1 + P–1

Since Elements C–1, D–1, and F–1 are all dependent on the variables volume and length/diameter, their values may be combined before plotting on coordinate paper. The combined values for these three elements for the 17 time studies are:

Study no.	$C–1 + D–1 + F–1$
S–13......................	1.045
S–111.....................	1.017
S–45......................	1.235
S–46......................	0.647
S–47......................	1.325
S–32......................	1.110
S–76......................	0.500
S–70......................	1.362
S–17......................	1.932
S–18......................	0.659
S–59......................	0.988
S–60......................	1.181
S–50......................	0.584
S–51......................	1.277
S–52......................	0.888
S–53......................	1.481
S–54......................	2.065

FIGURE 20-11

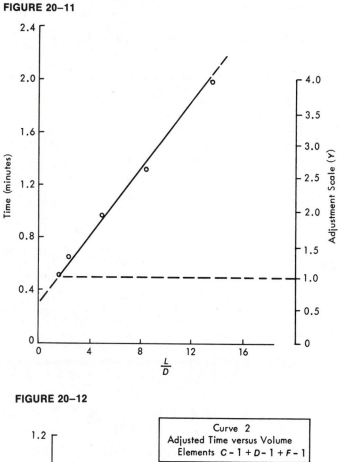

FIGURE 20-12

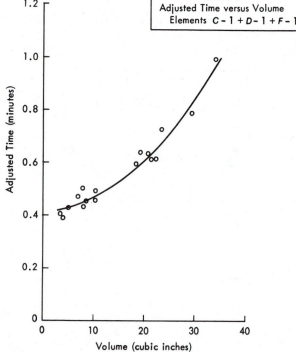

Curve 2
Adjusted Time versus Volume
Elements $C-1 + D-1 + F-1$

FIGURE 20-13

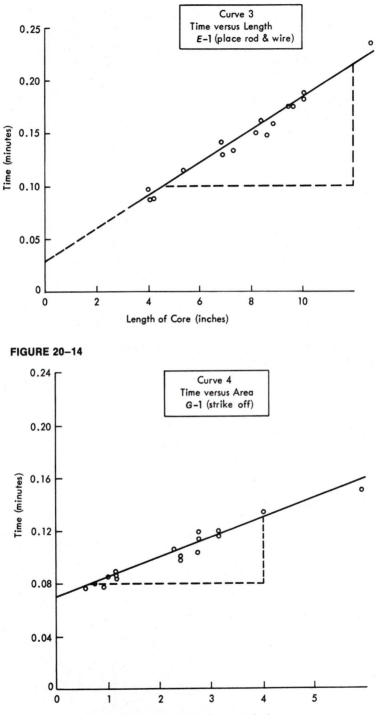

Curve 3
Time versus Length
E-1 (place rod & wire)

Time (minutes)

Length of Core (inches)

FIGURE 20-14

Curve 4
Time versus Area
G-1 (strike off)

Time (minutes)

Cross-Sectional Area (square inches)

For reference purposes, the combination of C–1 + D–1 + F–1 is designated R–1 (see Figures 20–11 and 20–12). Curves 1 and 2 are used to solve for the allowed time for these three elements by first getting the adjustment factor from curve 1 and then getting the adjusted time from curve 2. The product of these two values gives the allowed time for C–1 + D–1 + F–1.

The time for element E–1 (place rod and wire) has been plotted on curve 3 (see Figure 20–13), where its relationship to core length is shown.

This curve can be expressed algebraically by the equation:

$$T = fv + C$$

or

$$\text{Time} = (\text{Slope})(\text{length}) + y \text{ intercept}$$

Solving graphically:

$$f = \frac{0.210 - 0.100}{12 - 4.8} = \frac{0.110}{7.20}$$
$$= 0.0153$$
$$C = 0.03 \text{ (from graph)}$$

Thus:

$$\text{Time} = 0.0153L + 0.03$$

Element G–1 (strike off) has also been solved as a straight-line relationship. Here a cross-sectional area of the core has been plotted against time (see curve 4) (Figure 20–14). In this case

$$\text{Time} = (f)(\text{area}) + 0.07$$

and

$$f = \frac{0.132 - 0.08}{4 - 0.7} = \frac{0.052}{3.3}$$
$$f = 0.0157$$

Then:

$$\text{Time} = 0.0157A + 0.07$$

Element N–1 (roll out core) is shown on curve 5 (see Figure 20–15). Here

$$\text{Time} = f(\text{length}) + 0.045$$

and

$$f = \frac{0.125 - 0.068}{14 - 4}$$
$$= 0.0057$$

Then:

$$\text{Time} = 0.0057L + 0.045$$

FIGURE 20–15

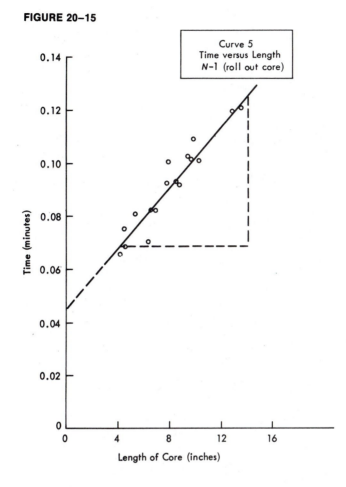

The time for element P–1 (clean box) when plotted against the volume of the core, gives a hyperbolic relationship as shown in curve 6 (Figure 20–16). This relationship with time can be expressed algebraically as follows:

$$\text{Time} = \sqrt{a^2 + \frac{v^2 a^2}{k^2}}$$

$$= \sqrt{0.082^2 + \frac{(v^2)(0.082)^2}{400}}$$

$$= \sqrt{0.0067 + 0.000016v^2}$$

Elements A–1, H–1, J–1, M–1 were classified as constants with the following respective allowed times determined by taking their average values: 0.046, 0.047, 0.043, 0.046. These constants are added to the sum of the y intercept values of equations (3), (4), and (5), to determine a total constant time of 0.327 minute:

FIGURE 20–16

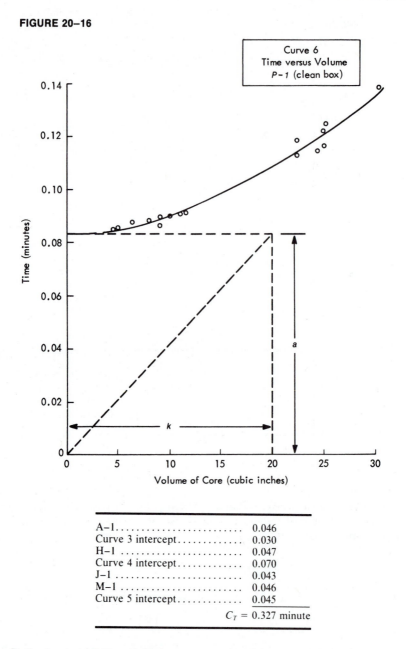

A–1..........................	0.046
Curve 3 intercept.............	0.030
H–1	0.047
Curve 4 intercept.............	0.070
J–1	0.043
M–1	0.046
Curve 5 intercept.............	0.045
C_T =	0.327 minute

Both elements B–1 and K–1 are proportional to the number of clamps N.
Thus, these two elements may be combined for simplicity as follows:

$$B\text{–}1 = 0.112N$$
$$K\text{–}1 = 0.061N$$
$$S\text{–}1 = 0.173N$$

Since elements E–1 (place rod and wire) and N–1 (roll out core) are proportional to length of core, we may combine their equated relationships with length and designate the combined total by the symbol T–1.
Thus:

$$E\text{–}1 = 0.0153L$$
$$\underline{N\text{–}1 = 0.0057L}$$
$$T\text{–}1 = 0.0210L$$

Simplifying our initial equation:

$$\text{Normal time} = \text{S–1} + \text{T–1} + \text{P–1} + \text{R–1} + \text{G-1} + C_T$$

$$= 0.173N + 0.0210L + \sqrt{0.0067 + 0.000016V^2}$$

$$+ Y \times CT + 0.0157A + 0.327$$

INSPECTION

The only inspection requirement on the cores covered by this formula is visual inspection done by the operator at the time the core is rolled free from the core box.

USE OF THE COMPUTER

Many special purpose work measurement software programs can be purchased. For example, programs for application of predetermined motion data (MTM as well as high-task bases), for the analysis of machining speeds for minimum time and cost of production, for work sampling analysis, and for learning curve or manufacture progress functions are available. For a modest cost the Institute of Industrial Engineers offers four microcomputer programs for applying work measurement methods. These are for stopwatch time study analysis, cumulative average hours per unit and labor cost based on learning curve theory, a work sampling accuracy analysis, and a work sampling observation generator.

Minicomputers are quite satisfactory for performing the following work measurement functions: line balancing, labor efficiency reporting, applying time study formulas, analyzing time study data, developing new standards based on fundamental motion data, processing work sampling data, and revising standards.

One software system—ADAM—is typical. It is designed to apply predetermined elemental time systems and to maintain standard data elements created by the predetermined time values.[1]

The commands the program executes include checking and accepting or rejecting a sequence of motion codes that comprise an operation,

---

[1] Douglas M. Towne, "ADAM—A Computer-based System for Generating and Maintaining Labor Standards and Standard Data," *The MTM Journal of Methods—Time Measurement* VII, no. 3.

storing standard data elements, evaluating formulas, starting a new study, and printing completed studies.

Initially users must input the element codes and their descriptive comments and frequencies to the core program. Then they can give a command to either display the resulting analysis on the cathode ray screen or print the output. At this time users can detect any input codes that were incorrect but not recognizable.

Editing features allow users to correct errors or adjust a previous study for a new application. Thus editing permits users to modify, insert, or delete a line at the current location at any time.

This system accepts six types of input: descriptive comments, element codes, frequencies, formulas, standard data elements, and grouping control.

A typical user's descriptive comment—designated by "C"—might be: C PLACE FORGING IN LEAF DRILL JIG. Element codes could include typical MTM data such as: M16C.

Users assign frequencies if an element occurs less or more than once per cycle. They enter this as "X" followed by the appropriate occurrence. An example would be M16C X 2.

Formulas, designated by "F," include the formula number. Then users input the values to be applied to the formula. For example, if formula number 126 were to be used where two variables were included, the input may be: F 126 0.42 179. The 0.42 and 179 would be the values of the two variables to be substituted in formula 126.

Standard data elements, designated by "Z," are followed by the particular reference code and then the mathematical manipulation of the standard data element. For example: Z 11 X 0.3. The system searches for and finds standard data element 11, multiplies its time value by 0.3, and places the result in the study.

Grouping control is symbolized by "/". This implies that a new group of elements follow.

Such computer-based systems not only save time and improve on the accuracy in the development of new standards, but also allow the economic revision of existing standards after methods engineering changes. The high-speed and mathematical accuracy of the computer gives its use significant advantages in locating, retrieving, and revising previously stored standards and standard data elements.

Although this software can compute new and maintain existing standards, the hardware has application in many other computation and display procedures discussed in this text. These include: manufacturing progress (learning curve) computations, work sampling analysis, queuing and worker and machine analyses, multiple regression analysis, optimization analysis, performance reporting, and production control.

The following simple example demonstrates the application of the com-

FIGURE 20–17
Program flowchart symbols

General Operation Symbol: Used for any operation which creates, alters, transfers or erases data, or any other operation for which no specific symbol has been defined in the Standard.

Subroutine (Predefined Process) Symbol: Used when a section of program is considered as a single operation for the purpose of this flowchart.

Generalized Input/Output Symbol: Used to represent the input/output (I/O) function of making information available.

Branch Symbol: Has one entry line and more than one exit. The symbol contains a description of the test on which the selection of an exit is based. The various possible results of this test are shown against the corresponding exits.

Offpage Connector Symbol: Used as a linkage between two blocks of logic that are to be found on separate pages of the flowchart. The symbol is only used on the "exit" page, on the "entry" page an onpage symbol is used.

Onpage Connector Symbol: Used as a linkage between two blocks on the same page, when it is not desirable to connect them using a linkage line. The label of the block to which the connection is being made is written inside the symbols.

Terminal Symbol: Used as the beginning or end of a flowline (e.g., start or end of a program).

Annotation Symbol: Used to add additional information to a symbol or block of program.

Flowlines (Linkage Lines): Used to show the flow between blocks of a flowchart. The normal flow is from top to bottom and left to right of the page. The programmer may dispense with the use of the direction arrows when the chart follows the normal flow. They must be used, however, for any portion of the diagram which does not follow the normal flow.

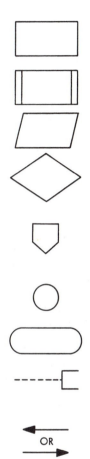

puter in performing arithmetic operations and in flowcharting (see Figure 20–17). Let the equation for knurling times be given by:

$$T = 0.1A + 1.5$$

where:

T = Normal time
A = Surface area of knurled shaft

Assume you want to generate a table of times (T) for areas (A) where A varies from 3 to 80 square inches by increments of ¼ inch. The flowchart for this very simple example is given in Figure 20–18. The parenthetical

FIGURE 20-18
Knurling times flowchart

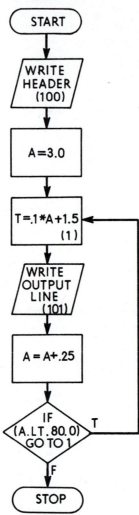

numbers relate to program statement numbers as they appear in the program.

Programming

Let's continue with the knurling times example which has been flow charted. It is easy to translate it into FORTRAN. For the achievement of repetitive formula computation, use the branching IF statement. Every READ and WRITE statement requires a FORMAT statement, the number of which is given by the second of the couple in the parenthetic expression after each; the first number represents the device: six being the printer. The source program is as follows:

```
C        PRØGRAM TØ PRØDUCE TABLE ØF KNURLING
C        TIMES
C        EQUATION--T = 0.1A + 1.5
         WRITE (6,100)
         A = 3.0
      1  T = .1*A + 1.5
         WRITE (6,101) A,T
         A = A + .25
         IF (A.LT.80.0) GO TO 1
         STØP
    100  FØRMAT (1H1,10X,14HKNURLING TIMES/1H,12X,9HT =
         .1A + 1.5/1H0,8X,4AREA1,10X,4HTIME/)
    101  FØRMAT (1H ,F12.2,F15.3)
         END
```

The job control language input is not included in the source program, as often this is unique to the computer installation. Generally, this requires less than 8 seconds of compile and execute time on a System/370 model 168 or above. The result of this program appears in Figure 20–19.

CONCLUSION

A time study formula can establish standards in a fraction of the time required to take individual studies. However, before a formula is released for use, the mathematics used in its development should be carefully checked to assure that the expression is correct. Then, too, several test cases should be tried to be certain that the formula establishes true, consistent standards. Clearly identify the formula's range of application, and in no case establish standards with data beyond the scope of the formula.

In summary, the following steps represent the chronological procedure in time study formula design:

1. Collect data.
    a. Use time studies with standardized elements that are already available.
    b. Use standard data based on standardized elements that are already available.
    c. Establish elemental values from basic motion data for standardized elements.
    d. Take new time studies. Be careful to break new studies into standardized elements.
2. Compile master work sheet and identify formula.

FIGURE 20–19

KNURLING TIMES $T=.1A+1.5$	
AREA	TIME
3.00	1.800
3.25	1.825
3.50	1.850
3.75	1.875
4.00	1.900
4.25	1.925
4.50	1.950
4.75	1.975
5.00	2.000
5.25	2.025
5.50	2.050
5.75	2.075
6.00	2.100
6.25	2.125
6.50	2.150
6.75	2.175
7.00	2.200
7.25	2.225
7.50	2.250
7.75	2.275
8.00	2.300
8.25	2.325
8.50	2.350
8.75	2.375
9.00	2.400
9.25	2.425
9.50	2.450
9.75	2.475
10.00	2.500
10.25	2.525
10.50	2.550
10.75	2.575
11.00	2.600
11.25	2.625
11.50	2.650
11.75	2.675
12.00	2.700
12.25	2.725
12.50	2.750
12.75	2.775
13.00	2.800
13.25	2.825

3. Analyze and classify elements.
   a. Constants.
   b. Variables.
4. Develop synthesis.
5. Compute the final expression.
6. Check mathematics of developed formula.
7. Test formula.
8. Write formula report.
9. Use formula.

By systematically following these nine steps, analysts have little difficulty in designing reliable time study formulas. Once a formula has been designed, it can be programmed on a digital computer and solved for all possible variations within the scope of the formula. Thus, tabulated standards can be made available for broad variations in the work assignment.

TEXT QUESTIONS

1. How can MTM and Work-Factor be used in formula development?
2. What advantages does the formula offer over standard data in establishing time standards?
3. Is the use of time study formulas restricted to machine shop operations where feeds and speeds influence allowed times? Explain.
4. What are the characteristics of a sound time study formula?
5. What is the danger of using too few studies in the derivation of a formula?
6. What is the function of the synthesis in the formula report?
7. Write the equation of the ellipse with its center at the origin and axes along the coordinate axes and passing through $(2, 3)$ and $(-1, 4)$.
8. Find the equation of the hyperbola with its center at $(0, 0)$ and $a = 4$, $b = 5$, foci on the y axis.
9. List the 14 sections that make up the formula report.
10. Which nine steps represent the chronological procedure in the design of time study formulas?
11. Explain in detail how it is possible to solve graphically for time when two variables are influential and they cannot be combined.
12. Develop an algebraic expression for the relationship between time and area from the following data:

Study no.	1	2	3	4	5
Time	4	7	11	15	21
Area	28.6	79.4	182	318	589

13. If the data shown in Figure 20–1 were plotted on logarithmic paper, would the plotting be a straight line? Why or why not?
14. Which types of problems lend themselves to solution on the digital computer?

GENERAL QUESTIONS

1. Would a union prefer standards to be set by formulas or standard data? Why or why not?
2. Would it be necessary for the "chart and formula" designer to have a background in time study work? Why or why not?
3. If an operator objected strongly to a rate established through a formula, explain in detail how you would endeavor to prove to him or her that the rate was fair.

PROBLEMS

1. In the Dorben Company, the work measurement analyst planned to develop standard data on a new milling machine that was recently installed. The material being cut in one group of studies involved a 1½-inch width of cut,

with lengths varying from 4 inches to 30 inches. For this work, a plain carbide-tipped milling cutter 3 inches in diameter and a width of face of 2 inches was used. Depths of cut ranged from 3/16 inch to 7/16 inch.

Give the equation that can be programmed to provide cutting time in terms of d (depth of cut) and l (length of casting being milled). In all cases, the feed per tooth is 0.010 inch and the 16-tooth cutter is running at a surface speed of 80 feet per minute.

2. The work measurement analyst in the Dorben Company was studying the hand filing and polishing of external radii. Six studies provided the following information:

Study	Size of radii	Minutes per inch
1	3/8	0.24
2	1/2	0.37
3	5/8	0.59
4	11/16	0.80
5	3/4	0.93
6	1	1.52

These data plotted as a straight line on semilog paper where time (the dependent variable) was the logarithmetic scale. Develop an algebraic equation for estimating the time of filing and polishing various radii.

3. In the assembly department of the Dorben Company, various hardware is bagged for shipment. The time study analyst desires to design a formula or a system of curves to establish standards on this work. The master table of detailed time studies contains the following information for the second element, "place hardware components in bag":

Study no.	Time (min.)	Weight of components (lb.)	Bag size	No. of components
1	0.264	6.62	4	11
2	0.130	1.15	2	6
3	0.186	5.61	3	8
4	0.169	2.91	2	6
5	0.126	4.01	2	4
6	0.220	6.14	3	9
7	0.200	5.50	3	6
8	0.332	7.25	4	14
9	0.222	7.02	4	6
10	0.155	1.75	2	8
11	0.345	5.45	4	17
12	0.256	6.91	4	10

From these data, design a formula or a system of curves for establishing the standard for this element.

4. In the blanking of various leather components from animal skins, the analyst noted a relationship between standard time and the area of the component. After taking five independent time studies, the analyst observed the following:

Study no.	Area of leather component (sq. in.)	Standard time (min.)
1.........................	5.0	0.07
2.........................	7.5	0.10
3.........................	15.5	0.13
4.........................	25.0	0.20
5.........................	34.0	0.24

Derive an algebraic expression to preprice the blanking of the various leather components.

5. The analyst in the Dorben Company decides to develop a formula for prepricing a particular assembly operation involving different sizes of work. The assembly operation involves three constant elements and one variable element. The constant elements are determined from MTM data. The classifications of the fundamental motions on the constant elements are as follows:

Element 1: One R10C, one G1B, one P2SSD, one AF (max), one M20C with 2# weight, and one RL (case 1)
Element 2: One eye travel with T = 20 and D = 10, one R12B, one G1A, one M10C, one P1SSE, ten T 30° 2#
Element 3: One R10A, one G1B, one M20B, one RL2

The variable element is based on the following data:

Study number	Standard time (min.)	Surface area (sq. in.)
1.................	0.282	7
2.................	0.163	5
3.................	0.022	2
4.................	0.120	4.5
5.................	0.227	6

Develop the algebraic expression for establishing standards for this operation for parts with surface areas of up to seven square inches.

6. The industrial engineer in the Dorben Company is in the process of building a formula to preprice the manufacture of a line of specialty forgings. The data gathered indicated a nonlinear relationship between forging volume and standard time for the variable elements related to positioning in the die and the actual forging. The data taken were as follows:

Study number	Time (min.)	Forging volume (cu. in.)
1..............	0.130	30
2..............	0.110	24
3..............	0.103	20
4..............	0.088	10
5..............	0.083	5
6..............	0.120	27

Develop an algebraic expression for the computation of standard time for any forging having a volume of between 5 and 30 cubic inches.

SELECTED REFERENCES

Denbow, Carl H., and Victor Goedicke. *Foundations of Mathematics*. New York: Harper & Row, 1959.

Lowry, Stewart M.; Harold B. Maynard; and B. J. Stegemerten. *Motion and Time Study*. 3rd ed. New York: McGraw-Hill, 1940.

Rotroff, Virgil H. *Work Measurement*. New York: Reinhold Publishing, 1959.

Whitehouse, Gary E. *Work Measurement IIE Microsoftware*. Atlanta, Ga.: Industrial Engineering and Management Press, 1985.

SELECTED SOFTWARE

Nicks, J. E. Regression Analysis, Basic Programming Solutions for Manufacturing, Dearborn, Mich.: Society of Manufacturing Engineers, 1982.

Work Sampling Studies

21

Work sampling is a technique used to investigate the proportions of total time devoted to the various activities that constitute a job or work situation. The results of work sampling are effective for determining allowances applicable to the job, for determining machine and personnel utilization, and for establishing standards of production. This same information can be obtained by time study procedures. Work sampling is a method that frequently provides the information faster and at considerably less cost than stopwatch techniques.

In conducting work sampling studies, analysts take a comparatively large number of observations at random intervals. The ratio of observations of a given activity to the total observations approximates the percentage of time that the process is in that state of activity. For example, if 1,000 observations at random intervals over several weeks showed that an automatic screw machine was turning out work in 700 instances, and that in 300 instances it was idle for miscellaneous reasons, then the downtime of the machine would be 30 percent of the working day, or 2.4 hours, and the effective output of the machine would be 5.6 hours per day. The application of work sampling was first made by L. H. C. Tippett in the British textile industry. Later, under the name "ratio-delay" study, it received considerable attention in this country.[1] The accuracy of the data determined by work sampling depends on the number of observations and the period over which the random observations are taken. Unless the sample size is of sufficient quantity, and the data are taken over a period of time that represents typical conditions, inaccurate results occur.

---

[1] Robert Lee Morrow, *Time Study and Motion Economy* (New York: Ronald Press, 1946), pp. 177–99.

The work sampling method has several advantages over that of acquiring data by the conventional time study procedure. These are:

1. It does not require continuous observation by an analyst over a long period of time.
2. Clerical time is diminished.
3. The total work-hours expended by the analyst are usually much fewer.[2]
4. The operator is not subjected to long-period stopwatch observations.
5. Crew operations can be readily studied by a single analyst.

THE THEORY OF WORK SAMPLING

The theory of work sampling is based on the fundamental laws of probability: at a given instant an event can either be present or absent. Statisticians have derived the following expression, which shows the probability of x occurrences of an event in n observations:

$$(p + q)^n = 1$$

p = Probability of a single occurrence
$q = (1 - p)$ the probability of an absence of occurrence
n = Number of observations

If the preceding expression, $(p + q)^n = 1$, is expanded according to the binomial theorem, the first term of the expansion gives the probability that $x = 0$, the second term $x = 1$, and so on. The distribution of these probabilities is known as the binomial distribution. Statisticians have also shown that the mean of this distribution is equal to np and that the variance is equal to npq. The standard deviation is, of course, equal to the square root of the variance.

Of what value is a distribution that allows only one event to either occur or not occur? For the answer to this, consider the possibility of taking one condition of the work sampling study at a time. All the other conditions can then be considered as nonoccurrences of this one event. Using this approach, we proceed now with the discussion of binomial theory.

According to elementary statistics, as n becomes large, the binomial distribution approaches the normal distribution. Since work sampling studies involve quite large sample sizes, the normal distribution is a satisfactory approximation of the binomial distribution. Rather than use the binomial distribution, it is more convenient to use the distribution of a proportion with a mean of p, that is, $\dfrac{np}{n}$, and a standard deviation of

---

[2] Ibid., p. 334.

$$\sqrt{\frac{pq}{n}}\ \left(\text{i.e.,}\ \frac{\sqrt{npq}}{n}\right)$$

as the approximately normally distributed random variable.

In work sampling studies, we take a sample of size n in an attempt to estimate p. We know from elementary sampling theory that we cannot expect the $\hat{p}$ ($\hat{p}$ = the proportion based on a sample) of each sample to be the true value of p. We do, however, expect the $\hat{p}$ of any sample to fall within the range of $p \pm 2$ sigma approximately 95 percent of the time. In other words, if p is the true percentage of a given condition, we can expect the $\hat{p}$ of any sample to fall outside the limits $p \pm 2$ sigma only about 5 times in 100 due to chance alone. This theory can produce the total sample size required to give a certain degree of accuracy. It is also used for subsample sizes a little later.

Illustrative Example

To clarify the fundamental theory of work sampling, it would be helpful to interpret the results of an experiment. Assume the following circumstances: we have observed one machine that has random breakdowns for a 100-day period. During this period, we have taken eight random observations per day.

Let:

n = Number of observations per day

k = Total number of days that observations are taken

x_i = Number of breakdown observations observed in n random observations on day i ($i = 1, 2, \ldots . k$)

N = Total number of random observations

N_x = Number of days that the experiment showed the number of breakdowns equal x ($x = 0, 1, 2, \ldots . n$)

$P(x)$ = The probability that the machine is down x times in n observations is given by the binomial distribution

$$P(x) = \frac{n!}{x!(n-x)!}\, p^x q^{n-x}$$

where:

p = Probability of machine being down

q = Probability of machine running

$$p + q = 1$$

For our example, $n = 8$ observations per day, $k = 100$ days of observations, and $N = 800$ total observations. An all-day time study for several days revealed that $p = 0.5$. Table 21–1 shows the number of days in which x breakdowns were observed in the work sampling study ($x = 0, 1, 2, 3$

TABLE 21–1

x	Nx	$P(x)$	$100P(x)$
0	0	0.0039	0.39
1	4	0.0312	3.12
2	11	0.1050	10.5
3	23	0.2190	21.9
4	27	0.2730	27.3
5	22	0.2190	21.9
6	10	0.1050	10.5
7	3	0.0312	3.12
8	0	0.0039	0.39
	100	1.00*	100*

* Approximately.

. . . . . n), and the expected number of breakdowns given by our binomial model, using $p = 0.5$ from the all-day time study.

There is close agreement between the observed days that a specified number of breakdowns occurred (N_x) and the expected number computed theoretically as $kP(x)$.

$$\bar{P}_i = \frac{x_i}{n} = \text{Observed proportion of downtime on day}$$

$$i \ (i = 1, 2, 3, \ldots . k)$$

$$\hat{P} = \frac{\sum_{i=1}^{k} \bar{P}_i}{k} = \frac{\sum_{i=1}^{k} x_i}{n \cdot k}$$

$$= \frac{\sum_{i=1}^{k} x_i}{N} = \text{Estimated proportion of machine downtime}$$
based on a work sampling experiment

A hypothesis that the theoretical information shows close enough agreement to the observed information for the theoretical binomial to be accepted, may be tested using the chi-square (χ^2) distribution. The (χ^2) distribution tests whether the observed frequencies in a distribution differ significantly from the expected frequencies.

In the example, where the observed frequency is N_x and the expected frequency is $kP(x)$:

$$\chi^2 = \sum_{k=0}^{k} \frac{[N_x - 100P(x)]^2}{100P(x)}$$

That is, the quantity under the summation is distributed approximately as χ^2 for k degrees of freedom. In this example $\chi^2 = 0.206$.

FIGURE 21–1

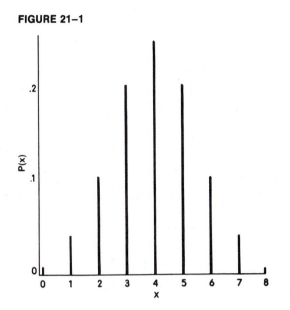

Analysts must determine whether the calculated value of χ^2 is suffi-ciently large to refute a null hypothesis that the difference between the observed frequencies and the computed frequencies is due to chance alone. This experimental value of χ^2 is so small that it could easily have occurred through chance. Therefore, accept the hypothesis that the ex-perimental data "fits" the theoretical binomial distribution.

In typical industrial situations, p (which was known to have a value of 0.5) is unknown to analysts. The best estimate of p is $\hat{p}$, which may be computed as $\dfrac{\sum\limits_{i=1}^{k} x_i}{N}$. As the number of observations taken at random per day (n) increases and/or the number of days the study is made in-creases, $\hat{p}$ will approach p. However, with limited observations, analysts are concerned with the accuracy of $\hat{p}$.

If a plot of $P(x)$ versus x were made from the above example, it would appear as shown in Figure 21–1.

When n is sufficiently large, regardless of the actual value of p, the binomial distribution very closely approximates the normal distribution. This tendency can be seen in the example when p is approximately 0.5. When p is near 0.5, n may be small and the normal can be a good approxi-mation to the binomial.

When using the normal approximation, set

$$u = p$$

and

$$\sigma_p = \sqrt{\frac{pq}{n}}$$

To approximate the binomial distribution the variable z used for entry in the normal distribution (see Table A3–2, Appendix 3) takes the following form:

$$z = \frac{\hat{p} - p}{\sqrt{\frac{pq}{n}}}$$

Although p is unknown in the practical case, estimate p from $\hat{p}$, and determine the interval within which p lies, using confidence limits. For example, imagine that the interval defined

$$\hat{p} - 2 \sqrt{\frac{\hat{p}\hat{q}}{n}}$$

and

$$\hat{p} + 2 \sqrt{\frac{\hat{p}\hat{q}}{n}}$$

contains p 95 percent of the time.

Graphically, this may be represented as:

$$\hat{p} - 2 \sqrt{\frac{\hat{p}\hat{q}}{n}} \qquad \hat{p} \qquad \hat{p} + 2 \sqrt{\frac{\hat{p}\hat{q}}{n}}$$

Derive the expression for finding a confidence interval for p as follows: suppose that we want an interval which contains p 95 percent of the time, that is, a 95 percent confidence interval. For n sufficiently large, the expression,

$$z = \frac{\hat{p} - p}{\sqrt{\frac{\hat{p}\hat{q}}{n}}}$$

is approximately a standard normal variable. Therefore, set the probability

$$P\left(z_{.025} < \frac{\hat{p} - p}{\sqrt{\frac{\hat{p}\hat{q}}{n}}} < z_{.975}\right) = 0.95$$

Rearranging the inequalities then gives:

$$P \left(\hat{p} - z_{.975} \sqrt{\frac{\hat{p}\hat{q}}{n}} < p < \hat{p} + z_{.975} \sqrt{\frac{\hat{p}\hat{q}}{n}} \right) = 0.95$$

remembering that $-z_{.025} = z_{.975} = 1.96$, or approximately 2. The interval with approximately a 95 percent chance of containing p is then

$$\hat{p} - 2 \sqrt{\frac{\hat{p}\hat{q}}{n}} < p < \hat{p} + 2 \sqrt{\frac{\hat{p}\hat{q}}{n}}$$

These limits imply that the interval defined contains p with 95 percent confidence since z has been selected as having a value of 2.

The underlying assumptions of the binomial are that p, the probability of a success (the occurrence of downtime), is constant each random instant that we observe the process. Therefore, it is always necessary to take random observations when taking a work sampling study. Thus, bias introduced by worker anticipation of observation times is reduced.

Another principle considered in the determination of the observation time is the stratification of the data collected. For example, an analyst who decides to gather 600 random observations by taking 10 random observations a day over the next 60 working days is stratifying by days while randomizing within days. By not allowing all 600 random observations to occur within two or three days observers get better results.

SELLING WORK SAMPLING

Before beginning a program of work sampling, sell the use and the realiability of the tool to all members of the organization who will be affected by the results. If the program will establish allowances, it should be sold to the union and the supervisor, as well as company management. This can be done by having several short sessions with representatives of the various interested parties and explaining examples of the law of probability, thus illustrating why ratio delay procedures work. Unions as well as workers favor work sampling techniques once the procedure is fully explained, since work sampling is completely impersonal, does not utilize the stopwatch, and is based on accepted mathematical and statistical methods.

In the initial session to explain work sampling, create a simple study by tossing unbiased coins. All participants should readily recognize that a single coin toss stands a 50–50 chance of being heads. When asked how they would determine the probability of heads versus tails, they will undoubtedly propose tossing a coin a few times to find out. When asked whether two times is adequate, they will say no. Ten times may be suggested, and the response will be: "That may not be adequate." When 100 times is suggested, the group will agree: "That should do it with some

FIGURE 21–2
Distribution of number of heads with infinite
number of tosses using four unbiased coins

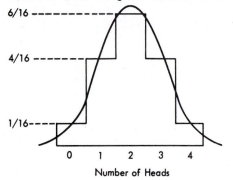

FIGURE 21–3

Operation	Observation	Total	Percent
Running	|||| |||| |||| |||| |||| |||| |||| |||| |||| |||| ‖ |||| |||| |||| |||| |||| |||| |||| |||| |||| |||| |||| |||| |||| |||| |||| |||| |||| |||| |||| |||| |||| |||| |||| |||| |||| |||| |||| |||| |||| |||| |||| |||| |||| |||| |||| |||| |||| |||| |||| ||||	252	70
Idle	|||| |||| |||| |||| |||| |||| |||| |||| |||| ||| |||| |||| |||| |||| |||| |||| |||| |||| |||| |||| |||| ||||	108	30
		Total 360	100

degree of assurance." This example firmly implants the principal requisite
of work sampling: adequate sample size to ensure statistical significance.

Next, discuss the probable results of tossing four unbiased coins. Explain that there is only one arrangement in which the coins can fall showing no heads, and only one arrangement that permits all heads. However,
three heads or one head can result from four possible arrangements. Six
possible arrangements can give two heads.

With all 16 possibilities thus accounted for, tell the group that if four
unbiased coins are tossed continually, they distribute themselves as in
Figure 21–2.

After this explanation and the demonstration of this distribution by
making several tosses and recording the results, your audience should
agree that 100 tosses could demonstrate a normal distribution. Further-

FIGURE 21-4
Cumulative percentage of running time

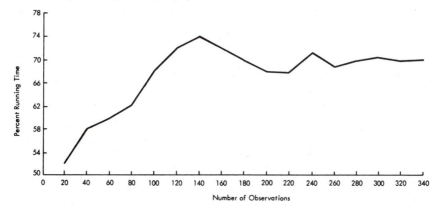

more, point out a thousand tosses would probably approach a normal distribution more closely and 100,000 would give a nearly perfect distribution. Such a distribution is not sufficiently more accurate than the 1,000-toss distribution, however, and not economically worth the extra effort. This establishes the idea of approaching significant accuracy rapidly at first, and then at a diminishing rate.

Next point out that a machine or operator could be in a heads or tails state. For example, a machine could be running (heads) or idle (tails). Demonstrate an example showing 360 observations that have been made on a turret lathe. These observations might give data like those in Figure 21-3.

A cumulative plot of "running" would level off, giving an indication of when it would be safe to stop taking readings (see Figure 21-4). Explain how "idle" machine time could be broken down into the various interruptions and delays and could be accounted for and evaluated.

Once the validity of work sampling has been sold up and down the line, clearly define the problem. If you wish to establish allowance data, make a summary of all elements customarily included in various allowance categories. To do this, make a preliminary survey of the class of work for which you are determining the allowances. Spend several hours on the production floor observing all the delays encountered, so that the element listing can be complete.

PLANNING THE WORK SAMPLING STUDY

After explaining the method and obtaining the approval of the supervisor involved, detailed planning must be done before making actual observations.

To begin, make a preliminary estimate of the activities on which you

are seeking information. This estimate may involve one or more activities. Frequently the estimate can be made from historical data. If you
cannot make a reasonable estimate, work sample the area or areas involved for two or three days and use the information obtained as the basis
of these estimates (see Figure 21–3).

Once the preliminary estimates have been made, determine the desired
accuracy of the results. This can best be expressed as a tolerance within a
stated confidence level.

Having made a preliminary estimate of the percentage occurrences of
the elements being studied, and determined an accuracy desired at a given
confidence level, estimate the observations to be made. Knowing how
many observations need to be taken and the time that is available to
conduct the study, you can determine the frequency of observations.

Next design the work sampling form or card on which to tabulate the
data and the control charts used in conjunction with the study.

Determining the Observations Needed

To determine the number of observations needed, you must know how
accurate your results must be. The more observations, the more valid the
final answer is. Three thousand observations give considerably more
realiable results than 300. However, if the accuracy of the result is not the
prime consideration, 300 observations may be ample.

In random sampling procedures, there is always the chance that the
final result of the observations will be beyond the acceptable tolerance.
However, sampling errors diminish as the size of the sample increases.
The standard error σ_p of a sample proportion or percentage is expressed
by the equation:

$$\sigma_p = \sqrt{\frac{pq}{n}} = \sqrt{\frac{p(1-p)}{n}}$$

where:

σ_p = Standard deviation of a percentage
p = True percentage occurrence of the element being sought, ex-
pressed as a decimal
n = Total number of random observations upon which p is based

By approximating the true percentage occurrence of the element being
sought (designated as $\hat{p}$), and by knowing the allowable standard error,
substitute in the above expression and compute n:

$$n = \frac{\hat{p}(1-\hat{p})}{\sigma_p^2}$$

For example, assume you want to determine the number of observations required with 95 percent confidence so that the true proportion of

FIGURE 21–5
The tolerance range of the percentage of unavoidable delay
allowance required within a given section of a plant

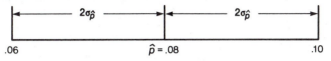

personal and unavoidable delay time is within the interval 6–10 percent.
The unavoidable and personal delay time encountered in the section of
the plant under study is expected to be 8 percent. These assumptions are
expressed graphically in Figure 21–5.

In this case, $\hat{p}$ would equal 0.08, and $\sigma_{\hat{p}}$ would equal 1 percent, or 0.01.
Using these values, solve for n as follows:

$$n = \frac{0.08(1 - 0.08)}{(0.01)^2}$$

$$= 736 \text{ observations}$$

After obtaining the initial estimate of observations compute a more
accurate estimation of p by recalculating n, based on the values of $\hat{p}$ and
σ_p calculated from the first few days of observations. If you do not get the
desired accuracy, take more observations and repeat the process.

Microcomputer software programs for determining the observations
required for a work sampling study are readily available today.[3] These
programs perform all the statistical calculations to determine sample sizes
and confidence intervals. For example, they can calculate the 90 percent,
95 percent, and 99 percent confidence intervals for a sample. They can
also provide the number of samples necessary to achieve 90 percent, 95
percent, and 99 percent confidence for a specified degree of accuracy.

This example illustrates the use of work sampling software. An analyst
wishes to determine the amount of downtime due to tool problems in an
area involving 10 CNC machining centers where very fine drilling is taking
place. After 500 random trips to the machining area (5,000 random observations), there were 243 instances where a facility was down because of
tooling. The analyst plans to use software to determine the confidence
intervals for the estimated downtime due to tooling and the number of
random observations to estimate within ±5 percent of the true value.

To calculate the confidence intervals, he or she would input the total
observations (5,000) and the number of downtimes (243). The program
reports the 90 percent, 95 percent, and 99 percent confidence limits on the
estimate (0.0486) of the downtime.

---

[3] For example, the program developed by Gary E. Whitehouse and David A. Washburn of
the University of Central Florida.

The analyst may want to determine if the sample size taken is adequate. To do this, he or she inputs a best estimate of p (.0486) and the degree of accuracy desired. The program then reports the requisite observations to be 90 percent, 95 percent, or 99 percent confident.

Determining Observation Frequency

The frequency of the observations depends for the most part on the observations required and the time limit placed on the development of the data. For example, for 3,600 observations during a study to be completed in 30 calendar days, we should need to obtain approximately:

$$\frac{3,600 \text{ observations}}{20 \text{ working days}} = 180 \text{ observations per working day}$$

Of course, the number of analysts available and the nature of the work being studied also influence the frequency of the observations. For example, if only one analyst is available who is accumulating allowance data on a limited battery of facilities, it may be impractical for that person to take 180 observations during one working day. After determining the number of observations per day, then select the actual time to record observations. To obtain a representative sample, take observations at all times of the working day.

For this example, assume that one analyst is available to study a battery of 20 turret lathes, completely independent of each other, to determine the personal and unavoidable delay allowance. The analyst has calculated that 180 observations per day will be required. Since there are 20 machines to observe, the observer must make nine random trips to the machine floor each day for 20 days. The time of day selected for these nine observations is chosen at random daily. Thus, the analyst establishes no set observation pattern from day to day on the production floor.

Frequently, work sampling analysts take observations at fixed intervals. This is referred to as systematic sampling. When sampling sizes are adequate and data is procured over the entire span of data, systematic sampling will give valid results. The fact that most indirect work does not have a fixed rhythm that could coincide with the sampling interval and that sampling times will not occur with great precision, systematic sampling is sufficiently random to produce reliable information. This is especially true when studying relatively long-cycle nonrepetitive work such as maintenance.

The analyst may select nine numbers daily from a statistical table of random numbers, ranging from 1 to 48 (see Appendix 3, Table A3–4). By letting each number carry a value in minutes equivalent to 10 times its size, the numbers selected can then set the time in minutes from the beginning of the working day to the time for taking the observations. For example, the random number 20 would mean that the analyst should make a series of observations 200 minutes after the beginning of the shift. If the working day began at 8 A.M., then at 20 minutes after 11 A.M. an inspection of the 20 turret lathe operators would begin.

Microcomputers can also determine the schedule of daily observations. For example, work sampling programs for the DataMyte collector described in Chapter 13 can print random time schedules. Computer programs are available (or may be developed) to provide the time of day to take each of a series of random observations of the workstation. Such programs generate a different schedule for each day of the work sampling study. For example, one program that has been developed is based on a 24-hour clock, so that random observations may be taken at any time over three shifts. The program asks four questions:

1. What time to start (in 24-hour clock)?
2. What time to stop (in 24-hour clock)?
3. How many random observations are to be taken between the start and stop times.
4. The minimum time between observations.

The software also allows the inputting of special conditions, for example, periods in which no observations are to be taken. It may not be desirable to take observations during a 10 A.M. to 10:15 A.M. coffee break and the 12 noon to 1 P.M. lunch period.

If the input were based on an 8 A.M. to 5 P.M. working day, 20 random observations were desired, and the minimum time between observations is 10 minutes, the four inputs would be: 8, 17, 20, and 10. Thus the schedule of observations to be generated would begin at 8 A.M. and end at 5 P.M. The observation schedule would then be displayed similarly to the following:

Observation 1 at 8:1	Observation 11 at 11:25
Observation 2 at 8:29	Observation 12 at 11:41
Observation 3 at 8:54	Observation 13 at 13:14
Observation 4 at 9:11	Observation 14 at 13:41
Observation 5 at 9:30	Observation 15 at 14:15
Observation 6 at 9:41	Observation 16 at 14:54
Observation 7 at 9:58	Observation 17 at 15:15
Observation 8 at 10:22	Observation 18 at 15:32
Observation 9 at 10:47	Observation 19 at 16:16
Observation 10 at 11:00	Observation 20 at 16:57

Here 9:58 means 58 minutes after 9 A.M.; 15:15 means 15 minutes after 3 P.M.

The study should be long enough to include normal fluctuations in production. The longer the overall study, the better the chance to observe average conditions. Usually work sampling studies are made over a block of time ranging from two to four weeks.

FIGURE 21–6
The analyst is performing her regular work while taking
a self-analysis work sampling study. Note the random
reminder instrument attached to her jacket. The analyst
records the activity she is engaged in when this
instrument beeps.

USING A RANDOM REMINDER.   Another alternative to help analysts
decide when to make daily observations is a random reminder. This
pocket-sized instrument beeps at random times letting analysts know
when to make the next observation. The user preselects an average sampling rate (observations per hour or observations per day) and responds
with a trip to the data collection area upon hearing the beep. Typically,
the instrument (Divilbiss Electronics, Model JD-7) can be preset at any of
the following average beeps per hour: 0.64, 0.80, 1.0, 1.3, 1.6, 2.0, 2.5,
3.2, 4.0, 5.0, 6.4, and 8. (See Figure 21–6.)

This instrument is especially useful for self-observation. A table with
times prepared in advance can occupy a disproportionate share of the
analyst's time when attempting to record data conscientiously at the listed
times.

Using an Electronic Work Measurement Recorder

Electronic work measurement machines with optional work sampling
software can be extremely helpful to the analyst. For example, the OS-3
Plus facility (see Chapter 13) is available with a work sampling program
that permits scheduling of random observations, performance rating of
individual readings, summary statistics and formatted printed reports.
Figure 21–7 illustrates a printout where the random schedule of tours for
the current shift are shown (at top) and a portion of a work sampling
summary report (at bottom).

FIGURE 21-7

Sample printout—schedule printout (The schedule printout lists the random schedule of tours for the current shift when the OS-3's scheduler is used.)

```
Your Company Name Here          OS-3 WORK SAMPLING   03/22/92   18:24

SHIFT#   START      END        BREAK    START      END
  1      08:00:00   17:00:00      1      10:00:00   10:15:00
                                  2      12:00:00   12:30:00
                                  3      14:00:00   14:15:00

TOUR TIME    MIN START TO START          MAX START TO START    SEED
00:30:00     00:30:00                     00:50:00             28348

sample   08:17:43
sample   08:55:29
break             10:00:00
sample   10:15:45
sample   10:56:55
sample   00:00:00
```

FIGURE 21-7 (continued)

Sample printout—schedule printout (The schedule printout lists the random schedule of tours for the current shift when the OS-3's scheduler is used.)

```
Your Company Name Here        OS-3 WORK SAMPLING  03/22/83 18:57    Page 1


OBSERVER        PLANT #          MO/DY/YR        HR:MN
John Anderson   1255             03/22/92        18:24


FRED


    MACHINE            # = 4   36%
    WELD               # = 2   18%
    FIT PIPE           # = 2   18%
    CLEAN UP           # = 1    9%
    productive TOTAL               # = 9   82%

    GET TOOLS          # = 4    9%
    WAIT JOB
```

COMBINED

MACHINE	# =	5	15%
WELD	# =	10	30%
FIT PIPE	# =	6	18%
CLEAN UP	# =	4	12%
productive TOTAL	# =	25	76%
GET TOOLS	# =	4	12%
WAIT JOB	# =	0	0%
CONFIR FOREMAN	# =	2	6%
IDLE	# =	1	3%
PERSONAL	# =	1	3%
nonproductive TOTAL	# =	8	24%
ABSENT	# =	0	0%
not coded TOTAL	# =	0	0%

Courtesy of GageTalker Corporation formerly Observational Systems

Designing the Work Sampling Form

The analyst should design an observation form to best record the data to be gathered during the course of the work sampling study. Do not use a standard form, since each work sampling study is unique from the standpoint of the total observations needed, the random times that observations are made, and the information being sought. The best form is the one tailored to the objectives of the study.

Figure 21–8 is a work sampling study form. An analyst designed this form to determine the time utilized for various productive and nonproductive states in a maintenance repair shop. It accommodates 20 random observations during the working day.

Some analysts prefer to use a specially designed card to gather work sampling data that allows observations without the attention caused by carrying a clipboard. The cards can be of a size that can be carried conveniently in the shirt or coat pocket. For instance, the form shown in Figure 21–8 could be split easily into two sections with one section on each side of a three inch by five inch card that could be carried in the shirt pocket.

Using Control Charts

The control chart techniques used so extensively in statistical quality control work can be applied readily to work sampling studies. Since work sampling studies deal exclusively with percentages or proportions, analysts use the "p" chart most frequently.

To understand how the "p" chart can be of value in a work sampling study, you must understand the theory behind control charting: While a complete discussion of control chart theory is impracticable, a brief discussion will explain the logic in using control charts.

The first problem encountered in setting up a control chart is the choice of limits. In general, seek a balance between the cost of looking for an assignable cause when none is present and not looking for an assignable cause when one is present. As an arbitrary choice, use the three-sigma limits as control limits on the "p" chart.

Suppose that p for a given condition is 0.10 and samples of size 180 are taken each day. Substitute in the equation for n (see page 638) to obtain control limits of ± 0.07. The nomogram in Figure 21–9 gives the correct three-sigma limits for various sample sizes and various values of p. Then construct a control chart similar to Figure 21–10. The p' values for each day would be plotted on the chart. (Note that $\frac{\Sigma p'}{N} = \hat{p}$, where N equals the number of days the work sampling study was made, and approximately the same number of observations were made each day of the study.)

In quality control work, the control chart indicates whether or not the process is in control. In a similar manner, in work sampling the analyst considers points beyond three-sigma limits of p as out of control. Thus, a

FIGURE 21–8

WORK SAMPLING STUDY

MAIN REPAIR SHOP

Remarks _____

Number Working This Study _____ Date _____ By _____

Obs. Nos.	Random Time	Productive Occurrences							Nonproductive Occurrences							Total Observations	Percentage Productive	Percentage Nonproductive
		Mch	Weld	Pipe Fit	Gen. Labor	Elect.	Carpen.	Janitor	Get Tools	Grind Tools	Wait Job	Wait Crane	Confer Foreman	Personal	Idle			
1																		
2																		
3																		
4																		
5																		
6																		
7																		
8																		
9																		
10																		
11																		
12																		
13																		
14																		
15																		
16																		
17																		
18																		
19																		
20																		
TOTAL																		

FIGURE 21–9
Nomogram for finding control limits on the results of work
sampling

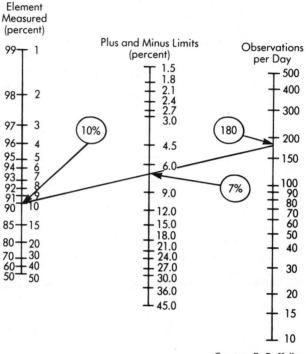

Courtesy R. P. Heller

FIGURE 21–10

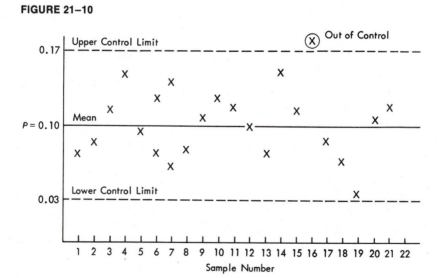

FIGURE 21–11

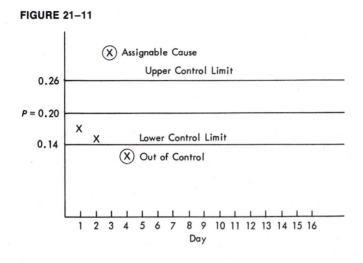

certain sample that yields a value of p' is assumed to have been drawn from a population with an expected value of p if p' fall within the plus or minus three-sigma limits of p. Expressed another way, if a sample has a value p' that falls outside the three-sigma limits, it is assumed that the sample is from some different population or that the original population has been changed.

$$3\sigma_p = 3\sqrt{\frac{p(1-p)}{n}} \tag{1}$$

As in quality control work, points other than those out of control may be of some statistical significance. For example, it is more likely that a point will fall outside the three-sigma limits than that two successive points will fall between the two- and three-sigma limits. Hence, two successive points between the two- and three-sigma limits would indicate that the population had changed. Series of significant sets of points have been derived. This idea is discussed in most statistical quality control texts under the heading "Theory of Runs."

The following hypothetical example shows how control charts can facilitate a work sampling study: Company XYZ wished to measure the percentage of machine downtime in the lathe department. An original estimate showed downtime to be approximately 0.20. The desired results were to be within ±5 percent of p with a level of significance of 0.95. Analysts took 6,400 readings over 16 days at the rate of 400 readings per day. They computed a p' value for each daily sample of 400 and set up a p chart for $p = 0.20$ and subsample size $N = 400$ (see Figure 21–11).

Each day, they took readings and plotted p'. On the third day, the point for p' went above the upper control limit. An investigation revealed that

FIGURE 21–12

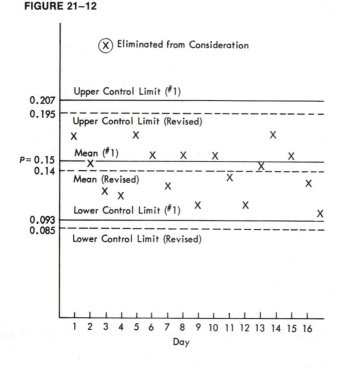

following an accident in the plant several workers left their machines to assist the injured employee to the plant hospital. Since an assignable cause of error was discovered, they discarded this point from the study. If they had not used a control chart, these observations would have been included in the final estimate of *p*.

On the fourth day the point for *p'* fell below the lower control limit. No assignable cause could be found for this occurrence. The industrial engineer in charge of the project also noted that the *p'* values for the first two days were below the mean *p* and decided to compute a new value for *p* using the values from days 1, 2, and 4. The new estimate of *p* turned out to be 0.15. To obtain the desired accuracy, *n* is now 8,830 observations. The control limits also changed, see Figure 21–12. They took observations for 12 more days, and plotted the individual *p'* values on the new chart. As can be seen, all the points fell within the control limits. Then they calculated a more accurate value of *p*, using all 6,000 observations. They determined the new estimate of *p* was 0.14. A recalculation of achieved accuracy showed it to be slightly better than the desired accuracy. As a final check, they computed new control limits using *p* equal to 0.14. The dashed lines superimposed on Figure 21–12 show that all points were still in control using the new limits. If a point had fallen out of control, analysts would have eliminated it and computed a new value of *p*. They would have repeated this process until the desired accuracy was achieved and all p' values were in control.

Do not assume that since the percentage downtime is 0.14 today, it will be 0.14 a year from now. Improvement should be a continuing process, and percentage downtime should diminish. One purpose of work sampling is to determine areas of work that might be improved. After discovering such areas, make an attempt to improve the situation. Control charts can show the progressive improvement of work areas. This idea is especially important if work sampling studies establish standard times, for such standards must change whenever conditions change if they are to remain realistic.

OBSERVING AND RECORDING THE DATA

A representative sample form for a shift study appears in Figure 21–13. Here analysts made six random observations of each facility per shift. A digit in the space provided for the state of each facility under study designates the particular observation. Since 14 facilities were being studied, they made a total of 84 observations per shift.

In approaching the work area, do not anticipate the recording you expect to make. Walk to a point a given distance from the facility, make your observation, and record the facts. It might be helpful to make an actual mark on the floor to show where you must stand before observing. If the operator or machine being studied is idle, determine the reason for idleness, confirming the reason with the line supervisor before making the proper entry. Learn to take visual observations, and make written entries after leaving the work area. This minimizes the shop workers' feeling of being watched and they will perform in their accustomed manner.

To assure that the operators perform in their usual fashion, inform them of the purpose of the study. The fact that you are not using a watch tends to relieve the operators of a certain mental tension; you should have little difficulty in getting their full cooperation.

Using a Random Activity Analysis Camera

Even if you observe the requisites of work sampling, the data tend to be biased when the technique is used for studying people only. This is because the arrival of an observer at the work center immediately influences the activity of the operator. The operator becomes productively engaged as soon as he or she sees the analyst approaching the work center. Then, too, there is a natural tendency for the observer to record what *has just* happened or what *will be* happening rather than what *is* actually happening at the exact moment of the observation.

Using a random activity analysis camera similar to that described in Chapter 8 permits unbiased work sampling studies of problems involving people. A work sampling study I made to determine the attentiveness of

WORK SAMPLING OBSERVATION RECORD

STUDY NO. ___46___

PLANT ___DuBois___ STUDY AREA ___Section C of Heavy Machine Shop___

DATE ___5/13___ REMARKS ___Were Producing Three Cylinder Turbines___

OBSERVER ___R. Guild___

MACHINE	DWG.	CUTTING	SETUP	MACHINE IDLE	CRANE WAIT	WAIT-INSPECTION	AID INSPECTION	WAIT-TOOLS NOT AVAILABLE	WAIT-TOOL TROUBLE	CONFER WITH OTHER SHIFT	TOOL HANDLING	GET OR GRIND TOOLS	CONFER WITH FOREMAN, INSP.	WAIT FOR JOB	REMOVE CHIPS CLEAN TABLE	MISCELLANEOUS	NO OPERATOR
20' VBM	93J967	3 5	4					1		6		2					
16' VBM ⑤	98J469 90J384	2 5 2 5	2 1	3									4 6 1				
28' VBM	34E9	2 6	4	3 5													
12' VBM	96J961	2 5	4							3 6 2							
16' PLANER	98J184	3 4 5 6	1					1									
8' IMM	97J686	3 4								2		5					
16' VBM	96J939	1 2 3 4 5 6															
14' PLANER	92J518	1 2 3 4 5 6	1							3							
72" E. LATHE	97J43	2 4 5 6	1							5							
96" E. LATHE	E-1810 30F307	2 3 4	1 6														
96" E. LATHE	36F463	2 3 5 6	1 4		5												
160" E. LATHE	93J771	2 3				5		1		4							
11-1/2' PLNR ⑤	90J158 96J798	2 4 5 6	3							1							
32' VBM	96J739	1 2 3 4 5 6															
		40	13	10	3	1		3		9		2	3				= 84

TABLE 21–2
Percentage of absent and not working time for workers engaged in data processing activities

Day	1	2	3	4	5	6	7	8	9
Camera study	42.9	47.1	48.9	47.9	45.7	53.2	50.7	49.3	43.6
Personal observation study	36.8	40.7	37.1	32.1	32.1	31.9	35.2	34.5	42.0

TABLE 21–3

Day	Number of observations with camera	Number of personal observations
1	280	280
2	280	140
3	280	294
4	280	280
5	280	280
6	280	210
7	280	210
8	280	252
9	280	224
Total	2,520	2,170

each worker significantly demonstrated the value of the random activity analysis camera. I made work sampling studies of the same activity by conventional, manual means and with the random activity analysis camera. Then I compared the results of the two methods. The work center was a data processing installation, where the workers keypunched and verified various quantities of data processing cards. The study was concerned only with the elements "working" and "not working." Working time included the elements adjusting cards, removing cards punching cards, and so on, while not working involved absence from the workstation and idleness.

The data collected with the random activity analysis camera revealed a 12.3 percent greater "not working" average than the personal observation method indicated. Table 21–2 summarizes the daily "Absent and Not Working" percentages for the two work sampling methods.

During a nine-day period, the random activity analysis camera, made 2,520 observations and during the next nine-day period, 2,170 personal observations were made. Table 21–3 summarizes the daily observations made by each technique.

TABLE 21–4
Difference between the results of a camera study of a personal observation
study in determining the percentage of absent and not working times

Day	1	2	3	4	5	6	7	8	9
Difference	6.1	6.4	11.8	15.8	13.6	21.3	15.5	14.8	1.6

A "*t*"-test compared the data gathered by both techniques to determine whether the difference in the data was significant. Accordingly, the null hypothesis established that the work sampling results obtained by personal observations and those obtained by memo-motion camera studies are the same.

Table 21–4 represents the difference in the percentage of "Absent and Not Working" times observed during the nine-day study by the two methods.

The mean difference is $\frac{\Sigma X}{n} = \frac{106.9}{9} = 11.9$

If there were no bias between the two methods, the mean difference would not differ significantly from 0.

With the assumed value of the population mean discrepancy of 0, we can compute a "*t*" value from the expression:

$$t = \frac{(\bar{X} - \bar{x})\sqrt{n-1}}{s}$$

where:

s = The standard deviation of the differences

$$s^2 = \frac{\Sigma d^2}{n} - \bar{d}^2$$

$$= 33.04$$

And

$$s = 5.75$$

Thus

$$t = \frac{(11.9 - 0)\sqrt{8}}{5.75}$$

$$= 5.86$$

By comparing the computed value of "*t*" (5.86) with the tabularized value at the 0.001 probability level for eight degrees of freedom (see Appendix A3–3), 5.041, we can conclude that the difference in the data obtained from the two methods is highly significant.

ESTABLISHING OF ALLOWANCES

One of the most extensive uses of work sampling has been establishing allowances for normal times to determine allowed times. However, the technique is also being used for establishing standards of production, determining machine utilization, allocating work assignments, and improving methods.

The determination of time allowances must be correct if fair standards are to be realized. Prior to the introduction of work sampling, analysts frequently determined allowances for personal reasons and unavoidable delays by taking a series of all-day studies on several operations, and then averaging the results. Thus, they recorded, timed, and analyzed trips to the rest room, trips to the drinking fountain, interruptions, and so forth, and determined a fair allowance. Although this method gave the answer, it was a costly, time-consuming operation that was fatiguing to both the analyst and the operator. Through work sampling study, analysts take a great number of observations (usually over 2,000) at different times of the day, and of different operators. Then, they divide the total number of legitimate occurrences other than work that involve normal operators by the total number of working observations. The result equals the percentage allowance that should be given the operator for the class of work being studied. The different elements that enter into personal and unavoidable delays can be kept separate, and an equitable allowance determined for each class or category. Figure 21–14 illustrates a summary of a work sampling study for determining unavoidable delay allowances on bench, bench machine, machine, and spray operations. In 26 cases out of 2,895 observations made on bench operations, there were interferences. This indicated an unavoidable delay allowance of 0.95 percent on this class of work.

DETERMINING MACHINE UTILIZATION

Observers determine machine utilization by the work sampling technique in the same manner that is used in establishing allowances. The following case history details the steps involved.

Analysts were told to gather information on machine utilization in a section of a heavy machine shop. Management estimated that the actual cutting time in this section should be 60 percent of the working day to comply with the quotations being submitted. There were 14 facilities involved in this section of the shop, and they had to take approximately 3,000 observations to get the accuracy desired.

Observers designed a work sampling form (see Figure 21–13) to accommodate the 16 possible states that each of the facilities under study might be in at the time of an observation.

To assure random observations, they set up a random pattern of visitation to the shop area. They made six observations of the 14 facilities

FIGURE 21–14

Summary of interruptions on various classes of work taken from a work sampling study for determining unavoidable delay allowance

OPERATION	Engineer	Supply	Quality	Mechanic	Supervision	Turning on Light	Miscellaneous	Working		Number of Interferences	TOTAL OBSER.	PERCENT ALLOWANCE
Bench	1	11	1	0	12	1	119	2,750		26	2,895	.95
Bench Machine	0	2	0	0	5	0	69	984		7	1,060	.71
Machine	0	1	6	11	9	0	29	1,172		27	1,228	2.30
Spray	0	0	0	7	36	0	262	1,407		43	1,712	3.06

XYZ ELECTRIC PRODUCTS, INC.

PLANT 26
DEPARTMENT 4
SUMMARY SHEET WORK SAMPLING
DATE 4/22/

NUMBER OF OBSERVATIONS

REMARKS: SUMMARY FOR INTERFERENCE ALLOWANCES. SEE OBSERVATION SHEET FOR DETAILS OF MISCELLANEOUS. MISCELLANEOUS WAS INCLUDED IN WORKING CATEGORY BECAUSE IN ALL CASES DOWNTIME WAS CREDITED TO THE OPERATORS.

NOTE: ALL OBSERVATIONS ARE TO BE TAKEN AT RANDOM.

during each shift. To get the required total number of observations, they observed 36 separate shifts—84 observations on each shift.

Since there were 14 machines and 6 observations of each per shift, a quick check for 84 separate readings per sheet assured complete coverage of the shift (see Figure 21–14). Each trip took one analyst about 15 to 20 minutes. This work occupied only about two hours per shift, leaving that analyst free to perform other work during the remaining six hours.

Since the principal purpose of the study was to learn the status of the actual cutting time in this section, an analyst kept a cumulative percentage machine cutting chart (see Figure 21–15). At the beginning of each day's study, the analysts took all previous cutting observations cumulatively as a ratio to the total observations to date. By the end of the 10th day of study, the percentage of machine cutting time began to level off at 50.5 percent.

After studying 36 shifts, they divided the sums of all observations in each category by the total number of observations. This resulted in percentages that represented the distribution of the cutting time, setup time,

FIGURE 21–15
Cumulative percentage machine cutting

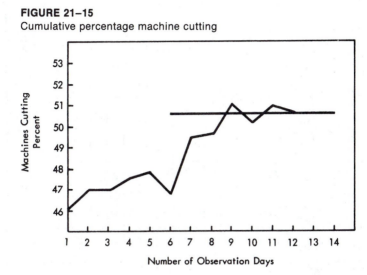

Number of Observation Days

and the various delay times listed. Figure 21–16 illustrates the summary sheet of this study. Cutting time amounted to 50.7 percent. The percentage of time required by the various delays indicated areas for methods improvement that would help increase the cutting time.

ESTABLISHING INDIRECT AND DIRECT LABOR STANDARDS

Some companies are finding work sampling applicable for establishing incentive standards on indirect as well as direct labor operations.[4] The technique is the same as that used for allowance determination. Take a large number of random observations; then the percentage of the total observations that the facility or operation is working approximates the percentage of total time that it is truly in that state.

One company's technique for taking systematic work sampling studies on clerical operations is to take observations on each operator at one-minute intervals. Observers assign the various elements of work in which the operator may be engaged identifying numbers. At the time of the observation, the analyst merely checks the appropriate number in the space provided on the form. Every five minutes, the analyst rates the performance of the operator. At the end of the study, the analyst determines a selective average rating factor.

---

[4] Aluminum Company of America and Douglas Aircraft Co.

FIGURE 21-16
Work sampling summary sheet

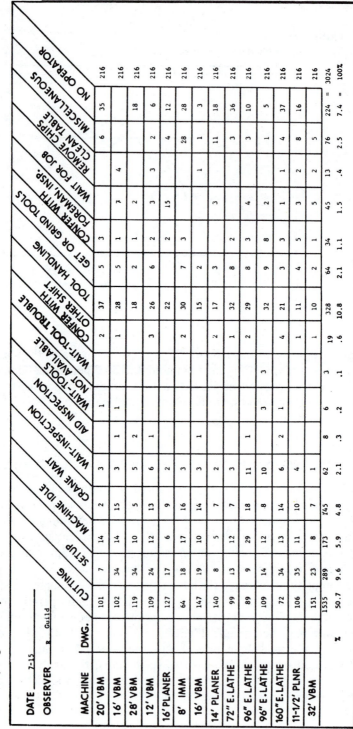

The normal minutes required for a given element of work may be expressed:

$$T_n = \frac{P \times N}{P_a}$$

where:

T_n = Normal minutes to perform the element
P = Selective average rating factor
N = Total observations for the given element
P_a = Total production for the period studied

For example, to determine the normal time for the element "write report heading on one BM report," the analyst may have prepared a work sampling study showing 84 observations during which the employee was actually engaged in performing this work element. During the study period, 12 reports were made, and the selective average rating factor was 110 percent. The normal time would then equal:

$$\frac{(1.10)(84)}{12 \text{ reports}} = 7.7 \text{ minutes normal time per report heading}$$

Figure 21–17 illustrates the study form developed for taking a 40-minute study. The columns "I," "P," and "M" accommodate idle, personal delays, and miscellaneous observations. Under the column headed "Lev." the analyst recorded the performance factor of the operator every five minutes. The table of standard data on page 504 illustrates representative standard data developed by the technique just outlined.

A more generalized expression for determining the normal minutes required to perform the element under study where observations are taken at intervals of fixed duration would be:

$$T_n = \frac{\dfrac{\Sigma \text{ Ratings}}{100} \times I}{P_a}$$

where:

I = Established interval between observations

For example, assume that analysts make 30 observations at 0.50-minute intervals on a work assignment involving three elements, and that workers produce 12 products. The tabular data involved appear in Table 21–5. The normal minutes for the three elements would be as follows:

$$T_1 = \frac{8.60 \times 0.50}{12} = 0.358 \text{ minutes}$$

$$T_2 = \frac{7.05 \times 0.50}{12} = 0.293 \text{ minutes}$$

$$T_3 = \frac{11.80 \times 0.50}{12} = 0.490 \text{ minutes}$$

FIGURE 21–17

Work sampling time study form for clerical operations

1. SIGHT—VERIFY REPORT TOTAL	11. OBTAIN INFORMATION (DISCUSS, PHONE, ETC.)
2. HEADINGS	12. OBTAIN SUPPLIES (PLUS TRIPS)
3. POSTING FROM REPORT	13. FILING
4. DELIVERY	14.
5. MANUAL REPORT PREPARATION	15.
6. EDITING—ASSIGN NO'S (RECAP NO'S ETC.)	16.
7. CORRECT ERROR CARDS	17.
8. ADDING	18. IDLE TIME
9. TYPING	19. PERSONAL TIME
10. TELEGRAMS—RECEIVING—SENDING	20. MISCELLANEOUS

TABULATING—CLERICAL

FROM ___8:00___ TO ___8:40___   DATE ___1/12/___

Cavenaugh

MI.	1	2	3	4	5	6	7	8	9	10	11	12	13	14	15	16	17	I	P	M	LEV.	REMARKS
1																				✓		NOT HERE
2																				✓		NOT HERE
3						✓																
4						✓																
5						✓															110	
6						✓																
7					✓																	
8					✓																	
9						✓																
10						✓															115	
11						✓																
12						✓																
13						✓																
14						✓																
15						✓															115	
16						✓																
17						✓																
18						✓																
19						✓																
20						✓															100	
21						✓																
22						✓																
23						✓																
24						✓																
25																			✓		105	SEARCH FOR STAMP
26																			✓			DATE STAMP
27					✓																	
28					✓																	
29					✓																	
30																			✓		110	OBTAIN CLIPS
31					✓																	
32					✓																	
33					✓																	
34					✓																	
35					✓																110	
36					✓																	
37					✓																	
38					✓																	
39					✓																	
40					✓																115	
TOTAL				2		20	13												5			

This company's technique can be criticized in that they have not taken random observations. Consequently, biased results could occur. For example, some short-cyclic element may be completely omitted by observing at regular one-minute intervals. This does not happen if analysts take sufficient random observations. The expression for establishing standards on office work can be modified to be applicable on work sampling studies involving random observations rather than regular ones a minute apart. This may be expressed:

TABLE 21–5
Tabular data of a three-element work sampling study

Observation number	Performance rating observed			
	Element 1	Element 2	Element 3	Idle
1	90			
2				100
3		110		
4	95			
5	100			
6		100		
7			105	
8	90			
9			110	
10	85			
11			95	
12		90		
13			100	
14			95	
15	80			
16			110	
17		105		
18			90	
19	100			
20			85	
21			90	
22			90	
23	110			
24			100	
25		95		
26				100
27		105		
28		100		
29			110	
30	110			
Σ Rating	860	705	1180	100
Σ Rating / 100	8.60	7.05	11.80	1.00

$$T_n = \frac{(n)(T)(P)}{(P_a)(N)}$$

$$T_a = T_n + \text{Allowances}$$

where:

T_n = Normal elemental time
T_a = Allowed elemental time
P = Performance rating factor
P_a = Total production for period studied
n = Total observations of element under study
N = Total observations of study
T = Total operator time represented by study

For example, assume that a standard must be established on the maintenance operation of lubricating fractional horsepower motors. A work sampling study of 120 hours revealed that, after 3,600 observations, lubrication of fractional horsepower motors was taking place in 392 cases and that a total of 180 facilities using fractional horsepower motors were maintained. The average performance factor was 0.90, and the normal time for lubricating a fractional horsepower motor would be:

$$\frac{(392 \text{ observations})(7{,}200 \text{ minutes})(0.90 \text{ performance factor})}{(180 \text{ total production})(3{,}600 \text{ total observations})}$$

$$= \frac{(392)(7{,}200)(0.90)}{(180)(3{,}600)}$$

$$= 3.92 \text{ minutes to lubricate one fractional horsepower motor}$$

This tool has many applications: determining machine downtime, calculating the relative amount of setup and put-away elements for all classes of work; and studying the relative amount of manual, mental, and delay times for clerical work, direct labor, and administrative work. These then form a basis for establishing the ideal work assignments and methods procedures.

SELF-OBSERVATION

Conscientious administrators periodically take work samples of their work to evaluate the effectiveness of their time usage. In the majority of cases, administrators spend less time on those aspects of their jobs that they deem important than they think they are spending. And, they spend more time on unimportant aspects, such as personal and avoidable delays, than they believe they are spending. Once administrators learn how much time is being taken by functions that could be handled readily by subordinates and clerical personnel, they can take positive action.

For example, a university professor may choose to work sample himself to learn just how he is spending his time. He decides to take random samples over an eight-week period during the academic year. This period should supply typical data not subject to seasonal variation. He sets his random reminder to provide, on the average 2.0 samples per hour. Thus, over the eight-week study period he would have 640 observations (8 weeks × 40 hours per week × 2 observations per hour). He could have chosen to take samples at a higher rate on randomly selected days within the study interval. For example, he may have chosen to take 4 samples an hour over 20 days randomly selected from the eight-week period. This would also provide 640 observations.

To record his data, the professor designed a form similar to that shown in Figure 21–18. Each time his random reminder beeped, he recorded the code letter for applicable category and the time. The form was designed to contain one week of daily random observations.

FIGURE 21–18
A specially designed work sampling form

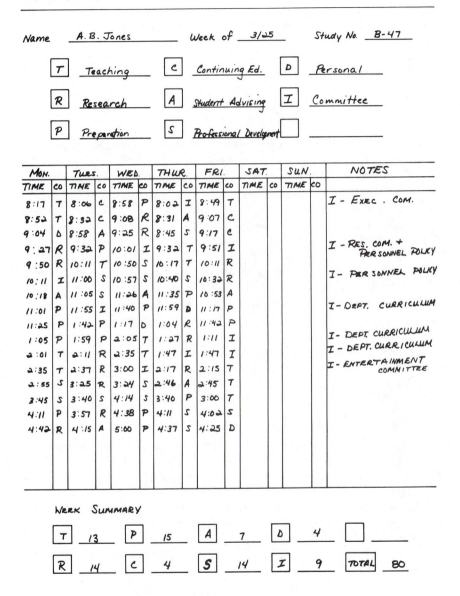

At the end of the eight-week study, he summarized the weekly data sheets. For example, the professor found that 80 of the total 640 observations were code I (committee participations). Thus about 12.5 percent of his work time is spent in committee participation. The 95 percent confidence interval would be:

$$\pm 1.96 \ \sqrt{\frac{0.125(1 - 0.125)}{640}} = \pm 0.013$$

The professor is 95 percent confident that committee work is occupying 12.5 ± 1.3 percent of his time.

He followed the same procedure for all the coded activities. Using the results of his study, he altered his calendar to utilize his potential in a more positive manner.

COMPUTERIZED WORK SAMPLING

By using a computer you can save an estimated 35 percent of the total work sampling study cost. This can be done because of the high percentage of clerical effort in relation to actual observation time. The majority of the effort involved in summarizing work sampling data is clerical: calculating percentages and accuracies, plotting data on control charts, determining the number of observations required, determining daily observations required, determining the number of trips to the area being studied per day and the time of day for each trip, and so on.

By mechanizing the processing of repetitive calculations, computers can calculate not only daily results but also cumulative results, such as the maintenance of control charts. For example, one company[5] developed a system referred to as MAST (Mechanized Activity Sampling Technique) which automates the clerical work, including mathematical calculations associated with the recording of observations, computation of element percentages, performance ratings, statistical accuracies, preparation and maintenance of control charts, and extrapolation of the data into equivalent staffing needs and/or machines and annual costs.

Users of MAST claim the following benefits:

1. The amount of industrial engineering time is increased through the reduction of clerical routines.
2. Results of the study are achieved more rapidly and the data is presented in a professional manner.
3. Cost of conducting work sampling studies is reduced significantly.
4. Accuracy of computations is improved.
5. Fewer errors are committed by analysts.
6. An incentive to make greater use of the work sampling technique is provided by the automated system.

CONCLUSION

The work sampling method is another tool that allows time and methods study analysts to get the facts in an easier, faster way.

---

[5] Applied Research Laboratories Division of Bausch & Lomb Inc.

FIGURE 21–19

Distribution of work measurement methodologies to process time, effort time, second work stoppages second delays under shop and mass production in the metal trades, and the approximate distribution of improvement from engineering and employees

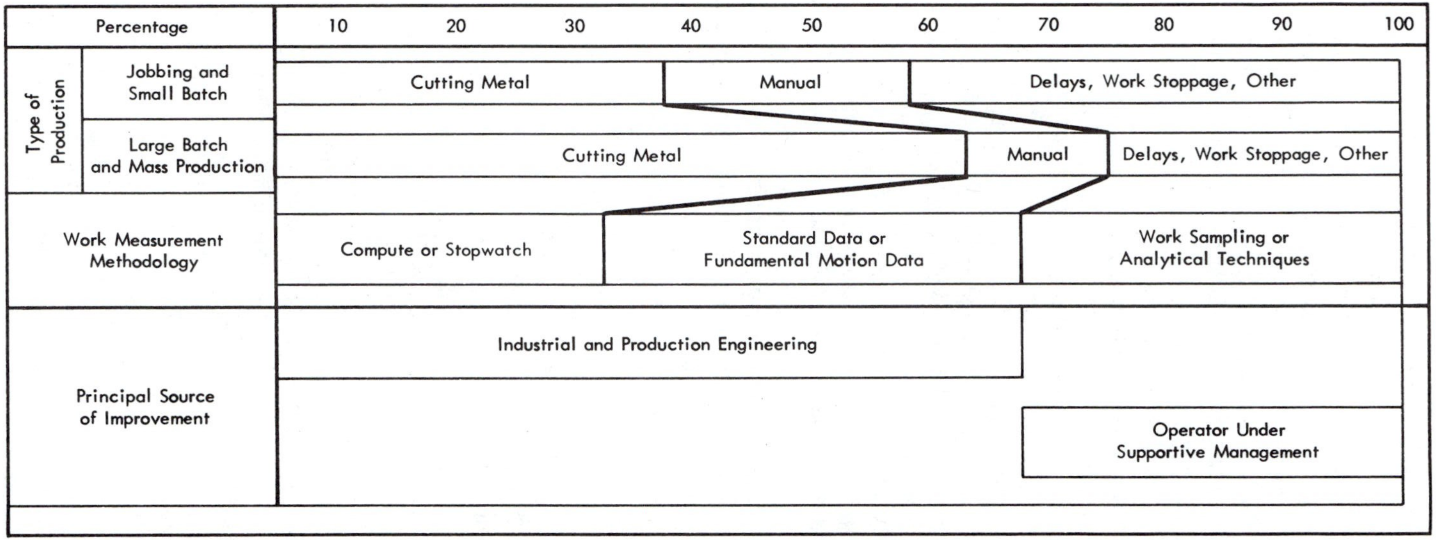

Data furnished by Professor K. Knott

Performance rated work sampling is especially useful in determining the amount of time that should be allocated for unavoidable delays, work stoppages, and the like. The extent of these interruptions is a suitable area for study to improve productivity. Typically in the machining industry these interruptions amount to between 24 and 45 percent of the total time. Figure 12–19 illustrates the magnitude of this time in both jobbing and mass production operations.

Certainly, computerized work sampling will become an increasingly popular method for uncovering unproductive labor, staffing imbalances, machine or facility downtime, and similar problem areas. It will also be used more heavily for establishing standards on production support labor, maintenance, and service labor.

Every person in the field of methods, time study, and wage payment should become familiar with the advantages, limitations, uses, and application of this technique. In summary, the following considerations should be kept in mind:

1. Explain and sell the work sampling method before using it.
2. Confine individual studies to similar groups of machines or operations.
3. Use as large a sample size as is practicable.
4. Take individual observations at random times so that observations are recorded for all hours of the day.
5. Take the observations over two weeks or more.

TEXT QUESTIONS

1. Where was work sampling first used?
2. What advantages are claimed for the work sampling procedure?
3. In which areas does work sampling have application?
4. How many observations should be recorded in determining the allowance for personal delays in a forge shop if it is expected that a 5 percent personal allowance will suffice, and if this value is to remain between 4 and 6 percent 95 percent of the time?
5. How is it possible to determine the time of day to make the various observations so that biased results do not occur?
6. What considerations should be kept in mind when taking work sampling studies?
7. What is meant by stratifying the data collected? Explain when it would be desirable to stratify data.
8. What are the principal advantages of using a random reminder in connection with the gathering of data for a work sampling study?
9. To get ±5 percent precision on work that is estimated to take 80 percent of the workers' time, how many random observations are required at the 95 percent confidence level?
10. If the average handling activity during a 10-day study is 82 percent, and the

number of daily observations is 48, how much tolerance can be allowed on each day's percentage activity?

11. If it should develop that four seconds of time would be desirable for each random observation, and if a sample size of 2,000 is necessary, how often would the analyst need to service the random activity analysis camera?

12. Over how long a period is it desirable to continue to acquire sampling data?

13. How biased can we expect work sampling data to be? Will this bias vary with the work situation? Explain.

14. In preparing for a work sampling study, why do some analysts prefer to use a form the size of a card that can fit into a shirt or coat pocket?

15. What is meant by systematic sampling? When would you recommend its use?

GENERAL QUESTIONS

1. Is there an application for work sampling studies in determining fatigue? Explain.

2. How can the validity of work sampling be sold to the employee not familiar with probability and statistical procedure?

3. What are the pros and cons for using work sampling to establish standards of performance?

PROBLEMS

1. The analyst in the Dorben Reference Library decides to use the work sampling technique to establish standards. Twenty employees are involved. The operations include cataloging, charging books out, returning books to their proper location, cleaning books, record keeping, packing books for shipment, and handling correspondence.

    A preliminary investigation resulted in the estimate that 30 percent of the time of the group was spent in cataloging. How many work sampling observations would be made if it were desirable to be 95 percent confident that the observed data were within a tolerance of ±10 percent of the population data? Describe how the random observations should be made.

    The following table illustrates some of the data gathered from 6 of the 20 employees. From this data, determine a standard in hours per hundred for cataloging:

	Operators					
Item	*Smith*	*Apple*	*Brown*	*Green*	*Baird*	*Thomas*
Total hours worked	78	80	80	65	72	75
Total observations (all elements)................	152	170	181	114	143	158
Observations involving cataloging..................	50	55	48	29	40	45
Average rating	90	95	105	85	90	100

The number of volumes cataloged equals 14,612.

Design a control chart based on three-sigma limits for the daily observations.

2. The work measurement analyst in the Dorben Company is planning to establish standards on indirect labor by using the work sampling technique. This study will provide the following information:

    T = Total operator time represented by the study
    N = Total number of observations involved in the study
    n = Total observations of the element under study
    P = Production for the period under study
    R = Average performance rating factor during the study

    With the above information, derive the equation for estimating the normal elemental time for an operation (t_n).

3. The analyst in the Dorben Company wishes to measure the percentage of downtime in the drop hammer section of the forge shop. The superintendent estimated the downtime to be about 30 percent. The desired results, using a work sampling study, are to be within ±5 percent of "p" with a level of significance of 0.95.

    The analyst decides to take 300 random readings a day for three weeks. Develop a "p" chart for "p" = 0.30 and subsample size N = 300. Explain the use of this "p" chart.

4. The Dorben Company is using the work sampling technique to establish standards in its typing pool section. This pool has varied responsibilities, including typing from tape recordings, filing, Kardex posting, and copying.

    The pool has six typists who work a 40-hour week. Seventeen hundred random observations were made over a four-week period. During the period, the typists produced 1,852 pages of routine typing. Of the random observations, 1,225 showed that typing was taking place. Assuming a 20 percent P.D. & F. allowance and an adjusted performance rating factor of 0.85, calculate the hourly standard per page of typing.

SELECTED REFERENCES

Barnes, R. M.   *Work Sampling,* 2nd ed. New York: John Wiley, & Sons, 1957.

Heiland, R. E., and W. J. Richardson.   *Work Sampling.* New York: McGraw-Hill, 1957.

Monks, Joseph G.   *Operations Management Theory and Problems,* New York: McGraw-Hill, 1977.

Pape, Elinor S. "Work Sampling." In *Handbook of Industrial Engineering,* 2nd ed., edited by Gavriel Salvendy, Chap. 64. New York: John Wiley & Sons, 1992

Richardson, W. J.   *Cost Improvement, Work Sampling and Short Interval Scheduling.* Reston, Va.: Reston Publishing, 1976.

SELECTED SOFTWARE

Whitehouse, Gary E. Work Sampling Accuracy and Work Sampling Observation Generator, Work Measurement IIE Microsoftware, Industrial Engineering and Management Press. Norcross, GA: Institute of Industrial Engineers, 1985.

Establishing Standards on Indirect and Expense Work

22

Since 1900 the percentage increase of indirect and expense workers has more than doubled that of direct labor workers. Groups usually classified as indirect labor include shipping and receiving, trucking, stores, inspection, material handling, toolroom, janitorial, and maintenance. Expense personnel include all positions not coming under direct or indirect—office clerical, accounting, sales, management, engineering, and so on.

The rapid growth in office workers, maintenance workers, and other indirect and expense employees is due to several reasons: first, the increased mechanization of industry and the complete automation of many processes, including the use of robots, have decreased the need for craftspeople and operators. This trend toward increased mechanization has resulted in a huge demand for electronics specialists, electricians, instrument makers, fluid mechanics technicians, and other service people. Also, the design of complicated machines and controls has resulted in a greater demand for engineers, designers, and draftspersons.

Second, the tremendous increase in paperwork brought on by federal, state, and local legislation is responsible, to a large extent, for an ever increasing number of clerical employees.

Third, office and maintenance work have not been subjected to the methods study and the technical advances that have been applied so effectively to direct labor in connection with industrial processes.

With a large share of most payrolls being earmarked for indirect and expense labor, progressive management is beginning to realize the opportunities for the application of methods and standards in this area.

AUTOMATION

The term *automation* may be defined as "increased mechanization." A completely automated manufacturing process is capable of operating for prolonged periods without human effort. Automation implies the use of robots. Today we are using robots for such mundane operations as screwing light bulbs on dashboards, tightening bolts, spot welding, and spray painting. Few industries or even processes within an industry have been completely automated, However, there has been a pronounced tendency toward semiautomation in American, Japanese, and European industry. With the increasing demand for greater productivity, more and more industries are automating. Frequently, robots perform routine boring operations as well as operations that require a dangerous environment. Such areas may be contaminated with radiation, have excess noise, or be extremely hot or cold.

A program of automation begins with the integration of a fully automatic machine, such as the automatic screw machine, with automatic transfer handling devices, so that a series of operations may be performed automatically. To determine the extent of automation justified, two factors should be considered: (1) the quantity requirements of the product, and (2) the nature or design of the product itself.

If the quantity requirements of the product are large, the design engineer endeavors to design the product so that it lends itself to automation. Frequently, adding something to a part, such as a lug, fin, extension, or hole, creates a means of mechanical handling to and from the workstation. Such a redesign can accommodate indexing at a workstation for successive production operations. For example, a designer added a holding fin to the die casting for producing a compressor piston. This permitted mechanical fingers to hold the work while it was being processed, as well as to automatically transport the work between production stations (see Figure 22–1).

Complete automation is feasible for the continuous processing of chemical products, such as gasoline, oil, and detergents. Partial automation is now widespread for mass-produced items—food products (for example, cookies, cereals, and pretzels), light bulbs, transistors, automobile parts, cigarettes, and so forth. Figure 22–2 illustrates the conversion of the assembly of terminals on wire leads to an automated process.

Factors that encourage plant automation include:

1. The increasing cost of labor.
2. Increasing foreign and domestic competition, which reduces selling prices and decreases profits.
3. The prospects for market expansion through cost and price reduction.

Among the principal factors that may discourage automation in a particular plant are:

FIGURE 22–1
Die casting with a fin so that work can be handled in automated equipment

1. The large capital investment necessary that must, subsequently, be absorbed in operating profit.
2. A market potential presently inadequate to absorb the increased output.
3. An existing technology unable to provide automation to produce a particular design.
4. Opposition by production workers, and adverse community relations caused by a reduction in the labor force.

However, there is no doubt that the present trend toward automation will continue. As automation equipment is developed, the need for effective preventive maintenance becomes apparent. The failure of a single minor component may cause the shutdown of a complete process or even an entire plant. This fact, together with the complexity of automated equipment using pneumatic, hydraulic, and electronic controls, brings out the reasons for the growth of indirect workers not only in numbers but in diversification of occupation. Companies must ensure the efficient use of machinery as they modernize. Workers in today's automated plants not only need problem solving skills but also must be able to work proficiently under stress as members of dedicated manufacturing teams.

FIGURE 22–2

A. Terminal assembly to wire leads were converted to an auto-mated process in view of the volume of work.

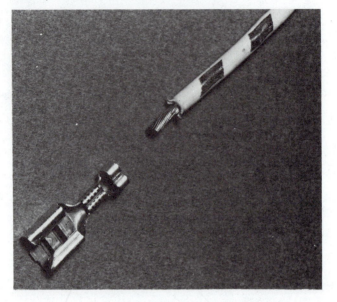

B. Hand-feeding of prestripped leads into bench press die. The press was activated by a foot pedal. The standard for this manual termination was 475 pieces per hour.

FIGURE 22–2 *(concluded)*
C. Automated equipment which strips and terminates the wire at a rate of 2,470 per hour.

General Electric Co., Louisville, Kentucky

As companies automate, they usually find it advantageous to compress management hierarchy and convert traditional direct labor to involved workers capable of making advanced manufacturing systems a success. There should be an improvement in productivity from a more motivated and knowledgeable work force.

METHODS IMPROVEMENTS ON INDIRECT AND EXPENSE WORK

The systematic approach to methods standards and wage payment outlined in Chapter 1 is just as applicable to indirect and expense areas as it is to direct labor. Careful fact-finding, analysis, development of the proposed method, presentation, installation, and development of job analysis should precede a program for establishing standards on indirect and expense labor. The methods analysis procedure in itself introduces economies.

Work sampling is a good technique to determine the severity of the problem and the potential savings in the indirect and expense areas. It is not unusual to find that the work force is productively engaged only 40 to 50 percent of the time or even less. For example, in maintenance work—

which represents a large share of the total indirect cost—analysts may find the following reasons for much of the time lost during the working day:

1. *Inadequate communication.* It is quite common to find incomplete and even incorrect job instructions on work orders. This necessitates additional trips to the toolroom and supply room to obtain parts and tools that should have been available when the craftspeople first started to work on the job. The work order that merely states "Repair leak in oil system" is indicative of poor planning and inadequate communication. The worker needs to know whether a new valve or pipe is needed, or whether a new gasket can do the job, or if the valve needs to be repacked.

2. *Unavailability of parts, tools, or equipment.* If the craftspeople do not have the facilities and parts to do the job, they are obliged to improvise in a manner that is usually wasteful of time and frequently results in inferior work. Proper planning assures that the correct tools and equipment to perform the job are available and that suitable spares are supplied to the job site.

3. *Interference of production employees.* With improper scheduling, maintenance employees may find that they are unable to begin a repair, service, or overhaul operation because the facility is still being used by production employees. This can result in the craftspeople waiting idly by until the production department is ready to turn over the equipment.

4. *Overstaffing the maintenance job.* This has been one of the principal causes of lost time in maintenance work. Too often, a crew of three or four is supplied when two or three workers are really needed.

5. *The work is unsatisfactory and must be redone.* Poor planning frequently results in an attitude of "this will get by" on the part of the mechanic. This results in the repair work having to be redone.

6. *The craftsperson waiting for instructions to begin the next job.* With improper planning, "wait for work" idle time can be significant. Good planning assures that there is ample work ahead of the maintenance person, who seldom if ever has to wait for the next job.

After developing good methods, standardize them and teach them to the workers who will be using them. This obvious step is too often neglected by management. By giving as much attention to operation improvement in indirect work as has been given to direct work, companies can decrease costs.

Indirect Labor Standards

When establishing standards in indirect labor departments, such as clerical, maintenance, and toolmaking, develop them from standard data or formulas. Time would not permit using stopwatch techniques for each and every standard developed. However, both stopwatch studies and fundamental motion data can help develop indirect labor standards.

Analysts can calculate time standards on any operation or group of operations that can be quantified and measured. If the elements of the work performed by the electrician, blacksmith, boilermaker, sheet metal worker, painter, carpenter, millwright, welder, pipe fitter, material handler, toolmaker, and others doing indirect work, are broken down and studied, it is usually possible to evolve equitable standards by summarizing the direct, transportation, and indirect elements. The tools used for establishing standards for indirect and expense work are identical with those used in establishing standards for direct work: time study, predetermined motion time standards, standard data, time formulas, and work sampling.

Thus, analysts can establish standards for such tasks as hanging a door, rewinding a 1-horsepower motor, painting a centerless grinder, sweeping the chips from a department, or delivering a skid of 200 forgings. For each of these operations analysts establish standard times by measuring the time required for the operator to perform the job. Next, they performance rate the study and apply an appropriate allowance.

Careful study and analysis often reveal that crew balance and interference cause more unavoidable delays in indirect work than in direct work. Crew balance is the delay time encountered by one member of a crew while waiting for other members of the crew to perform elements of the job. Interference time is the time that a worker waits for others to do necessary work. Both crew balance and interference delays are unavoidable delays; however, they are usually characteristic of indirect labor operations only, such as those performed by maintenance workers. Queuing theory—explained later in this chapter—is a useful tool to estimate the magnitude of waiting time.

Because of the high degree of variability characteristic of most maintenance and material handling operations, it is necessary to conduct sufficient independent time studies of each operation. This assures that the resulting standard is representative of the time needed for the normal operator to do the job under average conditions. For example, if a study indicates that it takes 47 minutes to sweep a machine floor 60 feet wide by 80 feet long, observers must assure that average conditions prevailed while taking the study. Obviously, the work of the sweeper would be considerably more time consuming if the shop were machining cast iron rather than an alloy steel. Not only is alloy steel much cleaner, but alloy steel chips are easier to handle; the use of slower speeds and feeds results in fewer chips. A 47-minute standard established when the shop was working with alloy steel, would be inadequate to produce cast-iron parts. Additional time studies would assure that average conditions prevailed and that the resulting standards represented those conditions.

In a similar manner, analysts can establish standards for painting. Thus, a standard for overhead painting is based on square footage. Likewise, vertical and floor painting can be measured, and time standards developed on a square-footage basis.

Just as the automative industry has established time standards for common repair jobs, such as grinding valves, replacing piston rings, and adjusting brakes, so can observers establish standards for typical maintenance and repair operations performed on the machine tools within a plant. Thus, they can determine a set of time standards for rewinding a ⅛-HP motor, a ¼-HP motor, a ½-HP motor, a ¾-HP motor, and so on.

Toolroom work is very similar to the work done in job shops. Analysts can predetermine the method by which such tools as a drill jig, milling fixture, form tool, or die can be made. Consequently, analysts use time study and/or predetermined elemental times to establish a sequence of elements and to measure the normal time required for each element. Work sampling provides an adequate tool to determine the allowances that must be added for fatigue, personal, unavoidable, and special delays. The standard elemental times thus developed are tabularized in the form of standard data and used to design time formulas for pricing future work.

FACTORS AFFECTING INDIRECT AND EXPENSE STANDARDS. All indirect and/or expense work is a combination of four divisions: (1) direct work, (2) transportation, (3) indirect work, (4) and unnecessary work and delays.

The direct work portion of indirect and/or expense labor is that segment of the operation which discernibly advances the progress of the work. For example, in the installation of a door, the work elements may include: cut door to rough size, plane to finish size, locate and mark hinge areas, chisel out hinge areas, mark for screws, install screws, mark for lock, drill out for lock, and install lock. Such direct work can be measured quite easily by using conventional techniques, such as stopwatch time study, standard data, or fundamental motion data.

Transportation refers to the work performed in movements during the course of the job or from job to job. Transportation may take place horizontally or vertically, or both. Typical transportation elements include: walk up and down stairs, ride elevator, walk, carry load, push truck, and ride on motor truck. Transportation elements similar to direct elements are easy to measure, and establish as standard data. For example, one company uses 0.50 minutes per 100-foot zone as its standard for horizontal travel time and 0.30 minutes as its standard time for 10 feet of vertical travel.

As a general rule, analysts cannot evaluate the indirect portion of indirect and/or expense labor by physical evidence in the completed job or at any stage during the work except by deductive inferences from certain characteristic features of the job. Indirect work elements may be separated into three divisions: *(a)* tooling, *(b)* material, and *(c)* planning.

Tooling work elements include the acquisition, disposition, and maintenance of all the tools needed to perform an operation. Typical elements under this category would include: getting and checking tools and equipment; returning tools and equipment at the completion of the job; cleaning tools; and repairing, adjusting, and sharpening tools. The tooling work elements are easy to measure by conventional means; statistical records provide data on the frequency of their occurrence. Waiting line or queuing theory provides information on the expected waiting time at supply centers.

Material work elements have to do with acquiring and checking the material used in an operation and disposing of scrap. Making minor repairs to materials, picking up and disposing of scrap, and getting and checking materials, are characteristic of the work elements in this division. As with tooling elements, the material elements are readily measured, and their frequency accurately determined through historical records. In acquiring material from storerooms, queuing theory provides the best estimate of waiting time.

The planning elements represent the most difficult area in which to establish standards. These elements include consultation with the supervisor, planning work procedures, inspection, checking, and testing. Work sampling techniques provide a basis for determining the time required to perform the planning elements. Analysts also measure planning elements by stopwatch procedures, but their frequency of occurrence is difficult to determine; consequently, work sampling is the more practical tool to use for this class of elements.

Planning and methods improvement can eliminate unnecessary work and delays. It is not unusual to find that unnecessary work and delays represent as much as 40 percent of the indirect and expense payroll. Much of this wasted time is management oriented; for this reason analysts should follow the systematic procedure recommended in Chapter 1 prior to establishing standards. In this work, analysts should get and analyze the facts, and develop and install the method, before establishing the standard.

Much of the delay time encountered in indirect and expense work is due to queues. Thus, workers are obliged to stand in line at the tool crib, the storeroom, or the stockroom, waiting for a forklift truck, a desk calculator, or some other piece of office equipment. Through the application of queuing theory analysts frequently can determine the optimum number of servicing units required in given circumstances. Such studies should be made to optimize the delay time in indirect and expense work.

BASIC QUEUING THEORY EQUATIONS. Queuing system problems may occur when a flow of arriving traffic (people, facilities, and so on) establishes a random demand for service at facilities of limited service capacity. The time interval between arrival and service obviously varies inversely with the level of the service capacity. The greater the number of service stations and the faster the rate of servicing, the smaller the time interval between arrival and service.

Methods and work measurement analysts should select an operating procedure that minimizes the total cost of operation. There must be an economic balance between waiting times and service capacity.

The following four characteristics define queuing problems:

1. *The pattern of arrival rates.* The arrival rate (for example, a machine breaking down for repair) may be constant or random. If random, the pattern is a probability distribution of the various values of the intervals between successive arrivals. A probability distribution of the random pattern may be definable or nondefinable.

2. *The pattern of the servicing rate.* The servicing time may also be constant or random. If random, analysts should define the probability distribution that fits the service time random pattern.

3. *The number of servicing units.* In general, multiple-service queuing problems are more complex than the problems of single-service systems. However, most problems are of the multiple-service type, such as the number of mechanics required to keep a battery of machines in operation.

4. *The pattern of selection for service.* Service is usually on a "first-come, first-served" basis; however, in some cases, the selection may be completely random or according to priorities.

The solutions to queuing problems fall into two broad categories: analytic and simulation. The analytic category covers a wide range of problems for which mathematical probability and analytic techniques have provided equations representing systems with various assumptions about queuing characteristics. One of the most common assumptions about the arrival pattern or arrivals per unit time interval is that they follow the Poisson probability distribution:

$$p(k) = \frac{a^k e^{-a}}{k!}$$

where a is the mean arrival rate, and k is the number of arrivals per time interval. A helpful graphic presentation of the cumulative Poisson probabilities appears in Figure 22–3.

A further consequence of a Poisson-type arrival pattern is that the random variable time between arrivals obeys an exponential distribution with the same parameter a. Being a continuous distribution, the exponential distribution has a density function $f(x) = ae^{-ax}$. The exponential distribution has a mean $\mu = \frac{1}{a}$, and a variance $= \frac{1}{a^2}$. μ can be recognized as the mean interval between arrivals. In some queuing systems the number of services per unit time may follow the Poisson pattern, and subsequently the service times follow the exponential distribution. Figure 22–4 illustrates the exponential curve $F(x) = e^{-x}$ which shows the probabilities of exceeding various multiples of any specified service time.

The basic equations governing the queue applicable to Poisson arrivals and service in arrival order fall into these five categories:

1. Any service time distribution and a single server.
2. Exponential service time and a single server.
3. Exponential service time and finite servers.
4. Constant service time and a single server.
5. Constant service time and finite servers.

Equations have been developed for each of the above categories. These provide quantitative answers to such problems as mean delay time in the waiting line and mean number of arrivals in the waiting line.

This typical example shows how to apply queuing theory. The analyst wishes to determine a standard for inspecting the hardness of large motor armatures. The time is composed of two distinct quantities: the time the inspector takes to make his Rockwell observations and the time the operator must wait until the next armature shaft is made available for inspection. The following assumptions apply: (1) single server; (2) Poisson arrivals; (3) arbitrary servicing time; and (4) first-come, first-served discipline.

This is a situation which fits into the first of the five categories. The equations which apply are:

$$a. \quad P > 0 = u$$

$$b. \quad w = \left[\frac{uh}{2(1-u)}\right]\left[1 + \left(\frac{\sigma}{h}\right)^2\right]$$

$$c. \quad m = \frac{w}{P > 0} = \frac{w}{u}$$

where:

a = Average number of arrivals per unit of time
h = Mean servicing time
w = Mean waiting time of all arrivals
m = Mean waiting time of delayed arrivals
n = Number of arrivals present (both waiting and being served) at any given time
s = Number of servers
u = Servers' occupancy ratio = $\frac{ah}{s}$
σ = Standard deviation of the servicing time
$P(n)$ = Probability of n arrivals being present at any random time
$P(\geq n)$ = Probability of at least n arrivals being present at any random time
t = Unit of time
$P > t/h$ = Probability of a delay greater than t/h multiples of the mean holding time

FIGURE 22–3
Poisson distribution of arrivals

$$P(c, a) = \sum_{x=c}^{\infty} \frac{e^{-a}a^x}{x!}$$

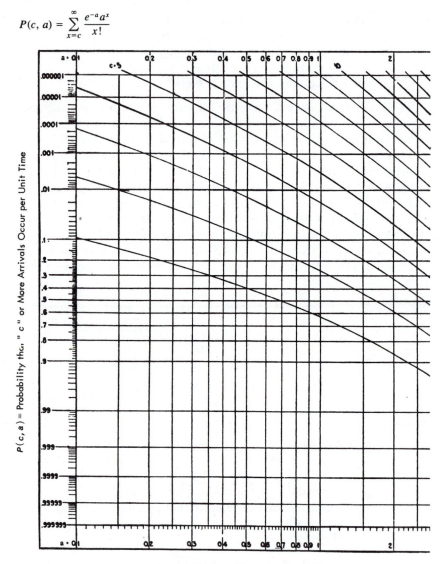

a = Average Number of Arrivals per Unit Time

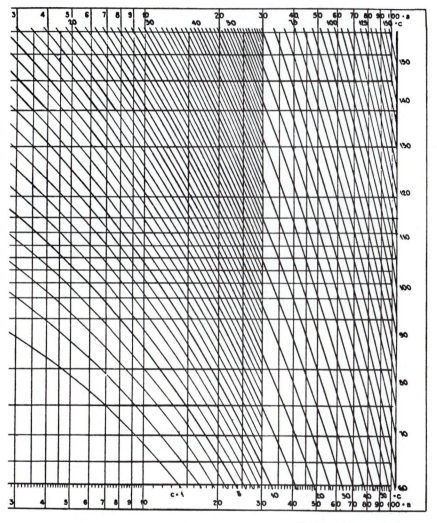

The Port of New York Authority

FIGURE 22–4
The exponential distribution

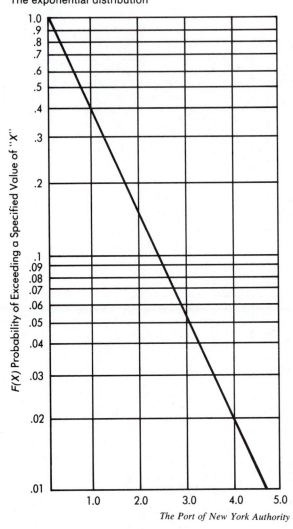

The Port of New York Authority

$P(d > 0)$ = Probability of any delay (delay greater than 0)
L = Mean number of waiting individuals among all individuals

Stopwatch time study establishes a normal time of 4.58 minutes per piece for the actual hardness testing. The standard deviation of the service time is 0.82 minutes, and 75 tests are made in every eight-hour working day. From the above data, we have:

$s = 1$ $u = (0.156)(4.58) = 0.714$

$a = \dfrac{75}{480} = 0.156$ $w = \left[\dfrac{(0.714)(4.58)}{2(1 - 0.714)}\right]\left[1 + \left(\dfrac{0.82}{4.58}\right)^2\right]$

$h = 4.58$ $= 5.95$ minutes mean waiting time of an arrival

$\sigma = 0.82$

Thus, the analyst is able to determine a total time of 10.53 minutes per shaft.

Many industrial problems fit into the second category; that is, exponential service and a single server with Poisson arrivals. The equations that apply here are:

a. $P > 0 = u$

b. $P > (t/h) = ue^{(u-1)(t/h)}$

c. $P(n) = (1 - u)(u)^n$

d. $P(\geq n) = u^n$ f. $m = \dfrac{w}{(P > 0)} = \dfrac{h}{(1 - u)}$

e. $w = \dfrac{h(P > 0)}{(1 - u)} = \dfrac{uh}{1 - u}$ g. $L = \dfrac{m}{h} = \dfrac{1}{(1 - u)}$

For example, the arrivals at a tool crib are considered to be Poisson, with an average time of seven minutes between one arrival and the next. The length of the service time at the crib window is distributed exponentially, with a mean of 2.52 minutes determined through stopwatch time study. The analyst wishes to determine the probability that a person arriving at the crib will have to wait, and the average length of the queues that form from time to time. Using this information, the analyst can evaluate the practicality of opening a second tool-dispensing window. The preceding equations can determine this information.

$a = 0.14$ average number of arrivals per minute

$h = 2.52$ minutes mean service time

$s = 1$ server

$P > 0 = u$

$u = \dfrac{ah}{s} = 0.35$ probability of a person arriving at crib having to wait

$L = \dfrac{1}{(1 - u)} = 1.52$ average length of the queue formed

MONTE CARLO SIMULATION.   The method of simulation used to solve queuing system problems does not employ formal mathematical models. Such models may be too complex or the systems too unique in light of the mathematical models in existence. Instead, a series of numerical statements creates a model of a queuing system. The analyst introduces a sample set of input values drawn from specified arrival and service time distributions. These input data generate a sample output distribution of waiting line results for the period.

Monte Carlo simulation estimates the expected waiting time and service times and develops an optimum solution through a proper balance of service stations, servicing rates, and arrival rates. This technique is most helpful to analyze the waiting line problem involved in the centralization-decentralization storage location of tools, supplies, and service facilities.

The Monte Carlo simulation technique is also applicable to problems with a Poisson arrival pattern. However, Monte Carlo is generally used for problems from which neither standard nor empirical formulas are obtainable.

For example, the analyst may wish to determine the minimum number of operators needed to set up machines in an automatic screw machine section. Analysis of past records in the department reveals the following probability distributions of work stoppages per hour and the time required to service a machine.

Labor rate......................	$12.00 per hour
Machine rate....................	$48.00 per hour
Number of operators	3
Number of automatics	15

Work stoppages per hour	Probability
0	0.108
1	0.193
2	0.361
3	0.186
4	0.082
5	0.040
6	0.018
7	0.009
8	0.003
	1.000

Hours to get machine into operation	Probability
0.100	0.111
0.200	0.254
0.300	0.009
0.400	0.007
0.500	0.005
0.600	0.008
0.700	0.105
0.800	0.122
0.900	0.170
1.000	0.131
1.100	0.075
1.200	0.003
	1.000

Work stoppages per hour	Probability	Numbers assigned
0	0.108	000–107
1	0.193	108–300
2	0.361	301–661
3	0.186	662–847
4	0.082	848–929
5	0.040	930–969
6	0.018	970–987
7	0.009	988–996
8	0.003	997–999

Hours to get machine into operation	Probability	Numbers assigned
0.100.	0.111	000–110
0.200.	0.254	111–364
0.300.	0.009	365–373
0.400.	0.007	374–380
0.500.	0.005	381–385
0.600.	0.008	386–393
0.700.	0.105	394–498
0.800.	0.122	499–620
0.900.	0.170	621–790
1.000.	0.131	791–921
1.100.	0.075	922–996
1.200.	0.003	997–999

The time to get a machine into operation results in a bimodal distribution and does not conform to any standard distribution. The analyst now assigns blocks of three-digit numbers, in direct proportion to the probabilities associated with the data for both the arrival and the service rates.

So 1,000 three-digit numbers have been assigned to both the arrival and the service distributions. By using a table of random numbers, analysts can simulate the expected behavior in the screw machine section over a period. Let us take 80 random observations, to simulate the work stoppages occurring during two weeks of activity on the floor. These 80 random numbers applied to work stoppages give the following:

Hour	Random number	Work stoppages	Hour	Random number	Work stoppages
1	221	1	41	028	0
2	193	1	42	871	4
3	167	1	43	988	7
4	784	3	44	100	0
5	032	0	45	479	2
6	932	5	46	228	1
7	787	3	47	678	2

Hour	Random number	Work stoppages	Hour	Random number	Work stoppages
8	236	1	48	276	1
9	153	1	49	337	2
10	587	2	50	131	1
11	573	2	51	689	3
12	485	2	52	137	1
13	619	2	53	096	0
14	369	2	54	204	1
15	188	1	55	733	3
16	885	4	56	071	0
17	097	0	57	923	4
18	129	1	58	995	7
19	859	4	59	938	5
20	386	2	60	184	1
21	534	2	61	245	1
22	407	2	62	220	1
23	021	0	63	071	0
24	951	5	64	297	1
25	357	2	65	579	2
26	262	1	66	333	2
27	778	3	67	494	2
28	464	2	68	651	2
29	375	2	69	920	4
30	616	2	70	987	6
31	934	5	71	004	0
32	219	1	72	579	2
33	952	5	73	125	1
34	978	6	74	315	2
35	699	3	75	961	5
36	043	0	76	854	4
37	610	2	77	728	3
38	859	4	78	913	4
39	217	1	79	775	3
40	156	1	80	372	2

To estimate the time required to get a machine into operation after it has stopped, analysts should select a different group of random numbers as input for each work stoppage. These values indicate the following:

Hour	Random number	Hours to get machine into operation	Hour	Random number	Hours to get machine into operation
1.............	341	0.200	41.............	684	0.900
2.............	112	0.200	42.............	366	0.300
3.............	273	0.200	43.............	220	0.200
4.............	106	0.100	44.............	920	1.000
5.............	597	0.800	45.............	346	0.200
6.............	337	0.200	46.............	347	0.200
7.............	871	1.000	47.............	639	0.900
8.............	728	0.900	48.............	308	0.200
9.............	739	0.900	49.............	319	0.200
10.............	799	1.000	50.............	848	1.000
11.............	202	0.200	51.............	562	0.800

Hour	Random number	Hours to get machine into operation	Hour	Random number	Hours to get machine into operation
12............	854	1.000	52............	481	0.700
13............	599	0.800	53............	000	0.100
14............	726	0.900	54............	587	0.800
15............	880	1.000	55............	275	0.200
16............	496	0.700	56............	267	0.200
17............	128	0.200	57............	430	0.700
18............	794	1.000	58............	160	0.200
19............	383	0.500	59............	096	0.100
20............	472	0.700	60............	653	0.900
21............	870	1.000	61............	876	1.000
22............	397	0.700	62............	086	0.100
23............	280	0.200	63............	085	0.100
24............	976	1.100	64............	898	1.000
25............	693	0.900	65............	422	0.700
26............	870	1.000	66............	164	0.200
27............	520	0.800	67............	255	0.200
28............	527	0.800	68............	145	0.200
29............	153	0.200	69............	144	0.200
30............	851	1.000	70............	322	0.200
31............	411	0.700	71............	610	0.800
32............	822	1.000	72............	855	1.000
33............	985	1.100	73............	167	0.200
34............	997	1.200	74............	463	0.700
35............	234	0.200	75............	886	1.000
36............	773	0.900	76............	081	0.100
37............	423	0.700	77............	826	1.000
38............	607	0.800	78............	566	0.800
39............	220	0.200	79............	228	0.200
40............	915	1.000	80............	491	0.700

In light of the indicated outcomes for 80 hours of operation in the automatic screw machine department, the following table shows the machine downtime predicted because of insufficient operators.

The simulated model indicates 19.6 hours of machine downtime every two weeks because of the lack of an operator. At a machine rate of $48 per hour, this amounts to a weekly cost of $470.40. Since the cost of a fourth operator would result in an added weekly direct labor cost of $480 (40 × 12), the present setup of three workers to service the 15 automatic screw machines appears to be the optimum solution.

Expense Standards

More and more, management is recognizing its responsibility to determine accurately the appropriate office forces for a given volume of work. To control office payrolls, management must develop time standards, since they are the only reliable "yardsticks" for evaluating the size of any task.

Hour	Random number	Work stoppages	Random number	Hours to get machine into operation	Operators available for next work stoppages	Downtime hours because of lack of operators for servicing
1 221		1	341	0.200	2	
2 193		1	112	0.200	2	
3 167		1	273	0.200	2	
4 784		3	106	0.100	2	
			597	0.800	1	
			337	0.200	0	
5 032		0	–	–	3	
6 932		5	871	1.000	2	
			728	0.900	1	
			739	0.900	0	
			799	1.000	0	0.9
			202	0.200	0	0.9
7 787		3	854	1.000	0	
			599	0.800	0	0.9
			726	0.900	0	0.1
8 236		1	880	1.000	1	
9 153		1	495	0.700	2	
10 587		2	128	0.200	2	
			794	1.000	1	
11 573		2	383	0.500	2	
			472	0.700	1	
12 485		2	870	1.000	2	
			397	0.700	1	
13 619		2	280	0.200	2	
			976	1.100	1	
14 369		2	693	0.900	2	
			870	1.000	1	
15 188		1	520	0.800	2	
16 885		4	527	0.800	2	
			153	0.200	1	
			851	1.000	0	
			411	0.700	0	0.200
17 097		0	–	–	3	
18 129		1	985	1.100	2	
19 859		4	997	1.200	1	
			234	0.200	1	0.100
			773	0.900	0	
			423	0.700	0	0.30
20 386		2	607	0.800	1	
			220	0.200	0	
21 534		2	915	1.000	2	
			684	0.900	1	
22 407		2	366	0.300	2	
			220	0.200	1	
23 021		0	–	–	3	
24 951		5	346	0.200	2	
			347	0.200	1	
			639	0.900	0	
			308	0.200	1	0.200
			319	0.200	0	0.200
25 357		2	898	1.000	2	
			562	0.800	1	

Hour	Random number	Work stoppages	Random number	Hours to get machine into operation	Operators available for next work stoppages	Downtime hours because of lack of operators for servicing
26 262	1	481	0.700	2	
27 778	3	000	0.100	2	
		587	0.800	1	
		275	0.200	0	
28 464	2	267	0.200	2	
		430	0.700	1	
29 375	2	160	0.200	2	
		096	0.100	1	
30 616	2	653	0.900	2	
		876	1.000	1	
31 934	5	086	0.100	2	
		085	0.100	1	
		898	1.000	0	
		422	0.700	0	0.100
		920	1.000	0	0.100
32 219	1	164	0.200	1	
33 952	5	255	0.200	2	
		145	0.200	1	
		144	0.200	0	
		322	0.200		0.200
		610	0.800		0.200
34 978	6	855	1.000	2	
		167	0.200	1	
		463	0.700	0	
		886	1.000	0	0.200
		081	0.100	0	0.700
		826	1.000	0	0.800
35 699	3	566	0.800	0	
		228	0.200	0	0.2
		491	0.700	0	0.8
36 043	0	623	0.900	1	
37 610	2	188	0.200	2	
		702	0.900	1	
38 859	4	323	0.200	2	
		354	0.200	1	
		464	0.700	0	
		806	1.000	0	0.200
39 217	1	494	0.700	1	
40 156	1	050	0.100	2	
41 871	0	–	–	3	
42 871	4	441	0.700	2	
		390	0.600	1	
		543	0.800	0	
		077	0.100	0	0.600
43 988	7	941	1.100	2	
		667	0.900	1	
		317	0.200	0	
		336	0.200	0	0.20
		650	0.900	0	0.40
		651	0.900	0	0.90
		370	0.300	0	1.40
44 100	0	–	–	0	

Hour	Random number	Work stoppages	Random number	Hours to get machine into operation	Operators available for next work stoppages	Downtime hours because of lack of operators for servicing
45 479		2	011	0.100	2	
			976	1.100	1	
46 228		1	229	0.200	1	
47 678		3	072	0.100	2	
			167	0.200	1	
			144	0.200	0	
48 276		1	843	1.000	2	
49 337		2	768	0.900	2	
			404	0.700	1	
50 131		1	636	0.900	2	
51 689		3	970	1.100	2	
			954	1.110	1	
			951	0.110	0	
52 137		1	562	0.800	0	0.100
53 096		0	–	–	0	
54 204		1	866	1.000	2	
55 733		3	348	0.200	2	
			327	0.200	1	
			995	1.100	0	
56 071		0	998	1.200	1	
			682	0.900	0	
			005	0.100	0	0.100
57 923		4	539	0.800	1	0.200
			616	0.800	0	
			181	0.200	0	
			647	0.900	0	0.200
58 995		7	273	0.200	2	
			683	0.900	1	
			623	0.900	0	
			018	0.100	0	0.200
			748	0.900	0	0.300
			408	0.700	0	0.900
			688	0.900	0	0.900
59 938		5	297	0.200	0	0.600
			985	1.100	0	0.600
			777	0.900	0	0.800
			294	0.200	0	0.800
			417	0.700	0	1.000
60 184		1	509	0.800	0	0.700
61 245		1	571	0.800	1	
62 220		1	925	1.100	2	
63 071		0	–	–	2	
64 297		1	753	0.900	2	
65 579		2	448	0.700	2	
			060	0.100	1	
66 333		2	899	1.000	2	
			238	0.200	1	
67 494		2	872	1.000	2	
			051	0.100	1	
68 651		2	123	0.200	2	
			886	1.000	1	
69 920		4	001	0.100	2	

Hour	Random number	Work stoppages	Random number	Hours to get machine into operation	Operators available for next work stoppages	Downtime hours because of lack of operators for servicing
			914	1.000	1	
			049	0.100	0	
			150	0.200	0	0.100
70 987		6	128	0.200	2	
			229	0.200	1	
			356	0.200	0	
			507	0.800	0	0.200
			114	0.200	0	0.200
			342	0.200	0	0.200
71 004		0	–	–		
72 579		2	513	0.800	2	
			361	0.200	1	
73 125		1	711	0.900	2	
74 315		2	619	0.800	2	
			558	0.800	1	
75 961		5	630	0.900	2	
			694	0.900	1	
			482	0.700	1	
			059	0.100	0	0.700
			321	0.200	0	0.800
76 854		4	027	0.100	2	
			426	0.700	1	
			161	0.200	0	
			524	0.800	0	0.100
77 728		3	572	0.800	2	
			780	0.900	1	
			298	0.200	0	
78 913		4	556	0.800	2	
			098	0.100	1	
			896	1.000	0	
			509	0.800	0	0.100
79 775		3	654	0.900	2	
			340	0.200	1	
			600	0.800	0	
80 372		2	996	1.100	2	
			651	1.100	1	
						Total 19.6

As for other work, methods analysis should precede work measurement in all expense operations. The flow process chart is the ideal tool for presenting the facts of the present method. Once presented, review the present method critically in all its details. Using the primary approaches to operation analysis, consider such factors as the purpose of the operation, design of forms, office layout, elimination of delays resulting from poor planning and scheduling, and the adequacy of existing equipment.

After the completion of a thorough methods program, standards development can commence. Many office jobs are repetitive; consequently, it is not particularly difficult to set fair standards. Word processing centers, billing groups, file clerks, and duplicating machine operators are representative groups that readily lend themselves to work measurement by stopwatch, standard data, and basic motion data techniques. In studying office work, analysts should carefully identify element end points, so that standard data may be established for pricing future work. For example, in typing production orders, the following elements of work normally occur on each page of each order typed:

1. Pick up production order from pile and position in typewriter.
2. Pick up sheet to be copied from, and place in Copy Right.
3. Read product order instructions.
4. Type heading on order:
    a. Date.
    b. Number of pieces.
    c. Material.
    d. Department.

Once analysts have developed standard data for most of the common elements used in an office, they can calculate time standards quite rapidly and economically. Of course, many clerical positions comprise a series of diversified activities that do not readily lend themselves to measurement. Such work is not made up of a series of standard cycles that continually repeat themselves and, consequently, is more difficult to measure than are direct labor operations. Because of this characteristic of some office routines, it is necessary to take many time studies, each of which may be but one cycle in duration. Then, by calculating all the studies taken, analysts develop a standard for typical or average conditions. Thus, they can calculate a time standard based on a page of copy typing. Granted, some pages of technical typing requiring symbols, radicals, fractions, formulas, and other special characters or spacing take considerably longer than routine pages. But if the technical typing is not representative of average conditions, it does not result in an unfair influence on the operator's performance over a period of time; simple typing and shorter than standard pages balance out the extra time needed to type complex letters.

It is usually not practical to establish standards on office positions that require creative thinking. Thus, such jobs as tool or product designing should be carefully considered before trying to establish time standards on the work done by the employees filling those jobs. If standards are established on work of this nature, they should be used for scheduling, control, or labor budgeting, but not for incentive wage payment. Putting pressure on these employees retards creative thinking. The result may be inferior designs that can be more costly to the business than the amount saved through the greater productivity of the designer.

In setting standards for office workers, analysts have found that white-collar workers do not like the practice, since they are not accustomed to having their work measured. Therefore, the same observance of good human relations practiced in the shop should be adhered to in the office.

SUPERVISORY STANDARDS. By establishing standards for supervisory work, it is possible to determine equitable supervisor loads and to maintain a proper balance between supervision, facilities, clerical employees, and direct labor.

The work sampling technique is the tool for developing supervisory standards. Observers could obtain the same information through all-day stopwatch studies, but the cost of reliable data would usually be prohibitive. Supervisory standards can be expressed in "effective machine running hours" or some other benchmark.

For example, one study of a manufacturer of vacuum tubes revealed that 0.223 supervisory hours were required per machine running hour in a given department (Figure 22–5). The work sampling study showed that, out of 616 observations, the supervisor was working with the grid

FIGURE 22–5

INDIRECT LABOR STANDARD

JOB- SUPERVISION   DEPT.- GRID   DATE- 4-16

Cost Center	Number of Obser-vations	Percent of Obser-vations	Prorated Hours	Base Indirect Hours	Effective Machine Running Hours (EMRH)	Direct Labor Hours	Base Indirect Hours per Machine Running Hour (includes 6% personal)
Grid Machines	129	21	130	130	2,461		0.223
Inspect Grids	161	26	160	160			
Desk Work	54	9	56	56			
Supply Material	18	3	18	18			
Misc. Allow.	150	24	148	148			
Walking	7	1	6	6			
Out of Dept.	11	2	12				
Idle	86	14	86				
Total	616	100	616	518	2,461		0.223

machines, inspecting grids, doing desk work, supplying material, walking, or engaging in activities classified as miscellaneous allowances a total of 519 times. Converted to prorated hours, this figure revealed that 518 indirect hours were required while 2,461 machine running hours took place. Adding a 6 percent personal allowance, analysts computed a standard of 0.223 supervisory hours per machine running hour.

$$\frac{518}{(2,461)(1.00 - 0.06)} = 0.223$$

Thus, in a department operating 192 machine running hours a week, a supervisor's efficiency would be:

$$\frac{192 \times 0.223}{40 \text{ (hrs./week)}} = 107 \text{ percent}$$

STANDARD DATA ON INDIRECT AND EXPENSE LABOR

Developing standard data to establish standards on indirect and expense labor operations is quite feasible. In fact, in view of the diversification of indirect labor operations, standard data are, if anything, more appropriate on office, maintenance, and other indirect work than on standardized production operations.

As individual time standards are calculated, tabularize the elements and their respective allowed times for future reference. As such an inventory of standard data is built up, the cost of developing new time standards declines proportionally.

For example, tabularizing standard data for forklift truck operations can be based on six different elements (travel, brake, raise fork, lower fork, tilt fork) and the manual elements required to operate the truck. Once standard data have been accumulated for each of these elements (through the required range), analysts may determine the standard time required to perform any fork truck operation by summarizing the applicable elements. In a similar manner, standard data can readily be established on janitorial work elements, such as sweep floor; wax and buff floor; dry mop; wet mop; vacuum rugs; or clean, dust, and mop lounge.

The maintenance job of "inspecting seven fire doors in a plant and making minor adjustments" can readily be estimated from standard data. For example, the Department of the Navy developed this standard:

Operation	Unit time (hours)	No. units	Total time (hours)
Inspection of fire door, roll-up type (manual chain, crankshaft, or electrically operated) fusible link. Includes minor adjustment............................	0.170	7	1.190
Walk 100 feet between each door, obstructed walking ...	0.00009	600	0.054
			1.24

The standard data time of 0.00009 hours per foot of obstructed walking was established from fundamental motion data time, and the inspection time of 0.170 hours per fire door from stopwatch time study.

Several manufacturers of material handling equipment have taken detailed studies of their product, and provide standard data applicable to their equipment when it is purchased. This information saves analysts many hours in developing material handling time standards.

Universal Indirect Labor Standards (UILS)

Where maintenance and other indirect operations are numerous and diversified, developing standard data and/or formulas to preprice all indirect operations may appear to be more costly than the expected savings brought about by introducing time standards. To reduce the number of different time standards for indirect operations, some engineers have sought to develop universal indirect standards.[1]

The principle behind universal standards is assigning the major proportion of indirect operations (perhaps as much as 90 percent) to appropriate groups. Each group has its own standard, which is the average time for all indirect operations assigned to the group. For example, group A may include these indirect operations: replacing defective union, repairing door (replace two hinges), replacing limit switch, and replacing two sections (14 feet of 1-inch pipe). The standard time for any indirect operation performed in group A may be 48 minutes. This time represents the mean $(\bar{x})$ of all jobs within the group, and the dispersion of the jobs within the group for $\pm 2\sigma$ is some predetermined percentage of $\bar{x}$ (perhaps ± 10 percent).

The three principal considerations in introducing a universal indirect labor system are:

1. Determination of the number of standards (groups or slots) to do a satisfactory job. (Twenty slots should be used when the range is up to 40 hours.)
2. Determination of the numerical standard representative of each group of operations contained in each slot.
3. Assignment to the appropriate slot of indirect labor work as it occurs.

The initial step is to determine good benchmark standards, based on measurement of an adequate sample of the indirect labor for which the UILS system is being developed. This is the most time-consuming and costly step in the installation of a universal maintenance standards program. To do this, establish a relatively large number of standards (200 or more) that are representative of the entire population of indirect work. Competent analysts can develop these measured benchmark standards by

---

[1] Universal maintenance standards were first developed by H. B. Maynard and G. J. Stegmerten in 1955.

using proven industrial engineering tools including stopwatch time study, standard data, formulas, fundamental motion data, and work sampling.

Once established, arrange the benchmark standards in numerical sequence. Thus, assuming 200 benchmark standards, the shortest would be listed first, the next shortest second, and so on, until ending with the longest. If there were 20 slots, and if a uniform distribution is used, compute the time standard for the first slot (UILS One) by calculating the mean of the first ten benchmark standards. Similarly calculate the value of UILS Two by taking the mean of benchmark standards 11 through 20. The last UILS (20) would be equal to the average of benchmark standards 191 through 200. Engineers have used this procedure extensively during the past in the development of UILS.

More reliable Universal Indirect Labor Standards result from using the normal rather than the uniform distribution. Using 20 slots, the assigning of the 200 standards would *not* be 10 per slot. Instead divide the standard normal variable into 20 equal intervals (truncation of the two tails allows this). For example, the standard normal variable may have a truncated range of

$$-3.0 \leq Z \leq +3.0$$

which accounts for 99.87 percent of the area under the curve. The range of each interval would be 0.3. The benchmark standards used in the compilation of the mean of each of the 20 slots (intervals) would equal:

$$P(Z \in \text{interval})(200)/.9987$$

Slot number 1 and 20 (because of symmetry) would have

$$\frac{P(-3.0 \geq Z \leq -2.7)(200)}{.9987} = \frac{P(2.7 \leq Z \leq 3.0)(200)}{.9987}$$

$$= \frac{(0.9987 - 0.9965)(200)}{.9987} = 0.4406 \text{ standards}$$

and slot number 10 and 11 would have

$$\frac{P(-0.3 \geq Z\ 0.0)(200)}{.9987} = \frac{P(0.0 \leq Z\ 0.3)(200)}{.9987}$$

$$= \frac{(.6179 - .5000)(200)}{.9987} = 23.61 \text{ standards}$$

Rounding off the fractions is dictated by the fact that all 200 jobs must be assigned to a slot. Calculate the universal standard time for each slot as the average of the benchmark standards assigned to the slot.

The normal approximation outperforms the uniform approximation when applied to a given set of data. To confirm this, analysts tabulated 270 maintenance standards. The data were taken from the *Engineering Performance Standards Machine Shop and Machine Repair Handbook*

FIGURE 22–6
Distribution of 20 Universal Indirect Labor Standards ranging in time from 0.625 hours to 24.38 hours

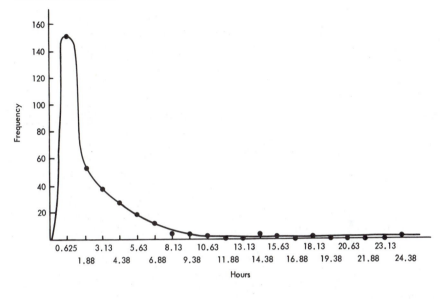

for Public Works Maintenance of the Department of the Navy, Bureau of Yards and Docks. For example, the following three jobs carry a standard of 4.4 hours:

1. Fabricate eight anchor plates, 1 × 10 × 6 inches. Cut off eight pieces from milled steel stock with power hacksaw. Drill four 1-inch holes in corners, using Cincinnati 21-inch vertical single-spindle drill press.
2. Fabricate one 8-inch flat belt pulley with 2-inch hole through center, ¼-inch keyway, and four 1½-inch stress holes. Material used: 9 inches diameter × 3 inches wide cast-iron blank. Machines used: engine lathe, horizontal mill, and radial drill.
3. Fabricate one shaft to dimensions from 4½ inches outside diameter × 25 inches milled steel. Turn four diameters; face, thread, mill keyway; and drill center pin holes. Machines used: power hacksaw, lathe, horizontal mill, and single-spindle vertical drill press.

When studying a new part to be made, analysts can fit jobs to categories where similar jobs have been studied and standards established. Although the normal distribution is superior to the uniform distribution, a skewed distribution may outperform the normal in view of the plotting of the data (see Figure 22–6).

The gamma distribution is a positively skewed distribution with a probability density function given by

$$f(x) = \begin{cases} \dfrac{1}{\beta^2\,\Gamma(\alpha)}\, x^{\alpha-1} e^{-x\beta} & \text{for } x,\,\alpha,\,\beta > 0 \\[2mm] 0 & \text{elsewhere} \end{cases}$$

where $\Gamma(\alpha)$ is a value of the gamma function given by

$$\Gamma(\alpha) = \int_0^{\infty} x^{\alpha-1} e^{-x}\, dx$$

$$= (\alpha - 1)\,\Gamma(\alpha - 1)$$

The skewness of the gamma distribution decreases as α increases for a fixed β.

The mean and variance of this distribution equal:

$$\mu = \alpha\beta$$
$$\sigma^2 = \alpha\beta^2$$

For a given set of data estimates of α and β, determine $\hat{\alpha}$ and $\hat{\beta}$ by first obtaining the mean and variance of the data. Then calculate estimates:

$$\hat{\alpha} = \mu^2/\sigma^2$$
$$\hat{\beta} = \mu/\hat{\alpha}$$

To determine the relative accuracy of the uniform, normal, and gamma techniques, analysts used the 270 maintenance standards developed by the Department of the Navy as benchmark standards. Next, they divided these benchmark standards into 20 slots using the three approaches. The gamma distribution and expected frequencies for the 20 slots appear in Table 22–1.

To compare the results of the uniform, normal, and gamma techniques, a simulation was done. For each of 25 weeks, analysts selected jobs at random until the sum of the actual standard times exceeded or equaled 40 hours. Then they determined the universal maintenance standard for each job and calculated the weekly sum. They assumed that each job was properly slotted.

For each week an error was calculated as

$$\left| \frac{\text{Actual Standard Time} - \text{Universal Standard Time}}{\text{Actual Standard Time}} \right| 100 \text{ percent}$$

The results of the simulations for the uniform, normal, and gamma distributions are given in Table 22–2.

This study confirms that the gamma distribution offers some improvement over the normal and that the normal gives better results than the uniform.

Increasing the pay period from one week (40 hours) to two weeks (80 hours) would markedly reduce the cumulative error per pay period. The

TABLE 22–1

The gamma distribution probabilities and expected frequencies for the 20 slots

Slot (hours)	Cumulative probabilities	Probabilities	Actual frequency	Expected frequency
0.0 — 1.25	.5404	.5405	150	145.94
1.26— 2.50	.7146	.1741	50	47.00
2.51— 3.75	.8108	.0962	29	25.97
3.76— 5.00	.8698	.0590	13	15.93
5.01— 6.25	.9077	.0379	11	10.23
6.26— 7.50	.9327	.0250	6	6.75
7.51— 8.75	.9495	.0168	2	4.54
8.76—10.00	.9610	.0115	2	3.11
10.01—11.25	.9688	.0078	2	2.11
11.26—12.50	.9743	.0055	0	1.49
12.51—13.75	.9780	.0037	0	1.00
13.76—15.00	.9807	.0027	2	.73
15.01—16.25	.9825	.0018	1	.49
16.26—17.50	.9838	.0013	0	.35
17.51—18.75	.9847	.0009	1	.24
18.76—20.00	.9854	.0007	0	.19
20.01—21.25	.9859	.0005	0	.13
21.26—22.50	.9862	.0003	0	.08
22.51—23.75	.9864	.0002	0	.05
23.76—25.00	.9866	.0002	1	.05
25.01—x..............	1.000	.0134	0	3.62
Total			270	270.00

magnitude of the error would also decrease as the number of groups (slots) increases.

UILS offers an opportunity to introduce standards for a majority of indirect operations at a moderate cost, and also minimizes the cost of maintaining the indirect standards system.

PROFESSIONAL PERFORMANCE STANDARDS

The cost of professionals is a sizable proportion of the total expense budget. In most manufacturing and business operations the professional salaries of employees in engineering, accounting, purchasing, sales, and general management represent a significant proportion of total cost. If the productivity of these employees can be improved by as little as a few percent, the overall impact on the firm's business is consequential. Establishing standards for professional employees and utilizing these standards as achievable goals, inevitably enhances productivity.

The difficulty in developing professional standards is due, first, to the determination of what to count and second, to the method for counting these outputs. In determination of what to count, begin by stating the

TABLE 22–2
Results of 25-Week Simulation

Week number	Absolute percent error		
	Uniform	*Normal*	*Gamma*
1	5.97	7.18	2.57
2	16.01	6.93	0.27
3	8.49	6.42	5.74
4	10.94	4.03	3.32
5	25.78	1.67	1.85
6	2.61	0.47	5.25
7	4.79	6.08	3.90
8	0.88	3.37	3.21
9	4.51	5.34	5.36
10	0.05	6.45	3.07
11	30.78	0.32	1.79
12	21.93	1.75	0.64
13	8.23	4.24	1.62
14	6.67	7.55	5.59
15	2.37	2.37	1.53
16	0.06	0.87	1.24
17	12.53	2.88	1.79
18	3.73	5.21	5.86
19	6.85	1.52	5.35
20	11.50	2.29	0.58
21	20.18	2.48	0.05
22	6.44	8.31	0.92
23	3.46	6.72	5.11
24	2.96	0.45	3.09
25	11.74	1.01	1.13
Mean	9.18	3.84	2.83
Variance	151.78	21.62	11.34
Std. Dev.	12.32	4.65	3.37

objectives of the professional employees' positions. For example, buyers in the purchasing department might have the objective "to procure quality components and raw materials at the lowest price, in time to meet company production and delivery schedules." To be effective, a count of buyers' outputs needs to consider five things; (1) proportion of deliveries made to schedule; (2) proportion of deliveries that meet or exceed quality requirements; (3) proportion of shipments that represent the lowest available price, (4) the number of orders placed during some interval of time, such as one month; (5) a total dollar value of the purchases made during a period of time.

The next problem is establishing measures to use as achievable goals. In such instances, using historical records supplemented with work sampling analyses to determine how time is utilized can serve as the basis for the development of professional standards.

Returning to the example of establishing buyers' standards, it would not be difficult to identify the purchases made by the various buyers and review what proportion of these orders were delivered on schedule over a six-month period. This historical data study may reveal something analogous to the following:

Buyer	Proportion of orders delivered on or before schedule (percent)
A	70
B	82
C	75
D	50
E	80

Based on this record, skilled buyers in this particular organization should be able to procure at least 72 percent of their purchases on schedule (the mean of their performance).

Similarly, a quality review of the purchases made by the five buyers may disclose the following:

Buyer	Proportion of orders delivered with less than 5 percent rejects (percent)
A	85
B	90
C	80
D	95
E	80

Here the quality standard could be 86 percent of orders received have less than 5 percent rejects (the mean of the past performance of the five buyers).

The proportion of procurements purchased at the lowest available price is another important measure of buyers. Once again, historical records allow comparison of the performance of the five buyers. Assume the following performance record applied:

Buyer	Proportion of orders procured at the lowest available price (percent)
A	45
B	50
C	60
D	47
E	40

The average of these values, 48.4 percent of the orders placed at the lowest available price, could reflect normal performance.

Historical records might indicate that on the average a buyer placed 120 orders per month having a total monetary worth of $120,840. These five criteria could then be used to develop an overall performance standard: delivery, quality, price, number of orders, and order value. For example, one method would be to add the means of the first three criteria (0.72 + 0.86 + 0.484) plus .002 times the mean of the order placed plus .000001 times the average monetary worth of purchases. The buyers standard in this fictitious operation would be

$$0.72 + 0.86 + 0.484 + 0.24 + 0.12 = 2.424$$

Another example will help clarify how performance standards may be achieved for managerial personnel. Consider the position of director of personnel administration. An analysis may suggest four specific objectives of this position:

1. Establish a methodology for identifying both the quantity and quality of the company's human resources.
2. Establish a procedure for attracting, employing, and retaining kinds and numbers of employees for the successful operation of the company.
3. Establish policies, programs, and practices that facilitate the achievement of departmental objectives and maintain employee morale.
4. Administer and maintain the company's benefit program.

Now that the objectives have been stated, it is relatively easy to develop a performance standard in terms of time. For example, the standard for objective 1 above may be: "Train, within the next three months, staff representatives to conduct an audit of the company's personnel, to determine projected needs from the standpoint of both numbers and type."

The performance standard for objective 2 might be: "Employ within the next 12 months *(a)* 2 Ph.D. Chemists; *(b)* 7 M.S. degrees in industrial and/or mechanical engineering; and *(c)* 35 B.S. degrees with a distribution of 10 in business, 20 in engineering, and 5 in liberal arts. Employ (based on anticipated turnover and expansion) 75 hourly employees. Investigate the turnover rate of professional employees in the past year and prepare a report showing how turnover may be reduced."

For objective 3, the performance standard may be as follows: "Within the next three months, update the current management handbook, bringing the salary administration program up-to-date. Within the next six months develop and distribute a booklet for all hourly employees describing the new grievance procedure established in our new labor contract. The booklet should explain not only the importance of reducing the number of grievances, but also how this can be done."

The performance standard for objective 4 could be: "Review within the next 12 months the company's entire fringe benefit program and compare our benefits to those of similar-size companies in this area. Make any appropriate recommendations to management."

These objectives identify performance standards for finite time periods. Thus, the standards may change as time passes since each standard is result based. The standards established for succeeding periods may include different work assignments to meet the stated objectives.

In the establishment of professional performance standards, professionals should assist in identifying the objectives of the position, gathering the historical performance records, and developing the standards. Performance standards developed without the complete involvement of the professionals are seldom realistic.

When gathering historical data to facilitate the development of professional standards, take a work sampling study during the period that serves as the basis of the historical record data. This work sampling study can reveal how much working time was spent on the various necessary work routines or on work assignments that would better be handled by clerical or semiprofessional employees; and how much time was literally wasted. After reviewing the work sampling study, performance rate the average data gathered over the historical period to attain a standard that is more representative of normal professional experience.

In the development of professional standards, observe the following guidelines:

1. Each manager should be involved with the setting of standards for his or her professional subordinates. Thus professional standards should be jointly developed by employees and their supervisors.
2. Standards should be result based and worded to include references to measurement.
3. Standards must be realistically attainable by at least one half of the group concerned.
4. Standards should be periodically audited and revised if necessary.
5. It is helpful to work sample managers to assure that they have adequate clerical and administrative aid support and are using their time judiciously.

ADVANTAGES OF WORK STANDARDS ON INDIRECT WORK

Standards on indirect work offer distinct advantages to both the employer and the employee. Some of these advantages are:

1. The installation of standards leads to many operating improvements.
2. The mere fact that standards are established results in better performance.

3. Indirect labor costs are related to the work load regardless of fluctuations in the overall work load.
4. Labor loads can be budgeted.
5. The efficiency of various indirect labor departments can be determined.
6. The costs of such items as specific repairs, reports, and documents are allocated. This frequently results in the elimination of needless reports and procedures.
7. System improvements can be evaluated prior to installation. Thus, it is possible to avoid costly mistakes by choosing the right procedure.
8. It is possible to install incentive wage payment plans on indirect work, thus allowing employees to increase their earnings.
9. Accurate planning and scheduling of all indirect labor work gets the jobs done on time.
10. Employees require less supervision, as a program of work standards tends to enforce itself. Employees who know what is required, do not arbitrarily waste time.

CONCLUSION

It is more difficult to study and determine representative standard times for nonrepetitive tasks characteristic of most indirect labor operations than for repetitive tasks. Since indirect labor operations are difficult to standardize and study, only infrequently are they subjected to methods analysis. Consequently, this area usually offers a greater percentage potential for reducing costs and increasing profits through methods and time study than does any other.

After good methods have been introduced and operator training has taken place, it is both readily possible and practical to establish standards on indirect labor operations. The usual procedure is to take a sufficiently large sample of stopwatch time studies to assure the representation of average conditions, and then to tabularize the allowed elemental times in the form of standard data. Fundamental motion data also have a wide application for establishing standards on indirect work. This is especially true of those systems that utilize larger blocks of fundamental motions, such as Brief Work-Factor, MTM-2, and MOST.

Once standards are developed, store them in a computer as standard data for future retrieval. Computers make calculations faster than electronic hand-held calculators. They establish allowed operation times consistently, accurately, and economically.

Companies representing the fundamental motion data systems, such as the Work-Factor Company, H. B. Maynard and Company, Inc., and MODAPTS, and Standards, International, have developed software to accompany their systems for standards development using the computer. In such cases, analysts only need to dictate work place and method data, which may be transcribed by a typist for input into the computer. The results may be displayed almost immediately on the cathode ray tube monitor. Software programs also produce hard copy printouts of the developed standard, which can include both allowances and process time, along with a sketch of the workplace, which serves as a record of the method used to develop the standard.

Analysts can estimate indirect elements involving waiting time accurately by using waiting line or queuing theory. Analysts must understand elementary waiting line theory to establish the mathematical model that fits the parameters of the problem. Where the problem does not fit established waiting line equations, analysts can use Monte Carlo simulation as a tool for determining the extent of the waiting line problem in the work area.

There are no automatic means for determining a standard job time for a maintenance or other type of indirect or expense job. Analysts must review each job request and determine both the labor and material requirements. They usually go through the following steps:

1. Examining the work request in detail; consulting with the initiator of the work request or even visiting the jobsite to determine the exact requirements of the job.
2. Preparing a material estimate for the job.
3. Studying the job from the standpoint of the work elements performed.
4. Selecting the appropriate direct work or task times and the applicable tooling, material, and planning times from standard data tables or universal time standards.
5. Assigning an estimated time from similar work shown on the master table of universal standard times or spread sheet if no specifically applicable times exist.
6. Adding standard times to cover the transportation times for the work order being analyzed and the necessary allowance to cover unavoidable delays, personal delays, and fatigue.

In establishing standards on indirect and expense work, the following table can be a guide:

Indirect and expense type work	Recommended method of establishing standards
Routine maintenance. Work standards 0.5 to 3 hrs.	Standard data, MTM-2, MTM-3, Ready Work-Factor, Brief Work-Factor, MOST Macro Motion Analyses, MODAPTS
Complicated maintenance, standards 3 hrs. to 40 hrs.	Slotting based on universal indirect labor standards
Shipping and receiving	Standard data MTM-2, MTM-3, Ready Work-Factor, Brief Work-Factor, MOST Macro Motion Analyses, MODAPTS
Toolroom	Slotting based on universal indirect labor standards

Indirect and expense type work	Recommended method of establishing standards
Inspection	Standard data, MTM-2, Ready Work-Factor, MOST, Macro Motion Analyses MODAPTS
Tool Design	Slotting based on universal indirect labor standards
Buying	Standards based on historical records, analysis, and work sampling
Accounting	Standards based on historical records, analysis, and work sampling
Plant engineering	Standards based on historical records, analysis, and work sampling
Clerical	Standards based on standard data, Ready Work-Factor, MTM-2, Brief Work-Factor, MOST, Micro and Macro Motion Analyses, MODAPTS
Janitorial	Standard data, slotting based on universal indirect labor standards
General management	Standards based on historical records, analysis, and work sampling

TEXT QUESTIONS

1. What would be the expected waiting time per shipment if a stopwatch time study established that the normal time to prepare a shipment was 15.6 minutes? Twenty-one shipments are made every shift (eight hours). The standard deviation of the service time has been estimated as 1.75 minutes. It is assumed that the arrivals are Poisson distributed and that the servicing time is arbitrary.

2. Using Monte Carlo methods, what would be the expected downtime hours because of the lack of an operator for servicing if four operators were assigned to the work situation described in the text?

3. Differentiate between indirect labor and expense labor.

4. Explain "queuing theory."

5. Which four divisions constitute indirect and expense work?

6. How are standards established on the "unnecessary and delays" portion of indirect and expense work?

7. The arrivals at the company cafeteria are Poisson, with an average time between arrivals of 1.75 minutes during the lunch period. The average time for a customer to obtain lunch is 2.81 minutes, and this service time is distributed exponentially. What is the probability that a person arriving at the cafeteria will have to wait? How long?

8. Why has there been a marked increase in indirect workers?

9. Why do more unavoidable delays occur in maintenance operations than on production work?

10. What is meant by crew balance? By interference time?

11. Explain how time standards would be established on janitorial operations.
12. Which office operations are readily time studied?
13. Why are standard data especially applicable to indirect labor operations?
14. Summarize the advantages of standards established on indirect work.
15. When were Universal Maintenance Standards first developed? Who first developed them?
16. Why is work sampling the best technique for establishing supervisory standards?
17. Explain the application of "slotting" under a universal standards system.
18. Why will a universal standards system involving as few as 20 benchmark standards work in a large maintenance department where thousands of different jobs are performed each year?

PROBLEMS

1. Work measurement procedures establish an average time of 6.24 minutes per piece on the inspection of a complex forging. The standard deviation of the inspection time is 0.71 minutes. Usually, 60 forgings are delivered to the inspection station on the line every eight-hour turn. One operator performs this inspection.

   Assuming that the castings arrive in Poisson fashion and that the service time is exponential, what would be the mean waiting time of a casting at the inspection station? What would be the average length of the casting queue?

2. In the tool and die room of the Dorben Company, the work measurement analyst wishes to determine a standard for the jig boring of holes on a variety of molds. The standard will be used to estimate mold costs only, and it will be based on operator wait time for molds coming from a surface grinding section and on operator machining time. The wait time is based on: a single server, Poisson arrivals, exponential service time, and first-come, first-served discipline.

   A study revealed the average time between arrivals was 58 minutes. The average jig-boring time was 46 minutes. What is the possibility of a delay of a mold at the jig borer? What is the average number of molds in back of the jig borer?

SELECTED REFERENCES

Crossman, Richard M., and Harold W. Nance. *Master Standard Data: The Economic Approach to Work Measurement*. Rev. ed. New York: McGraw-Hill, 1972.

Knott, Kenneth "Indirect Operations: Measurement and Control." In *Handbook of Industrial Engineering*, 2nd ed., edited by Gavriel Salvendy, Chap. 65. John Wiley & Sons, 1992.

Lewis, Bernard T. *Developing Maintenance Time Standards*. Boston: Cahners, 1967.

Nance, Harold W., and Robert E. Nolan. *Office Work Measurement*. New York: McGraw-Hill, 1971.

Newbrough, E. T. *Effective Maintenance Management.* New York: McGraw-Hill, 1967.

Pappas, Frank G., and Robert A. Dimberg. *Practical Work Standards.* New York: McGraw-Hill, 1962.

Raghavachari, M. "Queuing Theory." In *Handbook of Industrial Engineering,* 2nd ed., edited by Gavriel Salvendy, Chap. 94. New York: John Wiley & Sons, 1992

SELECTED VIDEOTAPES

Manufacturing Insights Videotape Series. Simulation ½″ VHS VT 253-1368 and ¾″ U-Matic VT 253U-1368 Dearborn, Mich.: Society of Manufacturing Engineers, 1987.

Work Measurement and Computers

23

Today, computers are the principal work measurement analysis tool. There are five reasons for this: (1) increasing use of fundamental motion data systems; (2) increasing memory capacities and lower prices of personal computers; (3) development of appropriate software by end users and educational institutions; (4) broader dissemination and use of statistical and mathematical techniques, including multiple regression, linear programming, and work sampling; and (5) increasing use of automation in data gathering.

When computers are utilized for methods and standards work, it is axiomatic that standard data are being tabularized from either, or both, stopwatch time studies and fundamental motion data. With these data available, digital computers offer a means of storing and recovering detailed standard data and elemental standard data for application on jobs being performed for the first time and for the development of proposals.

COMPUTERIZING THE ESTABLISHMENT OF STANDARDS

The conventional procedure to establish a new standard without data processing equipment, but utilizing standard data, is illustrated in Figure 23-1. Here methods analysts develop a workstation layout and motion pattern based on knowledge of motion economy and shop operations. From this proposed method, they make an elemental breakdown and look up the appropriate standard data times. If fundamental motion data are being used, they further subdivide the elemental breakdown so that basic motion times may be assigned. They develop the time standard for the operation by extending elemental time values by their frequency, totaling the times for each element, applying the correct allowance, and finally

FIGURE 23–1

The development of time standards through the application of standard data

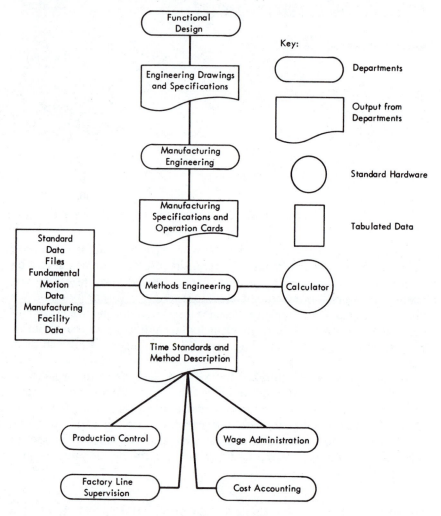

summarizing the allowed elemental times to determine the allowed operation time.

Much of the preceding procedure represents burdensome clerical activities prone to human error. Computers relieve analysts of much clerical
and computational work in the establishment of time standards.

Computer-based systems operate as follows:

1. Methods engineering develops a workstation layout and motion pattern.
2. Methods engineering also identifies the proposed method in detail by
   an elemental breakdown.

3. A computer program retrieves the description of each element, identifies the normal elemental times, extends the elemental times by frequencies, applies allowances, and computes the allowed time for the operation.
4. The system prepares all associated reports.
5. A computer operator enters the operation time and description into a permanent file for future use and maintenance.

The only way the data processing system can accomplish step 3 is by standard data coding. The coding can take any form a company may wish to use: numeric, alphabetic, or mnemonic. Generally, coding similar to that used in the fundamental motion data systems (MTM, Work-Factor, MODAPTS, MOST, Micro and Macro Motion Analyses) is appropriate.

Advantages of Computerization of Work Measurement

The principal advantages of using computers to develop standards include increased coverage, more accurate standards, and improved standards maintenance. Since standards are developed much more rapidly with computers, from the standpoint of both time and money, they make it feasible to increase any plant's coverage of measured work. The smaller the amount of unmeasured work, the greater the opportunity for effective control and efficient operation.

Standards developed with computer software programs are inevitably more error-free, since manual arithmetic and lookup techniques are subject to human error. Manual techniques prohibit accurate standard maintenance for no other reason than the avalanche of clerical work involved. For example, if basic research revealed that 2.2 TMU was a more valid value for a GIA grasp than 2.0 TMU, think of the clerical work involved to adjust every existing standard in the plant by increasing all GIA grasps by 10 percent! With the correct software, such a change could be economically effected within a short time on a computer.

Approach to Computerization for Work Measurement

To computerize work measurement records, it is necessary to develop files of all the existing standard data. These files must be identified by an acceptable code. For example, a standard data element characteristic of such a file might be:

Code	Element description	Standard time
PSP3A	Pick up small piece, place in three-jaw air chuck and clamp	0.062

Here the code "PSP3A" represents the input to the system and the element description "pick up small piece, place in three-jaw air chuck and clamp," along with the standard time of 0.062, the output.

FIGURE 23-2
Automated work measurement systems for processing new operations and
changing existing operations (dotted line applies to change operations only,
and solid lines apply to both change and new operations)

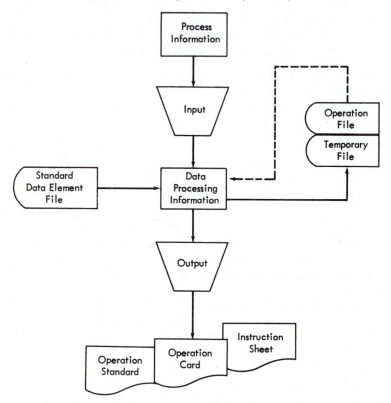

Analysts can program computers so that greater amounts of detail
appear on a specific output document, such as a detailed instruction
sheet. This is done by placing special symbols at appropriate locations
within the element description.

For example, the lozenge ($\Diamond$) symbol might be inserted between the
words "piece" and "place" as follows: "pick up small piece $\Diamond$ place in
three-jaw air chuck and clamp." One output document could show the
element as stated, and another could show additional detail, such as "pick
up small piece, casting #437106, place in three-jaw air chuck and clamp."

Figure 23-2 shows the flow of information in the development of a
detailed instruction sheet, operation card, and operation standard. Analysts enter process information—such as cutting feeds, cutting speeds,
depths of cuts, along with the codes of the elements and their frequencies—into the system. The program retrieves the element description and
its standard time from the standard data element file for each element

involved in the operation. Analysts store records involved in the development of the operation standard in a temporary file until completing a final review of the output. Upon completion of the operation review, they transfer the operation records to a permanent operation file, which may be on either tape or disk.

The output may take many forms. Typically included in output are the operation standard, which lists the applicable elements and their standard times; an operation card, which gives a brief description of the operation and its standard in minutes per piece and pieces per hour (this can serve as a move card, pay voucher, and so on); and an instruction sheet, which may be used by operating personnel.

Computers make it relatively easy to introduce a change that affects one or several operations. With details of operations stored on magnetic tapes or disks, retrieval techniques allow changes to be made through "add" and/or "delete" instructions entered for the sections of the operation to which the changes should be applied. When the operation file is initially generated, each element of an operation is assigned a line number identification. Spaces signifying line numbers allow element additions. When part of an operation must be deleted, it is referred to by the line number. When an element is added, it is assigned a line number reserved for future additions in the proper sequence.

The Work Measurement System

Figure 23–3 shows the general flow of information through a typical work measurement system.

Through visual motion study or micromotion, time study, and process information, analysts can make a methods study that results in the development of a workstation utilizing the basic principles of motion economy and operation analysis. The proposed method's appropriate element coding is then input, after edited input records, any obvious errors are corrected. Analysts retrieve information from the major input files, that is, the element file and the operation file. The program produces a temporary file for all output records, including operations, and lists any error conditions. Then it processes the output records and prints various reports.

After analysts review the output reports, the operation file is updated, using the temporary file.

The installation of a work measurement system, requires two major input files: the element file and the operation file. The element file is more basic than the operation file, since it contains the description and the normal time of all existing elements. The system retrieves this information when it processes a new operation. After the new operation has been processed, its records are stored in the operation file.

To create an operation file on an existing operation, analysts enter the information on the analysis input sheet in the system. The computer

FIGURE 23–3

The flow of information in a typical work measurement system

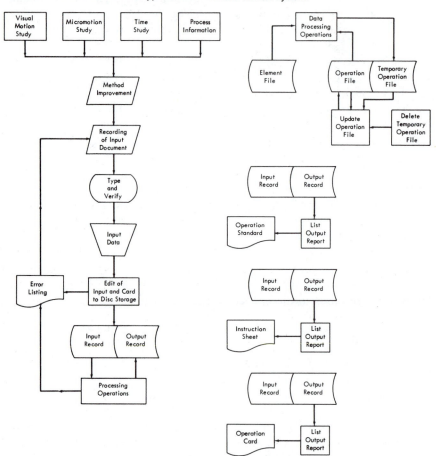

program then retrieves element record data from its element file, applies the data to each recorded element, performs the required frequency time multiplications and accumulations, applies the appropriate allowances, and stores the entire operation on a temporary storage tape or disk. On approval of the operation standard, the temporary operation records are transferred to the operation file.

The operation record is normally composed of two sections: (1) heading information, where the operation is described and its standard time is recorded; and (2) element information, where the elements involved in the operation are described and assigned a time value. Each element in an operation is usually given a "line item" number automatically by the computer to provide for an easy method of updating a particular element.

Both the element and operation files must be adequately maintained if the work measurement system is to be reliable. All operations must be processed using recent elemental descriptions and time values.

Figure 23–4 shows a typical work measurement input document. The computer program compares the element code with its element master file, retrieves element times, and computes the time standard based on the element time value, the element's frequency, and the appropriate allowance.

Changes in element description and/or time may have an appreciable effect on existing standards and operating instructions. Under a manual system, reviewing all existing standards to introduce every elemental change is not feasible. However, a computerized methods and standards system can readily incorporate element changes that would result in a significant alteration of the operation standard. This can be handled by generating a special file to obtain the impact of standard data changes. Those operations that have a significant change can then be reprocessed and a new standard developed.

In addition to the rapid development and maintenance of time standards, a computerized work measurement system renders several helpful reports. For example, it is possible to develop a report that shows the old and new standard for the operation, the variance, and the percentage of variance. Thus, the effect of change by area or department is readily available. Another helpful report is the "where-used index." This summary can identify where specific components, tools, and so on are used. For example, a particular drill jig may be used for the production of several components. The "where-used" index can show the study numbers, as well as the specific lines referring to the drill jig.

Computers can generate a workplace layout that can be used by line supervision to help assure that the prescribed method is followed. Figures 23–5 and 23–6 represent typical input and output documents.

A computerized work measurement system can be extremely helpful in the rapid development of time estimates and quotations. Many businesses today are experiencing substantial costs in the preparation of quotations, only a small fraction of which result in contracts. Assume for example, that a small job shop automatic screw machine business prepares approximately 100 quotations per week. The request for the quotation, along with a drawing of the part, is studied by a methods analyst, who prepares a planning sheet that includes the following information:

1. Alloy of material required, size (diameter and so on), material weight, cost per pound of material.
2. Length of material required per piece (length of part plus cutoff).
3. Recommended cutting speed in surface feet per minute.
4. Recommended feed in inches per revolution.
5. Money rate of the automatic to be employed.

FIGURE 23-4
Work measurement input document

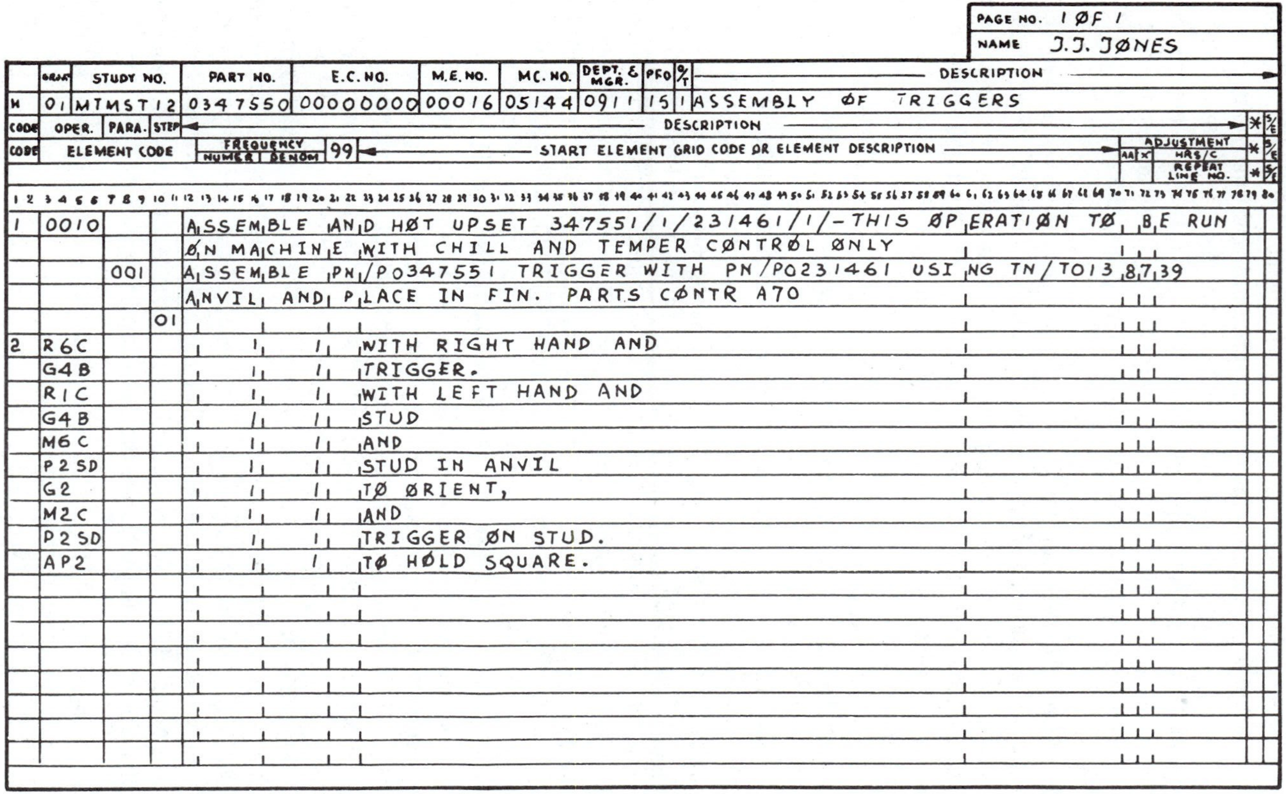

CHAR	STUDY NO.	PART NO.	E.C. NO.	M.E. NO.	M.C. NO.	DEPT. & MGR.	PFO %	DESCRIPTION
M	01 M TM ST 12	0347550	0000000000	00016	05144	09	11	15	ASSEMBLY OF TRIGGERS

PAGE NO. 1 OF 1
NAME J.J. JONES

CODE	OPER.	PARA.	STEP	FREQUENCY NUMBER	DENOM	START ELEMENT GRID CODE OR ELEMENT DESCRIPTION / DESCRIPTION
CODE	ELEMENT CODE			99		

ADJUSTMENT — MISC / MISC / REPAY LINE NO.

	Description
1 0010	ASSEMBLE AND HOT UPSET 347551/1/23146I/1/-THIS OPERATION TO BE RUN
	ON MACHINE WITH CHILL AND TEMPER CONTROL ONLY
001	ASSEMBLE PN/PO34755I TRIGGER WITH PN/PO23146I USING TN/TO13,87,39
	ANVIL AND PLACE IN FIN. PARTS CONTR A70
01	
2 R6C	WITH RIGHT HAND AND
G4B	TRIGGER.
R1C	WITH LEFT HAND AND
G4B	STUD
M6C	AND
P2SD	STUD IN ANVIL
G2	TO ORIENT,
M2C	AND
P2SD	TRIGGER ON STUD.
AP2	TO HOLD SQUARE.

IBM Corporation, Data Processing Division

716

FIGURE 23–5
Input document calling for a workplace layout

PAGE NO. 1
NAME A.W.

SRM	STUDY NO.	PART NO.	E.C. NO.	M.E. NO.	M.C. NO.	DEPT.&MGR	PFB	DESCRIPTION
01	MTMST02	0347550	00000000	00016	05144	0911	151	ASSEMBLY OF TRIGGERS

N	CODE	OPER	PARA	STEP	ELEMENT CODE	FREQUENCY NUMBER/DENOM	START ELEMENT GRID CODE OR ELEMENT DESCRIPTION	ADJUSTMENT	REPEAT LINE NO.
1		0010					ASSEMBLE AND HOT UPSET 34755/1/231461//-THIS OPERATION		
							ON MACHINE WITH CHILL AND TEMPER CONTROL ONLY		
	WP	WORK ARE	M		A31	12,TO,138739	ANVIL//STD.ELECTRODE/1	00004	
	WP	OPER SEAT			B31	00,80,000123	OPERATOR SEAT	00004	
	WP	PART LOC			A21	18,PO,231461	STUDS	00500	
	WP	PART LOC			A41	06,PO,347551	TRIGGERS	00100	
	WP	PART LOC			A42	06,PO,347550	FIN. PARTS IN SMALL CNTR.	00050	
	WP	CNTR 405	S		A11	18,PO,231461	STUD SUPPLY	00500	
	WP	CNTR 705			B11	18,PO,347551	TRIGGER SUPPLY	00500	
	WP	CNTR 150			B41	12,PO,347550	FIN. PARTS IN CNTR. A70	00200	
1		001					ASSEMBLE PN/PO34755I TRIGGER WITH PN/PO231461 USING TN/TO		
							ANVIL, AND, PLACE IN FIN PARTS CNTR A70		
				01					
2					R6C	/1	.WITH RIGHT HAND AND		
					G4B	/1	.TRIGGER		
					R1C	/1	.WITH LEFT HAND AND		
					G4B	/1	.STUD		
					M6C	/1	.AND		
					P2SD	/1	.STUD IN ANVIL		
					G2	/1	.TO ORIENT.		
					M2C	/1	.AND		
					P2SD	/1	.TRIGGER ON STUD.		
					AP2	/1	.TO HOLD SQUARE.		

IBM Corporation, Data Processing Division

717

FIGURE 23–6
Output document showing workplace layout

GROUP 01   ASSEMBLY OF TRIGGERS   PART NUMBER 0347550
STUDY MTMST02   MAY 20, 19--
EC 00000000 ME 00016   OCN 05-20-4 DEPT 091-1

0010

```
XXXXXXXXXXXXXXXXXXXXXXXXXXXXXXXXXXXXXXXXXXXXXXXXXXX
X               X           X               X               X
X   VENDORS     X    PART    X               X  PART LOC 2   X
X   CONTAINER   X           X       WORK     X               X
X   --------    X  LOCATION  X               X  XXXXXXXXXXXXXXX
X               X           X       AREA     X               X
X   SEALED      X     1      X               X  PART LOC 1   X
X               X           X               X               X
XXXXX -A1- XXXXXXXXX -A2- XXXXXXXXX -A3- XXXXXXXXX -A4- XXXXX
XXXXXXXXXXXXXX                              XXXXXXXXXXXXXX
X               X                           X               X
X               X               XXXXXXX     X               X
X     705       X               X     X  X  X  PLASTIC      X
X    STACK      X               X  X  X  X  TOTE BOX     X
X     BIN       X                 XXX     X  WITH PROT.   X
X               X               OPERATOR  X               X
X               X                           X               X
XXXXX -B1- XXXXX                  -B3-      XXXXX -B4- XXXXX
```

PARTS LOCATION TABLE

FROM LOC.	TO LOC.	PART/TOOL NUMBER	PART DIST	PART QTY.	PART/TOOL DESCRIPTION	REP. QTY.
A1-1		P0231461	18	0500	STUD SUPPLY	
A2-1		P0231461	18	0500	STUDS	
A3-1		T0138739	12	0000	ANVIL/1/STD. ELECTRODE/1/	
A4-1		P0347551	06	0100	TRIGGERS	
A4-2		P0000000	06	0050	FIN. PARTS IN SMALL CONTR.	
B1-1		P0347551	18	0500	TRIGGER SUPPLY	
B3-1			00			
B4-1			12	0200	FIN. PARTS IN CONTR. A70	

IBM Corporation, Data Processing Division

6. Coding of the setup elements.
7. Dollar rate of setup person.
8. Percent efficiency of the production machine employed.
9. Any miscellaneous costs and one-time charges.

The planning sheet, along with the request for the quotation, is then routed to the computer operator who prepares the quotation. The computer operator's responsibilities are to type in:

1. The information which will read and print out on the printer. This includes the customer's name and address, the part name, part number, and the customer request number.
2. The quantities being quoted on, as column heads.
3. The size, shape, and kind of material required, as shown on the planning sheet.
4. The part length and cutoff length.

5. The material weight (pounds per foot).
6. The cost per hundredweight according to the price, as read from the planning sheet.
7. The major diameter and recommended surface feet, as read from the planning sheet.
8. The throw and feed for each operation.
9. The percent efficiency.

The computer calculates the amount of material needed per hundred pieces (or some other unit of production) and prints this information on the quotation. It also cranks out such information as the pounds of material per hundred pieces, weight per quantity quoted on, cost per hundredweight, gross and net production, and spindle speed required for each operation.

The computer operator types in a description of the operation as read from the planning sheet, as well as the gross and net production, and the dollar rate of the facility. The computer automatically calculates the dollar amount per hundred pieces and prints out this information on the quotation. This procedure is repeated for each line of charges on the quotation. To total the quotation, the operator depresses a command key, and the total price per 100 pieces prints out in the appropriate quantity columns on the quotation. Such information as miscellaneous or nonrecurring tooling charges, or delivery schedule can be independently entered.

Upon completion of the quotation preparation, the computer automatically tabs to the quotation journal in the primary printer, and posts a recap of the quotation for future reference. This recap contains the customer name, customer request number, quotation number, prices per quantity quoted on, and delivery time. As this journal is printing, the printer posts the same information to the customer account file to provide a reference of quotations outstanding, by customer.

Computers following this procedure process about 12 quotations per hour. Figure 23–7 illustrates the flow of information on this estimating and quotation writing procedure.

ELECTRONIC DATA GATHERING. Wherever electronic data is gathered, the hardware utilized must be portable, have sufficient storage capacity, and be able to interface with a computer for data transfer, storage, and analysis. Of course, the hardware must be able to accept the analysts' inputs, prompt the observer, and have the ability to audit the data being gathered.

TYPICAL COMPUTER PROGRAMS. The Automated Data Application and Maintenance (ADAM) system is a computer program written in PASCAL, and designed to simplify the application of various fundamental motion data systems. Figure 23–8 illustrates the total ADAM system. The core program accepts element codes belonging to the MTM-C,

FIGURE 23–7
Estimating and quotation-writing procedure

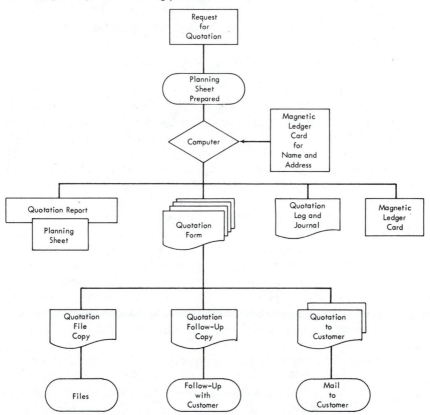

MTM-V, and MTM-2 families described in Chapters 19 and 20. After entering the appropriate element codes and their frequencies along with descriptive comments, analysts can edit the inputted data on the screen or obtain a hard copy printout. To edit the displayed input codes on the screen, analysts can delete, insert, or modify lines by moving the cursor.

ADAM accepts these six types of input:

1. Descriptive comments designated by C and followed by user-supplied comments.
2. Element codes. These are one or more of the codes belonging to the MTM–C or MTM–V data banks.
3. Frequencies. The input here is for element frequencies of more or less than one. In such cases, an "X" followed by the frequency is entered. Elements may be grouped within parentheses to indicate a common frequency.

FIGURE 23–8
Diagram of the ADAM system

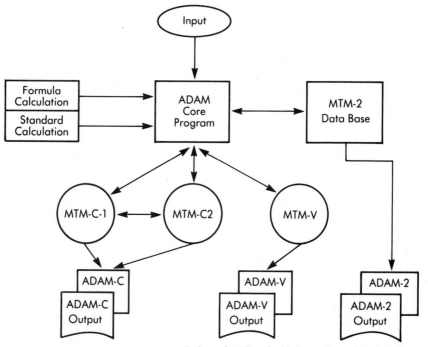

Redrawn from Douglas M. Towne General Analysis Corp.

4. Formulas. Formulas are designated by "F" and the designation is followed by the identifying formula number. Values for the variables in the formula are then inputted.
5. Standard data elements. These are designated by Z and this code is followed by the appropriate standard data reference number.
6. Grouping control. When a new group of elements follows, a "/" is inputted.

ADAM has two output formats: One of these locates the descriptive comment beside the first element code in the group. The other format assigns user-supplied descriptions immediately above all element codes.

ADAM prints a heading page for each analysis using any of these twenty heading entries:

1. Company
2. By
3. Date
4. Revision date
5. Department

11. Remarks
12. Typewriter (M or E)
13. Pitch (E or P)
14. PF&D allowance
15. Calculations

6.	Functional area	16.	Unit frequency
7.	Task name	17.	Misc. allowance #1
8.	Task code	18.	Value of misc. allowance #1
9.	Unit of measure	19.	Misc. allowance #2
10.	Approved	20.	Value of misc. allowance #2

Computations that can be printed on the heading page are:

Total TMU	Hours/hundred units
PF&D allowance	Units/minute
Standard TMU	Hours/1000 units
Standard minutes	Misc. allowance 1
Standard hours	Misc. allowance 2
Units/hour	Total allowances

Analysts can select the headings, computations, and the order of appearance on the heading page. ADAM-C utilizes both MTM-C level 1 and

FIGURE 23–9

GENERALIZED AUTOMATIC STANDARDS PROGRAM (GASP)   3916   Date:

Standard 01--For shearing flat stock, straight cut--up to & including 7" wide, 1¼" thick.
Standard 02--For shearing round stock--up to & including 2¼" diameter (EXCLUDE 3/4" diameter ABOVE C1025 H.R. & C1040 C.R.)
Standard 03--For combination cuts (1 R.E.)--(2 R.E.)--up to & including 2¼" wide & ½" thick.
Standard 04--For cut on curved knife (1 R.E.)--up to & including 3 3/4" wide & 7/8" thick.

STD. NO.	PART NUMBER	OPER. NO.	ORIGINAL BAR				SHEARED PIECE		SHEAR	ALDADB	OLD PRICE 2d Hlp	OLD PRICE 1st Hlp	OLD PRICE Oper.	RSONDE	ACCT
			THICK. -or--- GAUGE	WIDTH	LENGTH	FIN. PCS.	LENGTH	FIN. PCS.							
01	467335 R1	0 1 0	0.7500	6.0000	144.000	12	11.5000	0 1					0237	A	5
01	578036 R1	0 1 0	1.2500	25.000	912.000	0072	72.500	0 1				0259	0267	F	8
		0 1 0													
		0 1 0													
		0 1 0													
		0 1 0													
		0 1 0													
		0 1 0													
		0 1 0													
		0 1 0													
		0 1 0													
		0 1 0													

Col. 23 12 zone=cards only
         11 zone=forms only
         no zone=cards & forms

ALL DIMENSIONS MUST BE GIVEN IN INCHES AND DECIMAL PARTS OF AN INCH. (ORIGINAL BAR LENGTH MUST BE ROUNDED TO 3 DEC.)

If gauge is given, col 27 must contain # followed by gauge. Leave col 30 & 31 blank.

Equal Length=E
Extra stroke
  for C/K =1
2 R.E. (C/C) =2

Repairs =R   Package= P   Exports= E
Machine= M   Interworks= I   Special Labor= X

REASON CODE
10 A= A   Ex M   =E
10 B= B   Prc Adq =F
 9 C= C   Mtl Ch =G
10 D= D   Mtd Ch =H
          Eqp Ch =I

501=5
501=8

FIGURE 23–10

PIECEWORK PRICE AUTHORIZATION

DP NO. 20	ST NO. 01	EFFECTV. DY MN YR 03│09│64	CHANGE REASON 10–A A	TAG # 4737	GROUP TYPE SHR.PRS.	NO. 1	MACHINE TAG #│TYPE	PART NUMBER 467335│R1	M/P 01	USED	ON MODEL NO.

OPER NO.010	METHOD DESCRIPTION	SHEAR 12 PCS./BAR

MATERIAL	FORM	IN PER	OUT CYC	FIN PCS/ IN	OUT	DIMENSIONS				SET ANALYSIS BY
STEEL	FLAT BAR	1	12	12	1	.7500 THICKNS	6.0000 WIDTH	144.000 LENGTH	11.5000 SHR-LNG	G&P CANTN 062664 STND. MINUTES PER

CL. NO.	ELEMENT DESCRIPTION	FREQ/CY	CYCLE	FIN. PCE.
V 1	HANDLE ORIGINAL BAR/S TO SHEAR	1.000	.3214	
V 2	SHIFT BAR/S TO END GAUGE	10.000	.5693	
V 3	SHIFT BAR/S TO END GAUGE- END OF BAR/S ON TABLE	1.000	.0579	
V 4	BACK GAUGE TO MARK	1.000	.0693	
V 6	LAST PIECE/S AWAY	1.000	.0314	
V 8	LEVEL LOAD.	1.000	.1450	
V 9	GAUGE ONE PIECE	.060	.0110	
V10	TRUCKING	.060	.2400	
V11	BUNDLE CHANGE	.024	.0734	

DOWNTIME PERSONAL 24 CLEAN–UP 3 OTHER – TOTAL 27 MIN/DAY   TOTAL 1.5189

POSTED BY DEPT		ACCT #	OC. CLS	LG	O	H	STANDARD PER HUNDRED PIECES			PRODUCTION	CLASS	SUB TOTAL
20	COST STND.	501	SX0508	06	1	0	$ .2370	$ .2350	$	STD. 473.933 TOT. 447.427	VARIABLE MACHINE CONSTANT	

NEW PIECEWORK PRICE APPROVED BY		OLD PRICE	NEW PRICE	EXCS. COST	PIECES/HOUR	IDLE	
STANDARDS ENGINEER	FOREMAN	NEW PRICE RETROACTIVE TO EFFECTIVE DATE AS PER LABOR CONTRACT ARTICLE XII, SEC. 10 VALID ONLY WHEN PROPERLY APPROVED				STANDARD .1266 DOWN .0075 TOTAL .1341 HRS.CPS. .211	

MTM-C level 2 element codes which may be mixed. ADAM-V recognizes MTM-V element codes and the code "PT" (process time). ADAM-2 implements MTM-2 and is usable both independently and from within ADAM-C and ADAM-V.

One company has developed a modular program capable of processing a sequence of the known elements necessary to perform an operation. The input involves a description of the work location, equipment, raw material or stock, and desired piece dimensions.

Figure 23–9 shows an example of the input data for the development of a shearing standard. Figures 23–10 and 23–11 reflect the program output.

Where group technology is practiced, the use of the computer may be especially timesaving in the development of work standards. A computer program can describe all the operations for making a part characteristic of a product group. For example, gears of similar types may be classified by a family or group number. The input information for a specific gear includes the pertinent dimensions, the part name and number, the material,

FIGURE 23–11

PIECEWORK PRICE AUTHORIZATION

DP NO. 20	ST NO. 01	EFFECTV. DY MN YR 03 09 64	CHANGE REASON PRCADQ F	TAG # 4230	GROUP TYPE SHR. PRS.	NO. 1	MACHINE TAG # TYPE	PART NUMBER 518036 R1	M/P 01	USED	ON MODEL NO.

OPER. NO.010	METHOD DESCRIPTION	SHEAR 7 PCS./BAR

MATERIAL STEEL	FORM FLAT BAR	IN PER 1	OUT CYC 7	FIN PCS/ IN 7	OUT 1	DIMENSIONS					SET ANALYSIS BY G&P CANTN 062664
						1.2500 THICKNS	2.5000 WIDTH	192.000 LENGTH		27.2500 SHR-LNG	STND. MINUTES PER

CL. NO.	ELEMENT DESCRIPTION	FREQ/CY	CYCLE	FIN. PCE.
V 1	HANDLE ORIGINAL BAR/S TO SHEAR	1.000	.2961	
V 2	SHIFT BAR/S TO END GAUGE	6.000	.3906	
V 4	BACK GAUGE TO MARK	1.000	.0693	
V 7	LAST PIECE THRU KNIFE TO HELPER	1.000	.0310	
V 8	TRUCKING	.049	.1997	
V 9	BUNDLE CHANGE	.022	.0679	
	DOWNTIME PERSONAL 24 CLEAN-UP 3 OTHER - TOTAL 27 MIN/DAY	TOTAL	1.0548	

POSTED BY DEPT		ACCT #	OC. CLS.	LG	O	H	STANDARD PER HUNDRED PIECES			PRODUCTION		CLASS	SUB TOTAL
20	COST STND.	801	SX050A SX245	05 04	1 0	0 1	S .2670 .2590	S .2690 .2610		STD. 398.142 TOT. 375.704		VARIABLE MACHINE CONSTANT IDLE	

NEW PIECEWORK PRICE APPROVED BY		OLD PRICE	NEW PRICE	EXCS. COST	PIECES PER HR.	STANDARD	.1507
STANDARDS ENGINEER	FOREMAN	NEW PRICE RETROACTIVE TO EFFECTIVE DATE AS PER LABOR CONTRACT (ARTICLE XII, SEC. 10) VALID ONLY WHEN PROPERLY APPROVED				DOWN	.0090
						TOTAL	.1597
						HRS/CPS.	.251

and other descriptive details. The computer output includes a tabulation showing a sequence of the operations; the tooling involved; the feeds, speeds, and depths of cuts; and the standard times.

The United California Bank at Los Angeles stores its standards in a computer file. These standards are regularly recalled and used to evaluate work centers and to determine staffing policies.

Chapter 19 discusses the computerized work measurement system using Detailed and Ready Work-Factor developed by WOFAC Company, a division of Science Management Corporation, and the computerized Maynard Operation Sequence Technique (MOST) developed by H. B. Maynard and Company. Two other computerized work measurement systems: the Micro-Matic Methods and Measurement (4M Data) system, based on MTM-1, and available through MTM Association for Standards and Research and, the Automated Advanced Office Controls (Auto-AOC) system for clerical standards developed by the Robert E. Nolan Company.

Electronic data gathering and computers in standards development have significantly improved the output of the work measurement analysts. This trend will continue and result in the development of more refined work methods and equitable standards.

TEXT QUESTIONS

1. Explain in detail how a computerized methods and standards system can minimize standards clerical and computational work.
2. Outline how a change that affects several operations is introduced where a computerized data system is in place.
3. Which two major input files are needed in the installation of a work measurement system?
4. Which helpful management reports can be readily generated if an automated work measurement system is used?

GENERAL QUESTION

1. What disadvantages, if any, can you visualize if a computerized work measurement system is installed?

PROBLEMS

1. Write a computer program to provide output data for the painting of a line of castings from the equation:

$$T = [C_1 + 0.162S + n(v_1 + 0.0820)]1 + \alpha$$

where:

T = Time in minutes per casting
C_1 = 0.16 or 0.25 or 0.38, depending upon the grade of paint used
S = Surface area of casting in square inches
n = Number of holes that need to be plugged prior to painting
v_1 = 0.04 or 0.07 or 0.13, depending upon the plug diameter
α = 0.12 or 0.15 or 0.18, depending upon which paint booth is used

2. Prepare a flow diagram and write a computer program to estimate standard times for producing a variety of jigs and fixtures from the equation:

$$T = \sqrt{2.5 + 0.14W^2} + 0.75N + 1.82Q + 0.51S + 2.42B + 6.14 + t$$

where:

T = Time in hours to produce one jig or fixture
W = Gross weight of jig or fixture in pounds
N = Number of holes jig bored in the tool
Q = Number of clamps contained on the tool
S = Number of stops contained on the tool
B = Number of bushings contained on the tool
t = 0.25 when tolerances are closer than ±0.001 inch

= 0.10 when tolerances are between ±0.001 and ±0.005
= 0 when tolerances are greater than ±0.005

SELECTED REFERENCES

Bedworth, David D. *Industrial Systems: Planning, Analysis, Control.* New York: Ronald Press, 1973.

Lientz, Bennet P. *Computer Applications in Operations Analysis.* Englewood Cliffs, N.J.: Prentice-Hall, 1975.

Maisel, Herbert, and Guiliano Gnugnoli. *Simulation of Discrete Stochastic Systems.* Chicago: Science Research Associates, 1972.

Mishra, Davendra. "Computerized Work Measurement," In *Handbook of Industrial Engineering,* edited by Gavriel Salvendy. New York: John Wiley and Sons, 1982.

Towne, Douglas M. "ADAM-A Computer-Based System for Generating and Maintaining Labor Standards and Standard Data." *The Journal of Methods-Time Measurement* 7, no. 3.

Follow-Up Method and Uses of Time Standards

24

Follow-up is the last of the systematic steps in installing a methods improvement program. Although follow-up is as important as any of the other steps, it is the most frequently neglected step. Analysts have a natural tendency to consider the methods improvement program complete after developing time standards. However, a methods installation should never be considered complete. Follow-up is necessary to assure that the proposed method is being followed; that the established standards are being realized; and that the new method is being supported by labor, supervision, the union, and management. Follow-up usually results in additional benefits accruing from new ideas and new approaches that eventually stimulate the desire to improve a methods engineering program for the existing design or process. The procedure is to repeat the methods improvement cycle shortly after it is completed, so that each process and each design is continually being scrutinized for possible further improvement.

Without follow-up, it is easy for the proposed methods to revert to the original procedures. I have made innumerable methods studies where follow-up revealed that the method under study was slowly reverting to or had reverted to the original method. Humans are creatures of habit, and a work force must develop the habit of the proposed method if it is to be preserved. Continual follow-up is the only way to assure that the new method is maintained long enough to have all those associated with its details become completely familiar with its routines.

MAKING THE FOLLOW-UP

Although it is the normal function of production supervision to spot check and monitor the newly installed method, the extensiveness of this job seldom permits adequate time for completely effective follow-ups. Consequently, the methods and standards department should schedule regular follow-up.

The initial follow-up should take place approximately one month after the development of time standards for production jobs. A second follow-up should be made 2 months later, with the third follow-up 3 to 9 months after that. As brought out on page 473, the frequency of the audit should be based on the expected hours of application per year.

On each follow-up analysts should review the original method report and the development of the standard to be certain that all aspects of the proposed method are being followed. At times they find that portions of the proposed method are being neglected and that workers have reverted to the old ways. When this has happened, contact the supervisor immediately and determine why the unauthorized change has taken place. If no satisfactory reasons can be given for going back to the old method, insist that the correct procedure be followed. Tact coupled with firmness is essential; analysts need to display sales ability and technical competence.

Not only should the method be followed up, but also the performance of the operator. The worker's daily efficiency should be checked. Verify that the worker's performance is greater than standard. Performance should be evaluated with typical learning curves for the class of work. If the operator is not making the progress anticipated, make a careful study, including a conference with the operator to discover if any unforeseen difficulties have been encountered.

Review all factory layouts to ensure that ideal flow of materials and product is taking place. It is a good idea for the industrial engineering department to maintain detailed flow process charts for the major products being produced. If new equipment has been acquired in conjunction with the method, its capability should be audited regularly to assure that the anticipated productivity and performance are being realized.

Also, audit the job evaluation after the worker has performed the new method for six months. This review should assure that the compensation of all employees associated with the developed method is competitive with equivalent jobs in the area. Absentee rates, too, should be audited to obtain an additional measure of operator acceptance. A thorough and regular follow-up system can assure the expected benefits from the proposed method.

METHODS OF ESTABLISHING STANDARDS

Time standards may be determined in several ways:

1. By estimate (see Chapter 12).
2. By performance records (see Chapter 11).
3. By stopwatch time study (see Chapter 14).
4. By standard data (see Chapter 18).
5. By time study formulas (see Chapter 20).
6. By work sampling studies (see Chapter 21).
7. By queuing theory (see Chapter 22).

Methods 3, 4, 5, 6, and 7 give considerably more reliable results than either method 1 or 2. If standards are used for wage payment, they should be determined as accurately as possible. Consequently, standards determined by estimate and by performance records do not suffice. Of course, standards developed by performance records and estimates are better than no standards at all, and frequently can be used to exercise controls throughout an organization.

All of these methods have application under certain conditions, and all have limitations on accuracy and cost of installation. With regard to the more reliable methods of measurement, the following summary is helpful.

Stopwatch Time Study

Advantages.
1. Enables analysts to observe the complete cycle, providing an opportunity to suggest and initiate methods improvements.
2. Is the only method that actually measures and records the actual time taken by an operator.
3. Is more likely to provide coverage of those elements that occur less than once per cycle.
4. Quickly provides accurate values for machine-controlled elements.
5. Is relatively simple to learn and explain.

Disadvantages.
1. Requires the performance rating of a worker's skill and effort.
2. Does not force a detailed record of the total method being employed, including workplace layout, motion patterns, condition of materials, tools, and so forth.
3. May not provide accurate evaluation of noncyclic elements.
4. Bases the standard on a small sample since the standard is determined by one analyst studying one worker who is following one method.
5. Requires that the work be performed before the standard is established.

Predetermined Motion Time Data Systems

Advantages.
1. Force detailed and accurate descriptions of the workplace layout; motion patterns; and shape, size, and fit of components and tools.
2. Encourage work simplification to reduce standard times.
3. Eliminate performance rating.
4. Permit establishing methods and standards in advance of actual production.
5. Permit easy and accurate adjustments of time standards to accommodate minor changes in method.
6. Provide more consistent standards.

Disadvantages.
1. Depend on complete and accurate descriptions of the required methods for the accuracy of the time standard.
2. Require more time for the training of competent analysts.
3. Are more difficult to explain to workers, supervisors, and union officials.
4. May require more work-hours to establish standards for long-cycle operations.
5. Must use stopwatch, standard data, or formulas for process-controlled and machine-controlled elements.

Standard Data, Formulas, and Queuing Methods

Advantages.
1. Eliminate performance rating.
2. Establish consistent standards.
3. Permit establishing methods and standards in advance of actual production.
4. Allow standards to be established rapidly and inexpensively.
5. Permit an easy adjustment of time standards to accommodate minor changes in method.

Disadvantages.
1. May not accommodate small variations in method.
2. May require more skilled technicians for complex formulas.
3. Are more difficult to explain to workers than stopwatch procedure.
4. May result in significant inaccuracies if extended beyond the scope of the data used in their development.

Work Sampling

Advantages.
1. Eliminates tension caused by constant observation of the worker (when using stopwatch time study).
2. Represents typical or average conditions over a period of time where conditions change from hour to hour or day to day.
3. Permits the simultaneous development of standards for a variety of operations.
4. Is ideally suited to studies of machine utilization, activity analysis, and unavoidable and personal delays.
5. Can be used with performance rating to determine standard times.

Disadvantages.
1. Assumes that the workers uses an acceptable and standard method.
2. Requires that the observer identify and classify a wide range of delays and work activities.
3. Should be confined to constant populations.
4. Makes it more difficult to apply a correct performance rating factor than does stopwatch time study.

5. Accuracy of time standard is dependent on the number of random observations made, as well as upon the accuracy of the classification and recording of individual observations.
6. Requires accurate records of the hours worked and the units produced.

A further summary of the situations best suited to the various techniques follows:

Stopwatch time study.
1. Where there are repetitive work cycles of short to long duration.
2. Where new operations can be performed without standards until a study is performed.
3. Where a wide variety of dissimilar work is performed.
4. Where process control elements constitute a part of the cycle.

Predetermined motion-time data systems.
1. Where the work is predominantly operator controlled.
2. Where there are repetitive work cycles of short to medium durations.
3. Where it is necessary to plan work methods, including line balancing, in advance of production.
4. Where there has been controversy over the performance rating procedure.
5. Where there has been controversy over the consistency of standards.

Standard data, formulas, and queuing methods.
1. Where there is similar work of short to long duration.
2. Where there has been controversy over the performance rating procedure.
3. Where there has been controversy over the consistency of standards.

Work sampling.
1. Where it is necessary to establish delay allowances for various processes or departments.
2. Where there is considerable difference in the work content from cycle to cycle, as in certain shipping, material handling, and clerical activities.
3. Where activity studies are needed to show machine or space utilization, or the percentage of time spent on various activities.
4. Where standards are needed for crew activities which vary from cycle to cycle.
5. Where there is objection to stopwatch time study.

PURPOSES OF STANDARDS

Time standards are fundamental in the operation of any manufacturing enterprise or business. Time is the one common denominator from which all elements of cost evolve. In fact, everyone uses time standards for practically everything they do or want anyone else to do. Examples abound in everyday life: a man rising in the morning allows himself one hour to wash, shave, dress, eat breakfast, and get to work. This surely represents a time standard that has a bearing on the hours of sleep he gets. A football game involves but 60 minutes of playing time. The student reading this chapter allows herself so many minutes to cover the assignment. This standard affects the student's program for the remainder of the day.

We are particularly interested in time standards used in the effective operation of a manufacturing enterprise or business and the results of time study.

A Basis for Wage Incentive Plans

Time standards are usually thought of in their relation to wage payment. As important as this is, standards have many other uses in the operation of an enterprise. However, the need for reliable and consistent standards is most pronounced in the wage payment area. Without equitable standards, no incentive plan that compensates in proportion to output can possibly succeed. Without a yardstick, how can anyone measure individual performance? Standardized methods and standard times comprise the yardstick that forms a basis for wage incentive application.

A Common Denominator in Comparing Various Methods

Since time is a common measure for all jobs, time standards are a basis for comparing various methods of doing the same piece of work. For example, assume an operator thought it might be advantageous to install broaching on a close tolerance inside diameter rather than to ream the part to size as is currently being done. To make a sound decision on the practicality of the change, time standards are necessary. Without reliable standards, analysts would be groping in the dark.

A Means for Securing an Efficient Layout of the Available Space

Time is the basis for determining how much of each kind of equipment is needed. Only by knowing the exact requirements for facilities, can management make the best possible utilization of space. If a company requires 10 milling machines, 20 drill presses, 30 turret lathes, and 6 grinders in a particular machining department, then the manager can plan for the layout of this equipment to best advantage. Without time standards, the company could over-provide for one facility and underprovide for another, inefficiently utilizing the space available.

In determining the size of storage and inventory areas managers consider the length of time a part is in storage and the demand for the part. Again, time standards are the basis for determining the size of such areas.

FIGURE 24–1

Time standards allow the determination of projected direct labor requirements

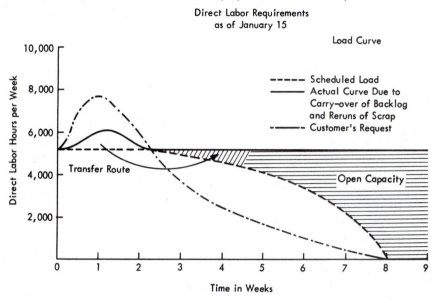

A Means for Determining Plant Capacity

Through time standards, not only is machine capacity determined, but also department and plant capacity. Once the available facility hours and the time required to produce a unit of product are known, it is a matter of simple arithmetic to estimate product potential. For example, if the bottleneck operation in the processing of a given product required 15 minutes per piece, and if 10 facilities for this operation existed, then the plant capacity based on a 40-hour per week operation on this product would be:

$$\frac{40 \text{ hours} \times 10}{0.25 \text{ hours}} = 1,600 \text{ pieces per week}$$

Figure 24–1 illustrates a weekly graphic analysis of the direct labor requirements for a specific industrial plant. This chart clearly indicates when the plant is in a position to produce new customer orders.

A Basis for Purchasing New Equipment

Since time standards allow analysts to determine machine, department, and plant capacity, they also provide the necessary information for determining how many of which facilities are necessary for a given volume of production. Accurate comparative time standards also highlight the advantages of one facility over its competitors. For example, a plant may find it necessary to purchase three additional single-spindle, bench-type drill presses. By reviewing available standards, plant managers can procure the style and design of drill press that produces the most favorable output per unit of time.

A Basis for Balancing the Working Force with the Available Work

Having concrete information on the required volume of production, as well as the time needed to produce a unit, enables analysts to determine the required labor force. For example, if the production load for a given week is evaluated as 4,420 hours, then the company needs 4,420/40, or 111 operators. This use of standards is especially important in a retrenching market, where the volume of production is going down. When overall volume diminishes, without a yardstick to determine the actual number of people needed to perform the reduced load, the entire working force may slow down so that the available work will last. Unless the working force balances with the available volume of work, unit costs rise progressively. It is only a matter of time, under these circumstances, until production operations are performed at a substantial loss, thus necessitating increased selling prices and further reductions in volume. The cycle repeats until it becomes necessary to close the plant.

In an expanding market, it is equally important to be able to budget labor. Rising customer demands necessitate a greater volume of personnel. Companies must determine the exact number and type of personnel to add to the payroll so that these workers can be recruited in sufficient time to meet customer schedules. If accurate time standards prevail, it is a matter of simple arithmetic to convert product requirements to departmental work-hours.

Figure 24–2 illustrates how overall plant capacity may be increased in an expanding market. Here the plant anticipates doubling its work-hour capacity between January and November. This budget projects the scheduled contracts in terms of work-hours and allows a reasonable cushion (crosshatched section) for receiving additional orders.

Besides allocating plant labor requirements, time standards serve in budgeting the labor needs of specific departments. Figure 24–3 illustrates the budgeting of spindle hours in the multiple winding department of one of the Owens-Corning Fiberglas plants. Four products areas are served: "streamline," "serving," "straight edge," and "fishing rod." In view of present customer requirements, 75 spindles have been allotted for the multiple winding of "streamline" products, 42 spindles for "serving," 14 for "straight edge," and 15 for "fishing rod." Based on this budgeting of spindles and work-hours, the load for "streamline" extends to the end of the first week in November, "serving" to the middle of October, "straight edge" to the middle of November, and "fishing rod" to the last week in October. Although customer requirements fluctuate, time stan-

FIGURE 24–2

Chart illustrating actual projected work-hour load and budgeted work-hour load

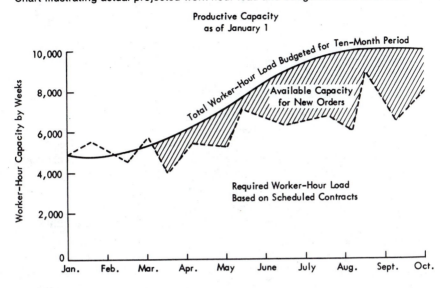

FIGURE 24–3

Chart illustrating the budgeting of the spindle hours of a specific department to best meet customer needs

Load chart, Huntingdon Plant, Owens–Corning Fiberglas Corp. multiple winding department

FACILITY	SPS.	SP. HR.			SEPTEMBER					OCTOBER					NOVEMBER			
Streamline	75	54,000			108	216	324	432	540	648	756	864	472	1080	1188	1276	1404	1512
Serving	43	30,960			62	124	186	248	307	372	434	496	558	620	682	744	806	888
Straight edge	14	10,080			20	40	60	80	100	120	140	160	180	200	220	240	260	280
Fishing rod	15	10,800			22	44	66	88	108	130	152	178						

Note: Individual orders scheduled by week in area
immediately above allocated spindle hours graph.

KEY: ▨ Ordered spindle hours
▦ Allocated spindle hours

dards make it possible to adjust the labor budget of this department to best meet all customer requirements. The solid bar on this chart shows the material that has been allocated to the various product classes as of September 1.

Valid time standards keep the working force in proportion to the volume of production required, thus controlling costs and maintaining operation in a competitive market.

IMPROVING PRODUCTION CONTROL. Production control is the phase of an operation that schedules, routes, expedites, and follows up production orders in an effort to achieve operating economies and satisfy customer requirements. The whole function of controlling production is based on determining where and when the work can be done. This obviously cannot be achieved without a concrete idea of "how long."

Scheduling, one of the major functions of production control, is usually handled in three degrees of refinement: (1) long-range or master scheduling; (2) firm order scheduling; and (3) detailed operation scheduling, or machine loading.

Long-range scheduling is based on the existing volume and the anticipated volume of production. In this case, specific orders are not given any particular sequence, but are merely lumped together and scheduled in appropriate time periods. Firm order scheduling involves scheduling existing orders to meet customer demands and still operating in an economical fashion. Here, workers assign degrees of priority to specific orders, and anticipated shipping promises evolve from this schedule. Detailed operation scheduling, or machine loading, is assigning specific operations day by day to individual machines. This scheduling is planned to minimize setup time and machine downtime while meeting firm order schedules. Figure 24–4 illustrates the machine loading of a specific department for one week. It documents that considerable capacity exists on milling machines, drill presses, and internal thread grinders.

No matter what the degree of refinement in the scheduling procedure, scheduling would be utterly impossible without time standards. The success of any schedule is in direct relation to the accuracy of the time values used in determining the schedule. If time standards do not exist, schedules formulated only on judgment cannot be expected to be reliable. Time standards help predetermine the flow of materials and of work in progress, thus forming a basis for accurate scheduling.

Work center routings provide process information to the shop floor, and convey time data to the shop floor control system. They are the preferred means of disseminating job standards to the employees.

Expediting and follow-up involve performance reporting. Modern production control systems utilize time standards from a variety of sources to generate performance reports. Today in many plants, shop floor time data collection devices are computers that allow line supervisors or staff to review the status of any job based on the most recent data submitted.

FIGURE 24–4

Machine loading of a machining department for one week (notice that several schedules depend on receiving additional raw material)

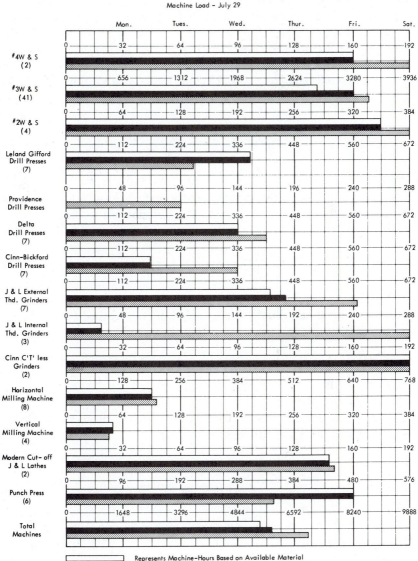

DORBEN MANUFACTURING COMPANY

Machine Load – July 29

Represents Machine-Hours Based on Available Material

Represents Machine-Hours Based on Prospective Material

Represents Worker-Hours

These modern follow-up devices help assure quality production control with improved edit checks and less paperwork.

THE ACCURATE CONTROL AND DETERMINATION OF LABOR COSTS. With reliable time standards, a plant does not have to be on incentive wage payment to determine and control its labor costs. The ratio of departmental earned production hours to departmental clock hours reveals the efficiency of a specific department. The reciprocal of the efficiency multiplied by the average hourly rate gives the hourly cost of standard production. For example, the finishing department in a plant using straight daywork may have had 812 clock hours of labor time, and in this period it has earned 876 hours of production. The departmental efficiency would then be:

$$E = \frac{He}{Hc} = \frac{876}{812} = 108 \text{ percent}$$

If the average daywork hourly rate in the department was \$16.80, then the hourly direct labor cost based on standard production would be:

$$\frac{1}{1.08} \times \$16.80 = \$15.56$$

In a second example assume that in another department the clock hours were 2,840 and that the hours of production earned for the period were only 2,760. In this case, the efficiency would be:

$$\frac{2,760}{2,840} = 97 \text{ percent}$$

and the hourly direct labor cost based on standard production with an average daywork rate of \$16.80 would equal:

$$\frac{1}{0.97} \times \$16.80 = \$17.32$$

In the latter case, management would realize that its labor costs were running \$0.52 per hour more than base rates and could increase supervision to bring total labor costs into line. In the first example, labor costs were running less than standard, which would allow a downward price revision, increasing the volume of production, or making some other adjustment suitable to both management and labor. Figure 24–5 illustrates a direct labor variance report indicating departmental performance above and below standard.

FIGURE 24–5

Weekly report illustrating departmental performance in a specific manufacturing plant (regular print indicates hours and percentages earned over standard; italics indicate to what degree standard has not been achieved)

DIRECT LABOR VARIANCE HOURS

Week Ending June 3

			Efficiency variance					Percent total variance over-under standards				
			Week ending			Weekly average		Week ending				
No.	Name	Allowed direct labor	6/3	5/26	5/19	First qtr.	Apr.	6/5	5/26	5/19	4 weeks, April	First qtr.*
11	Machine shop	892	204	29	110	33	3	22.9	2.5	9.2	0.2	2.2
12	Wire brush	178	–	–	–	–	–	–	–	–	–	–
19	Punch press	41	18	8	–	6	3	43.9	9.5	–	4.5	9.1
20	Rubber milling	21	101	43	124	21	51	481.0	18.1	172.2	37.4	13.8
31	Rubber fabricating	1,183	36	29	12	116	59	3.0	1.5	0.7	3.2	5.2
35	Pilot plant	53	–	–	–	–	–	–	–	–	–	–
39F	Finishing	339	60	107	27	42	50	17.7	18.5	5.8	10.0	6.6
39P	Paint	23	1	9	12	8	3	4.3	12.7	23.1	11.0	26.4
40	Assembly	13	1	15	15	14	4	7.7	28.3	25.9	6.0	19.9
50	Reclaim	20	–	–	–	–	–	–	–	–	–	–
65	Toolroom	–	–	–	–	–	–	–	–	–	–	–
	Total—This week	2,763	217	148	192	104	59	7.9%	3.7%	4.5%	1.2%	1.9%
	Last week	4,462										

Note: The latest planning and method changes are reflected in all groups.
* First quarter includes the 13 weeks from January 1 through March 31.

Requisite for Standard Cost Methods

Standard cost methods refer to accurately determining cost in advance of production. The advantage of being able to predetermine cost is obvious. Most contracts today are let on a firm cost basis, necessitating that the producer predetermine costs. By having time standards on direct labor operations, producers can preprice those elements entering into the prime cost of the product. (Prime cost is usually thought of as the sum of the direct material and direct labor costs.)

As a Basis for Budgetary Control

Budgeting is the establishment of a course of procedure: the majority of budgets are based on the allocation of money for a specific center or area of work. Thus, for a given period a company may establish a sales budget, a production budget, and so forth. Since money and time are definitely related, any budget is a result of standard times regardless of how the standards were determined.

As a Basis for the Supervisory Bonus

Wage incentives are discussed at length in Chapter 25. At this time, it will suffice to point out that any type of supervisory bonus keyed to productivity depends directly on having equitable time standards. Since workers receive more and better supervisory attention under a plan where the supervisory bonus is related to output, the majority of supervisory plans give consideration to worker productivity as the principal criterion for the supervisory bonus. Other factors usually considered in the supervisory bonus are indirect labor costs, scrap cost, product quality, and methods improvements.

Quality Standards Are Enforced

Establishing time standards forces the maintenance of quality requirements. Since production standards are based on the quantity of acceptable pieces produced in a unit of time, and since no credit is given for defective work turned out, there is a constant intense effort by all workers to produce only good parts. If an incentive wage payment plan is in effect, operators are compensated for good parts only; to keep their earnings up they keep their scrap down. Sampling inspection is invariably more effective under incentive conditions. The operator has already assured that the quality of each piece turned out is satisfactory before he/she releases it.

When some of the pieces are defective, either the operator who produced the parts is held responsible for their salvage or the worker's earnings are adjusted so that he or she is compensated only for satisfactory parts.

Personnel Standards Are Raised

Where standards are used, there is a natural tendency to "put the right person on the right job," so that the standards established are either met or exceeded. Placing employees on work for which they are best suited goes a long way toward keeping them satisfied. Workers are motivated when they know the established goals and understand how these goals dovetail with organizational objectives.

Problems of Management Are Simplified

Time standards are accompanied by many control measures, such as scheduling, routing, material control, budgeting, forecasting, planning, and standard costs. Having controls on practically every phase of an enterprise, including production, engineering, sales, and cost, minimizes the problems of management. By exercising the "exception principle," in which attention is given only to the items deviating from the planned course of events, management can confine its efforts to only a small segment of the total activity of the enterprise.

For example, Figure 24–6 illustrates a Weekly Lost Time Analysis developed so that management can take positive action when scheduled hours are not attained as planned. Note that a goal of 9.50 percent hours lost of scheduled time was established for 1991 as compared to 10.6 percent lost hours in 1990.

Figure 24–7 illustrates helpful management information in connection with failure analysis for seven major subsystems of a large compressor system. The data obtained from a work sampling study illustrates the Pareto distribution. (A distribution that reflects that the major part of an activity is accounted for by a minority.)

Governmental operations have found time standards extremely helpful. Added emphasis was given to the need for a standards program in all governmental agencies when President Harry S. Truman, by Executive Order 10072 in July 1949, and the 81st Congress, by Public Law 429 passed the same year, emphasized the need for the continuous examination and review of governmental operations to ensure the achievement of planned programs in each department. Title X of Public Law 429 makes specific provisions for establishing an efficiency awards system in each governmental agency. Time standards provide a means for evaluating the individual or group of individuals submitting proposals, and also enable those submitting entries to examine the potential of their ideas. As mentioned in Chapter 2, MIL–STD 1567A Work Measurement Guidance Appendix was finalized in 1986. This military standard requires the application of a disciplined work measurement program as a management tool to improve productivity.

FIGURE 24–6
ETF weekly lost time analysis. July 14–20, 1991

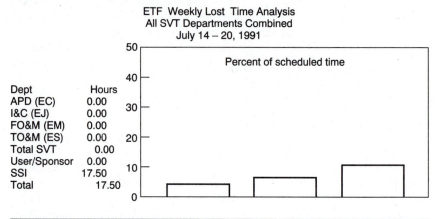

ETF Weekly Lost Time Analysis
All SVT Departments Combined
July 14 – 20, 1991

Dept	Hours
APD (EC)	0.00
I&C (EJ)	0.00
FO&M (EM)	0.00
TO&M (ES)	0.00
Total SVT	0.00
User/Sponsor	0.00
SSI	17.50
Total	17.50

		This Week	Last Week	FY 1991 TO-DATE	FY 1990
Scheduled hours	Total	45.67	55.91	1666.08	3062.24
Hours lost	All SVT	0.00	2.25	101.60	324.56
Percent	All SVT	0.00	4.02	6.10	10.60

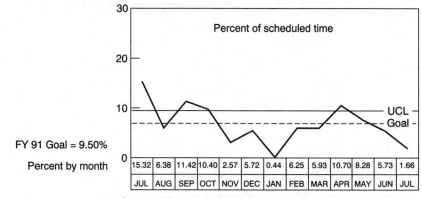

FY 91 Goal = 9.50%

Percent by month

	JUL	AUG	SEP	OCT	NOV	DEC	JAN	FEB	MAR	APR	MAY	JUN	JUL
	15.32	6.38	11.42	10.40	2.57	5.72	0.44	6.25	5.93	10.70	8.28	5.73	1.66

Courtesy of Ramesh C. Gulati, Sverdrup Technology, Inc. AEDC Group.

Service to Customers is Improved

Experience has proven that those companies that have developed sound standards based upon measurement are more likely to meet scheduled delivery dates of their products. The use of time standards allows the introduction of up-to-date production control procedures, with the result-

FIGURE 24–7

Failure analysis for seven major subsystems of a large compressor system and human error

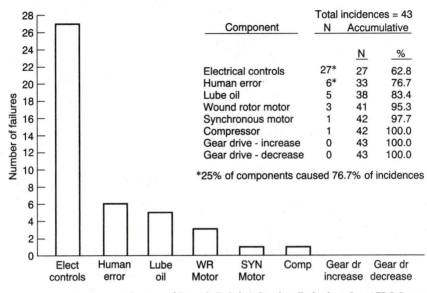

Component	Total incidences = 43		
	N	Accumulative	
		N	%
Electrical controls	27*	27	62.8
Human error	6*	33	76.7
Lube oil	5	38	83.4
Wound rotor motor	3	41	95.3
Synchronous motor	1	42	97.7
Compressor	1	42	100.0
Gear drive - increase	0	43	100.0
Gear drive - decrease	0	43	100.0

*25% of components caused 76.7% of incidences

Courtesy of Ramesh C. Gulati, Sverdrup Technology, Inc. AEDC Group.

ing advantage to customers who get their merchandise when they want and need it. Also, time standards tend to make any company more time and cost conscious; this usually results in lower selling prices. As has been explained, quality is maintained under a work standards plan, thus assuring the customers of more parts made to required specifications.

CONCLUSION

Thorough and regular follow-up assures the expected benefits from the proposed method.

There are, of course, many uses of time standards in all areas of any enterprise. Probably the most significant result of time standards is the maintenance of overall plant efficiency. If efficiency cannot be measured, it cannot be controlled, and without control it will diminish markedly.

Once efficiency goes down, labor costs rapidly rise, and the result is eventual loss of competitive position in the market. Figure 24–8 illustrates the relationship of labor costs to efficiency in one leading manufacturer's automobile accessories business. By establishing and maintaining effective standards, a business can standardize direct labor costs and control overall costs.

FIGURE 24–8
Relationship of labor costs to efficiency

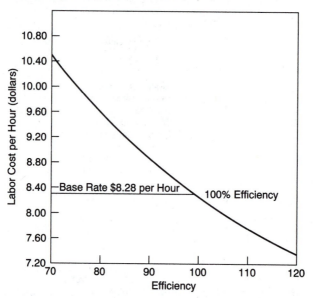

Developed from data furnished by AC Spark Plug Division, General Motors Corp.

TEXT QUESTIONS

1. Explain the use of learning curves in the "follow-up" step of the systematic approach to motion and time study.
2. List the different ways time standards may be determined.
3. How can valid time standards help develop an ideal plant layout?
4. Explain the relationship between time standards and plant capacity.
5. In what way are time standards used for effective production control?
6. How do time standards allow the accurate determination of labor costs?
7. How does developing time standards help maintain the quality of products?
8. In what way is customer service improved through valid time standards?
9. What is the relationship between labor cost and efficiency?
10. If a daywork shop was paying an average rate of $12.75 per hour and had 250 direct labor employees working, what would be the true direct labor cost per hour if during a normal month 40,000 hours of work were produced?
11. How are management problems simplified through the application of time standards?
12. In what way did President Harry S. Truman increase the use of time standards in governmental operations?

13. Explain how inventory and storage areas can be predicted accurately.
14. Explain what is meant by the Pareto distribution. Give some examples of this distribution with which you are familiar.

GENERAL QUESTIONS

1. Name several other uses of time standards not mentioned in this chapter.
2. What is the relationship between the accuracy of time standards and production control? Does the law of diminishing returns apply?
3. How does work measurement improve the selection and placement of personnel?
4. When is it no longer necessary to follow up the installed method?
5. Do you believe that the typical worker today is motivated if the company goals and objectives are clearly presented to him or her? Discuss and give examples.

PROBLEMS

1. In the XYZ Manufacturing Company, direct labor cost is based on the efficiency relationship illustrated in Figure 24–8. For a given product line, the selling price was based on the company's running at 95 percent efficiency. How much additional profit did the company realize if the actual efficiency turned out to be 110 percent? Profit was originally estimated at 10 percent of the total cost. On this product line, the total overhead was estimated to be 100 percent of prime cost (direct labor plus direct material). Material cost averaged $5 per unit of output, and direct labor averaged 0.50 hours per unit of output.
2. What would be the costing rate in the finishing department of the XYZ Company, where 32 employees work? Twenty-seven of these employees are on standard and have been averaging 1,310 hours earned per standard 40-hour week. The daywork hourly rate for these workers is $13.00 per hour. The remaining five workers are line supervisors who are not on standard. Their daywork hourly rate is $15.20 per hour.
3. In the XYZ Company, management is considering going from two 8-hour shifts per day to three 8-hour shifts per day or two 10-hour shifts per day in order to increase capacity.

    Management realizes that shift start-up results in a loss of productivity that averages 0.5 hour per employee. The premium for the the third shift is 15 percent per hour. Time over eight hours worked per day gives the operator 50 percent more pay. To meet projected demands, it is necessary to increase the work-hours of production by 25 percent. In view of insufficient space and capital equipment, this increase cannot be accommodated by increasing the employees on either the first or the second shift.

    How should management proceed?

SELECTED REFERENCES

Buffa, Elwood S. *Modern Production Operations Management,* 6th ed. New York: John Wiley & Sons, 1980.

Graham, C. F. *Work Measurement and Cost Control.* Oxford, England: Pergamon Press, 1965.

Greene, J. H. *Production and Inventory Control: Systems and Decisions.* Rev. ed. Homewood, Ill.: Richard D. Irwin, 1967.

Magee, J. F., and D. M. Boodman. *Production Planning and Inventory Control,* 2nd ed. New York: McGraw-Hill, 1967.

Moore, F. G., and R. Jablonski. *Production Control.* 3rd ed. New York: McGraw-Hill, 1969.

Mundel, Marvin E. *A Conceptual Framework for the Management Sciences.* New York: McGraw-Hill, 1967.

SELECTED SOFTWARE

Nicks, J. E. *Cost Estimating, Basic Programming Solutions for Manufacturing.* Dearborn, Mich.: Society of Manufacturing Engineers, 1982.

Wage Payment

25

The principal factors in creating highly productive and satisfied workers are reward and recognition for effective performance. The reward must be meaningful to employees whether it is financial, psychological, or both. Experience has proved that workers do not give extra or sustained effort unless some incentive, either direct or indirect, is in the offing. In one form or another, incentives have been used in business and industry for many years. Today, with the increasing need for American business and industry to improve productivity to retard inflation and maintain or improve their position in the world market, management should not overlook the advantages of wage incentives. Only about 25 percent of manufacturing employees are now on incentive. If this figure were doubled in the next decade, the improvement in our national productivity could be phenomenal.

With fringe benefits becoming increasingly significant (today, they average about 30 percent of direct labor), these costs must be spread over more units of output. At present, fringe benefits average approximately 19 vacation days a year, 9 paid holidays a year, $13,000 in life insurance, disability insurance at 72 percent of salary, and sick pay up to 85 days. See the representative paternalistic plan on page 748.

In a broad sense, all incentive plans that increase the employees' production may be referred to as flexible wage plans. Four types of flexible plans will be discussed briefly: (1) piece rates and standard labor hour plans; (2) gainsharing plans (Scanlon, Rucker, Improshare); (3) employee stock ownership plans (ESOPs); (4) profit-sharing plans.

Before analysts design wage payment plans for specific plants, they should review the strengths and weaknesses of past plans including sound day rate plans and plans other than financial.

Benefit	Approximate cost
Health insurance	$ 400.00
Health insurance (2 dependents)	700.00
Vision insurance	55.00
Dental insurance	90.00
Vacations (2 weeks per year)	600.00
Personal leave (5 days per year)	300.00
Holidays (10 per year)	600.00
Term life insurance ($35.00)	250.00
Accidental death & dismemberment	20.00
Long term disability (60%)	150.00
Pension (9%)	1,500.00
Educational expense reimbursement	60.00
Total	$5,025.00

SOUND DAY RATE PLANS

Sound day rate plans compensate the employee on the basis of number of hours worked times an established hourly base rate. Company policies that stimulate employee morale and result in good productivity without directly relating compensation to production fall into the sound day rate classification. Such overall company policies are fair and relatively high base rates based upon job evaluation, merit rating, sound suggestion systems, a guaranteed annual wage, and relatively high fringe benefits build healthy employee attitudes, which tend to stimulate and increase productivity.

All day rate plans have the weakness of allowing too broad a gap between employee benefits and productivity. After a period of time, employees take for granted the benefits bestowed on them and fail to realize that the means for their continuance must result entirely from their productivity. The theories, philosophies, and techniques of day rate plans are beyond the scope of this text; for added information in this area, the student is referred to books on personnel administration.

PLANS OTHER THAN FINANCIAL

Nonfinancial incentives include any rewards that have no relation to pay, and yet improve employees' spirit to such an extent that added effort is evident. Under this category would come such company policies as periodic shop conferences, quality control circles, frequent talks between the supervisor and employee, proper employee placement, job enrichment, job enlargement, nonfinancial suggestion plans, the maintenance of ideal working conditions, and the posting of individual production records. Effective supervisors and capable, conscientious managers also use many other techniques such as having the employee take his or her spouse out for dinner and send the manager the bill, providing tickets to sporting events or theater, arranging special trips to trade shows or other companies for exposure to "state of the art" technology. All of these approaches seek to motivate by improving the work environment. They frequently are referred to as quality of work life plans.

FLEXIBLE COMPENSATION PLANS

Flexible compensation plans include all plans in which the worker's compensation is related to output. This category includes both simple individual incentive plans and group incentive plans. In simple individual plans, each employee's performance for the period governs compensation. Group plans are applicable to two or more persons working as a team on operations that are dependent on one another. In these plans, each employee's compensation within the group is based on his or her base rate and on the performance of the group for the period.

The incentive for high or prolonged individual effort is not nearly as great in group plans as in individual plans. Hence, industry favors individual incentive methods. In addition to lower overall productivity, group plans have other drawbacks: (1) personnel problems brought about by nonuniformity of production coupled with uniformity of pay, and (2) difficulties in justifying base rate differentials for the various opportunities within the group.

Of course, group plans do offer some decided advantages over individual incentive: (1) ease of installation brought about through ease of measuring group rather than individual output; and (2) reduction of cost in administration through reduced paperwork, less verification of inventory in process, and less in-process inspection.

In general, individual incentive plans foster higher rates of production and lower the unit cost of products. If practical to install, the individual incentive plan should be given preference over group systems. On the other hand, the group approach works well where it is difficult to measure individual output and where individual work is variable and frequently performed in cooperation with another employee. For example, where four men are jointly working together in the operation of an extrusion press for extruding brass rods, it would be virtually impossible to install an individual incentive system, but a group plan would be applicable.

Plans in Which Employees are Rewarded in Direct Proportion to Output

PIECEWORK. Piecework implies that all standards are expressed in money and that operators are rewarded in direct proportion to output. Under piecework, the day rate is not guaranteed. Today, piecework is not used, since federal law requires a minimum guaranteed hourly rate. Prior to World War II, piecework was used more extensively than any other

FIGURE 25–1
Operator earnings and unit direct labor cost under piecework

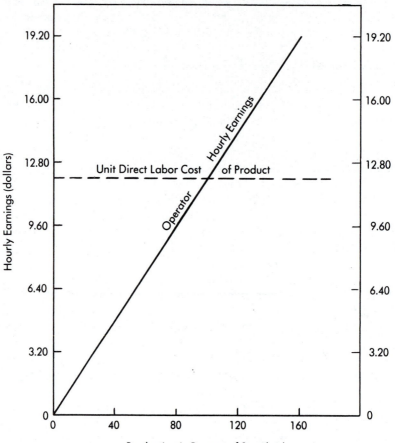

incentive plan. The reasons for the popularity of piecework are that it is easily understood by the worker, easily applied, and one of the oldest wage incentive plans. Figure 25–1 illustrates graphically the relationship between an operator's earnings and unit direct labor costs under a piecework plan.

THE STANDARD HOUR PLAN.   The standard hour plan with a guaranteed base rate, established by job evaluation, is by far the most popular incentive plan in use today. The fundamental difference between the standard hour plan and the piecework plan is that under the former, standards are expressed in time rather than money. Operators are rewarded in direct proportion to output.

Graphically, the relationship between operator earnings and unit direct labor cost, when plotted against production and money, would be identical to piecework. For example, a standard may be expressed as 2.142 hours per 100 pieces. It is an easy operation to calculate either the money rate or the operator's earnings, once the base rate is known. If the operator had a base rate of $12, then the money rate of this job would be: (12.00)(2.142) = $25.70 per hundred, or $0.257 per piece. Assume that an operator produced 412 pieces in an eight-hour working day; earnings for the day would be: ($12.00)(2.142)(4.12) = $106.01, and hourly earnings would be: $106.01/8, or $13.25. The operator's efficiency for the day, in this case, would then be: (2.142)(4.12)/8, or 110 percent.

The standard hour plan offers all the advantages of piecework and eliminates the major disadvantages. However, it is somewhat more difficult for workers to compute earnings under this plan than if standards were expressed in money. The principal advantage is, of course, that standards are not changed when base rates are altered. Thus, this plan reduces clerical work over a period of time when compared to the piecework plan. Moreover, the term *standard hour* is more palatable to workers than is the term *piecework,* and with standards expressed in time, the money earned by workers is not quite so closely linked with time study practice. For these reasons, there has been a marked growth in the popularity of standard hour plans where the base rate is guaranteed.

MEASURED DAYWORK.   During the early 1930s, shortly after the era of efficiency experts, organized labor tried to get away from time study practice and, in particular, piece rates. At this time, measured daywork became popular as an incentive system that broadened the gap between the establishment of a standard and worker's earnings. Many modifications of measured daywork installations are in operation today, and the majority of them follow a specific pattern. First, job evaluations establish base rates for all opportunities falling under the plan. Second, some form of work measurement determines standards for all operations. Third, analysts keep a progressive record of each employee's efficiency for usually one to three months. This efficiency, multiplied by the base rate, forms the basis of a guaranteed base rate for the next period. For example, the base rate of a given operator may be $12.00 per hour. Assume that the governing performance period is one month, or 173 working hours. If, during the month, the operator earned 190 standard hours, his or her efficiency for the period would be 190/173, or 110 percent. Then, in view of the performance, the operator would receive a base rate of (1.10)(12.00), or $13.20, for every hour worked during the next period, regardless of performance. However, achievement during this period would govern the base rate for the succeeding period.

In all measured daywork plans today, the base rate is guaranteed; thus, an operator falling below standard (100 percent) for any given period would receive the base rate for the following period.

The length of time used in determining performance usually runs three months to diminish the clerical work of calculating and installing new guaranteed base rates. Of course, the longer the period, the less incentive effort can be expected. When the spread between performance and realization is too great, the effect of incentive performance diminishes.

The principal advantage of measured daywork is that it takes the immediate pressure off workers. They know what their base rates are; they realize that, regardless of performance, they will receive that amount for the period.

The limitations of measured daywork are apparent. First, because of the length of the performance period, the incentive feature is not particularly strong. Second, to be effective, it places a heavy responsibility on supervisors for the maintenance of production above standard. Otherwise, the employee's performance drops, thus lowering the base rate for the following period and causing employee dissatisfaction. Third, keeping detailed rate records and making periodic adjustments is costly in all base rates. In fact, as much clerical work is involved under measured daywork as under any straight incentive plan where employees are rewarded according to output.

Although measured daywork is classified as a wage payment plan in which employees participate in all the gain above standard, this method of wage payment is really a hybrid of the other "one-for-one" plans. As workers' seniority increases, total earnings approximate the amount they would have earned if paid in direct proportion to output. Therefore, the plan is, in effect, similar to other types discussed that fall under this category.

GAIN SHARING PLANS.   Gain sharing plans, also known as productivity sharing plans, are those plans characterized by sharing the benefits of improved productivity or cost reduction and/or quality improvement. About 1,000 firms throughout the United States today have some form of gain sharing plan. In many of these plants gain sharing supplements rather than replaces existing compensation systems. However, in one survey of 508 establishments, 28 percent stated they would not need gain sharing if they had profit sharing. Most progressive managements accept the principle of rewarding employees for improvements in productivity and/or cost whether or not the improvements are due to performance above normal or to improvements in work methods. Proponents of cost saving sharing plans point out that, from the motivational viewpoint, direct incentives miss many improvement opportunities. These include savings in material, both direct and factory supplies, and myriad methods changes that result in improvements. Most of these plans tie the incentive to a "production value" above standard. Production value is the monetary difference between sales and purchases, and consequently represents the value that all employees have contributed. Previous performance, for at least a year, is considered standard. Future "value added" can be improved by:

1. Savings in raw materials, purchased parts, supplies, fuel, and power.
2. A reduction in scrap and less rework.
3. A reduction in customer allowances including less warranty service.
4. An increase in the volume of output.
5. Greater productivity without increases in time inputs by hourly and/ or salaried staff.

Under plans of this type, management computes incentives on a monthly basis. Customarily, only two thirds of the incentive earned in a given pay period is distributed. The remaining third is placed in a reserve fund to be used any month that performance falls below standard.

Cost saving sharing plans are relatively simple to establish since they do not require the development of engineered time standards. These types of plans warrant consideration as a means of stimulating workers' performance and enhancing the competitive advantage of the business. The three productivity sharing plans discussed here—Scanlon, Rucker, and Improshare—differ in the formula used to compute productivity savings and in the implementation method. The Scanlon and Rucker plans measure the payroll of the firm against total dollar sales and compare it to the past average of several years. The Improshare plan measures output against total hours worked. Thus, the Scanlon and Rucker plans use dollars as the measurement unit while Improshare uses hours. All three of these productivity plans are flexible regarding the personnel included in the plan. Direct and indirect workers, as well as all levels of management, may be included.

SCANLON PLAN. During the Great Depression, Joseph Scanlon developed the Scanlon plan to save a failing company. Three fundamental principles form the basis of this plan: bonus payment, identity with the company or firm, and employee involvement. Scanlon plans recognize the value and contribution of each member of the firm, encourage decentralized decision making, and seek to get each employee to identify with the organization's objectives through financial participation.

Before calculating the bonus, a base ratio must be computed. This traditionally is:

$$\text{Base ratio} = \frac{\text{Payroll costs to be included}}{\text{Value of production}}$$

Analysts make a historical study of approximately one year to gather data prior to calculating a proper base ratio. For example, if the base ratio is 15 percent and during the past month the value of production (sales plus or minus inventory) equals $2 million, then allowed labor equals $300,000 (0.15 × 2,000,000). An actual labor cost of $270,000 generates a bonus pool of $30,000. Typically, the company keeps a portion of this pool to provide for capital expenditures. The remainder is distributed to the employees as a monthly bonus based on a percentage of their wages.

To stimulate identity with the company, the Scanlon plan recommends a continuing program of management development where all employees, through effective communication, learn about the goals, objectives, opportunities, and problem areas characteristic of the plan. The Scanlon plan incorporates most quality of work life variables including job enlargement, job enrichment, feeling of achievement, and recognition.

Employee involvement is typically accomplished through a formalized suggestion system and two overlapping committee systems. Elected employee representatives meet at least monthly with their departmental supervisors to review productivity, cost reductions, and quality improvement suggestions. These committees frequently have certain decision-making authority for less costly suggestions. More costly suggestions or those affecting another department, are referred to a higher level committee.

RUCKER PLAN. This plan, too, came into being during the Great Depression (during the early 1940s). It was conceived by Allen W. Rucker, who noted the relationship between payroll costs and actual net sales plus or minus inventory changes minus purchased materials and services.

Like the Scanlon plan, the Rucker plan emphasizes identity with the company and employee involvement through the establishment of a suggestion system, Rucker committees, and good labor-management communications. The Rucker plan provides a bonus where everyone, excluding top administration, shares a percentage of the gains. In the evaluation of the bonus, a historical relationship between labor and value added must be established. For example:

Net sales (for period of one year)	$1,500,000
Inventory change (decrease)	200,000
	$1,300,000
Less materials and supplies used	700,000
Production value added	$ 600,000

$$\text{Rucker standard} = \frac{\text{Payroll costs included in group}}{\text{Production value}}$$

Assuming that the labor costs in the base one year period were $350,000, the Rucker standard becomes:

$$\frac{\$350,000}{\$600,000} = 0.583$$

Thus in any future period (usually one month) that the actual labor costs are less than 0.583 of production value, employees earn bonuses. Typically, 30 percent of this bonus is reserved for deficit months, a portion is kept by the company to provide improvements, and the remainder (often 50 percent) is distributed to employees.

Since materials and supplies used are deducted from net sales, the Rucker plan calculation partially accounts for variables, such as product mix. This plan also encourages employees to conserve supplies and materials since they would benefit from these savings.

IMPROSHARE. The IMproved PROductivity through SHARing plan was developed by Mitchell Fein in 1974. Its goal is producing more products in fewer hours of direct and indirect labor. Unlike the Scanlon and Rucker plans, IMPROSHARE does not emphasize employee involvement. IMPROSHARE measures performance rather than dollar savings, and includes the time of direct and indirect workers.

IMPROSHARE compares the work hours saved for a given number of units produced to the hours required to produce the same number of units during a base period. The savings are shared by the company and the direct and indirect employees involved with the production of the product.

Base productivity is measured by comparing the labor hour value of completed production to the total labor input for this production. Only acceptable products are counted. Thus:

$$\text{Work hour standard} = \frac{\text{Total production work hours}}{\text{Units produced}}$$

For example, assume that in a single product plant 122 employees produced 65,500 units over a 50-week period. If the total hours worked were 244,000, the work hour standard would be:

$$\frac{244,000}{65,500} = 3.725 \text{ hrs./unit}$$

If in a week 125 employees worked a total of 4,908 hours and produced 1,650 units, the value of the output would be $1,650 \times 3.725 = 6,146.25$ hours. The gain would be $6,146.25 - 4,908 = 1,238.25$ hours. Typically, one-half of this amount, or 619.125 hours, goes to the employees. This would be a 12.6 percent (619.125/4,908) bonus or additional pay to each employee.

The company, too, benefits since labor costs have been reduced. The unit labor cost of 3.725 hours established for the base period has been reduced to 3.350 hours per unit $\frac{(4,908 + 619.25)}{1,650}$.

Under IMPROSHARE management and the employees share the same goals: to improve productivity and reduce the costs of production.

EMPLOYEE STOCK-OWNERSHIP PLANS. Employee stock-ownership plans have become increasingly popular in the past decade. The Bureau of National Affairs' 1984 survey of 195 employers for type of productivity improvement programs administered indicated 37, or 19 percent, had installed employee stock-ownership plans. These plans involved the creation of a trust that holds company stock for its employees. Although 100 percent worker ownership plants are rare, ESOP can be used to develop such an organization.

PROFIT SHARING. The Council of Profit Sharing Industries has defined profit sharing as "any procedure under which an employer pays to all employees, in addition to good rates of regular pay, special current or deferred sums based not only upon individual or group performance, but on the prosperity of the business as a whole." [1]

No one specific type of profit sharing has received general industrial acceptance. In fact, just about every installation has certain "tailor-made" features that distinguish it from others. However, the majority of profit sharing systems fall into one of the following broad categories: (1) cash plans, (2) deferred plans, and (3) combined plans.

As the name implies, the straight cash plan involves the periodic distribution of money from the profits of the business to the employees. The payment is not included with the regular pay envelope, but is made separately, to identify it as an extra reward, brought about by the individual and combined efforts of the entire operating force. The amount of the cash distribution is based on the degree of financial success of the enterprise for the bonus period. The period varies in different companies. In some installations it is one month, and in others, one year. There is general agreement, however, that the shorter the period, the closer the connection between effort and financial reward to the employees. Longer periods are selected because they reflect more truly the status of the business over that particular length of time. For example, if a period of one month is used, employees might give excellent performance for the entire period in anticipation of a sizable bonus. However, due to conditions beyond their control, such as high inventories or slow-moving products, there may be no profit for the period. With no additional reward, employees would quickly lose confidence in the plan and reduce their efforts appreciably.

Another factor in selecting the period for the distribution of profits is the amount of cash to be shared. If the period is too short, the amount may be so insignificant in size that the plan itself may backfire. But, if a longer period, such as six months or a year, is used, the amount is large enough to accomplish the purpose of the profit sharing system.

Deferred profit sharing plans feature the periodic investment of portions of the profits for employees. On retirement or separation from the company, employees have a source of income at a time when their needs may be more pronounced.

Deferred profit sharing plans obviously do not provide the incentive stimulus to the degree that cash plans do. However, deferred profit shar-

---

[1] Council of Profit Sharing Industries, *Profit Sharing Manual* (Akron, Ohio: Council of Profit Sharing Industries, 1949), p. 3.

ing plans do offer the advantage of being easier to install and administer. Also, plans of this type offer more security than do the cash reward plans. This makes them especially appealing to stable workers.

Combined plans arrange to have some of the profits invested for retirement and similar benefits, and some distributed as cash rewards. This class of plans can realize the advantages of both the deferred plans and those employing the straight cash system. A representative installation might provide for sharing half the profits with the employees. Of this amount, one third may be distributed to employees as extra bonus checks, one third may be held in reserve to be given out during a less successful financial period, and the remaining third may be placed with a trustee for deferred distribution.

METHODS OF DISTRIBUTION UNDER PROFIT SHARING. Three methods for determining the amount of money to be given individual employees from the company's profits are common: The first and least used is the "share and share alike" plan. Here each employee, regardless of job class, participates in an equal amount of the profits, after attaining the prescribed amount of company service. Proponents of this method believe that individual base rates have already taken care of the relative importance of the different workers to the company. The share and share alike plan supplies a feeling of teamwork and importance to each employee, no matter what his or her position in the plant may be. Just as a complex mechanism is made inoperative by the removal of an insignificant pin, so an enterprise is dependent upon each and every employee for its efficient operation.

The most commonly used method of distribution under profit sharing is based on the regular compensation paid to workers. The theory is that the employee who was paid the most during the period contributed to the greatest extent to the company's profits and consequently should share in them to the greatest extent. For example, a toolmaker earning $15,000 during a six-month period would receive a greater share of the company's profits than a chip-hauler paid $7,000 during the same period.

Another popular means of profit distribution is through the allocation of points. Points are given for each year of seniority and each $100 of pay. Some point plans also evaluate such factors as attendance, the worker's cooperation, and standards of production. The number of points accumulated for the period determines each employee's share of the profits. This method takes into consideration the factors that influence the company's profits and then distributes the profits on a basis that gives each employee a just share. Perhaps the principal disadvantage of the point method is the difficulty of maintenance and administration brought about through complex and detailed records.

CRITERIA FOR A SUCCESSFUL PROFIT SHARING PLAN. The Council for Profit Sharing Industries has summarized these 10 fundamental principles a profit sharing plan must follow to succeed:

1. Management must have a compelling desire to install the plan to enhance the team spirit of the organization. If there is a union, its cooperation is essential.
2. The plan should be generous enough to forestall any feeling by employees that the lion's share of the results of their extra effort will go to management and the stockholders.
3. The employees must understand that there is no benevolence involved, that they are merely receiving their fair share of the profits they have helped to create.
4. The emphasis should be placed on partnership, not on the amount of money involved. When financial return is emphasized, it is more difficult to retain the employees' loyalty and interest in loss years.
5. The employees must feel that it is their plan as well as management's and not something that is being done for them by management. The employees should be well represented on any committee set up to administer the plan.
6. Profit sharing is inconsistent with arbitrary management—it functions best in companies operating under a democratic system. This does not mean that management relinquishes its right, and, in fact, its obligation, to manage, but rather that management functions by leadership instead of arbitrary command.
7. Under no circumstances can profit sharing be used as an excuse for paying lower than prevailing wages.
8. Whatever its technical details, the plan must be adapted to the particular situation, and should be simple enough so that all can readily understand it.
9. The plan should be "dynamic," in both its technical details and its administration. Management and employees—and in profit sharing companies it is often difficult to find the demarcation line—should give constant thought to ways of improving it.
10. Management should recognize that profit sharing is no panacea. No policy or plan in the industrial relations field can succeed unless it is well adapted and unless it evidences the faith of management in the importance, dignity, and response of the human individual.

ATTITUDES TOWARD PROFIT SHARING.   For the most part, union officials have not supported profit sharing. Their principal criticism is that the technique is a "hard-times, wage cutting method."[2] There can be no doubt that profit sharing, when practiced with perfect harmony between labor and management, minimizes the necessity of a union in the eyes of the employees. Therefore, union leaders can hardly be expected to uphold and propagate something that diminishes their personal prestige,

---

[2] Charles E. Britton, *Incentives in Industry* (New York: Esso Standard Oil Co., 1953), p. 30.

power, and income. It is interesting that in 1960 about 30 percent of the work force belonged to labor unions, while today this figure is only approximately 17 percent.

Where profit sharing has been practiced by a sincere, fair, and competent management team, individual workers are enthusiastic, and wholeheartedly support the plan. Attempts to unionize plants where profit sharing is flourishing, such as in the Lincoln Electric Company, have resulted in violent worker reaction against the organizers.

A successful profit sharing program depends on the profits of the company, which frequently are not under the control of the direct labor force. In periods of low profits or of losses, the plan may actually weaken rather than strengthen employee morale. Also, unless the team spirit is enthusiastically supported by *all* members of supervision, the plan does not provide the incentive to continuous effort characteristic of wage incentive installations. This is a result of the length of time between performance and reward; most companies wait 12 months and use the year-end to declare their profits. In large companies, it becomes virtually impossible to instill the team idea throughout the plant, and unless this is done generally, the plan does not succeed completely.

Perhaps the greatest objection to profit sharing is the taking for granted that an "extra" check will be received at the end of the year. Employees expect to receive these checks, and even make substantial purchases on a time basis in anticipation of the added remuneration. If the company has experienced a lean year and workers get no extra remuneration, they do not see any relation between their productivity for the past year and the fact that no bonus was received. Employees feel that they have been cheated. Thus, the whole profit sharing plan fails to engender the spirit intended.

For these reasons, any employer should be very cautious before embarking on a profit sharing program. On the other hand, many companies today are experiencing high worker efficiency, decreased costs, reduction of scrap, and better worker morale as a result of profit sharing. James F. Lincoln, past president and founder of the Lincoln Electric Company, who developed one of the most successful profit sharing installations, offers the following suggestions to companies contemplating a profit sharing system:

1. Determine that the system is going to be adopted and decide that whatever needs to be done to install it will be done.

2. Determine what plan and products the company will make that will carry out the philosophy of "more and more for less and less."

3. Get the complete acceptance of the board of directors and all management involved in the plan, together with their assurance that they will continue to take whatever steps are necessary for a successful application of it.

4. Arrange a means whereby management can talk to the men and the men can talk back. That means full discussion by all.

5. Make sure of co-operative action on the agreed plan of operation. This will include the plan for progressively better manufacturing by all people in the organization and the proper distribution of the savings that result from it.

6. Set your sights high enough. Do not try to get just a little better efficiency with the expectation that such gain will be to the good and expect to leave the matter there.

7. Remember, this plan for industry is a fundamental change in philosophy. From it new satisfactions will flow to all involved. There is not only more money for all concerned, there is also the much more important reward—the satisfaction of doing a better job in the world. There is that greatest of all satisfactions, the becoming a more useful man.[3]

The Unions' Attitudes toward Wage Incentives

The subject of wage incentives has always been controversial to employees, unions, and management. Industries that have assured a good living wage and then applied easily-calculated incentive earnings for extra or prolonged effort, find that their employees are receptive to wage incentives. In fact, where a successful installation of this nature has been made, considerable labor unrest would result if any attempt were made to do away with the plan. On the other hand, in industries or businesses where workers find it necessary to work at an incentive pace to earn the necessities of life, they can hardly be expected to be enthusiastic about any form of wage incentive payment.

The majority of union officials with whom the writer has been in contact oppose the installation of incentive wage payment in plants where incentives do not exist. However, where incentives do exist, most unions not only insist on their continuance but endeavor to expand the coverage of their membership under incentives. For example, the United Steelworkers of America summary of the August 1, 1969, incentive arbitration award stated:

> The first issue tackled at the arbitration hearing was the Union's demand that a rational system of coverage be provided in place of the present chaos. It was the Union's position that all jobs should be covered by incentives. The Companies, on the other hand, argued unsuccessfully that in many instances already too many jobs are covered by incentives, and that the Companies should be given the right to remove coverage from many jobs which now enjoy incentives.

---

[3] James F. Lincoln, *Lincoln's Incentive System* (New York: McGraw-Hill, 1946), pp. 171–72.

On the other hand, in an AFL–CIO Collective Bargaining Report that is still being used as the union's position paper about incentives, past President George Meany stated:

> Wage incentive plans, that is, plans which offer more wages for more production, present a host of special problems which normally far outweigh any possible benefits.
>
> With few exceptions, unions are opposed to wage incentive systems, both because of the damaging past experience with the abuses under such plans, and because of the difficulties and ill effects inherent in incentive plans.
>
> Unions which have accepted them or permitted them to continue have usually done so only with reluctance and misgivings. It simply has not always been practical or expedient actively to oppose or to eliminate such plans.
>
> A few unions, primarily in the rubber and needle trade industries where wage incentives are most firmly entrenched, have at least temporarily accepted incentives as part of their collective bargaining programs.
>
> Many industrial engineers agree that wage incentives have been abused "in the past," but maintain that this is no longer true. In fact, although some of the worst abuses have been toned down, primarily through union action, a variety of ill effects and strains on workers are still very much the rule.
>
> The presence of an incentive system invariably means special problems of education, representation, and protection of workers. It puts a strain on the entire collective bargaining process, making it more difficult, more complex, and most costly.
>
> The value of any such plan is also highly questionable because, even though it may initially yield increased earnings, it inevitably requires a speedup of work efforts, creates friction between workers, and produces continual wrangling over production standards.

The unions' principal objections to incentive plans arise from a fear that a reduction in personnel will be brought about by high effort, given only a fixed production of goods and services. Their leaders also believe that incentive installations de-emphasize the necessity of unions since the unions' major role is to achieve higher wages, which the incentive plan automatically does. Union officials have stated that they object to incentives because these "pit worker against worker." They state that when one worker makes high earnings and another low, a feeling of distrust and suspicion permeates the working group and disrupts the partnership relations among workers.

The Grand Lodge constitution of the International Association of Machinists (Section 6 of Article J), clearly expressed the policy of this labor group toward incentives in the statement that "any member guilty of advocating or encouraging any of these systems where they are not in existence is liable to expulsion." Carl Huhndorff, as past director of research of this labor group, stated that the aforementioned clause had never been invoked against an individual member to the best of his knowledge; however, the Grand Lodge policy "has always opposed these [incentive] plans whenever possible as we do not feel that they work to the best interests of working men and women."

There are, of course, unions which approve of incentives. In fact, the late Philip Murray, a past president of the CIO, favored incentive wage payment and believed that practically any sound system of wage payment can be made to work when a harmonious relationship prevails between labor and management.

In summary, most unions fight to keep incentives where they already exist. Where incentives do not exist, unions resist any attempt to install them.

PREREQUISITES FOR A SOUND WAGE INCENTIVE PLAN

Certainly the majority of companies that have incentive installations favor their continuance and believe that their plans are (1) increasing the rate of production, (2) lowering overall unit costs, (3) reducing supervision costs, and (4) promoting increased earnings of their employees.

However, in a survey of 160 companies practicing incentive wage payment, 84 replies from managers implied that they felt their plans to be only fair, and that additional improvement could be made.[4] Five plant managers felt that their incentive systems were poor, and that some changes must be made to warrant continuance.

Before installing a wage incentive program, management should survey its plant to be sure that the plant is ready for an incentive plan. Initially, a policy of methods standardization must be introduced so that valid work measurement can be accomplished. If operators each follow different patterns while performing their work, and the sequence of elements has not been standardized, then the organization is not yet ready for the installation of wage incentives.

Work schedules should create a backlog of orders for each operator, so that the chances of running out of work are minimized. Of course, this implies that adequate inventories of material are available, and that machines and tools are properly maintained. Also, established base rates should be fair, and should provide for a sufficient spread between job classes to recognize the positions that demand more skill, effort, and responsibility. Preferably, management has established base rates through a sound job evaluation program.

Last, fair standards of performance must be developed before wage incentive installation can take place. In no case should these rates be set by judgment or past performance records. To be sure that they are correct, some form of work measurement based on time study, fundamental motion data, standard data, formula, or work sampling procedure should be used.

---

[4] National Metal Trades Association.

Once these prerequisites are complete, and management is fully sold on incentive wage payment, the company is in a position to design the system.

DESIGN FOR A SOUND WAGE INCENTIVE PLAN

To be successful, an incentive plan must be fair to both the company and its operators. The plan should give operators the opportunity to earn approximately 20 to 35 percent above base rate if they are normally skilled and execute high effort continuously. Management benefits through the added productivity by being able to prorate fixed costs over a larger number of pieces, thus reducing total cost.

Perhaps next to fairness, the most important qualification of a good incentive plan is simplicity. To be successful, the plan must be completely sold to the employee, to the union, and to management itself. The simpler the plan is, the more easily it is understood by all parties; and understanding enhances the chances for approval. Individual incentive plans are more easily understood and work the best—if individual output can be measured.

The plan should guarantee the basic hourly rate set by job evaluation; the rate should be a good living wage comparable to the prevailing wage rate of the area for each job in question.

There should be a range of pay rates for each job, and these should be related to total performance. Total performance encompasses quality, reliability, safety, and attendance, as well as output. At periodic intervals, such as every six months or every year, management should review employees' rate steps in relation to their total performance. For performance greater than standard, operators should be compensated in direct proportion to output, thus discouraging any restriction of production.

To help employees associate effort with compensation, paycheck stubs should clearly show both the regular and the incentive earnings. It is also advisable to indicate on a separate form, placed in the pay envelope, the efficiency of the operator for the past pay period. This is calculated as the ratio of the standard hours produced during the period to the hours worked during the period.

In an efficient manner, management should provide for unavoidable loss of time not included in the standard. Also, techniques must be established to ensure accurate piece counts meeting quality requirements. The plan should entail the control of indirect labor through measuring the productivity of those involved. Indirect labor should not be given a bonus equivalent to that of the direct workers unless some yardstick has been determined to justify their incentive earnings. Once the plan has been installed, management must accept the responsibility for maintaining it. Administration of the plan calls for keen judgment in making decisions, and for close analysis of all grievances submitted. Management must exercise its right to change the standards when the methods and/or equipment are changed. Employees must be guaranteed an opportunity to present their suggestions, but the advisability of their requests must be proved before any changes are made. Compromising on standards must be avoided. A liberalizing of standards already based on facts leads to complete failure of the plan. No incentive plan can continue to operate unless it is effectively maintained by a watchful, competent management team.

The Motivation for Incentive Effort

Methods analysts should recognize that unless operators perform at good effort, the most favorably designed workstation cannot result in levels of productivity synonymous with company objectives. To achieve high levels of productivity, in addition to good physical work center design, the conditions surrounding the work should encourage all employees to do their best to help realize company objectives.

Some hypotheses related to motivation include:

1. Work is and always has been a natural human activity. Today, as always, most people basically want to work and to achieve.
2. Almost all people want to be involved in the achievement of the goals established by their group or organization.
3. Most people perform better if given both independence and control in their work situation.
4. Workers expect to see a relationship between their contributions and the magnitude of the resulting rewards.

To introduce an effective wage incentive plan, all four of the above hypotheses should be observed. It is insufficient to provide a one-for-one incentive plan based on fair standards. Even when people feel equitably paid, there is no guarantee of high performance. A climate of motivation must accompany any formal incentive plan.

Perhaps the first requirement in establishing the proper climate of motivation is developing a management style which emphasizes a supporting role rather than a directive role. Thus, the goal should be to have all the workers feel that it is their responsibility to meet the objectives of the business and that it is the supervisors' responsibility to assist the workers as best they can.

Second, the goals of the enterprise should be clearly established, and should be broken down into division, department, work center, and individual goals. It is important that established goals be realistic, that they emphasize not just quantity but also quality, reliability, and any other characteristic essential to the success of the business. All workers should understand the objectives of the company and the goals related to their work. These goals should be quantified in such a manner that workers are cognizant of their achievement in relation to established goals.

Third, there should be regular feedback to all employees. Timely reporting should inform workers about the results of their efforts and the impact these efforts have had on the established goals.

Fourth, every work situation should be designed so that operators are in a position to control to a large extent the assignments they are given. There is a considerable variation in different jobs from the standpoint of the amount of control that can be given to the operator. However, experience has verified that positive results are achieved where management has adopted the attitude that "job enrichment is important." A sense of responsibility is an important source of motivation, as is recognition for achievement.

Reasons for Incentive Plan Failures

An incentive plan may be classified as a failure when it costs more for its maintenance than it actually saves, and thus must be discontinued. Usually it is not possible to put a finger on the precise immediate cause of failure for a given incentive installation. If the facts were completely known, numerous reasons would be found for the plan's lack of success. One survey listed the principal causes of plan failure as poor employee attitude; excessive cost; an insufficiently stable product resulting in high standards costs; and too liberal allowances, resulting in too costly a program.

Certainly, companies should discontinue any plan when its cost of maintenance exceeds the benefits derived through its use. The reasons given in the survey are, for the most part, symptoms of a sick plan, of a plan doomed to failure—they are not really causes. The actual cause of the failure of any plan is incompetent management: A management that permits the installation of a plan with poor scheduling, unsatisfactory methods, a lack of standardization or loose standards, and the compromising of standards, has no one but itself to blame for the plan's failure.

Of course, all the requisites of a sound incentive system may be met and the plan may still be unsatisfactory because of failure to promote good industrial relations relative to the program. The complete cooperation of the employees, the union, and supervision must be won to foster the team spirit so necessary to attain ultimate success with the incentive installation.

Bruce Payne, president of Bruce Payne & Associates, Inc., reported on 246 companies where incentive plans had either failed or developed weaknesses that necessitated revision. Three factors were responsible for the failures:[6]

---

[6] Britton, *Incentives*, p. 60.

		Percent
1.	Fundamental deficiencies in the plan	41.5
2.	Inept human relations	32.5
3.	Poor technical administration	26.0

The reasons falling under each of these divisions and their relative importance were summarized as follows:

Fundamental deficiencies	*Percent*
Poor standards	11.0
Low incentive coverage of direct productive work	8.6
Ceiling on earnings	7.0
No indirect incentives	6.8
No supervisory incentives	6.1
Complicated pay formula	2.0

Inept human relations	
Insufficient supervisor training	6.9
No guarantee of standards	5.7
A fair day's work not required	5.0
Standards negotiated with the union	4.8
Plan not understood	4.1
Lack of top-management support	3.6
Poorly trained operators	2.4

Poor technical administration	
Method changes not coordinated with standards	7.8
Faulty base rates	5.1
Poor administration, i.e., poor grievance procedure	4.9
Poor production planning	3.2
Large group on incentive	2.8
Poor quality control	2.2

ADMINISTRATION OF THE WAGE INCENTIVE SYSTEM

To be successful, an incentive system must be adequately maintained; it cannot maintain itself. To maintain a plan effectively, management must keep all employees aware of how the plan works and of any changes introduced into the plan. One technique frequently used is to distribute to all employees an "Operating Instruction" manual outlining in detail not only company policy relative to the plan, but also all its working details with examples. This manual should thoroughly explain the basis of job classifications, time standards, performance rating procedure, allowances, and grievance procedure. It should also describe the technique of handling any unusual situation. Finally, it should present the objectives of the organization and the role of each employee in the fulfillment of those objectives.

Administrators of the plan should make a daily check of low performance and excessively high performance in an effort to determine their causes. Low performance is not only costly to management in view of the guaranteed hourly rate, but it leads to employee unrest and dissatisfaction. Unduly high performance is a symptom of loose standards, or the introduction of a methods change for which no standard revision has been made. In any case, a loose rate leads to the dissatisfaction of employees in the immediate vicinity of the operator who is working on the job carrying the low standard. A sufficient number of such poor standards can cause the whole incentive plan to fail. Frequently, operators who have the loose rate restrict their daily production in fear that management will adjust the standard. This restriction of output is costly to the operators and the company, and results in dissatisfaction among neighboring workers who see fellow employees on a soft job.

Management should make a continuing effort to include a greater share of the employees in the incentive plan. When only a portion of the plant is on standard, there will be a lack of harmony among operating personnel because of significant differentials in take-home pay. However, work generally should not be put on incentive unless:

1. It can readily be measured.
2. The volume of available work is sufficient to economically justify an incentive installation.
3. The cost of measuring the output is not excessive.

Periodically management should review old standards to assure their validity. On standards that have proved to be satisfactory, elemental values should be recapped for standard data purposes, so that even greater utilization may be made of the time values. Thus, analysts can achieve greater coverage of the plant relative to the use of standards.

Fundamental in the administration of any wage incentive plan that is keyed to production is the constant adjustment of standards to changes in the work. No matter how insignificant a methods change may be, it is well to review the standard for possible adjustment. Several minor methods improvements, in aggregate, can amount to a sufficient time differential to bring about a loose rate if the standard is not changed. When revising time standards due to methods changes, it is necessary to study only those elements affected by the changes.

To keep the incentive plan healthy, the company should arrange periodic meetings with operating supervisors to discuss fundamental weaknesses of the plan, and possible improvements in the installation. At these meetings, departmental performance should be compared, and specific standards that appear unsatisfactory should be brought to light and discussed. Employees expect and should receive an equitable working climate that assures a relationship between their contributions and their rewards.

The company should keep progress reports showing such pertinent information as departmental efficiency, overall plant efficiency, the number of workers not achieving standard performance, and the highest individual performance. These reports provide information on areas that need attention as well as areas where the plan is working satisfactorily.

Effective administration of the plan requires a continuing effort to minimize the nonproductive hours of direct labor. This nonproductive time for which allowance must be given represents lost time due to machine breakdowns, material shortages, tool difficulties, and long interruptions of any sort not covered in the allowances applied to the individual time standards. Managers must carefully watch this time, frequently referred to as "blue ticket time" or "extra allowance time," or it will destroy the purpose of the entire plan.

For example, assume that on a certain job the production rate is 10 pieces per hour, and that an hourly rate of $12 is in effect under a straight daywork operation. Thus, the unit direct labor cost is $1.20. Now this shop changes over to incentive wage payment where the day rate of $12 per hour is guaranteed, and, above task, the operator is compensated in direct proportion to his or her output. Let us assume that the standard developed through time study is 12 pieces per hour, and that for the first five hours of the working day, a certain operator averages 14 pieces per hour. His earnings for this period would then be:

$$(\$12.00)(5) \left(\frac{14}{12}\right) = \$70.00$$

Now assume that for the remainder of the working day, due to a material shortage, the operator could not be productively engaged in work. The worker would then expect at least the base rate or:

$$(3)(\$12.00) = \$36.00$$

which would give the earnings for the day of

$$\$70.00 + \$36.00 = \$106.00$$

This would result in a unit direct labor cost of:

$$\frac{\$106.00}{70} = \$1.514$$

Under daywork, even with the low performance, the operator would have produced the 70 pieces in less than the working day. Here the earnings would have been: $8 \times \$12.00$, or $96.00, and the unit direct labor cost would have been: $96.00/70, or $1.371.

Under incentive effort, production performance is considerably higher than under daywork operation. With the shorter accompanying in-process time of materials, there needs to be very careful inventory control to prevent material shortages. Likewise, management should introduce a program of preventive maintenance to assure the continuous operation of all machine tools. Equally important to material control is the control of all nondurable tools, so that shortages with resulting operator delays do not develop.

An effective technique often employed to control the "extra allowance" is to key the supervisor's bonus to the amount of this nonproductive time credited to the operator. The more of this time turned in for the pay period, the less would be the supervisor's compensation. Since supervisors are in an ideal position to watch schedules and material inventories, and to maintain facilities, they can control nonproductive downtime better than anyone else in the plant.

In addition to controlling the "extra allowance" or daywork time, it is essential that exact piece counts be recorded at each workstation. The piece count which determines the operator's earnings is usually done by the operator. Controls must be established to prevent operators from falsifying their production output.

Where the work is small (several pieces can be held in one hand), operators make a "weigh" count of their production at the end of the day or the end of the production run, whichever period is shorter. This "weigh" count is verified by the immediate supervisor, who initials the production report.

On larger work, one technique that is frequently employed is to have a tray or box with built-in compartments to hold the work. The tote box holds round numbers of the work, such as 10, 20, or 50. Thus, at the end of the shift, it is a simple matter for the operators' supervisors to authenticate production reports by merely counting the boxes and multiplying by whatever number each box holds.

Basically, management establishes wage incentive plans to increase productivity. In a sound and properly maintained installation, the percentage of incentive earnings of those workers on incentive would remain relatively constant over time. If analysis shows that incentive earnings continue to rise over a period of years, the installation has problems that will ultimately erode the effectiveness of the plan. If, for example, the average incentive earnings increased from 17 percent to 40 percent in a period of 10 years, most likely the 23 percent rise was not due to a proportionate increase in productivity, but to a creeping looseness in standards.

CONCLUSION

The only acceptable wage incentive plan applied to individual workers today is the standard hour plan with a guaranteed day rate. Similarly, group plans must guarantee their respective day rates to all members of the group, as well as reward members of the group in direct proportion to their productivity once standard performance is achieved.

TABLE 25–1

Respondent type	Profit sharing		ESOP	
	Have plan (%)	*Do not have plan (%)*	*Have plan (%)*	*Do not have plan (%)*
All respondents	52%	48%	25%	75%
With ESOP	54	46	—	—
With profit sharing	—	—	25	75
With gain sharing	48	52	16	84
With simple incentive	58	42	28	72

Source: Based on 508 surveys returned 10/15/86. Adapted from data by R. Broderick and D. J. B. Mitchell. "Who Has Flexible Wage Plans and Why Aren't There More of Them?" *IRRA 29th Annual Proceedings*, pp. 163–64.

TABLE 25–2
Characterization of flexible compensation plans by survey respondents

	Profit sharing (%)	*Employee stock ownership plans (%)*	*Gain sharing Scanlon Rucker improshare (%)*	*Simple std. hr. pc. rate%*
Best for:				
Raising productivity	28%	5%	26%	41%
Increasing loyalty	48	17	19	14
Providing for retirement	80	13	n.a.	n.a.
Linking labor cost to performance	53	n.a.	28	19
Easiest to:				
Explain to employees	32	9	4	49
Administer	40	7	4	38

* Refers to tax-deferred profit sharing plans. n.a. not asked.
Source: Adapted from data by R. Broderick and D. J. B. Mitchell. "Who Has Flexible Wage Plans and Why Aren't There More of Them?" *IRRA 29th Annual Proceedings*, pp. 163–64.

Profit sharing, employee stock ownership, and other related cost improvement savings plans have met with success in many cases. In general, they tend to be more effective when they are installed in addition to, rather than instead of, a simple incentive installation. See Table 25–1.

Incentive principles have been applied in job shops and production shops; in the manufacture of both hard goods and soft goods; in manufacturing and service industries; and in direct and indirect labor operations. Incentives have been used to increase productivity, improve the quality and reliability of the product, reduce waste, improve safety, and stimulate good working habits, such as punctuality and regularity of attendance.

Table 25–2 illustrates the thinking of 508 personnel/industrial relations managers as to characterization of flexible compensation plans. The majority feel that simple incentives—piece work, standard hour plans, and measured day work—are the best from the standpoint of raising productivity and ease of explanation.

Soundly administered incentive systems possess important advantages, both for workers and management. The chief benefit to employees is that these plans make it possible for them to increase their total wages, not at some indefinite time in the future, but immediately—in their next paycheck. Management obtains a greater output and, assuming that some profit is being made on each unit produced, a greater volume of profits. Normally, profits increase, not in proportion to production, but when a higher rate of production takes place, so that overhead costs per unit decrease. Then, too, the higher wages that result from incentive plans improve employee morale and tend to reduce labor turnover, absenteeism, and tardiness.

Since the proper functioning of incentive systems implies the existence of many prerequisites, such as good methods, standards, scheduling, and management practices, the installation of incentives normally results in important improvements in production and supervisory methods. The activities that bring about these improvements should be performed even though incentives are not introduced; the improvements, therefore, are not necessarily attributable to the employment of the incentive plan or plans.

In general, the harder the work is to measure, the more difficult it will be to install a successful wage incentive plan. Usually, it is not advantageous to install incentives unless the work is subject to reasonably accurate measurement. Furthermore, it is usually not advantageous to introduce incentives if the availability of work is limited to less than 120 percent of normal.

The following 16 fundamental principles are a guide for sound wage incentive installation and administration:

1. *Agreement on General Principles.* Management and labor should be in real agreement on the general principles involved in the relationship between work and wages.

2. *A Foundation of Job Evaluation.* There should be a sound wage rate structure, based upon an evaluation of the skill, responsibility, and working conditions inherent in the various jobs.

3. *Individual, Group, or Plant-wide Incentives.* It is generally conceded that standards, applied to individuals or to small integrated groups, are most effective. Such standards need to be set with the utmost care, and undoubtedly tend toward the lowest unit cost. At times, due to difficulties in recording individual production, or due to the possibilities of teamwork, group standards may be advisable. The larger the group, the less the individual response. With plant-wide incentives, some of the jealousies and transfer difficulties often inherent in group plans are eliminated, but without an unusual degree of leadership and co-operation, the incentive effect is greatly diluted.

4. *The Production-Incentive Relationship.* When production standards are properly set, and based upon well-engineered conditions, good practice has demonstrated the desirability of adopting an incentive payment in which earnings above the established standard are in direct proportion to the increased production.

5. *Simplicity.* The plan should be as simple as possible, without causing inequities. Workers should be able to understand the effect of their own efforts on their earnings.

6. *Quality Control and Improvement.* The desirable and economical degree of quality should be determined and maintained, tied in with bonus payment where advisable.

7. *Improved Methods and Procedures.* To secure the lowest costs and to prevent uneven standards and inequitable earnings, which lead to poor labor relations, the establishment of production standards should be preceded by basic engineering improvements in design, equipment, methods, scheduling, and material handling.

8. *Based on Proven Industrial Engineering Techniques.* Standards should be developed from detailed time studies, basic motion data, standard data, and formulae. A permanent record of standard elemental times for each unit of an operation eliminates the occasion for many arguments. A table of basic standard times prepares the way for proper introduction of technological improvements.

9. *Based on Normal Operation under Normal Conditions.* In general, the production standard should be established by management setting up the amount of work performed per unit of time by a normal qualified operator under normal conditions.

10. *Changes in Standards.* The plan should provide for the changing of production standards whenever changes in methods, materials, equipment, or other controlling conditions are made in the operations represented by the standards. In order to avoid misunderstandings, the nature of such changes and the logic of making them should be made clear to labor or its representatives who should have the opportunity to appeal through the grievance machinery.

11. *Considerations in Changing Standards.* Except to correspond properly with changed conditions, production standards once established should not be altered unless by mutual agreement between management and labor representatives.

12. *Keep Temporary Standards at Minimum.* The practice of establishing temporary standards on new operations should be kept at a minimum. It should, in any event, be made clear to all that the standards are for a reasonably short period only.

13. *Guarantee of Hourly Rates.* Under ordinary circumstances, the employees' basic hourly rates should become guaranteed rates.

14. *Incentives for Indirect Workers.* Effective standards may be established for most indirect jobs in the same manner as for direct jobs. If the exigencies of a situation demand that some form of incentive payment be applied to indirect workers as a whole, or in groups, then the indirect man-hours should be correlated to some measurable unit, such as production or direct employee hours, so that indirect labor cost may be kept under control.

15. *Counting.* There must be accurate control of piece counts, unmeasured work, setup, and downtime.

16. *Thorough Understanding of Human Relations Involved.* It should be emphasized that unless management is prepared to work on the problem with a thorough understanding of the human relations involved, it had better have no incentive plan. Whereas such a plan may be a progressively constructive force for increased production, it may also be a means of disrupting labor relations and of actually lowering production. While necessarily retaining its functions, management should take into account labor's point of view. It should impart to labor a complete understanding of the plan and patiently consider grievances, in whatever manner may be agreed upon.[7]

Due to recent research that has produced more reliable fundamental motion data, and to the advent of the hand-held calculator and the personal computer, the reliability of standards is assured. Furthermore, we now are able to cover the vast majority of work opportunities within industry or business with good standards. Poor standards and insufficient coverage are probably the principal reasons for failure of incentive systems in the past. Now that we have the capability to avoid these pitfalls, we are more conscious of the absolute necessity for good human relations and technical administration, recognizing that sound wage incentives can do much to stimulate productivity and reduce inflation. Wages incentives permit workers to increase their standard of living, as opposed to being beaten by inflation even though receiving sizable annual increases in compensation for doing the same amount of, or less, work.

In the immediate future, industry, in order to maintain a competitive posture, will need to be innovative in connection with rewards. In addition to a sound one-for-one payment procedure based upon fair base rates, it will need to consider a profit sharing and/or employee stock ownership plan so as to emphasize the importance of the team approach and group performance and the rewarding of systems embracing improvements in quality, methods, and technology. Management must be specific in letting all workers know:

---

[7] From a talk by John W. Nickerson, "The Importance of Incentives," given at the Second Annual Time Study and Methods Conference, New York.

1. That the profit sharing portion of employees' income is a reward for their part in the success of the company and is not to be thought of as added compensation.
2. The company's current performance in connection with established objectives.
3. The potential for each worker advancing in the company.
4. Its desires to provide work that is both interesting and challenging.
5. That it endeavors to provide opportunity for its employees to learn new skills and develop personally.
6. That the company will seek advice from employees in connection with issues that influence their lives at work.
7. That the company is committed to continuous improvement.

Well-planned and skillfully administered incentives increase production and decrease total unit cost. Usually, they more than compensate for the increased costs of industrial engineering, quality control, and time-keeping which may result from their use.

TEXT QUESTIONS

1. Under which three general categories may the majority of wage incentive plans be classified?
2. Differentiate between individual wage payment plans and group-type plans.
3. What is meant by the term *fringe benefits?*
4. Which company policies are included under nonfinancial incentives?
5. What are the characteristics of piecework? Plot the unit cost curve and operator earning curve for daywork and piecework on the same set of coordinates.
6. Why did measured daywork become popular in the 1930s?
7. What advantages does the Rucker plan have over the Scanlon plan?
8. How does IMPROSHARE differ from the Rucker and the Scanlon plans?
9. Define "profit sharing."
10. Which specific type of profit sharing plan has received general acceptance?
11. Which three broad categories cover the majority of profit sharing installations?
12. What does the amount of money distributed depend on under the cash plan?
13. What determines the length of the period between bonus payments under the cash plan? Why is it poor practice to have the period too long? What disadvantages are there to the short period?
14. What are the characteristic features of the deferred profit sharing plan?
15. Why is the "share and share alike" method of distribution not particularly common? On what basis is this technique advocated by its proponents?
16. List the 10 principles summarized by the Council for Profit Sharing Industries as fundamental for a successful profit sharing plan.

17. Why have many unions shown antagonistic attitudes toward profit sharing?
18. What suggestions has James Lincoln offered for those embarking on an incentive installation?
19. How does job enrichment differ from job enlargement?
20. Why is it usually advisable to have a range of pay rates that are applicable to each job?
21. What four important hypotheses related to motivation should the work measurement analyst be cognizant of?
22. Why do many union officials prefer straight daywork as a form of wage payment?
23. What are the fundamental prerequisites of a successful wage incentive plan?
24. What are the requisites of a sound wage incentive system?
25. What is extra allowance or "blue ticket" time? How may it be controlled?
26. Why is it fundamental to keep time standards up-to-date if a wage incentive plan is to succeed?
27. What does unduly high performance indicate?
28. What responsibilities of the supervisor, time study department, quality control department, production department, accounting department, and industrial relations department are essential to the effective administration of a sound wage incentive system?
29. List 15 fundamental principles to follow to assure a successful wage payment plan.
30. When would it be inadvisable to put an indirect labor activity on incentive?
31. How would you go about establishing a climate to increase worker motivation?

GENERAL QUESTIONS

1. Why do many unions object to management posting incentive earnings?
2. Should indirect labor that is not on incentive carry a higher base rate than direct labor on incentive when both jobs carry the same evaluation? Why or why not?
3. What is the major reason that many union officials dislike any form of incentive wage payment?
4. How can the piece count be controlled within the plant?
5. How can a yardstick of performance be established in the following jobs: tool crib attendant, tool grinder, floor sweeper, chip-hauler?
6. Explain how you would enrich the position of a punch press operator.
7. Interview three employees in three different companies. Are they satisfied with the wage payment program in the companies in which they are employed? Why or why not?
8. Explain why profit sharing increases loyalty to a much higher degree than simple incentives. See Table 25–2.

PROBLEMS

1. In a single product plant where IMPROSHARE was installed, 411 employees produced 14,762 product units over a one-year period, and recorded 802,000 clock hours. In a given week, 425 employees worked a total of 16,150 hours and produced 348 units. What would be the hourly value of this output? What percentage bonus would each of these 425 workers receive? What would be the unit labor cost in hours for this week's production?

2. Analysts established an allowed time of 0.0125 hours/piece for machining a small component. A setup time of 0.32 hour is established also, as the operator performs the necessary setup work on "incentive." Compute:
   a. Total time allowed to complete an order of 860 pieces.
   b. Operator efficiency, if job is completed in an eight-hour day.
   c. Efficiency of the operator who requires 12 hours to complete the job.

3. A "one-for-one" or 100 percent time premium plan for incentive payment is in operation. The operator base rate for this class of work is $10.40. The base rate is guaranteed. Compute:
   a. Total earnings for the job at the efficiency determined in problem 2(*b*).
   b. Hourly earnings, from above.
   c. Total earnings for job at the efficiency determined in problem 2(*c*).
   d. Direct labor cost per piece from (*a*), excluding setup.
   e. Direct labor cost per piece from (*c*), excluding setup.

4. A forging operation is studied, and a rate of 0.42 minute per piece is set. The operator works on the job for a full eight-hour day and produces 1,500 pieces.
   a. How many standard hours does the operator earn?
   b. What is the operator's efficiency for the day?
   c. If the base rate is $9.80 per hour, compute the earnings for the day. (Use a 100 percent time premium plan.)
   d. What is the direct labor cost per piece at this efficiency?
   e. What would be the proper piece rate (rate expressed in money) for this job, assuming that the above time standard is correct?

5. A 60–40 gain sharing plan is in operation in a plant (base rate is quaranteed and operator receives 60 percent of proportional gain after exceeding 100 percent). The established time value on a certain job is 0.75 minute, and the base rate is $8.80. What is the direct labor cost per piece when operator efficiency is:
   a. 50 percent of standard?
   b. 80 percent of standard?
   c. 100 percent of standard?
   d. 120 percent of standard?
   e. 160 percent of standard?

6. In a plant where all the rates are set on a money basis (piece rates) a worker is regularly employed at a job where the guaranteed base rate is $8.80. This worker's regular earnings are in excess of $88 per day. Due to the pressure of work, the operator is asked to help out on another job, classified to pay $10 per hour. This employee works three days on this job and earns $80 each day.
   a. How much should the operator be paid for each day's work on this new job? Why?

    b. Would it make any difference if the operator had worked on a new job where the base rate was $8 per hour and had earned $72? Explain.

7. An incentive plan employing a "low-rate high-rate" differential is in use. A certain class of work has the guaranteed "low-rate" of $6 per hour and the "high-rate" for work on standard of $9.20 per hour. A job is studied and a rate of 0.036 hours per piece is set. What is the direct labor cost per piece at the following efficiencies:

    a. 50 percent?
    b. 80 percent?
    c. 98 percent?
    d. 105 percent?
    e. 150 percent?

SELECTED REFERENCES

Campion, Michael A., and Gina J. Medsker. "Job Design." In *Handbook of Industrial Engineering*, 2nd ed., edited by Gavriel Salvendy, Chap. 32. New York: John Wiley & Sons, 1992.

Dingus, Victor R., and Justice, Russell E. "Celebrating Quality," *Quality Progress,* November 1989, p. 74.

Eden, Dov, and Shlomo Globerson. "Financial and Nonfinancial Motivation." In *Handbook of Industrial Engineering*, 2nd ed., edited by Gavriel Salvendy, Chap. 31. New York: John Wiley & Sons, 1992.

Fay, Charles H., and Beatty, Richard W. *The Compensation Source Book.* Amherst, Mass.: Human Resource Development Press, 1988.

Hill, John. "Employees Blunder into Rewards." *USA Today,* May 22, 1990.

Lokiec, Mitchell. *Productivity and Incentives.* Columbia, S.C.: Bobbin Publications, 1977.

U.S. General Accounting Office. *Productivity Sharing Programs: Can They Contribute to Productivity Improvement?* Gaithersburg, Md.: U.S. Printing Office, 1981.

Von Kaas, H. K. *Making Wage Incentives Work.* New York: American Management Associations, 1971.

Zollitsch, Herbert G., and Adolph Langsner. *Wage and Salary Administration.* 2nd ed. Cincinnati: South-Western Publishing, 1970.

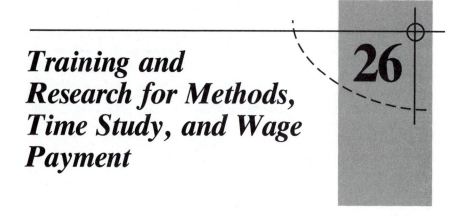

Training and Research for Methods, Time Study, and Wage Payment

26

In a survey conducted by Ralph E. Balycat on subjects found in industrial engineering curriculas and reported in the May 1954 *Journal of Industrial Engineering,* educators ranked motion and time study first in a listing of 41 subject areas. A somewhat similar survey made 10 years later and reported in the January 1964 *Factory* found that industrial engineers still spent most of their time on work measurement. This information was based on the response of 250 large U.S. manufacturing companies and was indicative of more than 8,700 nonclerical employees in industrial engineering. In 1977 R. S. Rice reported in the July issue of *Industrial Engineering* that approximately 89 percent of American and Canadian companies used work measurement techniques while only 71 percent used such techniques in 1960. Then in 1979 Neville Harris, Management Services in the United Kingdom, reported on a survey of industrial engineering practices in 667 firms there. Out of 32 management techniques, the three most frequently used were methods study, work measurement (direct), and incentive application.

Today, most of a practicing industrial engineer's work still centers on methods, standards, and wage payment. Table 26–1 shows how much time an industrial engineer spends on various major functions.

My own independent surveys indicate that most businesses and industries expect at least half of the industrial engineering effort to be directed toward work methods and standards. The utilization of industrial engineering techniques is spreading to all areas of the modern business, including marketing, finance, sales, and top management, yet the data input for almost all optimization techniques includes standard times. The importance of work measurement in such indirect areas as office activities,

TABLE 26–1
Percent of industrial engineer's time spent on major functions

		Percent
1.	Work measurement	33.4
2.	Work methods	21.1
3.	Production engineering	13.0
4.	Manufacturing analysis and control	9.9
5.	Facilities planning	8.6
6.	Wage administration	5.6
7.	Safety	2.6
8.	Production and inventory planning	2.0
9.	Quality control	1.1
10.	Other	2.7
		100.0

maintenance, shipping and receiving, sales work, inspection, and the tool-room will continue to grow.

To meet the demand, and to more quickly reap the benefits of training in this field, many industries have embarked on education programs conducted in their own plants on company time. For example, an extensive survey of over 5,300 U.S. companies revealed that 80 percent were providing formal training programs for first-line supervisors, and that 42 percent of these training programs dealt with work simplification and methods.

Production under a typical daywork environment is about 80 percent of normal productivity. Thus, by establishing standards after methods have been studied, improved, and standardized under daywork operation, an improvement of 25 percent in productivity can be expected.

Of course, companies must consider several cost items in introducing such a program. For example, if the annual wages of 1 operator represents 1 unit of direct labor payroll, 100 direct labor employees represent 100 units. Usually one capable methods and standards analyst can establish good methods and maintain reliable standards for approximately 100 employees. The cost of the methods and standards analyst can be estimated as three units of cost. To keep production records, make routine calculations, assist with the taking of studies, and so forth, a junior standards analyst engineering technician may be required at a salary of two units of cost. Let us estimate that inspection costs rise by two units in view of the added productivity. (This often is not the case—in fact, inspection costs frequently lessen because of methods improvement.) Secretarial help requires one unit of cost.

Thus, the total extra labor cost to introduce a methods and standards program per 100 direct labor employees is estimated at:

	Units
Methods and standards analyst	3
Junior standards analyst engineering technician	2
Secretarial assistance	1
Increased inspection	2
Total	8

This cost estimate should be adequate for most jobbing shops with a wide variety of operations and frequent changes of products. Costs should be considerably lower for well-standardized processes or mass-production plants.

With gross direct labor savings of 20 units and an additional administrative expense of 8 units, a good methods and work measurement program should effect a net savings of 12 units or 12 percent of a daywork direct labor payroll.

The savings in a fixed factory burden resulting from a work measurement program may be even more significant. Since a methods and work measurement program would increase productivity by an estimated 25 percent, then each dollar of fixed burden is distributed over 125 units, where formerly it was distributed over only 100 units. Thus, the fixed burden cost per unit of output is reduced to 100/125, or 80 percent, of the former rate. For example, if a company's fixed burden rate is 100 percent of the direct labor cost (this is a low estimate), the reduction in burden cost is equivalent to 20 percent of the direct labor payroll.

INDIVIDUAL PLANT METHODS TRAINING PROGRAMS

When properly undertaken, any methods training program is self-supporting. Several concerns that have introduced and gone ahead with training in methods have almost immediately realized substantial savings in all areas of their plants. One manufacturer of farm machinery gave a 64-hour course in methods analysis over 21 weeks. In attendance were 42 supervisors, assistant supervisors, time study analysts, and other key operating personnel. At the termination of the course, 28 methods projects were turned in, representing methods improvement solutions that could be introduced in the plant. These ideas included machine coupling; redesign of tools and products; paperwork simplification; improved layout; better means of material handling; improved dies, jigs, and fixtures; elimination of operations; adjusting of tolerances and specifications; and many other usable improvements that resulted in saving thousands of dollars annually.

In addition to the immediate gains mentioned, the training in operation analysis and work simplification developed an analytic approach on the part of the operating personnel so that in the future they were continually on the alert to find a "better way." They developed an appreciation of the cost of manufacture, and, at the completion of the course, were more cognizant of the relationship between output and selling price. By providing the means for on-the-job training in the field of methods and related areas, this company has gone a long way toward assuring its place in an extremely competitive market.

Training in Methods and Time Study

The lack of success of some time and methods study programs is in part due to a lack of understanding of the techniques by both management and operating personnel. One of the easiest ways to assure the success of any practical innovation is to inform all affected parties as to how and why it operates. When the theories, techniques, and economic necessity of methods, work measurement, and employee motivation are understood by all parties, little difficulty is encountered in their application. In unionized shops, it is especially important that the officers of the union at the local level understand the need for and the steps involved in establishing standards of performance. Training in the areas of performance rating, application of allowances, standard data methods, and job evaluation are especially essential. Companies that have provided training in the elements of time study for union officials and stewards, as well as representatives of management, have had harmonious relationships regarding methods, standards, and wage payment.

Management should sponsor training to acquaint various operating and supervisory members of the plant with the philosophies and techniques of time and motion study. In addition, industry must provide training for those of its personnel who plan to make time and motion study their life work. Not only is it necessary to train neophytes, but also experienced analysts should be continually checked to make certain that their conception of normal is not deviating from standard. New developments are constantly being made. As they are recognized, personnel in the methods, time study, and wage payment sections should be trained accordingly.

Any company that has, or plans to have, a program of work simplification or methods analysis, time study, work measurement, and incentive wage payment, should include as part of its installation a continuing training program. A two-hour training period once a week for supervisors, union stewards, direct labor, and management, can be well worth the time and money spent. Periodic verification of the rating ability of the time study staff is fundamental.

Developing Creativity

Creative work is not confined to a particular field or to a few individuals, but is carried on in varying degrees by people in many occupations: the artist sketches, the newspaper writer promotes an idea, the teacher encourages student development, the scientist experiments with a theory, and the methods and time study analyst develops improved methods of doing work.

Creativity implies newness, but often it is just as concerned with the improvement of old products as with the creation of new ones. A "how to produce something better" attitude, tempered with good judgment, is an important characteristic of effective methods and time study analysts.

Developing creativity in practicing methods and time study analysts is a continuing problem. Knowledge of the fundamental principles of physics, chemistry, mathematics, and engineering subjects is a good foundation for creative thinking. If practicing analysts do not have this basic background, they should acquire it either by education programs or else through study alone. Of course, knowledge is only a basis for creative thinking, and does not necessarily stimulate it.

The inherent personal characteristics of curiosity, intuition, perception, ingenuity, initiative, and persistence produce an effective creative thinker. Curiosity seems to stimulate more ideas than does any other personal characteristic. One aid to the development or restoration of curiosity is to train oneself to be observant. Methods and time study analysts should get into the habit of asking how a particular object was made, of what materials it was constructed, why it was designed to be of a particular size and shape, why and how it was finished as it was, and how much it cost. If they cannot answer these questions, they should find the answers, either through analysis or by referring to source materials and consulting others. These observations lead creative thinkers to see ways in which products or processes can be improved through cost reduction, quality improvement, ease of maintenance, or improved aesthetic appeal.

One significant creative idea usually opens up fields of activities that lead to many new ideas. Frequently one idea that has application on a given product or process has equal application on other products and similar processes.

DECISION-MAKING METHODS AND PROCESSES

Methods and time study analysts should become trained in the various decision-making processes. Judgment alone does not always provide the best answer. Analysts are continually confronted with such questions as: "Would it be economically wise to go ahead with this process?" "Should this tool be redesigned with the probability of 0.50 that production requirements will be increased by between 500 and 1,000 pieces?" "Will it be necessary to take 5,000 work sampling observations if we are seeking an element that occurs approximately 12 percent of the time and if tolerance limits are so much?" These are typical questions that must be answered; analytic means provide a much better method than does mere judgment.

To be able to handle problems similar to these, analysts should have a working knowledge of engineering economy, statistical analysis, and the theory of probability. They should also become familiar with the various decision-making processes for the evaluation of alternatives. For example, assume that an analyst has four alternatives to consider (a_1, a_2, a_3, a_4) which would be applied to four possible states of the product or market (S_1, S_2, S_3, S_4). Also assume that this analyst estimates the following outcomes for the various alternatives and states of the market:

Alternatives	States of product or market			
	S_1	S_2	S_3	S_4
a_1	0.30	0.15	0.10	0.06
a_2	0.10	0.14	0.18	0.20
a_3	0.05	0.12	0.20	0.25
a_4	0.01	0.12	0.35	0.25

If the outcomes represent profits or returns, and the state of the market will be S_2, then the analyst would definitely decide on alternative a_1. If the outcomes represent scrap or some other factor that the analyst wishes to minimize, then alternative a_3 would be chosen. (Although a_4 also has an outcome of 0.12, he or she chooses a_3, since under this alternative there is less variability as to outcome than with a_4.)

Seldom should decisions be made under an assumed certainty. Usually, some risk is involved in predicting the future state of the market. Assume that the analyst is able to estimate the following probability values associated with each of the four states of the market:

S_1	0.10
S_2	0.70
S_3	0.15
S_4	0.05
	1.00

A logical decision-making strategy would be to calculate the expected return under each decision alternative and then select the largest value to maximize or the smallest to minimize. Here

$$E(a) = \sum_{j=1}^{n} P_j C_{ij}$$

$$E(a_1) = 0.153$$
$$E(a_2) = 0.145$$
$$E(a_3) = 0.132$$
$$E(a_4) = 0.15$$

Thus, alternative a_1 would be selected to maximize.

A different decision-making strategy would be to consider the state of the market that has the greatest chance of occurring. This state of the market would, of course, be S_2, since S_2 carries a probability value of 0.70. The choice again, based on the most probable future, would be alternative a_1, with a 0.15 return.

A third decision-making strategy under risk is based on a "level of aspiration." Here, we assign an outcome value (C_{ij}) which represents the consequence of what we are willing to settle for if we are reasonably sure that we will get at least this consequence most of the time. This assigned value may be referred to as representing a level of aspiration which we shall denote as "A." Now determine the probability for each a_j that the C_{ij} in connection with each decision alternative is greater than or equal to "A." Select the alternative with the greatest $P(C_{ij} \gtrless A)$.

For example, assign the consequence of 0.10 to A to have the following:

$$(C_{ij} \gtrless 0.10)$$
$$a_1 = 0.95$$
$$a_2 = 1.00$$
$$a_3 = 0.90$$
$$a_4 = 0.90$$

Since decision alternative a_2 has the greatest $P(C_{ij} \gtrless A)$, it would be recommended.

Analysts may be unable to assign probability values to various states of the market with confidence and may, therefore, want to consider any one of them as being equally likely.

A decision-making strategy that may be used under these circumstances is based on the "principle of insufficient reason," since there is no reason to expect that any state is more likely than any other state. Here we compute the various expected values based on:

$$E(a) = \frac{\sum_{j=1}^{n} C_{ij}}{n}$$

In our example, this would result in:

$$E(a_1) = 0.153$$
$$E(a_2) = 0.155$$
$$E(a_3) = 0.155$$
$$E(a_4) = 0.183$$

and, based on this choice, alternative four would be proposed.

A second strategy analysts may consider when making decisions under uncertainty is based on the criterion of pessimism. When one is pessimistic, the worst is anticipated. Under a problem of maximization, the minimum consequence is selected for each decision alternative. The analysts compare these minimum values, and select the alternative that has the maximum of these minimum values. Thus, in our example:

Alternative	Min. C_{ij}
a_1	0.06
a_2	0.10
a_3	0.05
a_4	0.01

Here, alternative a_2 would be recommended since its minimum value of 0.10 is a maximum when compared with the minimum values of the other alternatives.

The plunger criterion is a third decision-making strategy that analysts may want to consider. This criterion is based on an optimistic approach. If one is optimistic, the best is expected, regardless of the alternative chosen. In a maximizing problem, analysts select the maximum C_{ij} for each alternative, and then select the alternative having the largest of these maximum values. Thus:

Alternative	Max. C_{ij}
a_1	0.30
a_2	0.20
a_3	0.25
a_4	0.35

Here, decision alternative a_4 would be recommended because of its maximum value of 0.35.

Most decision makers are neither completely optimistic nor completely pessimistic. Instead, a coefficient of optimism, X, is established where:

$$0 \le X \le 1$$

Then, a Q_i is determined for each alternative, where:

$$Q_i = (X)(Max. \ C_{ij}) + (1 - X)(Min. \ C_{ij})$$

The alternative recommended is the one associated with the maximum Q_i for maximization and with the minimum Q_i for minimization.

A final decision-making approach based on uncertainty is the minimax regret criterion. This criterion involves the calculation of a regret matrix. For each alternative, based on a state of the market, analysts calculate a regret value. This regret value is the difference between the payoff actually received and the payoff that could have been received if the decision maker had been able to foresee the state of the market.

To construct the regret matrix, select the maximum C_{ij} for each state S_j and then subtract from this maximum the C_{ij} value of each alternative associated with that state. In our example, the regret matrix would be:

| Alter- | States | | | |
natives	S_1	S_2	S_3	S_4
a_1	0	0	0.25	0.19
a_2	0.20	0.01	0.17	0.05
a_3	0.25	0.03	0.15	0
a_4	0.29	0.03	0	0

Now select the alternative associated with the minimum of the maximum regrets (minimax).

Alternative	Max. r_{ij}
a_1	0.25
a_2	0.20
a_3	0.25
a_4	0.29

Based on the minimax regret criterion, a_2 would be selected in view of its minimum regret of 0.20.

Different decision-making processes may very well give different answers. Analysts should become familiar with these decision making strategies and make use of those that are most appropriate to their organizations.

LABOR RELATIONS AND WORK MEASUREMENT

Every business owner recognizes the importance of harmonious labor-management relations. Sound work measurement philosophies and practices do a great deal to promote good relations between labor and management. Lack of consideration of the human element in work measurement procedures causes sufficient turmoil to make the profitable operation of a business impossible. Management should identify and implement conditions most likely to enable employees to achieve the organization's objectives.

Union Objectives

To understand the relationship between work measurement and labor relations, analysts must understand the objectives of the typical labor union. Briefly, the principal objectives of the typical union are to secure for its members higher wage levels, decreased working hours per workweek, increased social and fringe benefits, improved working conditions, and job security. The philosophy underlying the union movement has in the past had much to do with the opposition of organized labor to incentive systems. Unions formerly looked upon themselves primarily as fighting units that united workers by seeking ends common to all members. It was not to the advantage of the early unions to emphasize differences in workers' abilities and interests; to do this would increase rivalries and jealousies among their members and potential members. Consequently, organized labor usually sought percentage wage increases for all members of a group rather than means by which remuneration would be adjusted to the worth of individual workers. The work of methods standards and wage payment analysts came to be looked on as means by which management sought to destroy the solidarity of workers by stressing the differences in their capabilities.

The enactment of government legislation, however, has changed the status of labor unions. Management today recognizes the union as a bargaining agent for its employees. Unions, therefore, are less fighting units than bodies concerned with the orderly negotiation of wage contracts for all their members. Also many of their members are not content with wage negotiation concerned only with obtaining high minimum wages for the entire group, and leaving the determination of extra rewards for more valuable workers entirely up to management. To satisfy most of their members, unions must obtain equitable wages (recognizing workers' different skills and qualities) as well as high wages for all; in fact, unions have already done this in many instances. Methods, time standards, job evaluation, merit rating, and incentive systems are tools for assuring equitable wages and good working conditions. Soon they may be as important to organized labor as to management.

Labor executives, in bargaining over wage contracts, are in an excellent position to obtain safeguards for the proper handling of the development of work methods and standards and for fair wage payment practices. They may, for example, demand that provisions be inserted in contracts: (1) forbidding the reduction of standard times without a methods change; (2) requiring minimum hourly rates and payment for time lost due to no fault of the workers; (3) establishing grievance procedures for handling all workers' complaints arising out of the functioning of job evaluation, merit rating, and wage incentives; and (4) even giving labor the right to participate in job- and worker-rating activities, the setting of standard times, and the determination of piece rates.

Many unions today train their own work measurement personnel. In most instances, however, these time study people are employed to check standard times and to explain them to workers, rather than taking part in their initial establishment. However, the training the union time study analysts receive usually treats the concepts, philosophies, and techniques of methods, work measurement, and wage payment from a different point of view than that of management's time study persons. For example, in a recent collective bargaining report published by the AFL–CIO for the training of union stewards in motion and time study, the subject of time study is introduced as follows:

> Time study, which is widely used for determining work loads and wage incentive standards, is an imprecise tool and lends itself to easy abuse. Unions confronted with it must be consistently on guard against the use of arbitrary, unreasonable and unrealistic time study results.
>
> Of the whole field of so-called "scientific management," time study is the area in which most of labor's distrust and suspicions are centered. Ever since its introduction in the 1880's most unions have opposed the use of stopwatch time study.
>
> Labor's distrust stems from its practical experiences with time study. Problems arise both because of the inherent shortcomings of the time study process itself as well as the application of the technique in industry.
>
> Time study is usually represented to unions as "scientific." But time study produces results which are simply judgments. They are not, and cannot be, scientific or accurate: at best they represent approximations, and at worst they are no better than wild guesses.
>
> Time study is supposed to be a method of determining the time which should be allowed for a worker to perform a defined job according to a specific method and under prescribed conditions.
>
> Time studies are usually made by timing workers with a stopwatch while they are doing a certain job. This time is then adjusted for such factors as delays, fatigue, personal needs and incentive factors. The result is frequently called a standard. The job or time standard may be in terms of units per hour, standard hours per 100 units or time per unit. If an incentive plans exists, the standard may also be expressed in monetary terms, such as 1 cent per piece produced.

Since most time studies are taken of an individual worker or a small group of workers, disagreements over time studies are usually expressed as grievances on specific job standards at the local union level.

While international unions can and do provide expert assistance and information to their locals, the investigation and processing of time study disputes remain primarily a local union problem.

Based on their own experiences, local unions have devised approaches to time study which fall into four general categories:

1. Some locals prevent the use of time study altogether.
2. Some locals allow management to use any method of setting job standards it desires, but the locals reserve the right to bargain on the results.
3. Some locals participate directly with management in making time studies and in setting standards.
4. A majority of locals allow management to make time studies but insist on bargaining on both the methods used and their applications.

Unions faced with time study must be certain that they have complete information on how it is used by their company. In order to provide essential protection to members, this information must not be limited in any respect. It must include not only the results of time study, that is, the individual job standards, but also the plan in use by the company and the exact procedures followed.

That this information is essential to collective bargaining is apparent. Without it, a union would be unable to sensibly process grievances or discuss pertinent contract clauses.

The union's legal right to such information has been definitely established by decisions of arbitrators and the National Labor Relations Board.

In spite of this, some unions are still finding it difficult to secure all the necessary data. Some managements still claim that such information is confidential. When union pressure forces compliance, some managements attempt to restrict the information they will provide or the means of providing it.

To insure immediate availability of time study data, without question or limitation, most unions insist on a contract clause such as:

The company shall furnish to the union a copy of the time study plan presently being used. It shall make available for inspection by the union any and all records pertaining to time study and the setting of production standards including original time study observation sheets. Upon request by any shop steward or union officer, the company will furnish copies of any of the above information including copies of time studies.

Some companies, by contract, have been required to furnish the union with photostatic copies of the time study observation sheet each time a study is made.

Accuracy of Time Studies

We do not bargain standards!! is a frequent assertion by management when faced with union time study grievances.

Management claims that time studies produce facts and, of course, facts are not subject to bargaining or compromise.

If time study did result in facts, then this management position might be sound and the union might be able to bargain only on whether or not time study should be used in the plant.

But such is not the case with time study. At best, standards developed from time study are only approximations. They involve the use of considerable judgment by the time study man at every step of the time study procedure.

The actual recording of the time involves the least judgment of all, but even here a 10 percent error is expected and recognized by time study experts.

Among some of the variable factors which will significantly affect a time study result are:

1. The selection of the worker to be studied.
2. The conditions under which the work is performed during the time study.
3. The manner in which the operation is broken down into parts or elements.
4. The method of reading the stopwatch.
5. The duration of the time study.
6. The rating of the worker's performance.
7. The amount of allowances for personal needs, fatigue and delays.
8. The method of applying the allowances.
9. The method of computing the job standard from the timing data.

Inaccuracies of "Rating" Process

After a time study man has completed the stopwatch timing, he has obtained a figure which represents the time, on the average, that it took the particular worker he observed to perform the job.

This time could be used to set a standard only if the worker who has been time studied could be considered a qualified, average worker, working at a normal place exhibiting a normal amount of skill.

If the worker observed does not meet these specifications, then the observed time must be adjusted to make it conform to the normal time. This adjusting procedure most frequently is called "rating," but is sometimes called "leveling" or "normalizing."

Rating has been defined as the process whereby the time study man compared the actual performance which he observed in a concept or idea of what normal work performance would be on the job being studied. This concept of "normal" must by its nature be carried around in the time study man's head.

During the rating process, the time study man must make two distinctly personal judgments. First, he must formulate in his mind the concept of what a normal performance should be on the job being studied and second, he must numerically compare the observed performance with his mentally conceived normal performance.

The very nature of rating opens it to abuse. By manipulating this rating factor, it is easy for the time study man to end up with practically any result he chooses.

In fact, as many unionists know, the rating factor is often used to enable the time study man to end up with a standard determined before the time study is taken. In other words, time study is often used to "prove" to the workers that a workload or standard set by the company is fair.

Time Study Men "Adjust" Findings

For example, a company may decide, without measurement of any kind, that a certain item should be produced at the rate of 60 pieces an hour or one piece per minute. Its time study man then studies the job and finds that the average time it took the operator to make one piece was 1¼ minutes. This means that only 48 pieces would be produced in an hour. Obviously, this falls far short of management's set goal of 60 pieces per hour.

So the time study man simply decides that the operator was working below "normal" during the time study and that a "normal" operator would have worked faster. The time study man then adjusts his findings; he cuts the 1¼ minutes that it took the "slow" operator by 20 percent—to come up with the standard that the company wanted of 1 piece per minute or 60 pieces per hour.

It is practically impossible for a union to prove this kind of deliberate deceit since the "normal" operator is only a figment of the time study man's imagination.

So it is easy to see that unscrupulous time study men can readily manipulate time studies.

But the situation unfortunately is not much better even when management and its time study men are trying to be completely honest and objective.

Test of Time Study Men's Estimates

Most time study men claim that they can judge worker pace within 5 percent; some have even claimed that with experience this could be reduced to an average error of 2 percent. But these claims of precision have been proved false by many studies by university researchers and management groups to determine the ability of time study men to rate workers' performance.

These studies show that time study men will, in more than half their ratings, misjudge by more than 10 percent variations in work pace. Such errors of as much as 40 percent are not at all uncommon among qualified and experienced time study men.

Specifically, the Society for the Advancement of Management recently conducted one of the most extensive rating studies ever undertaken.

The study used films of workers performing a variety of industrial operations. Time study men rated different paces of these operations. (From the calibration of the film, it was possible to get an accurate measurement of variations in pace.)

Trade unionists were not surprised at the published results of the SAM study. They showed, for example, that for 599 time study men:

1. The average error in estimating variations in work pace was 10.57 percent.
2. Fifty-nine percent of the time study men had average errors larger than 10 percent; 41 percent averaged less than 10 percent error; and less than 12 percent had errors averaging below 5 percent.

One man's errors averaged as high as 22 percent. These, of course, are average errors. Some individual ratings were lower and some were higher than the averages. The SAM did not publish the range of errors, but similar studies have shown errors higher than 40 percent.

Burden on Grievance Procedure

Time study has become a major issue in union-management relations. Unions have found that grievance problems are greatly increased in plants where time study exists. Not only are there more grievances, but a greater amount of time must be spent in investigating and processing time study grievances.

One local union reported that in the first five months of a year, it had carried 254 grievances to the fourth step of the grievance procedure. (That's when the grievance committee meets with the plant superintendent.) Of these 254 grievances, 221 or 87 percent were time study cases.

Local Union Representatives Can Handle Time Study Problems

Some unionists have felt that time study grievances could not be handled in the same manner as other grievances. Some have felt that time study grievances could not be handled by local union representatives but that "outside" experts were needed.

While local unions may seek help in special cases, the majority of time study grievances not only should, but can be successfully handled by shop stewards. See the listing starting below under "Union Handling of Time Study Grievances," for points to be considered in handling disputes over production standards.

The most important factor to remember is that no time study is accurate. Union judgment is on a par with management's. The worker on the job is as accurate a judge as anyone of the propriety of any job standards.

Because management does not like to bargain on standards and because there is so little that is factual about time study, a large proportion of time study grievances are taken to arbitration.

Unfortunately, while many arbitrators may not be biased in favor of the company or union, they all too often accept time study as being scientific and mathematically precise. They do not recognize the shortcomings of time study and therefore are more likely to accept its results.

Despite their difficulties and shortcomings, local unions can do a good job of protecting workers against unfair management time studies. To do an effective job, however, locals need informed and alert officers and stewards. Above all union representatives should not be "snowed under" by the so-called scientific procedures and arguments of management time study men.

In locals where trouble with time study persists, the international union should be requested to give advice and assistance. Various international unions have staff members with specialized experience in handling time study difficulties.

Union Handling of Time Study Grievances

Time study grievances should be handled the same as any other grievances. The most important factor is that of getting facts.

While some knowledge of time study is helpful, it is not necessary for the shop steward or other union representative to be a time study man in order to process a time study grievance.

The union representative should:

1. Secure a copy of the company's record of the operation in dispute.
2. Make certain that the records of job conditions and job description are complete. If either is incomplete, it will be impossible to reproduce the job as it was when the time study was made, and therefore the company's time study cannot be checked. This alone is grounds for rejecting the study.
3. If the time study sheet does contain sufficient information as to how and under what conditions the job was being performed when the time study was made, then it is necessary to determine whether the job is still being performed in exactly the same way now.

   Check to see if there has been any change in machines, materials, tools, equipment, tolerances, job layout, etc. Any change should be checked to see if it affects the ability of the operator to produce at the same rate as that achieved during the time study.
4. Usually the total operation or job cycle is broken down for timing purposes into parts which are called elements. Check the elemental breakdown of the job on the time study sheet. See that the beginning and ending point of each element is clearly defined. If it is not, any attempt to measure elements and give them a time value was pure guesswork.
5. Check the descriptions of each element to see if they describe what the operator is presently required to do. Any change invalidates the original study.
6. Make sure that everything the operator is required to do as part of the job has been recorded and timed on the time study sheet. Watch for tasks which are not part of every cycle.

   Such things as getting stock, adjusting machines, changing tools, reading blueprints, and waiting for materials are examples of items most frequently missed.
7. Check for strike-outs. A time study finding is based on a number of different timings of the same job. The time study man may discard some of his timings as abnormal.

   If the time study man has discarded any of his recorded times, he must record his reasons for doing so. This enables the union intelligently to determine if the strike-out is valid. The fact that a particular time is larger or smaller than other times for the same element is not sufficient reason for discarding it.
8. Determine if the time study was long enough to accurately reflect all of the variations and conditions which the operator can be expected to face during the job. Was it a proper sample of the whole job? If not, the time study should be rejected, as its results are meaningless.
9. See that only a simple average, and not the median, mode or other arithmetic device was used in calculating the elemental times. The average is the only proper method for time study purposes.
10. Check the rating factor on the time study sheet. Try to find out if the time study man recorded his rating factor before leaving the job or after he computed the observed times. Ask the operator who was time studied if he feels the rating factor is a proper one.

> Watch the worker who was timed work at the pace he considers proper and then at the pace required to produce the company's workload. The judgment of the worker and steward are as valid as that of the time study man.
> 11. Make sure that allowances for personal time, fatigue and delays have been provided in proper amounts.
> 12. And finally check all the arithmetic for errors.

A final note:

> Most union representatives have found it unwise to take additional time studies themselves as a check, except as a last resort. It is more effective to show the errors in management's study than to try to prove a new union time study is a proper one.
> A time study taken by a union is still only the result of judgment. Even when proper methods are used, they tend merely to reduce the inconsistencies of time study, not eliminate them.[1]

Company training of union time study representatives has been quite successful, in many instances, as a means of promoting a more cooperative atmosphere in installing and maintaining methods, standards, and wage payment systems. This procedure provides for the joint training of company and union personnel. Having received this training, union representatives are much more qualified to evaluate the fairness and accuracy of the technique and to discuss any technical points relative to a specific case.

Employee Reactions

In addition to having an understanding of the unions' objectives and its attitudes toward the methods, standards, and wage payment approach, analysts must have a vivid understanding of the psychological and sociological reactions of operators.

These three points should always be recognized: (1) most people do not respond favorably to change; (2) job security is uppermost in most workers' minds; and (3) people have a need for affiliation and consequently are influenced by the group to which they belong.

Most people, regardless of their positions, have an inherent resistance to changing anything associated with their work patterns or work centers. This is due to several psychological factors. First, change indicates dissatisfaction with the present situation; there is a natural tendency to defend the present way since it is associated intimately with the individual. No one likes others to be dissatisfied with their work; if a change is even suggested, the immediate reaction is to expound on why the proposed change will not work.

---

[1] Collective Bargaining Report prepared by Department of Research, AFL–CIO, *American Federationist,* November 1965.

Second, people tend to be creatures of habit. Once a habit is acquired, not only is it difficult to give up, but also there is resentment if someone endeavors to alter the habit. Anyone in the habit of eating at a certain place is reluctant to change to another restaurant even though the food may be better and less expensive.

Third, it is only natural that people desire security. The desire to feel secure in one's position is just as basic as the instinct for self-preservation. In fact, security and self-preservation are related. Most workers prefer job security over high wages when choosing a place to work.

Fourth, to the worker, all work relating to methods and standards results in either a change or an effort to increase productivity. The immediate and understandable reaction is to believe that if production goes up, the demand will be filled in a shorter period; and that without demand, there will be fewer jobs.

The solution to the problem of job security lies principally in the sincerity of the management leadership. When methods improvement results in job displacement, management has the responsibility to make an honest effort to relocate those who have been displaced. This may include provision for retraining. Some companies have gone so far as to guarantee that no one will lose employment as a result of methods improvement. Since the labor turnover rate is usually greater than the improvement rate, the natural attrition through resignation and retirement can usually absorb any displacement of people as a result of improvement.

Fifth, the sociological need for affiliation and the resulting impact of "behaving as the group wants everyone to behave" also influence change. Frequently the worker, as a union member, feels that he or she is expected to resist any change that has been instituted by management; accordingly, the worker is reluctant to cooperate with any contemplated changes resulting from methods and standards work. Another factor is the resistance toward anybody who is not part of one's own group. Just as any parent would take offense to a neighbor suggesting that his son part his hair on the other side, so does a member of a group take offense when a member of another group suggests changes. A company represents a "group" which has several groups within its major boundaries. These individual groups respond to basic sociological laws. Change proposed by someone outside one's own group is often received with open hostility. The worker is associated with a different group than that of the practitioner of methods and standards, and tends to resist any effort analysts make that might interfere with the usual performance within the group.

The Human Approach

The approach and behavior of practitioners of methods engineering and work measurement are the real key to good management-labor relations and the success of the whole production system. Too many people in this field do not appreciate the importance of the human approach. The human approach recognizes three fundamental needs of all people: achievement, affiliation, and power.

When doing methods work, analysts should take time to talk to the operators and get their ideas and reactions. The work progresses so much more smoothly and effectively if operators become part of the team. However, they must be asked to "join the team." The operators are closer to their job situations than anyone else and usually have more specific knowledge of details than anyone else. This knowledge should be realized, respected, and utilized. Accept operators' suggestions gratefully; if they are practical and worthwhile, put them into effect as soon as possible. If they are used, be sure that the operators are appropriately rewarded. If they cannot be used at present, give a complete explanation as to why they cannot be used. This approach responds to operators' three basic needs.

Analysts should take time to explain all facets of the methods-standards work undertaken, and point out how this work improves the competitive situation of the company. Emphasize that both the security of a business and the job security of its employees depend on the business's earning a profit.

As the work progresses, analysts should keep the operator informed on the probable results of the analysis; take time to answer questions related to any problems the operator may have in relation to work or to contemplated changes in work.

At all times analysts should imagine themselves in the workers' place and then use the approach they would like used toward themselves. Friendliness, courtesy, cheerfulness, and respect tempered with firmness are the human characteristics that must be practiced to be successful in this work. In short the golden rule must be applied.

To summarize the "how" of the human approach, consider Dale Carnegie's advice on handling people, making people like you, influencing the thinking of people, and changing people. Carnegie summarizes as follows:

Fundamental techniques in handling people.
1. Instead of condemning (criticizing) people, try to understand them.
2. Remember that all people *need* to feel important; therefore, try to figure out the other person's good points. Forget flattery; give honest, sincere appreciation.
3. Remember that all people are interested in *their* own needs; therefore, talk about what they want and show them how to get it.

Six ways to make people like you.
1. Become genuinely interested in other people.
2. Smile.
3. Remember that a person's name is to him or her the sweetest and most important sound in the English language.
4. Be a good listener. Encourage others to talk about themselves.

5. Talk in terms of the other person's interest.
6. Make the other person feel important—and do it sincerely.

Twelve ways to win people to your way of thinking.
1. The only way to get the best of an argument is to avoid it.
2. Show respect for the other person's opinions. Never tell anyone they are wrong.
3. If you are wrong, admit it quickly and emphatically.
4. Begin in a friendly way.
5. Get the other person saying yes, immediately.
6. Let the other person feel that the idea is his or hers.
7. Let the other person do a great deal of the talking.
8. Try honestly to see things from the other person's point of view.
9. Be sympathetic with the other person's ideas and desires.
10. Appeal to the nobler motives.
11. Dramatize your ideas.
12. Throw down a challenge.

Nine ways to change people without giving offense or arousing resentment.
1. Begin with praise and honest appreciation.
2. Call attention to people's mistakes indirectly.
3. Talk about your own mistakes before criticizing the other person.
4. Ask questions instead of giving direct orders.
5. Let the other person save face.
6. Praise the slightest improvement and praise every improvement. Be hearty in your approbation and lavish in your praise.
7. Give the other person a fine reputation to live up to.
8. Use encouragement. Make the fault seem easy to correct.
9. Make the other person happy about doing the thing you suggest.

RESEARCH IN METHODS, TIME STUDY, AND WAGE PAYMENT

Many of the techniques presented here are not based on a scientific method, but on judgment, experience, and iterative methods. This is not because of choice on the part of the industrial engineer associated with methods, time study, and wage payment, but because the technology does not exist today to provide a scientific method for many aspects of this work. The techniques described in this text have been proven in countless instances; they can and do produce acceptable solutions, and they have contributed significantly to the economy of the country. However, there is no doubt that many of these techniques can be improved. In particular, the following areas need more research so that objective, quantitative methods can assure that judgment, bias, and favoritism cannot influence solutions to problems.

Methods

1. Worker and machine relationships where arrival and service times are not characteristic of typical distributions.

2. Production and assembly line balancing.
3. Human capabilities—mental and physical: The effect of age, sex, physical structure, intelligence, and manual dexterity on workers' performance.
4. Human behavior patterns: The effect of education, home life, social groups, and the like on workers' output.
5. Environmental factors: The effect of working conditions—such as noise, temperature, humidity—on workers' output.
6. The digitizing of video information for computer analysis to develop ideal work centers.

Time Study
1. Allowances: The influence of environmental factors and different work situations on fatigue.
2. The appropriate time to take the time study.
3. Learning curves: Their shape for different work situations.
4. Performance rating: Development of an objective method that gives consistent, reliable results.
5. Fundamental motion data:
   a. The additivity of fundamental data.
   b. The time required for combined motions.
   c. More data on body motions.
6. Automated data gathering and analysis employing electronic devices coupled with the digital computer.

Wage Payment
1. Job evaluation: More objective means of determining job worth.
2. Nonfinancial incentives: Human wants, reasons for motivation, and so on.

Methods

Although simulation techniques, such as Monte Carlo, provide a solution to worker and machine studies of systems too unique for existing mathematical models, the solutions may not be optimal. Past records become a basis for probability distributions; this procedure alone can introduce data whose reliability may be questioned. The assignment of a random number to a block of time can also be questioned. The size of the block of time and the point within the block of time in which the event (work stoppage, delivery, and so forth) occurs are points of concern.

The classical assembly line balancing problem does not provide usable solutions in most real-life situations. Some of our most advanced industries still sort decks of cards, each card representing a standard for an operation, until a satisfactory balance has been determined by manual means. The General Electric Company publication, "Assembly Line Balancing," states: "There is no question but what a capable methods and time standards technician can improve on any computer-generated assignment by a few percentage points. Our goal should be to give him a very good base, much better on the average than he could do manually, from which he can make intelligent and effective improvements.'' This statement speaks for itself as to the need for more analytic work in this area.

The standardization of the established method advocated by practitioners of methods, standards, and wage payment is certainly not in agreement with the views held by many psychologists. A question arises concerning the extent to which the operations of the worker-machine system can be divided, and the job resynthesized and standardized for use by every individual. The motivation of the worker, as affected by the sociological system of the community and the plant, may influence a constant cause system.

Much research needs to be done in job placement, where job requirements are matched with the inherent capabilities of employees. We all know of excellent milers who would come in last in a respectable hundred-yard dash. We also know first-class baseball players who would be most awkward on the basketball court. Many industrial assignments bear no information as to which human capabilities are suited for the work.

Time Study

With the technology available today, it is not feasible to develop a practical, deterministic, mathematical structure of the work measurement problem. The complexity of the mathematical relationships, involving variables as yet unmeasured, is so great that mathematics has not yet developed the power to treat a phenomenon as complex as work measurement.

In William Gomberg's text *A Trade-union Analysis of Time Study,* four sources of variation that influence the output of the operator are mentioned: mechanical, physiological, psychological, and sociological.[2] It is Gomberg's contention that these sources of variation result in a chance cause system that cannot be stabilized sufficiently to be measured over an extended period. Since present-day work measurement is a sampling approach in which the time for a unit of work is assumed to be a random variable, it may well be that the effect of all sources of variation results in a variable chance cause system which cannot be stabilized sufficiently to be measured accurately over an extended period.

For example, such mechanical effects as temperature, light, heat, humidity, and noise are suspected of affecting the fatigue accumulation of the worker, but little is known of the effect of fatigue on the constant chance cause system. Fatigue, a primary source of physiological

[2] William Gomberg, *A Trade-union Analysis of Time Study* (Chicago: Science Research Associates, 1948).

variation, has been a stumbling block, as there is no satisfactory method of measuring it. Much experimentation has tried to relate fatigue to physical output, but with incomplete success. Some researchers have tried to relate fatigue with Kefosteriod excretion, with some degree of success.[3] Some manufacturing concerns (Du Pont and Eli Lilly) are correlating the fatigue of different jobs through the measurement of energy consumed. This is done by measuring the heartbeat, and correlating it with oxygen consumption. Studies have indicated that the average man should expend no more than 2,500 calories during an eight-hour working day.

In practice, a fatigue allowance is often set by collective bargaining for the development of plantwide standards. This practice results in realistic allowances in only a portion of the developed standards, in that fatigue undoubtedly varies with the type of operation and the conditions surrounding it. Surely, a different fatigue allowance should be applied for different classes of work.

Selecting the appropriate time to take a time study is a realistic problem. Abruzzi recommends that a criterion of "local and grand stability" be satisfied before an operation is ready for measurement.[4] He suggests using the Shewhart control chart technique. The local stability criterion covers continuous observations taken over a few hours during a day, while the grand stability criterion includes small samples of observations taken over an extended time interval considered to be under homogeneous conditions. In applying Abruzzi's criterion, consider using the learning curve. This curve is generally conceded to be exponential in form and, consequently, it would be most appropriate to apply the stability criterion after the knee of the curve has been reached.

Assuming the validity of the statistical formulation, analysts should recognize that an unknown mean performance time for productive work is to be estimated by work measurement. In general, measurement systems have three major components: a sensing mechanism, a translation mechanism, and a readout system. Each of these sources is capable of contributing bias error (positive or negative) and a variable error to the measurements, resulting in the total error for the system.

Probably the most fertile area of research lies in performance rating. All performance rating systems establish a "within company" concept of a "normal" rate of work. Arbitrary times for benchmark jobs, rated motion-picture and/or videotape films of representative jobs, and the concept of normal embodied in predetermined time systems or in the views of experienced time study technicians may be bases for adopting the concept of normal for a company. Research needs to develop an objective means of rating that eliminates or minimizes the need of judgment on the part of the analyst. Perhaps the synthetic rating method using the nomogram

---

[3] Hoagling and Pincas.

[4] Adam Abruzzi, *Work Measurement* (New York: Columbia University Press, 1952).

suggested by the author can point the way toward more valid performance rating. In this connection, research needs to determine the variability in performance of successive elements within an operation.

The development of fundamental motion data certainly represents one of the most significant contributions of industrial engineering. MTM, Work-Factor, MOST, MODAPTS and other systems provide means for establishing realistic standards in a short time by semitechnical personnel after a relatively short (three to six months) training period. However, much research still needs to be done in this area.

The fact that fundamental motion data is not additive on certain motion patterns has been established in several research papers. Thus, more studies need to be made on combined motion patterns, and data need to be established for these patterns.

Existing fundamental motion data on body motions is inadequate. For example, data need to be established for body motions (walking and other) under load and when performed at different angles of inclination.

Job Evaluation

In the establishment of equitable base rates, more research needs to be done. Job evaluation removed some of the bias and favoritism characteristic of earlier wage payment programs. However, even with job evaluation, job descriptions can be manipulated so that more or less points are assigned and, consequently, the base rate is affected. There is still a real need to develop a completely fair method for determining base rates.

There is much need of research in the whole area of nonfinancial incentives. What do most people really want out of life after earning a satisfactory wage? How can these objectives be incorporated as part of every job? What barriers are inadvertently introduced in every job through poor supervision, poor communication, and poor leadership?

Wage Payment

As industry and business become increasingly automated, the responsibilities of workers become more diverse making precise measurement of work more difficult. Consequently, one-for-one type incentive plans such as the standard hour plan are difficult to install and maintain. Yet it is generally conceded that, given the opportunity and the due rewards, employees, in an environment of mutual trust and cooperation, willingly maintain above-normal productivity levels. Incentive plans characteristic of the Scanlon Plan, Rucker Plan, IMPROSHARE, profit sharing, and others that are based on the productivity-sharing concept allow employees to create an environment where human ingenuity and commitment to quality as well as performance are a way of life. Such innovative plans, when properly installed and maintained by competent management may do much to unlock and channel the productive and quality potential of a work force.

CAD/CAM

With the increasing use of computers in connection with computer-aided design and computer-aided manufacturing, there is the opportunity to design for producibility thus assuring ideal methods, and also to develop work standards as part of the CAD/CAM process. However more research and development needs to take place in identifying and storing the necessary information in the computer so that the best sequence of the correct operations can be selected. For example, dynamic decision equations need to be developed to evaluate competing processes to determine the most advantageous way to perform a given operation. Also needed are parameters identified and quantified for computer computation including the following: quantity to be produced, material being processed, size of part being processed, geometrical configuration desired by the operation, tolerance needed, and so on.

Industry has already proven that for complete, successful CAD/CAM installation there must be cooperation between the functional designer, manufacturing engineer, industrial engineer, quality assurance personnel, and data processing personnel. As each of these entities learns about the others' problems and responsibilities, the computer system can increase its effectiveness in the complete planning for producibility, which involves good methods and standards determination in the planning stage.

Today, there is wide use of electronic data gathering devices with storage capability. This trend will continue and in time should replace the stopwatch in the development of most work standards.

Much greater use will be made of videotape recorders in the future. Compact, portable, reliable 8 mm video systems provide analysts with factual data for the development of ideal work centers and input to computers for standards development based on standard data and/or fundamental motion data.

SUMMARY

To a large extent, the work of motion and time study analysts influences labor relations within an enterprise. Therefore, analysts must understand the objectives of the union that represents workers in their plants. Analysts should know the nature of the training that officers in their companys' locals are receiving. With this information they can understand the attitudes and problems of workers on the bench. Today, and in the immediate future, quality of output must be paramount in the eyes of both labor and management.

At all times analysts must be cognizant of the necessity of using the human approach. They must always ask for and develop methods, procedures and standards that are fair to both the company and the operator. Improvement in both quality and output needs to be a way of life.

Analysts must recognize that much of the work relating to methods, standards, and wage payment is not based on science, but on judgment and experience. Consequently, they should monitor current research in the field. This will enable them to introduce quantitative techniques to supplement or supplant those that can result in inequity. They should also take an active part in professional societies that are continually exploring the field to develop more objective techniques.

TEXT QUESTIONS

1. Outline the objectives of a typical labor union.
2. Why have unions in the past sought flat "across-the-board" wage increases for their members?
3. Explain why the union training program says that time study is an "imprecise tool and lends itself to easy abuse."
4. Which four approaches to time study are suggested by the typical union?
5. Which three points related to the psychological and sociological reactions of the operator should be recognized by the analyst?
6. What do we mean by "the human approach"?
7. Name 12 ways you can get people to agree with your ideas.
8. Is methods training conducted within a plant self-supporting?
9. Why is plantwide training in the areas of methods and time study a healthy management step?
10. Why should training in time study be looked on as a continuing project?
11. Why is American industry more receptive to time study analysts today than it was prior to World War II?
12. How does diversified industry rank time study in the development of an industrial engineering curriculum?
13. What intangible benefits can be achieved through a methods training program on a supervisory level?
14. Why should experienced analysts be continually checked on their ability to performance rate?
15. How can a person develop creative ability?
16. What are some of the areas where research needs to be done in methods? In standards? In job evaluation?
17. What is meant by "local and grand stability"?
18. Explain how videotape recorders can be used with computers.
19. Will the typical stopwatch be obsolete by the year 2000? Why?

GENERAL QUESTIONS

1. Who should handle in-plant training in methods, time study, and wage payment?
2. Is it advisable to provide training in the area of methods, time study, and wage payment down to the operator level? Why or why not?

3. To which positions within an industry can experienced analysts in methods and time study expect to be promoted?
4. What is the relation between work measurement and operations research?
5. Will "automation" diminish the need for the methods and time study analysts?
6. Which government legislation has changed the status of labor unions?
7. Why do unions often train their own time study analysts?
8. Would a young engineer find the work in the time study division of the CIO a challenging opportunity? Why or why not?

PROBLEMS

1. Based on the cost relationships presented in this chapter, what would be the total annual dollar savings (estimated) of a company having an annual direct labor payroll of $2,500,000 and a fixed burden rate of 150 percent of direct labor if it initiated a methods and standards program?
2. A company employing straight daywork as a method of wage payment is compensating its employees an average of $18.00 per hour. In addition, the cost of fringe benefits is running 30 percent of direct labor. Overhead in this company is 125 percent of direct labor. A methods, standards, and incentive plan is being contemplated in which the average incentive earnings have been estimated as equaling 20 percent of base wages. What payoff will the proposed plan yield?

SELECTED REFERENCES

Majchrzak, Ann. "Management of Technological and Organizational Change." In *Handbook of Industrial Engineering,* 2nd ed., edited by Gavriel Salvendy, Chap. 29. New York: John Wiley & Sons, 1992.

Glossary of Terms Used in Methods, Time Study, and Wage Payment

1

Abnormal time. Elemental time values that are either considerably higher or lower than the mean of the majority of observations taken during a time study.

Access time. Time required to retrieve information from a system's memory.

Accumulative timing. The application of three watches with linkage so that at element termination one watch is read for recording the latest element time, one watch is timing the element currently being observed, and one watch is cocked to time the next element.

Activity sampling. (See Work sampling.)

Actual time. The average elemental time actually taken by the operator during a time study.

Address. A designation indicating where a specific piece of information can be found in the memory storage.

Aerobic efficiency. The efficiency of the body during moderate work where the oxygen intake is adequate.

Algol. An early high-level programming language.

Algorithm. Step-by-step specifications of the solution to a problem, usually represented by a flowchart, which eventually is translated into a program.

Alignment chart. (see Nomogram.)

Allowance. The time added to normal time to provide for personal delays, unavoidable delays, and fatigue.

Allowed time. The time the normal operator takes to perform an operation while working at a standard rate of performance with due allowance for personal and unavoidable delays and fatigue.

Alphanumeric. Set of all machine-processable alphabetic letters (a–z), numeric digits (0–9), and special characters (such as those that appear on a typewriter).

Anaerobic efficiency. The efficiency (ratio of work done in calories to net energy used in calories) of the body during heavy work.

Analog. Data that has the characteristic of being continuous in form, as opposed to digital, which is characterized by discreet levels.

Arithmetic unit. That part of the computer processing section that does the adding, subtracting, multiplying, dividing, and computing.

Assemble. The act of bringing two mating parts together.

Assignable cause. A source of variation which can be isolated in a process or operation.

Automation. Increased mechanization to produce goods and services.

Auxiliary storage. Data stored in either floppy or hard disk drives external to the computer that is on-line to the computer and can be accessed by a program.

Available machine time. That portion of a time cycle during which a machine could be performing useful work.

Average cycle time. The sum of all average elemental times divided by the number of cycle observations.

Average elemental time. The mean elemental time taken by the operator to perform the task during a time study.

Average hourly earnings. The mean dollar-and-cent moneys paid to an operator on an hourly basis. They are determined by dividing the hours worked per period into the total wages paid for the period.

Avoidable delay. A cessation of productive work entirely due to the operator and not occurring in the regular work cycle.

Balanced motion pattern. A sequence of motions made simultaneously by both the right and the left hand in directions which facilitate rhythm and coordination.

Balancing delay. The cessation of productive work by one body member as a result of orienting another body member in the process of useful work.

Ballistic movement. The motion of arms (usually) or legs with smooth, flowing, rapid muscle action from the start to the termination.

Base time. The time required to perform a task by a normal operator working at a standard pace with no allowance for personal delays, unavoidable delays, or fatigue.

Base wage rate. The hourly money rate paid for a given work assignment performed at a standard pace by a normal operator.

Basic. High-level language frequently used with microcomputers.

Basic motion. A fundamental motion related to primary physiological and/or biomechanical performance capabilities of body members.

Baud. Rate with which bits of information can be transmitted per second between two devices. One baud is one bit per second.

Bedaux plan. A constant sharing wage incentive plan based on a high task rate, with the provision that the bonus for incentive effort be distributed between the employee and management.

Benchmark. A standard that is identified with characteristics in sufficient detail so that other classifications can be compared as being above, below, or comparable to the identified standard.

Binary digits. The numbers 1 and 0 (on and off) as used internally by computers.

Binomial distribution. A discrete probability distribution with mean $= np$ and variance $= np(1 - p)$ having a probability function $= C_{n,k}\, p^k\, (1 - p)^{n-k}$.

Biomechanics. The application of mechanical principles, such as levers, mechanical advantage, and forces, to the analysis of body part structure and movement.

Bit. Smallest possible unit of information represented as a binary digit.

Bonus earnings. Those moneys paid in addition to the regular wage or salary.

Breakpoint. A readily distinguishable point in the work cycle selected as the boundary between the completion of one element and the beginning of another element.

Buffer. Intermediate or temporary holding area for data that may be located in the computer, in an input or an output device, or its controller.

Byte. Group of eight bits, the amount of computer memory needed to hold a single numeral or alphabet letter. Represents one character, usually expressed in thousands (k) or millions (M).

CAAD. Computer-aided analysis/design.

CAD. Computer-aided design.

Candelas. A measure of luminance (emitted or reflected light) obtained by means of a photometer.

Cathode-ray tube (CRT). Electronic vacuum tub used to display text on graphic images.

Change direction. A basic division of accomplishment characterized by a slight hesitation when the hand alters its directional course while reaching or moving.

Changeover time. The time required to modify or replace an existing workplace. This time includes both the teardown time for the existing condition and the setup of the new condition.

Check study. A review of a job with either a stopwatch or a regular wristwatch to determine the appropriateness of a standard.

Chronocyclegraph. A photographic record of body motion that may be used to determine both the speed and the direction of body motion patterns.

Chronograph. A time-recording

device that operates by marking a tape driven at a constant speed. Time is determined by measuring the distance between successive markings on the tape.

Clo unit. A measure of the thermal insulation provided by clothing. One clo is 0.16 degrees Celsius per watt per square meter of body surface area.

Coding. The translation of a data processing machine program from descriptive, symbolic, or diagram form into machine language (code) or into an explicit symbolic language that may be translated directly into machine language by means of an assembly program or compiler.

Combined motions. Two or more nonconsecutive elemental motions performed during the same time interval by the same body member.

Consistency. The absence of noticeable or significant variation in behavioral or numerical data.

Constant element. An element whose performance time does not vary significantly when changes in the process or dimensional changes in the product occur.

Continuous timing method. A method of studying an operation in which the stopwatch is kept running continuously during the course of the study and is not snapped back at elemental termination.

Control system. A system that has as its primary function the collection and analysis of feedback from a given set of functions for the purpose of controlling the functions.

Controlled time. Elapsed elemental time that depends entirely on the facility or process.

Coverage. The number of jobs that have been assigned a standard during the reporting period or the number of personnel whose jobs have been assigned a standard during the reporting period.

Curve. A graphic representation of the relation between two factors, one of which is usually time.

Cycle. A series of elements that occur in regular order and make an operation possible. These elements repeat themselves as the operation is repeated.

Cyclegraph timing. The use of small lights on the hands or other body members to indicate their motion patterns. The lights are recorded by a still camera in a darkened room with an exposure time equal to at least one motion cycle.

Cycle timing. The measurement of the time for a complete work cycle rather than for the individual elements of the cycle.

Data base. A collection of data item that can be processed by a variety of applications.

Data control. The management of a data base to provide growth, service, utilization, security, recovery, and accuracy.

Daywork. Any work for which the operator is compensated on the basis of time rather than output.

dBA. A measure of sound pressure level, most commonly used to assess the noise exposure of workers.

Decimal hour stopwatch. A stop watch used for work measurement, the dial of which is graduated in 0.0001 of an hour.

Decimal minute stopwatch. A stop watch used for work measurement, the dial of which is graduated in 0.01 of a minute.

Delay. Any cessation in the work routine that does not occur in the typical work cycle.

Differential piecework. Compensation of labor in which the money rate per piece is based on the total pieces produced during the period (usually one day).

Differential timing. Timing an element by combining it with preceding and/or succeeding elements and then determining the elemental times by solving the simultaneous collective time equations.

Direct labor. Labor performed on each piece that advances the piece toward its ultimate specifications.

Disassemble. The basic division of accomplishment that takes place when two mating parts are separated.

Division of labor. The separation of jobs or tasks into less complex jobs or tasks usually to allow use of workers possessing less skill than that required by the overall job or task, or to make use of special skills.

Downtime. The time that is represented by operation cessation due to machine or tool breakdown, lack of material, and so on.

Drop delivery. The disposal of a part by dropping it on a conveyor or a gravity chute, thus minimizing move and position therbligs.

Earned hours. The standard hours credited to a worker or a work force as a result of the completion of a job or a group of jobs.

Effectiveness. The ratio of earned hours to actual hours spent on prescribed tasks.

Efficiency. The ratio of actual output to standard output.

Effort. The will to perform either mental or manual productive work.

Effort time. The portion of the cycle time that depends on the skill and effort of the operator.

Elapsed time. The actual time that has transpired during the course of a study or an operation.

Element. A division of work that can be measured with stopwatch equipment and that has readily identified terminal points.

Engineered work standards. Time standards based on measurement of work content (as opposed to historical standards) performed in the most productive way.

Ergonomics. The analysis of a work situation to optimize the physiological cost of performing an operation.

Exponential distribution. A continuous probability distribution with mean $= \dfrac{1}{a}$ and variance $=$ $\dfrac{1}{a^2}$, and having a density function $= ae^{-ax}$.

External time. The time required to perform elements of work when the machine or process is not in operation.

External transport. Transport between different plants or companies.

Extra allowance. An allowance to compensate for required work in addition to that which is specified in the standard method.

Fair day's work. The amount of work performed by an operator that is fair to both the company and the operator, considering the wages paid. It is the "amount of work that can be produced by a qualified employee when working at a normal pace and effectively utilizing his time where work is not restricted by process limitations."

Fatigue. A lessening in the capacity of the will to work.

Fatigue allowance. An amount of time added to the normal time to compensate for fatigue.

Feed. The speed at which the cutting tool is moved into the work, as in drilling and turning, or the rate that the work is moved past the cutting tool.

Film analysis. The frame-by-frame observation and study of a film of an operation or process with the objective of improvement.

First-piece time. The time allowed to produce the first piece of an order. This time has been adjusted to allow for the operator's unfamiliarity with the method and to accommodate for minor delays resulting from the newness of the work. It does not include time for setting up the workstation.

Fixture. A tool that is usually clamped to the workstation and that holds the material being worked on.

Float. The amount of material not directly employed or worked on in a system or process at a given point in time.

Flow analysis. The detailed examination of the progressive travel, either of personnel or material, from place to place and/or from operation to operation.

Flow diagram. A pictorial representation of the layout of a process showing the location of all activities appearing on the flow process chart and the travel paths of the work.

Flow process chart. A graphic representation of all operations, transportations, inspections, delays, and storages occurring during a process or procedure. The chart includes information considered desirable for analysis, such as the time required and the distance moved.

Footcandle. The measure of light falling on a surface. One footcandle equals 10.8 lumens per square meter.

Foot lambert. A unit of luminance (emitted or reflected light).

One foot lambert is equal to 3.43 candelas per square meter.

Foreign element. An interruption in the regular work cycle.

Frame. The space occupied by a single picture on a motion-picture film.

Frame counter. A device that automatically tabulates how many frames have passed the lens of the projector.

Frequency function. The complete listing of the values of a random variable together with their probabilities of occurrence.

Fringe benefits. The portion of tangible compensation that is not paid in wages, salaries, or bonuses given by the employer to employees. These include insurance, retirement funds, and other employee services. They exclude benefits paid for by employees by pay deductions, such as their participating portions of insurance premiums and retirement funds.

Gain sharing. Any method of wage payment in which the worker participates in all or a portion of the added earnings resulting from above standard production.

Gang process chart. A chart of the simultaneous activities of one or more machines and/or one or more workers.

Gantt chart. A series of graphs consisting of horizontal lines or bars in positions and lengths that show schedules or quotas and progress plotted on a common time scale.

Get. The act of picking up and gaining control of an object. It consists of the therbligs reach and grasp, and move; it also sometimes includes search and select.

Grasp. The elemental hand motion of closing the fingers around a part in an operation.

Gravity feed. Conveying materials either to or away from the workstation by using the force of gravity.

Hand time. That part of the work cycle controlled by manual elements exclusive of power or mechanized pacing elements.

Hertz. The unit of frequency, in cycles per second. One Hz equals one cycle per second.

High task. The production standards based on incentive conditions.

Human factors. Those axioms and postulates concerned with the physical, mental, and emotional constraints affecting operators' performance.

Idle time. Time the worker is not working.

Incentive. Reward, financial or other, that compensates the worker for high and/or continued performance above standard.

Incentive pace. A performance that is above normal or standard.

Indirect labor. Labor that does not directly enter into transforming the material used in making the product, but that is necessary to support the manufacture of the product.

Input storage. The temporary storage of a group of facts by a computing machine until the time that this group of facts should be processed.

Instruction. A coded program step that tells the computer what to do for a single operation in a program.

Interference time. Idle machine time due to insufficient operator time to service one or more machines that need servicing, because of operator engagement on other assigned work.

Internal transport. Transport within a company, plant, etc.

Internal work. Work performed by the operator during the operation of the machine or equipment.

Irregular element. An element which occurs randomly and can be statistically determined.

Jig. A tool that may or may not be clamped to the workstation and is used both to hold the work and to guide the tool.

Job analysis. A procedure for making a careful appraisal of each job and then recording the details of the work so that it can be equitably evaluated.

Job evaluation. A procedure for determining the relative worth of various work assignments.

Key job. A job representative of similar jobs or classes of work in the same plant or industry.

Kymograph. An electronic time study device used to measure extremely short work time intervals.

Learning curve. A graphic presentation of the progress in production effectiveness as time passes.

Leveling. (see Performance rating.)

Line equipment. Equipment guided along a predetermined transport route.

Loose rate. An established allowed time permitting the normal operator to achieve standard performance with less than average effort.

Low task. The production standards based on typical daywork levels.

Luminous flux. The total light output of a source or the amount of incident light on a surface expressed in lumens.

Luminous intensity. Light intensity within a small angle in a specific direction measured in candela.

Lux. The unit of illuminance equal to one lumen per square meter or 0.093 footcandle.

Machine attention time. That time during the work cycle in which the operator must devote attention to the machine or process.

Machine cycle time. The time required for the machine in process to complete one cycle.

Machine downtime. That time when the machine or process is inoperative because of some breakdown or because of a material shortage.

Machine idle time. That time when the machine or process is inoperative.

Machine pacing. The machine or mechanical control over the rate at which the work progresses.

Manufacturing progress function. The progress in production effectiveness with the passing of time.

Marsto-chron. A time study instrument that records elapsed time on a tape driven by a synchronous motor. Elapsed time is determined by measuring the distance on the tape between parallel markings recorded at element terminal points.

Maximum performance. That performance resulting in the highest obtainable production.

Maximum working area. The area readily reached by the operator when the arms are fully extended, while in a normal working position.

Mean of *x*. The expected value of *x*.

Measured daywork. Work for which performance standards have been established but where the operator is compensated on an hourly basis with no provision for incentive earnings. (First choice.)
An incentive system in which hourly rates are periodically adjusted on the basis of operator performance during the previous period. (Second choice.)

Memomotion study. The division of a work assignment into elements by analyzing motion pictures taken at speeds of 50, 60, or 100 frames per minute, and then improving the operation.

Memory storage. The section of

the computer that files or holds facts.

Merit rating. A method of evaluating an employee's worth to a company in terms of quantity and quality of work, dependability, and general contribution to the company.

Merrick differential piece rate. An incentive wage payment plan with three different piece rates based on operator performance.

Method. The technique employed to perform an operation.

Methods study. Analysis of an operation to increase the production per unit of time and consequently reduce the unit cost.

Microchronometer. A specially designed clock devised by Frank B. Gilbreth; it is capable of measuring elapsed time in "winks" (0.0005 minute).

Micromotion study. The division of a work assignment into therbligs by analyzing motion pictures frame by frame and then improving the operation by eliminating unnecessary movements and simplifying the necessary movements.

Minimum time. The least amount of time taken by the operator to perform a given element during a time study.

Modal time. The elapsed elemental time value that occurs most frequently during a time study. Occasionally used in preference to the average elemental time.

Motion study. The analysis and study of the motions constituting an operation to improve the motion pattern by eliminating ineffective motions and shortening the effective motions.

Move. A hand movement with a load.

MTM. A procedure analyzing any manual operation or method as to the basic motions required to perform it, and assigning to each motion a predetermined time standard based on the nature of the motion and the conditions under which it is made.

Network analysis. A planning technique used to analyze the sequence of activities and their interrelationships within a project.

Noise. Unwanted sounds that interfere with the detection of desired signals.

Nomogram. A graph, usually containing three parallel scales graduated for different variables so that when a straight line connects values of any two, the related value may be read directly from the third at the point intersected by the line.

Normal distribution. A continuous probability distribution with mean = m and variance = σ^2 and having a density function =

$$\frac{1}{\sigma \sqrt{2\pi}} \exp \left[\frac{-(x - m)^2}{2\sigma^2} \right]$$

Normal operator. An operator who can achieve the established standard of performance when following the prescribed method and working at an average pace.

Normal performance. The performance expected from the average trained operator when fol-

lowing the prescribed method and working at an average pace.

Normal time. The time required for the standard operator to perform the operation when working at a standard pace without delay for personal reasons or unavoidable circumstances.

Normal working area. That space at the work area which can be reached by either the left hand or the right hand when both elbows are pivoted on the edge of the workstation.

Numerical control. A method of controlling a machine or facility whereby either a binary or decimal digit system is programmed to carry out operations through electronic circuits and related activating mechanisms.

Observation. The gathering and recording of the time required to perform an element, or one watch reading.

Observation board. A convenient board used to support the stopwatch and hold the observation form during a time study.

Observation form. A form designed to accommodate the elements of a given time study with space for recording their duration.

Observer. The analyst taking a time study of a given operation.

Occupational physiology. Scientific study of the worker and environment.

Occurrence. An incident or event happening during a time study.

Operation. The intentional changing of a part toward its ultimate desired shape, size, form, and characteristics.

Operation analysis. An investigative process dealing with operations in factory or office work. Usually the process leading to operation standardization, including motion and time study.

Operation card. A form outlining the sequence of operations, the time allowed, and the special tools required in manufacturing a part.

Operation process chart. A graphic representation of an operation showing all methods, inspections, time allowances, and materials used in a manufacturing process.

Operator process chart. A graphic representation of all movements and delays made by both the right and the left hand, and of the relationship between the relative basic divisions of accomplishment performed by the two hands.

Output. The total production of a machine, process, or worker for a specified unit of time.

Output. (applied to computer). Computer results, such as answers to mathematical problems.

Output devices. Computer hardware that converts the results of a problem solved on the computer into the desired form, such as magnetic tape or a printed report.

Overall study. Recording cycle time as a verification of a developed time study standard.

Pallet. A load carrier, usually having a rectangular standardized load carrier.

Pareto's distribution. A distribution that reflects the fact that the major part of an activity (usually 80–85 percent) is accounted for by a minority (usually 15–20 percent). For example, 15 percent of the employees account for 85 percent of the absenteeism.

Performance. The ratio of the operator's actual production to the standard production.

Performance rating. The assignment of a percentage to the operator's average observed time, based on the actual performance of the operator compared to the observer's conception of normal.

Personal allowance. A percentage added to the normal time to accommodate the personal needs of the operator.

Picking rate. The rate at which a pallet, or some other transport unit, is completely picked.

Piece rate. A standard of performance expressed in money per unit of production.

Plan. A basic division of accomplishment involving the mental process of determining the next action.

Point. A unit of output identified as the production of one standard operator in one minute. Used as a basis for establishing standards under the Bedaux system.

Point system. A method of job evaluation in which the relative worth of different jobs is determined by totaling the points assigned to the various factors applicable to the different jobs.

Poisson distribution. A discrete probability distribution with mean = λ and variance = λ, and having a probability function =
$$\frac{\lambda^k e^{-\lambda}}{k!}$$

Position. An element of work which consists of locating an object so that it will be properly oriented in a specific location.

Pre-position. An element of work which consists of positioning an object in a predetermined place so that it may be grasped in the position in which it is to be held when needed.

Process. A series of operations that advance the product toward its ultimate size, shape, and specifications.

Process chart. A graphic representation of a manufacturing process.

Productive time. Any time spent in advancing the progress of a product toward its ultimate specifications.

Profit sharing. Any procedure in which an employer pays to employees, in addition to good rates of regular pay, special current or deferred sums, based not only on individual or group performance but also on the prosperity of the business as a whole.

Program. A set of instructions providing the information needed by the computer to handle a complete program.

Progress chart. A graphical representation of the status or extent of completion of work in process.

Qualified operator. An employee with sufficient training and education who has demonstrated an adequate level of skill and effort; this person is expected to perform at an acceptable level with respect to both quantity and quality.

Queuing theory. (See Waiting line theory.)

Random variable. A chance number resulting from a trial from among the set of numbers $x_1, x_2,$ and so on.

Rate. A standard expressed in dollars and cents.

Rate setting. The act of establishing money rates or time values on any operation.

Ratio-delay study. A work sampling study in which a large number of observations are taken at random intervals.

Reflectance. Percentage of light reflected from a surface.

Right- and left-hand chart. A chart showing the motions made by one hand in relation to those made by the other hand, and using standard therblig abbreviations or symbols.

Runout time. That time required by machine tools after cutting is completed so that the tool can be cleared from the work in preparation for the next sequence of work elements.

Selected time. An elemental time value chosen as representative of the expected performance of the operator being studied.

Setup. The preparation of a workstation or a work center to accomplish an operation or a series of operations.

Simo chart. (See Right- and left-hand chart.)

Simultaneous motions. Two or more nonconsecutive elemental motions performed during the same time interval by different body members.

Skill. Proficiency at following a prescribed method.

Sorting. A transport terminal activity by which goods are divided into groups.

Standard data. A structured collection of normal time values for work elements codified in tabular or graphic form.

Standard time. A unit time value for a work task as determined by the proper application of appropriate work measurement techniques by qualified personnel.

Standby time. That time in which the worker is not actively engaged, but is prepared to take action when needed.

Stowage. Handling for the purpose of positioning and/or securing goods in the space intended.

Synthetic basic motion times. A collection of time standards assigned to fundamental motions and groups of motions.

Temporary standard. A standard established to apply for a limited number of pieces or a limited time to account for the newness of the work or some unusual job condition.

Therblig. An abbreviated segment of a work element that describes the sensorimotor activities.

Tight rate. A time standard that allows less time than that required by a normal operator working at a normal pace to do the work.

Time study board. (See Observation board.)

Time study form. (See Observation form.)

Transmission. Facility of transmission of nerve impulses across the motor end plate in the muscle fiber.

Travel chart. A table providing distances travelled between points in a manufacturing or business facility.

Unavoidable delay. An interruption in the continuity of an operation beyond the control of an operator.

Unit load. A material in a packed state. Frequently, a standardized size transport unit.

Use. An objective therblig occurring when either or both hands have control of an object during that part of the cycle when productive work is being performed.

Value analysis. A review of product costs to evaluate contribution to product value. The review includes operations analysis and motion study to reduce costs.

Variable element. An element whose time is affected by one or more characteristics, such as size, shape, hardness, or tolerance, so that as these conditions change, the time required to perform the element changes.

Variance of x. A measure of the expected dispersion of the values of x about its mean.

Wage incentive. Providing a financial inducement for effort above normal.

Wage rate. The money rate expressed in dollars and cents paid to the employee per hour.

Waiting line theory. Mathematical analysis of the laws governing arrivals, service times, and the order in which arriving units are taken into service.

Waiting time. That time when the operator is unable to do useful work because of the nature of the process or because of the immediate lack of material.

Warehouse. An installation for storing products during long gaps between production stages or for storing finished products.

Wild value. (See Abnormal time.)

Wink. One division on the microchronometer equal to $1/2,000$ (0.0005) minute.

Wink counter. A mechanically or electrically driven timing device that records elapsed time in winks.

Work cycle. The total sequence of motions and events that comprise a single operation.

Worker-machine process chart. A chart showing the exact relationship in time between the working cycle of the operator and the operating cycle of the machine or machines.

Work factor. Index of the additional time required over and above the basic time as established by the Work-Factor system of synthetic basic motion times.

Work-hour. The standard amount of work performed by one worker in one hour.

Work pace. The rate at which an operation or activity is done.

Work physiology. The specification of the physiological and psychological factors characteristic of a work environment.

Work sampling. A method of analyzing work by taking a large number of observations at random intervals, to establish standards and improve methods.

Work station. The area where the worker performs the elements of work in a specific operation.

Collection of Helpful Formulas

(1) *Quadratic*

$$Ax^2 + Bx + C = 0$$

$$x = \frac{-B \pm \sqrt{B^2 - 4AC}}{2A}$$

(2) *Logarithms*

$$\log ab = \log a + \log b$$

$$\log \frac{a}{b} = \log a - \log b$$

$$\log a^n = n \log a$$

$$\log \sqrt[n]{a} = \frac{1}{n} \log a$$

$$\log 1 = 0$$

$$\log {}_a a = 1$$

(3) *Binomial theorem*

$$(a + b)^n = a^n + na^{n-1}b + \frac{n(n-1)}{2!} a^{n-2}b^2$$

$$+ \frac{n(n-1)(n-2)}{3!} a^{n-3}b^3 + \cdots$$

(4) *Circle*

$$\text{Circumference} = 2\pi r$$
$$\text{Area} = \pi r^2$$

(5) *Prism*

$$\text{Volume} = Ba$$

(6) *Pyramid*

$$\text{Volume} = \tfrac{1}{3}Ba$$

(7) *Right circular cylinder*

$$\text{Volume} = \pi r^2 a$$
$$\text{Lateral surface} = 2\pi ra$$
$$\text{Total surface} = 2\pi r(r + a)$$

(8) *Right circular cone*

$$\text{Volume} = \tfrac{1}{3}\pi r^2 a$$
$$\text{Lateral surface} = \pi rs$$
$$\text{Total surface} = \pi r(r + s)$$

(9) *Sphere*

$$\text{Volume} = \tfrac{4}{3}\pi r^3$$
$$\text{Surface} = 4\pi r^2$$

(10) *Frustrum of a right circular cone*

$$\text{Volume} = \tfrac{1}{3}\pi a(R^2 + r^2 + Rr)$$
$$\text{Lateral surface} = \pi s(R + r)$$

(11) *Measurement of angles*

$$1 \text{ degree} = \frac{\pi}{180} = .0174 \text{ radians}$$
$$1 \text{ radian} = 57.29 \text{ degrees}$$

(12) *Trigonometric functions*

Right triangles:

The sine of the angle A is the quotient of the opposite side divided by the hypotenuse. Sin $A = \dfrac{a}{c}$.

The tangent of the angle A is the quotient of the opposite side divided by the adjacent side. Tan $A = \dfrac{a}{b}$.

The secant of the angle A is the quotient of the hypotenuse divided by the adjacent side. Sec $A = \dfrac{c}{b}$.

The cosine, cotangent, and cosecant of an angle are, respectively, the sine, tangent, and secant of the complement of that angle.

Law of sines:

$$\frac{a}{\sin A} = \frac{b}{\sin B} = \frac{c}{\sin C}$$

Law of cosines:

$$a^2 = b^2 + c^2 - 2bc \cos A$$

(13) *Equations of straight lines*
Slope—intercept form

$$y = mx + b$$

intercept form

$$\frac{x}{a} + \frac{y}{b} = 1$$

APPENDIX

Special Tables

TABLE A3–1
Natural sines and tangents

Angle	Sin	Tan	Cot	Cos	
0	0.0000	0.0000	∞	1.0000	90
1	0.0175	0.0175	57.2900	0.9998	89
2	0.0349	0.0349	28.6363	0.9994	88
3	0.0523	0.0524	19.0811	0.9986	87
4	0.0698	0.0699	14.3007	0.9976	86
5	0.0872	0.0875	11.4301	0.9962	85
6	0.1045	0.1051	9.5144	0.9945	84
7	0.1219	0.1228	8.1443	0.9925	83
8	0.1392	0.1405	7.1154	0.9903	82
9	0.1564	0.1584	6.3138	0.9877	81
10	0.1736	0.1763	5.6713	0.9848	80
11	0.1908	0.1944	5.1446	0.9816	79
12	0.2079	0.2126	4.7046	0.9781	78
13	0.2250	0.2309	4.3315	0.9744	77
14	0.2419	0.2493	4.0108	0.9703	76
15	0.2588	0.2679	3.7321	0.9659	75
16	0.2756	0.2867	3.4874	0.9613	74
17	0.2924	0.3057	3.2709	0.9563	73
18	0.3090	0.3249	3.0777	0.9511	72
19	0.3256	0.3443	2.9042	0.9455	71
20	0.3420	0.3640	2.7475	0.9397	70
21	0.3584	0.3839	2.6051	0.9336	69
22	0.3746	0.4040	2.4751	0.9272	68
23	0.3907	0.4245	2.3559	0.9205	67
24	0.4067	0.4452	2.2460	0.9135	66
25	0.4226	0.4663	2.1445	0.9063	65
26	0.4384	0.4877	2.0503	0.8988	64
27	0.4540	0.5095	1.9626	0.8910	63
28	0.4695	0.5317	1.8807	0.8829	62
29	0.4848	0.5543	1.8040	0.8746	61
30	0.5000	0.5774	1.7321	0.8660	60
31	0.5150	0.6009	1.6643	0.8572	59
32	0.5299	0.6249	1.6003	0.8480	58
33	0.5446	0.6494	1.5399	0.8387	57
34	0.5592	0.6745	1.4826	0.8290	56
35	0.5736	0.7002	1.4281	0.8192	55
36	0.5878	0.7265	1.3764	0.8090	54
37	0.6018	0.7536	1.3270	0.7986	53
38	0.6157	0.7813	1.2799	0.7880	52
39	0.6293	0.8098	1.2349	0.7771	51
40	0.6428	0.8391	1.1918	0.7660	50
41	0.6561	0.8693	1.1504	0.7547	49
42	0.6691	0.9004	1.1106	0.7431	48
43	0.6820	0.9325	1 0724	0.7314	47
44	0.6947	0.9657	1.0355	0.7193	46
45	0.7071	1.0000	1.0000	0.7071	45
	Cos	Cot	Tan	Sin	Angle

823

TABLE A3–2
Areas of the normal curve

z	*Area*	*z*	*Area*
−3.0	.0013	0.1	.5398
−2.9	.0019	0.2	.5793
−2.8	.0026	0.3	.6179
−2.7	.0035	0.4	.6554
−2.6	.0047	0.5	.6915
−2.5	.0062	0.6	.7257
−2.4	.0082	0.7	.7580
−2.3	.0107	0.8	.7881
−2.2	.0139	0.9	.8159
−2.1	.0179	1.0	.8413
−2.0	.0228	1.1	.8643
−1.9	.0287	1.2	.8849
−1.8	.0359	1.3	.9032
−1.7	.0446	1.4	.9192
−1.6	.0548	1.5	.9332
−1.5	.0668	1.6	.9452
−1.4	.0808	1.7	.9554
−1.3	.0968	1.8	.9641
−1.2	.1151	1.9	.9713
−1.1	.1357	2.0	.9772
−1.0	.1587	2.1	.9821
−0.9	.1841	2.2	.9861
−0.8	.2119	2.3	.9893
−0.7	.2420	2.4	.9918
−0.6	.2741	2.5	.9938
−0.5	.3085	2.6	.9953
−0.4	.3446	2.7	.9965
−0.3	.3821	2.8	.9974
−0.2	.4207	2.9	.9981
−0.1	.4602	3.0	.9987
0.0	.5000		

Above table tabularizes $P(z)$ where

$$P(z) = \int_{-\infty}^{z} \frac{1}{\sqrt{2\pi}} e^{-\frac{z^2}{2}} dz$$

$$\text{and } z = \frac{x - u}{\sigma}$$

TABLE A3–3

Percentage points of the *t* distribution (probabilities refer to the sum of the two tail areas; for a single tail, divide the probability by 2)

Probability (*P*).

n	·9	·8	·7	·6	·5	·4	·3	·2	·1	·05	·02	·01	·001
1	·158	·325	·510	·727	1·000	1·376	1·963	3·078	6·314	12·706	31·821	63·657	636·619
2	·142	·289	·445	·617	·816	1·061	1·386	1·886	2·920	4·303	6·965	9·925	31·598
3	·137	·277	·424	·584	·765	·978	1·250	1·638	2·353	3·182	4·541	5·841	12·941
4	·134	·271	·414	·569	·741	·941	1·190	1·533	2·132	2·776	3·747	4·604	8·610
5	·132	·267	·408	·559	·727	·920	1·156	1·476	2·015	2·571	3·365	4·032	6·859
6	·131	·265	·404	·553	·718	·906	1·134	1·440	1·943	2·447	3·143	3·707	5·959
7	·130	·263	·402	·549	·711	·896	1·119	1·415	1·895	2·365	2·998	3·499	5·405
8	·130	·262	·399	·546	·706	·889	1·108	1·397	1·860	2·306	2·896	3·355	5·041
9	·129	·261	·398	·543	·703	·883	1·100	1·383	1·833	2·262	2·821	3·250	4·781
10	·129	·260	·397	·542	·700	·879	1·093	1·372	1·812	2·228	2·764	3·169	4·587
11	·129	·260	·396	·540	·697	·876	1·088	1·363	1·796	2·201	2·718	3·106	4·437
12	·128	·259	·395	·539	·695	·873	1·083	1·356	1·782	2·179	2·681	3·055	4·318
13	·128	·259	·394	·538	·694	·870	1·079	1·350	1·771	2·160	2·650	3·012	4·221
14	·128	·258	·393	·537	·692	·868	1·076	1·345	1·761	2·145	2·624	2·977	4·140
15	·128	·258	·393	·536	·691	·866	1·074	1·341	1·753	2·131	2·602	2·947	4·073
16	·128	·258	·392	·535	·690	·865	1·071	1·337	1·746	2·120	2·583	2·921	4·015
17	·128	·257	·392	·534	·689	·863	1·069	1·333	1·740	2·110	2·567	2·898	3·965
18	·127.	·257	·392	·534	·688	·862	1·067	1·330	1·734	2·101	2·552	2·878	3·922
19	·127	·257	·391	·533	·688	·861	1·066	1·328	1·729	2·093	2·539	2·861	3·883
20	·127	·257	·391	·533	·687	·860	1·064	1·325	1·725	2·086	2·528	2·845	3·850
21	·127	·257	·391	·532	·686	·859	1·063	1·323	1·721	2·080	2·518	2·831	3·819
22	·127	·256	·390	·532	·686	·858	1·061	1·321	1·717	2·074	2·508	2·819	3·792
23	·127	·256	·390	·532	·685	·858	1·060	1·319	1·714	2·069	2·500	2·807	3·767
24	·127	·256	·390	·531	·685	·857	1·059	1·318	1·711	2·064	2·492	2·797	3·745
25	·127	·256	·390	·531	·684	·856	1·058	1·316	1·708	2·060	2·485	2·787	3·725
26	·127	·256	·390	·531	·684	·856	1·058	1·315	1·706	2·056	2·479	2·779	3·707
27	·127	·256	·389	·531	·684	·855	1·057	1·314	1·703	2·052	2·473	2·771	3·690
28	·127	·256	·389	·530	·683	·855	1·056	1·313	1·701	2·048	2·467	2·763	3·674
29	·127	·256	·389	·530	·683	·854	1·055	1·311	1·699	2·045	2·462	2·756	3·659
30	·127	·256	·389	·530	·683	·854	1·055	1·310	1·697	2·042	2·457	2·750	3·646
40	·126	·255	·388	·529	·681	·851	1·050	1·303	1·684	2·021	2·423	2·704	3·551
60	·126	·254	·387	·527	·679	·848	1·046	1·296	1·671	2·000	2·390	2·660	3·460
120	·126	·254	·386	·526	·677	·845	1·041	1·289	1·658	1·980	2·358	2·617	3·373
∞	·126	·253	·385	·524	·674	·842	1·036	1·282	1·645	1·960	2·326	2·576	3·291

Reprinted from Table III of R. A. Fisher and F. Yates, *Statistical Tables for Biological, Agricultural, and Medical Research* (Edinburgh: Oliver & Boyd, Ltd.), by permission of the authors and publishers.

TABLE A3–4

Random numbers III

```
22 17 68 65 84    68 95 23 92 35    87 02 22 57 51    61 09 43 95 06    58 24 82 03 47
19 36 27 59 46    13 79 93 37 55    39 77 32 77 09    85 52 05 30 62    47 83 51 62 74
16 77 23 02 77    09 61 87 25 21    28 06 24 25 93    16 71 13 59 78    23 05 47 47 25
78 43 76 71 61    20 44 90 32 64    97 67 63 99 61    46 38 03 93 22    69 81 21 99 21
03 28 28 26 08    73 37 32 04 05    69 30 16 09 05    88 69 58 28 99    35 07 44 75 47

93 22 53 64 39    07 10 63 76 35    87 03 04 79 88    08 13 13 85 51    55 34 57 72 69
78 76 58 54 74    92 38 70 96 92    52 06 79 79 45    82 63 18 27 44    69 66 92 19 09
23 68 35 26 00    99 53 93 61 28    52 70 05 48 34    56 65 05 61 86    90 92 10 70 80
15 39 25 70 99    93 86 52 77 65    15 33 59 05 28    22 87 26 07 47    86 96 98 29 06
58 71 96 30 24    18 46 23 34 27    85 13 99 24 44    49 18 09 79 49    74 16 32 23 02

57 35 27 33 72    24 53 63 94 09    41 10 76 47 91    44 04 95 49 66    39 60 04 59 81
48 50 86 54 48    22 06 34 72 52    82 21 15 65 20    33 29 94 71 11    15 91 29 12 03
61 96 48 95 03    07 16 39 33 66    98 56 10 56 79    77 21 30 27 12    90 49 22 23 62
36 93 89 41 26    29 70 83 63 51    99 74 20 52 36    87 09 41 15 09    98 60 16 03 03
18 87 00 42 31    57 90 12 02 07    23 47 37 17 31    54 08 01 88 63    39 41 88 92 10

88 56 53 27 59    33 35 72 67 47    77 34 55 45 70    08 18 27 38 90    16 95 86 70 75
09 72 95 84 29    49 41 31 06 70    42 38 06 45 18    64 84 73 31 65    52 53 37 97 15
12 96 88 17 31    65 19 69 02 83    60 75 86 90 68    24 64 19 35 51    56 61 87 39 12
85 94 57 24 16    92 09 84 38 76    22 00 27 69 85    29 81 94 78 70    21 94 47 90 12
38 64 43 59 98    98 77 87 68 07    91 51 67 62 44    40 98 05 93 78    23 32 65 41 18

53 44 09 42 72    00 41 86 79 79    68 47 22 00 20    35 55 31 51 51    00 83 63 22 55
40 76 66 26 84    57 99 99 90 37    36 63 32 08 58    37 40 13 68 97    87 64 81 07 83
02 17 79 18 05    12 59 52 57 02    22 07 90 47 03    28 14 11 30 79    20 69 22 40 98
95 17 82 06 53    31 51 10 96 46    92 06 88 07 77    56 11 50 81 69    40 23 72 51 39
35 76 22 42 92    96 11 83 44 80    34 68 35 48 77    33 42 40 90 60    73 96 53 97 86

26 29 13 56 41    85 47 04 66 08    34 72 57 59 13    82 43 80 46 15    38 26 61 70 04
77 80 20 75 82    72 82 32 99 90    63 95 73 76 63    89 73 44 99 05    48 67 26 43 18
46 40 66 44 52    91 36 74 43 53    30 82 13 54 00    78 45 63 98 35    55 03 36 67 68
37 56 08 18 09    77 53 84 46 47    31 91 18 95 58    24 16 74 11 53    44 10 13 85 57
61 65 61 68 66    37 27 47 39 19    84 83 70 07 48    53 21 40 06 71    95 06 79 88 54

93 43 69 64 07    34 18 04 52 35    56 27 09 24 86    61 85 53 83 45    19 90 70 99 00
21 96 60 12 99    11 20 99 45 18    48 13 93 55 34    18 37 79 49 90    65 97 38 20 46
95 20 47 97 97    27 37 83 28 71    00 06 41 41 74    45 89 09 39 84    51 67 11 52 49
97 86 21 78 73    10 65 81 92 59    58 76 17 14 97    04 76 62 16 17    17 95 70 45 80
69 92 06 34 13    59 71 74 17 32    27 55 10 24 19    23 71 82 13 74    63 52 52 01 41

04 31 17 21 56    33 73 99 19 87    26 72 39 27 67    53 77 57 68 93    60 61 97 22 61
61 06 98 03 91    87 14 77 43 96    43 00 65 98 50    45 60 33 01 07    98 99 46 50 47
85 93 85 86 88    72 87 08 62 40    16 06 10 89 20    23 21 34 74 97    76 38 03 29 63
21 74 32 47 45    73 96 07 94 52    09 65 90 77 47    25 76 16 19 33    53 05 70 53 30
15 69 53 82 80    79 96 23 53 10    65 39 07 16 29    45 33 02 43 70    02 87 40 41 45

02 89 08 04 49    20 21 14 68 86    87 63 93 95 17    11 29 01 95 80    35 14 97 35 33
87 18 15 89 79    85 43 01 72 73    08 61 74 51 69    89 74 39 82 15    94 51 33 41 67
98 83 71 94 22    59 97 50 99 52    08 52 85 08 40    87 80 61 65 31    91 51 80 32 44
10 08 58 21 66    72 68 49 29 31    89 85 84 46 06    59 73 19 85 23    65 09 29 75 63
47 90 56 10 08    88 02 84 27 83    42 29 72 23 19    66 56 45 65 79    20 71 53 20 25

22 85 61 68 90    49 64 92 85 44    16 40 12 89 88    50 14 49 81 06    01 82 77 45 12
67 80 43 79 33    12 83 11 41 16    25 58 19 68 70    77 02 54 00 52    53 43 37 15 26
27 62 50 96 72    79 44 61 40 15    14 53 40 65 39    27 31 58 50 28    11 39 03 34 25
33 78 80 87 15    38 30 06 38 21    14 47 47 07 26    54 96 87 53 32    40 36 40 96 76
13 13 92 66 99    47 24 49 57 74    32 25 43 62 17    10 97 11 69 84    99 63 22 32 98
```

* Reprinted with permission from Random Numbers III of Table XXXIII of R. A. Fisher and F. Yates, *Statistical Tables for Biological, Agricultural and Medical Research* (Edinburgh: Oliver & Boyd, Ltd.).

TABLE A3–4 (concluded)

Random numbers IV

10 27 53 96 23	71 50 54 36 23	54 31 04 82 98	04 14 12 15 09	26 78 25 47 47
28 41 50 61 88	64 85 27 20 18	83 36 36 05 56	39 71 65 09 62	94 76 62 11 89
34 21 42 57 02	59 19 18 97 48	80 30 03 30 98	05 24 67 70 07	84 97 50 87 46
61 81 77 23 23	82 82 11 54 08	53 28 70 58 96	44 07 39 55 43	42 34 43 39 28
61 15 18 13 54	16 86 20 26 88	90 74 80 55 09	14 53 90 51 17	52 01 63 01 59
91 76 21 64 64	44 91 13 32 97	75 31 62 66 54	84 80 32 75 77	56 08 25 70 29
00 97 79 08 06	37 30 28 59 85	53 56 68 53 40	01 74 39 59 73	30 19 99 85 48
36 46 18 34 94	75 20 80 27 77	78 91 69 16 00	08 43 18 73 68	67 69 61 34 25
88 98 99 60 50	65 95 79 42 94	93 62 40 89 96	43 56 47 71 66	46 76 29 67 02
04 37 59 87 21	05 02 03 24 17	47 97 81 56 51	92 34 86 01 82	55 51 33 12 91
63 62 06 34 41	94 21 78 55 09	72 76 45 16 94	29 95 81 83 83	79 88 01 97 30
78 47 23 53 90	34 41 92 45 71	09 23 70 70 07	12 38 92 79 43	14 85 11 47 23
87 68 62 15 43	53 14 36 59 25	54 47 33 70 15	59 24 48 40 35	50 03 42 99 36
47 60 92 10 77	88 59 53 11 52	66 25 69 07 04	48 68 64 71 06	61 65 70 22 12
56 88 87 59 41	65 28 04 67 53	95 79 88 37 31	50 41 06 94 76	81 83 17 16 33
02 57 45 86 67	73 43 07 34 48	44 26 87 93 29	77 09 61 67 84	06 69 44 77 75
31 54 14 13 17	48 62 11 90 60	68 12 93 64 28	46 24 79 16 76	14 60 25 51 01
28 50 16 43 36	28 97 85 58 99	67 22 52 76 23	24 70 36 54 54	59 28 61 71 96
63 29 62 66 50	02 63 45 52 38	67 63 47 54 75	83 24 78 43 20	92 63 13 47 48
45 65 58 26 51	76 96 59 38 72	86 57 45 71 46	44 67 76 14 55	44 88 01 62 12
39 65 36 63 70	77 45 85 50 51	74 13 39 35 22	30 53 36 02 95	49 34 88 73 61
73 71 98 16 04	29 18 94 51 23	76 51 94 84 86	79 93 96 38 63	08 58 25 58 94
72 20 56 20 11	72 65 71 08 86	79 57 95 13 91	97 48 72 66 48	09 71 17 24 89
75 17 26 99 76	89 37 20 70 01	77 31 61 95 46	26 97 05 73 51	53 33 18 72 87
37 48 60 82 29	81 30 15 39 14	48 38 75 93 29	06 87 37 78 48	45 56 00 84 47
68 08 02 80 72	83 71 46 30 49	89 17 95 88 29	02 39 56 03 46	97 74 06 56 17
14 23 98 61 67	70 52 85 01 50	01 84 02 78 43	10 62 98 19 41	18 83 99 47 99
49 08 96 21 44	25 27 99 41 28	07 41 08 34 66	19 42 74 39 91	41 96 53 78 72
78 37 06 08 43	63 61 62 42 29	39 68 95 10 96	09 24 23 00 62	56 12 80 73 16
37 21 34 17 68	68 96 83 23 56	32 84 60 15 31	44 73 67 34 77	91 15 79 74 58
14 29 09 34 04	87 83 07 55 07	76 58 30 83 64	87 29 25 58 84	86 50 60 00 25
58 43 28 06 36	49 52 83 51 14	47 56 91 29 34	05 87 31 06 95	12 45 57 09 09
10 43 67 29 70	80 62 80 03 42	10 80 21 38 84	90 56 35 03 09	43 12 74 49 14
44 38 88 39 54	86 97 37 44 22	00 95 01 31 76	17 16 29 56 63	38 78 94 49 81
90 69 59 19 51	85 39 52 85 13	07 28 37 07 61	11 16 36 27 03	78 86 72 04 95
41 47 10 25 62	97 05 31 03 61	20 26 36 31 62	68 69 86 95 44	84 95 48 46 45
91 94 14 63 19	75 89 11 47 11	31 56 34 19 09	79 57 92 36 59	14 93 87 81 40
80 06 54 18 66	09 18 94 06 19	98 40 07 17 81	22 45 44 84 11	24 62 20 42 31
67 72 77 63 48	84 08 31 55 58	24 33 45 77 58	80 45 67 93 82	75 70 16 08 24
59 40 24 13 27	79 26 88 86 30	01 31 60 10 39	53 58 47 70 93	85 81 56 39 38
05 90 35 89 95	01 61 16 96 94	50 78 13 69 36	37 68 53 37 31	71 26 35 03 71
44 43 80 69 98	46 68 05 14 82	90 78 50 05 62	77 79 13 57 44	59 60 10 39 66
61 81 31 96 82	00 57 25 60 59	46 72 60 18 77	55 66 12 62 11	08 99 55 64 57
42 88 07 10 05	24 98 65 63 21	47 21 61 88 32	27 80 30 21 60	10 92 35 36 12
77 94 30 05 39	28 10 99 00 27	12 73 73 99 12	49 99 57 94 82	96 88 57 17 91
78 83 19 76 16	94 11 68 84 26	23 54 20 86 85	23 86 66 99 07	36 37 34 92 09
87 76 59 61 81	43 63 64 61 61	65 76 36 95 90	18 48 27 45 68	27 23 65 30 72
91 43 05 96 47	55 78 99 95 24	37 55 85 78 78	01 48 41 19 10	35 19 54 07 73
84 97 77 72 73	09 62 06 65 72	87 12 49 03 60	41 15 20 76 27	50 47 02 29 16
87 41 60 76 83	44 88 96 07 80	83 05 83 38 96	73 70 66 81 90	30 56 10 48 59

Reprinted with permission from Random Numbers IV of Table XXXIII of R. A. Fisher and F. Yates, *Statistical Tables for Biological, Agricultural, and Medical Research* (Edinburgh: Oliver & Boyd, Ltd.).

TABLE A3–5

Useful Information

To find the circumference of a circle, multiply the diameter by 3.1416.

To find the diameter of a circle, multiply the circumference by .31831.

To find the area of a circle, multiply the square of the diameter by .7854.

The radius of a circle $\times$ 6.283185 = the circumference.

The square of the circumference of a circle $\times$ .07958 = the area.

Half the circumference of a circle $\times$ half its diameter = the area.

The circumference of a circle $\times$ .159155 = the radius.

The square root of the area of a circle $\times$ .56419 = the radius.

The square root of the area of a circle $\times$ 1.12838 = the diameter.

To find the diameter of a circle equal in area to a given square, multiply a side of the square by 1.12838.

To find the side of a square equal in area to a given circle, multiply the diameter by .8862.

To find the side of a square inscribed in a circle, multiply the diameter by .7071.

To find the side of a hexagon inscribed in a circle, multiply the diameter of the circle by .500.

To find the diameter of a circle inscribed in a hexagon, multiply a side of the hexagon by 1.7321.

To find the side of an equilateral triangle inscribed in a circle, multiply the diameter of the circle by .866.

To find the diameter of a circle inscribed in an equilateral triangle, multiply a side of the triangle by .57735.

To find the area of the surface of a ball (sphere), multiply the square of the diameter by 3.1416.

To find the volume of a ball (sphere), multiply the cube of the diameter by .5236.

Doubling the diameter of a pipe increases its capacity four times.

To find the pressure in pounds per square inch at the base of a column of water, multiply the height of the column in feet by .433.

A gallon of water (U. S. Standard) weighs 8.336 pounds and contains 231 cubic inches. A cubic foot of water contains 7½ gallons, 1728 cubic inches, and weighs 62.425 pounds at a temperature of about 39° F.

These weights change slightly above and below this temperature.

TABLE A3–6
15% compound interest factors

	SINGLE PAYMENT		UNIFORM SERIES				
n	Compound Amount Factor caf'	Present Worth Factor pwf'	Sinking Fund Factor sff	Capital Recovery Factor crf	Compound Amount Factor caf	Present Worth Factor pwf	**n**
	Given P To find S $(1+i)^n$	Given S To find P $\dfrac{1}{(1+i)^n}$	Given S To find R $\dfrac{i}{(1+i)^n-1}$	Given P To find R $\dfrac{i(1+i)^n}{(1+i)^n-1}$	Given R To find S $\dfrac{(1+i)^n-1}{i}$	Given R To find P $\dfrac{(1+i)^n-1}{i(1+i)^n}$	
1	1.150	0.8696	1.00000	1.15000	1.000	0.870	1
2	1.322	0.7561	0.46512	0.61512	2.150	1.626	2
3	1.521	0.6575	0.28798	0.43798	3.472	2.283	3
4	1.749	0.5718	0.20026	0.35027	4.993	2.855	4
5	2.011	0.4972	0.14832	0.29832	6.742	3.352	5
6	2.313	0.4323	0.11424	0.26424	8.754	3.784	6
7	2.660	0.3759	0.09036	0.24036	11.067	4.160	7
8	3.059	0.3269	0.07285	0.22285	13.727	4.487	8
9	3.518	0.2843	0.05957	0.20957	16.786	4.772	9
10	4.046	0.2472	0.04925	0.19925	20.304	5.019	10
11	4.652	0.2149	0.04107	0.19107	24.349	5.234	11
12	5.350	0.1869	0.03448	0.18448	29.002	5.421	12
13	6.153	0.1625	0.02911	0.17911	34.352	5.583	13
14	7.076	0.1413	0.02469	0.17469	40.505	5.724	14
15	8.137	0.1229	0.02102	0.17102	47.580	5.847	15
16	9.358	0.1069	0.01795	0.16795	55.717	5.954	16
17	10.761	0.0929	0.01537	0.16537	65.075	6.047	17
18	12.375	0.0808	0.01319	0.16319	75.836	6.128	18
19	14.232	0.0703	0.01134	0.16134	88.212	6.198	19
20	16.367	0.0611	0.00976	0.15976	102.443	6.259	20
21	18.821	0.0531	0.00842	0.15842	118.810	6.312	21
22	21.645	0.0462	0.00727	0.15727	137.631	6.359	22
23	24.891	0.0402	0.00628	0.15628	159.276	6.399	23
24	28.625	0.0349	0.00543	0.15543	184.167	6.434	24
25	32.919	0.0304	0.00470	0.15470	212.793	6.464	25
26	37.857	0.0264	0.00407	0.15407	245.711	6.491	26
27	43.535	0.0230	0.00353	0.15353	283.568	6.514	27
28	50.065	0.0200	0.00306	0.15306	327.103	6.534	28
29	57.575	0.0174	0.00265	0.15265	377.169	6.551	29
30	66.212	0.0151	0.00230	0.15230	434.744	6.566	30
31	76.143	0.0131	0.00200	0.15200	500.956	6.579	31
32	87.565	0.0114	0.00173	0.15173	577.099	6.591	32
33	100.700	0.0099	0.00150	0.15150	664.664	6.600	33
34	115.805	0.0086	0.00131	0.15131	765.364	6.609	34
35	133.175	0.0075	0.00113	0.15113	881.168	6.617	35
40	267.862	0.0037	0.00056	0.15056	1779.1	6.642	40
45	538.767	0.0019	0.00028	0.15028	3585.1	6.654	45
50	1083.652	0.0009	0.00014	0.15014	7217.7	6.661	50
∞				0.15000		6.667	∞

TABLE A3–7

HOURLY PRODUCTION TABLE
Showing 60% to 80% Efficiency

Sec per Piece	Gross Prod. Per Hr	60%	65%	70%	75%	80%
1/2	7200	4320	4680	5040	5400	5760
5/8	5760	3456	3744	4032	4320	4608
3/4	4800	2880	3120	3360	3600	3840
7/8	4114	2468	2674	2880	3086	3291
1	3600	2160	2340	2520	2700	2880
1 1/4	2880	1728	1872	2016	2160	2304
1 1/2	2400	1440	1560	1680	1800	1920
1 3/4	2057	1234	1337	1440	1543	1646
2	1800	1080	1170	1260	1350	1440
2 1/4	1600	960	1040	1120	1200	1280
2 1/2	1440	864	936	1008	1080	1152
2 3/4	1309	785	851	916	982	1047
3	1200	720	780	840	900	960
3 1/4	1107	664	720	775	830	886
3 1/2	1028	617	668	720	771	822
3 3/4	960	576	624	672	720	768
4	900	540	585	630	675	720
4 1/4	847	508	551	593	635	678
4 1/2	800	480	520	560	600	640

Sec per Piece	Gross Prod. Per Hr	60%	65%	70%	75%	80%
12 1/2	288	173	187	202	216	230
13	276	166	179	193	207	221
13 1/2	267	160	174	187	200	214
14	257	154	167	180	193	206
14 1/2	248	149	161	174	186	198
15	240	144	156	168	180	192
15 1/2	232	139	151	162	174	186
16	225	135	146	158	169	180
16 1/2	218	131	142	153	164	174
17	212	127	138	148	159	170
17 1/2	206	124	134	144	155	165
18	200	120	130	140	150	160
18 1/2	195	117	127	137	146	156
19	189	113	123	132	142	151
19 1/2	185	111	120	130	139	148
20	180	108	117	126	135	144
21	171	103	111	120	128	137
22	164	98	107	115	123	131
23	156	94	101	109	117	125

Sec per Piece	Gross Prod. Per Hr	60%	65%	70%	75%	80%
50	72	43	47	50	54	58
52	69	41	45	48	52	55
54	67	40	44	47	50	54
56	64	38	42	45	48	51
58	62	37	40	43	47	50
60	60	36	39	42	45	48
62	58	35	38	41	44	46
64	56	34	36	39	42	45
66	54	32	35	38	41	43
68	53	32	34	37	40	42
70	51	31	33	36	38	41
72	50	30	33	35	38	40
74	49	29	32	34	37	39
76	47	28	31	33	35	38
78	46	28	30	32	35	37
80	45	27	29	32	34	36
82	44	26	29	31	33	35
84	43	26	28	30	32	34
86	42	25	27	29	32	34

4 3/4	757	454	492	530	568	606	24	150	90	98	105	113	120	88	41	25	27	29	31	33
5	720	432	468	504	540	576	25	144	86	94	101	108	115	90	40	24	26	28	30	32
5 1/4	686	412	446	480	515	549	26	138	83	90	97	104	110	92	39	23	25	27	30	31
5 1/2	654	392	425	458	491	523	27	133	80	86	93	100	106	94	38	23	24	27	29	30
5 3/4	626	376	407	438	470	501	28	128	77	83	90	96	102	96	37	22	24	26	28	30
6	600	360	390	420	450	480	29	124	74	81	87	93	99	98	37	22	23	26	28	30
6 1/4	576	346	374	403	432	461	30	120	72	78	84	90	96	100	36	20	22	25	27	29
6 1/2	553	332	359	387	415	442	31	116	70	75	81	87	93	105	34	20	22	24	26	27
6 3/4	533	320	346	373	400	426	32	112	67	73	78	84	90	110	33	19	20	23	25	26
7	514	308	334	360	386	411	33	109	65	71	76	82	87	115	31	18	20	22	25	25
7 1/4	497	298	323	348	373	398	34	106	64	69	74	80	85	120	30	17	19	21	23	24
7 1/2	480	288	312	336	360	384	35	103	62	67	72	77	82	125	29	17	18	20	23	23
7 3/4	465	279	302	326	349	372	36	100	60	65	70	75	80	130	28	16	18	20	22	23
8	450	270	293	315	338	360	37	97	58	63	68	73	78	135	27	16	17	19	21	22
8 1/4	436	262	285	305	327	349	38	95	57	62	67	71	76	140	26	15	16	18	20	22
8 1/2	425	254	275	296	317	338	39	92	55	60	64	69	74	145	25	14	16	18	20	21
8 3/4	411	247	267	288	308	329	40	90	54	59	65	68	72	150	24	14	15	17	19	20
9	400	240	260	280	300	320	41	88	53	57	63	66	70	155	23	13	15	16	18	19
9 1/4	389	233	253	272	292	311	42	86	52	56	62	65	69	160	22	13	14	15	18	19
9 1/2	379	227	246	265	284	303	43	84	50	55	60	63	67	165	22	12	14	15	17	18
9 3/4	369	221	240	258	277	295	44	82	49	53	59	62	66	170	21	12	14	15	17	18
10	360	216	234	252	270	288	45	80	48	52	57	60	64	175	21	12	13	15	17	18
10 1/2	342	205	222	239	257	274	46	78	47	51	56	59	62	180	20	12	13	14	16	17
11	327	196	213	229	245	262	47	77	46	50	55	58	62	185	20	11	13	14	15	17
11 1/2	313	188	205	219	235	250	48	75	45	49	54	56	60	190	19	11	12	13	15	16
12	300	180	195	210	225	240	49	73	44	47	51	55	58	195	18	11	12	13	14	14

Source: National Twist Drill & Tool Co.

TABLE A3–8

Speed and feed calculations for milling cutters and other rotating tools

Ft. per Min.	30	40	50	60	70	80	90	100	110	120	130	140	150
Diam. In.						Revolutions per Minute							
1/16	1833	2445	3056	3667	4278	4889	....	....	....	....	....	....	....
1/8	917	1222	1528	1833	2139	2445	2750	3056	3361	3667	3973	4278	4584
3/16	611	815	1019	1222	1426	1630	1833	2037	2241	2445	2648	2852	3056
1/4	458	611	764	917	1070	1222	1375	1528	1681	1833	1986	2139	2292
5/16	367	489	611	733	856	978	1100	1222	1345	1467	1589	1711	1833
3/8	306	407	509	611	713	815	917	1019	1120	1222	1324	1426	1528
7/16	262	349	437	524	611	698	786	873	960	1048	1135	1222	1310
1/2	229	306	382	458	535	611	688	764	840	917	993	1070	1146
5/8	183	244	306	367	428	489	550	611	672	733	794	856	917
3/4	153	204	255	306	357	407	458	509	560	611	662	713	764
7/8	131	175	218	262	306	349	393	437	480	524	568	611	655
1	115	153	191	229	267	306	344	382	420	458	497	535	573
1 1/8	102	136	170	204	238	272	306	340	373	407	441	475	509
1 1/4	91.7	122	153	183	214	244	275	306	336	367	397	428	458
1 3/8	83.3	111	139	167	194	222	250	278	306	333	361	389	417
1 1/2	76.4	102	127	153	178	204	229	255	280	306	331	357	382
1 5/8	70.5	94.0	118	141	165	188	212	235	259	282	306	329	353
1 3/4	65.5	87.3	109	131	153	175	196	218	240	262	284	306	327
1 7/8	61.1	81.5	102	122	143	163	183	204	224	244	265	285	306
2	57.3	76.4	95.5	115	134	153	172	191	210	229	248	267	287
2 1/4	50.9	67.9	84.9	102	119	136	153	170	187	204	221	238	255
2 1/2	45.8	61.1	76.4	91.7	107	122	138	153	168	183	199	214	229
2 3/4	41.7	55.6	69.5	83.3	97.2	111	125	139	153	167	181	194	208
3	38.2	50.9	63.7	76.4	89.1	102	115	127	140	153	166	178	191
3 1/4	35.3	47.0	58.8	70.5	82.3	94.0	106	118	129	141	153	165	176
3 1/2	32.7	43.7	54.6	65.5	76.4	87.3	98.2	109	120	131	142	153	164
3 3/4	30.6	40.7	50.9	61.1	71.3	81.5	91.7	102	112	122	132	143	153
4	28.7	38.2	47.7	57.3	66.8	76.4	85.9	95.5	105	115	124	134	143
4 1/2	25.5	34.0	42.4	50.9	59.4	67.9	76.4	84.9	93.4	102	110	119	127
5	22.9	30.6	38.2	45.8	53.5	61.1	68.8	76.4	84.0	91.7	99.3	107	115
5 1/2	20.8	27.8	34.7	41.7	48.6	55.6	62.5	69.5	76.4	83.3	90.3	97.2	104
6	19.1	25.5	31.8	38.2	44.6	50.9	57.3	63.7	70.0	76.4	82.8	89.1	95.5
6 1/2	17.6	23.5	29.4	35.3	41.1	47.0	52.9	58.8	64.6	70.5	76.4	82.3	88.2
7	16.4	21.8	27.3	32.7	38.2	43.7	49.1	54.6	60.0	65.5	70.9	76.4	81.9
7 1/2	15.3	20.4	25.5	30.6	35.7	40.7	45.8	50.9	56.0	61.1	66.2	71.3	76.4
8	14.3	19.1	23.9	28.7	33.4	38.2	43.0	47.7	52.5	57.3	62.1	66.8	71.6
8 1/2	13.5	18.0	22.5	27.0	31.5	36.0	40.4	44.9	49.4	53.9	58.4	62.9	67.4
9	12.7	17.0	21.2	25.5	29.7	34.0	38.2	42.4	46.7	50.9	55.2	59.4	63.6
9 1/2	12.1	16.1	20.1	24.1	28.2	32.2	36.2	40.2	44.2	48.2	52.3	56.3	60.3
10	11.5	15.3	19.1	22.9	26.7	30.6	34.4	38.2	42.0	45.8	49.7	53.5	57.3
11	10.4	13.9	17.4	20.8	24.3	27.8	31.3	34.7	38.2	41.7	45.1	48.6	52.1
12	9.5	12.7	15.9	19.1	22.3	25.5	28.6	31.8	35.0	38.2	41.4	44.6	47.8
Ft. per Min.	30	40	50	60	70	80	90	100	110	120	130	140	150

Source: National Twist Drill & Tool Co.

TABLE A3–9
Table of cutting speeds for fractional sizes

TO FIND	HAVING	FORMULA
Surface (or Periphery) Speed in Feet per Minute = S.F.M.	Diameter of Tool in Inches $= D$ and Revolutions per Minute $= R.P.M.$	$S.F.M. = \dfrac{D \times 3.1416 \times R.P.M.}{12}$
Revolutions per Minute = R.P.M.	Surface Speed In Feet per Minute = S.F.M. and Diameter of Tool in Inches $= D$	$R.P.M. = \dfrac{S.F.M. \times 12}{D \times 3.1416}$
Feed per Revolution in Inches = F.R.	Feed in Inches per Minute $= F.M.$ and Revolutions per Minute $= R.P.M.$	$F.R. = \dfrac{F.M.}{R.P.M.}$
Feed in Inches per Minute = F.M.	Feed per Revolution in Inches = F.R. and Revolutions per Minute $= R.P.M.$	$F.M. = F.R. \times R.P.M.$
Number of Cutting Teeth per Minute = T.M.	Number of Teeth in Tool $= T$ and Revolutions per Minute $= R.P.M.$	$T.M. = T \times R.P.M.$
Feed per Tooth = F.T.	Number of Teeth in Tool $= T$ and Feed per Revolution in Inches $= F.R.$	$F.T. = \dfrac{F.R.}{T}$
Feed per Tooth = F.T.	Number of Teeth in Tool $= T$ Feed in Inches per Minute $= F.M.$ and Speed in Revolutions per Minute $= R.P.M.$	$F.T. = \dfrac{F.M.}{T \times R.P.M.}$

Source: National Twist Drill & Tool Co.

TABLE A3–10

Comparative Weights of Steel and Brass Bars

Steel—Weights cover hot worked steel about .50% carbon. One cubic inch weighs .2833 lbs. High speed steel 10% heavier.
Brass—One cubic inch weighs .3074 lbs.
Actual weight of stock may be expected to vary somewhat from these figures because of variations in manufacturing processes.

Size, Inches	Weight of Bar One Foot Long, Lbs.					
	Steel			Brass		
	○	□	⬡	○	□	⬡
1/16	.0104	.013	.0115	.0113	.0144	.0125
1/8	.042	.05	.046	.045	.058	.050
3/16	.09	.12	.10	.102	.130	.112
1/4	.17	.21	.19	.18	.23	.20
5/16	.26	.33	.29	.28	.36	.31
3/8	.38	.48	.42	.41	.52	.45
7/16	.51	.65	.56	.55	.71	.61
1/2	.67	.85	.74	.72	.92	.80
9/16	.85	1.08	.94	.92	1.17	1.01
5/8	1.04	1.33	1.15	1.13	1.44	1.25
11/16	1.27	1.61	1.40	1.37	1.74	1.51
3/4	1.50	1.92	1.66	1.63	2.07	1.80
13/16	1.76	2.24	1.94	1.91	2.43	2.11
7/8	2.04	2.60	2.25	2.22	2.82	2.45
15/16	2.35	2.99	2.59	2.55	3.24	2.81
1	2.67	3.40	2.94	2.90	3.69	3.19
1 1/16	3.01	3.84	3.32	3.27	4.16	3.61
1 1/8	3.38	4.30	3.73	3.67	4.67	4.04
1 3/16	3.77	4.80	4.16	4.08	5.20	4.51
1 1/4	4.17	5.31	4.60	4.53	5.76	4.99
1 5/16	4.60	5.86	5.07	4.99	6.35	5.50
1 3/8	5.04	6.43	5.56	5.48	6.97	6.04
1 7/16	5.52	7.03	6.08	5.99	7.62	6.60
1 1/2	6.01	7.65	6.63	6.52	8.30	7.19
1 9/16	6.52	8.30	7.19	7.07	9.01	7.80
1 5/8	7.05	8.98	7.77	7.65	9.74	8.44
1 11/16	7.60	9.68	8.38	8.25	10.51	9.10
1 3/4	8.18	10.41	9.02	8.87	11.30	9.78
1 13/16	8.77	11.17	9.67	9.52	12.12	10.49
1 7/8	9.39	11.95	10.35	10.19	12.97	11.24
1 15/16	10.02	12.76	11.05	10.88	13.85	12.00
2	10.68	13.60	11.78	11.59	14.76	12.78
2 1/16	11.36	14.46	12.53	12.33	15.69	13.60
2 1/8	12.06	15.35	13.30	13.08	16.66	14.42
2 3/16	12.78	16.27	14.09	13.87	17.65	15.29
2 1/4	13.52	17.22	14.91	14.67	18.68	16.17
2 5/16	14.28	18.19	15.75	15.50	19.73	17.09
2 3/8	15.06	19.18	16.62	16.34	20.81	18.02

Source: Brown & Sharpe Manufacturing Co.

TABLE A3–11

S. A. E. Standard Specifications for Steels

S. A. E. STEEL NUMBERING SYSTEM

A numerical index system is used to identify compositions of S. A. E. steels, which makes it possible to use numerals that are partially descriptive of the composition of materials covered by such numbers. The first digit indicates the type to which the steel belongs. The second digit, in the case of the simple alloy steels, generally indicates the approximate percentage of the predominant alloying element and the last two or three digits indicate the average carbon content in points, or hundredths of 1 per cent. Thus, 2340 indicates a nickel steel of approximately 3 per cent nickel (3.25 to 3.75) and 0.40 per cent carbon (0.35 to 0.45).

In some instances, it is necessary to use the second and third digits of the number to identify the approximate alloy composition of a steel. An instance of such departure is the steel numbers selected for several of the High Speed Steels and corrosion and heat resisting alloys. Thus, 71360 indicates a Tungsten Steel of about 13 per cent Tungsten (12 to 15) and 0.60 per cent carbon (0.50 to 0.70).

The basic numerals for the various types of S. A. E. steel are listed below:

Type of Steel	*Numerals (and Digits)*
Carbon Steels	1xxx
Plain Carbon	10xx
Free Cutting, (Screw Stock)	11xx
Free Cutting, Manganese	X13xx
High Manganese	T13xx
Nickel Steels	2xxx
0.50 Per Cent Nickel	20xx
1.50 Per Cent Nickel	21xx
3.50 Per Cent Nickel	23xx
5.00 Per Cent Nickel	25xx
Nickel Chromium Steels	3xxx
1.25 Per Cent Nickel, 0.60 Per Cent Chromium	31xx
1.75 Per Cent Niekel, 1.00 Per Cent Chromium	32xx
3.50 Per Cent Nickel, 1.50 Per Cent Chromium	33xx
3.00 Per Cent Nickel, 0.80 Per Cent Chromium	34xx
Corrosion and Heat Resisting Steels	30xxx
Molybdenum Steels	4xxx
Chromium	41xx
Chromium Nickel	43xx
Nickel	46xx and 48xx
Chromium Steels	5xxx
Low Chromium	51xx
Medium Chromium	52xxx
Corrosion and Heat Resisting	51xxx
Chromium Vanadium Steels	6xxx
Tungsten Steels	7xxx and 7xxxx
Silicon Manganese Steels	9xxx

Source: Brown & Sharpe Manufacturing Co.

TABLE A3–12

HORSEPOWER REQUIREMENTS

FOR TURNING

When metal is cut in a lathe there is a downward pressure on the tool. This pressure, called chip pressure, depends on the material cut, shape and sharpness of the tool, and the size and shape of the chip.

For average conditions, a simple formula which will give sufficiently accurate results for power estimating purposes is as follows:

$$P = CA, \text{ where}$$

A = cross sectional area of the chip in square inches, which is the product of depth of cut and feed per revolution of the work.

C = a constant depending on material being cut.

P = chip pressure on the tool in pounds.

Values of C

MATERIAL CUT	CONSTANT C
Low alloy steel	270,000
High alloy steel	350,000
High carbon steel	340,000
Medium carbon steel	300,000
Mild Steel	270,000
Cast iron—soft	132,000
Wrought iron	198,000
Malleable iron	170,000
Brass and bronze	110,000

Horsepower may be figured by using the following formula:

$$H.P. = \frac{P \times S}{33,000} \text{ where}$$

Example 2

Example 2: Assuming that SAE 4140 is heat treated so that its strength is 100,000 pounds per square inch, the horsepower necessary to cut it, when other conditions remain the same as in Example 1, is:

$$H.P. = \frac{3.25 \times 100,000 \times \frac{3.14 \times 4}{12} \times 200}{33,000} = 7.8 \text{ as before.}$$

Multiplying the result by 1.25, we get 10 horsepower.

FOR MILLING

Generally accepted approximate values of power for steel cutting is one horsepower per ¾ cubic inches of material removed per minute, although 1 cubic inch can be used for rapid estimating purposes. The horsepower is figured using the following formula:

$$H.P. = KdfNnw, \text{ in which}$$

d = depth of cut taken in inches

f = feed per tooth in inches

$H.P.$ = horsepower necessary to cut

K = a constant depending on material cut

n = number of teeth in the cutter

N = number of revolutions per minute the cutter makes

w = the width of cut in inches

For estimating the horsepower, approximate values of constant K are given below:

MATERIAL CUT	CONSTANT K
Bakelite	0.2
Brass	0.4
Cast Iron, soft	0.5
Cast Iron, medium hard	0.7
Cast Iron, hard	1.0
Steel: 120 Brinell	1.2
150 Brinell	1.4

H. P. = the horsepower necessary to revolve the work against the cutting pressure.

P = pressure on the tool in pounds.

S = cutting speed in feet per minute, and is equal to $\dfrac{3.14 \times D \times N}{12}$

in which D is the diameter of the work, and N is revolutions per minute

Example 1: Determine the horsepower necessary to take a cut ¼ inch deep, with feed of ⅛ inch per revolution, on SAE 4140 steel bar 4 inches in diameter turning 200 times per minute.

Solution: Using C = 325,000

P = 325,000 x ¼ x ⅛ = 1220 pounds

The H. P. = $\dfrac{1220 \times \dfrac{3.14 \times 4}{12} \times 200}{33,000} = 7.8$

This should be multiplied by 1.25 to allow for the efficiency of the machine, thus:

H. P. = 7.8 x 1.25 = 9.7 or 10.

When the tensile strength of the material is known, the following formula may be used for computing the horsepower necessary to cut the material:

H. P. = $\dfrac{3.25 \ ATS}{33,000}$ where

A = the cross section area of the chip in square inches, and is equal to the product of depth of cut and feed per revolution.

H. P. = the horsepower necessary to cut the metal.

S = the cutting speed in f.p.m.

T = the ultimate strength of the material cut.

Source: *Vascoloy-Ramet Corp.*

175 Brinell	1.5
250 Brinell	1.7
300 Brinell	1.9
400 Brinell	2.0
500 Brinell	2.3
600 Brinell	2.5

It is to be noted that for a given material cut, fixed width of cut, and fixed number of teeth, the horsepower will vary with the depth of cut, the feed per tooth and the r.p.m.

Example: Assuming a width of cut 2 inches, the depth ⅛ inch, the feed 0.004" per tooth, what horsepower will be required to mill with a 3-inch 6-tooth cutter running at 600 r.p.m. and cutting steel 250 Brinell hardness.

Solution: K = 1.7 from Table; d = ⅛" or 0.125"; f = 0.004"; n = 6; N = 600, and w = 2".

Substituting these values in the formula, we get:

H. P. = 1.7 x 0.125 x 0.004 x 6 x 600 x 2 = 6.14

If the machine was powered with a 5 horsepower motor, we could reduce the r.p.m. and come within the capacity of the machine, using formula: N = $\dfrac{\text{H. P.}}{Kdfnw}$ in which the symbols have the same meaning as before.

Substituting the known values in this formula, we get:

N = $\dfrac{5}{1.7 \times 0.125 \times 0.004 \times 6 \times 2} = 490$ (approx.)

The speed of the machine should not be allowed to drop more than 50 per cent from that recommended on page 25, since this would impair the performance of the cutter. 490 r.p.m. for a 3-inch cutter will give us a cutting speed

S = $\dfrac{3.14 \times 3 \times 490}{12} = 385$ f.p.m.(approx.)

TABLE A3–13

+00000000 01	+00000000 01	+40000000 51	+10000000 51	+78539750 50	+31415900 51	+00000000 00	0001 000
+00000000 40	+00000003 14	+15000000 51	+42500000 50	+00000000 01	+00000000 00	+00000000 00	0002 000
+40000000 51	+10000000 51	+00000000 01	+13166432 51	+00000000 00	+00000000 00	+00000000 00	0003 015
+00000000 01	+00000000 02	+40000000 51	+12500000 51	+12271836 51	+49087344 51	+00000000 00	0004 000
+00000000 32	+00000004 90	+12500000 51	+42500000 50	+00000000 01	+00000000 00	+00000000 00	0005 000
+40000000 51	+12500000 51	+00000000 01	+12186924 51	+00000000 00	+00000000 00	+00000000 00	0006 015
+00000000 01	+00000000 03	+40000000 51	+15000000 51	+17671444 51	+70685776 51	+00000000 00	0007 000
+00000000 26	+00000000 06	+12500000 51	+47500000 50	+00000000 01	+00000000 00	+00000000 00	0008 000
+40000000 51	+15000000 51	+00000000 01	+12920935 51	+24052798 51	+96211192 51	+00000000 00	0009 015
+00000000 01	+00000000 04	+10000000 51	+17500000 51	+00000000 00	+00000000 00	+00000000 00	0010 000
+00000000 22	+00000009 62	+10000000 51	+50000000 50	+00000000 01	+00000000 00	+00000000 00	0011 000
+40000000 51	+17500000 51	+00000000 01	+12122121 51	+00000000 00	+00000000 00	+00000000 00	0012 015
+00000000 01	+00000000 05	+40000000 51	+20000000 51	+31415900 51	+12566360 52	+00000000 00	0013 000
+40000000 51	+00000012 56	+10000000 51	+55000000 50	+00000000 01	+00000000 00	+00000000 00	0014 000
+40000000 51	+20000000 51	+00000000 01	+12793782 51	+00000000 00	+00000000 00	+00000000 00	0015 015
+00000000 02	+00000000 01	+41250000 51	+10000000 51	+78539750 50	+32397647 51	+00000000 00	0016 000
+00000000 41	+00000003 23	+15000000 51	+42500000 50	+00000000 01	+00000000 00	+00000000 00	0017 000
+41250000 51	+10000000 51	+00000000 01	+13193287 51	+00000000 00	+00000000 00	+00000000 00	0018 015
+00000000 02	+00000000 02	+41250000 51	+12500000 51	+12271836 51	+50621324 51	+00000000 00	0019 000
+00000000 33	+00000005 06	+12500000 51	+47500000 50	+00000000 01	+00000000 00	+00000000 00	0020 000
+41250000 51	+12500000 51	+00000000 01	+12839626 51	+00000000 00	+00000000 00	+00000000 00	0021 015
+00000000 02	+00000000 03	+41250000 51	+15000000 51	+17671444 51	+72894707 51	+00000000 00	0022 000
+00000000 27	+00000007 28	+12500000 51	+47500000 50	+00000000 01	+00000000 00	+00000000 00	0023 000
+41250000 51	+15000000 51	+00000000 01	+12950110 51	+00000000 00	+00000000 00	+00000000 00	0024 015
+00000000 02	+00000000 04	+41250000 51	+17500000 51	+24052798 51	+99217792 51	+00000000 00	0025 000
+00000000 23	+00000009 92	+10000000 51	+50000000 50	+00000000 01	+00000000 00	+00000000 00	0026 000

23AUG

This is a FOR TRANSIT coding of the oil-sand mix core problem specifying a single clamp for lengths 9 inches or less and two clamps for longer lengths. The following definitions are listed:

CLO: Initial length

DL: Increment length

NL: Number of lengths to be used in calculation

CDO: Initial diameter

DD: Increment diameter

ND: Number of diameters to be used in calculation

$Y = f \dfrac{L}{D}$ was stored in tables as whole numbers and handled as subroutine function ADJYF, which was separately programmed for "table lookup." CT was stored in tables as whole numbers and handled as subroutine function CORTF, which was separately programmed for "table lookup."

TABLE A3–13 (*concluded*)

0000001500+	6164718866+	9595916572+	ADJYF	551EK
0000001500+	6376798366+	9595926572+	CORTF	552EK
0000001500+	8278798366+	9190916572+	SQRTF	101EK

```
        C           CYLINDRICAL OIL-SAND CORES
        C           FORMULA M-11-NO.15 - NIEBEL
        C           AUGUST 24, 1961
        C
        1           READ, CLO, DL, NL
        2           READ, CDO, DD, ND
        3           PI = 3.14159
        4           DO 15 I=1,NL
        5           DO 15 J=1,ND
        6           CL = CLO + (I-1) * DL
        7           CD = CDO + (J-1) * DD
        8           A = PI * CD**2 / 4.0
        9           V = A * CL
       10           LD = (CL / CD) * 10.0
       11           KV = V * 100.0
       12           Y = ADJYF(LD) / 100.0
       13           CT = CORTF(KV) / 1000.0
                    PUNCH, I,J,CL,CD,A,V,LD,KV,
                  1 Y,CT
       14           AT = 0.173*V + 0.021*CL
                  1 + 0.0157*A
                  2 + SQRTF(0.0067 + 16.0E=6*V**2)
                  3 + Y*CT + 0.327
       15           PUNCH, CL, CD, AT
       16           STOP
       17           END
08/21/
```

TABLE A3–14

Tables of waiting time and machine availability for selected servicing constants *†
(values expressed as percentages of total time, where $T_1 + T_2 + T_3 = 100$ percent)

k = 0.01

n	(a) T_s	(a) T_1	(b) T_s	(b) T_1
1	0.0	99.0	0.0	99.0
10	0.1	99.0	0.1	98.9
20	0.1	98.9	0.2	98.8
30	0.2	98.8	0.4	98.6
40			0.6	98.4
50			0.9	98.1
60			1.3	97.8
70			1.8	97.2
80			2.7	96.3
85			3.4	95.7
90			4.2	94.9
95			5.2	93.8
100			6.7	92.4
105			8.5	90.6
110			10.7	88.4
115			13.4	85.8
120			16.3	82.9
121			16.9	82.3
122			17.5	81.7
123			18.1	81.1
124			18.8	80.4
125			19.4	79.8
126			20.0	79.2
127			20.6	78.6
128			21.2	78.1
129			21.8	77.5
130			22.4	76.9
131			22.9	76.3
132			23.5	75.7
133			24.1	75.2
134			24.6	74.6
135			25.2	74.1
136			25.7	73.5
137			26.3	73.0
138			26.8	72.5
139			27.3	71.9
140			27.9	71.4
141			28.4	70.9
142			28.9	70.4
143			29.4	69.9
144			29.9	69.4

k = 0.02

n	(a) T_s	(a) T_1	(b) T_s	(b) T_1
1	0.0	98.0	0.0	98.0
5	0.1	98.0	0.2	97.9
10	0.2	97.8	0.4	97.6
15	0.4	97.7	0.7	97.4
20	0.6	97.5	1.1	97.0
25	0.8	97.2	1.6	96.5
30	1.2	96.9	2.2	95.9
35			3.1	95.0
40			4.3	93.8
45			6.1	92.0
50			8.7	89.5

k = 0.02 (cont.)

n	(a) T_s	(a) T_1	(b) T_s	(b) T_1
51			9.3	88.9
52			10.0	88.3
53			10.7	87.6
54			11.5	86.8
55			12.3	86.0
56			13.1	85.2
57			14.0	84.3
58			14.9	83.4
59			15.9	82.5
60			16.8	81.5
61			17.9	80.5
62			18.9	79.5
63			19.9	78.5
64			21.0	77.5
65			22.0	76.4
66			23.1	75.4
67			24.2	74.4
68			25.2	73.3
69			26.2	72.3
70			27.2	71.3
71			28.2	70.4
72			29.2	69.4

k = 0.03

n	(a) T_s	(a) T_1	(b) T_s	(b) T_1
1	0.0	97.1	0.0	97.1
5	0.2	96.9	0.4	96.7
10	0.5	96.6	1.0	96.2
15	1.0	96.2	1.8	95.4
20	1.6	95.5	3.0	94.2
25	2.8	94.4	4.7	92.5
26	3.1	94.1	5.2	92.1
27	3.4	93.7	5.7	91.6
28	3.8	93.4	6.2	91.1
29	4.3	92.9	6.8	90.5
30	4.8	92.4	7.4	89.9
31			8.1	89.2
32			8.9	88.5
33			9.7	87.7
34			10.6	86.8
35			11.6	85.9
36			12.6	84.9
37			13.7	83.8
38			14.9	82.6
39			16.1	81.4
40			17.4	80.2
41			18.8	78.9
42			20.1	77.5
43			21.6	76.2
44			23.0	74.8
45			24.4	73.4
46			25.9	72.0
47			27.3	70.6
48			28.7	69.2

k = 0.04

n	(a) T_s	(a) T_1	(b) T_s	(b) T_1
1	0.0	96.2	0.0	96.2
2	0.1	96.1	0.2	96.0
3	0.2	96.0	0.3	95.9
4	0.2	95.9	0.5	95.7
5	0.3	95.8	0.7	95.5
6	0.5	95.7	0.9	95.3
7	0.6	95.6	1.1	95.1
8	0.7	95.5	1.3	94.9
9	0.8	95.4	1.5	94.7
10	1.0	95.2	1.8	94.4
11	1.1	95.1	2.1	94.1
12	1.3	94.9	2.4	93.8
13	1.5	94.7	2.8	93.5
14	1.8	94.5	3.2	93.1
15	2.0	94.2	3.6	92.7
16	2.3	94.0	4.0	92.3
17	2.6	93.6	4.5	91.8
18	3.0	93.3	5.1	91.3
19	3.4	92.9	5.7	90.7
20	3.9	92.4	6.4	90.0
21	4.5	91.8	7.1	89.3
22	5.2	91.2	8.0	88.5
23	6.0	90.4	8.9	87.6
24	6.8	89.6	9.9	86.7
25	7.9	88.6	11.0	85.6
26	9.0	87.5	12.2	84.5
27	10.4	86.2	13.4	83.2
28	11.9	84.7	14.8	81.9
29	13.6	83.0	16.3	80.5
30	15.5	81.3	17.9	79.0
31			19.6	77.4
32			21.3	75.7
33			23.0	74.0
34			24.8	72.3
35			26.6	70.6
36			28.4	68.9
37			30.1	67.2

k = 0.05

n	(a) T_s	(a) T_1	(b) T_s	(b) T_1
1	0.0	95.2	0.0	95.2
2	0.1	95.1	0.2	95.0
3	0.2	95.0	0.5	94.8
4	0.4	94.9	0.7	94.5
5	0.5	94.7	1.0	94.3
6	0.7	94.6	1.4	94.0
7	0.9	94.4	1.7	93.6
8	1.1	94.2	2.1	93.3
9	1.4	93.9	2.5	92.9
10	1.6	93.7	3.0	92.4
11	2.0	93.4	3.5	91.9
12	2.3	93.0	4.1	91.4
13	2.7	92.6	4.7	90.8
14	3.2	92.2	5.4	90.1
15	3.8	91.7	6.2	89.3

k = 0.05 (cont.)

n	(a) T_s	(a) T_1	(b) T_s	(b) T_1
16	4.4	91.0	7.1	88.5
17	5.2	90.3	8.1	87.6
18	6.1	89.5	9.1	86.5
19	7.1	88.5	10.4	85.4
20	8.4	87.3	11.7	84.1
21	9.8	85.9	13.1	82.7
22	11.5	84.3	14.7	81.2
23	13.4	82.5	16.5	79.6
24	15.5	80.5	18.3	77.8
25	17.8	78.2	20.2	76.0
26	20.3	75.9	22.2	74.1
27	22.8	73.6	24.3	72.1
28	25.3	71.2	26.4	70.1
29	27.9	68.8	28.5	68.1

k = 0.06

n	(a) T_s	(a) T_1	(b) T_s	(b) T_1
1	0.0	94.3	0.0	94.3
2	0.2	94.2	0.3	94.0
3	0.4	94.0	0.7	93.7
4	0.6	93.8	1.1	93.3
5	0.8	93.6	1.5	92.9
6	1.1	93.3	2.0	92.5
7	1.4	93.1	2.5	92.0
8	1.7	92.7	3.1	91.4
9	2.1	92.4	3.7	90.8
10	2.6	91.9	4.5	90.1
11	3.1	91.4	5.3	89.4
12	3.8	90.8	6.2	88.5
13	4.5	90.1	7.3	87.5
14	5.4	89.2	8.4	86.4
15	6.5	88.2	9.7	85.2
16	7.8	87.0	11.2	83.8
17	9.3	85.6	12.8	82.3
18	11.1	83.9	14.6	80.6
19	13.2	81.9	16.5	78.8
20	15.6	79.7	18.6	76.8
21			20.8	74.7
22			23.1	72.5
23			25.5	70.3
24			27.9	68.0
25			30.3	65.8

k = 0.07

n	(a) T_s	(a) T_1	(b) T_s	(b) T_1
1	0.0	93.5	0.0	93.5
2	0.2	93.2	0.4	93.1
3	0.5	93.0	0.9	92.6
4	0.8	92.7	1.4	92.1
5	1.1	92.4	2.0	91.6
6	1.5	92.1	2.7	91.0
7	1.9	91.7	3.4	90.3
8	2.4	91.2	4.3	89.5
9	3.1	90.6	5.2	88.6
10	3.8	89.9	6.3	87.6

* All tables assume random calls for service. Column (a) is for constant servicing time and column (b) for an exponential distribution of servicing times. It is hoped that the missing values in column (a) can be secured by approximation in the near future.

† Where no entry appears in column, the figures were not available.

TABLE A3–14 (concluded)

n	(a) T_s	T_1	(b) T_s	T_1
	k = 0.07 (*cont.*)			
11	4.7	89.1	7.5	86.4
12	5.7	88.1	8.9	85.1
13	7.0	86.9	10.4	83.7
14	8.6	85.4	12.2	82.1
15	10.4	83.7	14.1	80.3
16	12.6	81.6	16.2	78.3
17	15.2	79.3	18.5	76.2
18	18.1	76.6	21.0	73.9
19	21.1	73.7	23.5	71.5
20	24.4	70.7	26.2	69.0
21			28.9	66.5
	k = 0.08			
1	0.0	92.6	0.0	92.6
2	0.3	92.3	0.5	92.1
3	0.6	92.0	1.2	91.5
4	1.0	91.7	1.9	90.9
5	1.4	91.2	2.7	90.1
6	2.0	90.8	3.5	89.3
7	2.6	.90.2	4.5	88.4
8	3.4	89.5	5.7	87.3
9	4.3	88.6	7.0	86.1
10	5.4	87.6	8.5	84.8
11	6.7	86.4	10.1	83.2
12	8.4	84.8	12.0	81.4

n	(a) T_s	T_1	(b) T_s	T_1
	k = 0.08 (*cont.*)			
13	10.4	83.0	14.2	79.5
14	12.8	80.8	16.5	77.3
15	15.6	78.2	19.0	75.0
16	18.8	75.2	21.8	72.4
17	22.2	72.0	24.6	69.8
18	25.7	68.8	27.6	67.1
19	28.2	66.5	30.5	64.4
	k = 0.09			
1	0.0	91.5	0.0	91.7
2	0.4	91.4	0.7	91.1
3	0.8	91.0	1.4	90.4
4	1.3	90.6	2.3	89.6
5	1.9	90.0	3.3	88.7
6	2.6	89.4	4.5	87.7
7	3.4	88.6	5.8	86.5
8	4.5	87.6	7.3	85.1
9	5.7	86.5	9.0	83.5
10	7.3	85.0	10.9	81.7
11	9.3	83.2	13.1	79.7
12	11.7	81.0	15.6	77.5
13	14.5	78.4	18.3	75.0
14	17.8	75.4	21.2	72.3
15	21.5	72.0	24.2	69.5
16	25.3	68.5	27.4	66.6
17	29.2	65.0	30.6	63.7

n	(a) T_s	T_1	(b) T_s	T_1
	k = 0.10			
1	0.0	90.9	0.0	90.9
2	0.4	90.5	0.8	90.2
3	1.0	90.0	1.8	89.3
4	1.6	89.5	2.8	88.3
5	2.3	88.8	4.1	87.2
6	2.2	88.0	5.5	85.9
7	4.4	86.9	7.1	84.4
8	5.8	85.7	9.0	82.7
9	7.5	84.1	11.2	80.8
10	9.7	82.1	13.6	78.5
11	12.4	79.8	16.3	76.1
12	15.6	76.8	19.3	73.4
13	19.2	73.4	22.5	70.4
14	23.3	69.8	25.9	67.4
15	27.4	66.0	29.4	64.2
16	31.5	62.0		
	k = 0.15			
1	0.0	87.0	0.0	87.0
2	0.9	86.2	1.7	85.5
3	2.1	85.1	3.6	83.8
4	3.9	83.8	6.0	81.8
5	5.5	82.2	8.7	79.4
6	8.0	80.0	11.8	76.7
7	11.2	72.2	15.4	73.5
8	15.2	73.7	19.5	70.0
9	20.1	69.5	23.8	66.2
10	25.5	64.8	28.4	62.3
11	31.0	60.0		

n	(a) T_s	T_1	(b) T_s	T_1
	k = 0.20			
1	0.0	83.3	0.0	83.3
2	1.5	82.0	2.7	81.1
3	3.6	80.4	5.9	78.4
4	6.3	78.1	9.8	75.2
5	10.0	75.0	14.2	71.5
6	14.7	71.1	19.2	67.4
7	20.6	66.2	24.6	62.8
8	27.3	60.6	30.3	58.1
9	32.6	56.1		
	k = 0.30			
1	0.0	76.9	0.0	76.9
2	3.0	74.6	5.1	73.0
3	7.4	71.3	11.1	68.4
4	13.3	66.7	18.0	63.1
5	21.1	60.7	25.4	57.4
6	29.9	53.9	33.0	51.6
	k = 0.40			
1	0.0	71.4	0.0	71.4
2	4.8	68.0	7.5	66.0
3	11.8	63.0	16.3	59.8
4	21.2	56.3	25.6	53.1
5	31.9	48.6	34.9	46.5

TABLE A3-15
Metric system conversion chart

LENGTH

U.S.	METRIC
1 inch = 25.4 millimeters
1 foot = 30.48 centimeters
1 yard = 0.914 meter

METRIC	U.S.
1 millimeter = 0.039 inch
1 centimeter = 0.394 inch
1 meter = 39.37 inches

THICKNESS

1 mil = .025 millimeter

1 millimeter = 39.37 mils

AREA

1 sq. foot = 929.03 sq. centimeters

1 sq. centimeter = 0.155 sq. inch

VOLUME

1 gallon = 3.785 liters

1 liter = 0.264 gallon

VOLUME

1 cu. inch = 16.387 cu. centimeters   1 cu. centimeter = 0.061 cu. inch

CC

CU. IN.

TEMPERATURE

$$°Fahrenheit = \frac{(°Celcius)\ 9}{5} + 32 \qquad °Celcius = \frac{(°Fahrenheit - 32)\ 5}{9}$$

°C

°F

WEIGHT

1 ounce (dry) = 28.35 grams   1 gram = .035 ounce

GRAMS

OUNCES

WEIGHT

1 pound = .454 kilogram   1 kilogram = 2.204 pounds

KILOGRAMS

POUNDS

PRESSURE

1 pound/sq. inch = 0.703 kilogram/sq. centimeter   1 kilogram/sq. centimeter = 14.22 pounds/sq. inch

KG./CM²

LBS./SQ. IN.

With the compliments of McGraw-Hill Book Company, College Division. Publishing for the engineer's diversity. 1221 Avenue of the Americas, New York, NY 10020.

843

844 *Motion and Time Study*

TABLE A3–16
Program Name: "LEARN"

The memory required for this program is 4136 bytes.

This program has four separate parts combined in one program. When "booted up" the user selects the module or routine he wishes to work with. The four sections of this program are:

1. Calculate the exponent for a curve.
2. Calculate time for the first unit.
3. Calculate time for the Nth unit.
4. Calculate launching costs.

Some microcomputers calculate log base 10 directly; the TRS-80[1] microcomputer works in *e* based logs and it is necessary to write the calculations using log (*x*)/log(10). The program is presented in its entirety on the next several pages.

```
3 REM PROGRAM NAME LEARN
4 REM WRITTEN BY J.E. NICKS
5 REM COMPUTE APPLICATIONS FOR THE MANUFACTURING ENGINEER
6 REM COPYRIGHT 1981 ALL RIGHTS RESERVED
10 CLS
20 PRINT"THIS PROGRAM CALCULATES"
21 PRINT"LEARNING CURVES AND LAUNCHING COSTS"
22 PRINT"INPUT THE SUB PROGRAM YOU WISH TO USE"
23 PRINT:PRINT
30 PRINT"CALCULATE THE EXPONENT FOR A CURVE------------1"
40 PRINT"CALCULATE THE TIME FOR THE FIRST UNIT---------2"
50 PRINT"CALCULATE THE TIME FOR THE N TH. UNIT---------3"
60 PRINT"CALCULATE LAUNCHING COST----------------------4"
70 INPUTN

80 ONNGOTO90,250,410,600
81 CLS
90 PRINT"THIS ROUTINE CALCULATES EXPONENTS FOR A LEARNING CURVE"
91 PRINT
92 PRINT"THE NUMBER OF UNITS PRODUCED (LARGER) MUST BE"
93 PRINT"2 TIMES THE NUMBER OF UNITS PRODUCED (SMALLER)"
94 PRINT
100 INPUT"TIME OR PER CENT (LARGER)";Y1
110 INPUT"NUMBER OF UNITS PRODUCED (LARGER)";N1
120 INPUT"TIME OR PER CENT (SMALLER)";Y2
130 INPUT"NUMBER OF UNITS PRODUCED (SMALLER)";N2    Lines 140 and 150
140 A=Y1/Y2:A1=LOG(A)/LOG(10) ←──────────────────  convert natural logs to
150 C=N1/N2:C1=LOG(C)/LOG(10)                       base 10 logs
160 B=A1/C1:B1=Y2/Y1
170 LPRINT CHR$(27);CHR$(14);"LEARNING CURVE EXPONENTS"
171 LPRINT" "
180 LPRINT"THE LARGER TIME OR % WAS";Y1;"THE SMALLER WAS";Y2
181 LPRINT" "
190 LPRINT"LARGER UNITS PRODUCED WAS";N1;"THE SMALLER WAS";N2
191 LPRINT" "
200 LPRINT"THIS IS A ";B1;"% LEARNING CURVE"
201 LPRINT" "
210 LPRINT"THE EXPONENT FOR THIS CURVE IS";B
220 CLS:INPUT"FOR ANOTHER CALCULATION TYPE 1 or 2 to EXIT";X
230 ONXGOTO10,240
240 END

241 CLS
250 PRINT"THIS ROUTINE CALCULATES TIME OF THE FIRST UNIT"
```

[1] TRS-80 is a trademark of the Radio Shack Division of Tandy Corporation.

TABLE A3–16 (continued)

```
260 INPUT"THE AVERAGE ACCUM QUANTITY IS ";Q
270 PRINT"THE AVERAGE ACCUM TIME FOR ";Q;"QUANTITY IS"
280 INPUTY
290 PRINT"TO CHOOSE THE LEARNING CURVE YOU WISH TO WORK WITH"
300 INPUT"TYPE 1";Z
310 IFZ=1THEN320
320 GOSUB2000
330 A=(QCB)*Y
335 LPRINT CHR$(27);CHR$(14);"TIME FOR THE FIRST UNIT"
336 LPRINT" "
340 LPRINT"THE AVERAGE ACCUM TIME FOR THE ";Q;"UNIT"
350 LPRINT"WAS";Y
355 LPRINT" "
360 LPRINT"TIME FOR THE FIRST UNIT USING A ";C;"% CURVE"
370 LPRINT"IS ";A
380 INPUT"FOR ANOTHER CALCULATION TYPE 1 OR 2 TO EXIT";X
390 ONXGOTO10,400
400 END

401 CLS
410 PRINT"TIME FOR THE N TH UNIT"
420 INPUT"NUMBER OF UNITS PRODUCED, THE N TH UNIT IS";P
430 INPUT"TIME FOR THE FIRST UNIT";A
440 PRINT"TO CHOOSE THE LEARNING CURVE YOU WISH TO WORK WITH"
450 INPUT"TYPE 1";Z
460 IFZ=1THEN470
470 GOSUB2000
480 Y=(P[-B)*A
490 Y1=(1-B)*Y
500 LPRINT CHR$(27);CHR$(14);"TIME FOR THE N TH UNIT"
505 LPRINT" "
510 LPRINT"WHERE TIME FOR THE FIRST UNIT IS ";A
520 LPRINT"USING A";C;"% LEARNING CURVE"
525 LPRINT" "
530 LPRINT"THE AVERAGE ACCUM TIME FOR THE N TH UNIT";P
540 LPRINT"IS";Y

545 LPRINT" "
550 LPRINT"AND THE TIME FOR THE N TH ";P;" UNIT ONLY"
560 LPRINT"IS";Y1
570 PRINT"FOR ANOTHER CALCULATION TYPE 1 OR 2 TO EXIT"
571 INPUTX
580 ONXGOTO10,590
590 END

600 CLS
601 PRINT"THIS ROUTINE CALCULATES LAUNCHING COSTS"
610 INPUT"TIME IN HOURS FOR THE FIRST UNIT IS";A
620 INPUT"STANDARD TIME IN HOURS PER UNIT ";S
630 INPUT"SHOP RATE DOLLARS IS";R
640 PRINT"TO CHOOSE THE LEARNING CURVE YOU WISH TO WORK WITH"
650 INPUT"TYPE 1";Z
660 IFZ=1THEN670
670 GOSUB2000
680 B1=-1/B:B2=1-B
690 P=((S/B2)/A)[B1
700 Y=(P[-B)*A
```

TABLE A3–16 (concluded)

```
710  O=((Y/S)*100)-100
720  L=(Y-S)*P*R
730  LPRINT CHR$(27);CHR$(14);"LAUNCHING COSTS"
731  LPRINT" "
740  LPRINT"TIME FOR THE FIRST UNIT IS";A;"HOURS"
750  LPRINT"STANDARD TIME IS";S;"HOURS PER UNIT"
760  LPRINT"SHOP RATE IS $";R;" DOLLARS"
761  LPRINT" "
770  LPRINT"USING A ";C;"% LEARNING CURVE"
780  LPRINT"STANDARD IS REACHED AT THE ";P;"UNIT"
781  LPRINT" "
790  LPRINT"WHEN THE ";P;"UNIT IS REACHED AS A UNIT"
800  LPRINT"THE AVERAGE ACCUM TIME UP TO THE ";P;"UNIT"
810  LPRINT"IS";Y
811  LPRINT" "
820  LPRINT"AND THE OFF STANDARD IS";O;"PER CENT"
821  LPRINT" "
822  LPRINT" "
830  LPRINT CHR$(27);CHR($14);"LAUNCHING COSTS ARE $";L;"DOLLARS"
840  INPUT"FOR ANOTHER CALCULATION TYPE 1 OR 2 TO EXIT";X
850  ONXGOTO10,860
860  END

2000  CLS
2010  PRINT"THE FOLLOWING CURVES ARE AVAILABLE TO WORK WITH"
2030  PRINT"TYPE IN THE NUMBER YOU WISH TO USE
2040  PRINT"1    = 98%--------------2   = 96%"
2050  PRINT"3    = 95%--------------4   = 94%"
2060  PRINT"5    = 92%--------------6   = 90%"
2070  PRINT"7    = 88%--------------8   = 96%"
2080  PRINT"9    = 85%-------------10   = 84%"
2090  PRINT"11   = 82%-------------12   = 80%"
2100  PRINT"13   = 78%-------------14   = 76%"
2110  PRINT"15   = 75%"
2120  INPUTN
2130  ONNGOTO2160,2170,2180,2190,2200
2140  ONN-5GOTO2210,2220,2230,2240,2250
2150  ONN-10GOTO2260,2270,2290,2300
2160  B=.029146:C=98:RETURN
2170  B=.058894:C=96:RETURN
2180  B=.074:C=95:RETURN
2190  B=.089267:C=94:RETURN
2200  B=.120294:C=92:RETURN
2210  B=.152003:C=90:RETURN
2220  B=.184425:C=88:RETURN

2230  B=.217591:C=86:RETURN
2240  B=.234465:C=85:RETURN    <──── Sub-routine contains exponents for
2250  B=.251539:C=84:RETURN             the last three modules
2260  B=.286304:C=82:RETURN
2270  B=.321928:C=80:RETURN
2280  B=.358453:C=78:RETURN
2290  B=.395929:C=76:RETURN
2300  B=.415038:C=75:RETURN
```

Courtesy of: Dr. J. E. Hicks. Source code by: MiCapp, Inc. Reprinted courtesy: Society of Manufacturing Engineers, Copyright 1982.

MIL–STD–1567A

5.1.1 *Predetermined Time Systems.* It is not the intent of this Military Standard to challenge the accuracy of those predetermined time systems whose inherent accuracy meets the requirements of paragraph 5.1. However, when a predetermined time system is used, it shall be incumbent on the contractor to demonstrate to the Government that the accuracy of the original data base has not been compromised in application or standards development.

5.2 *Operations Analysis.* Operations analysis is considered an integral part of the development of a Type I Engineered Labor Standard. An operating analysis shall be accomplished and recorded prior to the determination of a Type I standard; and in the improvement of established labor standards.

5.3 *Standard Data.* The contractor shall take full advantage of available standard time data of known accuracy and traceability.

5.4 *Labor Standards Coverage.* The contractor shall develop and implement a Work Measurement Coverage Plan which provides a time-based schedule for achieving 80% coverage of all categories of touch labor hours with Type I standards. (See 3.9, Touch Labor.)

5.4.1 *Cost Trade-Off Analysis.* The Work Measurement Coverage Plan shall be based on cost trade-off analyses which consider the status and effectiveness of the contractor's existing work measurement program.

5.4.2 *Initial Coverage.* Type II Standards are acceptable for initial coverage. All Type II standards shall be approved by the organization(s) responsible for establishing and implementing work measurement standards and estimating when Type I Standards have not yet been developed.

5.4.3 *Upgrading*. The Work Measurement Touch Labor Coverage Plan shall provide a schedule for upgrading Type II to Type I Standards.

5.5 *Leveling/Performance Rating*. All time studies shall be rated using recognized techniques.

5.6 *Allowances*. Allowances for personal, fatigue, and unavoidable delays shall be developed and included as part of the labor standard. Allowances should not be excessive or inconsistent with those normally allowed for like work and conditions.

5.7 *Estimating*. The contractor's procedures shall describe how touch labor standards are utilized to develop price proposals.

5.8 *Use of Labor Standards*. Labor standards shall be used:

5.8.1 *Budgets, Plans and Schedules*. As an input to developing budgets, plans and schedules, when available.

5.8.2 *Touch Labor Hours*. As a basis for estimating touch labor hours when issuing changes to contracts and as a basis for estimating the prices of initial spares, replenishment spares and follow-on production buys, when available.

5.8.3 *Measuring Performance*. As a basis for measuring touch labor performance.

5.9 *Realization Factor*. When labor standards have been modified by realization factors, major elements which contribute to the total factor shall be identified. The analysis supporting each element shall be available to the Government for review.

5.10 *Labor Efficiency*. A forecast of anticipated touch labor efficiency shall be used in manpower planning, both on a long-range and current scheduling basis.

5.11 *Revisions*. Labor standards shall be reviewed for accuracy and appropriate system data revision made when changes occur to:

a. Methods or procedures
b. Tools, jigs, and fixtures
c. Work place and work layout
d. Specified materials
e. Work content of the job

5.12 *Production Count*. Work units shall be clearly and discretely defined so as to cause accurate measurement of the work completed and shall be expressed in terms of completed:

a. End items
b. Operations
c. Lots or batches of end items

5.12.1 *Partial Credit.* In those cases where partial production credit is appropriate, the work measurement procedures shall define the method to be used to permit a timely and current production measure.

5.13 *Labor Performance Reporting.* The contractor's work measurement program shall provide for periodic reporting of labor performance. The report shall be prepared at least weekly for each work center and be summarized at each appropriate management level; it shall indicate labor efficiency and compare current results with pre-established contractor goals. (When this report is required to be delivered, see 6.2.)

5.13.1 *Variance Analysis.* Labor performance reports shall be reviewed by supervisory and staff support functions. When a significant departure from projected performance goals occurs, a formal written analysis which addresses causes and corrective actions shall be prepared.

5.13.2 *Report Retention.* Performance reports and related variance trend analyses shall be retained for a six-month period.

5.14 *System Audit.* The contractor shall use an internal review process to monitor the work measurement system. This process shall be so designed that weaknesses or failures of the system are identified and brought to the attention of management to enable timely corrective action. Written procedures shall describe the audit techniques to be used in evaluating system compliance.

5.14.1 *Scope of Audit.* The audit shall cover compliance with the requirements of this standard at least annually. The audit, based upon a representative sample of all active labor standards and work measurement activities, shall determine:

a. The validity of the prescribed method and the accuracy of the labor standard time values as validated against the data baseline.
b. Percent of coverage by Type I and Type II labor standards.
c. Effectiveness of the use of labor standards for planning, esitmating, budgeting, and scheduling.
d. The timeliness, accuracy and traceability of production count reporting.
e. The accuracy of labor performance reports.
f. The reasonableness and attainment of efficiency goals established.
g. The effectiveness of corrective actions resulting from variance analyses.

5.14.2   *Audit Reports.*   A copy of the audit finding shall be retained in company files for at least a two-year period and shall be made available to the Government designated representative for review upon request.

6.   NOTES

6.1   *Intended Use.*   This standard is intended to promote the cost effective acquisition of systems and equipment by requiring the use of work measurement to increase productivity and efficiency.

6.2   *Data Requirements.*   The following data requirements should be considered when this standard is applied on a contract. The applicable Data Item Descriptions (DIDs) should be reviewed in conjunction with the specific acquisition to ensure that only essential data are requested/provided and that the DIDs are tailored to reflect the requirements of the specific acquisition. To ensure correct contractual applications of the data requirements, Contract Data Requirements Lists (DD 1423) must be prepared to obtain the data, except where DoD FAR Supplement 27.410-6 exempts the requirements for a DD 1423.

Paragraph No.	Data Requirement Title	Applicable DID No.
5.13	Work Measurement Labor Performance Report	DI-MISC-80295

Index